Bird Families of the World

A series of authoritative, illustrated handbooks of which this is the fourteenth to be published

Series editors

C. M. PERRINS, Chief editor
W. J. BOCK
J. KIKKAWA

THE AUTHORS studied, appreciated, and argued herons together for over 25 years starting when Hancock, then soon to retire as an oil company executive, met with Jim Kushlan at his heron study area in the Everglades of south Florida. Their collaboration continued through the writing of *Herons Handbook*, which has served as the definitive monograph on herons of the world since its publication in 1984, and *Storks, Ibis and Spoonbills of the World*, published in 1992. In the present book they update their collaboration in what will undoubtedly become the authoritative contemporary statement on this family of birds.

James A. Kushlan is a biologist, writer, and conservationist. He has studied colonial wading birds since the late1960s, particularly in Florida but also at locations throughout the world. He received his Ph.D. from the University of Miami (Florida) in 1974. Since then he has been awarded honorary doctorates from Thiel College and John Cabot University. In addition to his books with James Hancock, he is the co-editor of *Herons Conservation* (2001) and author of over 175 technical papers. He has held positions as Supervisory Biologist with the US National Park Service, Professor of biological sciences and Director of the Water Resources Institute at Texas A&M University—Commerce, Professor and Chair of Biology at the University of Mississippi, Director of the Patuxent Wildlife Research Center, Senior Science Adviser to the US Geological Survey, and Research Associate at the Smithsonian Environmental Research Center. He is formerly president of the Waterbird Society, editor of the journals *Colonial Waterbirds* and *Florida Field Naturalist*. He is a Fellow and president of the American Ornithologists' Union, and Chair of the Herons Specialist Group and of the Bird Conservation Alliance (USA). He serves on the governing boards of the American Bird Conservancy, Hawk Mountain Sanctuary, Waterbird Conservation for the Americas, and John Cabot University. His interests include waterbird and wetland conservation and education.

The late James Hancock was a writer, photographer, ornithologist, former business executive, and worldwide traveller. In addition to his co-authorship with James Kushlan of *Herons Handbook* and *Storks, Ibis and Spoonbills of the World*, he was co-author of the first heron monograph, *Herons of the World* (1978), author and principal photographer of the natural history books *Herons and Egrets of the World—A Photographic Journey* (1999), *Herons of North America—Their World in Focus* (2000), and author of *Birds of the Wetlands* (1999). He held an honorary doctorate from the University of Southampton and was formerly president of the British Trust for Ornithology, councillor for the Royal Society for the Protection of Birds, President of the Hampshire Ornithological Society, and member of the Royal Photographic Society. He was elected a Corresponding Fellow of the American Ornithologists' Union and awarded the Order of the British Empire (OBE) by Her Majesty the Queen for his services to Ornithology. His interests included writing, photography, heron taxonomy and conservation, fly fishing, and a collection of grandchildren. James Hancock passed away in February 2004.

THE ARTIST David Thelwell was born in March 1950 at Birkenhead, Merseyside, England. His interest in birds was sparked at a very early age, when his father showed him some garden birds and explained they all had different names. Watching his father, a well-known cartoonist and illustrator, at work in his studio, inspired David to become an artist himself. He studied at Southampton Art College, and gained his first job as a cartographic draughtsman. He later worked as a graphic designer and illustrator, and eventually became a fulltime freelance artist in 1989. As a keen ornithologist and field sketcher, David has travelled to many countries in search of birds and wildlife, and is particularly interested in waterbirds. He has contributed illustrations to many books, including *Birds of Prey of the World*, *The Encyclopedia of Birds*, and both the *New Atlas of Breeding Birds* and *Atlas of Wintering Birds in Britain and Ireland*. He has also illustrated Christmas cards for the British Trust for Ornithology and Royal Society for the Protection of Birds, and regularly contributes illustrations to a number of magazines.

Bird Families of the World

1. **The Hornbills**
   Alan Kemp
2. **The Penguins**
   Tony D. Williams
3. **The Megapodes**
   Darryl N. Jones, René W. R. J. Dekker, and Cees S. Roselaar
4. **Fairy-wrens and Grasswrens**
   Ian Rowley and Eleanor Russell
5. **The Auks**
   Anthony J. Gaston and Ian L. Jones
6. **The Birds of Paradise**
   Clifford B. Frith and Bruce Beehler
7. **The Nightjars and their Allies**
   D. T. Holyoak
8. **Toucans, Barbets and Honeyguides**
   Lester L. Short and Jennifer F. M. Horne
9. **Ratites and Tinamous**
   S. J. J. F. Davies
10. **The Bowerbirds**
    Clifford B. Frith and Dawn W. Frith
11. **Albatrosses and Petrels across the World**
    Michael Brooke
12. **The Grebes**
    Jon Fjeldså
13. **The Hawaiian Honeycreepers**
    H. Douglas Pratt
14. **Herons**
    James A. Kushlan and James A. Hancock
15. **The Cuckoos**
    Robert B. Payne
16. **Ducks, Geese and Swans**
    Edited by Janet Kear

Bird Families of the World

Herons

James A. Kushlan and
James A. Hancock

Illustrated by
David Thelwell

OXFORD
UNIVERSITY PRESS

OXFORD
UNIVERSITY PRESS

Great Clarendon Street, Oxford, OX2 6DP,
United Kingdom

Oxford University Press is a department of the University of Oxford. It furthers the University's objective of excellence in research, scholarship, and education by publishing worldwide. Oxford is a registered trade mark of Oxford University Press in the UK and in certain other countries

First Edition published in 2005

Published in the United States of America by Oxford University Press
198 Madison Avenue, New York, NY 10016, United States of America

British Library Cataloguing in Publication Data
Data available

Library of Congress Cataloging in Publication Data
Data available

ISBN 978-0-19-854981-9

Preface

Herons are an intriguing bunch of birds. Coming from different ornithological worlds—one the scientific ornithologist, the other the recognized photographer and natural historian—the two authors of this volume have shared a passion for herons. Twenty years ago we published a synthesis of the biology and conservation of the herons of the world, *Herons Handbook* (Hancock and Kushlan 1984). We were pleased with its reception among scientists, naturalists, and conservationists. In it we attempted to synthesize what was then known about herons. Drawing on the artistic talents of Robert Gillmor and Peter Hayman, the book offered a series of illustrative portraits, some of which are not likely to be surpassed. Although incorporating the scientific literature of the day, much of its content turned on our own observations and research, constituting as they did, then, an interesting portion of what was known. At that time we had 30 years of heron-watching experience between us and a contributor list composed of dozens of heron-watching colleagues. *Herons Handbook* turned out to be a successful enterprise, galvanizing interest in herons and apparently inspiring scientists to discover more about herons.

Unfathomably, together we now have 60 years of heron-watching experience, on all the heron-inhabited continents, having seen in the flesh all but a dozen of the species and now enjoying a colleague list of many hundreds of heron devotees. Despite our increased time in service, our own contribution has become an increasingly minor part of the extant knowledge base of these birds. This is because since our last book information and understanding of heron biology and conservation have advanced exponentially. We couldn't be more gratified. Information has come from across the globe, and has been published in books and in articles in the most distinguished of scientific journals. Scientists, conservationists, birders, and naturalists have all weighed in to fill in the knowledge gaps, correct our misimpressions, added new ideas and syntheses, and used herons as tools to address questions of basic biology. What we now know, compared with what we knew then, is staggering.

Given this explosion in knowledge, it is clearly time to update the biology and conservation material of the *Herons Handbook*. But we no longer could do this in one volume. The present book, *Herons*, a contribution in the distinguished Bird Families of the World series, aims to summarize the biology of the family of herons. Its companion volume, *Heron Conservation* (Kushlan and Hafner 2000), summarizes the conservation of herons. *Herons* particularly deals with descriptions, distribution, systematics, foraging, and breeding, as well as a short summary of conservation and research needs. We do not try to redo the species portraits of *Herons Handbook* but offer a fine new series of identification plates by David Thelwell. The distribution maps are totally updated, and hopefully as accurate as current knowledge allows. The senior author will be pleased to learn of errors, new distribution records, and range changes.

So what has happened in the intervening years? The White-eared Night-Heron was rediscovered after being seen for sale in a Chinese country market. It and herons such as the White-bellied Heron and the Japanese Night-Heron teeter on the brink of extinction, as do populations such as the Purple

Herons on Cape Verde Island, the Eastern Great Egrets in New Zealand, the southern African population of the Eurasian Bittern, and the Grey Heron in Comoros and Aldabra. The range contractions of large bitterns continue. But there are more Chinese Egrets and Zigzag Herons than we previously thought. Little Egrets continue their expansion, now even established as a nesting species in the Western Hemisphere. The Yellow Bittern's range is expanding following advances in rice farming. The Cattle Egret's exponential rate of increase in North America has stabilized. The Black-crowned Night-Heron has settled in the rice fields of the Po Valley in Italy forming the densest concentration in Europe. Great heronries are being discovered in China. The Great White Egret is now recognized as two species, a western one and an eastern one, a resolution initiated by nature photography. The cyclical splitting and combining of *Butorides* has taken a splitting turn. The little Egrets remain a great puzzle.

Particularly exciting have been many advances in heron systematics. Behavioural, morphological, and in particular, molecular studies are helping to resolve issues that have been confusing taxonomists for more than a century. In the case of DNA studies, although much has been done there is much left to do; indeed, material from many taxa has not yet been examined and some of the findings must continue to be regarded as provisional. Nevertheless, the presently available evidence is leading us to reappraise many long-established views. We have reflected some of these new ideas in the taxonomy of the current book. These changes include such findings that the Cattle Egret, Intermediate Egret, and the great egrets are all *Ardea*, that *Ardea* and *Egretta* are not closely related sister taxa, and that the bitterns are the most derived of the herons. The tiger herons, the Boat-billed Heron and perhaps the Agami Heron are more distinctive than were thought a few years ago, a distinctiveness we recognize by placing them in separate subfamilies.

Some of the most encouraging developments have been in the area of heron conservation, which is now maturing as a movement worldwide. The Heron Specialist Group (HSG) has taken the lead in evaluating, summarizing, organizing, and encouraging conservation action. Continent-wide water bird conservation initiatives in North America have heron conservation as a key component. Wetland conservationists and managers are coming to appreciate how their actions can positively affect herons. Wonderful community action and education programme based on heron conservation are developing. These and other advances are discussed in much more detail in *Heron Conservation*.

In *Herons*, it is our intention to summarize the status of biological knowledge of the herons of the world and to point out questions and areas where observations and research are still needed. It is hoped to encourage the next generation of heron observers and biologists to go forth and learn interesting things about these intriguing birds. Challenges and controversy are not avoided, so as to inspire additional study and debate. If 20 years from now our knowledge of herons has advanced to the same degree shown in the past 20, this book will have accomplished its goal.

James A. Kushlan
Annapolis,
Maryland, USA

James A. Hancock
Sparsholt,
United Kingdom

Acknowledgements

The authors express their thanks to our many colleagues and companions in the study and conservation of herons, whose work, published and unpublished, constitutes the core content of this book. This book is based on the hard work and astute observation of hundreds of colleagues and is a product of the Herons Specialist Group.

We thank especially the many colleagues have generously shared new information, provided critical evaluation, or reviewed early drafts. Their contributions were extremely helpful. These include: Ghisselle Alvarado, George Archibald, Simon Aspinall, Neil Baker, Richard Banks, Joanna Burger, Stuart Butchart, Robert Butler, Bosco Chan, Xiaulin Chen, Mike Crosby, Ian Davidson, William E. Davis, Robert Dickerman, David Diskin, Tim Dodman, Zhou Fang, Mauro Fasola, John Fellows, Jacques Franchimont, Justin Gerlach, Doug Harebottle, Floyd Hayes, Peter Holmes, Roger Jaensch, Michael Jennings, M. Kaiser, Richard Lansdown, Max Maddock, Loic Marion, Kevin McCracken, Heather McGuire, Eric Mellink, Tibor Mikuska, Ma Ming, Gonzalo Morales, Michael Moser, Lourdes Mugica, Taij Mundkur, Martin O'Brian, Fabio Olmos, Storrs Olsen, Richard Paul, Olivier Pineau, Hugo Rainey, Fredrick Sheldon, Subbu Subramanya, Melanie Steinkamp, Maris Strazds, Sephanie Tyler, Christophe Tourenq, Jack Tordoff, Hishey Tshering, Donald Turner, Janine van Vessem, Clair Voisin, Stefano Volponi, Captain Lun Cheong Wong, Robert Williams, Glyn Young, and Lew Young. We especially acknowledge the many contributions to this project of Heinz Hafner, who also co-edited the companion volume, *Heron Conservation*. Of course the authors, not our conferees, remain responsible for interpretations in this book.

We gratefully acknowledge those who assisted in the technical preparation of this volume: our editors Lisa Begley, Stephen Benaim, Abbie Headon, Cathy Kennedy, Judith May, and Christopher Perrins, whose guidance throughout this long project are much appreciated, administrative assistants Bonnie Knight and Marilyn Whitehead, who provided much essential help, and librarians Lynda Garret and Wanda Manning. Kinard Boone, created the range maps.

We acknowledge institutions that provided support during this project, including the University of Mississippi, the National Biological Service, The US Geological Survey, Smithsonian Environmental Research Center and U. S. National Museum of Natural History. We appreciate the support of Denny Fenn, Sue Haseltine and Ross Simons. The Heron Specialist Group has been sponsored by The Species Survival Commission of The Conservation Union (IUCN), The International Waterbird and Wetland Research Bureau now Wetlands International, and Station Biologique de la Tour du Valat. We appreciate the support of Michael Moser, Janine van Vessem, Tunde Ojei, Luc Hoffmann, and Jean-Paul Taris.

The identification plates and bird topography figure are the work of David Thelwell. Photographs are by James Hancock, except as follows: John Bildahl, 5.4, 5.5 a–c; Patricia Caruso, Figs. 5.1c; Sylvia Hancock, The Authors; James Kushlan, The Illustrator, Figs. 3.4, 5.1 except as otherwise noted, 5.2, 5.5d, Plate 13.3, Plate 18.1; Terry Master, Plate 14.4, Plate 16.4, Plate 17.5, Plate 18.3; Eric Mellink, Fig. 4.7; Christopher Perrins, Fig. 2.4; Rahav Segev, Fig. 5.5; Melanie Steinkamp, 5.1i; Eric Vidal, 2.5, Plate 18.5.

As this book went to press, James Hancock passed away at his home, Jollers in Sparsholt, England. This followed by only a few months the passing of Heinz Hafner, at his home Petit Badon, in the Camargue, France. These great colleagues were two of the persons most instrumental in guiding, over the past two decades, the advances in heron biology and conservation that are presented in this book and its companion volume, *Heron Conservation*. In gratitude for their years of friendship and collaboration, *Herons* is dedicated to their memory.

Contents

List of Colour Plates

Colour plates fall between pages 66 and 67.

Plan of the book and notes on the species accounts

PART I

This monograph is divided into two parts. Part I consists of an introduction to the herons followed by five chapters providing more details on special topics including systematics and evolution, feeding ecology, nesting biology, herons and people, and conservation. This part aims to provide generalized information on the family.

PART II

Part II consists of 62 species accounts. Information on each species is presented within a standardized format, further discussed below. The species accounts differ in the amount of detail provided, reflecting the information available on each species.

Nomenclature

The accounts begin with a recommended English name and the valid scientific name for the species. The English name was chosen to reflect standard names used internationally and is followed by a listing of other common names used in various languages. The bibliographic reference to the source of the original description and type locality follows. Finally, there is a listing of the scientific names of subspecies recognized in the present book. When subspecies are not recognized, the species is considered to be monotypic.

Description

This section begins with a short summary statement of the species' overall appearance. The adult male is then described in detail, starting with non-breeding plumage. Generally description starts at the head, bill, and face followed by the upper sides, undersides, legs, and feet. Our intention is to describe the normal and usual plumages of the species rather than provide detailed coverage of alternative plumages, moult patterns, or annual variations. Descriptions of plumage and skin tint use simple colours, compounded as needed, realizing that verbal descriptions of coloration can only be indicative of actual appearance. The plumage of some herons is highly complicated and virtually impossible to describe in the space available. Subtle colour differences, intricate shading, barring or spotting are described in general terms. Each species is illustrated in colour in the accompanying Plates. After the general description, appearance changes that occur annually are described, particularly those associated with courtship and breeding. Variation between genders is noted, especially in plumage or size. Then known variation due to geography is described, whether this variation has been recognized in subspecific taxonomy or not. The appearance of juveniles, particularly in plumage and soft-part colours, is described next followed by a description of the chick, if known. The voice is described next. Calls and songs are given names that are capitalized and placed in quotations at their first use. Call names are onomatopoetic, unless other sorts of names were previously used in the literature. When known or suspected, the functions of calls are given, such as advertising, flight, disturbance, aggression, or greeting. Other sounds, such as those made with the bill, are also noted. The result is the beginning of a catalogue of heron vocalizations, which we hope will be a useful start to further analysis. Our understanding of heron vocalizations is in its infancy. Recording and cataloguing calls and songs, determining their behavioural contexts, and exposing their evolutionary history are important to the advance of our understanding of heron biology. The Description section ends with weights and measures, given in metric units, where possible as a range. Much additional work on the cataloguing of heron size information remains to be done.

Field characters

This section gives the characteristics that distinguish a species from other herons or other birds with which it might be confused. The section begins with a statement of identifiable field characters including flight. This is followed by explanations of how the species may be distinguished from similar-looking birds.

Systematics

This section identifies the species' closest relatives. It also discusses outstanding or historic issues of generic, specific, or infra-specific systematics, especially geographic variation.

Range and status

This section starts with a short overall description of the range of the species, generally at continental and subcontinental scales. This is followed by a detailed description of the breeding range, generally described at a country scale, although for large countries it is sometimes described at the province/state level. Place names follow the *National Geographic Atlas of the World* (Revised sixth edition, 1996, National Geographic Society, Washington, DC). For polytypic species, the overall range is partitioned among the subspecies. This is followed by a description of how the range changes in the non-breeding season as a result of postbreeding dispersal, population shifts, or true migrations. Migration is then described, on a flyway basis, when this information is available. A range map is provided for each species. Much has been learned about heron distribution since the *Herons Handbook*, and this knowledge is reflected in the new range maps. The maps show the range divided into areas used during the breeding season (black), areas used during the non-breeding season (light grey), and areas used in both seasons (medium grey). However, it is that the case for many temperate-breeding species the non-breeding range is not necessarily synonymous with the wintering range. Extra-limital occurrences are not shown on the maps, but are noted in the text. Maps are based on the best information available. New and published information is cited in this section. Much of the detail of current heron distribution has not previously been available, and personal communications from regional experts are also cited in the text. Caution should be used in interpreting the maps. First, much remains to be known about the details of species occurrences. This is especially the case for species shown as having expansive, generalized ranges. Secondly, at the scale and grey-tone format at which the maps are presented many details are lost. Thirdly, herons are habitat dependent and may be expected to occur only in their restricted habitat within their overall range. It is hoped that corrections to the maps will be brought directly to the attention of the senior author so that future maps, especially those made for conservation purposes, may be updated. Finally, in the Range and Status section the range-wide status of the species is summarized. Where possible sizes of various populations are given quantitatively, and where this is not possible in descriptive terms. The trend of population change is noted where known or inferred.

Habitat

The habitats used by the species are noted briefly. Generally, the characteristic vegetation (e.g. forest, marsh, pasture), natural habitat feature (e.g. coast, islands, stream edge), and artificial habitat (e.g. rice field, fish pond) are used to describe habitat use. Species of plants used are noted where appropriate. Habitat used for feeding, breeding, roosting, and wintering are noted where these are known to differ from the general habitat requirements.

Foraging

Foraging ecology is described in some detail beginning with feeding behaviour. Names for specific feeding behaviours, as defined in Chapter 3, are used to list the behavioural repertoire and to characterize the feeding behaviour of each species. Such foraging characteristics as nocturnality, sociality, and feeding interactions are particularly described. Then food habits are briefly noted. These are generally listed by general type of prey using common names (e.g. fish, insects, crustaceans, dragonflies) and illustrated by representative genera of the prey listed.

Breeding

The account of the breeding biology of each species begins with identifying the nesting season. This may be in general terms, such as spring or the rainy season, or by months covered by breeding. When seasons are noted alone, they refer to the appropriate months in the northern or southern hemisphere, depending on the area being described. Then nesting habitat is described, generally with regard to physical structure, vegetation, or placement within the landscape. The account of actual nesting begins by noting the breeding sociality of the species followed by a description of the nest site and nest. Breeding displays are then described starting with the advertising displays of the male. The names for specific breeding behaviours, as defined in Chapter 4, are used to list and describe the behavioural repertoire. Attention is paid to how courtship or breeding behaviours have elements distinct from the general descriptions. The eggs are described relative to their colour, dimensions (range of length by width in millimetres), clutch (usual range and extreme range), and laying interval. Incubation period, hatching, hatching success, and early care for the young are outlined. Growth and maturation of young and fledging time are then noted.

Population dynamics

Little is known about the population dynamics of nearly all species. Data on the age of first breeding and mortality are provided when known. It is hoped that the paucity of information in this section will emphasize the need for such information to be collected.

Conservation

The conservation status of the species, including listing status, is noted. Important conservation issues are then discussed.

Research needs

Important deficiencies in the knowledge of each species are noted with recommendations for appropriate research emphases.

Overview

The species account finishes with a generalized comment on the species, emphasizing its overall way of life.

Bibliography

This volume is based explicitly on over 1500 citations culled from over 5000 references. In citing literature, we have concentrated on documenting the advances in our understanding since 1984 and so we cite few of the papers previously cited in the *Herons Handbook*. Relative to the literature, the *Herons Handbook* continues to stand alone as a primary and secondary reference up to its date of publication. We urge the reader to consult it as a starting point for earlier information. We also have made available via the Internet a more complete bibliography on herons at (Kushlan, J. A. and Garrett, L. 2004, A Bibliography of the Herons). This bibliography incorporates a selected core of the heron literature to date. We hope it proves a useful tool.

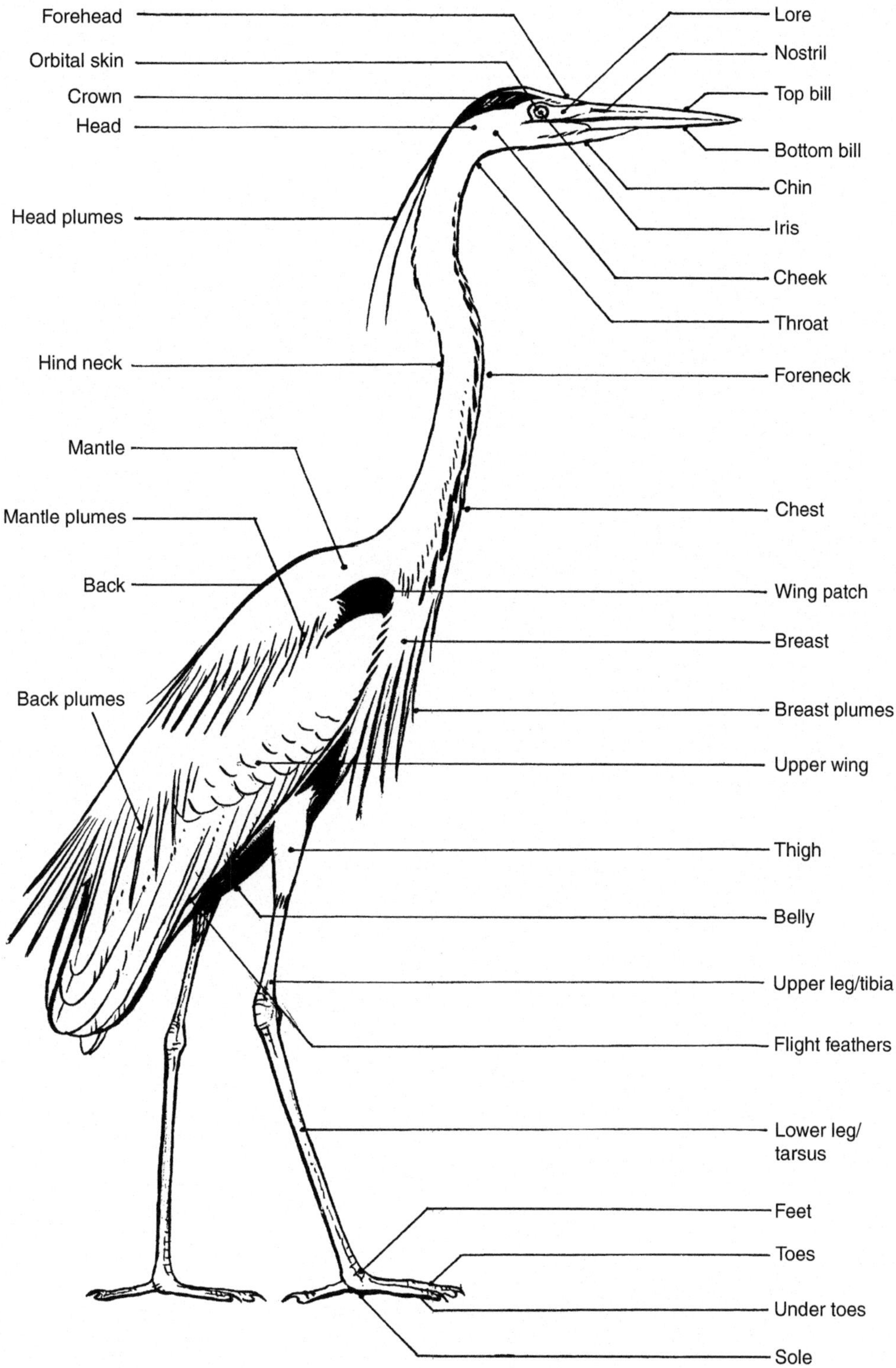

Drawing of a heron to show the topology of the bird identifying the body parts mentioned in the text.

PART I

General chapters

1

Introduction to the herons

Herons are recognized the world over. Their distinctive sleek figure, emphasized by their long legs and neck, standing erect against a wetland backdrop or perched in a tree at sunset, is a well-appreciated sight. Worldwide they are known for their tolerance of humans and, for many species, adoption of the human landscape. These birds are called by many vernacular names, some aligned with their taxonomy such as herons, egrets, or bitterns. Other names reflect a more local perspective, such as quawk or shitepoke. Local names such a 'The Heron' of England, the 'American Egret', the 'Tick Bird' of Africa, and the 'Blue Crane' of Australia show they are well integrated into their neighbouring human cultures.

These are birds of intriguing contrast—of remote and inaccessible swamps but also of backyard fishponds and motorway margins. Herons nest in deep forests and swamps, hidden away far from human contact, but also in neighbourhood parks and zoological gardens. They breed in massive colonies but also in solitary pairs. They are beloved by many as a comforting companion in the solitude of natural world. But commercial fishers and fish farmers despise them as a threat to their livelihoods.

In the contemporary world, the image of a heron increasingly is used as a symbol of wetland or coastal protection and of other conservation movements. Given their cosmopolitan distribution, the 62 heron species have a special potential to represent the value and role of wetlands and other wild places.

1.1 Goliath Heron, the largest of the living herons.

General appearance

What makes a heron? Familiar 'typical' herons all seem much alike. They are tall and slim, with long pointed bills, a long neck with a kink in the middle, broad slow-beating wings, long legs, long toes, coloured in basic black, white or grey. This in fact well describes the majority of the typical herons, and understanding these features goes far to understanding what herons are all about. Although this model heron does not do justice to the subtle diversity of morphological adaptations of herons, nor of the range of body types within the family, it does serve as a starting point for understanding the functional morphology of herons. A typical heron is shown on p. xvii, with the body parts labelled that are usually mentioned in the species accounts. Where possible these are kept non-technical.

The Bill

As is the case for many bird families, the heron's bill may be its defining morphological element. Most are at least modestly elongated; few bills are short and truly stubby. These are not seed-cracking, meat-tearing, or flying-insect-capturing birds. They are fundamentally fish-catching and invertebrate-gleaning birds, with bills of various species mostly falling along a continuum from highly specialized fish catchers to more generalized gleaners of terrestrial insects.

The Agami Heron is the ultimate expression of the long-billed heron. Interestingly, it may be representative of a very early lineage of herons. Tricoloured Herons are perhaps next in line. In general the longer and thinner the bill, the more the species is a fish eater (as shown in the excellent series of papers by Boev). This sort of bill seems an optimal complement to the rapid strike needed to catch fish, which are protected by their size,

1.2 Bill of the Green Heron, a typical heron bill used for fish catching.

1.3 The long and thin bill of the Tricoloured Heron, specialized for catching fish.

slipperiness, mobility, the refractory characteristics of water, and their own antipredator devices such as spines.

Shorter bills are better adapted to picking-off smaller, shorter, and terrestrial prey. The Cattle Egret is a good example of a heron with a relatively short bill. The *Gorsachius* night herons have the smallest relative bills, for reasons that are not clear. They probably indicate terrestrial habits. For catching resting insects it is not the long strike that is important but the pick, glean, or grab. These appear to be better done with a shorter feeding apparatus.

Heron bills can be similarly contrasted along a continuum between thin and thick. Thin bills are like daggers or even rapiers. They can penetrate the water while minimally disturbing the surface during the strike. Thinness is an adaptation for fish eating, not only capturing small fish but impaling larger ones. Birds that feed on small fish have thin bills (such as most of the medium *Egretta*). Birds that feed on large thick fish need to have stronger and therefore stouter bills (like most of the larger *Ardea*).

Herons grabbing insects out of the air, gleaning them from plants, or taking crustaceans off a stream bottom or submersed rock do not need a penetrating bill. So their bills can be thicker and this facilitates the handling of hard skinned or shelled prey. Insect-eating bills, such as the Cattle Egret's, are thicker than those of similarly sized aquatic herons.

1.4 The stout bill of the Yellow-crowned Night-Heron, used for catching and handling crabs.

Crustacean-eating herons have even thicker bills. Some populations have bill thickness attuned to the local diet. The best example may be the Yellow-crowned Night-Heron, a crab and crayfish specialist. Bill thicknesses of populations of this species differ depending on the size of crabs that predominate in the local diet. In the Little Egret complex of populations, coastal 'reef heron' populations have discernibly thicker, more down-turned bills than do inland populations. These differences in bill thickness tend to be subtle; no heron has a crushing bill of the sort found on birds that crack hard objects, such as seeds.

The thickest and widest bill, of course, is that of the Boat-billed Heron. For many years this outlandish bill was a great puzzle. Despite its massiveness, it is not particularly strongly designed. In this it differs from a similarly endowed species, the Shoebill Stork, which uses its huge bill to catch large fish. The Boat-billed Heron, at least today, does not eat large prey. The mystery of this bill may have been solved in the discovery that the bill is used for Ploughing feeding. The Boat-bill's bill increases the surface area to capture prey by touch, rather analogous to the bill of spoonbills (*Platalea*).

Herons capture prey only with their bill. There is no evidence of their use of feet for this purpose in the way a raptor would. Small fish and many larger fish are bitten not pierced. The bill of a fish-eating heron is best considered a very long, very sharp pair of forceps or tweezers. The bill edges and tongue typically are slightly serrated to help hold slippery, struggling prey. Large or laterally compressed prey may be stabbed through by one or both of the slightly opened mandibles.

Once caught, a small fish is swallowed immediately by a forward lunge of the head. The fish goes down the gullet as the bill releases its hold. However, if the prey is large, squirmy, long, or defensively attired, a fish may have to be subdued. It may be released and recaught in the air, bitten, or if a platform is available, placed there and stabbed. Large fish can take a long time to 'handle', which affects their profitability in two ways. While a heron is attending to bashing its prey, other birds—larger or smaller—can steal it. Also time spent handling is time not spent catching more food. So the capacity of specific bill morphology to handle prey is as important as its capacity to catch the prey in the first place.

For most of the year heron bills are usually yellow or brown, or a combination of brown/black above and at the tip and yellow below. During this time of year, colour of the bill has to support prey catching. Sombre tints and bicolour patterns probably help disguise the waiting bill from prey below. However, the large heron bill is too good of a signalling device not also to play a prominent role in communications. Heron bill colour is subject to seasonal variation, turning shinier, brighter, and often redder in the courtship and breeding season. In certain species, such as the Reddish Egret, bill colours are especially highly developed for signalling. Descriptions of bill colour in the literature are fraught with contradictions, because of individual, seasonal, and developmental changes and also because of observational variability—people see and report colours differently. Much remains to be done in documenting the range of heron bill colours photographically.

Eyes and vision

The eye and its positioning are critical elements of heron anatomy. For most species the binocular field of vision is long (about 170°), allowing vision over almost the full range of height in front of the head, and narrow (about 20°), providing panoramic coverage only in a narrow field in front of the bill (Martin and Katzir 1994, Katzir and Martin 1998). With its bill horizontal, the most typical feeding posture, the bird can see its own feet with binocular vision. This, of course, is functional in seeing fish below the line of the bill in the water. By redirecting its vision slightly a heron can see birds flying above it, a protective feature and one essential for aggregative species that need to care about incoming competitors. It also explains the visual functionality of the Bittern posture. Clearly the principal role of the heron's highly directed binocular vision is to precisely guide the rapid bill thrust toward mobile prey. The narrow binocular field explains the repertoire of head and neck feeding movements, such as Head Swaying and Peering Over, used to help calculate the precise location of prey.

Most of the heron's vision range is monocular. Furthermore, by moving their eyes, herons can abolish their binocular overlap, more so than would be expected from the retinal field alone. In doing so herons can gain a wide scope of vision around their head, lacking vision only in an arc of about 45° to the rear. Visual acuity would not be great, but precise vision is not required in these quadrants where the function is to detect the presence of threatening objects and movement.

A serious consideration for a bird standing with its head in the air but intending to catch fish under the water is refraction, which displaces the apparent position of an item underwater relative to its real position. The prey strike starts out of water and, at that moment, where the heron sees a prey item underwater is not where it actually is and certainly not where it is going. Aquatic herons are able fully to accommodate for refraction (Katzir and Intrator 1987, Lotem *et al.* 1991). Diurnal fish-eating herons appear to have the best abilities, whereas, as might be predicted, Cattle Egrets have less of an ability to make this accommodation (Katzir *et al.* 1999). Perhaps even more interesting, Black-crowned Night-Herons become less proficient with increased prey depth or distance. Feeding at night, it is likely that night herons can only see, and therefore only need to attempt to catch, prey that are rather near at hand and near the surface. To the extent that diurnal herons are evolutionarily conservative, secondary adaptation to nocturnal or terrestrial foraging appears to have resulted in degradation of the ability to accommodate to refraction.

Heron eyes are relatively large, more so in night herons and huge in Boat-billed Herons. Nocturnal species have better night vision than diurnal species, such as Cattle Egrets (Rojas *et al.* 1999). The large eyes of night herons appear to increase vision over a range of light conditions, although they do not, as might be expected, increase their binocular vision (Katzir and Martin 1998). Species that feed passively have a greater potential visual acuity than do active-feeding species (Lish 1983). Such standing-feeding birds need to see subtle differences in the appearance of objects and detect small movements of potential prey.

Eyes are also used for communication. Iris colour changes with growth, maturation, and season, in many species becoming part of the breeding coloration of herons. We are now learning that in some herons during courtship and intense interpersonal encounters, irises can quickly flash to a different colour indicating changes in the bird's mood and motivations. As observed so far, such iris changes—usually to a red or orange—indicate and communicate extreme excitement such as that involved in threats, close-hand courtship, and copulation. Transient eye colour changes represent a most interesting area for future research in documenting the correlation of rapid iris colour change with behaviour and motivation.

Head and neck

The face typically has unfeathered portions, the most usual being between the eye and the bill, called the lores. Unfeathered skin may also extend onto the lower bill, around the eye (orbital skin), or as a line diagonally from the bill. This skin is coloured in ways that are distinctive for species, population, gender, or season. Hancock and Kushlan (1984) established loral colour as one of the important biological and taxonomic landmarks of herons, and since then much has been written and thought about this feature.

Some colour change is due to pigmentation of the skin, which lasts on the order of weeks. Some is due to blood flow, the expression of which may be under autonomic control. Like the irises, in some species the lores can flash a colour change in seconds to minutes. The physiology of loral colour and colour change is not understood and quantitative data are few. The study of loral coloration suffers from differences in the way individual observers see, record, and communicate colour as well as how the colours can change with emotion and post-mortem. We remain convinced this is an important element of heron biology that needs study. Photography has been an important tool (see Hancock 1999 for an example). Discussion has particularly focused on the potential importance of loral colours in understanding and identifying the inherently difficult white herons, which lack the easy plumage identification characters of multicoloured species.

As with the bill, irises, and lores, other parts of the heron's head are hugely important for social signalling. Heads are often distinctively marked, such as in the Purple Heron. Many heron species, particularly *Egretta* and *Ardea*, have long head plumes. Especially important is the crest found on many species, so some extra care has been taken in attempting to describe the head, crest, and head plumes in the species accounts in this book. In some species, such as the Boat-billed Heron and Whistling Heron, the crest and plumes become truly impressive structures. But more subtle crests are also important for other species. Many species, even those lacking pronounced crests as an adult, have elongated crest down as nestlings, presumably to aid communication with parents and siblings. The crest is not a uniform structure. Studies of the Cattle Egret have shown that erection of different parts of the crown signify different motivations, particularly threat or sexual (see Voisin 1991 for good illustrations). The crest is used in both situations, but it probably fundamentally is part of the threat system. It is used in responding to disturbance and in communicating presence and threats, in part increasing the size and impression of the head and emphasizing the bill, which is the body part that will carry out the threat. Herons do not have featherless heads, as do many other large aquatic birds such as some storks, ibises, spoonbills, and cranes.

The heron's neck is one of its fundamental adaptations. Prominent in long-necked species, the neck kink is a distinctive feature of the family. The kink is due to the elongation of the sixth cervical vertebra. In a forward strike, the bill, head, and upper neck travel forward with this vertebra functioning as a hinge. The muscles of the neck are aligned to create the rapid fish-catching strike of some species. It is likely that the distinctive neck morphology was an element in the long-term success of the heron family. The rapid strike is one aspect of neck morphology, but the thin bill and elongated neck also need to withstand the impact of a strike. Herons have complex ligaments that provide locking

1.5 Pied Heron showing the variably elongated neck, characteristic of herons.

mechanisms that minimize torque during the strike (Hale and Homberger 1993). Herons also are able to hold their head still while moving the rest of their body, a feature important for stalking (Carmi *et al.* 1995).

Species with shorter necks, such as night herons and bitterns, do not make use of the lightning bill strike but rather lunge the bill forward—neck, body and all. In these species strength may be more important than the long strike since much of the weight of the bird is behind the lunge. And in these species, the neck is thick and muscular, and the heron has a fundamentally different appearance than do long-necked species.

Once caught, a prey item needs to be swallowed whole. Herons lack some of the more usual morphological adaptations for enlarging the gullet, such as a flexible mandible, loose mandibular articulations, or severely reduced tongue. Instead, to increase the size of the gullet, muscles protract and retract the hyoid apparatus together with the walls of the neck and gular region allowing a disproportionate expansion of the gular region (Cummins and Homberger 1986, Homberger 1989). That the tongue of herons is not reduced, as might be expected, is unexplained. Perhaps it is important in securing the prey in some way.

The long neck, although useful for feeding, would become a bother in flight. Rather than having strengthened neck bones and musculature to hold the head and neck extended, the long-necked heron flies with its neck coiled back in an 'S' shape, bringing the head inboard. This posture is so characteristic that it can be used to identify flying herons at a distance.

Plumes

Herons use their feathers as signalling devices. Plumes and other feather oddities that characterize the family are distinctive among species. The head (or crown), as already noted, the lower front of the neck (or breast), the base of the upper neck (or mantle), and the back (or scapulars) are the primary plume-bearing areas. Bitterns have tufts of feathers on either side of their breast. In some species, plumes can be totally erected, but in others they can be moved only modestly. Heron plumes are of distinctive structure. Lanceolate plumes are rather like typical feathers but have edges with few barbs, giving them a frayed appearance. Aigrettes have long shanks with long elements and minimal barbs, such that they are loose and flexible. Filoplumes are long and hairy.

Lanceolate plumes occur particularly on the breast and also on the crown. Such crest plumes are particularly characteristic and appear to function importantly in communication and probably assessment of mate quality (Krebs 1991). These plumes are characteristic of night herons and some *Ardea* and *Egretta*. Filoplumes are characteristic of pond herons and, especially, the Reddish Egret. Aigrettes are characteristic of the *Egretta* and *Ardea* and are especially developed in the Snowy Egret, Intermediate Egret, and Great Egret.

For many years it was known without fear of contradiction that these plumes were used for courtship. But as courtship of herons has become better understood, it gradually dawned on observers that the plumes were not playing a role in pair bonding displays commensurate with their extravagant development. Plumes are now understood to be used principally to enhance threat and other antagonistic displays. They accentuate the heron's aggressive and threatening appearance, particularly Forward displays. In retrospect, it should not be a surprise that herons use plumes to enlarge apparent body size, reflecting a common theme in animal behaviour.

1.6 Cocoi Heron showing distinctive bill and head colours of the courtship period.

In fact appreciating the role of plumes in aggressive display suggests just how important aggression, particularly territorial defence, is for herons. All species defend their nesting sites. If these sites are not sufficiently defended, reproduction is not possible. Herons have long, sharp bills, a well-developed stabbing ability, and also long sharp toes. They are known to have killed each other. The interactive behaviour of herons has evolved in a way to reduce actual fighting as they engage in one-on-one aggression. The defending bird is generally accorded its dominance by the intruding bird. Display plumes maximize the aggressive and defensive displays, allow shades of motivation to be communicated, and permit communication at a distance, sufficient to moderate interpersonal interactions in most circumstances.

Threat displays do function in courtship, variably among species. Likely, females can interpret threat-based displays as information on male quality, given that nest site defence is one of the primary responsibilities of the male. Furthermore, these displays are the initial communications between perspective mates, as the female attempts to enter the defended display area of the male. The female, having much the same display plumes as the male, approaches him deferentially with her feathers sleeked down, the contrary position to an aggressive bird. So plumes have a critical function to play in breeding and courtship, but not so prominently in displays that have predominantly sexual meanings.

Plumage

The body feathers of herons are aligned in tracts, with down occurring only between the tracks. One type of down, nearly unique to herons and a familial characteristic, is powder down. The down occurs in tracts on the lower back and breast, and in some species on the sides and along the upper legs. Typical for the family is three pairs of down patches. Bitterns have two; the Boat-billed Heron has four. This indeterminately growing down is friable, breaking off easily when rubbed, bitten, or combed. The preening heron will do all of these things, rubbing its head, nibbling with the bill, or

1.7 White and dark phases of the Eastern Reef-Heron.

combing with its toe. The latter is literally the case. The longest, middle, toe has a serrated nail, called a pectinate nail that is used to comb powder down into the feathers. The down likely functions in waterproofing, partially replacing the function of the oil gland. Perhaps the use of powder down is what allows herons to have retained fully feathered heads and necks, despite their aquatic, fish-eating way of life. In some species the down may alter apparent feather colour, at least during courtship, a possible function that requires additional study.

The general body coloration in herons is black, white, grey, and brown. Some are entirely one of these colours. Most species have a very complex plumage design; and, for some complex-plumaged species, geographic and individual variation can be extensive. It has been hypothesized that the primary correlate of base plumage colour in these birds is conspicuousness to other birds and to prey (Kushlan 1978a). Some species need to be seen by others, some would rather not. All herons need to not be seen by their prey.

The all-white herons are on the simple end of the plumage range. It has been theorized that a principal function of white dorsal plumage is to facilitate a high degree of sociality (Kushlan 1978a). Most white species are aggregative foragers, implicating plumage colour in local facilitation, i.e. the gathering of birds at feeding sites. White birds are easily seen from the air, and several studies have

1.8 White plumaged juvenile of the Little Blue Heron.

1.9 White necked juvenile of the Pied Heron.

shown that white birds form the core in multi-species feeding aggregations. That white birds are all social is not universally true, of course, the Intermediate Egret being the best counter-example. Additional studies are needed. An insufficiently explored hypothesis is that birds with both white and dark forms are differentially social. Adult and immature Little Blue Herons provide an age-related model; pond herons provide a seasonal model; the dimorphic herons provide geographic examples, all amenable to study.

Herons that are white on their under sides show less contrast from below than birds with dark ventral plumage. So these birds are less conspicuous to prey in the water, and have been shown to frighten fish less than dark-bellied herons.

There are several dark egrets. These birds are decidedly not highly visible and generally speaking are less social. These are often birds that feed alone by methods that are not helped by being near other birds. In mixed species aggregations, the dark herons tend to position themselves along the outer periphery. So they are attracted to a productive feeding site by the white birds, but tend to stay out of the intensely competitive and commensal relationships taking place in the aggregative core. The best example of an all-dark heron is the Black Heron, which feed typically well dispersed even when aggregated in good feeding sites, the birds tend to be well dispersed, as would be required by their mobile feeding behaviour. Likely the same situation holds, more subtly, for other species.

Mixing black, grey, and white colours, sometimes overlaid with buff, chestnut, or maroon, is a very typical pattern for herons, a number of which vaguely follow the pattern of a Grey Heron. This mixing tends to produce dark backs, variably distinctive usually cryptic necks, and light or cryptically striped undersides. It is likely that the plumage of such species is cryptic—sky colour when seen from below and substrate colour when seen from above. The Tricoloured Heron is the poster-heron for the dark-backed, white-bellied, striped-necked heron. Based on the hypothesis offered above, one would theorize that this would be a solitary species (the dark back) that feeds with a need to reduce contrast when viewed from below, i.e. likely not under deep cover (the white belly), and a predator on escapable prey with a need to disguise its neck and head (the cryptic throat).

There are many variations on the theme of black, grey, and white. Several forms are dark with white chins or throats, which feature in display. The White-faced and White-necked Herons have white as indicated by their names and are easily identifiable. More complexly, the Black-crowned Night-Heron is a study in black and white—the black on the back reduces the visibility of the solitarily feeding bird from above, whereas the white belly reduces the visibility against a backlighted night sky.

More typical camouflage coloration is also well developed in the family, derived from contrasting brown, black and light stripes, spots, and vermiculations. As noted above, some of the typical herons tend to be cryptically coloured below with vertical lines in brown, black, and white on the throat and foreneck extending down onto the belly. The function is clearly a disguise from prey below. This coloration is mostly found on herons that typically feed by Standing or Walking very slowly. Other species are entirely cryptically coloured. Herons coloured this way are solitary species using dense vegetation. Entire groups show this plumage, particularly the solitary marsh herons (bitterns) and the solitary forest herons (tiger herons). This cryptic brown, black, and white complex of feathers together creates patterns of bars, stripes, speckling, spots, and patches. The plumage is hard to describe and in shaded cover just as difficult to see.

One of the more interesting aspects of heron coloration is the existence of colour forms of the same species. In only a few cases, notably the small bitterns, are male and female plumage markedly distinct. In most species the female is merely a somewhat duller-coloured, smaller version of the male, the slight difference perhaps functioning as a way of further decreasing aggressive responses of males. More species have a distinctive juvenile plumage. Most of the night herons, the *Butorides* herons, tiger herons, and the Boat-billed Heron have distinctive juvenile plumage, which in all cases is more cryptic that that of the adult. The functionality of this phenomenon has not been well studied. It is possible that cryptic coloration reduces conflict with adults over feeding areas. The oddest juvenile plumage must be that of the Little Blue Heron. Its juvenile is entirely white. The function of the difference between juvenile and adult Little Blue Heron has not been adequately explained.

A number of species have adults of different colours. These populations occur mostly along ocean coasts. Dimorphic herons include a coastal race of the Great Blue Heron, Little Egrets along the coasts of Africa and Asia, Reddish Egrets, and Eastern Reef Herons. Colour morphs persist in populations at differing ratios, presumably owing to differing selective pressures. Some evidence suggests that the different colour morphs can use different feeding techniques or tend to choose different habitats (Rohwer 1990). It is likely the less

1.10 Little Egret (*Egretta garzetta dimorpha*) a bird intermediate between the white and dark phases.

contrast of white birds in the highly reflective tropical marine environment enhances feeding success. However, such plumage also reduces the costs of thermoregulation (Ellis 1980). Odd colour forms have been reported in other species. Individual and geographic variation in plumage also occurs. The most geographically diverse species is the Striated Heron, with subspecies scattered on continents and islands, including two morphs on a single island group.

A colour pattern not found in herons is white body and black flight feathers, a combination occurring in several other groups of large aquatic birds (cranes, storks, ibises, and pelicans). The lack of dark flight feathers is likely associated with the limited soaring done by herons. Apparently herons do not need to have exceptionally strong flight feathers (derived from the high melanin content).

Legs and feet

The legs of herons are thin but strong and typically long. Legs are among the earliest developing body parts, which enables nestlings to scramble from the nest and climb in trees at an early age. Leg length does differ; the Least Bittern hardly comes to the 'ankle' of a Goliath Heron. The relative size of upper and lower leg also differs. Birds that wade have relatively long lower leg bones. Species with relatively smaller lower legs (vs. the upper leg) tend to be those that fish from overhanging vegetation

1.11 Grey Heron, long legs and long toes are characteristic of typical herons.

1.12 Snowy Egret's contrasting yellow feet.

or riverbanks, such as the Agami Heron. They strike by lunging forward on the hinge formed by their upper and lower legs.

Heron feet also differ according to their way of life. The relatively shortest toes are found on species that habitually walk on dry land, such as the Cattle Egret. The relatively longest toes belong to the bitterns, which use them to clamber through and over reeds. The moderate length toes are in species that wade. Feet have light webbing between the toes. Herons are not swimmers by preference. Should they land in the water they can swim out, and sometimes choose to feed this way. The pectinate toenail has already been described, the toes being used as grooming tools and sometimes for fighting in the air.

Flight and movements

The wings of herons are relatively long and broad, with 10–11 flight feathers, except the Boat-billed Heron. The tail is rather short. For most species the wings and tails are built for long flight rather than for quick departures. Although perfectly capable of rising from land or water, large herons do so with apparent effort and fly with legs dangling and head stretched out for some time. Smaller herons can take flight easier. Bitterns, for example, take flight with more rapid wing beats and drop quickly back into cover. Large herons are designed for continuous, economical flights, using the tail for steering, whereas small herons are more designed for rapid, short flight (Boev 1986).

Once in straight flight, the broad wings come to their own and herons can fly long distances with little apparent effort. The flight of herons tends to be slow and deliberate with shallow, bowed wing beats, not descending much past horizontal on the down stroke. Some species do have faster beats, such as night herons and bitterns, and so have commensurately narrower wings. Flap rates vary from nearly 200 beats per minute in small herons, 120 in the large herons, and about 100 in the great herons. The large herons are capable of soaring, but tend not to. Larger herons glide easily. Somewhat surprisingly, herons are capable of aerobatics on occasion, including zigzaging and plummeting from heights. They also can feed on the wing, usually accomplished by medium-sized species and less

1.13 The Snowy Egret in full aggressive behaviour, crest raised, neck feathers expanded, and lores flushing red.

1.14 Snowy Egrets fighting in the air.

frequently by large and small herons. Feeding from the air is energetically costly, and so occurs when food is easily obtained by these methods.

Herons do fly long distances. Larger species may feed 40–50 km from the nesting colony, making daily round trips in order to feed their chicks. Flight distance to feeding grounds appears to be a critical limit to nesting location, and in some regions heron colonies are spaced the equivalent of normal foraging distances. Long flights from nesting to feeding sites affect not only energy expenditures but also pose time constraints during a period when efficient use of time is critical (Smith 1995a).

Most species move around at some time in their annual cycle. Many undertake post-breeding dispersals; temperate species migrate; tropical species shift seasonally in response to changes in weather and hydrology.

Post-breeding dispersal occurs as birds move from their breeding sites. Young birds especially seem to disperse, with distances totalling hundreds to over 3000 km (Geering *et al.* 1998). Dispersal gradually merges into a seasonal migration or to shorter distance population shifts.

Temperate herons tend to migrate just prior to the local winter, sometimes in response to the first frosts of the year. In other species migration can be prolonged, occurring gradually over three to four months. As far as is known, herons migrate alone or in loose flocks generally using flapping flight. Speed has been calculated for Cattle Egrets to be 40–50 km/h. They are seldom observed soaring on migration and can fly both during the day and at night. They often migrate during high pressure or at the edge of a high-pressure area aided by the passage of fronts and wind shifts (Maddock and Bridgman 1992, Bridgman *et al.* 1998). They avoid flying in the face of contrary wind shifts and during rainy and foggy weather. Autumn migration tends to proceed directly, with species following land bridges to minimize crossing ocean gaps and also island hop. But offshore coastwise migration and ocean crossings do occur for some species in some areas. Herons either go around deserts or over them in a broad front, at times stopping at oases. Return migration seems to be more in stages, although there are few documented staging locations of the sort observed for shorebirds. Much more remains to be understood about the details of heron migration.

Tropical populations appear to be more sedentary, and some species may be entirely residential. However, important seasonal shifts occur in the tropics, some of which are only now becoming appreciated. Birds mostly move to an area to breed, taking advantage of local conditions and then migrate to other areas when conditions change. The most common pattern is between inland and coastal areas. These population shifts should be recognized as true migrations. The only differences between these and longer movements is the distance and direction travelled (Maddock and Geering 1994). The migratory movement of tropical heron populations is an important area of future study.

1.15 The broad wings of an Intermediate Egret.

Both postbreeding dispersal and migration account for many of the out-of-range records of herons. Nearly all species, except for the most sedentary, show up on occasion in odd places, much to the delight of birders. These dispersals at times result in significant range expansions. The colonization of the New World by Cattle Egrets, undoubtedly following a cross-ocean dispersal, is one of the more intriguing bits of heron natural history. When in its new continents, the egret took up a typical temperate migratory pattern. Cattle Egrets still show up out of range throughout the world. Equally fascinating is the ongoing invasion of the New World by the Little Egret, which has in the past few years established itself as a breeding species in Barbados.

Vocalizations

Heron vocalization is one of the least studied aspects of heron biology. Long known for their raucous screams when disturbed and similar screeching when nesting or in flight, heron vocal communication has not been examined in sufficient technical detail. There has been little effort to document calls and songs, to provide common nomenclature, and to examine the evolutionary derivations or behavioural functions of heron calls. There is no study of geographic variation or individuality, and the functions of identified calls are only surmised. Cataloguing, interpretation, functional analysis, phylogenetic correlation, and understanding the ecological use of heron vocalization are needed.

As a suggestion of the usefulness of such study, McCracken and Sheldon (1997) have shown that call characteristics such as peak energy frequency and frequency range have ecological correlations. Number of syllables, syllable structure, and fundamental frequency carry phylogenetic information. This breakthrough study shows clearly the information available in heron vocalization.

2

Systematics and evolution

The early herons

The herons are a fairly ancient group of birds. Although bird fossils are rare, herons are exceptionally rare even by avian standards (Bochenski 1995), totalling fewer than 40 identified species. Even the number of existing fossils is a moving target, as the restudy of available material periodically reassigns various specimens to non-heron taxa. Fortunately the lower leg bone of herons, the tarsometatarsus, often the bone preserved in fossils, is quite distinctive (Becker 1985, Olson 1979).

As far as we know, herons first emerge in the fossil record some 60–38 million years ago in the Eocene. The oldest known heron, *Proardea*, is from the Eocene of France. Heron-like tracks, similar to those of Great Blue Herons, have been found in Eocene deposits (Mustoe 1993). Other now extinct heron genera come from the Miocene and Oligocene: Birds attributed to contemporary genera first are recognized in the Oligocene about 7 million years ago, including *Nycticorax, Ardea, Egretta*, and *Ardeola* (Bochenski 1995, Becker 1985). Thus by the Miocene, a period characterized by extensive aquatic habitats, herons rather closely related to modern forms had evolved and had radiated into day and night herons and also into large, medium, and small herons (Becker 1996). These are all birds we now classify as the typical herons (the subfamily Ardeinae). What is unfortunately missing from the fossil record are representatives of the more basal lineages, those that eventually led to such contemporary herons as the Boat-billed Heron, Agami Heron, and the tiger herons.

Fossils are commoner in more recent times. Those from the Pleistocene are often assignable to extant species. Subfossils from islands represent extinct forms, species, or subspecies, particularly of *Nycticorax*. These fossils are illustrative of the ever-changing face of heron distribution in the world and provide hints about contemporary heron distribution. For example, *Nycticorax* had already occupied both hemispheres 135 000 years ago. The widespread *Butorides* occurred much farther north in the Middle East, bridging the gap between extant African and Asian populations. Of interest is evidence for the existence in the Middle East of a heron larger than the Goliath Heron, today's giant among herons. What became of this beast, alive only 5000 years ago (Hoch 1977), is not known but it likely was eaten by humans.

The family of herons

Considering all the analyses and evaluations, there is a remarkable consensus that the herons form a taxonomically unified, rational, and monophyletic grouping among the birds of the world. The recognition of herons as a unified group originated at the beginning of scientific nomenclature when in 1788 Linneaus (1956) grouped all the herons he knew under his genus *Ardea*. With modifications related to what species might be included with the core herons, this monolithic view of herons prevailed over the next two hundred years, including the first definitive evolutionary classification of Huxley (1867)—the foundation of subsequent classic evolutionary classifications—and Wetmore's (1960) historically definitive classification. Some debate persisted around the edges of the family, focusing on oddities such as the Hamerkop, Shoebill, and Boat-billed Heron. It is now fairly well appreciated that the Hamerkop and Shoebill are not herons, but part of a group of birds that includes storks, pelicans, and maybe other water birds as well (Sibley and Monroe 1990, Hancock *et al.* 1992, Sheldon and Slikas 1997). The Boat-billed Heron continues to surprise. Although Wetmore recognized it as very distinctive, more recent morphologically based studies tended to consider it no more than a big-billed night heron. But ongoing molecular study is confirming Wetmore's insight. It is a heron, but basal to the heron family tree, representing a separate lineage from all or most of the other herons (Sheldon *et al.* 1995, McCracken pers. comm.), a position that, given the benefit of hindsight, can be supported by what is known of its behaviour. Considering the difficulty of resolving monotypic lineages at the base of the tree, the Boat-billed Heron is for now best left included within the heron family.

Technically speaking, the heron family, in general, is characterized by having relatively elongated legs, bodies, and necks, long spear-like bills, rounded wings, short tails, a pectinate (comb-like) middle toe nail and strongly differentiated powder down patches, narrow feather tracts, down present in apteria (the space between feather tracks), bare lower tibias (except the Zigzag Heron), (for some) a characteristic modification of the articulations of the 5th, 6th, and 7th cervical vertebrae that cause a 'kink' in the neck, highly developed neck muscles, four long, thin toes on the same plane with one directed backwards and the front three having short webs between them, relatively high capacity for eye movement, 10 well-developed primaries (nine in the Boat-billed Heron) and one small primary, 12 tail feathers (10 in bitterns), and long undertail coverts; many have elaborate display feathers, the aigrettes.

The near relations

Which bird groups are the closest relatives of the heron family have been long debated. The family is nearly always grouped with other wading birds in the family Ciconiiformes (called sometimes Ardeiformes), which traditionally also includes the storks, ibises, spoonbills, Shoebill, Hamerkop, and sometimes flamingos (Hancock *et al.* 1992). Rasmussen *et al.* (1987) described a new family, Xenerodiopidae, from the Oligocene, which is thought to be closely related to herons. It is by no means certain if the traditional ciconiiform order is monophyletic, since included birds have characteristics that differ among them. Some of these families were long thought to harbour affinities to waterfowl, shorebirds, or pelicans as well as to herons.

In their reclassification of the birds of the world, Sibley and Monroe (1990) recognized the apparent multiple connectivity of traditional ciconiiforms to other groups and proposed collecting them with other birds such as frigatebirds, penguins, loons, petrels, albatrosses, storm-petrels, and with the new world vultures within a subdivision of a much larger Order Ciconiiformes that also includes such birds as sandgrouse, shorebirds, gulls and terns, hawks, grebes, pelicans and relatives. This proposal clearly views the 'Order' as a much larger and more inclusive collection of related birds than is traditional. But it does suggest the existence of an as yet unfathomed complexity of the evolutionary tree of the waterbirds and identifies an array of birds to be considered as sister groups to the herons. This proposal has and will provide much food for thought and guidance for additional research on the superfamilial relationships of waterbirds, including herons. So, above the family level, much remains to be understood regarding the relatives of the herons. But it is not likely that such studies will undermine our view of the herons (perhaps eventually minus the Boat-billed Heron) as constituting a monophyletic assemblage of birds recognizable at the family level.

Within the family

The Boat-billed Heron aside, the limits of the family are commonly agreed. However, within the family identifying limits of subfamilies, genera, species, and subspecies has not been so easy. These have been matters of iterative debate through the last 50 years. But step-by step, these decades have seen significant advances in understanding infrafamilial relationships due to detailed studies of morphology, molecular structure, behaviour, and, most recently, vocalizations by Parkes, Payne, Dickerman, Sheldon, McCracken, and others.

The subfamilies and tribes

Previous morphological and behavioural study had resolved the family into four groupings, the day herons, the night herons (including the Boat-billed Heron), the bitterns (including the Zigzag Heron), and the tiger herons. These groups were represented as subfamilies in the classification scheme previously offered by Hancock and Kushlan (1984). Since then, these proposed groupings have been challenged by molecular results and by the simultaneous evaluation of data from several independent approaches (Payne and Risley 1976, Sheldon 1987, Sheldon *et al.* 1995, 2000, McCracken and Sheldon 1997, 1998, Chang *et al.* 2003, K. McCracken pers. comm.).

The results are exciting, intriguing, and puzzling but together represent a remarkable advance in our understanding of infrafamilial classification over that of Hancock and Kushlan (1984). These studies suggest exceptionally promising directions for further molecular and population studies. We are grateful indeed to our colleagues, especially Fred Sheldon and Kevin McCracken, for sharing their up-to-the-minute analyses and insights that allow the current book to base a higher-order classification on hypotheses derived from molecular studies. We can all look forward to continued progress in resolving the heron family tree as new material becomes available.

Although the birds collected in the major heron groupings of Hancock and Kushlan (1984) appear to share specific looks and ways of life—e.g. the typical diurnal herons, the nocturnal herons, the cryptic, marsh-dwelling bitterns, and cryptic forest-dwelling tiger herons—this view of the heron world is simplistic and incomplete, failing to account for convergences. Figure 2.1 and Table 2.1 shows the classification of herons proposed in this book, reflecting evolutionary relationships among herons as currently hypothesized. This classification, of course, is only a snapshot of an ongoing process of discovery, and is certain to change.

The most sweeping result of 15 years of molecular study is recognition that the vast preponderance of extant species belong to a single monophyletic group, a subfamily of 'typical herons', the Ardeinae. This subfamily includes the herons and egrets (*Ardea*, *Egretta*), the green herons (*Butorides*), the pond herons (*Ardeola*), a few monotypic genera, but also, perhaps surprisingly, the night herons (*Nycticorax*, *Gorsachius*). It now appears clear that the 'night herons' are not more

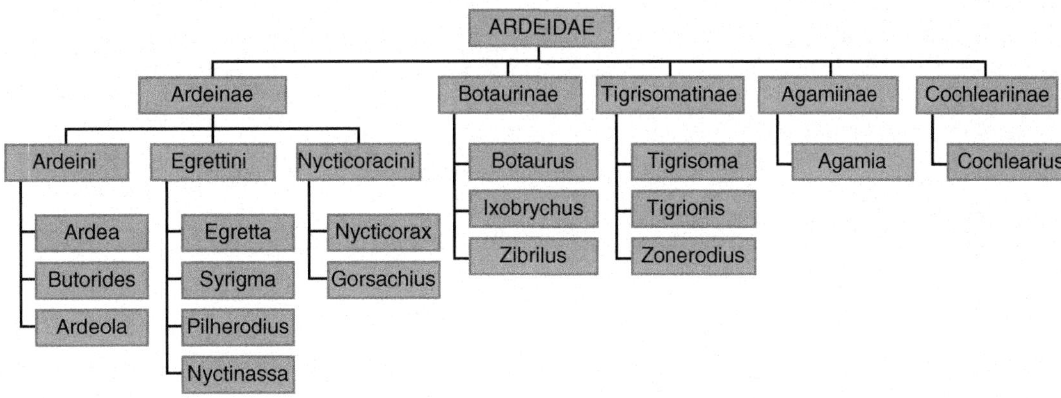

2.1 Diagramatic representation of the relationships of the herons of the world.

Table 2.1 Classification of the Herons of the World

Family ARDEIDAE
Subfamily ARDEINAE—The Typical Herons
Tribe—ARDEINI
Ardea cinerea Grey Heron
Ardea herodias Great Blue Heron
Ardea cocoi Cocoi Heron
Ardea pacifica White-necked Heron
Ardea alba Great Egret
Ardea modesta Eastern Great Egret
Ardea intermedia Intermediate Egret
Ardea melanocephala Black-headed Heron
Ardea humbloti Malagasy Heron
Ardea insignis White-bellied Heron
Ardea sumatrana Sumatran Heron
Ardea goliath Goliath Heron
Ardea purpurea Purple Heron
Ardea ibis Cattle Egret
Butorides virescens Green Heron
Butorides striatus Striated Heron
Ardeola ralloides Squacco-Heron
Ardeola grayii Indian Pond-Heron
Ardeola bacchus Chinese Pond-Heron
Ardeola speciosa Javan Pond-Heron
Ardeola idae Malagasy Pond-Heron
Ardeola rufiventris Rufous Bellied-Heron
Tribe—EGRETTINI
Egretta rufescens Reddish Egret
Egretta picata Pied Heron
Egretta vinaceigula Slaty Egret
Egretta ardesiaca Black Heron
Egretta tricolor Tricolored Heron
Egretta novaehollandiae White-faced Heron
Egretta caerulea Little Blue Heron
Egretta thula Snowy Egret
Egretta garzetta Little Egret
Egretta eulophotes Chinese Egret
Egretta sacra Eastern Reef-Heron
Syrigma sibilatrix Whistling Heron
Pilherodius pileatus Capped Heron
Nyctinassa violacea Yellow-crowned Night-Heron
Tribe—NYCTICORAXINI
Nycticorax nycticorax Black-crowned Night-Heron
Nycticorax caledonicus Rufous Night-Heron
Nycticorax leuconotus White-backed Night-Heron
Gorsachius magnificus White-eared Night-Heron
Gorsachius goisagi Japanese Night-Heron
Gorsachius melanolophus Malayan Night-Heron
Subfamily BOTAURINAE—The Bitterns
Botaurus pinnatus South American Bittern
Botaurus lentiginosus North American Bittern
Botaurus stellaris Eurasian Bittern
Botaurus poiciloptilus Australasian Bittern
Ixobrychus involucris Streaked Bittern
Ixobrychus exilis Least Bittern
Ixobrychus minutus Little Bittern
Ixobrychus sinensis Yellow Bittern
Ixobrychus eurhythmus Schrenck's Bittern
Ixobrychus cinnamomeus Cinnamon Bittern
Ixobrychus sturmii Dwarf Bittern
Ixobrychus flavicollis Black Bittern
Zebrilus undulatus Zigzag Heron
Subfamily TIGRISOMINAE—The Tiger Herons
Tigrisoma mexicanum Bare-throated Tiger-Heron
Tigrisoma fasciatum Fasciated Tiger-Heron
Tigrisoma lineatum Rufescent Tiger-Heron
Zonerodius heliosylus New Guinea Tiger-Heron
Tigriornis leucolophus White-crested Tiger-Heron
Subfamily AGAMIINAE—The Agami Heron
Agamia agami Agami Heron
Subfamily COCHLEARIINAE—The Boat-billed Heron
Cochlearius cochlearius Boat-billed Heron

distinctive from other typical herons than, say, *Ardea* is from *Egretta*. The latter two genera, usually thought to be closely related sister taxa, are as distinct from each other as each is from night herons. The sister group for *Ardea* is not *Egretta* but *Butorides/Ardeola*. The night herons are a sister group to both the *Ardea* group and *Egretta* group. Within the typical heron subfamily, these three groups are recognized as tribes, Ardeini, Egrettini, and Nycticoraxini. The subfamily Ardeinae also contains the oldest generally accepted fossils of herons, *Proardea amissa* and *Proardeola walkeri*.

The second subfamily, Botaurinae, encompasses the bitterns (*Ixobrychus, Botaurus, Zebrilus*). These birds share a number of morphological and behavioural characteristics that differ from the typical herons and are distinctive on the basis of molecular genetics (Chang *et al.* 2003).

Not included in the Ardeinae and Botaurinae are several intriguing species, the Boat-billed Heron, the tiger herons, and the Agami Heron. All recent results tend to show that the tiger herons are not particularly closely related to the bitterns but merely look somewhat similar in sharing a cryptic life style in common. Perhaps the most intriguing, most novel, and most tenuous of recent suggestions is that the Agami Heron is likely not closely related to the typical herons. This species has always been considered just an exaggerated typical heron, but once hypothesized to be from a different lineage, its many distinctive features become clearer and more interesting. The Boat-billed Heron is so genetically different from other heron that it may not be monophyletic with them (Sheldon *et al.* 1995).

Resolving these few more distantly related species that are basal to the main tree structure of extant herons is not, and will not be easy (K. McCracken pers. comm.). As more information becomes available and alternative analyses are run, these apparent relationships are likely to shift. Acknowledging this uncertainty, and in anticipation of further study, each is assigned to a subfamily, the Cochleariinae, the Tigrisominae, and the Agamiinae. Each of these has morphological and behavioural characteristics that differ from other extant herons. Intriguingly, they are all located on the southern continents (or former pieces of southern continents), with the most forms

2.2 Subfamily Ardeinae–Great Blue Heron with young in nest.

2.3 Subfamily Botaurinae–North American Bittern in reeds.

2.4 Subfamily Tigrisominae–Rufescent Tiger-Heron on nest.

2.5 Subfamily Agamiinae–Agami Heron.

2.6 Subfamily Cochleariinae–Boat-billed Heron.

in the Americas. Might these represent the remnants of a southern continent radiation of herons? Might the typical herons and bitterns derive from radiation on the northern continents? Much remains to be learned. For example the social behaviour of Boat-billed Herons (Mock 1976) is certainly better reinterpreted through its differences, rather than its similarity, to typical herons.

The genera

At the generic level, the situation has been in a state of flux for decades. Earlier authorities placed herons in numerous single-species (monotypic) genera based on the argument that herons, being of ancient origin, have evolved to a point where differences between the species should be substantial. Peters (1931) recognized 31 genera. The counter-argument is that maintaining numerous monotypic genera obscures infrafamilial relationships among the species. In recent years the tendency has been to classify birds in ways that show monophyletic relationships, which reduces the number of genera recognized. For this book, 17 genera are recognized (see Table 2.1). This is the same number recognized in 1984 (with two offsetting changes). Thus the list of genera has become relatively stabilized, except for the continual re-evaluation of problematic species. The birds remaining in monotypic genera, *Pilherodius, Syrigma, Agamia, Zebrilus, Zonerodius, Tigriornis, Nyctinassa*, and *Cochlearius*, seem properly so. But there may be other species that are not assigned appropriately to genus.

The species

At the species level, many herons are well characterized. But what is surprising is, that after a century or more of study, there remains so much uncertainty as to the species limits of so many forms. The green herons, the Little Egret group, Intermediate Egret, great egrets, pond herons, small bitterns, large bitterns, Boat-billed Heron, Great Blue Heron, and Grey Heron are some of the forms for which species limits remain unsettled. Even in this book, a new egret species is recognized. Of course uncertainty may reflect in part the difficulty in applying the species concept to widespread forms with isolated populations. But it more fundamentally reflects a lack of behavioural and molecular information on these birds. In this book a list of 62 species of extant herons is adopted (see Table 2.1).

Ardea includes the largest modern heron, *Ardea goliath*, which seems most closely related to *Ardea melanocephala*. As mentioned above, an even larger *Ardea* existed only a few thousand years ago (Hoch 1977). The three large *Ardea* (*cinerea, herodias*, and *cocoi*), based on different continents, are different but do appear to be closely related. Recent biochemical, vocalization, and composite evidence suggested that several problematic species are *Ardea*. The Cattle Egret (previously placed in *Bubulcus, Ardeola*, or *Egretta*), the Great Egret and the Eastern Great White Egret (previously referred to *Egretta* and *Casmerodius* as well as *Ardea*), and the Intermediate Egret (previously *Egretta*), all are properly *Ardea*. The Eastern Great Egret is recognized as a species separate from the Great Egret. There remain a number of interesting research questions within *Ardea*. These include the possible specific distinctiveness of certain populations, including *Ardea cinerea monicae* in Mauritania (Erard *et al.* 1986), *Ardea herodias occidentalis* in southern Florida and the Caribbean, and *Ardea herodias fannini* on the north Pacific coast of North America.

The *Ardeola* herons appear to be related to the *Ardea*. These are intriguing birds, characterized by interesting plumage changes. They appear also to be closely related to *Butorides*, a relationship that needs further elucidating.

As a whole the *Butorides* herons demonstrate an unmatched degree of geographic and individual variation in plumage. Considering these forms as a single polymorphic species, Hancock and Kushlan (1984) recognized and mapped 30 subspecies. The species limits within this genus have been a huge problem for decades. Monroe and Browning (1992) re-separated the North American *virescens* from the South American and Old World *striatus*, based on the same specimens used by Payne (1974) to come to the opposite conclusion. Hayes (2002) in a new re-examination demonstrates the existence of a substantial zone of intermediacy. By traditional criteria, the extent of intermediacy would call for recognition of a single species. Both field studies and an additional molecular study are needed, but in the meantime, we dutifully follow the trend in recognizing two species of *Butorides*. This likely is far from the end of the story. The appropriate relationship between New World and Old World *Butorides striatus* has not been thoroughly explicated; it seems they may be different species. And within the Old World, continental and island forms may be as distantly related to each other as they are to New World forms. We think it likely that rather than one or two, there may be several species of *Butorides*.

With respect to *Egretta*, having reassigned the previously puzzling great and intermediate egrets into *Ardea*, the remaining species in the genus *Egretta* seem more cohesive. The White-faced Heron is confirmed to be an *Egretta* as proposed by Payne and Risely (1976) and Hancock and Kushlan (1984). However, other *Egretta* species-level questions are far from resolved, particularly the taxonomy of large and intermediate-sized all-white egrets found throughout the world. These are birds that lack the useful plumage differences that allow other species and races to be readily recognized. It seems likely that these large white birds that differ in size and are relatively unstudied behaviourally actually include separate species. At the local level, the taxonomic status of *Egretta caerulea* and *Egretta thula* in west Mexico needs further elucidating.

Regretfully, also awaiting further resolution is the hypothesis posed by Hancock and Kushlan (1984) that several populations of closely related and similarly appearing medium-sized herons are best all considered to be Little Egrets, a hypothesis that has not been universally applauded since its introduction. This Little Egret group contains forms previously recognized as the Western or West African Reef Heron (*Egretta gularis gularis*), Arabian or Indian Reef Heron (*Egretta gularis schistacea*), the Dimorphic Heron or Mascarene Reef Heron (*Egretta dimorpha*), and several distinctive populations of the traditional Little Egret. It is clear to those who know these birds in real-life, that the typical forms are readily distinguished. But intermediate looking birds are often reported, and not only in zones of overlap. Atypical birds are often observed well within the range of another form. Coastal populations of several other heron species tend to be morphologically different from inland populations. When Hancock and Kushlan (1984) proposed this taxonomy, we had hoped that it would generate debate and new quantitative information. Our hopes were well met by the former, but disappointed in the latter. There has certainly been considerable discussion of opinions. But it is unfortunate that, at this time, no additional data have appeared to test the hypothesis. Opinions aside, the relationships inferred by combining these populations into a single species need to be subjected to additional scientific scrutiny, particularly detailed morphological study and molecular analysis. With some disappointment, we retain all these populations as Little Egrets and challenge others to do better.

Other species also within the typical heron subfamily are three New World species, *Syrigma*, *Pilherodius*, and *Nyctinassa*. Molecular study has suggested that the South American *Syrigma* and *Pilherodius* are sister taxa and that they are related to *Egretta* (K. McCracken pers. comm.). The Yellow-crowned Night-Heron appears not to be closely related to *Nycticorax*, but a distinctive form, now restored to the genus *Nyctinassa*. Additional study might reverse this subfamilial placement, however.

The idea that night herons (*Nycticorax, Gorsachius*) are not geneologically distinct from the *Ardea* and *Egretta* groups of herons is an important advance in our understanding of relationships among herons (Sheldon 1987). All these species are adapted for night foraging, although they are not by any means exclusively nocturnal, and it is this suite of adaptations rather than ancestry that distinguish them from the other typical herons. *Gorsachius* includes three species of solitary nesting Old World herons that are little understood. *Nycticorax* is a widely dispersed group of two extant colonial species, the Black-crowned and Rufous Night-Herons that are sometimes considered the same species. Within recent times, *Nycticorax* included additional recognized species, which have become extinct on islands in the Pacific, Indian and Atlantic oceans in the past 5000 to 300 years. Among these species are *Nycticorax kalavikai* from Niue (Polynesia) (5300–3600 years BP), an undescribed *Nycticorax* from Eua (Tonga), *Nycticorax mauritianus* of Mauritius, *Nycticorax megacephalus* of Rodrigues, *Nycticorax borbonensis* (previously *duboisi*) from Reunion, three possibly flightless species from the Mascarene Islands (500 years BP), and *Nycticorax olsoni* from Ascension Island (Cowles 1994, Worthy *et al.* 1998, Steadman *et al.* 2000, Bourne *et al.* 2003). The extinction of these species corresponded to the arrival of Polynesians and Europeans to these islands.

Many hybrids have been reported between species of typical herons in the wild and in captivity. Cross pairing and production of hybrid herons in the wild seem uncommon. Recent reported crosses and pairing in the wild include Grey Heron × Great Egret, Grey Heron × Purple Heron, Grey Heron × Little Egret, Pied Heron × Little Egret, Tricoloured Heron × Snowy Egret (Eigenhuis 1984, Lippens and Burggraeve 1983, 1985, Campos 1990, Fenyvesi 1992, (Meeks *et al.* 1996) Baumanis 1998, Passarella *et al.* 1999, J. Hancock pers. obs.). In captivity more distantly related pairings are reported with some regularity. Some of these seem rather puzzling based on the previous way of thinking about heron relationships, for example Pond Heron and Intermediate Egret (Parasharya 1985). The understanding that these species are more closely related than previously thought makes the situation less jarring, at least on an intellectual if not physical level. The typical herons appear to be sufficiently closely related to interbreed within currently recognized genera and sometimes across currently recognized genera. Heron distributions change, and this brings previously isolated species and populations into contact. The recent invasion of North America by Little Egrets has spawned observations of puzzling birds that may be are hybrids (Perkins 1995). The biological and evolutionary significance of hybridization in this adaptable group of birds has not been studied sufficiently.

Within the second heron subfamily, the bitterns, *Botaurus* includes four superficially similar species in North America, South America, Australasia, and Eurasia-Africa. They do however show distinctiveness in skeletal characters (Payne and Risley 1976), and they have been considered to be variously related. It appears likely that the two New World species and the two Old World species form related pairs of species.

Eight small bitterns are included in the genus *Ixobrychus*. It appears that *Ixobrychus* bitterns may be among the fastest evolving of the herons, quite a different point of view from the historic one that considered bitterns to be 'primitive'. It is possible, here too, that there may be much convergence within the group and the species limits are not as they appear to be. It may be, for example, that the Least Bittern is not as closely related as the others, and the disjunct ranges of the Little Bittern raise possibilities that these may be more than one species. However, it now seems clear that the puzzling, little known and morphologically distinctive Zigzag Heron belongs within the bittern group, rather than with the tiger herons (Sheldon *et al.* 1995).

The tiger heron subfamily includes five species. Members of the group have somewhat bittern-like plumage, but differ from bitterns in a number of skeletal characteristics and in behaviour. The distributions of the species are scattered in the tropics of

America (*Tigrisoma*), New Guinea (*Zonerodius*), and Africa (*Tigrionis*), but clearly constitute a monophyletic group (Sheldon *et al.* 1995).

The Agami Heron is a single unique species from the American tropics. Evidence emerging from molecular study is suggesting that this species is not monophyletic with the typical herons, as might be suggested from its general appearance. It is little understood but from what is known is clearly different. It appears to belong in a basal position in the heron family tree, but its exact position and relationships are not clear and may never be, given that there is only one species extant. Of the species assignments to subfamilies, this is the most tenuous. It is hoped that this hypothetical relationship will be further examined from multiple perspectives, including behavioural.

The Boat-billed Heron is a single polytypic species. Re-confirmation of the distinctiveness of this bird is an important advance in heron biology. The lesson here is to be wary of the convergence of morphology and plumage among herons of similar life style. Given its apparent distinctiveness from other herons, the relationships among the several populations should be re-examined. Perhaps they are more different from each other than now is appreciated.

The subspecies

While most herons are monotypic, geographic variation in others is recognized by the identification and description of subspecies. The issue of heron subspecies is not an easy one, especially considering the widespread distribution of some species, the potentially isolated situations of some island and coastal populations, the antiquity of some species, and the tendency of some species to disperse and undergo rapid range change. It is because of the complex forces at work that the issue of heron subspecies takes on interest and importance, both biologically and to ensure conservation of all genetically distinctive populations. Morphological and plumage differences in populations need to be better articulated, although these can easily be confounded by individual variation. Among promising characters are soft-part colours. Molecular characterization of subspecies is hoped for, but subspecies are by definition only modestly divergent. Subspecific characterization has not been applied consistently across species and a common taxonomic approach to the island forms of different herons is highly desirable. Characterization of geographically distinctive populations remains a high priority.

Future debates

The taxonomic nomenclature used in this book is based on a good faith effort to interpret available information, including the results of ongoing molecular study. The taxonomy of this book differs in many interesting ways from our previous best effort (Hancock and Kushlan 1984), reflecting the exciting advances in understanding heron systematics that have occurred in the interim. This is an encouraging outcome of the passage of 20 years of study. More intriguing results and even more surprises are anticipated as molecular studies progress covering more material and as additional behavioural observations are made on little known taxa. As was the case in 1984, the taxonomy presented is meant to suggest hypotheses to be scientifically tested. It continues to be our hope that resolutions of the interesting taxonomic issues will be forthcoming through new quantitative study and analysis.

3

Feeding ecology

Herons are, fundamentally, birds that wade in the water to capture aquatic prey with their bills. Many of the biological characteristics, described in Chapter 1, influence and reflect the way they feed. Specifically, the archetypical heron is a medium-sized bird, relative to the range of all birds, with long legs, long neck, and long bill. It uses its long legs to wade and stand in shallow water where it strikes at fish with its long neck, and bill. It feeds alone and also at times joins others aggregating in shallow water habitats having a supply of fish. To the extent that herons are a rather homogeneous group in their morphology, behaviour, and habitat choice, these generalizations tend to hold up surprisingly well across species. But even subtle deviations from the archetypical pattern open up different foraging opportunities. So within the theme of a long-legged wading bird, herons do differ substantially from each other in the details of habitat choice, behaviour, and feeding strategies.

Feeding habitats

Herons are water birds. Nearly all herons feed in water, which may be of almost any relatively shallow sort, ranging from potholes to lakes, gullies to rivers, marshes to coral reefs. Most herons do also feed on land when it is worthwhile, but only a few species do so customarily. The Black-headed Heron, the Cattle Egret, and the Whistling Heron are good examples of terrestrial herons. But even the most terrestrial of herons at times use damp places preferentially and may depend on rainy season flooding to nest. The most terrestrial, and currently most widespread heron, the Cattle Egret, is believed to have evolved using short grass margins of wetlands, particularly floodplains subject to regular annual inundation, where it became associated with the Cape buffalo. African Cattle Egrets still depend on frogs for feeding young late in nesting in order to secure minerals not provided by insects. All herons will tend to depend to some degree on aquatic habitats at some time in their annual cycle.

Herons most commonly feed in one of three situations–still water, flowing water, or shore. Herons very typically feed in shallow still waters, such as ponds, marshes, swamps, rice fields, and artificial fishponds. Large species can wade; smaller species can perch on vegetation over the water or float on mats. Herons feed in flowing waters, such as rivers, streams, and artificial canals and ditches. Only a few of the larger species are big enough to withstand fast flowing rivers, but many others can wade in slower water or feed from overhanging vegetation. And thirdly, herons feed along the shore. They stand on shoreline rocks or vegetation, but they also feed on beaches and reef crests. These situations cover the vast preponderance of the places herons feed. They also on occasion can feed effectively in deep water, either by standing on floating objects or feeding from the air or by swimming, but this is economical only under special circumstances.

Herons may be habitat generalists or habitat specialists. Specialists include the coastal species such as the Eastern Reef Heron and Reddish Egret, and obligatory marsh species, such as the bitterns. Generalists, on the other, hand choose among available habitats, presumably depending on their relative profitability. Herons observed choosing to switch habitats tend to achieve higher energy intake than those that do not (Richner 1986).

Herons are tropical by and large. Many of their fundamental adaptations seem highly suited for tropical environments, which are characterized by wet–dry rainfall cycles, constant day lengths, and a high diversity of competitors, predators, and prey. Tropical species either use many habitats or are extreme habitat specialists, such as the tiger herons, the Agami Heron, pond herons, and coastal herons. In temperate areas, herons tend to be linked to the fewer types of aquatic habitats available and also to the seasonal availability of that habitat, winter being rather inhospitable to herons.

Feeding behaviour

Flexibility in feeding is the rule. Each species of heron has a repertoire of feeding behaviours from which it may choose what seems most appropriate for the moment. Nearly all herons feed at times by standing and waiting, and by walking slowly after prey. Most also have other behaviours in their arsenal. They may use their feet, head, wings, or full body in various ways to access particular prey.

The usual foraging behaviours used by herons are familiar to observers, and less usual behaviours are also often reported. Heron feeding behaviours, even unusual ones, show a lot of commonality among species sufficient to support a common nomenclature (shown in bold here) (Meyerriecks 1960, Kushlan 1976a, 1978a, Hancock and Kushlan 1984, Voisin 1991). The following is the latest update of a nomenclature for heron feeding behaviour, superceding Kushlan (1978a) and Hancock and Kushlan (1984).

Prey capture

Herons usually catch prey with a **Bill Stab**, which is a downward or lateral strike involving fast, directed movement of the head and neck while the body remains still. This is the characteristic capture stroke of the long-necked herons, which also show full development of specialized neck vertebrae acting as a hinge for the forward strike. The Bill Stab is a fundamental foraging behaviour of most herons.

3.1 The Bill Stab, Grey Heron Grasping a fish in a forceps grip. This photo also shows the characteristic kink in the neck.

Herons can catch prey also by moving their body forward. One way to do this is with a **Bill Thrust**. The head, neck, and body are pushed forward as part of a headfirst leap, and the feet may ever leave the substrate. This is characteristic of species with relatively shorter, stouter necks. Another approach is the **Bill Lunge**. The bird lunges its body forward but its feet stay in place. This is a behaviour perfected by species with shortened lower legs, the 'knee' joint serving as the hinge allowing the longer upper leg to push forward as the neck and bill are extended. Herons also use Bill Thrusts and Bill Lunges in their threat behaviour. Herons feeding from perches have a couple of modifications of the typical prey-catching moves. In **Forward Throwing**, a bird grasping the branch throws its body toward a prey item but remains attached to the branch. The Green Heron **Hangs** upside down to catch fish beneath its perch.

A heron can use a **Probing** technique by quickly and repeatedly moving its bill into and out of the water or substrate, in a method more typical of ibises. Probing is a non-visual, tactile foraging technique. A more common behaviour is **Pecking**. The heron merely picks up an item from the substrate, often repeating the movement. The substrate may be the ground, a plant, or surface of the water. A specialized method is **Ploughing** used by the Boat-billed Heron (Willard 1979). For many years it was unclear in what special way the great bill of this species was used. By Ploughing, the individual walks forward with its bill partially submerged, thrusting forward and scooping with each step, thus using the wide bill.

Prey are caught in the bill either by **Impale Capture** or **Grasp Capture**. Impaling means the bill tip goes into or through the prey. This occurs mostly on large and wide prey, and mostly by herons with bills strong enough to withstand the stress of impaling and landing a larger prey item. Grasping is the more common capture method in which prey are caught in a tweezers grip.

Body postures

Herons may use several different postures to feed. Their choice of posture can enhance the effectiveness of foraging in a particular habitat or for a particular prey item. Posture also has a signalling function, indicating a heron's presence and perhaps its claim to territory. In **Crouched** posture, the heron stands legs bent with its body parallel to the ground or perch. The head is withdrawn or outstretched. This posture is used for concealment, stalking, and in preparation to strike. The Crouched posture is particularly used by species feeding in the open and in very shallow clear water and when feeding from along banks and from overhanging branches. The Crouched posture is widely used among herons and highly effective (Cezilly and Boy 1989). Using a Crouched posture is enhanced if a bird has short lower legs relative to the upper legs, such as is the case in the Eastern Reef Heron, the Agami Heron, and large bitterns. The Crouched posture is also used in interactions. A **Submissive Crouch** is assumed when deferring to a threatening bird.

In **Upright** posture, the heron stands straight, its body and neck angled away from the water above horizontal. This posture allows scanning of a larger feeding zone than it can in a Crouched posture. The feathers are often erected when used as a threat pose.

In **Erect** posture, the head and neck are held high, up to 90° from horizontal, and the bill is level or pointed upwards. An Erect posture offers the highest visibility, both to see and to be seen. The Erect posture has also been called the **Alert** posture, and it is often assumed in response to a disturbance or to conduct a reconnaissance of the area. In this mode, the body feathers are smoothed down and the head extends up almost vertically. These postures can differ subtly among species (McCanch and McCanch 1991).

An interesting posture characteristic of herons is the **Bittern Posture**. The bird not only puts its head straight up but its bill as well. The bird looks forward beneath the upturned bill. Given the lineated ventral coloration of most herons, especially bitterns, the posture is camouflaged. It is assumed usually in response to disturbance of a heron's normal behaviour. Birds also assume this posture without provocation. In this situation it is used for surveillance. The more ventrally cryptically coloured the heron, the more likely the species is to use this posture.

Head postures

Head position is an important component of feeding behaviour. The heron holds its head in one of several identifiable ways that influence its visual field, usually to make optimal use of its narrow binocular field of vision. It may extend its neck and tip its head so that its bill points straight down towards the ground. This **Peering Over** head posture reduces glare and distortion. It also provides a binocular view of a potential prey item by looking down the tip of the bill rather than beneath it. This head posture may be especially useful in finding prey where visibility is restricted by vegetation (Recher *et al.* 1983). **Head Cocking** is turning the head to one side to look straight down with one eye and straight up with the other. This is probably used to fix on a prey item with its monocular vision. In other birds, Head Cocking is used to observe aerial threats, and this possible function should not be overlooked in herons. **Head Tilting** is turning both its head and its neck to one side of its body. It is to reduce glare by shifting its head away from the strike zone. **Facing Down** is when a heron puts its bill, head, and neck straight down. It is usually used to probe or pick something up. This posture is also used as a display in some species. **Neck Craning** is looking about with head and neck outstretched. Herons also move their head and neck around to look more carefully for prey. While **Head Swaying**, a heron moves its head and neck from side to side. A head movement in the other plane is **Head Bobbing**, a sinuous forward and backward movement of the head and neck. Head Swaying and Head Bobbing are probably used to obtain parallax and gain a more accurate estimate of a prey item's location and distance. Head Swaying is usually used while Standing, and Head Bobbing while Walking. The terrestrial-feeding herons such as the Cattle Egret, Black-headed Heron, and Whistling Heron use these motions routinely. While **Neck Swaying**, the neck and sometimes the body is moved from side to side, the head tending to be rather stationary. Neck Swaying looks quite odd and may be a method of enticing small movements of prey, while the heron's head remains ready to strike. Neck Swaying is also used in the Bittern Posture, probably to increase camouflage value. While **Head Swinging**, the heron moves its bill from side to side in the water like a spoonbill (*Platalea*). All these neck and head movements seem to be used particularly for prey that are hard to catch.

Standing and walking

Standing is the simplest feeding behaviour. It is remaining in one place while feeding, waiting for a prey to become apparent. It is by far the most common behaviour among herons and one that characterizes best the fundamental adaptive suite of the group. Herons Stand in shallow water, on perches next to or over the water, on land, on emergent or floating plants, on a rock, or even on mobile platforms such as large animals. Many of the larger and the smaller sized herons specialize in Standing behaviour, notably the great herons, the pond herons, and *Butorides* herons. The large herons (Grey, Great Blue, Cocoi herons) and the night herons also are fundamentally Standing herons. Cryptically coloured species, such as the large bitterns and tiger herons, also primarily use this technique.

Various postures are used in Standing feeding. The Erect posture is particularly common in birds

3.3 Standing, Striated Heron (*Butorides striatus sundevalli*) Standing in Crouched posture on the lava shoreline of the Galapagos Islands.

3.2 Standing, Capped Heron Standing in Erect posture by the waterside.

3.4 White phase Eastern Reef-Heron feeding from a rock Peering Over.

feeding in groups, where it also has an antagonistic and distancing message. From an Upright posture, Standing herons can view a wide area for potential prey, competitors, and predators. Crouched posture is often used on land or from perches.

The Standing bird waits, avoiding detection until a potential prey item makes itself visible. While simplistic, the behaviour undoubtedly takes considerable skill and practice in that the heron has to hold its immobility until the prey is observed within reach of a Bill Stab. The difficulty of doing this successfully should not be underestimated. Species for which standing is characteristic tend to defend their feeding territories, no doubt to exclude intruders who would disturb their prey or their own concentration. Standing herons may tend to have delayed maturity, typically a sign that young birds are not effective foragers.

Walking is also a widespread feeding behaviour, used by all species, at least on occasion. Herons walk about looking for prey or stalk specific prey items. Herons walk in shallow water, on aquatic plants, on land, or along branches of trees. Walking allows the heron to cover more area than it could do by Standing but increases the risk of frightening potential prey. Walking slowly is defined as a rate less than 60 steps per minute (Rogers 1983); It merges with Standing behaviour as steps become more infrequent, at times as slow as one per minute. The transition is particularly difficult to determine in species such as large bitterns that may take minutes to move a leg.

The speed of Walking varies among species and situations. Typical Walking rates in water are 20–30

3.6 Goliath Heron Walking slowly in erect posture while wading in the shallow water.

3.7 Eastern Reef-Heron Walking slowly in Crouched posture while wading in shallow water.

3.5 Great Egret Walking with neck extended.

3.8 Reddish Egret Walking Quickly as part of a walk quickly, run, hop foraging sequence.

3.9 Intermediate Egret Walking, using Head Bobbing, which in this species also shows off its lower neck plumes.

3.10 Tricoloured Heron Running using wings for lift to skip lightly.

steps per minute but can be faster in very shallow water or on land. Slow Walking specialists include such species as the Little Blue Heron, which wades slowly and deliberately in shallow water, and the Cattle Egret, which walks on land, especially when following slow moving cattle.

Herons **Walk Quickly** when their steps become faster than 60 per minute. Walking Quickly covers more ground than intermittent Standing or Walking slowly and tends to disturb more prey. As a result, it is often considered to be a disturb and chase behaviour, but it may also be used merely to move quickly from one place to the next. Several species frequently use the Walking Quickly technique within their normal repertoire, including the Cattle Egret, Reddish Egret, and Tricoloured Heron.

Disturb and chase

Active feeding usually aims to disturb prey sufficiently to inspire them to move. Running from place to place or chasing after prey is a characteristic of certain herons. Actively feeding herons use their wings on the ground, while Standing, Walking, or Running. The functions appear primarily to disturb prey into revealing themselves and also to get the heron in a posture whereby it can better see into the water.

The simplest active behaviour is **Running**. A Running heron dashes about, usually to chase after specific prey items. Running involves steps in excess of 120 per minute (Rodgers 1983) but a bird often moves so quickly that it uses its wings for balance, to gain lift enabling faster movement, and for stopping or changing direction. Medium-sized marine-foraging species, especially the Reddish Egret and the Eastern Reef-Heron, often specialize in such behaviour. Cattle Egrets run frequently in terrestrial feeding habitats. In disturb and chase feeding, herons string a series of active behaviours together to disturb prey and then chase them down. Although Running is the principal component of this sort of behaviour, other behaviours, especially using wings and feet, are often part of the sequence. For example, a sequence for a Reddish Egret might be Walk Quickly–Run–Hop– Open Wing–Whirl–Stab.

Feeding using wings

There exists a wonderful progression in complexity of wing feeding behaviours, first deciphered by pioneering heron biologist Andrew Meyerriecks. The simplest is **Wing Flicking**. A heron quickly, partially extends and then retracts one or both wings. In **Open Wing Feeding**, a heron extends one wing and leaves it extended for a few seconds before retracting it. In **Underwing Feeding**, a heron puts its head under an open wing in order to strike at a prey item. **Whirling** and **Pirouetting** are variations of Underwing Feeding in which the bird turns rapidly. In Whirling, the heron spins fast with its wing out and peers under it. In Pirouetting

3.11 Black Heron in full Canopy Feeding, casting a shadow with its wings while feeding in shallow water.

3.12 Purple Heron showing Forward, threat display on the feeding ground.

a heron Walks and turns gracefully, not as fast as Whirling. While **Double Wing Feeding**, a heron brings both wings forward and holds them over its head. The fullest extent of wing use is **Canopy Feeding**. A heron brings both wings forward above its head so that the forward edges meet and feather tips dip in the water. This forms an umbrella over the heron's head and neck.

These several wing movement behaviours may function in different ways. Wing Flicking and Open Wing behaviours probably disturb prey, whose consequent movements increase the heron's chances of seeing them. Underwing, Double Wing, and Canopy Feeding reduce glare, making prey more visible, and may attract prey to the shade formed by the wings. A heron Canopy Feeding is often with its back to the sun. Open Wing Feeding is typical of the Reddish Egret, which often uses it before Running about, and is probably employed primarily on shoaling fish. Pied Herons, Tricoloured Herons, Snowy Egrets, and reef herons also use it. Tricoloured Herons, Reddish Egrets, and Snowy Egrets use Underwing Feeding. Double Wing Feeding is used especially by Reddish Egrets, but, so far as is known, only Black Herons use Canopy Feeding. This latter species walks or runs from place to place, stopping every few steps to form a Canopy over its head. It uses this behaviour almost obligatorily.

Herons also use their wings to fly while feeding. Aerial foraging is not a commonly used behaviour, probably because it uses so much energy. But it does seem to be widespread among herons of all sizes. While **Hopping**, a heron flies from a feeding site for a short distance before alighting again. This behaviour may be used merely to change place, or in response to seeing a potential prey. A variant of Hopping is called **Leap Frog Feeding**, in which a heron flies from the rear of a forward-moving feeding flock to the front. Cattle Egrets frequently do this, as do Black Herons in South African coastal waters and Whistling Herons prior to roosting. Another variant is **In place Hopping**, in which a heron lands in the same place as it began (Kent 1983). While **Hovering**, the heron flies in place over the water, stabbing at prey below usually for prey that do not try to escape. In a variant called **Hovering Stirring**, the heron pats, stirs, or rakes with its feet while hovering. While **Dipping**, a heron flies along, periodically reaching into the water to pick up prey. Dipping and Hovering are often associated (Davis and Jackson 2001). A heron may also trail its toes in the water while flying along, a behaviour called **Foot Dragging**. While in the air, a heron may catch a flying prey item; this is called **Aerial Flycatching**. Or it may dive into the water from the air—either head first by **Plunging**, or feet first by **Feet First Diving**. Herons may also launch themselves into the water from perches, either by **Diving** head first or by **Jumping** feet first. Green Herons, pond herons and the Great Blue Heron are known to use these behaviours.

So far as is known, the Snowy Egret possesses the largest repertoire of aerial feeding techniques.

Foot Dragging is a widespread aerial behaviour and has been reported in the Snowy Egret, Reddish Egret, Tricolored Heron, Little Blue Heron, and Little Egret. Hovering is used by many species, including, infrequently the large herons such as the Great Blue Heron, most of the medium-sized herons, and the Black-crowned Night-Heron. Aerial Flycatching is used especially by the smaller species, including the Green Heron, pond herons, and bitterns.

Feeding using feet

Herons also make special use of their feet in feeding. Foot movement behaviours include **Foot Stirring** in which the foot and leg are vibrated, **Foot Raking** in which the toes are scratched across the substrate, **Foot Probing** in which the toes are inserted into the substrate, and **Foot Paddling** in which the feet are moved up and down on the substrate. Several species frequently using Foot Stirring behaviour, such as Snowy and Little Egrets and Black Herons, have feet that are more brightly coloured than the rest of their legs. This suggests that the foot movement is used to startle prey and cause them to move. Foot movements are often used in mud or vegetation. The Snowy Egret is a foot movement specialist and uses each of these behaviours. Reddish Egrets, Tricoloured Herons' and Green Herons use both Foot Stirring and Foot Raking. Black Herons use foot movement to attract fish to under canopy (Vandewalle 1985).

Herons can use their feet to swim. Their feet have slight webbing, but it is probably more functional for standing in mud than for swimming. While **Swimming Feeding**, a heron floats or swims on the surface of the water stabbing at or picking up prey. The Great Blue Heron, Goliath Heron, Great Egret, and Black-crowned Night-Heron are known to swim with some frequency. Herons are not, however, swimming birds—it is not a behaviour often used.

Non-aquatic feeding

Most of the heron's feeding attention is directed downwards into the water. Herons also use several behaviours to catch prey that are not in the water. A heron may catch prey resting on objects above ground such as plants or other animals. This is called **Gleaning**. Or it may catch an airborne prey by **Flycatching**. Variations are **Standing Flycatching, Jumping Flycatching**, and Aerial Flycatching (as noted above).

Lures and tools

Herons not only disturb prey but also use behaviour that classifies as tool use. Herons place bait, either natural or artificial, in the water to lure prey. The behaviour is called **Baiting**. The Green Heron and Striated Heron have been most widely reported to use this behaviour. They can use real food, such as bread, maize, or a dead insect, or choose as a lure something

3.13 Snowy Egret attacking in a Supplanting Flight.

3.14 Little Blue Heron Following, a White Ibis.

that floats, such as a stick. Fish are attracted to the lure, coming within striking range of the Baiting bird.

Another way to attract prey is by **Bill Vibrating**, in which a heron places its bill in the water and rapidly opens and closes its bills. The disturbance caused by these bill movements can attract fish. This behaviour has been reported in the Snowy Egret, Black-crowned Night-Heron, and Cattle Egret. **Tongue Flicking** is similarly used. In this behaviour the heron places its open bill in the water and moves its tongue. As these behaviours have been variously described additional information is desirable (Kushlan 1973a, Mirick 2003).

Area defence

Herons defend their feeding areas. Depending on the species and the situation, the defended area may be as far as the bill reaches (their individual distance), or it may incorporate a large tract of habitat. Herons defend these areas to ensure their monopolization of the prey there. Defended areas can change constantly depending on the situation. Aggregative herons tend to defend the space near them. Residential and solitary herons defend larger territories that may be permanently held. The latter may nest within their feeding territories. The size of feeding territories depends mostly on their defensibility. When the prey base is scattered and the heron feeds in a solitary manner, territories may be large because they need to be and also because the competition for space is not great. At the opposite end of the spectrum, when prey is highly concentrated especially near roosts or nesting colonies, competition for space may be intense, and in this situation it is likely not economically feasible to defend space much larger than a heron can conveniently reach.

Herons defend their space with ritualistic displays, but when displays fail, also by fighting. The most common defensive display is the **Forward** threat, which is also used during nesting. The heron typically erects its head crest and the feathers on neck and back and makes a threatening lunge toward its opponent. The wings are slightly drooped, the head and neck are fully extended upwards with the bill horizontally aimed at the other bird. **Crest Raising** is an important threat and territorial display. **Twig Shaking** is a very obvious display that is primarily a threat. The bird extends its head out and down while leaning forward, and grasps and shakes a twig, slightly elevating the crest and neck feathers. This is mostly used in nesting but can be used at other times when a physical territory is being defended such as at roosts. Another threat display, which probably often has breeding implications, is the **Upright and Spread Wing**, a feeding ground interaction.

More intense defensive and aggressive behaviour involves actually engaging the opponent. The **Supplanting Run** occurs when distant threats fail and a bird runs after another bird usually driving it from the contested spot. The supplanting run is somewhat like a mobile Forward. The other way to displace a bird is from the air. In a **Supplanting Flight**, with its feathers erect and while calling, a heron flies at another bird. Unless repelled the attacker lands on the spot vacated by its opponent or on top of the opponent itself. In engaging the opponent at close quarters, herons use their bills in threats and in attacking other birds. As in feeding, herons project their bill towards their opponent using a **Bill Stab**, a **Bill Thrust**, or a **Bill Lunge**. They also use a **Bill Jab**, which does not have the extreme movement of the other attacks but is mostly the head and neck jabbing at the opponent. If contact is made, a fight may ensue. **Face to Face Fighting** is on the ground and **Aerial Fighting** is in the air. Two herons fly up from the ground facing each other, lunging bills and thrashing feet, often calling. The feet seem to be an important part of Arial Fighting, perhaps in fending off the other bird.

Commensalism and piracy

Beyond threats and fights, other kinds of interactions occur on the foraging ground. **Following** occurs when one bird trails behind another, or even a terrestrial animal. Following may be directed to any part of the followed animal, called the 'beater'. It has been shown in many studies that Following birds often derive a benefit from the behaviour, being more efficient or effective when following than when not. Leap Frog Feeding has been interpreted as a flock doing its own 'beating'. Birds also

3.15 Cattle Egrets following along side a grazing water buffalo.

3.16 Black-crowned Night-Heron in day roost.

indulge in **Pirating**. This involves robbing other birds of their food. Usually it is done in a Supplanting attack, which forces the victim to drop its prey. Herons rob other herons and also other species of birds foraging with them, such as ibises. In turn they are subject to robbing by both larger wading birds and by small agile birds as well. The energetics of prey robbing are not well understood, but it may not be an efficient foraging behaviour, and so may have additional functions such as dominance, or may be used when feeding efficiency is not required (Kushlan 1978c, Garrido *et al.* 2002).

Roosting

When not feeding, herons rest or repair to a roost, often with other birds. All the defensive behaviours used on the feeding ground are used at roost, if practicable. Threat and defensive behaviours are used between adjacent roosting individuals and between arriving and already-roosting individuals. Different species have different roosting patterns. Nocturnal species roost during the day. Diurnal species roost at night, but also during lull periods of the day. The large species tend to spend less time loafing than do the smaller species (DuBowy 1996). Some species that forage solitarily gather in single or multispecies roosts at night.

Roosting birds also conduct comfort activities such as sleeping, **Preening** and **Sunning**. Preening involves the bill, the toes, and rubbing. They use the bill for nibbling and stropping feathers. The pectinate middle toenail is used for scratching, combing, and spreading powder down. To sun, birds assume one of several wing postures, **Spread Wing** sunning, **Droop Wing** sunning, and **Delta Wing** sunning. The latter is an 'at ease' position, the wings drooped in a delta shape.

Food

The purpose and consequences of much of this behaviour, of course, are to get food. With a few exceptions for scavenging (e.g. Klapste 1987), herons are predators on living prey. Food taken depends on the interaction between food availability and the feeding techniques chosen, which are determined by the heron's size, shape of the bill, and the lengths of neck and leg. Within the limits thus imposed, each species' repertoire of behaviour is engaged to obtain its food. Prey differ in profitability, depending on their size, morphology, behaviour, and accessibility due to conditions of habitat, water, wind and sun, proximity to other birds, and the individual's particular skills and experience.

Taken as a whole herons eat almost all the available kinds of animal food within a size range of 1 cm to 1 m. Species and individuals can have a broad diet. The most important foods among all species are fish, crustaceans, and insects. Types of prey within these categories depend on the heron and the feeding site. The ranges of prey eaten by each species are noted in the species accounts. Food may include snails, bivalves, small mammals, young birds, reptiles, and all types of aquatic invertebrates and their larvae. But these are usually supplements to the basic diet of fish crustaceans or insects. The swallowing of small amounts of vegetable matter is occasionally reported, but this likely is only incidental to the capture of prey. However, scattered reports of herons deliberately eating vegetation (e.g. King 1985) should not be discounted without further investigation.

Fish are certainly the predominant prey of most herons in most situations. As discussed before, certain physical features correspond to increased ichthophagy, namely long thin bill, long neck, and slim stature. In any comparison among herons, the longer billed and thinner billed birds are likely to eat more fish. Goliath Herons are reported to take fish 1.4 kg, a weight larger than an average heron. It spends its life waiting patiently for a few very large fish. On the other extreme, small and medium herons tend to take small fish. There are many species eaten around the world, but one of universal interest is the mosquitofish (*Gambusia*). A native to North America, it has been introduced around the world for mosquito control. It is an important part of the diet of medium herons in southeast North America, and similarly now appears in and even dominates heron diets in Europe and Australia.

Fish come with a range of defensive mechanisms that the herons have to cope with. Small fish are swallowed quickly by the open-gullet lunge. But large, long, spiny or uncooperative fish need to be beaten, stabbed, and bitten, which influences foraging preferences (Werner *et al.* 2001). Herons have been described as repeatedly dropping a prey item rather than consuming it, a behaviour called **Prey Dropping** by Kelley (*et al.* 2003). Handling large prey takes time, may be dangerous (Forbes 1988), and opens the heron to piracy, but the energetic reward is high. Fish by their behaviour make themselves more or less available. Herons specialize in seeking out situations where fish are particularly vulnerable, such as drying pools, fish kills, tidal creeks, tide pools, and seasonal or daily periods of low oxygen.

Crustaceans, especially crayfish and crabs, are important prey in many situations. Some species and populations specialize in these prey. Thick, strong bills are associated with a cancivorous habit. The Yellow-crowned Night-Heron is the best example. The Striated Heron of the Galapagos is another example of a population showing adaptation in its bill morphology to crab eating.

Herons seem not to be particularly well designed to be insectivores, but these prey none the less constitute an important food source for many species. Aquatic insects, including insect larvae, are caught by most herons in the same way they would catch fish. Terrestrial insects are a bit different, requiring Flycatching, Gleaning, or Pecking types of behaviour. Of the many types of insects, orthopterans (grasshoppers, locust, crickets) appear to the prey most sought after and captured by herons. Following, Neck Swaying, and Head Swaying are behaviours that seem designed for effective grasshopper catching.

Foraging strategies

Many species can be seen to accommodate their foraging tactics to the local situation. Having a repertoire of behaviours to choose from, individuals feed in various ways. Herons will choose behaviours, strike rates, prey targets, and even the direction of movement based on relative foraging effectiveness (e.g. Davis 1986b, Itzkowitz and Makie 1986). The entire array of behaviours is available to day-feeding herons. Standing and Walking, of course, are the most frequently used. Behaviours that chase down fleeing prey are restricted to very shallow, and very clear water. Aerial feeding and other high-energy feeding techniques are used where the energy return is relatively high. Often studies find that herons have about the same success in different places using different behaviours, owing to their ability to match habitat with behaviour (Draulans and Hannon 1988). Conversely, different species may have very different success rates depending on behaviour. For example, Snowy Egrets using an active foraging behavioural repertoire had to forage twice as long as did Tricoloured Herons or Little Blue Herons (Kent 1986c).

The ultimate behaviour switchers are the Snowy and Little Egrets. For example, a typical Little Egret foraging sequence might be: Walk slowly with frequent halts, Stand and wait for a prey item to show itself, followed by Foot Stirring in the mud or plants aimed at hidden prey. In open situations, the egret might then revert to disturb and chase behaviours, such as pursuing swimming fish with Open Wing feeding, Running rapidly back and forth, and Hopping from place to place. In rice fields, Little Egrets used Foot Stirring with Walking, whereas in streams Walking alone consumed more feeding time (Fasola 1994). Foraging choices can translate into prey capture differences (Fasola 1994, Yamada 1994, Maccarone and Brzorad 2002). The Little Egret feeding with a Walking slowly—Foot Stirring repertoire takes small prey, averaging 4 cm. Feeding by Standing, it takes larger but fewer prey. By using behaviours best associated with the particular prey situation, a heron can maintain a sufficient capture rate (Kushlan 1978c).

Herons feed by both day and night. Some herons feed all day. Perhaps the most common feeding times are dawn and dusk, when energy needs are greatest. Also in the morning, prey may be maximally available. During the day, herons rely almost entirely on visual cues, seeming to use few if any tactile behaviours. Refraction, turbidity, vegetation, and substrate disturbance all have been shown to detrimentally affect feeding success. Herons use various approaches to accommodate to these problems. Probably the most common response is site selection. Through the course of the day herons continually explore their environment, trying out new places or new situations. They end up finding and choosing the sufficiently productive situations among those available.

Some herons specialize in night feeding. The typically night-feeding herons, such as night herons, feed by sight. Other herons also feed at night, including bitterns, tiger herons, and large *Ardea*. Some feed at night when energy demand is high, others when tidal conditions make food available during the night (McNeil *et al.* 1993). Some herons feed at night near artificial lights. Night feeding is not atypical among any of the herons and more and more we are discovering that few herons are obligatory day time foragers.

Fish that are caught visually often are caught near the surface. Efficiency of capture decreases with depth of prey in the water, which enhances their ability to escape. To capture surfacing fish, Standing is a good strategy, if the fish surface frequently enough. In the tropics many fish breath air or get oxygen from the highly oxygenated surface layer. These are particularly vulnerable to large Standing herons. On a diurnal basis, fish have to respond to varying oxygen levels in the water, which are lowest about dawn. This is when many fish tend to swim to the surface and there become susceptible to early-foraging herons.

Food requirements vary seasonally. During most of the year, most herons appear not to be food stressed. Although herons feed vigorously in the morning and the evening, they repair to day roosts during midday rather than continue to feed. During nesting, the situation is different and they are pressed to deliver sufficient food. It is then that they often choose to feed all day or at night, to fly long distances, or to use atypical behaviours.

Feeding dispersion

There is much variation among herons in the degree to which they are dispersed or concentrated while foraging. Some feed only solitarily, some in pairs or family groups, some in mixed species aggregations, and others in more-or-less coherent flocks. The consequences of feeding dispersion are profound.

Solitary species tend to be cryptically coloured and feed by Standing or Walking slowly in well-known habitat patches. Even highly social species, such as the Little Egret, may forage in solitary territories at times particularly when energy demands are not high, such as in the non-breeding season (Chu and Wang 1988). Solitary feeding especially occurs along the coast, inland where prey are not concentrated, in dense habitat. A solitary bird is usually at least temporarily resident and defends a feeding territory to which it returns daily. Advertising by long-distance calls, Supplanting Flights, Upright displays, and fights are not uncommon. Defence may be vigorous, and intruders may be killed. That the result can be so drastic may be due to solitary birds and solitary species not having developed the sophisticated inter-specific communications of more aggregative herons. Solitary birds also need to make trade-offs between site fidelity and exploring new feeding opportunities. The biology of solitary feeding in herons needs to be better explored.

3.17 Herons foraging socially.

Aggregative foraging occurs where prey are collected in high density, where prey are especially vulnerable, or where feeding habitat is scarce. Multispecies aggregations accumulate simply when birds see other birds foraging and join them. This is the reason white birds are usually the core species within an aggregation; they are easiest to see. The accumulation of birds at a location has been called local enhancement. Birds flying from a roost or colony, searching for a feeding site, are attracted to other herons and join them (Kushlan 1977, Fasola 1982). It is likely that patchy food distribution, which is effectively accessed through aggregative foraging, is a strong correlate of coloniality in herons (Toquenaga *et al.* 1995). The availability of highly concentrated food accessed by aggregative foraging can increase reproductive output compared with situations where such concentrations are not available (Hafner *et al.* 1993).

Aggregative groups may consist of up to 20 species and hundreds to thousands of individuals. Feeding in aggregations, those species that can catch stirred up and moving prey can benefit from joining others to feed. Several studies have shown that for these kind of herons there is a subtle commensal benefit to aggregative foraging. This is why it pays to be a white bird and have other birds join you at places where food is highly available. Herons, depending on species and circumstance, have been shown to expend less energy, have a higher striking rate, higher capture rate, or have less risk of failure by joining aggregations rather than feeding alone. Aggregations also increase the risks associated with a higher degree of social interaction leading to aggression, displacements, and piracy. Time expended in threats, defence, and other social interactions can be high and be distracting to feeding. So benefits of increased food intake, decreased energy used, or decreased variability need to exceed the costs of increased interactions. Egrets appear to make choices among feeding sites relative to these constraints (Gerharot and Taliaferro 2003).

A well-studied aggregative foraging species is the Little Egret. They form single and mixed species aggregations at sites of high food availability such as when low oxygen forces fish or prawns to the surface (Cezilly *et al.* 1990, Kersten *et al.* 1991, Hafner *et al.* 1993), where they are effectively captured. As the day goes on and dissolved oxygen is replenished, birds move on to more solitary foraging. Striking efficiency and capture rates are higher when birds are feeding in aggregations than alone, especially for juveniles. When foraging aggregatively different species can overlap broadly in prey consumption, undoubtedly due to the abundance of food at such sites (Fasola 1986). The variables affecting aggregative foraging are complex, involving, in addition to prey availability, such factors as the intensity of the foraging behaviours chosen, time of day, number of other herons, and age, sex and size of individual heron (Draulans *et al.* 1986, Nota 2003).

Heron aggregations tend to be composed of birds acting individually, but drawn to a common feeding site. Once there, as noted above, the interactions with other birds, of the same or different species, are intense and birds often must alter their foraging behaviour accordingly. Aggregating species show similar aggressive behaviours that are interpretable by other species with whom they are interacting. Some species, such as the Cattle Egret, may be considered to flock, a situation in which the group as a whole comes and goes more or less together, and shows its own internal coherency, such as by Leap Frog Feeding. Heron flocks tend to be temporary, as individuals join and leave independently.

Some species, in fact most of the typical herons, may choose to feed at times individually and at times in groups. Solitary feeding birds tend to defend territories while birds aggregating must adjust to the close proximity of others. Some species are territorial during some part of the year and aggregative at others. Grey Herons and Agami Herons feed individually even when nesting colonially. Great Blue Herons may alternate between territorial and nonterritorial feeding within the course of a day. The biology of individual choice of feeding sociality is poorly understood. In many cases, the choice is one of defensibility of resources, with birds feeding aggregatively when food is indefensibly concentrated and solitarily when it can be defended (Kushlan 1978b, Fasola 1982, Hafner *et al.* 1982, 1993, Kirsten *et al.* 1991). But the complexities are great. Nota (2003) demonstrated that Little Egrets in fact show three types of territorial dominance, highly exclusive territories to which a bird returns over a long period, moderately exclusive territories the integrity of which is sometimes compromised by intruders, and situations where exclusiveness is minimal. The body size of individuals correlates with the ability to hold the most exclusive type of territory. So not only food supply, but individual characteristics and choices affect feeding dispersion in herons.

Herons are remarkable in finding sites of temporarily high prey abundance. If there are herons nearby, it seems that few locations having abundant prey are overlooked. The dynamics of heron feeding at these patches is intriguing. Fishing success tends to increase with increasing prey density, to a point, due to higher encounter rates and the use of more intensive hunting methods. Aggregations generally form around highly visible, white plumaged, smaller species, such as the Snowy Egret or Little Egret. Birds join as they see other birds feeding. Intake is highest when birds first arrive at a feeding site, declining with time. Capture rates are high not only because of the high prey availability but also the commensal effects of the crowd. As feeding efficiency or intake wanes, herons move off to find another spot of high density. Herons tend to feed opportunistically on the prey that are available and are the most profitable (Fasola 1994) and different species have a different array of responses. So the success of aggregative foraging depend ultimately on the foraging behaviour a heron chooses to use (Smith 1995b).

Age and foraging

As shown by numerous studies, juvenile herons tend to be worse foragers than adults, suggesting there is a need for them to spend time perfecting the complexities of heron feeding. The training period needed for juveniles to attain proficiency may explain delayed maturity in some herons. It is particularly the case for large birds, Standing feeders, and those that specialize in complex disturb and chase behaviours. It is not clear how much of this ineffectiveness is due to juvenile inexperience or to dominance of adults at prime sites. It may be the case that juveniles of some species forage differently than adults, a possibility worth research. Juveniles appear to be less flexible than adults in use of time, but more flexible in use of space

Food supplies

Foraging success is linked intimately to food supplies. Similarly, feeding success and reproductive success are tightly linked. Herons tend to nest during the seasons of relatively high prey availability. In the temperate zone this is late spring and summer. In the tropics, nesting is more associated with the rain cycle. In coastal areas, it is geared to seasonal fluctuations in prey. In tropical coastal environments nesting may occur year round. In each case, it is ultimately food availability that determines the nesting cycle. The success of herons is accessing food determines their reproductive success (Hafner *et al.* 1986, 1993). Many aspects of heron reproduction can be interpreted as being accommodations to their food supply.

Energy demands in the non-breeding season are lower than when raising young, and so the imperative to maximize energy gain or efficiency is less. However in winter, birds can again be at risk in that poor weather conditions can disrupt foraging, leading to weight loss and even death (DuBowy 1996). Under very severe winter conditions, a large proportion of a local population can die, with persisting population consequences.

The foraging of herons may have substantial ecological consequences with regard to prey populations or energy flow. Their predation has been shown to benefit prey population by ensuring survival during drought and by culling diseased fish in aquacultural situations. Predation by a colony of herons over a nesting period involves a substantial movement of energy and materials around the landscape (Baxter and Fairweather 1994).

Most species of herons take advantage of man-induced food supplies. In some places more herons feed in man-altered habitats than natural habitats. In fact, this flexibility is probably the reason that so many heron species remain secure. Ponds, reservoirs, rice fields, aquacultural facilities, managed wetlands, ditches, pasture, farm fields, suburban and urban environments, and parks are used by one species or another. They also interact with humans more closely by appearing in backyards, at restaurants, on fishing boats, hauled nets, and garbage dumps.

Overall, the resource-use strategies of herons revolve around making use of diverse foraging opportunities as these present themselves. Other aspects of their population biology, such as seasonal migrations, intra-regional population movements, and dispersals, permit herons to make use of temporarily available food supplies. Herons use a wide variety of food sources and habitats. Especially during nesting, however, they often depend on superabundant but ephemerally available prey concentrations. With a diversity of behaviours, foods, and habitats to choose from, it is clear that herons have the potential to take advantage of many natural and artificial feeding opportunities.

Feeding and habitat

Herons as a group use a tremendous variety of habitats, from mountain forests to coral reefs, and most sorts of aquatic habitats in between. Additionally, terrestrial foraging sites from savannahs to road sides are used. And herons can be found in man-made habitats everywhere. The availability of rice fields and crayfish ponds have actually influenced population sizes and distribution. They are just as likely to turn up in a garden goldfish pool. The common denominator in habitat choice is a match among food supply, foraging conditions, heron morphology, and foraging styles. An appreciation of these species-specific factors is often sufficient to predict the presence or absence of herons in many situations.

Some species are true generalists. Many of the *Egretta* are good examples–Little Egrets and Snowy Egrets feed in many kinds of aquatic and terrestrial habitats. At some point it is likely that some Cattle Egret somewhere has attempted to feed in every sort of accessible spot. Large birds specialize in deep-water feeding, small birds in perching next to the water. Studies have repeatedly found quantitatively that long legged herons feed in deeper water than short legged herons. It actually is not a trivial question as to why long legged herons should choose to feed in deeper rather than shallower water.

Some species are more specialized. Terrestrial herons have feet, legs and behaviour specialized for dry land walking. Other species are even more specialized, the tiger herons and *Gorsachius* night herons being the best examples, most of them being forest herons–a heron way of life that we do not well understand. Zigzag, Agami, and Boat-billed Herons certainly are constrained by their morphology.

Herons will choose from the available habitats. It is often the case that a particular heron has the opportunity, on a daily basis, to choose between habitats-for example, inland vs coastal, lake vs stream, large wetland vs small wetland. Several species may choose the same place, especially if prey is superabundant, or species choose different habitats. For example, Great Egrets chose small ponds and Great Blue Herons chose larger waterbodies (Custer and Galli 2002). Studies tend to show they make such choices based on their ability to capture sufficient prey with sufficient efficiency to meet their, and perhaps their chick's, daily energy needs. On a typical day herons spend much time resting, and feeding in various habitats tends to result in similar food intake. These patterns suggest that energy is not normally in critically short supply. Given the potential for using a wide array of habitats, the choices that herons make determine in large part the outcome of their daily foraging.

4

Nesting biology

The archetypical heron strategy may be characterized as serially monogamous colonial breeding. Reproduction barring mishap, occurs once per year at a time of mild weather and seasonally abundant food supplies usually determined by annual temperature and rainfall cycles. Special breeding plumes and seasonal skin coloration, some of which persist longer into the nesting season, are characteristic. The male heron typically selects a nesting territory, which he defends and from which he advertises using calls and displays. The female heron selects a male by succeeding in gaining entrance to his territory and persisting there. Both members of the pair share nest building, usually with the male taking more responsibility for supplying material and the female for construction. They also share incubation duties and care and feeding of the young until newly fledged birds leave the colony site, but they do not tend them afterwards.

This general pattern is the norm, but not universal. As would be expected, species in the family display a range of breeding strategies. Herons may nest in huge, mixed-species colonies, small, intimate, single-species colonies, loose assemblages of nests, or alone. Pair bonds range from monogamous, to opportunistically promiscuous, to polygamous. Courtship ranges from extravagant to hardly existing, or at least not reported. Nest building and parental care may be the sole responsibility of the female. Parental care can extend for many months after nesting is completed.

Nesting habitat

As water birds, herons typically nest in wetland and other aquatic situations, usually on some sort of patch of ground that is protected from disturbance and ground predators. This can be a true island in a lake or offshore, an island of trees or reed within a wetland, or a patch of trees or reed some how set off from a larger expanse of swamp. Islands offer protection from terrestrial predators. Even the terrestrially feeding Cattle Egret tends to nest in colonies over water or surrounded by water, along with more aquatic herons. At the other extreme, some very aquatic herons, such as the Great Blue Heron and Grey Heron, may nest in tree stands over dry land. These nests are often very high, gaining protection from that feature. And the sites are often at least damp below. Herons nesting on offshore islands gain protection by their isolation.

Many of the marsh-nesting species seem to depend on reed beds and grass beds dominated by either cane (*Phragmites*), especially in Europe, or cattail (*Typha*), especially in the Americas (Barbraud *et al.* 2001b, Parsons 2003). The large bitterns are particularly well adapted to this habitat. Small bitterns nest in reeds but also in bushes within the wetland, a situation avoided by large bitterns. Nests are made of stems of the reeds or of small sticks inserted and woven into the reeds. Large reed-nesting herons start with a trampled spot at the water's surface and build up from there. The large bitterns start with a mass of reed stems. Small nests may have a cover of bent-over stems above while the large nests of the Purple Heron are exposed and can become huge, 1 m or more high.

Trees and bushes provide more support for nests than do reeds, and these are used by most heron species. These nests are made of twigs gathered from the surrounding area and inserted to create a platform or cup. Some nests are small; some are very large being added to year after year. Nests of

4.1 Mixed species heron colony in Indonesia. Birds are predominantly Cattle Egrets.

the great herons can reach 1.5 m in diameter. Nests may be placed near the trunk or out on the tangle of leafy branches and may be close to the ground or higher than 30 m. Larger herons usually take the upper layers where they can more easily come and go. Herons on islands may nest in low vegetation or on the ground. In some case, such as the Grey Heron of Coastal Mauritania, the ground nest may be nothing more than a few sticks and bones. In most places nests do not survive from one season to the next. Where they do, they will be reused, or at least the twigs will. Nesting is hard on the trees, both through breakage and the effects of nutrient build-up. Some colony sites are destroyed by heron nesting activity, causing the colony eventually to shift to another site (e.g. Julin 1986). One of the best documentation of nesting herons affecting vegetation is by Telfair and Bister (2004). Over a 28 year study period beginning at first occupancy, the nutrient buildup from nesting Cattle Egrets killed the native vegetation to be replaced by exotic plants, brought to the site by the egrets themselves as nest material. Egrets used the new vegetation but also began nesting on the ground.

In the tropics, mangroves are a universally important heron nesting habitat. Mangrove swamps line rivers as gallery forests even well into fresh water. They cover parts of islands, edge lagoons and bays, and occur as extensive coastal forests. Mangroves are hardy, physically complex trees offering good nest support. They achieve best development in places where nutrient accumulation is flushed by tidal or stream flow, and mangroves are well adapted to physical disturbance. So mangrove trees are somewhat resistant to the impact of the herons' activities. As a result they make an ideal nesting habitat. Supporting the diverse tropical heron fauna, mangrove colonies in some areas can become extremely large, involving tens of thousands of pairs and eight or more heron species.

Herons by and large are lowland species. Some are coastal, occurring only on or near the water to about 100–200 km inland. These nest on offshore islands, lagoon islands, mangrove swamps, coastal marshes and along coastal rivers. Species such as the Sumatran Heron, Eastern Reef-Heron, and Reddish Egret depend on coastal systems during nesting. Even apparently inland species are relatively coastal on a continental scale, examples are the Tricoloured Heron, Rufous Night-Heron, and Pied Heron. Other species, such as the Slaty Egret, nest exclusively in low-lying, inland swamps and marshes.

Exceptions to lowland nesting are of interest. Some species occur in low mountains and foothills, usually nesting in swamps or moderately flowing streams. The White-bellied Heron, White-eared and Japanese Night-Herons, and tiger herons nest in this habitat. The Rufescent Tiger-Heron and the Agami Heron are birds of forested mid elevation mountain streams nesting in trees and bushes overhanging the water. Sometimes herons nest at great altitudes along mountain streams or high plateaus, the champion probably being Black-crowned Night-Herons at nearly 4800 m in Chile. But herons also are reported from mountains in New Guinea, India and Nepal, and New Zealand.

Nesting season

Herons tend to nest at a time when prey is readily available and climatic conditions are mild. So temperature and water conditions in feeding areas are the two major factors determining the nesting season, principally because they affect prey abundance and availability. As a basically tropical group, herons have more species and more individuals in the tropics of all continents than in their corresponding temperate zones. Some species, such as the large bitterns, night herons, and large *Ardea*, nest far north. Seasonal patterns of prey availability differ substantially from the tropics to the high latitudes.

Herons in the tropics tend to nest in response to hydrological conditions. They may nest in the big rainy season, in the small rainy season, in the declining arm of the rainy season, or when rivers and river swamps are flooded or receding which may not be synchronized with the local rainy season. Thus it is generally rainfall's affect on foraging area rather than rainfall itself that is correlated with heron nesting. It is often the case that nesting begins at the peak or just past the peak of the rains at a time when the swamps are drying, thereby providing shallow water and aquatic prey that are concentrated by falling water levels. In other cases, particularly where the dry season is long, it is the wet season itself that provides more food. The nesting schedule in the tropics, even for a species, varies from place to place in response to these factors.

Nesting along the coast, is less constrained by hydrology and temperature. Under these more constant environmental conditions breeding, especially along tropical coasts is often protracted, and some birds may be nesting at any time of year. Coastal nesting may, however, be influenced by oceanic conditions that affect food supplies.

The farther removed from the tropics, the more temperature cycles tend to govern nesting season. A climate suitable for heron nesting needs to have sufficient time to allow herons to nest in freeze-free months when prey are available. In the temperate zones, nesting tends to be early, in the spring and early summer, ending by mid to late summer. The timing of frosts and freezes set breeding range limits for temperate and subarctic species.

Nesting sociality

Herons can nest solitarily, loosely colonially, or colonially. Nesting strategies tend to follow taxonomic lines. Most Ardeinae, Cochleariinae, and Agamiinae are colonial or loosely colonial. Botaurinae and Tigrisominae are solitary or infrequently loosely colonial. Given our understanding of the phylogeny of the family, this pattern is suggestive that coloniality may have been the primitive state for the family and that solitary nesting of cryptic species is derived.

Solitary breeders nest away from other birds, usually in defended territories that may incorporate their feeding area. This is typical for the small and large bitterns, and tiger herons. The larger *Ardea*, pond herons, and *Butorides*, although colonial, can also nest solitarily when appropriate. Bitterns and typical herons nest occasionally in loose colonies, in which the nesting birds are gathered at the landscape scale but are not usually in sufficient proximity to be in regular interaction with neighbours.

Colonial nesting herons gather in one location and nest near each other. Colonial species are also aggregative in the non-breeding season, roosting in small to large groups at night and other times when not feeding. Colonies may be composed of one species, several species, or many species of herons, and also of other colonial waterbirds such as pelicans, cormorants, darters, storks, ibises, and spoonbills. Non-waterbirds such as crows, hawks, and eagles are attracted to nest near colonies. Colonies can range from a few birds to huge numbers. In 1962 a colony in Tanzania supported 40 000–50 000 pairs of herons; and in 1986–87 colonies in the Inner Niger Delta together supported about 68 300–70 800 pairs of herons (Turner 2000). More typically colonies number in the hundreds to low thousands of nests.

Within a mixed-species colony, species behave differently. There tends to be a sorting of species due to dominance, body size, and nest site preferences (e.g. Burger 1978b, Burger and Gochfeld 1990, Fasola and Alieri 1992a). As noted, large species tend to take the higher positions. Smaller species tend to nest within the canopy. Studies on the role of aggression in determining nest position have not shown a consistent role for interspecies interactions. In general, large species dominate small species, diurnal species dominate nocturnal species, and some species such as Cattle Egrets are particularly aggressive. For marsh-nesting species the situation is reversed. Small bitterns can affix their nest higher in the reed than the large bitterns that need more support. Species sort themselves laterally as well. Some species tend to nest in single-species groups within a larger mixed colony. This usually is the result of displaying males of a species aggregating during the advertising period and then nesting in the same place. This phenomenon also results in colonies being temporally structured, with similarly aged nests together and earlier nests more central in the colony than later nests. Other species choose typically to nest near the periphery of the colony.

The function of coloniality in herons and other waterbirds has been well debated. Of the theories proposed, protection from predators, food finding, and social facilitation of the breeding cycle are three of the most prominent (Krebs 1974, 1978, Burger 1981, Wittenberger and Hunt 1995, van Vessem and Draulans 1986a, Draulans 1988b, Mock *et al.* 1988, Marion 1989). While these three ideas are discussed below with some sense of preference, there is not a firm conclusion, and may never be, as to a fundamental function of coloniality. Their relative roles differ in different species and different situations. Some studies have shown that colony size does not influence nesting success (Butler *et al.* 1995).

Colonial nesting provides protection because birds choose sites that are in some way insulated from predators, ground predators at least. However, herons nesting solitarily could achieve that sort of protection as well, perhaps more so because their nesting sites would not be so obvious. Heron colonies, rather than being a way of avoiding predators, attract them (Bellinato and Bogliani 1995). If that is the case, it is unlikely that coloniality developed as a way to lessen predation pressure.

Colonial nesting birds, then, have to succeed despite predation risks. In most places, predation on unattended chicks is a persistent threat. Large aerial predators, such as eagles, can threaten entire colonies. Birds of prey, Black-crowned Night-Herons from within the colony itself, crows, and other aggressive birds are predators on chicks and eggs. Night herons wander through the colony looking for unattended young. Aquatic predators are also a threat. Crocodilians and fish eat chicks that fall from the nest. And any predator that can swim, like lizards and small mammals, can access colonies. Owing to the higher diversity of predators, predation is more intense in the tropics. However, it is also intense in some locations in the temperate zone, for example in northwest North America where eagles affect Great Blue Heron nesting dispersal and success. Everywhere predators may be critical when previously isolated colonies be connected to the land by falling water levels.

Herons do achieve some additional security from predators because their nest is within a herd of other nests. But herons tend not to engage in active communal defence such as mobbing predators the way terns do. Beyond physically covering their eggs or young, adult herons are surprisingly passive when their young or eggs come under threat. There are two non-exclusive explanations for this. Long-lived herons may be trading off chick survival for the more sensitive demographic parameter of adult survival. And/or active defence is not well developed because it is not regularly needed; the heron's size and presence alone deter most predators. That colonial nesting birds are not more effected by predation than they are mostly seems to depend on individuals being protected by the herd.

If not predation, perhaps food is the primary correlate of coloniality. The function of coloniality in food gathering has been particularly well debated. If there were not sufficient food available within effective flight distances from a colony, coloniality could not occur. Given that food is available, coloniality must be in some way effective in allowing birds to access needed food sources. Few species tend to maintain regular feeding sites while nesting and still nest colonially. Those that do are mostly large or coastal species, such as the Eastern Reef-Heron and Tricoloured Heron. Rather, most colonial nesting species feed on ever-changing food resources, affected by tide, seasonal hydrology, daily availability patterns, and depletion. They fly from the colony to find food and usually feed in association with other birds from the colony. Flight distances vary with species, but generally are less than 20 km. Distance to feeding sites constrains colony development and size, as is suggested by the regular dispersion of colonies in uniform habitat. Despite much effort, it has not been shown that much information on food availability changes hands within or near the colony. However, it is totally clear that birds find food by seeing other birds from their colony already feeding at a site. This phenomenon is sufficient to explain how colonies function to support food acquisition when food resources are patchy and changeable (Toquenaga *et al.* 1995).

Coloniality also provides opportunities for mate selection, social stimulation to start and synchronize the nesting cycle, and increased choice for mate selection. Whether these factors alone are enough to explain the existence of coloniality is not clear. However, the many herons that are solitary or nest in very small colonies seem not to require the massive social stimulation of an expanded group. So why should coloniality be required for breeding initiation in some species and not in others. It seems likely that social facilitation is a secondary benefit of coloniality.

Life within a colony or roost depends on interactions and communication among the birds, including among different species. Herons that gather in colonies, perhaps unlike solitary residential birds, probably do not know the ever-changing line-up of neighbours. There is not an individually based hierarchy as exists in truly social birds. Nor are there group actions. Interactions are individualized, one on one. Disputes are usually won on the basis of residency within species, and by size and aggressiveness between species. Signalling in herons is sufficiently similar among species to allow several species to communicate. Interactions, both intraspecific and interspecific, are especially critical during the courtship phase of the nesting cycle. The interspecific aspects of heron communication has been little studied, and so is only marginally understood.

Courtship

The beginning of courtship is similar among herons. Courtship can be divided into three stages, Solo Male, Bachelor Stage, and Paired Stage. In the solo male stage, an unpaired heron selects and defends a display/nest site. He displays spontaneously, using behaviours and calls detectable from a distance to advertise to females and defend against males. When a displaying male has attracted potentially interested females, he enters the bachelor stage where he performs displays oriented towards a potential mate. Finally, upon pairing, the birds use close-contact behaviours to reinforce the pair bond and continue to reduce aggression.

As discussed previously, unusual feathers, plumes that moult into full development just before breeding and are featured in the display, postures, and movement characterize nesting herons. That these plumes are more used in threat than in sexual display is a major advancement in understanding heron biology. Why should such elaborate structures be used so prominently in threat displays? Clearly threat must be an important part of breeding. In fact, the role of territorial protection through threat displays can scarcely be overstated. Threat displays are the same between different sexes of a species and tend to be similar among heron species; and they are very unambiguous. This is probably not coincidence since, unlike courtship displays, threat displays must communicate across species. Gaining and holding a nesting territory is essential to reproduction for the security of the nest itself (from pilfering), of the female (from promiscuous advances), and of the eggs and young (from predation). A heronry by definition is full of predators, quite willing to eat small birds. The various threat and defensive behaviours described in herons serve not only their original territorial purpose but also as courtship displays to the extent they provide information on the defensive abilities of male to prospecting females.

The changeable soft part colours also are an important element in the breeding biology of herons. Colour changes in the face (lores), bill, legs, and feet carry information on the physiological and emotional status of the individual and likely species recognition information as well. These colours are featured in various displays. Changes that last days and weeks have long been documented, although this information is still incomplete for most species. That colours also change over very short time-spans during courtship adds additional complexity. For many years there have been reports of atypical eye and loral coloration in herons that did not fit into either the known non-breeding or breeding colours. Many of these now appear to be transitory colours that last from days during courtship to minutes during aggression. In bitterns the colours can come and go within a few minutes as the excitement level of the bird changes. These colour changes provide intense signals during courtship, defence, aggression, and other critical activities. Much more needs to be understood about this phenomenon.

A species-specific courtship repertoire is used to attract, pair, and maintain contact with a mate. They clearly serve as isolating mechanisms among the closely related species cohabiting in a colony. Some displays are gender restricted. The repertoire and the behaviours for most species are usually not

4.2 The back plumes and crest of the Reddish Egret.

clearly described or understood. This is because they are complex; birds are individuals; the behaviours tend not to occur when observers are nearby; and the observers themselves have to interpret what they see. So characterizing and understanding the behaviours listed below and others described in the species accounts will require much additional research. Courtship behaviours do have several elements in common (Voisin 1991). They tend to use the back plumes, but not always other plumes (which are used more for aggression). The bill is not directed toward the other bird, but either down or up (probably to avoid signalling a serious threat).

Courtship begins as males select a display territory and advertise from it. Solitary birds nesting in thick habitat, such as bitterns, advertise with long-carrying calls. Some species have been described as displaying and courting away from the nesting area, but the function of these interactions is not conclusively demonstrated. Boat-billed Herons appear to pair before arriving at the nesting site. For solitary nesters, the display site may be within an already established feeding territory. For colonial species, males may be in competition for prime display sites and vigorously defend them with aggressive displays that probably also serve an advertising function. Advertisement is abridged (and sometimes not yet reported) for species that are solitary, sedentary, and occupy the same ground year round. It is thought that these birds may be reconnecting with individuals they already know.

The male advertises using posture, display movements, plumes, advertising calls and songs, and his coloration. He advertises his presence by signalling from a place he selects, which in most species is near the eventual nest site. A female heron, attracted to his displays, chooses whether to engage him in mutual courtship. Using both acoustic and visual signals the male displays and defends his site. Advertisement and defence are intimately connected. Males initially use defensive displays against prospecting females as they would against any intruding bird. The postures and behaviours of females are characterized by appeasement. They have slicked plumage, approach surely and quickly, and stand in a **Withdrawn Crouch**. They are often attacked several times before being allowed to stay. Gradually, using appeasement, stealth, and perseverance, a female insinuates herself into the male's space, and she is gradually accepted. Advertise-ment and defensive displays are supplemented by mutual contact displays. But even into nesting, the defensive tendencies of the sitting bird need to be overcome by approaching birds through use of **Greeting Displays**.

The courtship behaviours of a number of species have been well described in studies such as those of Meyerriecks (1960), Blaker (1969a), Tomlinson (1974), Mock (1976), Wiese (1976), Rodgers (1978a, 1980c), and Viosin (1991). The repertoire of heron courtship displays is described below, with the terminology that is used in the species accounts.

Of all heron displays the **Stretch** is the most typical. Some form is used by most species. It is a primary male advertising display and used also between the pair. The Stretch has been observed in most typical herons so far studied. In its general form, it consists of the heron lifting its head and closed bill towards the vertical. At the peak, a call may be given; in the Great Blue Heron it is a moan or howl. As the head is brought down the legs are usually flexed, and in some species a bill snap is given. The crest and the plumage of the lower neck and scapulars are usually erected. This is a very conspicuous display, one that accentuates a heron's specialized plumes and the bright soft-part colours of the head and bill. In some species the head is down, in others it is placed all the way to the back, the Stretch of the Great Egret perhaps being among the more sensational with its head back and huge plumes erect and waved. The Stretch is especially characteristic of *Ardea* and *Egretta* and is used not only in advertisement but also in the Greeting Ceremony of the *Ardea*, including the Grey Heron, Purple Heron, and Cattle Egret—evidence of the affinity of Cattle Egret.

The **Snap** is also a typical heron display, although variable in form among species and individuals. It is primarily a defensive display, the last behaviour before an attack occurs. But it is also used for male advertisement in that it may convey a lot of information about the male and is the display that the female has to overcome to gain access to the display site. Generally the displaying heron, with its head and neck feathers erected, moves its head smoothly forwards and downwards; with the neck fully extended out, it snaps its mandibles shut,

4.3 Crest and plumes of the Snowy Egret used in aggression.

typically while bending its legs. Variability among species involves the direction of the neck movement (whether down or up), neck curvature, body orientation, and the extent of bending of the legs. Snaps by a displaying heron are generally not directed to their recipient. Some species do not have an actual snap component and some bite at or grab branches. In some species, such as the Reddish Egret, the Snap includes pronounced **Bowing**. In the Black-crowned Night-Heron the head is extended, feathers of head, back and neck are partially erected, and the bird snaps the bill or grasps a twig. Snap-type behaviour is so varied among species that the homologies are not always clear. Its presence in some species is undetermined. The Little Blue Heron has been described by Meyerriecks (1960) as using this display, but Rodgers (1980c) did not observe it during his studies.

The Snap can be connected to other displays, such as the **Stretch–Snap** . In the Tricoloured Heron and the Great Egret, the Stretch and Snap may be performed together in a single sequence (Rodgers 1977, Mock 1980b). The Black-crowned Night-Heron also uses a Stretch–Snap, which also has been called reverse stretch, snap–hiss, and song-and-dance. The bird stands erect, treading from one foot to the other, extends its head and neck forward and down, feathers of back, crown and breast fully erect, eyes bulging. With its bill down, it gives 'Plop' calls, which can go on for many minutes. Between songs it does Preening towards its belly and Twig Shaking.

One of the most conspicuous components of the courtship displays of herons is the **Advertising Call**. The sound of this long-distance signal varies among species, from a gurgle to a yell. The Green-backed Heron gives a 'Skeow', the Grey Heron a yelp. The **Boom Calls** of the large bitterns are remarkable. Such low-frequency sounds penetrate the thick surrounding vegetation.

The **Forward** is the primary defensive, territorial, and antagonistic display, found in most species. The heron erects its head crest or head feathers as well as any plumes on the neck or scapulars, the head and neck are fully extended upwards with bill horizontal, and wings slightly drooped. The heron lifts its wings, retracts its head and then rocks forward and lunges toward another bird, usually giving a call or a bill Snap. The Forward may be the final precursor to an attack. The display is variable among individuals and species. The Forward seems always to be used as an aggressive display in territorial defence. Even young chicks aggressively perform Forward displays while still in the nest. It has been described for nearly all species studied, although its presence in the Boat-billed Heron and, until recently, the large bitterns has not been well described. **Tail Flipping**, in which the tail is flipped up rapidly, also is an antagonistic display, preparatory to a strike at another bird. In some species it is part of Forward.

The **Twig Shake** is also an obvious aggressive display in which the heron partially extends its head and neck slowly out, grasps a twig with its bill, sometimes with a slight elevation of crest or neck plumes, and shakes it vigorously. There may be a flexion of legs causing a bounce, such as in the Great Egret where it has been called a Bow display. The intensity may become greater with the approach of females or other intruders. The Cattle Egret utters calls when it performs this display. The side-to-side movements resemble the behaviour used to insert twigs into the nest, from which the stereotyped behaviour probably originated. This aggressive or defensive display is aimed inter- or intraspecifically and used by both males and females. It is the heron equivalent of sabre-rattling.

Many displays are derived from normal comfort movements, and in fact continue also to function as such. Some of these named displays are no different than the usual behaviour but when done on the nest site carry such functions as advertisement, defence, and contact. In the **Body Shake** the bird vigorously shakes its body with feathers loosely ruffled. At the nest site, this display brings attention to the bird. **Standing** at the nest or display site also has a display component to it as the bird makes itself obvious and its presence claims the site or attracts interested birds. In the **Alert Posture**, the bird goes beyond Standing and stands very vertical with body feathers raised or smoothed, observing in a highly aware manner. The heron assumes the Alert Posture when it senses a disturbance or threat. This is also called Alarm Posture. In the **Upright**, the heron raises its head up and out until straight, body feathers are raised moderately, and a threat call is usually given. This is a common antagonistic display among herons, used in many contexts, and is also a feeding posture. At the nest sites, it is a mild form of territorial display.

Preening is the same as everyday comfort preening and is not ritualized in the strict sense. None the less it is a recognizable behaviour during courtship and serves a display function when done at the display or nest site. Other preening-like displays have lost much of their original function. In the **Wing Touch**, the bird brings head and neck forward arching it under the chest, to under the wings where it mock touches the flight feathers or front of wing. Given as a display at the potential nest site by the male, it often alternates with the Stretch. It is an advertising and contact display and also a conflict display between males and females in the presence of the other. A similarly derived display is the **Wing Preen**, in which the heron moves its bill smoothly along its primaries as if preening. In doing so it often orients towards the observing heron, but in many cases does not even touch its own feathers. This display has not been observed in all herons, but its subtlety allows it to be overlooked easily. It can be an aggressive display. In the **Back Bite** the bird nibbles at its mate's back feathers. Back Biting is a common form of **Allopreening**, in which one bird preens or mock preens the other aimed towards various spots. Back Biting and Allopreening may have a Bill Clappering component.

Given its obvious centrality to a heron's appearance, there should be little surprise that a number of displays involve neck and head positioning. In the **Fluffed Neck** display the heron raises its head and neck, erects its neck feathers, and opens its bill at a slight inclination. This behaviour is usually used in a low intensity **Greeting Ceremony** between mates, and has been seen in most species studied. The **Head Shake** involves, with neck fluffed and crest raised, the bird shaking its head gracefully. In the **Arched Neck** display the heron erects its neck feathers and head plumes and curves its neck in such a way that the closed bill is aimed downwards. It is usually used by a moving heron as a threat, and has so far been reported from the typical *Ardea* species. The **Wing Spread** involves the wings

4.4 Rufescent Tiger-Heron in Bittern Posture, on its nest.

being spread downward with the bird making rocking movements as if balancing and all plumes partially or fully erect. It is used after a Circle Flight, before a Forward, and while Walking.

Another common display for some species is **Crest Raising**, in which the heron erects its head plumes. This behaviour increases the apparent size of the bird and is used as a threat. The Cattle Egret shows the great complexity of Crest Raising in herons. Fear inhibits the erection of the anterior crest and aggression stimulates feather erection posterior of the crest. Long posterior crown plumes in many species are used in aggressive posturing. So the combinations of feather erection indicate emotional state. In **Bill Wiping** the bird wipes its bill on a branch or any other structure. In **Neck Craning** the bird looks far forward with head and neck outstretched. In **Straddling**, the bird stands hunched over, with back plumes erect, and prances from foot to foot. This is a characteristic display of the Cattle Egret. The most typical heron posture involving the neck is the **Bittern Posture**. Although best developed in bitterns, it is used by most short and moderately necked herons at one time or another. In the Bittern Posture the neck and bill are held vertical in response to disturbance or just for surveillance. In its full glory, the bitterning bird watches horizontally beneath it vertical bill, sways in the breeze along with nearby reeds, and turns slowly to continue to face a moving intruder. Chicks assume the Bittern Posture while in the nest.

Herons use their bills for display in several distinct ways. Newly paired birds engage in **Bill Duels**, which are ritualized attacks in which the male stabs while the female retracts her head, finally lowering it below that of the male. The female may instead seize the male's bill and hold it. This behaviour serves to strengthen the pair bond as the female tolerates the decreasingly intensive attempts of the male to drive her off his display or nest site. Over time, Bill Duels help to reduce the male's aggressiveness towards his new mate. Bill Dueling occurs in Great Blue Herons, Great White Egrets, and Cattle Egrets, and probably in other but not all species. **Bill Clappering** is the rapid chattering of the mandible tips (also called Bill Nibbling). If no contact is made with another bird, clicking noises can result. It may be performed while nibbling on the feathers of a mate, in pushing at another bird, or without making contact with the other bird (Non-contact Bill Clappering). Contact Bill Clappering is a very common interaction between paired birds and is used in making close, appeasing contact with a mate. All heron species studied use some form of Bill Clappering behaviour. The origin of Bill Clappering has been much discussed and is probably derived from preening behaviour.

Most species use flight for display. Aerial display behaviours are used more frequently by smaller herons than by larger ones. These have been variously described and named. **Circle Flights** start from the colony, travel around, through, over and beyond the colony sites and return to about the same place. Herons dangle their feet, call, raise their crest, cut about in curious curves and zigzags, and make species-characteristic flapping noises on slow wing beats. These sounds are rendered as 'whomp, whomp' or similar, and the display is frequently called a Flap Flight. It is principally a male advertisement display, although both sexes may do it in some species. In others several birds will be flying at once. In the **Tumbling Flight** display a bird, and perhaps also its mate, flies upward, and then drops toward the ground tumbling over and over until just prior to landing when it straightens up and alights. **Pursuit Flights** consist of one bird chasing the other around, usually in aggression. In the **Aerial Stretch** display a heron launches into flight while bringing its head and neck into the Stretch display position. It may sway its head from side to side and call. It begins and ends the flight with its plumage erected. A **Landing Call** is often given when alighting, the heron extending its head and neck when doing so; it is often part of a Greeting Ceremony.

4.5 The back plumes of the Great Egret. The displaying bird is showing the Bill Touch display.

Most of these displays are given at the nesting site. Displays given at other places, such as the feeding areas, are poorly documented. Some of these displays likely are purely aggressive, but dominance on the feeding ground might carry over into the nesting colony. The Reddish Egret display is the best described, although still little understood. In this **Dancing Ground** display, the displaying bird uses Head Shaking, rapidly raising and lowering its head with crest erected, Walking or Hopping toward other birds, giving a 'Crog' call. A similar observation of the Great Egret is of a bird advancing on the feeding party with bill up, head erect, stiff legged, and plumes erected. Another feeding ground display is **Peering Down**, in which standing birds bend over with bills touching the water for up to a minute or two. Birds on the feeding area also use **Jumping Over**. The jumping bird head tosses and erects its head, back, and chest plumes and leaps over the other. **Upright and Spread Wing** is also a feeding ground display shown by the Great Blue and Sumatran Herons. Two herons encounter each other, wings spread, fan their tail, droop and spread their wings, extend their neck and head high over back, and become more vertical as the display continues.

Fighting

When territorial displays fail, herons fight. After the Forward, the next level of near contact is **Bill Jabbing**, which is jabbing the bill at the opponent, perhaps making contact. At a longer distance, a **Supplanting Run** occurs when a bird takes off after another to chase it from a spot. When perched birds face off and duel to the point of attack and counter-attack, this is called **Face to Face Fighting**. Birds will stab, bite, and kick. In a **Supplanting Flight**, a bird with its feathers erect and while calling, flies at another bird. The attacker lands on the spot vacated by its opponent or on top of the opponent itself. The display carries a message of impending attack, and may end in a fight if the opponent does not withdraw. **Aerial Fighting**

4.6 Stretch display of the Great Egret.

occurs when two herons fly up from the ground facing each other, lunging bills and thrashing feet, often with the aggressive vocalization. The **Withdrawn Crouch** is a submissive posture given in response to aggressive birds.

Pairing

Given the intense territoriality of the male, pairing takes several days. As birds enter the **Paired Phase** they interact with each other in sophisticated ways. Standing, Preening, and Allopreening increase as the birds spend time together. At this stage and thereafter, **Greeting Ceremonies** occur. These are important because they allow a returning bird access to the nest site. Depending on the species, these ceremonies may be complex and involve such displays as Landing Call, or full mutual Stretches with heads and necks arched up, heads held horizontal, often with a call or Bill Clappering. The mate at the nest may answer the Landing Call. The Greeting Ceremony is first of all a display that allows individual recognition and therefore reengagement after a time apart. But it is fundamentally a sequence of behaviours that appease the intense nest defence of the guarding mate. Nest relief and twig passing are occasions when this behaviour is used.

Eventually **Copulations** begin, usually at the nest site. The female assumes a horizontal posture, without ceremony the male mounts by stepping on to her back, often waving or slowly flapping wings mostly for balance, and brings the cloacae into contact briefly. In some species the male grabs the female with his bill. The male steps off, often Wing Waving. Serial (seasonal) monogamy is the fundamental rule in herons. It is thought, but unproven, that species that live year round in the same place may have the same mate in successive years, but this remains to be demonstrated. Some colonial species engage in promiscuous extrapair copulations (Gladstone 1979). As a consequence, during the Paired Stage the male tightly guards the female. But males prowl through the colony seeking inattentive pairs. When approached, the female, which in many species is smaller than the male, is rather complacent in responding to the new male's advances. The impregnation success of extrapair copulations appears low overall, but probably serves as insurance for the female. It is thought that the large bitterns are fundamentally polygamous, although this has been shown conclusively only in some populations of the Eurasian Bittern and certainly is the case only in optimal habitat.

Nesting

Following courtship, the birds enter the next phases of breeding, **Nest Building**, **Laying**, **Incubation**, and **Hatching**. These have been well documented for a number of herons, as cited in the species accounts.

In some species the males begin to build a nest, but in nearly all, the female chooses the site of the nest and does most of the construction. The behaviour used is called **Tremble Shoving**. The female grasps the twig or reed and positions it in the nest by wiggling it up

4.7 Eggs of the Reddish Egret, nesting in mangroves.

4.8 Nest of Great Blue Heron in an unexpected location, a saguaro cactus.

and down and sideways. Sticks are usually gathered by the male and presented to the female. Greeting Ceremonies at this time often involve stick presentation. This division of labour allows the female to guard the

4.9 Incubating Pied Heron.

nest site, because in colonies if a nest is left unguarded its sticks are quickly stolen. In large bitterns, as far as is known, the male is not allowed near the nest and the female does the building and gathering. In the small bitterns, the male makes the nest alone.

Heron eggs typically are a shade of pale blue or blue green, unmarked, with some changes with time. Small bitterns are unmarked white. Large bitterns are brown. Tiger herons' have some mottling. Eggs are laid at intervals of one to several days, depending on the species. Small bitterns lay every day; large bitterns may lay only every 5–6 days. Most herons begin incubating before the last egg is laid. Since the incubation period is set physiologically, the staggered start has important consequences on hatching time.

Clutch sizes are between 1 and 10, but usually 3–5. Some tiger herons are reported to have a single egg, a situation so unusual among herons that it requires additional documentation. Within a species, clutch size tends to decrease in the tropics and increase in the temperate zone. Clutch size tends to be larger inland than on the coast and may also vary temporally, depending on habitat conditions (Hafner *et al.* 2001). Control of clutch size in herons is only beginning to be understood (Fasola 1998). Brood parasitism may occur, with one species dumping eggs in the nest of another (Gonzalez-Martin and Ruiz 1996).

For most species, both parents incubate; sometimes their schedules have the sexes incubating at different times of day. Incubating birds develop brood patches and some species become very broody. In large bitterns only the female incubates while the male guards

4.10 Nestling of Rufescent Tiger-Heron covered in natal down.

4.11 Asynchronous hatching in Great Egret. Chicks hatching first are older and larger and in large broods or under poor feeding conditions generally have a better chance of survival.

the territory. After egg laying, displays tend to become limited to Greeting Ceremonies given on exchanging nest duties. Generally, during incubation one parent is in attendance on the nest while the other feeds elsewhere. The brilliant soft-part colours used in courtship fade rapidly. Incubation is 2–4 weeks, generally dependent on body size. Small bitterns take as few as 12–14 days. Most herons tend to be single brooded, particularly in the temperate zones, unless replacing a clutch lost early in nesting. Multiple yearly broods are found mostly in the tropics, primarily small bitterns, pond herons, *Butorides*, and night herons.

Hatching is almost always asynchronous, meaning that the oldest egg hatches first followed in turn by younger eggs. If incubation begins with the first egg, all young are staggered. If it begins with a later egg, early chicks hatch together. In large bitterns, eggs may hatch 12–13 days apart, due both to staggered laying and the female having to leave the nest to feed the first young. As the chicks grow, the older are larger, stronger, and dominate the younger. Brood reduction often occurs, leading to the death of the younger chicks, sometimes at the bills of the older chicks, called sibilicide (Mock 1987, Creighton and Schnell 1996). The parents feed the most aggressive chick, not attending to all equally. Much has been written about the causes and consequences of asynchronous hatching. But for the most part, brood reduction brings the energy demand on the parents into alignment

4.12 A Great Blue Heron young begging by grasping bill of the adult to encourage regurgitation.

with its availability for the current year and ensures some chicks (the older ones) survive even if younger ones die. In obligatory sibilicidal species, the younger chicks are killed anyway, and in these cases they serve as insurance against the loss of older chicks. Hatch sequence is critical to brood-reducing species. First and second hatch chicks have higher andrgen levels, and presumably are more aggressive and therefore have higher survival while competing with siblings (Schwabl *et al.* 1997). Survival is correlated with hatch order (Telfair and Blister 2004) and hatch sequence has life-long consequences in that older chicks have been found to be more successful as subsequent nesters in the Little Egret. (Thomas *et al.* 1999). Overall it appears that parental favoritism for early-hatched chicks is advantageous to the parent's long term fitness.

There are several stages of post hatching: **Brooding Stage**, **Guard Stage**, **Brancher Stage**, and **Fledging Stage**. Generally speaking, the young are brooded or shaded for 1–2 weeks after hatching. They are then guarded for 1–2 weeks. A critical stage of development occurs when the chick's energy demands peak. The timing differs among species. The growth rates of the nestling are attuned to this period, and it is then that they are sufficiently developed (particularly legs and bills) to be safe if left alone. At this Brancher Stage, both adults begin to forage at the same time away from the nest and that the quantities of food returned to the nest significantly influence overall chick survival. Young then are independent of the nest but are still fed by the adults. The next stage is when they can fly (fledge). The final stage is when they are capable of independence and they leave the colony and the nesting area. Of course the actual timing differs among species. The young of small bitterns fledge in 2–3 weeks compared with 12–13 weeks for the large *Ardea* herons.

Parent herons feed their young by regurgitating food they have consumed. This food is usually only semi-digested or not at all. Species differ as to whether they put food in the nest or into the chick's bill or have it taken from their gullet by the aggressive chicks. In fact for most species, this behaviour needs to be explicitly documented Young beg with characteristic calls. Adults return to the nest site to feed. At the Brancher Stage, the young tend to encounter the adults before they reach the nest and so they are fed else-where.

Chicks grow quickly and differentially. They are altricial and semi-nidicolous at hatching. That means they cannot walk and have only a partial covering of down. Within a day or so they can stand and eat and their eyes are open. For most species the legs and feet develop relatively quickly and the wings relatively slowly. This allows the chicks to clamber out of the nest early and seek protection within the canopy, reeds, or with other young, with whom they form groups.

Both parents continue to feed the young by returning to the colony, but not necessarily to the nest site. As they grow, young begin to fly around and fly up to the arriving parents. At this stage the young take food directly from the gullet of the adult. Young disperse from the colony site, perhaps following adults to feeding areas. In only a few species, such as the Whistling Heron, has care of the young been recorded off the nesting area. So for most species the young are on their own once they leave the colony.

5

Herons and people

THE PEACE OF WILD THINGS

When despair for the world grows in me
and I wake in the night at the least sound
in fear of what my life and my children's life may be,
I go and lie down where the wood drake
rests in his beauty on the water, and the great heron feeds.
I come into the peace of wild things
who do not tax their lives with forethought
of grief. I come into the presence of still water.
And I feel above me the day-blind stars
waiting with their lights. For a time
I rest in the grace of the world, and am free.

Habitats in common

Humans have always sought out and settled on the edges of marshes, rivers, lakes, and ocean shores where food resources were often plentiful. Sharing wetlands, lakes, and ocean shores, people have always interacted with herons. Human civilization developed along the edges of great marshes and rivers. The Mesopotamian marshes of Iran and Iraq are the core of the fertile crescent of the ancient Mideast. Undoubtedly these marshes were then, as they are now, home to some of the most important concentrations of wintering herons in Eurasia. The Nile and its flood cycles set the annual calendar for Egyptian civilization. The Amazon, Congo, Yangtze, and Indus all supported their civilizations, and their heron fauna. Waterbirds migrate with the annual flood and dry cycles and come to represent these cycles to the people of the area.

The co-existence of humans and herons in a common habitat has probably always been fraught with potential conflict. Both humans and herons are dependent on water and these habitats are easily susceptible to alteration. Humans were certainly draining marshes thousands of years before the first millennium; the great cities of Rome and Tihuatican were built partially on drained marshes. Ditching, draining, and reclamation always have been high priority human activities, with adverse consequences to heron populations. Within the last century, drainage has been worldwide, with some of the largest wetlands in the world, such as the Everglades, being severely altered. The history of population fluctuations of herons in west Europe is certainly correlated with the history of land reclamation.

In the contemporary world, human land change is proceeding at an unprecedented pace. Wetlands critical to herons, such as the Pantanal in South America, the Everglades and Mississippi Delta in North America, the Okavango in Africa, the coastal wetlands of Bangladesh, and the Mesopotamian Marshes in the Middle East, are under threat. The Pantanal supports one of the greatest concentrations of waterbirds in South America, but channelization of the Parana–Paraguay River system would change the hydrology on which herons depend. Heron populations in the Florida Everglades, once the epicentre of heron diversity and abundance on the continent, have crashed in the past 40 years due to human management of the hydrology. The Mississippi Delta is disappearing as a result of channelization of the Mississippi River, diverting silt needed for its maintenance. Combined with local ditching, these critical freshwater marshes turn first to salt and then to open pools. The Slaty Egret depends for its existence almost entirely on the Okavango Delta system, which is under threat of damming. The floodplain wetlands and mangroves of Bangladesh support both waterbirds and over 100 million people. Despite both being attuned to the cycle of flood and drying, plans to control the flooding of the Ganges and other mighty rivers will change millennia-old cycles.

The common threat to humans and herons of wetland drainage is no better illustrated than in the great Mesopotamian marshes of Iraq and Iran. Water is the most precious of commodities in the Middle East, a fact that has put the Euphrates itself at risk. Capture of the river's water in Turkey and major water diversion in the marsh itself are killing this marsh, which has always served as a principal wintering area for east European herons. As much at risk as the herons are the indigenous people of the area, the Ma'dan or Marsh Arabs. These people lived isolated in the marshes for 5000 years, stretching back to the foundation of western civilization. They lived a culture little changed that depends entirely on the marshes from which they construct mats, houses, and boats, raise buffalo, and fish the waters. Following the Gulf War of 1990–1, these people came under continued attack both physically and by drainage schemes, driving them from the marshes. The future of herons in this 100 000 square kilometers of former habitat is in similar doubt.

Marshes are not the only heron habitat under devastating attack. The destruction of low elevation mature wet forests in Asia has brought the White-eared Night-Heron, the White-bellied Heron, and the Japanese Night-Heron to the brink of extinction. The survival of these herons requires that humans quickly find ways to continue to share the habitat they hold in common. Coastal development around the world is accelerating as industry, housing, forest harvesting, water quality degradation, and aquaculture take over. The Chinese Egret and Reddish Egret are specifically at risk.

The dependency of herons and humans on the same habitats has its positive aspects as well. It is surprising and encouraging how so many herons have accommodated to the contemporary human-dominated environment (Kushlan and Hafner 2000). With this accommodation has come an increasing level of dependency. For example, range expansion of the Cattle Egret was clearly in response to the spread of irrigated pastureland. Widespread use of human-made landscapes worldwide suggests that, for many species, herons now have a tight and increasingly obligatory commensal relationship with man.

Night herons nest in parks and zoos worldwide. Grey Herons and Great Blue Herons occupy neighbourhood woodlots. Several North American egrets have shifted populations to use crayfish and catfish production ponds. As herons become dependent on artificially maintained environments they also become at risk when those conditions are changed. The human landscape is continually changing with new economic realities, the spread of agriculture and intensive agricultural practices into marginal lands and newly reclaimed forests and marshes, and the spread of people into formerly agricultural lands due to increasing populations. As humans increasingly manage their shared habitats, the need to do this in ways benevolent to herons becomes even more important.

Perhaps the best example and the best symbol of herons' dependence on human habitat maintenance is rice culture. The importance of rice cultivation to contemporary heron populatizzon sustainability can hardly be overstated. The core of the west Europe night heron population depends on rice fields. In fact it is the case around the world that the dependence of herons on rice fields is critical to the health of extant populations (Fasola and Alieri 1992b, Fasola *et al.* 1996, Fujioka *et al.* 2001, Richardson and Taylor 2003). Not only is the existence of rice fields critical, but also their benefits depend in large measure on how they are farmed, and farming practices are always evolving in response to new markets, new varieties, and cost reductions. As rice farming goes so do heron populations around the tropical world.

Herons depend on people managing their habitat in more subtle ways. In most developed parts of the world, nesting sites are to some degree protected by an array of parks and refuges scattered across the landscape. Outside refuges, management practices for farming, irrigation, wetlands, ponds, lakes, rivers, and coasts affect herons. Herons, in their nesting and feeding habitats, are now dependent on humans throughout the world.

Heron hunting

Herons had been around for tens of millions of years before humans. Probably from the beginning, herons were undoubtedly hunted for food. They were an important food over 3000 years ago in Mexico (Steadman *et al.* 2003). Killing a few adults or, more likely, eggs and young, need not affect otherwise vigorous heron populations. However, there is mounting evidence of herons having been impacted by human hunting.

The effect of humans on the post Pleistocene and more recent island faunas is well appreciated. Invading humans are suspected of being responsible for several heron extinctions, and probably more than we know from scant subfossil and historical evidence. On the Pacific islands, first the immigrating Polynesian and Micronesians found birds that had limited defences and so became easy prey to the people, the domestic stock they brought with them, especially pigs, and to the diseases they introduced. Herons were among the species affected. A few centuries later, Europeans made their own invasions and impacted the remaining native fauna.

With scant fossils, we will never know the full extent of the lost island heron fauna. The best known are the several species of island night herons that went extinct coincident with the arrival of one or the other wave of human invaders. The Reunion Night-Heron, Mauritius Night-Heron, Rodrigues Night-Heron and a Yellow-crowned Night Heron on Bermuda all became extinct in relatively recent times. There are clearly more lost herons yet undiscovered on the isolated islands of the world. Less well understood is the fate of the already mentioned giant heron of southern Europe. This species became extinct about 8000 years ago, at the dawn of human civilization in the area. The coincidence is too great for the extinction of such a vulnerable bird not to be the result of humans. If any specific proof were needed, the recent loss of these heron forms suggests strongly that humans can drive herons to extinction.

Herons still are hunted. They are edible birds and are consumed in various cultures around the world. Much of the hunting has probably always occurred at nesting sites. Herons are hunted for food in China, Madagascar, India, Europe, and other places. In China, the White-eared Night-Heron was rediscovered when a young bird was spotted being offered for sale at a market. This is part of a trade in many wild species of animals for food and medicine that is endemic in the local society. Hunters target nests and take the young that are eaten or put up for sale. Because the young of night herons look much alike, rare and not so rare species are not distinguished.

Traditionally in Madagascar, eggs were eaten as a beauty enhancer. Hunting in heronries in Madagascar has altered the distribution and probably the viability of populations. The near endemic Malagasy Heron and Malagasy Pond-Heron are at risk.

In India, hunting along the coast has decimated colonies, forcing birds to move inland to nest (Naik and Parasharya 1987, Parasharya and Naik 1990). Recently it was reported that Cattle Egret chicks were being harvested for their livers, which are being used for medicinal purposes (Bharos 1998).

For species such as herons, in which annual reproduction is demographically less important than adult survival, some level of egg harvest may be sustainable. So hunting of eggs or young might be possibly sustainable, as appears to be the case in Borneo and the Americas. Indigenous people likely can indeed continue to take eggs and young herons for their use, if done in a managed way.

Long-lived birds seldom can stand up to intense hunting pressures on adults. In southern Europe and nations that are former colonies of Europe, birds such as the large *Ardea* still make inviting targets for sport hunting. In this culture, the use of herons as hunting quarry goes back centuries. The first great ornithologist, Frederick II (1194–1250), wrote about and illustrated herons in his manuscript on birds and hunting, *De arte venadi cu avibus*. Some of the earliest examples of protection for heronries in Europe were aimed at securing falconry targets for the local nobility. In what is now southwest Germany, the Baron of Crailsheim and the Margrave of Ansbach fought a heron war over ownership of a colony. Herons were hunted with falcons that would be sent in the air to attack the fleeing bird. While most herons were probably killed, this was not always the case as is shown by the story of a heron in the early 1700s that was caught, banded, and released six times by the Elector of Cologne. Modern hunting of herons for sport is less common, but not unknown in some parts of the world.

It was heron plumes that led to the greatest bout of heron hunting ever. Among birds, some of the most sought after feathers have always been the breeding plumes of egrets. Appreciation and use of heron feathers no doubt go back to the dawn of herons' interactions with people. For example, the decorative plumes of the Great and Little Egrets were used as expensive head decorations by the Hungarian nobility and also by the Turks, who had close economic and political ties with Central Europe in the Middle Ages.

By the late 1800s and early 1900s birds and their feathers were widely used in Europe and elsewhere for ornament, particularly on hats. The fashion became quite outlandish with whole birds being featured, but aigrettes were among the most highly admired feathers. As a result, a worldwide hunt was mounted for heron plumes. High prices were paid since demand was high and, at least initially, it was not easy to reach the breeding colonies in trackless swamps in order to obtain the plumes. In Europe, colonies on the Danube and Theiss Rivers were among those devastated and London became the centre of the European plume market. Merchants mounted plume-collecting expeditions resulting in the destruction of egret colonies in Europe and Asia. In the Americas, where the hunting ethic was a core value, backwoodsmen had little difficulty finding remote colonies. Colonies were hunted in southern North America and through tropical America as well.

All birds with useful plumes were killed in the colony, but birds of all species had their nesting disrupted. Not only the individuals killed for plumes, but their young and those of co-nesting species died, and after a few decades of hunting, known colonies were decimated. The feather trade involved millions of dollars. In 1902, 1608 packets of egret plumes were sold in London, each weighing about thirty ounces; hence the total weight was about 48 240 ounces. To get one ounce of plumes, four egrets had to be killed. This meant that to supply this demand for plumes, 192 960 egrets were killed.

Heron interactions with humans then took another turn, in that the final outcome of all this slaughter was the beginning of the animal conservation movement. Pressure groups came into existence, which developed into persistent bird conservation organizations, such as the National Audubon Society and the Royal Society for the Protection of Birds, among many more worldwide. Laws were passed, enforcement increased, and herons along with other birds were finally protected from wanton killing for their plumes. As the conservation ethic spreads worldwide through treaties such as the Ramsar and Biodiversity Conventions, herons continue to serve a role. Today the heron is a frequent image backing up conservation messages and the Great Egret remains today an emblem of the National Audubon Society of the United States.

The most potentially serious continuing hunting of herons may be antidepredation killing. Herons are killed at fish farms worldwide. The larger herons are perceived to be great threats to the economic viability of hatcheries and fishpond operations. The numbers killed are high. Herons were devastated in Europe in the late 1800s and early 1900s by aquacultural interests. Even in recent decades in England the take of Grey Herons has been estimated to approximate annual production. The legally permitted killing in the southern USA totals tens of thousands of birds. Despite much research showing minimal impacts and even benefits of heron predation and the existence of alternative management options (e.g. Glahn *et al.* 2002 but see e.g., Lekuona 2002a), the relations between herons and fish farmers remain poor.

Similarly, fishermen worldwide seem to believe herons are adversely affecting fishing activities, contrary evidence notwithstanding. Herons are harassed, shot at, and killed near fish traps and along rivers. Although not receiving quite the animosity reserved for cormorants, they continue to be badly thought of. There is much room for improvement in managing and influencing relationships between humans and herons in this arena.

Herons in society

Despite the many problems in the interactions between herons and people, it is more remarkable how well most heron populations have accommodated to humans and to the human-dominated

5.1 Herons as part of human society.

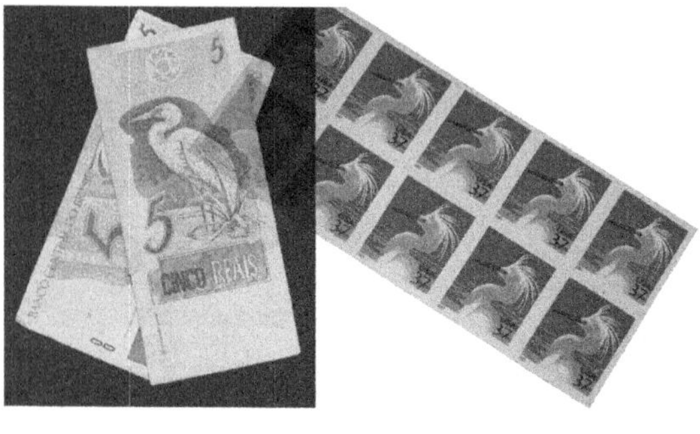

5.2 Correng & Stamps.

5.3 Friendly Great Blue Heron helping with laundry on the Galapagos Islands, Ecuador.

world. Nesting in towns, feeding in farm fields, ditches, and ponds, moving about in response to irrigation, and using many human habitats, herons are part of the landscape. Particularly along the coast and on islands, such as the Galapagos, herons tend to become exceptionally acclimatized to human activities. And herons are for the most part appreciated by their local human hosts.

It is hard not to miss a large wading bird, or their colourful plumage or idiosyncratic behaviour or a smelly neighbouring breeding colony. As people become more concerned with their environment, herons are increasingly noticed and appreciated. Magazines, children's books, field guides, and conservation events all use heron images to tell their stories. Magazine articles on the natural history and conservation of herons hold much interest. Herons also serve as a focal point for children's stories. Sometimes herons are only the name of a setting, such as Heron Creek, or a symbolic core of an entire region, such as the *World of the Great White Heron* by Marjory Bartlett Sanger. Or they become more, as in an essay written by a young boy in *The Valley of the Blue Heron* by Wilma Rugh Taylor: 'In flight his graceful curved neck was tucked back between bluish-gray shoulders, with flashes of downy white breast. The shadow of his huge wingspan spread over the water. His legs trailed straight out behind. For just a moment, he looked right at me, his yellow eye ringed with a wing-like feather of black. Not really at me, but into me'. At times the use of the heron in children's stories has risen to the level of literature, the most famous perhaps being Sarah Orne Jewett's *A White Heron*.

As noted above, herons are symbols of national and international conservation movements. Facing near extermination in the southern United States in the early 1900s, the fates of the Snowy and Great Egrets were capturing national attention. With the founding of the National Association of Audubon Societies, wardens were hired to protect the colonies. Two wardens, Guy Bradley and Columbus McLeod, were killed. The effect on fashion was immediate: plumage bills were introduced, and the conservation movement gained legal footing, with the Great Egret as its symbol.

The symbolic quality of herons is used frequently in local contexts, particularly in areas that are water oriented. An example is the Chesapeake Bay region of eastern North America. There the heron image, typically the Great Blue Heron, is heavily used to convey both commercial and conservation messages. Special automobile licence plates can be purchased that feature the Great Blue Heron. Extra funds raised by the sale of this plate are funnelled into environmental projects. Given the inherent conflict between development and the health of heron populations, it might seem odd that it is used as a corporate symbol for homebuilders and housing parks. Yet these examples are particularly telling in that the image being sold is the quality of life that a heron represents. Perhaps even more blatant is the use of the heron image to conjure up an environmentally benign appearance to an oil tank farm.

5.4 Herons in books, magazines and conservation.

A water-oriented society exists along coastal Florida in the southern United States. Herons have pull here. Their image is used to slow cars and to decorate municipal water tanks. This is the world of the Great White Heron noted above and the location of the Everglades, the mismanagement of which has led to the collapse of local heron populations. For decades reports, news stories, books, and lectures noting the decline in waterbird populations have been used to galvanize public support for changing water management practices in the Everglades, current proposals now putting the bill for this work in the billions of dollars.

Perhaps the most unlikely new use of the heron image is its finding a place in pop culture. In New York City a nightclub features as its primary backdrop for rock music an egret serenely surveying its marsh. While the artists playing in front of it are likely oblivious to its presence, the establishment clearly has a successful commercial goal in its use.

5.5 Backdrop mural in a night club in New York City.

In many cultures the knowledge of herons goes back to their beginning. At the birthplace of western civilization, herons merit mention in the Old Testament and so probably were appreciated in

5.6 Herons in art. Alexander Wilson from American Ornithology, 1808–14; John Gould from Bird of Europe, 1832–37; J. J. Audubon, from Birds of North America, 1827–38; M. Oake, Botswana, 1997.

preceding civilizations. Herons are noted in Egyptian pictographs, certainly a worthy precursor to the New York nightclub. The heron in hieroglyphics represented rebirth and morning. The Little Egret is illustrated with other waterbirds in the tomb of Menna in Thebes, Egypt, from about 1400 BC. Interestingly in the fresco, although attempting to fly away, two egrets are being held by their legs by a human hand. Herons were known by Aristotle and the Romans, as one is illustrated in a wall painting in Pompeii (House of Venus Marina), from the first century AD.

Some indigenous cultures are known to especially prize herons. In New Zealand, egrets were held in high stature and used to decorate chieftain regalia. In culture after culture, the feathers of birds, including egrets, have been used for such adornment. In Hindu culture, heronries often developed and were protected within temple grounds. Even today, despite universal habitat loss, heron status is stronger in Hindu than among neighbouring cultures. Herons were highly prized in Japan where some colonies are centuries old. The heron colony of Sagiyama near Tokyo persisted in the same location for at least 170 years protected by local peasant families, who were once paid for this service by the shogun generals who passed through the village on their way to the temples. Other colonies not protected were hunted. Herons have severely declined in Japan in recent years, but they remain much appreciated as shown in Shingi Itoh's (1988) magnificent photographic book, *The White Egret*.

Other cultures shun herons. The usual recipients of such reaction are the large bitterns. In Australia, the aborigine people thought the Boom Calls of the Australasian Bittern were bad tidings. In North America, as late as the 1700s, a community gathered on the Sabbath to rid the town's swamps of these evil creatures (Gibbs *et al.* 1992).

The use of heron images in illustration re-emerged in the Middle Ages. Although the heron has always been the most abundant of the large wading birds in Europe, it was not often figured in medieval manuscripts. This is probably due to its sombre colours, its lack of expressive posturing (as contrasted with the White Stork), and its similarity to more colourful cranes. Most of what might be herons in the illuminated manuscripts are likely some rendered combination of heron and crane (Yap 1981). The earliest possible heron is about 1170, although its leg colours are wrong. A definitive 'heyren' is in the Sherborne Missal from about 1400.

Despite this inauspicious start, the heron has developed a fine pedigree in the world of natural history art. One of the first natural history illustrations of herons was by Konrad von Gesner in his *Historia animalium* in 1555. Herons were illustrated in woodcuts over the next two hundred years. By the time of scientific exploration of North America, natural history illustration achieved new heights. Mark Catesby and Alexander Wilson painted the known North American herons in what today would be considered a primitive style. But it was John James Audubon and later John Gould and others in the 1800s that brought naturalistic bird illustration to some of its finest stages. Audubon's image of the Great White Heron, cardinal fish in bill and the Florida Keys in the background, is one of the most striking wildlife images of the era. Audubon in fact scientifically discovered and named this bird, then considered a separate species. John Gould, the most exceptional illustrator of this time and concentrating mostly on the more sensational looking birds, painted the herons of Europe. Thereafter to today, herons continue to be well-received subjects in drawing, painting, photography, sculpture, and other media.

What is it about herons that capture the human imagination? Beyond their size and obviousness, I suspect it is their lightness of being. As Sylvia in *A White Heron* found out, it is 'the wild, slender bird that floats and wavers'. It is a bird whose poetic message seems obvious, as 'it rises from the among the reeds and flies forever, . . . , towards the distant islands'.

People and herons epilogue

Of the many bonds between herons and people, one seems worth noting especially, namely that of the bird people of Manchar (Pelletier 1999). These people of the Kirthar Mountains of Pakistan have an enduring relationship with water birds. Each Mohana has his own heron or egret that acts as a guard and companion. This relationship, thought to go back to 2500–1500 BC, reflects the common bond of man to herons and to the wetland habitat needed by both. It is common bonds such as these, a person and a bird, that may by the single great hope for the herons of the world.

6

Conservation

As a group of birds so intimately engaged in the human-dominated environment, herons face conservation issues throughout the world. Some species and populations are at risk, a few are on the very brink of extinction. Other species have limited ranges or habitat scope. Many appear secure, but depend for their status on habitats supplied by human activities. Other species are vigorously expanding their range and population sizes. As noted in each of the species accounts, even the most common species require conservation attention directed to their habitat needs, whereas the very rare ones require immediate attention and action to prevent their extinction.

The conservation of herons has been discussed in detail in the companion volume to the present book, *Heron Conservation* (Kushlan and Hafner 2000). Conservation issues for each species are noted briefly in each species account. In this chapter we discuss very briefly the most threatened of the herons and examine the primary issues facing herons worldwide.

Herons at risk

The principal authority on the status of threatened species, including herons, is Birdlife International which conducts periodic evaluations for the IUCN (IUCN 2003). This evaluation is done on a species scale. The Heron Specialist Group-HSG has furthered the process by evaluating herons on a population scale (Kushlan and Hafner 2000). Based on these analyses, eight species of herons are seriously at risk: White-eared Night-Heron, Japanese Night-Heron, White-bellied Heron, Chinese Egret, Slaty Egret, Australasian Bittern, Malagasy Heron, and Malagasy Pond-Heron. All occur in Asia, Australasia, or Africa Madagascar.

The White-eared Night-Heron is the most critically endangered of all the herons. This is a species that has lost its primary habitat of lowland wet forest in south China. The principal cause has been deforestation followed by reforestation with pine plantations. More recently a reservoir was constructed on one of its remaining sites and gold mining and other activities are expanding. Also it is captured and eaten, a practice that threatens all the herons in the area. It is now known from three sites, which contain only a few birds each. There are likely other sites, so there remains hope that it will be found in other places. Its now uses remaining midlevel forests and is known to feed nocturnally in artificial sites. So much remains to be understood of the current status and ecology of the species. Its current habitat needs to be better understood. All its nesting and feeding sites need to be protected, enlisting the help and engagement of the local people. This is a species for which a captive breeding programme is needed if stock can be found.

The Japanese Night-Heron is similarly critically endangered, for much the same reasons as the White-eared Night-Heron. It is now known to nest in only six sites in Japan. Its hill forest habitat has been and continues to be converted to plantations. Remnant forests have developed scrubby undergrowth that may be incompatible to this forest-dwelling species. All the remaining forest patches used by this heron need to be protected in reserves. It winters in lowland primary forest in the Philippines, where similar habitat conversion is underway. Habitat here also needs to be protected if wintering sites can be located. This too is a species for which establishing a captive population is needed.

The White-bellied Heron also is an exceptionally rare heron of mature forests, and like the night herons is critically endangered. Also like the night herons, its habitat has been under threat in Asia for a century. It is a shy bird, occurring away from human activity. Most of its historically inaccessible wet forest habitat has been cleared or reforested. The very few places it is known to remain are now subject to disturbance and even hunting by local people. Its survival depends on the development of protected reserves where it is found, and engagement of the local populations in concern for its welfare. A captive breeding programme is desirable.

The Chinese Egret is a vulnerable coastal heron of Asia. Its conservation history reflects the two greatest historic challenges to heron conservation, namely hunting and habitat destruction. Like other plume-bearing egrets, it was impacted by hunting at the beginning of the 1900s. It continued to be hunted for its eggs. While it was always persistent in some areas, it did not experience a significant population rebound. More recently its habitat in east Asia has come under threat by the entire array of coastal changes—including harvest of the mangrove forest, drainage and reclamation of tidal mudflats and estuaries, industrializing the coastline, conversion to aquaculture ponds, and even an airport on its previously occupied habitat. It is now known to breed in only a few sites, although there is hope that others may be found. Protection of colony sites and its coastal feeding areas are needed on both breeding and wintering areas.

The Slaty Egret is a vulnerable heron of south central Africa. This is a heron with a small global population and highly specialized in that it depends on a limited area of inland floodplain marshes. Water management schemes in the Okavango River and Delta threaten the continued functionality of these wetlands. Livestock grazing and other immediate threats affect its reed bed habitat. Sustainable management of the Okavango is required for its conservation.

The Australasian Bittern is vulnerable due to habitat loss and alteration. Its population in Australia is small, restricted, and declining rapidly. In the marked wet–dry season of Australia, its wetland habitat is sensitive to alterations and such changes are occurring at alarming rates. Drainage and subsequent salinization, damming, grazing, clearing for agriculture, and peat extraction are threats. In New Zealand its swamps are being drained. The Australasian Bittern relies heavily on well-flooded fresh water swamps, and its future may depend on appropriate management of a single river system, the Murray–Darling.

The Malagasy Heron is a rare and vulnerable heron nearly endemic to Madagascar. Its once inaccessible habitat is being converted to rice and other developments. Also, it is still hunted by local people. Protected areas need to be established in areas where it still occurs. Local people, who tend not to distinguish it from the more abundant Grey Heron, need to be engaged in its conservation.

The Malagasy Pond-Heron also is vulnerable. Like the Malagasy Heron, its habitat in Madagascar has been lost through clearing, drainage, and conversion to rice fields, and it is hunted. Protecting the most important nesting areas in Madagascar needs to be the focus of conservation action. Its non-breeding status in Africa is very poorly known. Important sites in its wintering habitat need to be delineated and protected.

Other species and populations of herons are also of concern owing to small population sizes or habitat threats. These include the Sumatran Heron, Rufous-bellied Heron, Reddish Egret, Fasciated Tiger Heron, New Guinea Tiger-Heron and Zigzag Heron. Although all these are perceived to be at somewhat lower risk than the endangered and vulnerable species, these deserve conservation attention, and their issues are similar—loss of restricted habitat, population declines, conflict with humans, and lack of information.

Although the conservation of species has to be of extremely high concern, the conservation of distinctive populations is similarly important if the genetic diversity of herons is to be maintained. Several geographically distinctive populations are at risk. Those most of concern are the Purple Heron on Cape Verde Islands, the Eastern Great Egret in New Zealand, the Eurasian Bittern in southern Africa, the Grey Heron on the Comoros and Aldabra, and the Striated Heron on the Seychelles.

Other populations at risk include the Grey Heron on the coast of Mauritania, the Great Blue Heron of southern Florida and the Caribbean, the Great Blue Heron of the North American north Pacific coast, the Purple Heron on Madagascar, the Indian Pond-Heron on the Maldives Islands, Striated Herons on the Chagos Islands, the Maldives Islands (two subspecies), and Tahiti, the Malayan Night-Heron, and Little Bitterns in Madagascar and Australia.

Conservation issues

The species and populations of herons that are endangered or vulnerable have characteristics in common. They are suffering from habitat loss, perhaps compounded by human killing. Most herons at risk are found in Asia and on islands. So the main issues facing herons may be listed as: protection of their important areas; conservation and management of habitat; management of interactions with humans, particularly killing; defining population goals; and overcoming data deficiencies.

Herons, being for the most part colonial, gather into colony and roost sites and feed in distance-restricted areas around these sites. Those that are not colonial are not found randomly scattered over the landscape but confined to patches of suitable habitat. These characteristics make them particularly susceptible to being aided by conservation programmes that are based on protecting areas that are deemed most important to species or regional populations.

The important areas for herons are those that support significant portions of the population for some part of the year. Those areas that have been evaluated are called Important Bird Areas. Important heron habitat patches can be set aside, protected by agreement with local people and management agencies, or managed as part of regionwide conservation schemes. Herons now depend on humans, and the engagement of human institutions is required to secure and manage heron habitat in ways that are compatible with the sustainability of heron populations. Areas of importance for herons are scattered worldwide. Those locations that have been identified as important areas for herons are shown in the Figs. 6.1 and 6.2.

Within these important areas and at other locations, herons occupy habitat that often is vulnerable. Above all else the primary issue of heron conservation is to arrest the loss and degradation of this habitat. While many heron species have been successful in accommodating to the human-dominated landscape, as discussed previously, some have not. Although herons occur from Siberia to Tierra del Fuego and from low mountains to coral reefs, the habitats most at risk are generally swampy forests, large wetlands of the tropics, and coastlines everywhere.

Natural habitats, particularly the marshes and swamps that characterize the principal habitat for herons, have been under threat for centuries and continue to be at an accelerating rate. Drainage, conversion to agriculture, pollution, salinization, water diversions, fire, water withdrawal, and similar developments threaten the functioning of natural habitats. Heron conservation needs to be placed within the context of regional management of these wetland and coastal habitats for their multiple values and functions, one of which must be sustaining regional heron populations.

In specific sites, habitat will need to be managed for herons. Nesting colony sites often require protection, vegetation and water management, establishment of networks of colony sites and potential colony sites, and continued connectivity to feeding areas. Sites used for foraging may require specific management as well, or if other sorts of management is to occur, the needs of herons need to be accounted for. It is well documented that restoration or creation of wetlands habitat and prey base can result in its use by herons and development of nesting colonies (Mauchamp *et al.* 2002). Little is known of habitat use by most herons outside the breeding season. Identification and protection of stopover and non-breeding areas is essential, especially given the likelihood that non-breeding adult mortality may be the most critical demographic factor for some species. The habitat needs of each species in each region should be characterized for inclusion within regional planning efforts.

In many ways, as a group and for particular species, herons have thrived by making good use of human-made habitats. Feeding and nesting as they do in aquacultural ponds and raceways, rice fields, watercress beds, reed beds, pastures, parks, wildlife refuges, zoological parks, ditches, reservoirs, stock ponds, and many other aquatic and semiaquatic situations, they have proved adaptable. However, such

6.1 Map of the Americas showing areas important for heron conservation.

use means that they have become dependent on these features of the human landscape. In some cases traditional, low intensity, agricultural practices have benefitted herons, which are now threatened by modernization of those practices (e.g. Shrubb 2003).

Perhaps the most critical landscape features for herons across the tropical and subtemperate regions of the world are rice fields and fish ponds. The current status of the Black-crown Night-Heron of west Europe depends on rice cultivation in Italy. The herons of southeastern North American have come to depend on aquacultural ponds for winter foraging. Local heron populations have collapsed when aquacultural or agricultural practices change. The human-dominated agricultural landscape is a core aspect of the well-being of many herons nearly worldwide. There likely is no more important feature of heron conservation than the management of aquatic agriculture in ways that allow its sustainable use by herons.

Being in close proximity to humans carries risk. Disturbance to colonies, feral animals, pesticides and other contaminants, cultural practices that

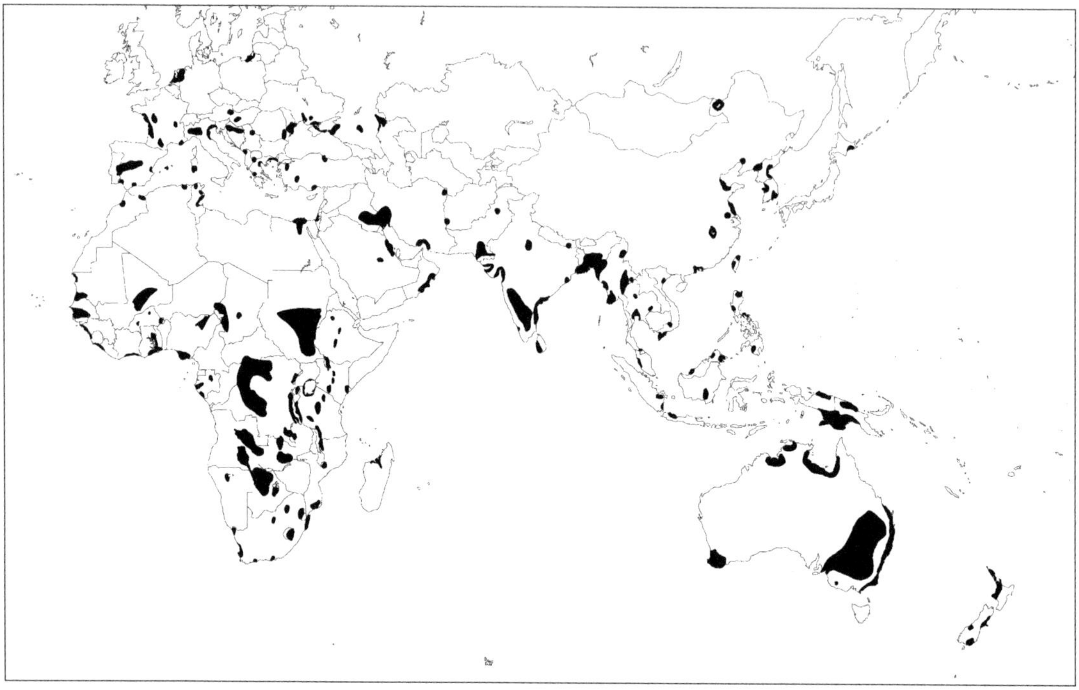

6.2 Map of the Europe, Asia, Africa, and Australia showing areas important for heron conservation.

undermine conservation, hunting, random killing, habitat alteration, and many more interactions occur across the landscape. Human–heron interactions need to be considered and managed where possible. Sociological or economic obstacles make action difficult in many areas. Certainly in the poorest parts of the world a heron is as good a protein source as any other. In the richest parts of the world, disturbance by recreationists and their play vehicles can have a devastating impact. Management requires legislation, enforcement, protective measures, and especially education programmes that enlist the local population in the care and survival of the species.

Conservation must have goals, and among the hardest are defining desired population and habitat outcomes of conservation planning and action. For only a few species are there sufficient population, demographic, and habitat-use data to allow the quantitative expression of conservation goals. Goals in general are the sustainability of the regional population. But these goals must be quantified in terms of population sizes and trends, and coverage and dispersion of habitat. But much additional information is needed to set such goals. Dedicated work is needed to bring them to fruition.

Kushlan and Hafner (2000), and especially Hafner *et al.* (2000), present specific conservation needs of the species and populations most at risk and discuss regional and topical issues underlying heron conservation. But in summary, worldwide, the following are needed: the distribution and status of little known populations and taxa need to be determined; population conservation goals need to be set; important areas for at risk populations need to be identified; management of important heron areas need to be linked through personal communication and institutional cooperation, habitat goals for populations at risk and also other species need to be determined regionally, including the location and status of nesting areas; methods for monitoring populations and habitat need to be developed and implemented; feeding sites and food needs have to be determined; human interactions, both positive and negative, need to be identified and their sociological and economic consequences determined; and environmental- and community-based programmes need to be developed and action undertaken.

Plates

Plate 1

Large herons (*Ardea*)

1. Grey Heron, *Ardea cinerea*, adult
2. Grey Heron, *Ardea cinerea*, non-breeding
3. Great Blue Heron, *Ardea herodias*, adult breeding
4. Great Blue Heron, *Ardea herodias*, white phase
5. Cocoi Heron, *Ardea cocoi*, non-breeding
6. Cocoi Heron, *Ardea cocoi*, breeding
7. Great Blue Heron, *Ardea herodias*, intermediate plumage

1
2
3
4
5
6
7

Plate 2

Large and great herons (*Ardea*)

1. White-necked Heron, *Ardea pacifica*
2. Black-headed Heron, *Ardea melanocephala*
3. Malagasy Heron, *Ardea humbloti*
4. White-bellied Heron, *Ardea insignis*
5. Goliath Heron, *Ardea goliath*
6. Sumatran Heron, *Ardea sumatrana*

1
2
3
4
5
6

Plate 3

Purple Heron and large egrets (*Ardea*)

1. Purple Heron, *Ardea purpurea*, juvenile
2. Purple Heron, *Ardea purpurea*, adult
3. Intermediate Egret, *Ardea intermedia brachyryncha*, breeding
4. Intermediate Egret, *Ardea intermedia intermedia*, breeding
5. Intermediate Egret, *Ardea intermedia intermedia*, non-breeding
6. Intermediate Egret, *Ardea intermedia plumifera*, non-breeding
7. Great Egret, *Ardea alba egretta*
8. Great Egret, *Ardea alba melanorhynchos*
9. Great Egret, *Ardea alba alba*
10. Great Egret, *Ardea alba*, juvenile

1
2
3
4
5
6
7
8
9
10

Plate 4

Little Egrets (*Egretta garzetta*)

1. Little Egret, *Egretta garzetta garzetta*
2. Little Egret, *Egretta garzetta nigripes*
3. Little Egret, *Egretta garzetta garzetta*, breeding
4. Little Egret, *Egretta garzetta immaculata*
5. Little Egret, *Egretta garzetta dimorpha*, dark phase
6. Little Egret, *Egretta garzetta dimorpha*, white phase
7. Little Egret, *Egretta garzetta schistacea*, white phase
8. Little Egret, *Egretta garzetta schistacea*, flying
9. Little Egret, *Egretta garzetta gularis*, dark phase
10. Little Egret, *Egretta garzetta gularis*, white phase
11. Little Egret, *Egretta garzetta schistacea*, dark phase

1
2
3
4
5
6
7
8
9
10
11

Plate 5

Eastern Great Egret and medium egrets (*Ardea* and *Egretta*)

1. Eastern Great Egret, *Ardea modesta*, non-breeding
2. Eastern Great Egret, *Ardea modesta*, breeding
3. Eastern Great Egret, *Ardea modesta*, Aerial Stretch Display
4. Pied Heron, *Egretta picata*, adult
5. Pied Heron, *Egretta picata*, juvenile
6. Reddish Egret, *Egretta rufescens*, white phase, juvenile
7. Reddish Egret, *Egretta rufescens*, dark phase, breeding
8. Reddish Egret, *Egretta rufescens*, white phase, breeding
9. Little Blue Heron, *Egretta caerulea*, adult
10. Little Blue Heron, *Egretta caerulea*, juvenile

1
2
3
4
5
6
7
8
9
10

Plate 6

Egrets (*Egretta*)

1. White-faced Heron, *Egretta novaehollandiae*, breeding
2. White-faced Heron, *Egretta novaehollandiae*, non-breeding
3. Slaty Egret, *Egretta vinaceigula*
4. Black Heron, *Egretta ardesiaca*
5. Black Heron, *Egretta ardesiaca*, Canopy Feeding
6. Tricoloured Heron, *Egretta tricolor*, breeding
7. Tricoloured Heron, *Egretta tricolor*, juvenile
8. Tricoloured Heron, *Egretta tricolor*, non-breeding
9. Snowy Egret, *Egretta thula*, non-breeding
10. Snowy Egret, *Egretta thula*, juvenile
11. Snowy Egret, *Egretta thula*, breeding
12. Chinese Egret, *Egretta eulophotes*, non-breeding
13. Chinese Egret, *Egretta eulophotes*, breeding
14. Eastern Reef-Heron, *Egretta sacra*, white phase
15. Eastern Reef-Heron, *Egretta sacra*, dark phase

1
2
3
4
5
6
7
8
9
10
11
12
13
14
15

Plate 7

Cattle Egret (*Ardea*), Striated herons (*Butorides*), and pond herons (*Ardeola*)

1. Cattle Egret, *Ardea ibis ibis*, non-breeding
2. Cattle Egret, *Ardea ibis ibis*, breeding
3. Cattle Egret, *Ardea ibis coromandus*, breeding
4. Green Heron, *Butorides virescens*
5. Striated Heron, *Butorides striata atricapilla*
6. Striated Heron, *Butorides striata sundevalli*
7. Indian Pond-Heron, *Ardeola grayii*, breeding
8. Indian Pond-Heron, *Ardeola grayii*, non-breeding
9. Chinese Pond-Heron, *Ardeola bacchus*
10. Javan Pond-Heron, *Ardeola speciosa*
11. Malagasy Pond-Heron, *Ardeola idae*, breeding
12. Malagasy Pond-Heron, *Ardeola idae*, non-breeding
13. Squacco Heron, *Ardeola ralloides*, juvenile
14. Squacco Heron, *Ardeola ralloides*, breeding
15. Squacco Heron, *Ardeola ralloides*, non-breeding
16. Squacco Heron, *Ardeola ralloides*, flying
17. Rufous-bellied Heron, *Ardeola rufiventrus*

1
2
3
4
5
6
7
8
9
10
11
12
13
14
15
16
17

Plate 8

Neotropical herons (*Zibrilus, Cochlearius, Pilherodius, Syrigma, Agamia*)

1. Zigzag Heron, *Zibrilus undulatus*
2. Zigzag Heron, *Zibrilus undulatus*, juvenile
3. Zigzag Heron, *Zibrilus undulatus*, adult
4. Boat-billed Heron, *Cochlearius cochlearius*
5. Boat-billed Heron, *Cochlearius cochlearius*, display
6. Boat-billed Heron, *Cochlearius cochlearius*
7. Capped Heron, *Pilherodius pileatus*
8. Whistling Heron, *Syrigma sibilatrix*, breeding
9. Whistling Heron, *Syrigma sibilatrix*, adult
10. Agami Heron, *Agamia agami*, adult
11. Agami Heron, *Agamia agami*, juvenile

1
2
3
4
5
6
7
8
9
10
11

Plate 9

Night herons (*Nyctinassa, Nycticorax, Gorsachius*)

1. Yellow-crowned Night-Heron, *Nyctinassa violacea*, juvenile
2. Yellow-crowned Night-Heron, *Nyctinassa violacea*, adult
3. Black-crowned Night-Heron, *Nycticorax nycticorax*, juvenile
4. Black-crowned Night-Heron, *Nycticorax nycticorax*, adult
5. Rufous Night-Heron, *Nycticorax caledonicus*, juvenile
6. Rufous Night-Heron, *Nycticorax caledonicus*, adult
7. White-backed Night-Heron, *Nycticorax leuconotus*
8. White-eared Night-Heron, *Gorsachius magnificus*
9. Japanese Night-Heron, *Gorsachius goisagi*
10. Malayan Night-Heron, *Gorsachius melanolophus*

1
2
3
4
5
6
7
8
9
10

Plate 10

Tiger herons (*Tigrisoma, Zonerodius, Tigriornis*) and small bitterns (*Ixobrychus*)

---

1. Bare-throated Tiger-Heron, *Tigrisoma mexicanum*
2. Rufescent Tiger-Heron, *Tigrisoma lineatum*
3. Fasciated Tiger-Heron, *Tigrisoma fasciatum*
4. New Guinea Tiger-Heron, *Zonerodius heliosylus*
5. White-crested Tiger-Heron, *Tigriornis leucolophus*
6. Least Bittern, *Ixobrychus exilis*, female
7. Least Bittern, *Ixobrychus exilis*, male
8. Little Bittern, *Ixobrychus minutus*, female
9. Little Bittern, *Ixobrychus minutus*, male

1
2
3
4
5
6
7
8
9

Plate 11

Small bitterns (*Ixobrychus*)

1. Black Bittern, *Ixobrychus flavicollis*, male
2. Black Bittern, *Ixobrychus flavicollis*, female
3. Streaked Bittern, *Ixobrychus involucris*, female
4. Streaked Bittern, *Ixobrychus involucris*, flying
5. Streaked Bittern, *Ixobrychus involucris*, male
6. Yellow Bittern, *Ixobrychus sinensis*, female
7. Yellow Bittern, *Ixobrychus sinensis*, male
8. Schrenck's Bittern, *Ixobrychus eurhythmus*, male
9. Schrenck's Bittern, *Ixobrychus eurhythmus*, female
10. Cinnamon Bittern, *Ixobrychus cinnamomeus*, male
11. Cinnamon Bittern, *Ixobrychus cinnamomeus*, female
12. Dwarf Bittern, *Ixobrychus sturmii*, female
13. Dwarf Bittern, *Ixobrychus sturmii*, male

1
2
3
4
5
6
7
8
9
10
11
12
13

Plate 12

Large bitterns (*Botaurus*)

1. Eurasian Bittern, *Botaurus stellaris*
2. North American Bittern, *Botaurus lentiginosus*
3. South American Bittern, *Botaurus pinnatus*
4. Eurasian Bittern, *Botaurus stellaris*, nestling
5. Australasian Bittern, *Botaurus poiciloptilus*

1
2
3
4
5

Plate 13

Large herons and large egrets (*Ardea*)

1. Cocoi Heron
2. Great Egret
3. Great Egret on boats
4. Intermediate Egret

1

2

3

4

Plate 14

Large herons and Cattle Egret (*Ardea*)

1. Cattle Egret
2. Black-headed Heron
3. Purple Heron
4. Cattle Egret on hippos

1

2

3

4

Plate 15

Small egrets (*Egretta*)

1. Little Blue Heron, dark adult
2. Little Egret, *Egretta garzetta gularis*, white phase
3. Little Egret, *Egretta garzetta dimorpha*, dark phase
4. Little Egret, *Egretta garzetta immaculate*

Plate 16

Eastern Reef-Heron, Whistling Heron, Capped Heron (*Egretta, Syrigma, Pilherodius*)

1. Eastern Reef-Heron, white phase
2. Eastern Reef-Heron, dark phase
3. Whistling Heron
4. Capped Heron

1

2

3

4

Plate 17

Green Heron, night herons, pond heron (*Butorides, Ardeola, Nycticorax, Nytinassa*)

1. Green Heron
2. Rufous Night-Heron
3. Indian Pond-Heron
4. Black-crowned Night-Heron (Adult)
5. Yellow-crowned Night-Heron (Juvenile)

1

2

3

4

5

Plate 18

Large bittern, Yellow-crowned Night-Heron, tiger heron, Agami Heron (*Botaurus, Tigrisoma, Agamia*)

1. North American Bittern
2. Rufescent Tiger-Heron, juvenile
3. Rufescent Tiger-Heron, adult
4. Agami Heron

1

2

3

4

PART II

Species accounts

Grey Heron *Ardea cinerea* Linnaeus

PLATES 1.1, 1.2

Other names: The Heron in English; Garza Real in Spanish; Héron cendré in French; Graureiher, Fischreiher in German; Blauwe Reiger in Dutch; Серая папjiя in Russian; Gråhäger in Swedish; Cangak abu in Indonesian; Ao-sagi in Japanese; Cang lu in Chinese.

Ardea Linnaeus, 1758. Syst. Nat. ed. 10, 1, 141. Type by subsequent designation *Ardea cinerea* Linnaeus (Gray, 1840, List Gen. Birds, p. 66).

Ardea cinerea Linnaeus, 1758. Syst. Nat. 10(1), 143: Sweden.

Subspecies: *Ardea cinerea jouyi* Clark, 1907, Proc. U.S. Natl. Mus. 32, 468: Korea; *Ardea cinerea firasa* Hartert, 1917, Bull. Brit. Ornith. Club. 38, 6: Antinosy country, Madagascar; *Ardea cinerea monicae* Jouanin and Roux, 1963, Oiseau, 33, 104: Ile Arel, Banc d'Arguin, Mauritania.

Description

The Grey Heron is a large grey heron having white and black accents, a white crown with black plumes, black belly, and white thighs.

ADULT: The adult has a white head (including crown, sides, throat) with a broad black eye stripe extending from above the eye to the back of the crown and continuing as a crest with several elongated, black plumes. The long and heavy bill is yellow with a dull brown suffusion along the lower bill and top of the upper bill. The irises are yellow and lores are yellow turning darker green around the eye. The chin, throat, and neck are light grey to white, with the neck being tinged buff at its base. The foreneck is grey-white with two distinct broken black streaks running parallel down the median. The upper back and hind neck are pale grey, the lower back and upper wing blue-grey. Pale grey lanceolate feathers occur along the back. The flight feathers are dark grey to black contrasting with the paler upper wings and uniform grey to white under the wing. At rest, a black 'shoulder' patch with a few white feathers is formed at the forward bend of the wing. The flanks are grey, the sides of the belly black, but the rest of the under parts are light grey to white, including the feathered thighs. The breast feathers are loose and elongated. The tail is grey. The legs and feet are green-grey to yellow-brown, varying in shade with age and season; the upper leg is paler (more yellow) than the lower leg. During breeding, the black crest plumes attain full development and long white lanceolate plumes develop along the back, lower foreneck, and breast. The iris, bill, and legs flush deep orange to red. The bill may retain some of this colour until after the young have hatched.

VARIATION: The sexes are identical in plumage, but the male averages larger (Boev 1987a, 1987b, 1987c). Considerable individual variation can exist, including birds that are various shades of grey, white, black, brown, or buff (e.g. Krotoski 1983, Sanders and Ebels 1998). Geographic patterns of variation are recognized taxonomically. The European–African race, *A. c. cinerea*, is as described above. The buff colour on the neck decreases geographically from east to west, and the eastern race *jouyi* typically lacks the buff colour and is markedly paler on the neck, upper wing feathers, and back plumes (Vaurie 1965). The island race *firasa* averages larger, especially in bill and legs than *cinerea*. The West African *monicae* is paler, looking sufficiently pale grey to white to be distinguishable from migrant *A. c. cinerea* (Erard *et al.* 1986). In the western Indian Ocean, Grey Herons breeding on various islands vary from *firasa*-type to *cinerea*-type (Penny 1974, Langrand 1990).

JUVENILE: Juvenile Grey Herons are more uniformly grey than adults and lack black plumage highlights and ornamental plumes. They have a grey to dark grey crown. The chin is white and foreneck appears brown grey. The upper parts are grey-brown and under parts grey with brown-grey streaking on the foreneck. The soft parts are duller than in the adult's. Legs are dark grey, the upper legs being paler and tinged yellow-green. In the first autumn and winter the upper parts become more blue-grey. The feathers of the back and breast are moderately elongated. From the second autumn subadult plumage is assumed, differing from the full adult in having the forehead and crown grey instead of white and the black of the sides of the belly less developed. Juvenile plumages are retained into the second year and vary among individuals from relatively more juvenile to more adult.

CHICK: The chick is brown-grey above, with notable head crest, and white below. Irises are yellow, bill grey, and legs green-grey.

VOICE: The Grey Heron is a rather vocal heron. Its distinctive vocalization is the 'Frarnk' call, a loud far-carrying flight call. The 'Go' call, rendered 'go, go, go', is an alarm call. The 'Oooo' call is an aggressive call, also given in the Forward display. The Oooo call is also used in the Stretch display, becoming a more gurgling 'oooo' when given by both sexes in the crouched part of the Stretch. At the breeding colony, the species is very vocal, uttering a variety of yelps, squawks and other, softer notes. The 'Rwo' call is the male's advertising call. The 'Arre' call is a landing call, rendered 'arre, arre', tending to a clucking on alighting. The Snap Display ends in a 'Clop'. A Greeting Display includes bill snaps. Bill Clappering occurs among pairs during formation and when in contact.

WEIGHTS AND MEASUREMENTS: Weight: 1020–2073 g; length: 90–98 cm.

Field characters

The Grey Heron is identified by its large size, grey coloration, distinctive eye stripes, black 'shoulder patch' (wing edge), black belly, and white thighs. At rest, the black shoulder patch is a useful field character; and in flight the leading edge of the wing and especially the carpal joint are noticeably white against the general blue-grey of the body and under wing. The flight action is slow, with wings bowed. When descending to a feeding area or to the nest, the Grey Heron can be very agile in parachuting downwards. These herons usually are seen alone or in small numbers either standing erect, or hunched over next to the water.

It is distinguished from the Purple Heron by its white head and lack of chestnut colour. Juveniles and darker than normal adults are distinguished from Purple Herons with difficulty, and wild hybrids are reported (Passarella *et al.* 1999), but Grey Herons are larger and lack rufous brown coloration. The Grey Heron is distinguished from the Black-headed Heron by its white head, larger size, grey (not black and white) under wing colour. Immature Grey Herons have lighter legs (skin and feathers) than immature Black-headed Herons. The Grey Heron is distinguished from the Goliath Heron by its smaller size and has a white or grey neck (not chestnut) in the juvenile. The Malagasy race of the Grey Heron is distinguished from the Malagasy Heron by being smaller and lighter and having a black shoulder patch. It is distinguished from the Imperial and Sumatran herons by being smaller and more brightly black and grey.

Distinguishing Grey from Great Blue and Cocoi herons, where each is vagrant, poses challenges (Gantlett 1998). Fundamentally, the Grey Heron is a smaller and slighter bird (about 60% of the Great Blue Heron and 40% of Cocoi by weight), with a smaller bill, shorter and light-coloured neck, and lacking chestnut thighs and chestnut edging to the 'shoulder patch' of the Great Blue. If a field comparison is possible, the Grey Heron is more similar in size and shape to the American Great White Egret, but heavier (the Great Egret being about 60% of the Grey Heron's weight). The general appearance of the Great Blue is of a larger, longer-legged, longer-necked, thicker-billed bird with "warmer" colour tones (derived from the venous to chestnut wash to the neck and chestnut thighs) contrasted with the flat grey of the Grey Heron. The Cocoi Heron is a starkly black and white bird, with the general appearance, size, and shape of the Great Blue, with white neck and thighs like the Grey Heron, but with an entirely black cap that may be paler on the forecrown.

Out of range identification of the three species is even more difficult among juveniles. All three have a brownish wash to their plumage, which is both variable and changeable, dark crowns, dark upper bill, and shorter plumes. Juvenile Grey Herons are quite pale, have grey thighs, strong black and white throat and breast streaking, light legs, and a white 'shoulder patch'. They tend to have a lighter crown than the other species, although this is variable among subspecies, *jouyi* being lighter than *cinerea*. Juvenile Great Blues are rather darker (except for some in southern Florida), have a pale to intense rufous tinge to the thighs and 'shoulder patch', and grey and rufous throat and breast streaking. Thigh colour, so important in identifying adult Great Blue Herons, is grey as in the other species. Juvenile Cocoi Herons are duskier than the others, with dusky streaked thighs and stripes on the throat; most importantly they have a distinct cap that is dark grey to black, paler on the forecrown, even in young juveniles. In all the species, adult plumage is assumed gradually over the first or second year with the grey of the crown diminishing very gradually, accounting for much of the variation to be seen, and much of the confusion in identification. The most important marks are the very dark and distinctively edged crown of the Cocoi Heron, the larger size and bill of the Great Blue Heron, and the thigh and 'shoulder patch' coloration.

Systematics

The Grey Heron is a typical heron related most closely the other *Ardea* and the *Egretta* herons. Indeed, in many ways it typifies the heron family—in nomenclature, dispersion, habitat, and behaviour. This is the Eurasian form of the large, solitary heron type, which is replaced by similar species in other biogeographic regions—the Great Blue Heron in the Nearctic, the Cocoi Heron in the Neotropics, and the White-necked Heron in Australasia.

Of the four subspecies recognized here, two distinguish extremes of clinal variation across Eurasia. The third subspecies represents an isolated island form, a taxonomic distinction with analogies in other herons. The pale Banc d'Arguin has a highly restricted distribution. Erard *et al.* (1986) proposed the interesting hypothesis that it is a remnant form, more closely related to populations of Asia and Madagascar, and worthy of specific rank. More typical Grey Herons nest in nearby Senegal, and an individual in Senegal was reported as showing intermediate features between *cinerea* and *monicae* (Baillon 1989, Dowsett and Dowsett-Lemaire 1993). Other described subspecies are problematic. Grey Herons on islands in the Asia-Oceania region include the heavy billed birds described as *A. c. altirostris*, occurring in Sumatra and Java (Voous and van Marle 1988). Those on Comores have been attributed to *A. c. cinerea*, whereas those on Aldabra have been considered intermediate between *cinerea* and *firasa* (Penny 1974).

Range and status

BREEDING RANGE: The breeding range of the Grey Heron covers most of the Old World south of the Arctic Circle, including Europe, Africa, Asia, and East Indies islands to Wallace's Line. *H. cinerea cinerea* occurs across northern Eurasia from Outer Hebrides, Great Britain (Bourne 1999), and Scandinavia, generally south of the Arctic Circle, through Russia (to Sakhalin). More to the south, it ranges from France, Spain, the Mediterranean, east Europe, Turkey, Iraq, possibly Israel (Ilani and Shalmon 1984), north Iran, north Afghanistan, Turkmenistan, Pakistan, India to Sri Lanka. Its status in Africa is intriguing. In north Africa, it nests rarely in isolated locations—Canary Islands, Cape Verde (Palacios and Barone 2001), Morocco, Algeria (van Dijk and Ledant 1983), and Tunisia. South of the Sahara and north of the Equator, small and sedentary breeding populations are known from Senegal, Mali, Niger, Côte d'Ivoire, Ghana, Nigeria, Sudan, and Ethiopia (not all mapped) (Turner 2000). Farther south, small breeding populations occur in Kenya, Tanzania, and Rwanda. They are resident throughout southern Africa, breeding patchily in suitable wet areas. In the east Palearctic, the race *cinerea* integrates with *jouyi* in east Russia. This subspecies ranges from India, Mongolia (Stephan 1988), China to Hainan, Korea, Japan (Matsunaga *et al.* 2000), Myanmar, Vietnam and Cambodia (Mekong Delta-J. Tordoff pers. comm.), Malaysia (peninsular), Indonesia to Java and Sumatra (Danielson and Skov 1985, Silvius 1986). The race *firasa* occurs in Madagascar. The race *monicae* breeds on islands of the Banc d'Arguin, Mauritania. It has been observed breeding elsewhere rarely, a few nests in the Aftout es Saheli located 100–200 km south of Banc d'Arguin (B. Lamarche through H. Hafner).

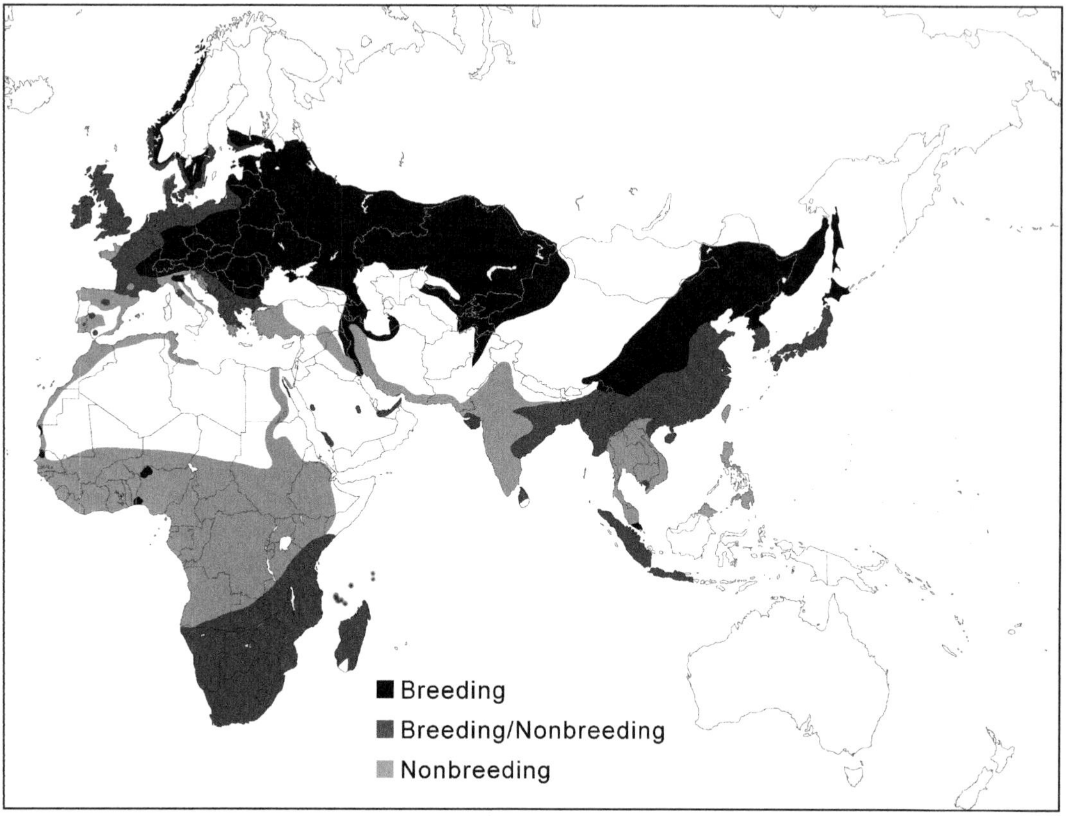

Outside of Mauritania it has been reported only as far as Senegal.

NON-BREEDING RANGE: European nesting birds occur in winter in Great Britain, west and central Europe (e.g. Draulans *et al.* 1986a), even increasingly in northern Europe, such as in coastal lakes in Latvia (M. Strazds pers. comm.), the Mediterranean, Africa, and the Middle East including Saudi Arabia, and Iran. Asian nesting birds (Biondi *et al.* 1994) occur in winter in Pakistan (Balochistan, Sindh), India, south China including Hong Kong and Taiwan, south east Asia, Japan, Thailand, Malaysia, Philippines, and Indonesia (Borneo, Sumatra). Birds occur only in winter in Hong Kong, Taiwan, Philippines, and Borneo (Sabah–Sheldon *et al.* 2001). Sub-Saharan Africa north of the equator is the important African wintering area for many populations (Turner 2000). This winter range generally is from Cape Verde Islands, Senegal and west Africa, south to 4° in Zaire, east to Somalia. Western European birds winter from Senegal, Guinea, Sierra Leone, Mali, Upper Volta, Togo, Nigeria, and Zaire. Eastern European birds winter in Mali, Upper Volta, Togo, Nigeria, and Ethiopia. Birds from Russia winter from Senegal, to south Egypt, to Kenya. There is no direct evidence of northern birds wintering south of the Equator but seasonal increases in birds in parts of Tanzania are suggestive (N. Baker and E. Baker in prep.).

MIGRATION: Grey Herons disperse widely after the breeding season, this movement beginning generally soon after the young can fly. Dispersal may be in almost any direction, although mainly to the southwest in Europe. In southern Europe, postbreeding migration is in September and October, and prebreeding migration in February and March (Lekuona and Campos 1996c). In northern Europe, autumn migration is from early September to late November and return is in February and April (Grishchenko and

Serebryakov (1993). Birds breeding in Scotland and England are generally non-migratory (Partridge 1986), although British birds have also been recovered in Spain and Gambia (Christmas 1994). In southern Europe, some birds remain in winter but others migrate (Lekuona and Campos 1996a, 1996c, Marion 2000). Migration is generally to the southwest in west and east Europe, to the southwest in west Asia, and to the south and southwest in east Asia (Hancock and Kushlan 1984, Li *et al.* 1989, Round 1995a). In the Mediterranean, migration is broad but corridors occur along both the eastern and western shores. Birds nesting in Africa, India, and south east Asia appear to be sedentary.

Individuals sometimes wander widely. The Grey Heron is regularly seen in Iceland, Faeroes, and Ascension islands. In the New World, it has been recorded in Greenland, United States (Burton and Smith 2001), Barbados (Smith and Smith 1990), Tobago (D. Finch 2002), French Guiana, Azores, Trinidad (F. Hayes pers. comm.), Montserrat, Martinique, off Bermuda, Brazil (Para), and Venezuela (Marion 1988). The last six records are of birds that originated in France. It has been reported offshore in South Africa (Kirsch 1998). There are also occasional records from New Guinea, Kalimatan (Van Balen 1999), Australia, and New Zealand (Parkes 1974). Early sight records from New Zealand have been discounted (Dawson 1974).

STATUS: The Grey Heron is the most widespread heron in Eurasia, with a breeding population in Europe between 150 000 and 180 000 pairs (Marion 2000). In Britain, where the longest running census occurs (Bourne 1999), the population was nearly 6000 pairs in 1991 (Carter 1992). The core of the European population is in France, Russia, Ukraine, Germany, the United Kingdom and Netherlands. With few exceptions, Grey Heron populations have increased markedly in numbers and range in Europe in this century. Northern expansion may be enhanced by post Pleistocene climate amelioration, but population increases are due principally to decreased hunting and depredation control (Moller and Olesen 1983, Marion and Marion 1987a, Bara 1988, Bakka 1986). In Britain, for example, the nesting population increased from 4000 to 6000 pairs between 1928 and 1991, rising at an annual rate of 8% per year during the 1980s (Tomlinson 1992). In France,the nesting population increased from 4500 to 27 000 pairs between 1974 and 1994, becoming the largest European population, due to protection (Marion 1997).

In West Africa, the nesting population outside Mauritania is small, perhaps less than 500 pairs (Turner 2000). The population in Niger averaged 8500 (1994–7) (Brouwer and Mullie 2001). In east Africa, it is widespread but scattered. The resident population in Tanzania may be about 15 000 birds (Baker and Baker 2004). In southern Africa, the Grey Heron is increasing in Zimbabwe and South Africa, due principally to irrigation and reservoirs. Aquaculture has provided substantial additional feeding opportunities to which populations have responded (Kurz and Schmidt 1994). The coastal Mauritania population is highly isolated, although the known population there has increased from 1100–1600 pairs in the 1960s to over 4188 in 1997 (Hafner *et al.* 1998b). The Malagasy subspecies is largely restricted to the western portion of Madagascar and is vulnerable owing to restricted range, exceedingly high levels of habitat alteration, hunting, and predation at the nesting colonies (Turner 2000). The Grey Heron is increasing on the Seychelles (J. Gerlach pers. comm.).

The Grey Heron is generally an abundant summer resident in China, Mongolia, Japan, Korea, and parts of Indonesia (Lansdown *et al.* 2000, Matsunaga *et al.* 2000). It is increasing in Japan (Kyoto) (Sasaki 2001). Breeding populations are small in Indochina and peninsular Malaysia.

Habitat

Grey Herons are generalists in habitat use. They are typically found in and around shallow water, generally along watercourses and shorelines, and usually in locations having roost trees nearby. They may occur in inland fresh waters, along estuaries, or in marine habitats. Shallow water, relatively large prey, and 4 or 5 months of an ice-free breeding season seem to be essential characteristics of suitable habitat. Among typical habitats are lakeshores, rivers and floodplains, reeds (Jeanmonod and Roulin 1989), swamps, ponds, mudflats, beaches, mangroves, and salt marshes. Dry or damp grasslands away from water are also used. Generally most abundant in the lowlands, Grey Herons can be found at surprising altitude, breeding to 2000 m in Armenia and occurring to 4000 m in northwest India. Use of man-made habitats is the rule over many parts of its range. Rice fields and fishponds provide important feeding sites nearly range wide. Individual birds and even local populations may become dependent on these habitats and so are affected by changes in agricultural management, as well as other anthropogenic factors (Fasola 1986, Biondi *et al.* 1993, Thomas and Hafner 2000). Grey Herons nest and roost in trees, with exceptions such as the Mauritanian population, so trees or other high vantage points are generally desirable. Some degree of isolation and protection is typical of places chosen for roosting and nesting.

Foraging

The Grey Heron usually hunts solitarily, but in situations where food is more concentrated, birds may feed in loose aggregations or even mixed species flocks. If conditions are especially favourable, quite large feeding aggregations may form, particularly after the breeding season. Herons feeding alone defend their feeding territories. Defence may be vigorous, and killing of intruders is known (Richner 1985). Aggression varies seasonally being most intense when young are being fed (Lekuona 1999). Sites may be near to the colony site or as far as 38 km away. When available feeding areas are poor and/or more distant from the colony, adults use two or three feeding areas and are no longer territorial (15% of breeders, Marion 1984, 1989).

Depending on prey availability and distance, herons use three types of feeding sites (L. Marion pers. comm.): (1) the individual defended feeding area composed of a single patch of about 20 ha, usually in wetlands or rice fields; (2) individual non-territorial feeding areas composed of two or three patches in poor habitats or in typical habitats more distant from the colony; and (3) neutral feeding sites used by many herons, where individual site appropriation is not possible. The latter are either ephemeral sites where high densities of food occur for short periods or sites too close to the colony and therefore attracting too many birds to be defended successfully. Herons appear to make trade-offs between nearer and distant sites, between using energy in territorial defence, or between site fidelity and exploring new feeding opportunities. Non-breeders and birds with little experience in an area spend more time exploring than do birds with local experience (van Vessem and Draulans 1987a).

Although considered in the literature to be typically a diurnal feeder, in fact Grey Herons feed at any time of the day. Generally, they feed most actively at dawn and at dusk and roost—usually in trees—during the middle of the day and at night. But they also feed at night, particularly during breeding when adults have to feed up to 23 hours a day to care for the young (Marion 1988). At fish farms, herons generally feed at early twilight or night, sometimes using artificial light but often feeding in the dark, where probably not coincidently they are out of sight of fish farmers who harass them during daylight (Draulans and van Vessem 1985a, Duquet 1987, Marion and Marion 1987a, 1987b, Amies 1990, Carss 1993). Time budgets actually observed depend on the interaction of food availability and food demand (van Vessem and Draulans 1987a, 1987b). As noted, during the peak of nesting energy demand, birds feed nearly all day. In tidal situations, birds forage according to how tides make food available, selecting high or low tides (Matsunaga 2000). In France, they fed predominantly during the four hours of low tide (Lekuona 1999).

By far the most common feeding technique (used half or more of the time) is Standing, often thigh-deep in water, but also in shallower water, on land next to water, or on dry land away from the water. Walking slowly is used to move between Standing bouts or for slow exploration. Walking is done at a very slow pace. When fishing from a bank, herons frequently use Peering Over. They also may use more active behaviours such as Running with wings half-open, Swimming Feeding (for many minutes at a time), Diving and Plunging, but these behaviours are rather exceptional (e.g. Brown 1985, Guntert 1986, Laurenti 1986, Craney *et al.* 1989, Taylor and Carss 1990, Husband and Siering

1991). Despite its size, it feeds while in flight by performing aerial manoeuvres of surprising dexterity (Keighley and Hall 1995, Bowey 1997, Barbour and Barbour 2001). Scavenging along the shore or in pastureland is not uncommon (Hiraldo *et al.* 1991), and it follows ploughs feeding on what is scared off or unearthed (Fox 1989a). In winter, it generally feeds in pastures on small mammals. The usual method of capturing prey is by a rapid Bill Thrust, preceded by retracting its head and neck. Pecking is used on dry ground, such as for earthworms. Food is nearly always dipped in the water, if available, before being swallowed. It is handled by biting, stabbing, and pounding to soften, to remove spines, to position for swallowing (fish are swallowed head first), or for breaking into smaller pieces.

Being a relatively large bird, it is not often harassed by other species on its feeding ground. In the air, it is usually able to avoid aerial attacks from other species by sideslipping, but is not always successful as there are instances of it being harried to its death by crows (Walters 1983). Being among the larger birds in mixed feeding situations, it can steal prey from other species such as gulls, cormorants, bitterns, and grebes (Cooper 1984, Sellin 1986, Kington 1991, Hume 1992). In turn it is subject to piracy and other aggressions from smaller, more agile birds such as crows and gulls (Walters 1983, Warner 1986, Todhunter 1987, Skeen 1988). Intraspecific robbing is also common, especially among juveniles (Lekuona 2002b). It also feeds symbiotically with species such as mergansers and osprey (Laurenti 1990, Jacobs 1994).

The Grey Heron's flexibility in diet and feeding behaviour suggests it can make use of a complex set of foraging tactics in the face of different or varying prey abundance (Draulans and van Vessem 1985a, 1985b, Draulans *et al.* 1986b, Draulans 1987, Draulans and Hannon 1988, Marion 1989, Peris *et al.* 1996). Evidence suggests that the species is highly opportunistic in prey selection. Prey do differ in profitability, depending on prey size, morphology, and behaviour (Moser 1986b), and Grey Herons rapidly respond to changes in prey availability (Adams and Mitchell 1995). Fishing success of Grey Herons increases with increasing prey density, to a point, due to higher encounter rates and the use of more intensive hunting methods. Prey intake is highest when birds first arrive at a feeding site, declining with time. The ability of herons to adjust their foraging intensity, behaviour, and expectations to ambiant conditions means that success rates turn out to be similar over different habitats. During the non-breeding season, herons can easily change feeding to more productive sites in response to environmental conditions (wind, rain, clouds) that alter the sites' relative value. However, during-breeding season the territorial feeding strategy induces differences in feeding success between territories and thus adults, and directly influences the breeding success (Marion and Marion 1987a, Marion 1989).

Individuals differ in their behavioural choices, juveniles especially being less effective in both feeding and selection of the site and time for feeding. It is not clear how much of this ineffectiveness is due to juvenile choice or to dominance of adults at prime sites (Draulans 1987, Draulans and Hannon 1988). Juveniles, being less experienced, improve their foraging efficiency at fish farms more than do adults. Overall, juveniles appear to be less flexible than adults in the use of time, but more flexible in the use of space (Draulans and van Vessem 1985b).

That prey capture success, energy intake, and frequency of feeding strikes of Grey Herons increase with higher prey densities indicates the importance of herons finding and feeding in locations that have high food abundance. At very high prey density, success plateaus out but the energy spent foraging decreases. Thus an important aspect of Grey Heron ecology is the ability to access sites supporting acceptable to high foraging efficiency. Diet variability follows, as herons change diet to take advantage of more catchable prey (Draulans and van Vessem 1987a, Adams and Mitchell 1995, Boldreghini *et al.* 1995, Peris *et al.* 1995). For example at a fish farm, herons focused on caged fish that were blind or in poor condition (Carss and Marquiss 1991, Carss 1993). Diet differs among colonies and among years (Draulans *et al.* 1987). Differential feeding success becomes exceptionally important during nesting. When feeding opportunities are distant and/or poor, for example if only frogs or voles are available, no more than two young can be produced, whereas if the feeding sites are rich, such as with eels, even if as far as 35 km away from the colony, herons can produce four young (L. Marion pers. comm.). Herons have more options for feeding site selection outside the nesting season, when energy demands are lessened.

Most studies have shown that Grey Herons do not seem to severely impact their prey supplies, taking only 6%–8% of fish standing stock available in ponds (Draulans 1988a, Feunteun and Marion 1994), although effects may be greater in trout streams (Kramer 1984). These are crucial findings showing the well-being of local heron populations are often tied to their interactions with fish farms.

The Grey Heron takes a very wide variety of prey, the diet varying according to habitat and season (Draulans *et al.* 1987, Marquiss and Leitch 1990, Sawara *et al.* 1994). It is something of a specialist on larger prey, up to 19–25 cm long. Some may be much larger or longer. The size of prey eaten or attempted is often remarkable and at times this leads to swallowing problems (Davis 1982, Bentzien 1991, Geroudet 1993a, Cuenoud 1994). The typical strategy is to catch a few large prey during the course of the day.

Depending on location, fish or crustaceans tend to dominate a diet; amphibians, small mammals, reptiles, crustaceans, worms, insects, and small birds are also eaten (Fox 1989b, Bentzien 1991, Greaves 1991, Mienis 1991, Camici and Zimmerli 1993, Besson 1994, Cuenoud 1994, Carruette 1995, Hewson 1985, 1995, Lekuona and Campos 1996c, Kreuziger and Achenbach 1998, McCanch 2003). It is remarkably versatile and is reported taking such prey as muskrats, rats, grebes, wood hoopoe, frogs in winter, scavenged fish, and mussels. It also swallows a good deal of vegetable matter and even other foreign material (Hill 1988), but it is unclear if this is functional. It can scavenge carrion, dead fish, and scraps from fishing boats (van der Kelen 1993, Dies 1999a, 1999b). Prey and size of prey varies seasonally, especially with respect to the food needs and food-handling capacity of young in the nest (Lekuona 1999).

Breeding

The time of nesting differs across the extensive range of this species. In temperate areas, the breeding season is restricted to spring and summer. In northern China, herons arrive in late March and in some years not until May. In coastal Russia (Maritime Province), herons arrive mid March and lay eggs in the first half of April (Litvinenko 1983). In Britain, nesting begins in February and March. In Spain nesting can be as early as January, corresponding to when minimum daily temperatures reach 8°C (Campos and Fernandez-Cruz 1991). Nesting in most situations is early relative to other herons, in Spain three months before other herons (Prosper and Hafner 1996). In the tropics, the nesting season is more flexible, often occurring in the rains, the timing of which may differ from place to place. For example, in north India nesting is in July–October, but in south India in November–March. Birds breed at the height of the rainy season in Africa, from Senegal through west Africa, Sudan, Ethiopia, Kenya, Tanzania, and Rwanda (Turner 2000). The coastal *monicae* nest from April to December peeking in April–May (H. Hafner pers. comm.) Even in monsoon areas, nesting may start relatively early, well before the rains actually begin. In some tropical areas, nesting can occur nearly year round, such as in coastal Kenya, where birds feed in the littoral zone.

Most frequently these herons nest in patches of trees located where they offer some degree of isolation and privacy. They generally choose the taller of the tress available, up to 50 m, and in mixed colonies are the highest of the nesting herons (Fasola and Alieri 1992a). However, the total range of nest substrates and colony sites used is very large, including wood lots, rocky islands and city parks, herbaceous marsh plants, rock cliff ledges, buildings, walls, and even the bare ground (e.g. Litvinenko 1983, Polasek 1991, Helm 1996). Nesting sites need to be within convenient flying distance of feeding areas. They fly 2–38 km to feed (Litvinenko 1983, Marion 1989).

The Grey Heron is primarily a colonial nester, although often in small colonies of 2–10 nests (e.g. Korvin 1987). The largest colony in Europe is 800–1300 pairs (Feunteun and Marion 1994). Being the first nesters, their colonies attract other species such as ibis, spoonbills, cormorants, other herons (e.g. Commecy 1989, Carpegna *et al.* 1990), and especially corvids (Tkachenko and Shakula 1983). Heron nests are attractive to and are taken over by other species such as hawks (Boonman and de Vrieslaan 1992).

The nest is generally a stick platform with a grass liner about 50 cm in diameter, constructed from the most easily available material at the site. As noted above, nest sites vary widely, but they tend to be built toward the top of the canopy. While new nests may be so flimsy that the eggs are visible through them, pre-existing nests are often reused, being repaired and enlarged annually, becoming bulky masses. Windstorms can destroy nests, so they tend to be situated on secure tree branches. On bare ground, herons use twigs, herbaceous plants, and seagrass for nesting material (e.g. Litvinenko 1983). Males do more stick gathering than females and females more of the nest building (van Vessem and Draulans 1986c).

The courtship of the Grey Heron has been intensively studied (Witherby *et al.* 1939, Lowe 1954, Bauer and Glutz von Blotzheim 1966, Milstein *et al.* 1970 and others). The male on arrival at the courtship site or the colony site claims an advertising location, an old nest if available. In small populations, males may gather and courtship occurs at display sites prior to the birds moving as a group to the colony site (L. Marion pers. comm.). In larger colonies, there are generally several waves of bird arrivals over two or three months, and each group of new birds tends to nest together forming subcolonies (Marion 1988). While displaying, through the day, the male repeatedly gives the Rwo call, its yelping advertising call. It also performs Upright displays with plumes erected, a bill-down version of the Stretch display, and the Twig-shake display. The male defends his territory against other males (and approaching females) mainly with Forward displays. The main pairing displays are the Stretch with up-pointed bill, combined with a milder version of the lunge movement or Bow–Snap display. Males Fly Around, and when alighting give a landing display, with powerful wing beats and arched neck with raised neck and head plumes. Females approach the male, who is at first responds aggressively toward the intrusion giving Forward, Upright, and Stretch displays. As aggression wanes, Snap displays increase. After pairing occurs, Snap displays decrease, Stretch displays continue, as do much mutual preening, billing, and Bill Clappering. Pairs may Fly Around, and when one lands with a Landing Display, the other responds with Upright and Stretch displays. The individual permutations of the display repertoire can be complex, varying a great deal with age, temperament, and the stage of the pair bond. As the pair bond strengthens, the intensity and elaboration of displays diminish, particularly for males. Copulation occurs with the female standing and continues well into egg laying. At larger colonies extrapair copulations can occur, usually with neighbouring females (Ramo 1993, Lekuona and Campos 1998). The pair bond appears to last a single season only.

The eggs are pale green-blue, relatively large but variable, 57–61 × 41–43 mm, laid at intervals of two days or longer. The clutch size differs with latitude. In temperate zones (Europe) a clutch is usually 3–5 eggs (3–6 in Belgium—van Vessem 1991). In the tropics, egg clutches are reported to be 3–4 in India, 3–5 in Africa, and 3 in Madagascar. Complete clutches of 1 egg to 10 eggs are on record, but the latter seem unlikely to be attributable to a single female. Replacement clutches are reported, but this is less common than has been assumed because birds more typically mate with another bird or both leave the colony rather than renest (Marion 1988). Nests from failed attempts may be reused by other birds. In Spain early season losses led to bimodal laying peaks, in March and again in April–May (e.g. Campos and Fernandez-Cruz 1991). Incubation takes 23–28 days, averaging about 26, but in tropical conditions may be only 21 days. Both parents incubate. They attend the nest for similar periods of time through the day (van Vessem and Draulans 1986b, 1986c). Young hatch asynchronously, depending on the laying date. Young are fed upon hatching, so the earliest chicks become much larger than later chicks. This leads to differential survival, influenced by sibling competition, aggression, and even cannibalism (Litvinenko 1983, Lekuona and Campos 1993). In years of food shortage or other difficulties the smaller chicks do not survive, and in larger broods young chicks rarely survive irrespective of food supply. The chick's eyes open at hatching, and begging begins soon after. Parents brood young for up to 18 days. One remains on guard until about day 30. Thereafter, both parents forage at the same time. Both parents feed the young, which take food directly from the adult's bill, except during the first days. When young cannot eat the sizes of prey most profitably caught by adults, the larger prey are removed from the nest by the adult, consumed, and predigested so that young are able to ingest them (Marion 1988).

Parents feed chicks first at the nest and then on nearby branches. Chicks that fall to the ground are not usually tended, although there are contrary reports (Lowe 1984). Feathers begin to grow about 7 days, and chicks are fully feathered at 28 days. Chicks remain in the nest for up to 10–20 days, where they spend most of the time standing, sleeping, begging, and tussling among sibs. They perform Upright, Forward and Snap displays, defending the nest. They can step away from the nest in 20–30 days, fly at about 50 days of age, and are independent at about 60–70 days. After fledging, generally juveniles are not fed or attended away from the nesting site.

Many factors affect the success of breeding. Egg failure can be high in some populations; in Belgium, egg failure was 17% (Campos and Fernandez-Cruz 1991, van Vessem 1991); nestling mortality was 33.3% (van Vessem 1991). In Spain, brood size averaged 2.2 nestlings from a clutch size averaging 3.5 (Fernandez-Cruz and Campos 1993). In Poland, brood size was 3.4 and 2.0 from a clutch size of 4.8 (Czapulak and Adamski 2002). Success differs among pairs within a colony. For example, brood size increased with increasing age of parents, suggesting higher production by older, experienced birds. Breeding success was not affected by the number of birds present or by synchronization or by location in the colony. The most critical factor in nesting success is availability of food through the nesting season. Young also can be killed by predators, especially large aerial predators such as eagles (Berthelot and Navizet 1987, Naoroji 1990, Ytraberg 1992). Breeding success can be higher for birds using fish farms (Carss and Marquiss 1996). Nesting high in trees puts nests and nestlings at risk from windstorms (Lekuona and Campos 1995).

Population dynamics

The population dynamics of the Grey Heron are among the best understood of all the herons (North 1979, Marion 1981, Marquiss *et al.* 1983, Marion and Marion 1987a, Campos and Fernandez-Cruz 1989, van Vessem 1991b, Tomlinson 1992). Grey Herons generally do not breed until after the second year, although some frequent colonies as young yearlings. Mortality is estimated to be 66.7% for yearlings (Fernandez-Cruz and Campos 1993), which none the less allowed a population increase of 34.6% for 8 years prior to stabilization. The demography of the species is driven by its longevity (Cordonnier 1985), suggesting that adult survival is the most critical factor in population growth and stability. In Spain, the breeding population is skewed toward older adult birds suggesting saturation of the breeding population (Campos and Fernandez-Cruz 1989, 1991, Campos and Fraile 1990). Killing, hunting, and accidental deaths of Grey Herons associated with humans can be a significant risk to stability (Kriedemann 1991). Century-scale population trends in Europe are attributable in large part to changes in its protective status (Marion 2000). Hunting is a particularly serious problem in some areas, such as Bavaria (Utschick 1983).

Fish farming has a conflicting influence. Populations can increase as birds, especially inexperienced young birds, feed at aquacultural facilities (Sawara *et al.* 1992, Samusenko 1993). However, mortality can also increase due to birds being killed at farms. In England water pollution has caused the deaths of breeding adults and eggshell thinning, but has had no obvious effects on total populations.

Winter survival is a critical factor for temperate populations. In Britain survival of both adult and first-year herons depends on severity of the winter (North 1978, Marquiss *et al.* 1983). The largest annual decreases in the British population occurred after the winters of 1947 and 1963. However, in Brittany, France, no difference in mortality rates was noted between hard and mild winters (Marion 1981). Young appear to be more susceptible than adults to winter kill (Tomlinson 1992). The interactive effects of humans killing birds and hard winters have been demonstrated by Utschick (1983) showing that the potential exists for hunting to decrease populations to levels that inhibit recovery following a severe winter.

Range wide, as populations have increased in the twentieth century, hunting has decreased as a factor, and density-dependent factors, such as territoriality, and also winter mortality have become more important factors in regulating population size (Marion 1984, Marion and Marion 1987a). It is likely that in much of Europe population size is becoming governed by habitat availability (Utschick 1983, Fasola and Alieri 1992b).

Conservation

The species is widespread, adaptable, and globally abundant. Overall, the species is secure. However, certain populations are of conservation concern. The Malagasy population (*firasa*) is threatened within its limited range by human activity and ongoing habitat destruction. It is considered globally Endangered by the Heron Specialist Group. Nesting sites require active protection and a local awareness campaign describing the critical nature of the bird's situation should be undertaken. The population of distinctive pale birds from islands in the Banc d'Arguin, Mauritania (*monicae*) also has a highly restricted range. Given the population increase and protection offered by the national park, its current situation seems secure at present. Surveys and monitoring should be continued and the birds of this colour form should be looked for elsewhere in west Africa. Populations on various Pacific Islands are small enough to warrant conservation concern. The Grey Herons of Sumatra (which have been called *altirostris*) are considered to be globally vulnerable by the Herons Specialist Group. A known population numbering about 700 birds occupies restricted habitats under threat. A conservation plan for this population is needed. Populations of concern include birds of the Comores, Aldabra, and North Marianas. Surveys are needed to determine the true numbers of these populations, seasonal distribution, and migrations and location of important areas as well as their appropriate taxonomic status. Other populations of concern include that in Indochina (Thailand, Malaysia), where few colony sites are known, and that in the Danube Delta, which appears to be decreasing.

Through the rest of its range, where known, the species is either stable or increasing, in some areas spectacularly so. In west Europe it is likely at carrying capacity, its populations now being influenced by winter conditions and the amount of habitat. The European population had reached an extremely low level by the mid twentieth century due to killing by hunters and fishermen for the preceding century (Knief and Drenckhahn 1984, Marion 1984, 2000, Marion and Marion 1987a). In central Europe, for example, there was an increase after 1880–90 and then a decrease from 1930 to 1950. Legal restrictions that began in the 1950s, along with the increase in artificial feeding situations and general warming of the climate have been beneficial. The increasing heron population has brought it into renewed conflict (or the perception of conflict) with the expanded aquacultural industry. Through its known history, the species' welfare has depended in large part on the outcome of its direct interactions with people (Cordonnier 1985, Marion 2000). These also are essential ingredients for management within a contemporary conservation strategy.

Among the factors under human control, the elimination of hunting and the killing of herons at fish farms are those that require immediate attention in conservation planning. For a long-lived bird demographically dependent on adult survival, it would be hard to make an argument that herons should be a huntable species for either cultural or food purposes. Renewed hunting poses a threat to the Bavarian heron populations (Utschick 1983). In Scotland 800 herons were estimated to have died in one year at fish farms (Carss 1994). Perhaps the only reason why the number of birds killed at aquaculture facilities has not reduced the overall populations so far is that it is probably young birds that are mostly killed. Such killing occurs despite excellent studies showing that the effects of herons on most fish farming operations are slight (relative to the full scope of mortality) or manageable through aquacultural practices (Utschick 1984a, 1984b, Marion and Marion 1987b, Perennou 1987, Carss and Marquiss 1991, 1996, Carss 1993 Marion 2000). For example, at a fish farm with caged trout, herons selected small, blind or poor condition fish, and losses were small compared with overall losses, yet these appeared serious to farmers. Furthermore, in France, it was found that impacts to aquaculture were not caused by the local nesting birds, so reducing the local heron population would not have any effect on depredation (Cordonnier 1985). Utschick (1994a) found that to reduce heron presence at fish farms in Bavaria from 0.6 to 0.1 herons per day would require eliminating 75% of the population of Central Europe. Fishery management plans and heron conservation plans need to develop ways to minimize both the damage by herons at fish farms and the damage to herons at fish farms.

Timber operations are a constant problem through most of the range (e.g. Bracko 1985, Kneif 1986, Tallone 1991). Unless sensitively controlled, harvesting can remove nesting trees and the disturbance of nearby operations, even if a colony site is preserved, can be devastating to the nesting herons. Population increases in South Africa and Zimbabwe correlate with the establishment of reservoirs, dams, and irrigation areas (Turner 2000). This is an example of how the maintenance of current population levels requires maintenance of these artificial situations.

Research needs

The Grey Heron is one of the best known of the herons. As a result, the knowledge base is sufficient to indicate the direction of conservation actions. Studies of Grey Herons have also contributed significantly to understanding heron behaviour and biology in a larger sense. Important gaps exist in the biology, taxonomy, and conservation needs of peripheral populations, particularly island populations, in west Africa and the Indo-Pacific. These populations require detailed examination. Additional study in both the field and of specimens of *monicae* and island forms is needed to determine the range of variation in these populations and relationships among recognized subspecies and other populations. Given the adaptability and vagrancy of the species, observers should search for breeding occurrences near where vagrants occur. Its potential for establishment in North America and the Caribbean islands should be monitored. The research base showing the limited impact of Grey Herons on fish farming is strong. What may now be needed are studies of ways to further reduce what impacts do occur, namely a socio-economic study of heron–fish farmer interaction, and sociological studies to find new approaches to achieve co-existence for herons and fish farmers. Demographic analysis is desirable to study conclusively the potential importance of adult mortality (especially from hunting and other killing) to population stability.

Overview

The Grey Heron is a highly successful species due to its biological flexibility. Its large size allows it to remain relatively far north in winter or to migrate as appropriate. Its large size also allows it to dominate other herons in defending feeding territories and to take a wide variety of prey types. It uses many habitats, from open seashore to secluded ponds and from fields to ditches and can take advantage of man-caused prey abundances, such as at fish farms. The Grey Heron tends to be a communal nester but a solitary feeder, a combination that takes advantage of both strategies. Herons can continue to feed in defended areas, but if superabundant food becomes available can shift to those sites and feed more aggregatively. The Grey Heron is one of the most abundant species over its large range, due primarily to its flexibility and adaptability.

Great Blue Heron *Ardea herodias* Linnaeus

PLATES 1.3, 1.4, 1.7

Other names: Blue Crane, Blue Heron, Great Blue, Great White Heron (white birds), Würdemann's Heron (dark-white intermediate), Blue Gaulin, Arsnicker in English; Garza Azulada, Garza Azul Mayor, Garzón, Garzón Azul, Garzón Cenizo, Garzón Azulado, Garcilote, Garcilote Ceniciento, Garzón Blanco (white birds) in Spanish; Grand héron, Krabye nwa, lajirond, Crabier bleu, Crabier radar, Crabier noir in French, Haiti; Amerikaanse blauwe reiger in Dutch; Kanadareiher in German.

Ardea herodias Linnaeus, 1758. Syst. Nat. ed. 10(1), p. 143: America = Hudson's Bay.

Subspecies: *Ardea herodias occidentalis* Audubon, 1853: Keys near Key West, Florida; *Ardea herodias wardi* Ridgway, 1882: Oyster = Estero Bay, Florida; *Ardea herodias fannini* Chapman, 1901: Skidegate, Graham Island, in Queen Charlotte Islands; *Ardea herodias cognata* Bangs, 1903: Indefatigable Island (Santa Cruz), Galapagos.

Description

The Great Blue Heron is a large, dimorphic heron. Dark birds are grey with chestnut thighs and a white cap over a black eye stripe. Light birds are all white.

ADULT: The dark form of the Great Blue Heron is basically a grey-blue bird. It has a white crown, cheeks, and throat and a black stripe running from the eye, along side of the head merging into long black plumes. The plumes and white crown together form an erectile crest. The irises are yellow, and in non-breeding the lores are a dark dull green, often surrounding a small yellow spot or line near the base of the bill. The long, deep bill is yellow, most often with a dusky top to the upper bill. The neck is grey, often with a chestnut to vinous tinge on the back and sides. The throat is vertically striped grey and white. The back is blue-grey. At rest, a patch at the apparent 'shoulder' (the epaulette) is black with chestnut toward the back. The undersides are light grey with black stripes. Thigh feathers are a distinctive chestnut. In flight, it has two-tone upper wings, darker behind, lighter in front, with patches of pale chestnut feathers at the bend in the wing (as field characters these are often called 'headlights'). The lower legs are green-brown, the upper legs often shading to pink. The white morph is all white. The legs are pale cream-yellow and lores very pale blue. The bill is yellow, often with a dusky top. During breeding the plumes achieve full development. The crest, incorporating several long black head plumes, when raised, is black with a white centre. The elongated throat and neck plumes are light grey with vinous wash. The similarly elongated back plumes also are light grey, contrasting with the darker back. The bill takes on an orange colour near the base and more yellow to the tip, turning darker after courtship. The legs also take on an orange-red colour. These colour changes may vary geographically as they are not reported in *fannini* (Butler 1992). The lores become blue-green to blue tending to yellow at the base of the bill. In the white form, the plumes are all white, and the bill and legs turn very pink. The lores turn blue.

VARIATION: The sexes have similar plumage but females are about 10% smaller. Some individuals, especially in southern Florida, appear to be intermediate between the dark and white morphs, having pale body plumage and a white head and light neck. These birds (the Würdemann's Heron) have been the subject of discussion for over a century (Cory 1886, 1887). They likely are the result of parental or ancestral pairings of white and dark birds. Subspecies of Great Blue Heron differ primarily in plumage shade and in size. The western race *fannini* has substantially darker dorsal coloration, longer wings, and a shorter bill than *herodias*. The southern race *wardi* is markedly larger and paler than *herodias*. The tropical races, Galapagos *cognata* and the dark form of *occidentalis*, are paler. *Occidentalis* has a high proportion of white morph birds, which are the largest Great Blue Herons, being up to 10% heavier and with distinctly heavier bills.

JUVENILE: Juveniles are darker than the adults and have a variably dark crown, lacking the white centre and plumes of the adult. Soft parts are duller; the upper bill is slaty and lower bill yellow. The irises are yellow and the lores are dark with yellow interior. Juveniles have much more ventral striping and mottling on the neck and chest than do adults. The back of the neck and the back are grey overall with a brownish tinge due to the upper wing feathers being tipped in chestnut. Legs are olive grey, tending to pink close to the body. Feathered thighs are light, grey to white. Juvenile white birds are off-white with buff grey legs. Yearlings more resemble adults but with small white crowns, shorter plumes, and remnants of chestnut-tipped wing feathers (Baker pers. comm.).

CHICK: Dark form chicks at hatching are covered with a pale grey down, especially exuberant on the crown. Legs and bill are pink. White form herons have white down.

VOICE: The harsh, deep, hoarse 'Frawnk' call is the alarm, disturbance, taking flight, and aggressive call (Bayer 1985b). This characteristic call is highly variable, ranging from a simple 'frawnk', 'frank', or 'braak' when taking flight after disturbance to an aggressive series—'fraawnk, fraawnk, fraawnk, taaauw, taaauw'. The 'Ark' call is given at the nest and part of the Greeting Ceremony. The 'Gooo' call is a bleat given at the end of a Forward display, in pursuit flights, and during aggressive encounters. The 'Roh' call, rendered 'roh, roh, roh', is used to advertise territory either while Standing in conjunction with Arched Neck or in Circle Flight displays. As a feeding territorial call, it is answered in kind by birds on neighbouring territories. It is also given when in active pursuit of another heron. The similar Landing Call is given at the nesting colony. The 'Go' call is a series of clucks, 'go, go, go', given at the nest and approached. The 'Ee' call serves as a social contact vocalization, given on arrival or departing. Birds also use a Bill Snap especially in defending a nest site and in other situations (Mock 1979). Bill Clappering is common among pairs.

WEIGHTS AND MEASUREMENTS: Weight: 2200–2500 g; length: 97–137 cm. It is the largest heron in North America, which lacks a representative of the giant heron group.

Field characters

The Great Blue Heron is identified by its large size, white crown, grey back and upper wings with black epaulettes, and chestnut thighs. It generally is seen alone, standing in an erect posture. When taking off and flying short distances, the Great Blue Heron often keeps its neck extended, but in level flight coils it back in typical heron fashion. Its wing beats are slow, powerful, and steady; it can soar to height before gliding downward again. The bicoloured dark and light grey wings with white spots on the leading edge aid identification.

It is distinguished from other dark herons in its range (Little Blue Heron, Tricoloured Heron) by its much larger size. It is distinguished from the Sandhill Crane (*Grus canadensis*) by its black (not red) head, lighter (not black) bill, and in flight by having its head pulled back (not extended). The dark form Great Blue Heron is similar to the other large *Ardea*, whose wanderings can bring individuals into the its otherwise allopatric range. It is distinguished from the Cocoi Heron by having a white crown with a black crest (not black crown), grey (not white) neck with complex vertical neck stripping (not a single black line), light grey (not black) belly, chestnut (not white) thighs, and chestnut edges to the 'shoulder' patches (not black epaulettes). It is distinguished from the Grey Heron by being larger (10%–40%), having chestnut edges (not black) epaulettes, and chestnut (not grey) thighs. Distinguishing the field characteristics of juveniles is discussed in the Grey Heron account. The white form of the Great Blue is identified by its size, its large, thick, straight bill, leg colour, and plumes. It is distinguished from the Great Egret by its large size, cream-yellow to buff-grey (not black) legs, relatively larger, thick and straight (not thinner and slightly down curved) bill, presence of head plumes, and less developed back plumes. It is distinguished from the medium-sized white herons (Snowy Egret, immature Little Blue Heron, white phase

Reddish Egret) by its much larger size and leg and bill colour.

Systematics

Subspeciation of the Great Blue Heron in North America, proper, is not as clear as one would hope. There is certainly clinal variation from north to south and east to west. However, clinal gradation, surprisingly, is not well studied due to the lack of specimens (Dickerman 1992). The number of subspecies described from North America range from seven to two (Oberhouser 1912, Payne 1979). Besides those recognized here, *lessoni* has been described as darker birds breeding in Mexico and *repens* a dimorphic form from coastal Venezuela described as light birds having short plumes and being smaller than *occidentalis.* These populations deserve closer scrutiny. The most interesting aspects of the Great Blue Heron's systematics are the situations at the edges of its range, in the Caribbean and south Florida (*occidentalis* and by extension the similar dimorphic birds of the Venezuela coast) and in the Pacific north west of North America (*fannini*). Dickerman (in press) reviewed variation in North America, recognizing the four subspecies in North America recognized here.

Occidentalis was erected by John James Audubon for the white birds of the Florida Keys. But these birds pair with dark birds and the subspecies name is now considered to apply to all the dark, white, and intermediate birds of south Florida and the Caribbean. Birds in this population are relatively larger than other populations, although likely this represents the end of a clinal variation in size down the Florida peninsula that remains to be quantitatively documented. For some time it has been known that there is a tendency for assortative pairing, by colour, among Florida Bay Great Blue Herons. The most likely explanation for this situation is that the white birds evolved as part of a dimorphic coastal population in the Caribbean and this population is now in secondary contact in Florida with birds derived from the continental population (*wardi*). This interpretation is supported by recent behavioural and molecular studies by Heather McGuire (pers. comm.). Based on several criteria (morphological, white–dark pairing, molecular), the geographic variation shown in the southern Florida and Caribbean population is recognizable as subspecific but not species level of divergence. The situation in *fannini* is similarly intriguing. This is a fairly isolated population of darker, shorter-billed birds possibly with differences in breeding soft part coloration. This population may have diverged farther from the eastern stock than is now appreciated (R. Butler pers. comm.).

Range and status

The Great Blue Heron breeds throughout much of North American except for high mountains and deserts, in Central America, and on certain islands in the Caribbean and Pacific.

BREEDING RANGE: Its North American breeding range is from southeast Alaska (Webster 1988, Butler 1997), British Columbia, across Canada from north Alberta, to central Manitoba, through south Ontario, south Quebec, New Brunswick, and Nova Scotia southward through the United States except the Rocky Mountains, into Mexico to Baja California (Palacios and Mellink 1993, Becerril and Carmona 1997), Sonora and Sinaloa on the west coast, and in Tamaulipas, Tabasco, and Yucatan on the east; also Cuba, St Johns Virgin Islands, Los Rocas Venezuela, and Galapagos. Breeding status in Central Mexico is unclear. The race *herodias* occurs over most of this North American range including Inland British Columbia (Gebauer and Moul 2001). The eastern race *wardi* occurs from Kansas and Oklahoma across the Mississippi river basin to Florida. The western race *fannini* occurs along the Pacific coast of North America from southeastern Alaska, Coastal British Columbia, to coastal Washington. The Caribbean race *occidentalis* occurs in extreme south Florida, from Tampa Bay south especially in Florida Bay, and on Cuba, St Thomas, and Los Roques, Yucatan (Lopez-Ornat and Ramo 1992). It occurs along the coast of Yucatan and Belize (Haas 1983) but has not been confirmed to be breeding there. It is also possible that it breeds in Venezuela (G. Morales pers. comm.). The race *cognata* occurs in the Galapagos Islands.

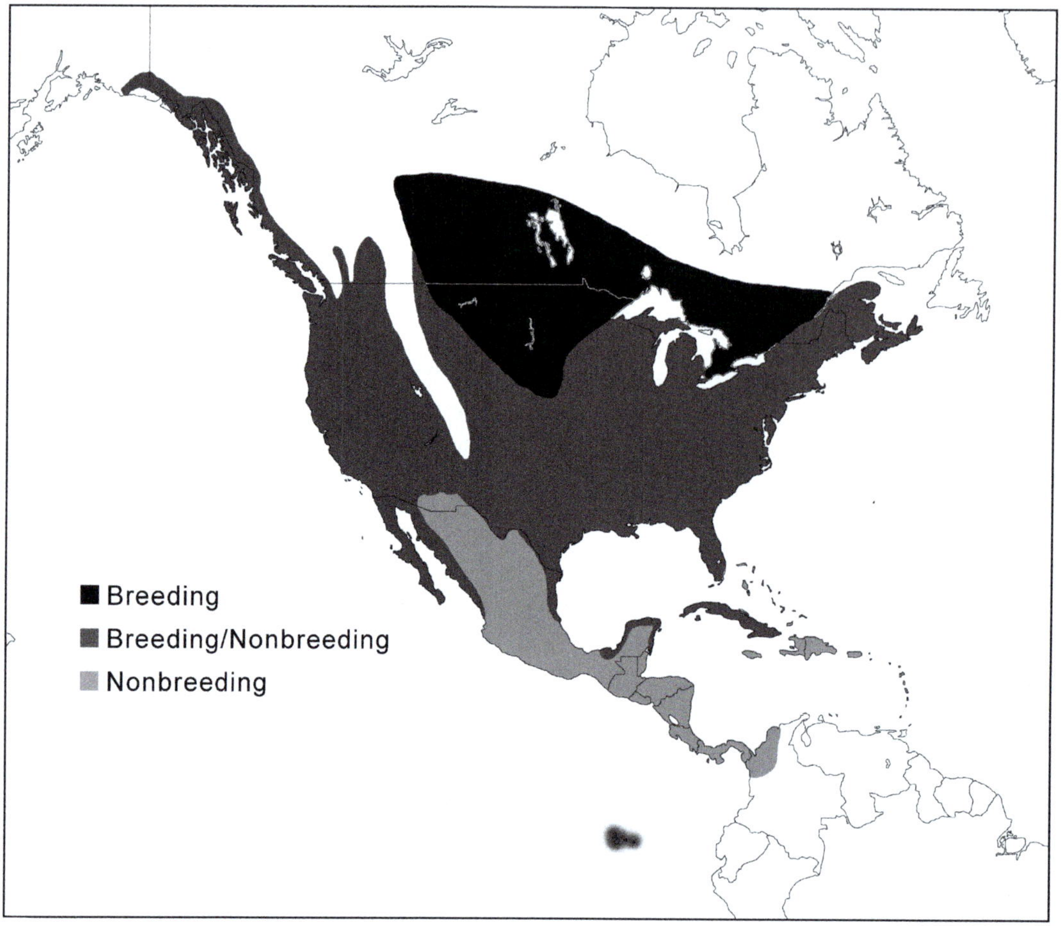

NON-BREEDING RANGE: Great Blue Herons have the largest non-breeding range of any North American heron (Mikuska *et al.* 1998). Many remain rather far north in the non-breeding season, from southeast Alaska, through its USA breeding range except in the north central prairies. Birds remain widely dispersed in winter, but important wintering areas include Strait of Georgia, Puget Sound central California, the USA midwest, Colorado River, Great Salt Lake, Gulf of Mexico and Pacific coast of Mexico, and Florida (Root 1988, Mikuska *et al.* 1998). Great Blue Herons occur year-round but with no evidence of breeding in the Bahamas. The non-breeding range includes the Greater Antilles, especially Hispaniola and Cuba, with fewer occurring in the Lesser Antilles south to Trinidad. It also occurs year-round in the Netherlands Antilles off the coast of Venezuela (Aruba, Bonaire). It is widely dispersed in northern winter in coastal Mexico to Guerrero on the west, Central America, rarely to Panama and northern South America as far south as the upper Cauca Valley in Colombia. It occurs rarely in northern Venezuela and even more rarely southward—but Great Blue Herons have been observed as far south as Rio Negro, Brazil (03 00 S) (Parkes 1998).

MIGRATION: Migratory tendency depends on location. Herons migrate south in September and October, in numbers of a few to 100, flying both day and night. Great Blue Herons from north central North America all are highly migratory in that their feeding habitats ice over. Eastern herons are more flexible, some shift southward, others remain rather far north, more in mild winters (George and Moore 1990). *Fannini* is non-migratory (R. Butler pers. comm.). There is no evidence for migration in *occidentalis*, although wandering occurs. Upon return migration, birds arrive early, usually in February to March (Butler 1992). Great Blue Herons disperse widely after nesting. Dispersal records include Newfoundland, Greenland, Hawaii (Pyle 1999), Bermuda, ships in the North Atlantic (on which they have arrived in Europe), near the Azores, Canary Islands, and France (Casement 1995, Herroelen 1995, Dobson 1996, Guermeur 1996, Clarke 1999). *Wardi* disperse northward (Dickerman 1992). Birds of the white form have wandered as far north as the Great Lakes and New York (Anonymous 2000, 2002c).

STATUS: Population overall appears to be at least stable over most of its range, increasing in eastern and southern North America, with some concern for a possible slow decline in the north Pacific population due to habitat losses and eagle predation (Butler 1997, Butler *et al.* 2000, pers. comm.). It is common to abundant throughout the rest of its breeding range to Mexico. It is rarer in the remainder of its breeding range, where in most cases its accurate status remains to be determined. Population sizes and quantitative trends are not possible to determine accurately, because nesting sites are so numerous and so highly dispersed (Butler *et al.* 2000). Available information includes over 35 000 birds along the south and east North American coast in the 1970s, but the nesting population in Louisiana alone now numbers over 10 000 birds and has been increasing (Fluery and Sherry 1995). Few data are available from its inland range, although over 12 000 pairs are reported in Quebec, 13 000 pairs in Ontario, and 5000 pairs in Illinois, so numbers inland are not inconsequential. Overall the population numbers in the 100 000– 250 000 range. The white morph in south Florida is thought to have been hugely reduced by plume hunting in the early part of the twentieth century but recovered to about 1500 individuals (Powell *et al.* 1989).

Habitats

The Great Blue Heron is a generalist in its habitat selection. Its long legs and large size allow it to feed from deep water to dry land. It uses freshwater and salt marshes, freshwater and mangrove swamps, estuaries, meadows, flooded agricultural fields and pastures, ditches, island shores, river banks, and lake edges. It can feed on dry land when profitable. In salt water, it feeds in coastal lagoons, mangroves, muddy and rocky seashores, tidal flats, sea grass flats, salt marshes, and also in the surf and along beaches. The white form is found almost entirely in marine habitats, particularly tidal grass flats. In the non-breeding season, outside the continental USA, birds tend to be marine using tidal flats, rocky shores, brackish lagoons, and mangrove forests (Morales 2000). Overall, natural habitats tend to be more important than man-made ones (Heitmeyer 1986). Birds require trees or bushes to nest in most locations. Although a species of the lowlands, it has been recorded up to 2600 m in South America. These herons are flexible, and can respond to changes in habitat availability (e.g. Elphick and Oring 1998).

Foraging

This heron usually feeds in the water or at its edge. Over 90% of the time it feeds by Standing or Walking slowly, although it also uses active behaviours such as Hovering, Plunging, Jumping, and Swimming Feeding to forage in even deeper water than its substantial leg length would otherwise allow. It also uses Standing Flycatching, Probing, Pecking, Running, Hopping, and Wing Flicking. Given the advantage of their size, they occasionally rob other birds of prey (e.g. Bildstein 1980, Bayer 1985a, Merrifield 1992, Squires 1998). Similarly, being obvious and not so agile as smaller birds, they are subject to piracy from other species (Johnson *et al.* 1996). Prey taken depends on feeding behaviours used. Forbes (1987a) found that by Standing, birds caught more active prey; whereas by Walking, birds caught more sedentary prey; by Feet First Diving, birds took dead and floating prey. Net return was lowest for Standing, higher for Walking, and highest for Feet First Diving. So Great Blue Herons choose foraging behaviour on the basis of its profitability, unusual or energy costly behaviours being used when the return is high (e.g. Morey and Smits 1987). Given their flexibility, Great Blues can take advantage of unusual foraging opportunities. For example, herons Dip behind fishing boats as nets are hauled (Ewins and Hennessey 1992), appear at grunnion runs (a beach-spawning fish) (Griem and Martin 1997) or catch fish attracted to floating bread (Zickefoose and Davis 1998).

Herons take advantage of food provided by aquaculture, such as pond-raised catfish, crawfish, and bait fish, and are considered to be significant nuisances to these industries. Local Great Blue Heron populations can obtain up to 44% of their diet from catfish ponds and up to 50% from trout hatcheries (Glahn *et al.* 1999a, 1999b). However, in a controlled experiment no difference was found in fish stocks between aquaculture facilities in which herons foraged and control situations (Glahn *et al.* 1999a, 1999b). Herons were found to be poor predators on healthy fish but good predators on sick fish (Glahn *et al.* 2000, 2002). Unhealthy fish and less desirable species, such as sunfish (*Lepomis*), were the preferred prey of Great Blues at catfish ponds.

Great Blue Herons feed by day or night. Especially in tidal situations, herons feed in relation to the tidal cycle, feeding at night at the proper tide (Bayer and McMahon 1981, Austin 1995). They also feed at night in non-tidal conditions, so nocturnality is a typical part of their behavioural pattern and their night vision is good (McNeil *et al.* 1993, Rojas *et al.* 1999). These herons typically defend their feeding sites, either a large territory held alone or individual areas when feeding in an aggregation. Territories are defended in both winter and summer. In Oregon, territories averaged 8.4 ha on the coast and 0.6 ha inland. In Canada, males defended territories while females and juveniles fed communally (Butler 1992). Pursuit and Supplanting Flights are common, as are standing displays such as Upright. They will advertise their territorial claim by their presence with Roh call and also by Arched Neck and Circle Flight displays (Bayer 1985b). Their most interesting display is given on the feeding grounds, the Upright and Spread Wing display: two herons encounter each other wings spread and drooped, with neck extended, head high over the back. The bill becomes more vertical and the back of the head further on the back as the display continues. Herons also use Forward and Upright Displays on the feeding site, the former merging into duels, which can be sufficiently severe that herons are injured.

The Great Blue Heron, being a large bird, can catch and use both small and large prey. Handling time depends on the size and the defences of the prey item. Herons tend to grab prey but can spear them as well (George 1941). They also probe and peck. They attack surprisingly large prey, but large and defensive prey can overwhelm herons, even causing their death (Wolf and Jones 1989). For a bird that catches prey infrequently, increased handling time seems not to be a great constraint and Great Blue Herons devote considerable time to subdue and prepare prey for swallowing (Forbes 1982, 1989). Herons can readily lose their relatively large prey to other birds (e.g. Squires 1998). Aquiring their foraging technique takes time and experience to learn. Juveniles are less effective than adults, and their skills increase with age (Quinney and Smith 1980, Davis 2001).

The Great Blue Heron typically eats fish, generally relatively large fish such as *Micropterus, Lepomis*, and *Esox*. But overall prey items include a wide range of insects, fishes, amphibians, birds, mammals, and reptiles. Birds are often reported; and in the west, mammals are frequently taken. In fact, seemingly atypical prey can in certain places be important in a heron's diet, such as is the case for hatching sea turtles (Sage 1995), ground squirrels (Hoffmann 1997), voles (Collazo 1985), and birds (Chapman and Forbes 1984, Vaniman and Luna 1998, Stolen 2001). Herons, especially on islands such as Florida Keys and the Galapagos, frequent human habitations for scraps of food provided for them, and these may become important dietary supplements.

Breeding

The Great Blue Heron is generally the first heron to begin nesting, due to its long nesting period. Nesting in most of its range, including continental USA, Cuba, and Mexico, begins in February through April continuing to June through August. In coastal southern Florida nesting is nearly year-round, more concentrated in winter, December–April.

As a generality, nesting Great Blue Herons need tall trees with some isolation from human disturbance and nearby aquatic feeding areas. These areas may be singular sites that the heron can use continually as food is renewed, or are a series of sites that can be used sequentially during the nesting season. Colonies tend to be well dispersed, but located so that the birds can fly 3 to 24 km to surrounding wetland complexes (Dowd and Flake 1984, 1985, Gibbs and Kinkel 1997). Colony size is correlated with the amount of wetlands within flying distance from the colony site (Gibbs *et al.* 1987, Gibbs 1991). Great Blue Herons tend to nest in relatively undisturbed sites, buffered from disturbance, with low road density, and surrounded by large forest stands (Scharf 1989, Gibbs and Kinkel 1997). Colony sites tend to shift around, sometimes for no apparent reason, but at other times due to tree mortality (Julin 1986, Dusi and Dusi 1996). They are flexible, such as when they rapidly occupy re-established beaver ponds (Potter *et al.* 1997). Great Blue Herons nest either solitarily or in colonies. Colony size can vary from year to year (Julin 1986). Regionally they tend to nest at a few large and many smaller colonies, the larger sites being more stable (Iverson 1993, Castrale 1994). Nests are stick platforms that vary from 0.5 to over 1 m across. Nests that survive the winter are rebuilt and enhanced, becoming larger over several years (Gretch 1989). Surviving nests also serve as display sites, and if unclaimed by herons can be used by other species such as owls (Burkholder and Smith 1988). Nests tend to be placed in the taller trees of those available (Carlson 1995). They do use bushes where these are more available, especially mangrove bushes in the tropics. They also nest on artificial structures (Henny and Kurtz 1978, Paton and Kneedy 1993), on the ground in protected sites, or on cliffs.

The courtship behaviour of the Great Blue Heron has been well studied for decades (Cottrille and Cottrille 1958, Meyerriecks 1960, Mock 1976). Herons change mates and nest sites from year to year *et al.* 1987) and so pair selection is a critical part of the annual cycle. Birds sometimes gather on foraging grounds to display. At a colony site, a displaying male usually first secures a display site, often an old nest site. During pair formation, males defend the nest and site with Forward displays, Jabbing, and Supplanting Flights. An Advertising Call is seldom given, the bird instead advertising with a highly stereotyped Stretch display, which includes much lateral swaying. The Snap display includes an audible mandible 'clack'. Other displays used include Wing Preen, Circle Flight, Twig Shake, and Fluffed Neck. Crest Raising is performed by both sexes throughout the breeding season. Aggressive displays include the Upright, Arched Neck, and Forward. These are not used in predictable ways, and there is much individual variation in their use. After pairing, Contact and Non-contact Bill Clappering are common.

Eggs are pale blue, having a size range of 61.3–65.6 × 41.9–46.5 mm (Schonwetter 1967). Eggs are laid at 2- to 3-day intervals. The clutch size varies from 2 to 7 eggs, increasing from south to north. Incubation takes about 25–29 days, most typically 28 days. Incubation is continuous after the first egg, consuming 54 minutes of each hour (Pratt 1970). Nest relief occurs once per day, and nests are seldom left unattended during incubation. Males tend to stay at the nest during the day, and females at night.

Both parents attend the nest and young, relieving each other two to five times per day. Hatchlings are not left alone until after 28 days old (Pratt 1970). Feeding activity peaks when chicks are about 4 weeks. Both parents feed the young by regurgitation, as many as six to ten times per day. Nesting success depends on food supplies, generally producing two to three young per successful nest. Chicks fledge at 60 days and leave the nest at 64–91 days (Pratt 1970). Mortality of chicks is often high, one to two usually being fledged. Among the highest success rates reported was about three per nest in Nova Scotia (Quinney 1983). Colony characteristics, food supplies, parental care, and predation affect nesting success. Herons in smaller colonies were more likely to fail completely but also more likely to raise three young than herons in larger colonies, overall there being no difference in fledging (Butler *et al.* 1995). Nesting success is limited by food supplies and by the ability of adults to supply sufficient food (Quinney 1982, Powell 1983, Mock 1985a, 1987, David and Berrill 1987, Sullivan 1988, Bennett *et al.* 1995). Chicks in competition for food can attack other chicks (David and Berrill 1987, Mock *et al.* 1987), but this trait is not as pronounced as in other species and chicks killing other chicks is not common. Becoming increasingly appreciated is the extent to which predators such as crows (*Corvis*), eagles (*Haliaeetus*), vultures (*Cathartes*), owls (*Bubo*), wolves (*Canis*), raccoons (*Procyon*), and bears (*Ursus*) attack nestlings (Gretch 1987, Norman *et al.* 1989, Robinson *et al.* 1991, Butler 1992). Heavy predation can affect nesting success (Seabolt 1984), and in some parts of the range the populations of these predators are increasing.

Population dynamics

The population dynamics of the Great Blue Heron are generally understood due to a long-term banding record and intensive studies at colonies (Henny and Bethers 1971, Henny 1972, Blus *et al.* 1980, Blus and Henny 1981, Butler 1997, Crawford and Neel 1997). However, no single colony or group of related colony sites have been subjected to long-term study of marked birds. They appear to begin to nest at two years, although younger birds do attempt nesting. Mortality rates are high for juveniles. In the first year mortality is 69–71%, decreasing thereafter and with regional differences (Henny 1972, Bayer 1981). Highest postfledging death rates are from August to December (Owen 1959), probably related to the difficulty of learning to feed (Quinney and Smith 1980). Considerable study has investigated contaminant burdens in Great Blue Herons, but there is no evidence of population impairment (e.g. Henny and Bethers 1971, Blus *et al.* 1980, 1985). For example, while nest attendance was inversely related to contaminent levels, there was no effect on pre-fledging nesting survival (Thomas and Anthony 2003). Mortality on the wintering grounds may be important (Blus and Henny 1981), particularly during severe winters for birds remaining far north. Overall, the Great Blue Heron is a relatively long-lived bird (Crawford and Neel 1997). Being so large, the adult Great Blue Heron is immune to most predators. However, they are harassed and their prey taken by eagles and it is likely that predation by eagles is more important than previously appreciated (Forbes 1987b). Great Blue Herons can kill each other in territorial disputes (Nehls 1992).

Conservation

This is a very adaptable species. Overall populations are stable or increasing, so there is no range-wide conservation concern. Certain coastal populations occupy relatively small areas. These need particular conservation attention. Colony site protection is the most critical conservation issue for Great Blue Herons. Range-wide, nesting sites are not limiting, but the inexorable loss of potential sites, particularly those most efficiently positioned near productive feeding areas, is continuing. Herons adversely affect nesting trees and feeding conditions change from time to time. As a result heron colonies naturally shift sites, particularly the smaller ones. So colonies need to be protected and managed on a landscape scale to preserve a matrix of potential sites. Research is clear that nesting Great Blue Herons prefer to avoid human disturbance. They nest away from disturbance and shift colony sites and breeding dispersion relative to disturbance factors (Scharf 1989, Gibbs and Kinkel 1997). In Nova Scotia human disturbance affects nesting success (Quinney 1983), probably also the case elsewhere. Colony sites need to be protected from disturbance, ranging from habitat destruction to intrusion by humans or their domestic pets. Buffer zones are required around active colony sites (Rodgers and Smith 1995b, Carlson and Mclean 1996). Colony site protection effort needs to be active, ongoing, flexible, and at a landscape scale.

The adaptability of the species can put it in conflict with human activities. Great Blue Herons are able to take advantage of abundant food supplies made available by aquaculture operations such as trout hatcheries, catfish ponds, and bait fish (Parkhurst *et al.* 1992, Hoy 1994, Stickley *et al.* 1995, Pitt and Conover 1996, Pitt *et al.* 1998). Studies have shown that, except in restricted situations, depredation caused by this species is not economically important, and where it is important it can be managed without harming the birds (Glahn *et al.* 1999a, 2000, 2002). Many birds are killed legally and illegally in the USA at aquaculture facilities. Over 12 000 herons were reported to be killed legally from 1987 to 1995 in southeast USA (Belant *et al.* 2000). Such killing of a long-lived species can be expected to affect population viability. There is little valid justification for killing Great Blue Herons because of their purported impact on aquaculture. In Louisiana, populations have increased due to the food made available by crawfish aquaculture (Fluery and Sherry 1995). In Florida Bay, white herons facing low reproduction attributable to declining environmental conditions affecting food supply are being maintained by taking handouts from people (Powell and Powell 1986). Being a top predator, Great Blue Herons accumulate persistent toxins, and many studies have been conducted of contaminant loads and possible effects. So far, population effects have not been shown and individual effects are subtle. It is possible that the versatility of the species includes its ability to deal with contaminants. As a result, this species can serve as an environmental sentinel, and monitoring contaminant loads should continue range wide. Great Blue Herons are impacted by catastrophic toxic events, such as occurred at the Salton Sea, USA. Management of such incidents is critical in that they can affect wintering birds concentrating in such areas.

Conservation of feeding habitat and food supplies is another critical issue. Throughout its range, natural situations are preferred over altered habitats. Conservation of the wetlands, rivers, streams, and other aquatic habitats supporting this species is essential. This is best done as part of landscape-scale wetland conservation programmes. In many areas the Great Blue Heron, as one the larger and more apparent water birds, can be used as a 'poster' species to encourage local and regional wetland and stream conservation action. Despite being continent wide in distribution, the species is principally a coastal bird in the winter. Conservation of coastal habitats in the southern USA, Mexico, and the Caribbean is essential.

Research needs

This is a well-studied species. Much of its basic biology is known and it has been used as a study model for behavioural and evolutionary questions. Infraspecies variation needs to be better understood, not only the validity and utility of described subspecies but also clinal and other patterns of geographic variation. Biochemical studies of these relationships are needed, especially in the Caribbean and Central America. It is possible that these forms represent a significantly divergent lineage. Nesting status of the species in Central Mexico, coastal Mexico and Belize, Central America, and the Caribbean (Bahamas especially Andros Island, Cuba, Lesser Antilles, and the islands off Venezuela) needs to be clarified. Migratory movements and wintering areas of various subpopulation need to be better understood. Colour marking, satellite telemetry, and biochemical studies can accomplish this. Because its nesting is highly dispersed over its North American range, the overall and seasonal population size of this species is poorly known and its trend is only guessed at. It is important to develop a regional and range-wide census and monitoring progamme for this species, giving due consideration to the error bias inherent in such counts (Gibbs *et al.* 1988, Dodd and Murphy 1995). Studies need to continue of the relationship of herons to aquaculture focusing on the populational risks of killing birds and ways to mitigate adverse impacts of the birds to aquaculture.

Overview

The Great Blue Heron is the North American representative of the large heron group. It is a specialist at patient feeding, tending to catch a few large prey items each day, seemingly being pressed for time only when feeding young. That success varies with age has important demographic implications, and may explain why maturation is delayed to a time when the birds have mastered the species' difficult feeding tactics. Birds tend to choose to forage in profitable sites, changing locations as needed. Birds from a single colony can feed in different situations (Parsons and McColpin 1995). Despite breeding in colonies, birds show a mixed social strategy tending to occupy and defend individual feeding areas. Although some colony sites are occupied for many years, it is more common for colony sites to shift slowly as the breeding habitat degrades or available food supplies shift over the long term. This variability in nesting site use allows the species to adapt to ever-changing conditions. Taking advantage of artificial food supplies can increase nesting success and population size. It is flexible in feeding timing, especially along the coast (Horvath and Moholt 1986, Austin 1995). So this adaptable species has an admirable ability to exploit available habitats and food resources. As a result it is abundant, widespread, and well-known throughout its extensive range.

Cocoi Heron *Ardea cocoi* Linnaeus

PLATES 1.5, 1.6, 13.1

Other names: White-necked Heron in English: Garza Cuca, Garza Morena, Garza Gris, Garza Mora, Garza Real, Garza Cenicienta, Garzón in Spanish; Socó-grande, João-grande, Maguari, Baguari in Portugese; Héron cocoi in French; Cocoireiher in German. *Ardea cocoi* Linnaeus, 1766. Syst. Nat. ed. 12(1), p. 237: Cayenne.

Description

The Cocoi Heron is a large grey heron with white neck and black cap.

ADULT: The Cocoi Heron is a starkly grey, black, and white bird. The crown is entirely black to below the eye, but may be paler on the forecrown. Head plumes trail from the cap. The bill is dull yellow, a little paler below and with a blackish base. The irises are normally yellow and the lores blue. The neck is markedly white. The bird's dominant colour is the grey of the back and wings, with a black 'shoulder' patch that shows when standing. The thighs and undertail also are white. The belly is black as are the legs and feet. In nuptial plumage, the bills turn bright yellow with a strong red tinge at the base, and the legs turn dusky pink.

VARIATION: The sexes are alike. No subspecies are described, but southern birds are larger than northern birds.

JUVENILE: In immature birds, the lower underparts are ashy streaked with buff, and the neck has an ashy tinge. The thighs and under tail are streaked dusky. The distinctive cap is present but is a duller black or grey, paler on the forecrown and lacks plumes.

CHICK: The downy chick is grey and shows the contrasting darker colour of the cap.

VOICE: The 'Raab' call is its principal call, rendered 'rraahb, rraabb'. It is similar in function to the 'Frannk' call of the Great Blue Heron. The rest of the vocal repertoire remains to be characterized.

WEIGHTS AND MEASUREMENTS: Weight: 1900–1980 g; length: 97–127 cm. This is the largest South American Heron.

Field characters

The Cocoi Heron is identified by its complete dark cap and by the combination of white neck and black thighs and shoulder patch. It is distinguished from the Great Blue Heron by its completely black cap, light neck, and black thighs. See the Grey Heron account for a discussion of juveniles.

Systematics

This is a distinctive bird, related to the Grey and Great Blue Herons. There is likely more geographic variation than has so far been described. The alternative English name, White-necked Heron, is used for another species.

Range and status

The Cocoi Heron a South American species.

BREEDING RANGE: The Cocoi Heron nests north as far as Columbia and Venezuela. It is a year-round resident in extreme southeast Panama, but breeding has not yet been documented. It nests south throughout most of South America south to Chile (Aisen) and Argentina (Chubut), excluding the Andes.

NON-BREEDING RANGE: Generally the species occurs in its breeding range during the non-breeding season. But it is found in Trinidad from August to June and more rarely in Tobago from January to June (ffrench 1973, Hancock 1999, F. Hayes pers. comm.).

MIGRATION: Although usually considered rather sedentary, it is expected that southern Cocoi Herons probably undertake movements to escape the southern winter and postnesting dispersion should also be expected. Evidence is scanty. Occasional appearances of this heron in the far south of

Chile, on the Falkland Islands, and Fernando de Noronha (Nacinovic and Teixeira 1989), suggest postbreeding dispersal occurs.

STATUS: The species is widespread and common throughout its range. Its population centre is the northern and central part of the continent (Morales 2000). It is scarce or absent along the west coast from Chile to Ecuador. It is patchily distributed in some areas and in some seasons, but the reasons are not clear (Morales 2000). It can be very common in certain habitats, such as in the Venezuela llanos.

Habitat

This heron is a bird of natural wetlands, fresh or salt, grass or forested. It feeds in shallow to deep water along lake shores, rivers, estuaries, beaches, forested rivers, small streams, grassy marshes, swamps, and other wetlands. Although a bird of lowland wetlands, it can be found as high as 2500 m (Borrero and Cruz-Millan 1982).

Foraging

This species feeds in typical large-heron fashion, primarily by Standing and by Walking slowly between standing sites. It uses Head Tilting for better vision. It feeds both day and night. Additional descriptive details of foraging are needed. It generally feeds alone and defends its feeding territory. It also joins large mixed species aggregations, especially during the dry season. In such situations it not only forages for itself but also steals prey from other species, such as Wood Storks (Gonzalez 1996). Its diet is primarily large fish, but it also eats frogs, the larvae of aquatic insects, and carrion. Fish can be up to 30 cm long (Borrero and Cruz-Millan 1982). It probably has a broad diet, for example, eating dead crabs.

Breeding

The nesting season starts in July in Surinam, October in Uruguay, and November in Buenos Aires. It nests in swamps and other protected sites, apparently preferring large trees 20–25 m tall (Borrero and Cruz-Millan 1982). But it may nest in bushes primarily, reeds, or even cactus. It may nest solitarily or in colonies of single or mixed species (Parera and Bosso 1996). As many as 400–800 breeding pairs used a coastal colony in Surinam (Spaans and de Jong 1982). Nests are made of twigs, dry thistles, and reeds lined with grass. Nests are roughly circular, 65 cm in diameter and very deep. In a reed swamp, they were supported by *Zizaniopsis* reeds and were only 30–40 cm above the water. It took about 7 days to build a nest (Borrero and Cruz-Millan 1982). Little is known about courtship behaviour. Although it is assumed to be similar to the Great Blue Heron, this needs to be documented. Soft-part colours brighten in courtship, the lores turning cobalt blue (Borrero and Cruz-Millan 1982). The eggs are sky blue with paler speckles and a few whitish traces. Clutches are 3 eggs in Argentina. Eggs size appears variable, 51–66 × 34–42 mm (Borrero and Cruz-Millan 1982). The incubation period is 20–23 days (Borrero and Cruz-Millan 1982). Generally, 2–4 chicks survive. Predators may take young, such as Caracara (*Caracara*).

Conservation

The Cocoi Heron is widespread and common over most of its range, more so in the north than the south of South America. There are no known compelling conservation issues at present. It is thought that contaminants, especially pesticides, could affect them, but this remains to be documented. Most critical is habitat alteration over much of its range. It appears to prefer natural over artificial habitat, and these natural habitats need to be maintained.

Research needs

Although this species is large, obvious, widespread, and common, surprisingly little is recorded on its general biology. Basic studies of its biology are needed. Given the extensive knowledge available for Great Blue and Grey Herons, studies should be relatively easy to design and carry out. Especially important is a better understanding of its foraging habitat requirements. The migratory or dispersive tendencies of various populations need to be understood.

Overview

The Cocoi Heron is a large, generally solitary bird that feeds principally in grassy marshes and other wetland situations. Typical of the large herons, it is a patient feeder, foraging alone and defending its territory from other herons. But when prey is abundant, it joins into aggregations and feeds socially. As is typical of the other large *Ardea*, it feeds on a wide variety of prey, some rather large. But much remains to be documented about this abundant and conspicuous species that could well serve as an emblematic species of South American wetlands.

White-necked Heron *Ardea pacifica* Latham

PLATE 2.1

Other names: Pacific Heron, Crane in English; Garza Cuelliblanca in Spanish; Héron à téte blanche in French; Weißhalsreiher in German. *Ardea pacifica* Latham, 180 1. Ind. Orn., Suppl. 1, p. 45: New South Wales.

Description

The White-necked Heron is a medium to large, robust looking, grey heron with a completely white neck and head.

ADULT: The head and neck are white, with black spotting running vertically down the foreneck. The irises are yellow to green; lores are dark grey to black, often with a cream-green to pale yellow streak from the bill to the eye. The bill is black. The upper parts of this heron are sooty to dark grey. The upper wings are dark slate-grey, with a white 'shoulder' patch visible in flight. The underparts are dark slate-grey with white streaking. The legs are black, but it can have a yellow-olive tinge on the lower leg. During breeding, plumage tends to become darker, dark grey to black on the back, tail, and upper wing having a greenish sheen. The head and neck are white, but the dark spotting on the throat may be more chestnut coloured. The bill is black and the irises are yellow to green. The lores become blue during courtship. Maroon lanceolate plumes develop at the base of the foreneck and along the back. The characteristic white patch on the wings becomes more prominent. The under parts are streaked with white. Legs are black.

VARIATION: The sexes are alike, but females average smaller than males. There is considerable variation in size among individuals, up to 30% in length.

JUVENILE: Immature birds have grey to buff-white heads and necks and lack plumes. Their lower bill is yellow to white. The irises are dark brown to bronze.

CHICK: Downy chicks are grey-white, with whiter head and neck apparent.

VOICE: This heron is characteristically a quiet bird, at least away from the nest site. It issues a hoarse guttural croaking alarm call when disturbed. At nest adults make guttural noises. The 'Oomph' call is given during Stretch and Wing Touch. Cackles are made during the Greeting Ceremony. Chicks give a rasping call to contact parents. Additional information is needed on the vocalizations of this species.

WEIGHTS AND MEASUREMENTS: Weight: 650–860 g; length: 76–107 cm.

Field characters

At all ages and stages this is an unmistakable heron, with its contrasting white neck and head and its grey body. It is identified by the combination of medium size, dark plumage, and white neck and head. The White-necked Heron flies with slow wing beats with head drawn back, looking heavy and large. In flight the white patch is highly visible on the leading edge of the otherwise dark wing and is a good field character.

It is distinguished from the White-faced Heron by its all white head. It is distinguished from adult Pied Herons by its larger size, white (not black) head, white neck with prominent ventral spotting, dark (not yellow) bill, and dark (not yellow) legs. Pied Herons, however, are variable and can have very white heads, but generally with less spotting and always with light rather than black legs and bills. The White-necked Heron can be confused with juvenile Pied Herons that have a white head and light to white underparts. It is distinguished by its larger size, neck spotting, dark bill and legs, and more solitary (not highly aggregative) behaviour.

Systematics

This species is related to the other medium to large *Ardea* herons, but much remains to be understood about interspecific relationships among these herons. Two English names are widely used. We suggest the name 'White-necked Heron' as being more descriptive whereas 'Pacific Heron' is derived from its scientific name and is not a good description of its range. White-necked could also be appropriate for the Cocoi Heron, but the Cocoi Heron's predominant characteristic is its black head, whereas the predominant characteristic of *pacifica* is its white neck and head.

Range and status

The White-necked Heron is endemic to Australia.

BREEDING RANGE: The species has bred throughout much of Australia outside of the deserts. Although some colony sites are permanently occupied, this is an eruptive, nomadic bird that can locate suitable feeding conditions and nest as conditions allow. Consequently its exact nesting range is highly flexible. It has expanded its range in northwestern Australia in the past 50 years.

NON-BREEDING RANGE: As a nomadic species, it can turn up almost anywhere outside the nesting season, occupying a site for as briefly as a few days. It is regularly reported in New Guinea (Transfly, Port Mosby, and elsewhere), Tasmania, and Bass Strait Islands.

MIGRATION: This is a mobile heron that undertakes both regular migrations and event-triggered dispersal. Migrations occur between breeding sites and seasonally suitable wetland feeding habitats. For example, inland from Darwin, herons nest from June to August and then most migrate away for January to May. Conversely they appear in the north of Australia during the non-breeding dry season (Morton *et al.* 1993). Irruptions take place when especially favourable feeding conditions occur and the herons move to take advantage of them (Dell 1985). The eruptive tendency of this species results in many dispersal records elsewhere in the nearby Pacific, for example New Zealand, New Guinea, Norfolk Island, and Tristan da Cunha.

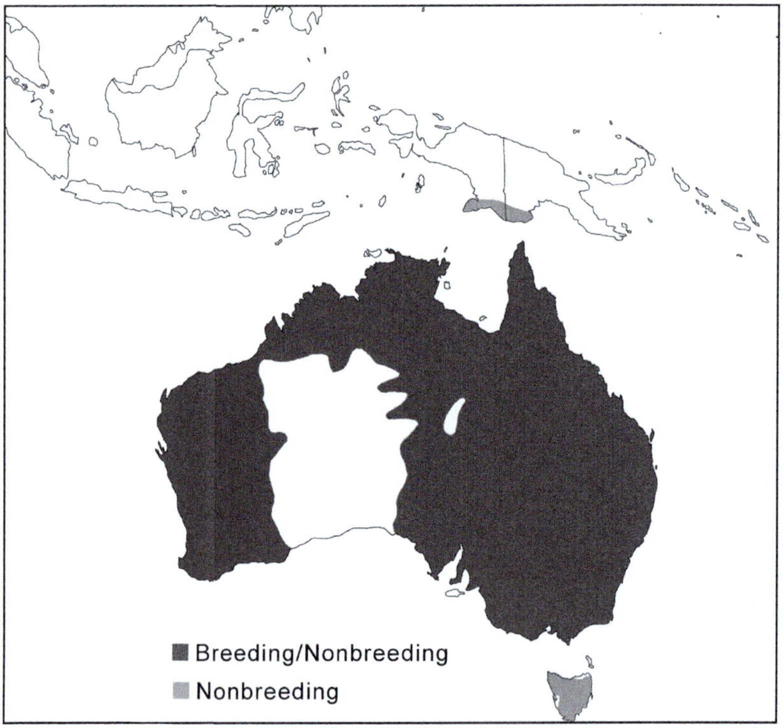

STATUS: This is an uncommon species but one that is widespread. All the signs are that the Australian population is stable. It is uncommon in Tasmania and New Guinea and a vagrant in New Zealand.

Habitats

The White-necked Heron uses a variety of inland wetlands, grasslands, permanent and temporary water bodies, freshwater swamps, lagoons, billabongs (waterholes), and river floodplains. It also uses reservoirs, dams, and roadside ditches. The heron most often chooses very shallow water sites having a depth of less than 7 cm, usually in marsh or wet grass where there is a clear view of its surroundings.

Foraging

The White-necked Heron typically feeds alone or in pairs in wet grasslands, shallowly flooded marshes, and other shallow sites. The species will defend its feeding site from incursions. At sites of abundant prey, it sometimes joins in small dispersed groups or in flocks up to several hundred. It feeds mostly by Standing and also by Walking Slowly from both Upright and Crouched posture travelling slowly along the margins of swamps, wet pasture, or along the water's edge. Its feeding rate is less than a stab a minute. It also uses Foot Raking, Diving into the water from overhanging limbs, and Hovering (Ley 1991) and will steal food from ibises and raptors. Diet is versatile, depending largely on availability. Food is typically small aquatic prey, tending to be less than 3 cm long, although a prey item over 10 cm long has been recorded (Recher *et al.* 1983). Its diet includes frogs and tadpoles, newts, lizards, mussels, shrimp, crayfish, insects such as dragonfly larvae, mantids, grasshoppers, beetles and beetle larvae, fish, small birds, and mammals. Although it eats fish less than other herons, it does include them in the diet when they are available (Ley 1991). Under appropriate conditions, it apparently will eat carrion.

Breeding

The nesting season is flexible, generally taking place during the rains; in wet years it can be year-round. The nesting season is typically in the spring and summer in the south and Tasmania, September to February and sometimes into April. Nesting in the north is December to March. They nest in freshwater swamps, marshes, and grasslands, especially areas that remain flooded for several months or permanently (Briggs *et al.* 1997). Within this habitat, they use tall trees, often dead, as high as 40 m, near of over standing water, and convenient to feeding areas. This heron is colonial, nesting in small, dispersed colonies of 2–30 nests, but also up to hundreds. They nest also in mixed colonies of other herons, ibis (*Threskiornis*), spoonbills (*Platalea*), and cormorants (*Phalacrocorax*). Compact nesting is possible with up to five nests having been observed in a single tree. Solitary nesting occurs such as at small water bodies, but contrary to some literature, isolated nesting is not typical of the species. Nests are loosely constructed stick platforms 30–60 cm across. Nests can be reused in successive years. Twigs are collected by males near nest sites, from other nests, or even in flight from the water.

Some aspects of nesting behaviour are described (McCulloch 1976, Marchant 1988, Lowe 1989, Marchant and Higgins 1990). Solitary males select and attend a future nest site, standing on or near it for long periods preening or displaying. Birds fly out in Circle Flights, without calling, but with deep wing beats and neck extended (McCulloch 1967). When occupying a potential nest site, the displaying male gives Wing Touch and Stretch displays, often repeated between bouts of Preening and Standing. The Stretch includes an Oomph call. Both birds occupy the nest site, Preening and Standing. Body Shake and Twig Shake displays are given. They may take mutual Circle Flights that include hovering and swooping toward each other. It should be noted that Marchant and Higgins (1990) consider this behaviour to be between adults and young. After pair formation, the nest exchange Greeting Ceremony consists of Upright display with raucous cackles. The White-necked Heron also appears to perform display rituals away from the nesting sites. The bird has been observed doing the Stretch, followed by a bobbing downwards of the legs, in a tree standing in water at a location well isolated from any known nesting site. The action is described as being by far less vigorous than the equivalent display of the Eastern Great Egret (M. Maddock pers. comm.).

Eggs are pale green-blue, slightly pitted, and average 58 × 39 mm. Clutch is 4 eggs, occasionally 2–6. Incubation is 28–30 days. Both adults feed young by regurgitation. Young give rasping calls in the nest. Parents continuously attend and guard the nest and small young until 3–4 weeks. Older siblings harass the smallest chick, which usually dies. In colonies, young form groups on branches. Generally only two chicks survive. Fledging occurs in 6–7 weeks.

Population dynamics

Little is known of the population dynamics of this species.

Conservation

As a widespread, adaptable, and still common bird in Australia, its population is considered to be stable and is not of immediate conservation concern. Its expansion into northwest Australia was due to its ability to exploit irrigation and reservoir development. However, this Australian endemic depends, for nesting on inland seasonal wetlands, which have been under threat for decades. Artificial wetlands and other aquatic sites can be managed to support herons by water control and maintenance of nesting trees (Briggs *et al.* 1997).

Research needs

Much additional study is needed to fill gaps in the basic biology of this species, particularly the movements of individuals and populations within Australia, and the relation of hydrologic conditions to breeding success, demographics, and population biology.

Overview

The White-necked Heron is an excellent example of a nomadic heron. It is highly mobile, moving around in response to seasonally and annually varying rainfall patterns that typify much of the Australian continent. The resulting flooding provides ephemeral food supplies. Finding and exploiting temporary hydrologic conditions are keys to this species' ecology and life history. Where conditions are stable, it breeds in successive years. In the comparatively reliable rainfall areas of eastern and southeastern Australia, fairly stable groups of 20–30 may congregate to nest and remain indefinitely. Possibly these are the small resident populations from which nomadic individuals radiate during droughts. Such movements result in distant dispersals. Its nomadism sets this species apart from many other medium to large herons. As a result of its versatility, its diet is broad. It can feed, nest, or roost alone or in groups. It can expand and contract its distribution as conditions dictate, finding and exploiting available food supplies for feeding and breeding. It is a species well adapted to the harsh climatic conditions of the Australian continent.

Great Egret *Ardea alba* Linnaeus

PLATES 3.7–3.10, 13.2, 13.3

Other names: Great White Egret, Common Egret, American Egret, Great White Heron in English; Garceta Grande, Garzón Blanco, Garceta Real, Garza Real, Graza grande, Garza Blanca in Spanish; Garça branca-grande in Portuguese; Grande aigrette in French; Silberreiher in German, Grote zilverreiger in Dutch; Airone bianco maggiore in Italian; Ъольшая белая цапля in Russian; Da bailu in Chinese.

Ardea alba Linnaeus, 1758. Syst. Nat. 10(1), p. 144: Europe.

Subspecies: *Ardea alba alba* Linnaeus. *Ardea alba egretta* (Gmelin), 1789: in insula S Dominici, insulis Falkland et America australi ad Louisianum usque; *Ardea alba melanorhynchos* (Wagler), 1827: Senegambia.

Description

The Great Egret is a large, slender, white heron, with long neck, dark legs, and long back plumes when breeding.

ADULT: The Great Egret is an all white bird. The feathers of the crown are slightly elongated but do not form a crest. The bill is yellow or yellow with a black tip. Irises are yellow. Lores are green-yellow to yellow. A black line of bare skin extends from the edge of the gape to beyond the eye. The head is without crown plumes. The legs are variably dark. Breeding season soft-part coloration appears to differ geographically. The back plumes reach full development during courtship when as many as 50 plumes, up to 50 cm long, extend as much as 10 cm past the tail. These aigrettes have stiff quills and are able to be fanned in courtship. Feathers of the lower neck and chest are thick and only slightly elongated, frayed but not fully lanceolate.

VARIATION: Sexes are alike in plumage, but males are larger than females. The races differ in soft-part coloration, particularly during breeding. *Egretta* has a dull, mottled, yellow bill, often tipped in black in the non-breeding season. Legs and feet are black. In courtship, the bill turns bright orange, then fading back to yellow. The iris has an outer red ring. The lores are a bright green. *Alba* is the larger race. In non-breeding season, the bill is yellow. During courtship the bill is red with a dark tip, turning to black with yellow base and red streaking. In breeding, lores turn bright green. In non-breeding, leg colour is variable, upper legs are generally black with a green or yellow tinge. In courtship, upper legs, and sometimes stripes on the lower leg, are pink. Postcourtship, during the remainder of nesting, the upper leg turns yellow and the lower leg dark. In *melanorhynchos*, the non-breeding bill is yellow, lores yellow, and legs and feet black. During courtship the bill turns black and the irises turn red. The lores turn bright blue-green. After pairing, the irises become yellow, the lores pale green then yellow-green, and the bill becomes yellow starting at the base leading to a bicolour condition, usually with the tip remaining black.

JUVENILE: Juveniles are similar to adults, but lack ornamental plumes. They have yellow bills with a blackish tip.

CHICK: Nestlings are covered in white down, elongated on the head forming a crest. The upper bill is grey with a black tip, the lower bill is yellow-grey, both turning yellow. The iris is straw to green. The lores are pale blue-grey becoming yellow. The throat is pink. The legs are pink becoming grey-green. In *melanorhynchos* the skin is green becoming yellow.

VOICE: A low-pitched, deep, 'Kraak' call is the flight, disturbance, and threat call. The high pitched 'Eeeiee' call, rendered 'eeeee, i, eeee', also is a threat call. The 'Cuk' call, a rapid 'cuk, cuk, cuk', is also a disturbance call. The characteristic advertising call is the soft 'Frawnk' call. The 'Rha' call is a courtship call. In Africa a 'Cra' call, rendered 'craa, craa, craa', has been reported as the nest greeting. A 'Rrrooo' call, used in the Greeting Ceremony call, is also called the 'Laughing Call'. Young beg with a 'Ket' call, rendered 'ket, ket, ket, ket'. The literature on vocalizations is not clear, suggesting additional geographic or racial differences likely exist.

WEIGHTS AND MEASUREMENTS: Weight: 930–1700 g; length: 85–102 cm.

Field characters

The Great Egret is identified by its relative size, its back plumes, dense breast plumes, lack of crest and head plumes, yellow and yellow-black bill, the extension of a dark line from the gape to beyond the eye, and dark legs. It has long legs and neck relative to the size of the body (Boev 1988a, 1989); the obviously kinked neck is 1.5 times the body length. It flies strongly with a down stroke that elevates the body. Legs extend far beyond the body. Owing to its relative length, the neck is deeply bowed when it is flying with its head pulled back.

It is distinguished from the white form of the Great Blue Heron by its smaller size, smaller, slighter and very slightly down curved (not straight) bill, dark (not pale) legs, lack of head plumes, and its substantial back plumes in breeding plumage. It is distinguished from the white form of the Reddish Egret by its larger size, yellow/yellow-black (not bicoloured) bill, very dark (not bluish-dark) legs, lack of crest and neck plumes. It is distinguished from the Snowy and Little Egrets by its larger size, yellow (not black) bill except in courtship, dark (not yellow) feet, relatively narrower wings. It is distinguished from the

Intermediate Egret by its somewhat larger size, relatively longer (not stubby) bill, relatively longer, thinner more kinked neck, relatively longer legs, lack of head plumes, and the line of its gape (extending beyond not ending at the eye) and by a more upright alert posture (McCanch and McCanch 1991). It is distinguished from the Cattle Egret by its larger size, height, long neck, and lack of buff colour. It is distinguished from the Eastern Great Egret by the latter's combination of two-tone brown to grey black legs and yellow bill and in breeding by a combination of black bill, bright green lores, and pink to purple red wash on dark legs.

Systematics

The taxonomic position of this species has long been debated. For some time this indecision was resolved by placing it in its own genus, *Casmerodius*, between *Ardea* and *Egretta*. Morphological, behavioural, and biochemical evidence all point to these birds being *Ardea* (Payne and Risley 1976, Sheldon 1987, Sheldon and Slikas 1997, Chang *et al.* 2003). Geographic variation among the large white egrets has not been thoroughly investigated using molecular techniques. Based on DNA–DNA hybridization, the eastern population is recognized as a species, having been found to be as different from the North American form as it is from the Intermediate Egret (Sheldon 1987, Sheldon and Whittingham 1997). Similar analyses are needed to compare all populations. Attention especially needs to be focused on the area of potential overlap in north China and Japan between *Ardea alba alba* and *Ardea modesta* (Amadon and Woolfenden 1952).

Range and status

The Great Egret occurs through most of North, Central and South America, east Europe, Africa, and north Asia, avoiding deserts.

BREEDING RANGE: In the Americas, *egretta* nests to the north USA and south Canada in central Washington, California, Nevada, and parts of Arizona, Texas, south Idaho, southeast Saskatchewan (formerly), southwest Manitoba, South Dakota, central Minnesota, Wisconsin, south Michigan, Ohio, south Ontario, southwest Quebec, to southern Maine, central and eastern USA (Pratt 1983, Scharf 1989, Sauer 1991). They occur north of their breeding range even in summer, as for example in Alaska (Gibson *et al.* 1987b). They nest in coastal Mexico, Bahamas, Greater Antilles, north Lesser Antilles, Central America, South America to south Chile and south Argentina (44°S) and the Galapagos Islands

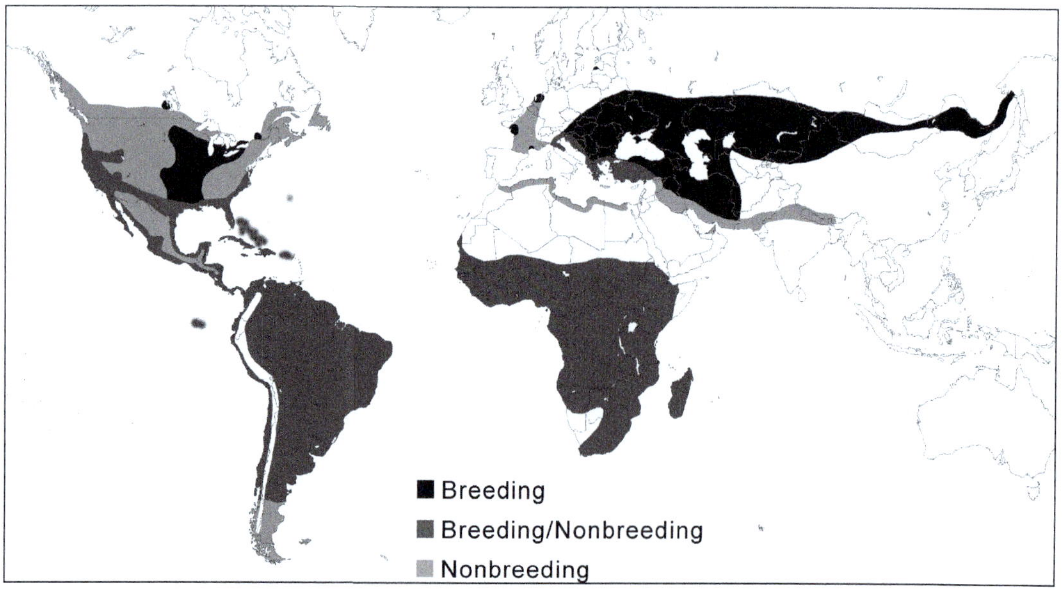

(Nores 1986). In Eurasia, *alba* nests in Netherlands, Latvia (Lake Engures-M. Strazds pers. comm.), France, Italy, Austria, Bulgaria, Romania, Poland, Ukraine, Hungary, Slovakia, Croatia, Albania, Greece, Moldova, Turkey, Iran, Turkmenistan, Uzbekistan, Afghanistan, Kazakhstan, south Russia (to Ussuriland), Mongolia, north China (northwest Heilungkiang, and north Simkiang (Voslamber 1992, Piacentini 1993, Gavris' 1994, Marion and Marion 1994, Shernazarov 1995, Volponi and Emiliani 1995, Schogolev 1996, Pugacewicz and Kowalski 1997, Van der Kooij and Voslamber 1997). In Africa, *melanorhynchos* nests from the southern tip of Mauritania, Senegal, Mali, Sudan, south Somalia, Ethiopia, Kenya, Tanzania, Zimbabwe, to South Africa (Baccetti 1983, Wallace *et al.* 1992, Turner 2000, O. Messaoud pers. comm.). Actual breeding records are scarce over much of this African range due in part to inaccessibility during the rainy season (Baker and Baker 2004). It also occurs on Madagascar, the Comoro Islands, and Pemba Island (Cheke and Diamond 1986).

NON-BREEDING RANGE: Depending on temperature and rainfall, the Great Egret spends the non-breeding season fairly far north in North America and south in South America. It occurs north to coastal Oregon, north California, central Nevada, central Utah, central New Mexico, central Texas, and sometimes farther (McMahon 1984). Most birds winter on the Gulf of Mexico coast and Florida, but are found on the Atlantic coast as far north as coastal Maryland and as far east as Bermuda. Birds occur in the non-breeding season in the Bahamas, through Mexico and the rest of Central and South American, south of the breeding range to the Straits of Magellan and Falkland Islands. European birds spend the non-breeding period around the Mediterranean Sea, Red Sea, Persian Gulf, in France, Italy, Slovenia, Croatia, Albania, Greece, Turkey, Egypt, Tunisia, Morocco (Biondi *et al.* 1994, Vogrin 1999), but increasingly also farther north (Bousquet 1998). West Asia birds winter in Iran, Iraq, and Pakistan, straggling to India. There is a record of a Russian bird in sub-Saharan central Africa, but there is no evidence of a regular influx of northern birds to the area and no evidence of a trans-Saharan migration by *alba* (Turner 2000). East Asia birds winter in south Tibet, north India, and south China. In Africa, *melanorhynchos* numbers fluctuate seasonally suggesting population movements (Brown *et al.* 1982). Great Egrets winter in west Africa (Gatter 1988) but the origin of these birds is not clear, likely not Europe. Birds spend the non-breeding season in South Africa and also occur in the Seychelles.

MIGRATION: Temperate populations of *egretta* disperse and migrate in complex ways. The two processes are not totally distinguishable. After breeding, egrets disperse in many directions and are regularly reported north of the usual breeding range. In South America they occur south of the breeding range. They also disperse to nearby and oceanic islands. Dispersal records include south Alaska, Northwest Territories, northwest and central British Columbia, south Alberta, south Saskatchewan, southeast Manitoba, south Quebec, Newfoundland, Great Britain, Hawaii, Tierra del Fuego, Falkland Islands, Tristan da Cuna, Hawaii, and Kuri Atoll (Byrd 1978, Pyle 1984, Gibson *et al.* 1987a, Sirois *et al.* 1991, Tobish and Isleib 1992, Anonymous 2002a). The Atlantic coast population moves along the coast to the Bahamas and Antilles. Egrets from interior and Midwest North America move to or through Florida. Western North American egrets migrate to Mexico and Central America, although some remain farther north. *Alba* disperses in all directions after nesting in July to September, followed by a southward migration from late September to early November (Voisin 1991). On dispersal, it occurs in England, Wales, Scotland, Ireland, Iceland, Belgium, Netherlands, Scandinavia, France, Spain, Malta, Germany, Switzerland, Poland, Czech Republic, Algeria, Morocco, Libya, Mauritania, Sao Tome, Cape Verde Islands, and Canary Islands (Stawarczyk 1984, Cotton and Murphy 1985, Manson 1985, Donovan 1988, Eccles 1988, Holmstrom 1989a, Dolton 1991, De Juana and Ferrer 1996, Hazevoet 1999, Hjartarson and Petursson 2003). The population is partially migratory with a few birds remaining in some breeding areas, at least in mild winters, a phenomenon that has been increasing in recent years (Festetics and Leisler 1999). Birds from Austria and Hungary winter in Italy, the Balkans, Albania, and Greece. Some birds move farther south to Turkey, Israel, Egypt, and north Africa. In west Asia, birds move south to Iran and Iraq. In east Asia, birds migrate from north China, to winter in south Tibet and south China. Birds from Korea and north Japan winter in the Philippines. Return migration is in late February to early April. In Africa, *melanorhynchos* has a substantial postbreeding dispersal or short distance migration (Brown *et al.* 1982). Some birds appear to migrate south to South Africa. Great Egrets winter in numbers in west Africa (Gatter 1988), although the origin of these birds is not clear.

STATUS: In North America, the species has been expanding its range northward in the past century, recently beginning breeding in southern Ontario (Butler *et al.* 2000). This followed population reductions and range contractions due to plume hunting in late 1800's. Nesting and wintering populations are high in North America and increasing in Louisiana, Texas, and California, although perhaps decreasing locally in the southeast USA (Fluery and Sherry 1995, Butler *et al.* 2000). In the mid 1970s over 110 000 nesting birds were censused in east and south USA. In the 1980s over 43 000 nested in Louisiana (Martin and Lester 1991). The species also is abundant farther south with over 27 000 pairs reported in Mexico (Tabasco) (Scott and Carbonell 1986). Around 2700 birds were counted in Honduras and Nicaragua (Frederick *et al.* 1997). In Costa Rica, 2650 pairs were counted in a colony, and 1000 pairs were reported along the coast of Suriname. It is common and probably increasing in South America (Morales 2000). Populations of *alba* in western Europe decreased in the nineteenth and early twentieth centuries due to plume hunting but have been recovering in recent decades. It has expanded its nesting range, and increased nesting and wintering population sizes in west Europe, nesting becoming established in Netherlands (1977), Latvia (1977), Italy (1992), France (1994) and Poland (1996) (Celmins 1992, Voslamber 1992, Kayser *et al.* 1994, Marion and Marion 1994, Volponi and Emiliani 1995, De Juana and Ferrer 1996, Marion 1997, Pugacewicz and Kowalski 1997, van der Kooij and Voslamber 1997). Its breeding numbers have increased rapidly to Netherlands in recent years (Anonymous 2002b). Total numbers in Europe are now 12 900–17 500 breeding pairs (Carpegna *et al.* 2000, Marion *et al.* 2000). It is common, although patchy, in Africa. At its northern limit in west Africa, about 150 pairs have been recorded in the Mauritanian portion of the Senegal delta (H. Hafner pers. comm.). About 3000 pairs were counted in Niger inner delta Mali, 1450 pairs in Senegal, 1000 birds in a colony in Kenya, up to 3800 birds in Zambia, 200 pairs in the Okavango in Botswana, and up to 500 birds in Transvaal South Africa (Skinner *et al.* 1987, Turner 2000). It has been common in Madagascar where large flocks of 1000 or more birds have been seen, but breeding colonies appear to be declining. The resident population on the Comoros Islands is 50–100 birds (Turner 2000).

Habitat

The Great Egret is a bird at home in a variety of wet habitats. It frequents marshes, damp meadows, swamps, river margins, lake shorelines, flooded grasslands, and saltpans. It is also found in marine habitats such as tidal marshes, seagrass flats, mangrove swamps, coastal lagoons, and offshore coral reefs. It uses human-altered environments and agricultural lands, including drainage ditches, rice fields, crawfish ponds, aquaculture ponds, pasture, and dry seashores. In some areas, fishponds, rice fields, crawfish ponds, and winter-rape fields have become critical habitats in the year-round status of the species (Ern 1990, Fluery and Sherry 1995, Gruell 1998). It breeds in a variety of situations in trees, bushes, bamboo, reeds, and other plants near water and on islands, sites that are protected from ground predators (Festetics and Leisler 1999). It occurs mainly in lowlands, but can occur up to 1800 m in Russia and 3000–4000 m in the Andes.

Foraging

The foraging ecology and diet of the Great Egret has been well studied through most of its range (Kilham 1980, 1984, Reichholf 1983, Rodgers 1983, Wiggins 1991, Ramo and Busto 1993, Austin 1996, Havens *et al.* 1996, Dimalexis *et al.* 1997, Miranda and Collazo 1997a, Smith 1997, Grull 1998, Kasoma 2000, Figueroa and Coreles Stappung

2003). The Great Egret feeds in groups, but more often alone. It feeds almost entirely during the day, most actively near dawn and dusk; and in tidal environments it feeds according to tide stage, principally on outgoing tides (Powell 1987, Bancroft *et al.* 1994, Austin 1996). Elsewhere water depth often determines foraging habitat choice (e.g., Smith *et al.* 1995). Egrets roost in trees when not feeding and communally at night.

The Great Egret principally feeds by Walking slowly, which may occupy 60%–90% of its time (Rodgers 1983). It feeds in shallow to moderately deep water, on the shore next to the water, or even on dry ground. It generally Walks in Erect posture, its head and neck well extended at about a 45° angle. It stops Walking when it sees prey. It also Stands next to places where prey is active. Its Crouched posture, usually assumed when a potential prey item is seen, involves holding its body horizontal and its head retracted. When examining a spot for prey, it may use Head Tilting, Peering Over, Head Swinging, Head Swaying, or Neck Swaying. It usually uses a Bill Stab to capture prey, but may probe or peck. It moves between feeding spots by Walking Quickly or Hopping. It uses Foot Stirring, Foot Paddling, Wing Flicking, Foot Probing, Bill Vibrating, Gleaning insects off plants, and aerial techniques including Inplace Hopping, Hovering, Dipping, Plunging, Diving and Foot Dragging (Shanley 1984, Gonzalez 1997). Sometimes behavioural sequences can be quite complex. Catley (1983) reported that an egret Standing in the water began Foot Paddling while flapping its wings violently and periodically Hopping from the water; it stabbed at prey apparently brought to the surface by this activity. When feeding solitarily it will vigorously defend its site, using Upright and Forward displays, Supplanting Flights, and actual attack (Ryan 1999). It also is highly aggressive in flocks, defending its feeding area, supplanting nearby birds (usually after seeing the other bird capture something), robbing other birds or being robbed. Robbing is common when birds attempt to handle relatively large prey (Kushlan 1978b, Kilham 1984, Smith 1989, Kasoma 1995). It also feeds by following other animals, using them to find feeding sites and Leapfrog feeds (Kilham 1980, Peter 1994, Tyler 1993, Berezovikov and Gistsov 1994, McCall 1996, Sueur 1998). In foraging, the Great Egret is not very efficient, having a fairly low rate of success per attempt, and also a not particularly high strike rate (Rodgers 1983). Immature birds are even less efficient, in Florida averaging a 9% success rate.

Overall diet is diverse, but its principal prey in most situations is fish. In other situations insects or shrimp predominate. Insects include water beetles, water bugs, dragonflies, crickets, and locusts. Its diet includes frogs, lizards, snakes, small mammals, and small birds. Small prey dominate the diet (Mock 1985a, Berezovikov and Gistsov 1994).

Breeding

The breeding biology of the Great Egret has been well studied (Palmer 1962, Pratt 1970, 1972, McCrimmon 1974, Tomlinson 1976, Wiese 1976, Maxwell and Kale 1977, Mock 1979, Voisin 1983, Pratt and Winkler 1985, Wallace *et al.* 1992, McCrimmon *et al.* 2001). Temperate birds breed in the early spring and summer, April–July in most situations. In tropical situations rains are more important than the solar season and breeding timing varies from place to place and even from year to year. The Great Egret usually nests in the part of the rain cycle in which food becomes maximally available. This can be during the rains (Kenya, Ethiopia), end of the rains (Nigeria), or dry season (south USA, Mexico, Nigeria). This egret nests in reed beds or in trees or bushes, usually isolated stands near water or on islands. Along tropical shores mangrove-lined trees are often used, as are willows (*Salix*) inland. Reed beds are particularly important habitats in some areas, such as east Europe (Schogolev 1996). The Great Egret may nest either solitarily or colonially with other species, sometimes in colonies of over 1000 nests. The race *alba* tends to nest alone, whereas other races tend to be more highly colonial. Nests may be close together, less than 1 m apart, or even touching. They are more spread out in reed colonies. The nest is 80–120 cm wide, 20 cm high, made of sticks and lined with finer material. Nests from previous years may be reused. Depending on substrate, nests may be placed 1 to 15 m above ground, tending to be placed high relative to smaller heron species.

The male, after a flying around period, chooses a display and nesting site. He occupies the site Standing (with hunched neck and aigrettes fanned over the back), Twig Shaking, giving the Advertising Call, and building the beginnings of a nest. The site is defended vigorously by the male and also later by the female using Upright (with sleeked plumage), Arched Neck and Forward displays, Supplanting Flights, and attacks. The Forward includes having the body horizontal with head and neck retracted into an 'S' shape, wings partly spread, rapid Tail Flick, extreme erection of the back plumes and head and breast feathers; Eeiee and Kraak calls may be given. The male advertises using the Stretch display, in which the egret bends its fully extended neck slightly (not markedly) backwards while pointing its bill skywards; back plumes (but not head and neck feathers) are erect and fanned; the neck is slowly retracted while bending legs. Upon completion of the retraction the male stands tall again. This is often followed directly by the Snap display, which is directed upwards. The Stretch–Snap combination is highly attractive to other birds. The male also advertises with Twig Shaking (also called Bow) in which it lowers its head and grabs a branch or nest sticks sways from side to side, and crouches shallowly. Advertisement also includes much flying around the colony, Standing, the Frawnk call, and Circle Flights. Preening, Back Biting, and Wing Touch occur after pair formation. The Greeting Ceremony includes erection of the display plumes and raising of the wings.

The eggs are pale blue. Averaging 61 × 43 mm and 56 × 40 mm in two studies. The clutch size is usually 3–5, range 1–6, being smaller in the tropics. Some averages are 2.97 in California, USA, 2.4 in Australia, 3.4 in Kenya, 3.7 in Zimbabwe, and 3.1 in South Africa (Brown *et al.* 1982, Pratt and Winkler 1985, Marchant and Higgins 1990). Incubation, by both parents, lasts about 25–26 days. Incubation starts with the first or second egg, perhaps irregularly until the third egg. Young are semialtricial and nidicolous and hatch asynchronously. They are brooded and attended by both parents for times that appear to vary, periods of 15 to 30 days have been recorded, possibly related to food availability (Voisin 1991). In hot climates, parents shade the young. Feathers begin to emerge at 7 days. They are feathered except for thighs by 15–21 days. The young start clambering from the nest in 21–30 days and leave the colony in 42–60 days. Young are fed by regurgitation in the nest; later the chicks grab the adult's bill, and eventually young come to the arriving parent to grab food. Brood reduction is the rule in Great Egrets (Mock 1985a, 1987, Godfray 1986, Mock *et al.* 1987). High levels of aggression occur between the older and the youngest chicks, with the younger generally dying. Aggression is less in two-chick broods. Aggression is not directly related to food availability but chick survival is related to the amount of food, and perhaps type of food, delivered. Chicks appear to have a critical growth point, which may be variable due to food supply, at 2–4 weeks. At this point the ability of the parents to supply food becomes crucial (Voisin 1983, Pratt and Winkler 1985). In Europe, family groups are reported to stay together after fledging.

Hatching success tends to be high, barring catastrophic events, the hatching rate in one study being 77.6% (Voisin 1983). Late hatched young typically die of starvation or are killed. Fledging success is annually variable, 0.03–2.04 young fledged/nest, and averaging 0.90 over one long-term study in California, USA. The deaths of older young prior to fledging are generally due to predation or starvation. Colony-wide reduced nesting success is usually due to adverse rainfall or conditions on the foraging grounds (Frederick and Collopy 1989a, Smith and Collopy 1995).

Population dynamics

Juveniles probably start to breed in their second year. First year morality is 76% and adult mortality is 26%. Longevity is 22 years. Mortality is high in the first year probably due to the difficulty of learning to feed and also initial migration and dispersal risks. Thereafter, annual risk decreases. As a long-lived bird, local population size does not depend strongly in the short-term on annual reproductive success (Pratt and Winkler 1985) and was able to recover quickly from population decreases due to hunting.

Conservation

The Great Egret, as currently recognized, is extremely widespread and has locally large populations in places throughout its range. In North

America, recent range expansion and high populations in the south suggest it is secure there. In South America, the species is well distributed and stable as far as is known. In west Europe the historic range is being reclaimed (and expanded). Great Egrets are common in Africa, where large flocks of 1000 or more birds have been seen. There is a concern in Madagascar that breeding colonies are declining due to human plundering, but impact on the egret's population is less than clear as large colonies are likely being abandoned for smaller ones (Turner 2000). Throughout the range, the most critical conservation issue is the identification, protection, and management of important nesting sites and associated feeding grounds, and the identification of important stopover and non-breeding feeding areas. Colony site conservation includes protection, control of disturbance, and vegetation management (Obradovic 1988, Rodgers and Smith 1995). Conservation of feeding areas includes management of hydrology, salt intrusion, contaminants, and disturbance (Pyrovetsi and Crivelli 1988, Maccarone and Parsons 1994, Puglisi *et al.* 1995a).

Artificially maintained sites are critical to many population segments. Appropriate conditions need to be provided in order that these habitats continue to support populations (Elphick 2000). The establishment of the Great Egret in the Netherlands was in response to the creation of polder, its declaration as a nature reserve, and subsequent active management in favour of the species (van der Kooij and Voslanber 1997). Northern wintering is attributed to an increase in agricultural land (winter-rape) (Grull 1998). Colony sites can be actively managed on artificial sites and reintroduction programmes can be successful (Pullin 1987, 1990). In aquacultural situations in North America and elsewhere, the Great Egret comes into conflict with humans. Resolution of these conflicts through changes in aquacultural practices is essential to the conservation of the species (Mott and Flynt 1995). Excluding Great Egrets from aquacultural and other man-made situations such as reservoirs and farm ponds may not be the best conservation strategy. Aquacultural sites have become important feeding sites in some areas, likely maintaining population stability (Fluery and Sherry 1995), probably by reducing winter adult mortality. Maintenance of current population levels may require continuation of these practices and access of egrets to them. Great Egrets accumulate and are affected by persistent contaminents (Ohlendorf and Marois 1990, Spalding *et al.* 2000a). Clutch size, for example, was smaller prior to the ban of DDT (Pratt and Winkler 1985).

Research needs

The geographic species and subspecies limits of the Great Egret are no longer clear. There is a need to study geographic variation with respect to morphological, behavioural, vocal, and genetic differences and similarities. Among the questions are: What are the relationships among populations of Great Egret? How different is the North American form from the European or African forms? What is the breeding distribution and taxonomic status of birds in east Russia, north China, Japan, and Korea? Little is known about dispersal and survivorship. Additional migration and survival studies should be undertaken in major colonies dispersed across the species range. Study is needed of the foraging ecology of the species relative to its overall energy capture strategy. It would be worthwhile to better understand how feeding behaviour, aggressive behaviour, and sociality interact within this strategy.

Overview

The Great Egret is a successful heron. It has a large range, successful dispersal behaviour, and broad habitat preference, including an ability to use man-made feeding sites, a broad diet, and locally large populations. Its relatively long legs and neck and slow wading feeding behaviour allow it to feed in deeper water than the medium herons, but when appropriate also in very shallow water, on banks, or even on dry land. However, this apparently diverse approach to feeding habitat, behaviour, and acceptable food disguises a much more narrow and, at times, energy-limited niche. Despite feeding in many types of places, it often uses artificial situations. Despite possessing a large behavioural repertoire, it uses Walk Slowly and Stand techniques nearly all the time. Despite its long prey list, it feeds principally on a small fish and small insects. Feeding on small prey using a restricted feeding repetoire means that it has to capture many items to meet its daily energy demands. To accomplish this end, it feeds for exceptionally long periods during the day. So fundamental is its dependence on small prey that its nesting behaviour is characterized by internestling aggression and the (nearly obligatory) killing of younger chicks (Mock 1985a). Because the abundance and availability of small fish is determined in large part by water conditions, the Great Egret depends on finding feeding spots having appropriate tides or seasonal drying or relatively high productivity (Powell 1987, Kelly *et al.* 1993, Master *et al.* 1993, Gruell 1998). If it can defend these sites and feed alone, it does so. When it cannot, it joins other birds in locations of relatively high food availability such as drying marshes and tidal situations.

Despite feeding aggregatively, it seems best adapted to feed alone. Through most of the year, the egret preferentially does so. Even when nesting colonially, it tends to fly to feeding areas alone, choose vacant sites, and land away from other feeding birds (Smith 1995b). Feeding in aggregations does not increase efficiency, decrease capture variability, or increase prey intake, and takes more of its time in aggressive interactions (Wiggins 1991, Master *et al.* 1993). And it tends to capture even smaller prey in aggregations than alone. It is likely that aggregative feeding takes place because of the high availability of prey and despite the presence of other birds. Its aggressiveness, supplanting, and prey robbing appear to compensate for the flock disrupting the effectiveness of its Wading/Stalking feeding behaviour.

While sufficiently successful for most of the year, during nesting the egret experiences a delicate balance between energy needs and its ability to meet these needs. Around mid nesting when there is an energy crisis and the young can easily begin arrested biomass growth that is usually fatal. At this time both parents forage intensively and it is then that brood reduction occurs, thereby determining reproductive performance for the year.

Eastern Great Egret *Ardea modesta* Gray

PLATES 5.1–5.3

Other names: Great Egret, Large Egret, White Egret, White Heron, Kotuku (of Maori origin in New Zealand), White Crane, Large Heron in English; Garceta Grande in Spanish; Grande aigrette in French; Большая белая цапля in Russian; Kuntul putih besar in Indonesian; Tagak, Talabong in Pilipino; Dai-sagi in Japanese; Da bailu in Chinese.

Ardea modesta J. E. Gray, 1831. Zool. Misc., p. 19: India.

Description

The Eastern Great Egret is a large white heron with yellow bill, dark legs, and, in breeding, long back plumes.

ADULT: The base plumage colour is entirely white. The head lacks plumes. The bill is yellow with a grey-black tip. The lores are yellow. The iris is yellow. A black line of bare skin on the gape extends beyond the eye. The back lacks breeding plumes. The legs and feet are brown to grey-black, two-toned, the upper leg being lighter than the lower leg. During breeding, the long straight (not recurved) aigrette plumes extend beyond the tail. The crest and neck have slightly long, loose feathers but not plumes. During courtship the bill turns grey-black. In courtship there appear to be short-term flushes of colour, especially in the irises that can turn red. The lores are variably pale green to olive green. The whole legs and sometimes feet can take on a reddish wash over the otherwise dark base colour, being variably pink, red, purple-red, or brown-red. After courtship, during incubation and chick rearing, the bill returns to yellow starting from the base leading to a transient bicolour condition; the irises return to yellow; the lores fade to dull green; legs and feet change to black.

VARIATION: Sexes have similar plumage. Males are larger than females. Birds from New Zealand (sometimes called *maoriana*) are reported to be larger than birds from China and India.

JUVENILE: Juveniles are similar to adults, but lack ornamental plumes. They have yellow bills with a blackish tip.

CHICK: Nestlings have cream-white down. Their bill is black tipped with yellow, and a gape that extends beyond the eye. The iris is straw coloured. The lores are green. Longer down on the head forms a crown. The legs are grey.

VOICE: The 'Cuk' call is the disturbance call given at the nest, rendered 'cuk, cuk, cuk, cuk', or 'glok, glok, glok, glok'. The 'Arrrr' call is the disturbance call uttered when taking flight, rendered 'ar, ar, ar, ar, aaar'. It is probably equivalent to the 'Kraak' call attributed to the Great Egret. The 'Gorork' call, the Greeting Ceremony call, is a 'gorork, gorork, gorork' followed by rapid, repeated croaking 'Grock'. This call seems very different from the 'Rrrooo' call of the Great Egret. Young beg with a 'Kek' call, rendered 'kek, kek, kek, kek'.

WEIGHTS AND MEASUREMENTS: Weight: males 970 g, females 760 g; length: 83–103 cm.

Field characters

The Eastern Great Egret can be easily confused with the other white herons. It is identified by its size, bill size and colour, back plumes (in breeding), long kinked neck, and extension of a dark line of the gape from the corner of the bill to beyond the eye.

It is distinguished from the Great Egret of the west by its slightly smaller size, two-tone brown to grey-black legs, and yellow bill and in breeding by a combination of black bill, green lores, and pink to purple-red wash on dark legs. It is distinguished from the Intermediate Egret by its somewhat larger size, relatively longer (not stubby) bill, relatively longer, thinner more kinked neck, relatively longer legs, lack of obvious head crest, the line of its gape (extending beyond not ending at the eye), and in breeding its back plumes and red-washed legs. It is distinguished from Little Egrets and Chinese Egrets by its larger size, yellow (not black) bill except in courtship, dark (not yellow) feet, and long neck. It is distinguished from Cattle Egret by its larger size, height, long neck, and lack of buff colour. It is distinguished from white Eastern Reef-Herons by its larger size, tall, slim appearance, long neck, and in breeding its back plumes.

Systematics

In this description we recognize the eastern form of the great egret, long classified as *Egretta alba modesta*, as a separate species in the genus *Ardea*. Treating first the issue of genus, one of the more critical recent findings about heron systematics using molecular techniques is the confirmation of morphological evidence that *Ardea* and *Egretta* are not necessarily sister taxa and that there are, in fact, character gaps between the two groups of birds (Payne and Risley 1976, Sheldon 1987, Sheldon and Slikas 1997, Sheldon and Whittingham 1997). In this light, the congruence of morphological, behavioural, and biochemical evidence leaves little doubt that the birds known as great egrets are not *Egretta* but are more closely allied to *Ardea*.

At the specific level using molecular evidence, Sheldon (1987) found clearly that the southeast Asia population of great egrets is distinct from the North American population. The degree of distinction is equivalent to that of the Intermediate Egret and so should be recognized equivalently, i.e. as separate species. Beyond molecular evidence, the Eastern Great Egret differs from the western Great Egret in having red legs during nesting (Hancock 1984b) and using an Aerial Stretch display. In courtship it uses the Stretch as an advertising display but not also as a greeting display and its Snap lacks a heavily aggressive component, a behavioural trait showing similarity to *Egretta* (Hancock 1984a, Voisin 1991).

Range and status

The Eastern Great Egret occurs in south and east Asia and Australasia.

BREEDING RANGE: It nests in Pakistan, India, Sri Lanka, Bangladesh, Myanmar, Thailand, China (Heilongjiarg, Shandong, south Yunnan and Fujian (historically), Guangxi (Zhou 2003)). Korea, north-east Russia (south Ussuriland), Japan (Kanai 1990, Narusue 1992, Ueda 1994), Indo-china, Indonesia (Sumatra, Java, Borneo, Sulawesi, Lesser Sundas, Moluccas, Iran Jaya), Papua-New Guinea, Solomon

Islands, Australia (except drier interior), and New Zealand (a single colony on South Island (moore 2001)). There is not now breeding known in Malaysia or Philippines (Lansdown *et al.* 2000).

NON-BREEDING RANGE: It is a summer visitor to Japan as far as Hokkaido and the Kurils (Iijima 1984, Kosugi 1993, Shimada *et al.* 1994). It occurs in the non-breeding season from extreme south Japan (Honsu, Ryukus Islands), southeast China, Hainan, and Taiwan, through the East Indies, Australia, and New Zealand.

MIGRATION: Postbreeding dispersal is characteristic of Eastern Great Egrets. Within Australia birds disperse in all directions but tagged birds have been recorded returning to the natal colony to breed (Geering *et al.* 1998). Dispersing birds appear on distant islands, such as Macquarie, Lord Howell, Chatham and Norfolk islands, and Fiji (Suva) (Mackereth 1992). Red-legged Great Egrets seen in Yugoslavia and in the Netherlands have been attributed to vagrant individuals of *modesta* (Gaasbeek 1982, 1985, Hancock 1984a). The question of leg colour in the great egrets clearly needs more study. Although birds in Indo-china through Indonesia are apparently resident, portions of Asian and Australian populations migrate. The northeast population migrates from Korea to Japan to the Philippines. In Australia, inland nesting birds migrate through Queensland (Geering *et al.* 1998) to New Guinea and southern birds to New Zealand.

STATUS: The species is widespread and common to abundant over much of its range including India, Australia, and New Guinea (Lansdown *et al.* 2000, Maddock 2000). Counts include almost 11 000 in India and 5000 in the Philippines. The single nesting site in New Zealand has increased four-fold since 1958 and now supports a stable nesting population of about 100 birds (Miller 2001), owing in part to the increase in fish ponds and protection of

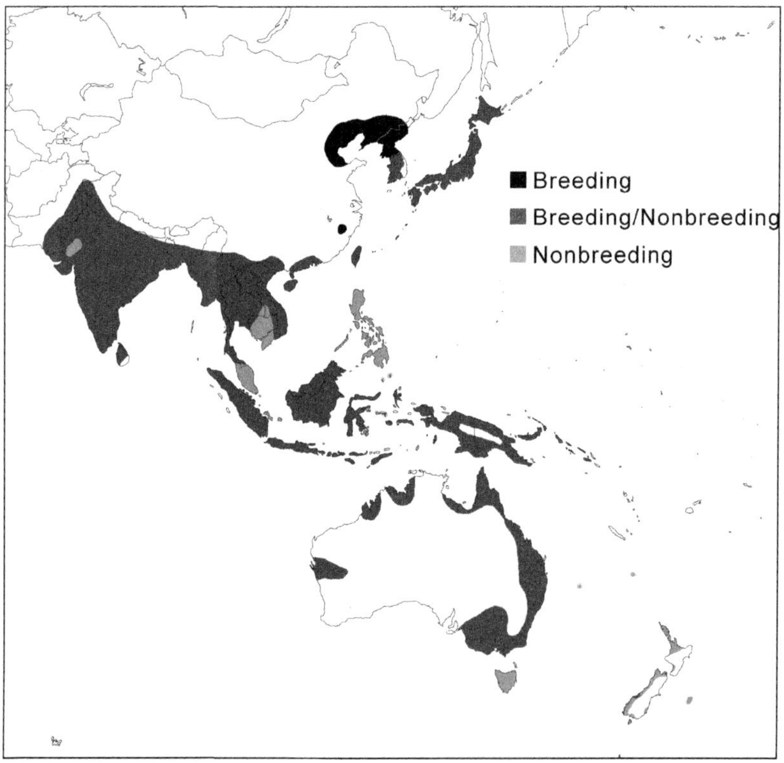

the site (Maddock 2000). Recent surveys of two areas suggest it is very scarce in China (M. Fasola pers. comm.).

Habitat

The Eastern Great Egret uses a variety of habitats, preferring permanent water of pools and streams in flood plains, lake and river margins, swamps and marshes (*Eleocharis, Typha, Phragmites, Scirpus*), and seasonal herbaceous marshes. Along the coast, it uses mud flats, lagoons, mangrove swamps, tidal streams, salt marsh, salt pans, and offshore reefs. Colony size has been related to the available area of salt marshes (Baxter and Fairweather 1998). It also uses human-altered habitats including reservoirs, sewage ponds, flooded fields, pasture, and ditches (Young and Chan 1997). For nesting it requires isolated trees or bushes, usually surrounded by or near water. Its favoured areas are wetlands with tall trees. It uses mangroves along the coast, inland swamps, river edge forests, and lake edges and islands. It also seems to require trees and islands for roosting.

Foraging

The Eastern Great Egret is a diurnal forager, its foraging schedule depending on daylight, tides, and weather (Min and Shiraishi 1984a). It feeds singly, in small groups, and occasionally in very large aggregations; it feeds by Walking slowly, stalking prey (Recher *et al.* 1983, Tojo 1996). It primarily Walks slowly in open shallow water about 30 cm deep, using an Erect posture. It also Stands motionless with head extended. It often uses Head Tilting, Peering Over, and Head Swaying. It feeds from the air by Plunging. Swimming has been reported (Viney 1988). When feeding as part of aggregations, it frequently engages in antagonistic behaviour to defend its feeding spot or take prey from other birds such as cormorants. It is also robbed of prey and feeds commensally by following other birds within the aggregation (Maddock 1992). Sexually oriented aggression has been reported on the feeding ground. One bird in Hong Kong in spring migration advanced on other feeding birds in stiff legged fashion, neck outstretched, bill pointing to 45°, and back feathers erected. The importance of feeding ground display is undetermined.

Prey taken is extremely diverse but principally fish. It also takes frogs, invertebrates, and even small birds. Invertebrates include crickets, dragonfly nymphs, flies, beetles, snails, bugs, crayfish, and prawns. Among the birds recorded are chicken, crake (*Porzana*), kingfisher (*Halcyon*), sparrow (*Passer*), white eye (*Zosterops*), swallows (*Hirundo*), and tern (*Sterna*) (Peter 1994, Darnell and Darnell 1995). Prey taken tends to be relatively small, under 12 cm. One value of this is to avoid being pirated by eagles (*Haliaeetus*), kites (*Haliastur*), or other wading birds.

Breeding

The breeding biology of the Eastern Great Egret has been well studied (Min and Shiraishi 1984a, 1984b, Min *et al.* 1984a, 1984b, Maddock 1986, Maddock and Baxter 1991). Nesting timing in Australia is variable depending in part on rainfall (Geering 1993), but generally from November to early May. In Korea, nesting is from April to August (Min *et al.* 1984b). This egret nests in reed beds, trees, or bushes, usually in isolated places or on islands. Trees used include larch (*Larix*), mangroves, paper bark (*Melaleuca*), eucalypt, and ironwood (*Casuarina*). The Eastern Great Egret is primarily a colonial nester, usually with other species including other herons such as night herons (*Nycticorax*), ibises, spoonbills, and cormorants. It places its nest relatively high in the vegetation. The nest, made of sticks, is a shallow platform and usually less than 1 m in diameter. It is built almost entirely by the female, using sticks brought by the male.

The bird is highly colonial. Large numbers of birds accumulate at the nesting site, and the male claims and defends its display territory. The Stretch is its principal advertising display. In the Stretch display, the head is thrust vigorously upwards, followed by withdrawal and equally vigorous strong downward bob of the legs (M. Maddock pers. comm.). Wing Preen, Twig Shake, Circle Flight, and Bill Clappering with bills in contact have been observed during and following pair formation. Long sequences of repeated Wing Preens and violent Twig Shakes, interspersed with Stretch displays, is a typical behavioural series (M. Maddock pers. comm.). An Aerial Stretch display has been documented in *modesta*, but has not been recorded in

Ardea alba (Hancock 1984b). Mock Preening and Allopreening occur after pair formation. In the Greeting Ceremony, the bird erects its display plumes and raises its wings giving the Gorork call.

The eggs are pale blue to pale blue-green, 52 × 36 mm. The clutch size is 2–5, averaging 3.8 in Korea. Incubation is accomplished by both parents but mainly by the female who incubates 80% of the time. Incubation begins after the first egg and lasts 23–29 days.

Young are semi-altricial and are tended by both parents. Feeding is first at the nest and later away. Older chicks grab the adult's bill to force regurgitation. In the early stages of chick rearing, one parent remains at the nest on guard. The elaborate Greeting Ceremony continues to be performed at the changeover. In the latter stages of nesting, the young are left unguarded (M. Maddock pers. comm.). Little information is available on the details of chick development. Nesting success is often related to hydrological conditions (Maddock 1986, Maddock and Baxter 1991). High chick mortality is also due to starvation following the death of parents and loss of chicks due to sibilicide during dry seasons has been observed (M. Maddock pers. comm.). Nests built in inappropriate trees are often loss (Min and Shiraishi 1984b) and only 40% of these hatched eggs. At a colony in Australia, 91% of nests fledged young averaging 2.1 chicks per successful nest.

Population dynamics

Tagged birds have been recorded returning to their nesting colony as two-year old birds (M. Maddock and D. Geering pers. comm.). Adult mortality during the breeding season was 5.8%–7.2% during April–June, attributed to energy deficits due to spring migration, stresses of breeding, and pesticide poisoning (Min *et al.* 1984b).

Conservation

The species, in Australia and New Zealand, has benefited from irrigation schemes creating foraging sites. Its tendency to feed in rice fields makes it dependent on this habitat is some areas, such as Japan. Changes in rice culture may adversely affect use it, and this important issue requires additional study (Narusue and Uchida 1993, Subramanya 1996, Lane and Fujioka 1998). Protection and management of feeding sites are essential. Its natural habitat in Australia remains under constant threat from drainage, grazing, clearing, burning, salinization, and exotic plants. As a result, there has been a serious decline (80%) in the breeding population at eastern NSW nesting colonies 1988–89 to 1997–98 (Maddock 2000), with no evidence of a recovery since (M. Maddock pers. comm.). Decline has also occurred in Victoria, where its Conservation Classification is endangered (NRE 2000).

It is flexible in its nesting sites and can even nest near residential and industrial settings. But nesting colonies, whether on natural or artificial sites, need to be protected and in some cases actively managed (Min and Shiraishi 1984b, Baxter and Fairweather 1998). For example, appropriate trees need to be planted for both short-term and long-term use. In Australia, current moves to provide environmental water flows from regulated rivers to breeding colonies in the Murray–Darling inland river system (Maddock 2000) have the potential to halt population declines there. However, the threats to the species are not only confined to the breeding colonies but also to habitat along the migration routes, which must also be addressed.

Research needs

Needless to say, interesting taxonomic issues remain within great egret populations. Our recognition of eastern populations of great egrets as a separate species deserves critical examination. A study of geographic variation in size and soft-part colours of all eastern great egret populations is needed to decipher the relationships among Australian egrets, and the egrets of mainland Asia. Attention needs to be focused on the breeding situation in the area of potential overlap in north China and in Japan, where birds considered to be intermediate between *albus* and *modesta* have been described (Amadon and Woolfenden 1952). Analysis is confounded by the presence of breeding birds, winter migrants, and summer dispersants in these areas. The question also remains as to whether the European and African forms are more closely related to the North American *egretta* or Australasian *modesta*, or even constitute separate species themselves.

The specific courtship behaviour of the Eastern Great Egret has not been studied in sufficient detail. Given the importance being placed in its own species, partially due to the existence of an Aerial Stretch display and breeding leg coloration, there is a need to better document courtship behaviour, nesting behaviour, colour changes, and vocalizations of *modesta.* Banding studies are needed to further decipher the range of the species, its migration patterns, and important non-breeding areas.

Overview

The Eastern Great Egret is a bird of shallow wetlands, where its slow, patient stalking and waiting behaviour can be effective. It catches relatively small fish and other prey, and so feeds for long periods during the day. It is primarily solitary in its feeding, which avoids disturbing its prey and competing with others. When feeding at concentrated food sources, it is highly aggressive, so much so that feeding appears to be interrupted. It can use a variety of feeding locations, from reefs to pasture including artificial situations. Despite its solitary preferences in feeding, it is highly colonial in nesting. Much remains to be observed about the basic biology of these species, especially in comparison with other large white egrets.

Intermediate Egret *Ardea intermedia* Wagler

PLATES 3.3–3.6, 13.4

Other names: Yellow-billed Egret, Plumed Egret, Median Egret, Smaller Egret, Lesser Egret in English; Garceta Intermedia in Spanish; Aigrette intermédiare in French; Mittlereiher in German; Middelste Zilverreiger in Dutch; Средняя Оеляя цапля in Russian; Kuntul perak in Indonesian; Tagak, Talabong in Pilipno; Chū-sagi in Japanese; Zhong bailu in Chinese.

Ardea intermedia Wagler, 1829. Ibis, vi, p. 659: Java.

Subspecies: *Ardea intermedia plumifera* (Gould), 1847: New South Wales; *Ardea intermedia brachyrhyncha* (Brehm), 1858: Blue Nile, eastern Sudan.

Description

The Intermediate Egret is a medium white heron with a relatively short yellow or bicoloured bill and, in breeding, exceptionally long back plumes.

ADULT: The plumage of this heron is all white with cream or offwhite tinge. The bill colour is basically yellow, but varies seasonally and geographically from yellow, black, to yellow and black, to yellow and brown. The irises are yellow and the lores are pale yellow. The legs are dark, sometimes two toned. In breeding, a train of long aigrette plumes extends along the back to 10 cm beyond the tail. Similar plumes extend from the lower neck forming a pronounced bib. The Head has a slight crest but no plumes. Soft parts change colour in the breeding season, in geographically variable ways (see below). Coloration changes rapidly during nesting, being intense during courtship and fading thereafter.

VARIATION: The sexes are alike. Geographic variation in size and soft-part colour is recognized in three subspecies. *Intermedia* is intermediate in size among the subspecies. Its bill is yellow, often with a black tip. Legs are slate grey, dark green, or brown. Feet are grey to black. During courtship, lores are yellow green, the bill is black, and legs are black. After courtship in breeding, the bill turns yellow below remaining black above. *Plumifera* is smaller than *B. i. intermedia.* Its bill is yellow-orange to buff. Upper legs are variable, pale brown, green, grey, or yellow. The rest of the lower legs and feet are dark brown. Lores are green-yellow to yellow. The irises are horn. During courtship, the bill is deep pink to red with a yellow tip. The base of the lower bill is green, lores bright green, and irises red. The Upper leg is ruby red, lower leg is variably ruby with or without black on the front, or black with red wash. Postcourtship, during incubation and chick rearing, the bill becomes dull red changing to orange-yellow and the lores become pale green changing to yellow-green to yellow. The irises turn yellow.

The upper legs become yellow and the lower leg, grey. *Brachyrhyncha* is larger than *A. i. intermedia*. Its bill is all yellow or yellow-orange. The upper legs are yellow. The lower legs and feet are dark brown to black. During breeding, the bill is pink-red. In courtship, lores turn bright green, irises red, and upper leg pink-red. The lower legs are black with a yellow line along the side (the line turning pink red in courtship), and the feet are black.

JUVENILE: Juveniles look like adults but may have a faint yellow wash to their plumage.

CHICK: The chick is covered in white down with a white crest. Its soft-part colours change with development (McKilligan 1990a). The bill colour is variable with time and among individuals. Generally it starts yellow, turns brown near the head, and becomes mostly brown in time. In a study of *plumifera* in Australia (Maddock 1988b, 1989b), chicks were found with several types of bill colour: beak all yellow (37%); black-tipped yellow (18%); variegated black and yellow (45%); and occasionally all black. The mouth colour was also variable. Eighty percent of mouths (hard palates) were all black, 2% were black in front and yellow behind, 1% all yellow, and 2% pink and black. The lores are bright yellow or yellow and black and the irises are yellow-white turning to pale yellow. Legs are olive green turning pale green then grey-green on the back, black on the front.

VOICE: This egret is normally fairly silent, Most vocalizations are quiet, and its soft buzzing call is unique among herons. The soft repeated 'Glock' call, rendered 'glock, glock', and harsh 'Kroo' call, rendered 'kroo, kroo', are alarm calls (heard in Australia but not Africa—Blaker 1969b, Marchant and Higgins 1990). The 'Kraa' call, rendered 'kraa, krr', is a flight call. The 'Kroh' call, rendered 'kroh, kroh', is a greeting call. The 'Grrk' call, rendered 'grrrk, grrrk', is a soft, two-syllable, rasping, staccato sound used in Greeting Ceremony (called a soft buzzing call by Blaker 1969b). Chicks beg with 'chi, chi, chi', which becomes 'khe, khe, khe' or 'cro, cro, cro' with age.

WEIGHTS AND MEASUREMENTS: Weight: 400–500 g; length: 65–72 cm.

Field characters

The Intermediate Egret is identified during the breeding season by its extensive plumes, especially the very long back plumes and also by a lack of head plumes. It is a sleek bird, with a bill relatively short for a medium aquatic heron. It has a long head and neck, often held in an S shape when standing or flying (McCanch and McCanch 1991), but relatively shorter than that of the Great Egret. In flight it resembles that of the Great Egret, rather buoyant, and with slow wing beats. Races are variably identifiable by courtship bill, and leg colour. In the non-breeding season it can be easily confused with other white herons. It is distinguished from the Eastern Great Egret in Asia and Great Egret in Africa by its smaller size, by its relatively shorter, deeper bill, shorter neck (relative to the exceptionally long-necked great white egrets), bicolour (not all black) legs, and the thin line of the mouth gape ending at the eye rather than continuing past the eye. It is distinguished from the Little Egret by its somewhat larger size, yellow to bicolour bill, yellow lores, black feet (not yellow), usually not all black legs, back and neck plumes and lack of head plumes in the breeding season, and less active feeding behaviour. It is distinguished from the Eastern Reef Heron by its slightly larger size, yellow or yellow and black bill (not horn upper), bicolour to dark legs and black (not yellow) feet, taller and more graceful less haunched appearance, and by its habitat choice (inland and protected coastal sites as opposed to open reef crests). It is distinguished from the Cattle Egret by its larger size, larger bill, relatively longer neck, dark (not yellow) legs, and slim appearance.

Systematics

The generic position of this species has long been a puzzle. Often assigned a genus intermediate between *Ardea* and *Egretta* (*Casmerodius, Mesophoyx*) or to *Egretta* due its extravagant plumes, it has been shown to be closer genetically to *Ardea* (Chang *et al.* 2003). The disjunct, continent-based distribution of the recognized subspecies and confusing individual and geographic variability in soft-part colour suggest that the taxonomic relationship among the African, Australian, and Asian forms should be re-examined. There may well be more than a single species within this cluster of medium white plumaged herons.

Range and status

This species occurs in apparently isolated populations in Africa, southeast and east Asia, and Australia.

BREEDING RANGE: In Africa, *brachyrhyncha* occurs south of the Sahara, its distribution avoiding deserts and jungle. In west Africa, it is in Senegal, Mali (Wallace *et al.* 1992), Guinea–Bissau, Nigeria, and Chad but is absent from desert and equatorial rain forest. In west Africa it occurs in north Sudan, northwest Ethiopia (avoiding the desert horn of Africa), Uganda, Kenya, Botswana, Tanzania, Zimbabwe, and Zambia. In southern Africa, it occurs in north central Angola and Namibia to South Africa. In Asia, *intermedia* breeds in Pakistan, India, Nepal, Sri Lanka, China (north to Szechwan, Hunan), Korea, Japan (Honshu, Shikoku), and Indonesia (Java, Sumatra). In Australia *plumifera* occurs in northeast Western Australia, east to Queensland, south to Victoria, with large breeding colonies throughout the Murray–Darling inland river system. It also occurs and breeds in temporary wetlands (Jaensch and Bellchambers 1997).

NON-BREEDING RANGE: North Asia populations spend the non-breeding season in Malaysia, Philippines, and Indonesia (Borneo, Sulawesi). Australian birds occur in south New Guinea, and Tasmania.

MIGRATION: North Asian birds undergo true migration, leaving Japan in September–October moving to and through peninsular and insular Malaysia and across the Luzan Strait to the Philippines and to Borneo (Sabah—Sheldon *et al.* 2001). Birds may winter in Oceania (Carolines, Marianas). There is clearly non-breeding seasonal movement between breeding areas in northeast Australia and non-breeding areas in New Guinea (Geering *et al.* 1998). In India and Africa, the species has been generally considered residential, but in most areas it undertakes partial or complete population shifts

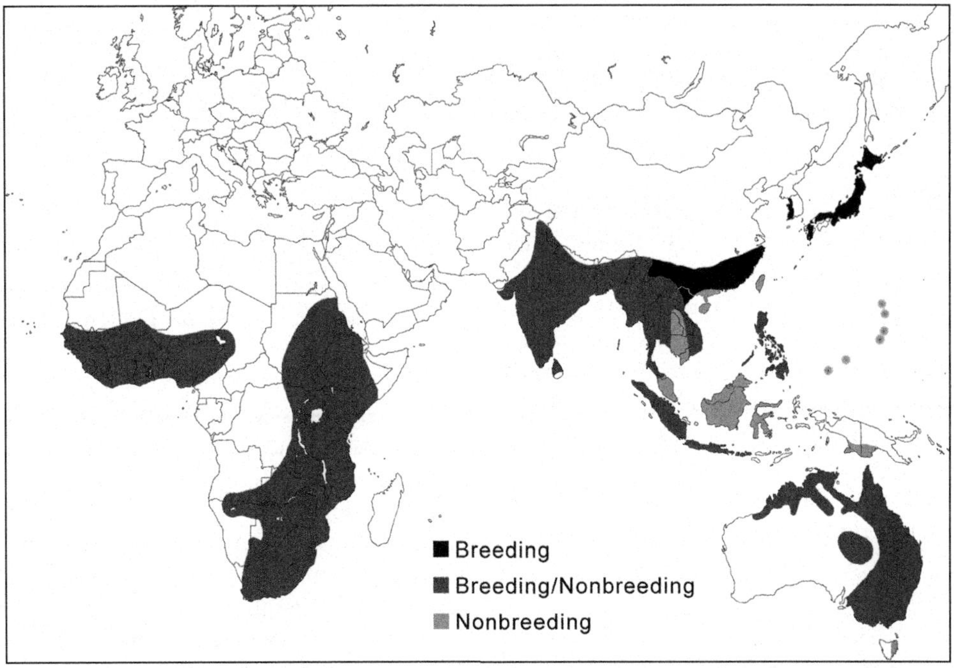

related to wet/dry conditions. Postbreeding, birds disperse and shift from wet season nesting areas to dry season non-breeding sites. In droughts, birds concentrate around permanent water. It is semimigratory in Australia. Australian birds disperse and shift and some migrate to Tasmania, and southern New Guinea, returning to their natal colonies during the breeding season (Geering *et al.* 1998). Dispersal results in occurrences outside its normal breeding range, including Mauritania, Cameroon (Martinez *et al.* 1996), Cape Verde Islands, Jordan (Dead Sea), Hong Kong, Tasmania, New Zealand, and Norfolk Island. Spring migrants overshoot, resulting in dispersal records occurring as far west as Italy (Taube and Grylle 2001), as far north as Russia (Ussuriland, Sakhalin Island), North Korea, and Midway Atoll (Richardson 1999).

STATUS: There are few available data on population sizes or trends, but the species is common to abundant in many places where it occurs and presumably it is stable in most areas. It is locally common to abundant in places throughout its range. Numbers in the hundreds to over a thousand have been recorded nesting in Kenya, Mali, Ethiopia, Zimbabwe, India, Pulau Dua Indonesia, and Australia. It has increased in numbers and range in west Africa (Turner 2000); a population estimate for Tanzania is 12 000 birds (Baker 1995, Baker and Baker 2004). It is abundant in non-breeding areas in Malaysia, Philippines, and Papua New Guinea (Lansdown *et al.* 2000, Maddock 2000). It is common in Bali. About 130 000 were counted in a regional survey in Papua New Guinea (Halse *et al.* 1996). Numbers in the hundreds of thousands were also found in north Australia (Morton *et al.* 1993). However, in some areas the species has suffered declines. Once the most common egret in Japan, it has declined markedly since the 1960s (e.g., Sasaki 2001). In east Australia decreases from year to year in colony size have caused concern (Maddock 2000), although fluctuations in colony numbers seem characteristic of the species.

Habitat

The Intermediate Egret uses a wide variety of floodplain wetlands but principally shallow water marshes and pastures (Blaker 1969b, Sato and Maruyama 1996, Tojo 1996, Takumi and Ezaki 1998). Habitats include ponds and bilabongs, fresh water marshes, swamp forests, water courses such as rivers and streams, fresh and alkaline lake margins, reservoirs, sewage ponds, rice fields, wet meadows, wet fields, and flooded and dry pasture. The less frequently used coastal habitats include mangrove swamps, salt marsh, intertidal flats, tidal streams, and rivers. It very commonly feeds in wet meadows and grasslands near water and is not characteristically a coastal bird. It prefers sheltered water less than 80 cm deep with emergent grasses. Egrets tend to avoid habitats where vegetation is too thick for feeding (Sato and Maruyama 1996).

Foraging

Intermediate Egrets are diurnal and typically solitary hunters. On occasion they do form small loose flocks or aggregations numbering in the hundreds. In Africa it has been suggested that these are old reports and that flocks are not seen today (N. Baker and E. Baker in prep.). They hunt primarily by Standing and by Walking slowly in shallow water or on damp ground in a slow and graceful manner (Recher and Holmes 1982, Recher *et al.* 1983, Sodhi and Khera 1986). When Walking they can exhibit a decided strut (Hancock 1999). They also can feed in deeper water by Walking on the matted vegetation. When Standing or Walking they use Gleaning, Head Tilting, and Peering Over. More active behaviours include Foot Stirring, Hopping, Hovering, and Plunging (Klaptste 1976, Sivasubramanian 1988). Head Swaying is used for feeding on insects in pasture. They have been reported to feed commensally, following spoonbills (Maddock 1992).

The diet of this heron is principally small fish in aquatic habitats and insects in terrestrial habitats, especially grasshoppers and crickets (Maddock 1986, Baxter and Fairweather 1989, Sodhi 1992). Fish include eels, perch (*Macquaria*), gudgeons (Eleotridae), and in Australia the introduced mosquitofish (*Gambusia*). It also eats frogs, snakes, and lizards. Invertebrates include crayfish, leeches, water bugs, and dragonfly larvae in aquatic habitats, and grasshoppers, mole crickets, bugs, beetles, and spiders in drier habitats.

Breeding

The Intermediate Egret breeds typically in the wet season, beginning just past the height of the rains or floods. In Africa it may be at the waning of the long rains or the short rains, depending on location. As a result, the months of breeding are highly variable across its range. For example, breeding has been reported in July–October in Senegal, March–August in Kenya, September–October in South Africa, July–September in north India, November–February in south India, February–July in west Java, December–February in east Java, and November–April in Australia. Timing in an area may change among years, due to drought or floods (Geering 1993).

The Intermediate Egret is highly colonial, nesting in mixed-species colonies with other herons, spoonbills, ibises, and cormorants, which may be very large. Pairs do nest singly as well. Colonies are typically sited in trees in and near water, in inland swamps, or mangrove swamps. Trees used in Australia are eucalyptus, mangroves (*Avicennia*), paper bark (*Melaleuca*), and ironwood (*Casuarina*). They also nest lower to the ground, in bushes and reeds. Colony sites may be used erratically, depending on water conditions. The nest is a shallow platform of loosely woven reeds, sticks, or twigs 60–80 cm across and 20–40 cm deep. It is placed from 1.5–20 m above the water or waterlogged ground, but there is a tendency to nest high in the trees (Ando 1993). Both sexes build the nest, with the female usually completing it with material brought by the male. Both males and females steal sticks from nearby nests.

As might be expected, courtship makes ample use of the extravagant plumes of the species. Birds claim a display and nesting territory. Males occupy, advertise, and defend the site using Standing, Preening, Upright, Twig Shake, Forward, Supplanting Flights, and Flap Fights. In the Forward, the body is fluffed out and all plumes are erected. Courtship displays are well described (Blaker 1969a, Marchant and Higgins 1990). Courtship displays reported are Stretch, Standing, Preening, Wing Preen, Head Toss, Snap, Back Biting, Bill Clappering, Flap Flights, and Circle Flights, which may have a neck-extended component. Stretch displays are rare. The predominant display is the Snap, in which the bird raises it back plumes and crest, flexes it legs, and bobs up and down rapidly. In Africa the birds are reported to have their head and neck horizontal, whereas in Australia they are reported to have the bill tucked into the chest (a behaviour called Bobbing). There is no reported Bill Snap or Twig Grasping component or calling. This display is most common around the time of pair formation. After pairing, the Snap continues to be used by both birds as the Greeting Ceremony in which plumes and crown feathers are raised, the ap-proached bird stands and does the Snap several times. The arriving bird may also give a Snap display and give the Kroo call or the quiet buzzing Grrk calls. Paired birds also spend considerable time at the nest site Standing, Preening, Wing Preening, Back Biting with Bill Clappering, and intertwining necks. Exchanging sticks includes neck arching and flaring of plumes.

The eggs are pale green, smooth, and slightly pitted. They average 47–49 × 35–40 mm. Eggs of the smallest race are 45 × 32 mm. The clutch is usually 2–3 eggs; range is 2–6. Incubation is by both sexes and takes 24–27 days. Hatching is asynchronous. Adults brood for 12 days. Nests and young are attended by both parents who defend eggs and young from ravens and hawks (McKilligan 1991), crouching over the nest, raising plumes, and pointing the bill to the predator (Marchant and Higgins 1990). The young are fed by regurgitation, first onto the nest floor and later chicks take food from the parent's mouth. Competition for food can occur among siblings (McKilligan 1990a). Young are semi-altricial. Pinfeathers appear at day 4. They can leave the nest at 24 days, but return for feeding. They fledge at about 40 days and leave the colony at about 70 days. Success is reported as relatively high in Africa and in Australia, where survival of hatchlings to fledging was 96%, and 88% of nests fledged at least one young (Maddock in Marchant and Higgins 1990). Nesting success is higher in wet than in dry years (Maddock and Baxter 1991, Baxter 1994).

Population dynamics

Birds in breeding plumage but not breeding coloration returned to the natal colony in the first year but did not nest (Maddock in Marchant and Higgins 1990). Survival has not been evaluated.

Conservation

This species is widespread over its range, locally abundant, and by all measures in an admirable conservation situation. As a bird that seasonally uses grasslands, it has benefited from the clearing of forests for agriculture and the extension of irrigation into dry areas. It uses human-altered situations, such as reservoirs, ponds, and rice fields. However, there are threats to these habitats. Changes to floodplain habitat on which it depends for nesting are caused by drainage, levee breaking and salt intrusion, grazing, and burning (Marchant and Higgins 1990, Maddock 2000). Changes in rice agriculture can affect the use of rice fields by the Intermediate Heron (Lane and Fujioka 1998). In coastal eastern Australia, there has been a significant decline in the breeding population in the period 1988 to 1998, exemplified by the 98% decrease shown by long-term monitoring at the Shortland colony (Maddock 2000), with no evidence since of a recovery (M. Maddock pers. comm.). Breeding at colonies in the inland Murray–Darling river system has been erratic, with long periods without nesting, such as in four of seven breeding seasons in the Macquarie Marshes 1990–91 to 1996–97 (Maddock 2000). The decline in Victoria, where its Conservation Classification is Critically Endangered (NRE 2000), has been especially severe, with no significant nesting event for about 20 years (O'Brien pers. comm. in Maddock 2000).

Despite its use of rural human environments, the Intermediate Egret is considered to be shier than other egrets and it is thought to be easily disturbed. Colony sites near potential disturbances need to be protected. Inland colonies, such as on regulated rivers in Australia, must continue to receive adequate water flow to ensure the long-term continuation of successful breeding (Maddock 2000). Close attention must be paid to the conservation of foraging and roosting habitats along migration routes.

Research needs

African, Asian, and Australian Intermediate Egrets appear to differ in size, breeding and courtship coloration, and courtship displays. This suggests that patterns of geographic variation and the taxonomic relationship among these forms should be re-examined, including by the use of molecular teechniques. The rapid change in breeding colours makes these difficult to document, and a careful comparative study needs to be undertaken. It is possible that these populations might be found to represent different species. Continuation of the study of dispersal patterns in Australia and the initiation of similar studies in other parts of the range are needed to better understand seasonal movements and provide data in support of conservation and habitat management efforts.

Overview

The Intermediate Egret appears intermediate not only in size, between large and small egrets, but also appears to be intermediate in its ecology, between being an aquatic heron and a terrestrial one. It has two distinct major feeding habitats (shallow water and pasture–prairie), two primary foods (fish and insects), and two seasonal ranges (wet season nesting and dry season non-nesting.) In general, during breeding it feeds in wetlands on fish during the wet season, and during non-breeding it feeds more broadly including in wet and dry grasslands on insects and lizards. It appears morphometrically attuned to its dichotomous life, with its bill being relatively shorter and thicker than that of the Great Egret yet relatively larger and thinner than a Cattle Egret's. Similarly its leg and neck proportions appear to be intermediate. It shows behavioural flexibility. It uses the same primary Stand and Walk slowly feeding techniques in both habitats. But it also is flexible in being able to use the Cattle Egret's Head Swaying and Gleaning and the typical middle-size egret's ability to use a range of active feeding behaviours such as feeding on the wing in deep water. Its most marked accommodation may be its migratory tendency, which allows populations to shift to accommodate seasonally and annually changing water conditions and its ability to alter its nesting season, geographically and annually, to nest when aquatic prey are available. It is a highly colonial species, nesting in small to large mixed-species aggregations. It is migratory and dispersive as well, moving in response to seasonal changes in weather and rainfall. The Intermediate Heron is adaptable in taking advantage of two alternative ecological situations. Its flexibility is clearly successful, given its relatively high nesting success rate and large population sizes.

Black-headed Heron *Ardea melanocephala* Vigors & Children

PLATE 2.2, 14.1

Other names: Garza Cabecinegra in Spanish; Héron mélanocéphale in French; Schwarzhalsreiher in German.

Ardea melanocephala Anon. = Vigors and Children, 1826, in Denham and Clapperton's Travels, Vol. 2, App. xxi, p. 20 1: near Lake Chad.

Description

The Black-headed Heron is a medium, black, grey, and white heron commonly found in the grasslands of Africa.

ADULT: The distinctive black of the head extends from the forehead down the back of the neck, including the few head plumes. The irises are yellow, and the lores are yellow and green. The upper bill is black, and the lower, yellow. The chin and throat are white, stippled with black, merging into a blue-grey on the foreneck. Most of the body is grey—the back, wings, and undersides, including lanceolate plumes on the neck and back. The underwing feathers are white and legs are black. During courtship the iris turns ruby red, reverting to yellow through an intermediate orange phase. Apparently there are no seasonal colour changes in the legs and bill.

VARIATION: The sexes are alike. There is no geographic variation described, but there is individual variation in the amount of white on the chin and throat. In some the lower throat is rufous; in others it is all black. Variation also occurs in the base grey colour, from light to dark. In lighter birds, the head stands out more.

JUVENILE: In the immature bird the throat and neck are dark grey to brown-grey, the breast buff, and the undersides are white with some buff.

CHICK: Chicks are covered with pale grey down and have a crest.

VOICE: The nasal 'Kvaak' call is its typical call. The 'Kow' call is used on landing. It is a double 'kow', followed by a growl and ending with a triple 'kut', rendered 'kow, owk, errr, kut, kut, kut'. A harsh 'Kaah' call is given in the Forward. During the Stretch display, it gives a soft 'How-oo' call.

WEIGHTS AND MEASUREMENTS: Weight: 710–1650 g; length: about 92 cm.

Field characters

The Black-headed Heron is identified by its black head and neck with white chin, grey belly, and in flight light under wings contrasting with black flight feathers. It is distinguished from the Grey Heron by its slightly smaller size, black head and neck, slimness, and contrasting wings in flight. It is distinguished from the Goliath Heron by its smaller size and grey colour and lack of chestnut. It is distinguished from the Purple Heron by its shorter bill and lack of chestnut.

Systematics

The Black-headed Heron is related to the other *Ardea*, but it is a distinctively terrestrial heron.

Range and status

The Black-headed Heron is an African species.

BREEDING RANGE: It occurs in Africa south of the Sahara. Its range extends west to east, from Senegal, The Gambia, Nigeria, and Chad, to Sudan and Ethiopia, south to Cape Province of South Africa. Its distribution is relatively patchy, depending on vegetation state and season.

MIGRATION: Birds migrate north during the rainy season to nest in the Sahel and Sudano–Guinean zones. In other areas of Africa, birds move to drier areas during the rainy season. In equatorial and south Africa, birds are more sedentary. Vagrancy seems uncommon, but dispersal records have been reported from Algeria, south west Europe, Oman, and Madagascar.

STATUS: This species is widespread throughout Africa. Generally it is one of the more numerous

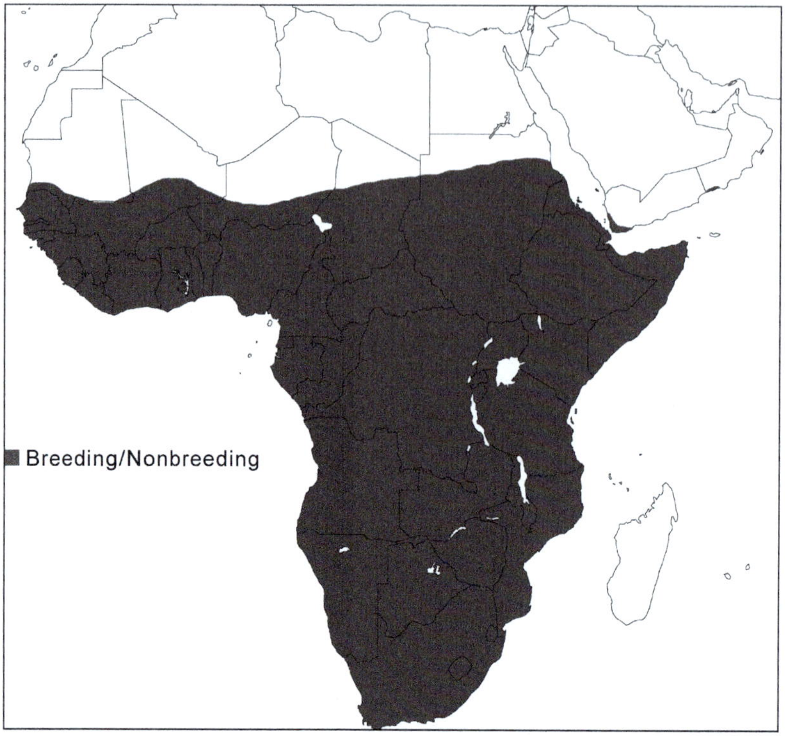

and commonly seen of the African large herons. The population in Niger was estimated to be 1648 (1994–7) (Brouwer and Mullie 2001). It has expanded its range and increased its population size benefiting from man-made pastures, dams, reservoirs, and urbanized habitats (Turner 2000).

Habitat

This is a heron of grasslands, inhabiting fairly open country either natural savannas or human-created grassland. It often is found in human-altered environments, even breeding in towns. It uses such diverse but uniformly open habitats as wet meadows, damp pasture, farmland, grassland, shallow marshes, edges of rivers, ponds and lakes, coastal flats, and clearings. It is attracted to fields flooded by rains.

Foraging

The Black-headed Heron is generally a solitary feeder, although on occasion it participates in loose flocks (Erwee 1989). At roosts and during non-nesting periods it congregates in the tens to hundreds of birds. It feeds primarily by Walking slowly, stalking through the grass. It lifts its feet high, keeps its head and neck erect and back, until it sights its prey. It frequently uses Neck Swaying, with increasing speed, before the final thrust. It also feeds by Standing. It feeds commonly at night.

This highly terrestrial heron feeds on a variety of invertebrates and vertebrates. Invertebrates include insects (especially grasshoppers and locusts), earthworms, blind worms, scorpions, crabs, and spiders. Vertebrates include fish, frogs, and lizards and small mammals such as rats, water voles, and musk-shrews (O'Connor 1984, Stuart and Dürk 1984). Ground-nesting and young birds are frequently taken (Forsyth 1984, Uys and Underhill 1995, Kahindi 1996, Cizek 2001, Tyler and Tyler 2001). Food varies with season—insects are most important during the rains and become less important after grasslands dry. It also scavenges.

Breeding

Black-headed Herons nest typically in the rainy season. In some areas they may nest year round, but peak in the rainy season. They typically nest high in trees of many sorts, including palms but also in reed beds or on the ground. They nest in mixed-species colonies of up to hundreds of pairs. The nest is rather flimsy, built of twigs. Males collect the material, and fresh twigs with leaves are added during early nesting. The bird may even soak twigs before using them (van Gass 1989). Building takes about 15 days. Both birds attend the nest. Both parents brood and shade the nestlings.

The Forward display and Kaah call are given toward intruding birds. In its Stretch display, the bird points its bill upward while inflating the chin and throat, accentuating their whiteness. The birds gives the How-oo call, followed by a gentle gurgling. Crest Raising and Fluffed Neck displays are used. In Contact Bill Clappering the heron runs its bill up and down the back and flank of its partner. Greeting Ceremony includes an exchange of loud cries. Until incubation begins the Greeting Ceremony tends to accompany every nest relief. Copulation occurs at the partly built nest.

The eggs are pale green-blue. The average is 60 × 43 mm. The clutch is 2 or 3 eggs; range of 2–6. Some pairs may have second or even third broods. Incubation takes 25 days. Adults feed young by regurgitation at the nest. Birds fledge at 40–50 days. Young return to the nest to solicit food. They are independent by about 60 days. Nesting success is not well documented. Heavy predation of the young occurs by eagles (*Hieraaetus, Haliaeetus, Aquila*) (Brown *et al.* 1982, Petzer 1990).

Population dynamics

Population biology and demography of the species are little known.

Conservation

The species is widespread, common, increasing and adaptable. It is not a major conservation concern. It has benefited from land clearing for pasture, irrigation, farming, and other human alterations. Protection of large colony sites and feeding habitat is needed throughout Africa.

Research needs

The basic biology of this species is fairly well known. But little is known about demographic parameters, such as survival and recruitment. There is a need to better delineate the actual breeding vs. non-breeding ranges and the important nesting sites. The movement patterns of individual populations should be better understood.

Overview

The Black-headed Heron is a terrestrial heron, a larger edition of the Cattle Egret. It uses open African grasslands but has been able to invade and uses many similar habitats created by humans. It consumes terrestrial prey, both insects and vertebrates, which it encounters by stalking slowly through the grass. Like the Cattle Egret it Walks slowly using Neck Swaying to encourage the prey to move. It is attracted to sites where food becomes ephemerally available, such as after rains, floods, fires, and locust outbreaks. It is highly adaptable, feeding in many different situations. It nests colonially with other herons, but feeds individually unless food is highly abundant. It is a species widespread in Africa, and well accommodated to the grassland habitat that typifies much of the continent.

Malagasy Heron *Ardea humbloti* Milne-Edwards & Grandidier

PLATE 2.3

Other names: Madagascar Heron, Humblot's Heron in English; Garza Malgache in Spanish; Héron de Humblot in French; Madagaskarreiher in German.

Ardea humbloti Milne-Edwards & Grandidier, 1885. Hist. phy. nat. et polit. de Madag., Ois., xii(i), p. 546: eastern Madagascar.

Description

The Malagasy Heron is a large grey heron occurring on the island of Madagascar.

ADULT: It is a dull grey bird having a distinctive black coloration on its crown, side of head, and chin. It is the only heron with a contrasting black chin. Its bill is yellow with a black base. Irises are yellow and legs brown. Soft-part colours likely change seasonally, but details are needed.

VARIATION: There is no information on sexual or other variation.

JUVENILE: Juveniles are very dark birds and have white to brownish chins.

CHICK: The chick is unrecorded.

VOICE: The voice is not reported, although it likely has similarities to that of the Grey Heron.

WEIGHTS AND MEASUREMENTS: Length: 92–100 cm.

Field characters

The Malagasy Heron is identified by its dark colour, black head and chin and very large, yellow bill. It is distinguished from the Grey Heron by its being larger, darker, with less contrast, and having a decidedly heavier bill. Interestingly, local people apparently do not distinguish between the two species although they do look quite different.

Systematics

This species is related to the Grey Heron and the other large *Ardea*.

Range and status

This heron occurs on the island of Madagascar and the nearby Comoro islands (Draulans 1986).

BREEDING RANGE: On Madagascar it is distributed especially along the west of the island. It may occur on the east coast, but recent records are few.

MIGRATION: There is no evidence of migration, but it may be prone to long-distance wandering, given the Comoro records.

STATUS: The Malagasy Heron is very rare and is considered to be decreasing, in some areas rapidly (Turner 2000, IUCN 2003). It occurs primarily in four protected areas in west Madagascar and in January 1983, 463 individuals were censused in the Antsalova region (Turner 2000). Confirmation of breeding status is needed both there and in the Comoro Islands. The global population is fewer than 5000 birds (Rose and Scott 1997, IUCN 2003).

Habitats

The Malagasy Heron appears preferentially to be a bird of coastal areas, particularly tidal mud flats and shallow water bays in mangrove swamps. However, it has been reported as using various aquatic situations including lakes, tidal flats, reefs, wetlands, and rice fields. At Lake Kinkony, a freshwater locality not far from the coast southwest of Majunga, the species used masses of floating vegetation (Rand 1936). In the Comoro Islands it fed in coastal tidal flats.

Foraging

The Malagasy Heron feeds by Standing, remaining motionless for long periods with its neck stretched out or in typical heron resting position (Draulans 1986). It is an extremely patient, passive forager, staying in one place, often in deep water, and then moves by flying or Walking slowly to another site. On the Comoros herons fed on the outgoing and incoming tides, and rest at high tide. It defends its feeding territory against intruders. The heron appears to specialize on medium to large fish. Its large dagger-like bill seems to be adapted to such prey. It has been reported to take eels, 48 cm long (Benson 1960), but probably observations on the Comoros are more typical, where they took fish about 1.5 times the size of their bill (Draulans 1986).

Breeding

The breeding season is not clear. Nests were found active in mid July, but females in breeding condition were collected in early December (Rand 1936). The herons have been recorded as nesting in a small patch of forest and also on the ground. As far as is known, the Malagasy Heron nests in very small groups in mixed-species colonies, but more needs to be known. Nothing is known of the nest or courtship behaviour of this species. Similarly, nearly nothing is known of the nesting biology. Only three nesting sites have been reported. One nest had three blue eggs.

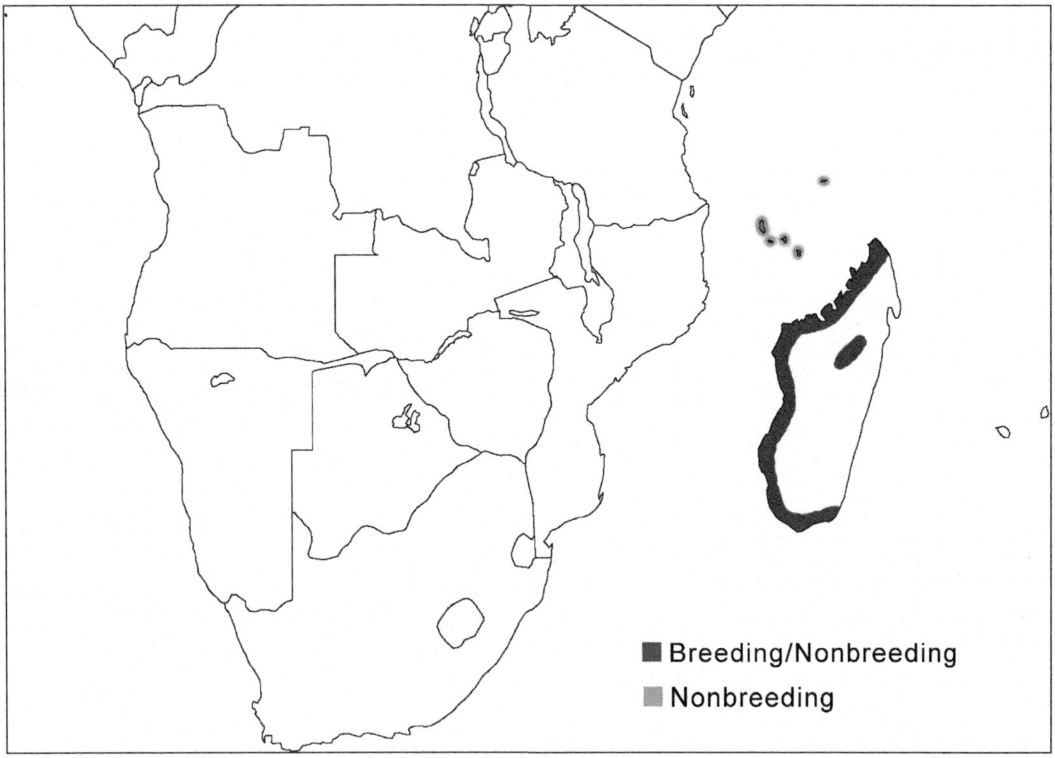

Population dynamics

Nothing is known of the population biology.

Conservation

Except for the Comoro Island observations, this species is endemic to Madagascar. It was probably never abundant, but is extremely rare now. It is globally vulnerable (Hafner *et al.* 2000, IUCN 2003). It has a patchy distribution along the west coast of Madagascar, fortunately including four protected areas (Turner 2000, Hafner *et al.* 2000). Until recently its habitat was inaccessible but is now threatened by the conversion of wetlands to rice, other development pressure, as well as hunting and nest predation by local people. The rarity of this species suggests that it is one of the herons of the world most in need of protection and conservation. Surveys for nesting and feeding sites, including along the east coast, protection of known sites, and legal and effective protection of individual birds and nests are needed (Turner 2000). These need to be part of a conservation action plan for the species, in concert with other conservation action ongoing on the island.

Research needs

So little is known about the basic biology and conservation status that detailed studies are required. Surveys of its distribution and abundance in Madagascar are needed, as is an understanding of its breeding status on Comoro Islands. Nesting sites and feeding areas need to be identified and protected. Further study of its highly patient foraging behaviour, diet, and habitat needs is needed.

Overview

Little is known about the species ecology, and studies are critically needed to understand its biology and conservation needs. This seems to be a heron of coastal tidal situations where it infrequently catches large fish. It is found solitarily or in small numbers. As a vulnerable species, conservation action is required immediately.

White-bellied Heron *Ardea insignis* Hume

PLATE 2.4

Other names: Imperial Heron, Great White-bellied Heron in English; Garza Ventriblanca in Spanish; Héron impérial in French; Kaiserreiher in German Baifu lu in Chinese.

Ardea insignis Hume 1878, 'Stray Feathers' vi, p. 470: Sikkim terai, Bhutan duars.

Description

The White-bellied Heron is a very large, brownish-grey heron. It is the second largest living species of heron.

ADULT: The crown appears blackish; crest feathers are pale tipped with grey. The massive bill, 152–176 mm long, appears blackish; but the lower bill is rather dull yellow-green at the tip, on the top edge and at the gape. The irises are yellow, tending to ochre, while the loral and orbital skin is green-yellow. The neck is mostly grey. The upper parts are slaty having a distinctly brown tinge, with the upper wing appearing solid grey. The chin and underparts are white. The underwing is grey with contrasting white wing linings. The legs and feet are dull grey.

VARIATION: The sexes are similar, but of a pair studied in Bhutan, one was lighter grey and slightly smaller than the other (RSNP 2003).

JUVENILE: The immature bird is browner with a white belly and lighter legs.

CHICK: The chick's body and head are white, bill black, undersides yellowish, legs pink. The wings and back turn darkish grey and legs whitish (RSNP 2003).

VOICE: The voice is a loud, deep croak.

WEIGHTS AND MEASUREMENTS: Length: 127 cm.

Field characters

The White-bellied Heron is identified by a combination of its large size, grey colour with white underparts and, in flight, grey upper wing and grey underwing contrasting with white linings.

It is distinguished from the Grey Heron by its large size, especially its large bill, and lack of a dark 'shoulder' patch. It is distinguished from the other great herons by its grey neck and brownish back. The larger Goliath Heron has a chestnut neck. Their historic ranges overlapped in Bangladesh. The smaller Sumatran Heron has grey and uniformly grey underwings. Their historic ranges overlapped in Burma. The White-bellied Heron occurs principally in forest, whereas the other great herons occur in more open wetland. It is distinguished from cranes in flight by flying with its neck crooked back rather than outstretched. Correct identification of this species in the wild is essential for its conservation.

Systematics

The White-bellied Heron is one of the great herons, related to the other *Ardea*, and likely most closely related to the Sumatran Heron. The two specific names *imperialis* and *insignis* are used in the literature as are two English names. The acceptable name has been a matter of prolonged debate (Baker 1932–1935, Biswas 1960, Walters 2001) resolved by Ripley (1982). Of the two oft-used English names, 'Imperial' is derived from the incorrect scientific name and is not particularly descriptive. White-bellied Heron is a more appropriate English name.

Range and status

The White-bellied Heron occurred historically in south Asia, from northern India through northern Myanmar.

BREEDING RANGE: This species historically had a small range in south to southeastern Asia including central and east India (Uttar Pradesh, Bihar, West Bengal, Assam, Arunachal Pradesh, Nagaland), Nepal, likely north Bangladesh, Bhutan, Myanmar, and China (south Tibet) (Sugathan *et al.* 1987,

Inskipp 1989, Choudhury 1992, Anon. 1994, BirdLife 2001, 2003). Records previously accepted from the Sundarbans (Scott 1989, Hancock and Kushlan 1984) are likely erroneous, given the habitat and potential for confusion with the Goliath Heron reported to be found there. The species has not been reported recently from much of this presumed historical range. In the past two decades it has been recorded only from west Uttar Pradesh in central India (in winter), Assam in east India, Bangladesh, Bhutan, and north Myanmar (Barua 1998, Choudhury 2000, Kazmierczak 2000, BirdLife 2003). Its breeding status over most of this range is unclear. It is confirmed to breed at present only in Bhutan (Wangduephodrang) (Inskipp *et al.* 1999, RSPN 2003, T. Norbu pers. comm.). The discovery of a nest and also of a juvenile bird (at Punaka) is an exceptionally important documentation of breeding status in Bhutan.

MIGRATION: Post breeding dispersion occurs, the extent of which is unclear. The birds studied in Bhutan in 2003 left the breeding area after fledging. There are suggestions of a tendency to move from higher breeding areas into marshy areas at lower altitudes. Reports of birds in the nonbreeding season have occurred in areas where they are not known to breed, which may indicate dispersal or reflect a lack of firm information on nesting status in the area.

STATUS: The species is very rare and rarely reported, even historically. Individuals have been seen recently primarily in Bangladesh, Assam,

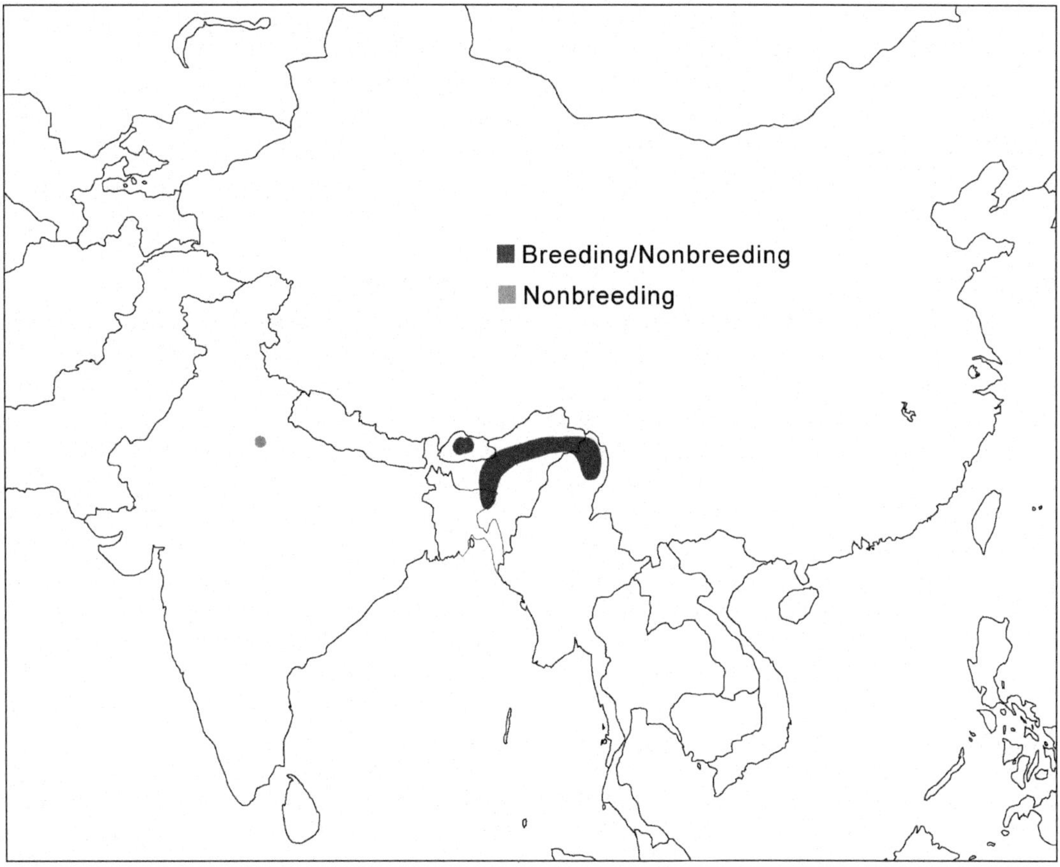

Myanmar, and Bhutan (Harvey 1990, Inskipp and Inskipp 1993a, 1993b, Thompson *et al.* 1993, BirdLife 2003). Overall it has very small population that gives every indication of continuing to decrease. The Indian population is no more than in the low hundreds. It is very rare in Bangladesh, regular in Bhutan, and rare in Myanmar (BirdLife 2001). The worldwide population is probably under 1000 individuals.

Habitat

This is a heron of large inland swamp forests and forested rivers and less frequently submontane grasslands. It appears to depend during nesting on mature forests having large trees for nesting associated with wetlands, rivers, or lakes. In the southern Himalayas its habitat includes the dense forest and rivers (duar) and the terai, an often-marshy area of elephant grass and dense forest patches dependent on soil washed down from the mountains. Elsewhere in its range it occurs in patches of remaining forest (Perennou *et al.* 2000). It primarily uses riverine sites for foraging, especially during nesting. These range from sandy and shingle beds in large rivers to small forested streams. It has occurred to 1500 m in India.

Foraging

It feeds solitarily, although members of a pair or family group may feed in the same area, but not closely together. It forages characteristically by Standing motionless 'knee-deep' in water with its neck held in an 'S' shape or Walking very slowly along a river shore, looking alternately to each side (RSPN 2003). When disturbed it assumes an Erect posture. It feeds most characteristically in and along fast-flowing mountain streams, but also uses slower flowing waters and even grassy marshes, the latter more commonly in the nonbreeding season. It feeds during daylight, flying to the feeding site in the early morning and returning to roost in the early evening. Its diet is poorly known but includes insects, large crayfish, and fish including large ones (T. Norbu in RSPN 2003).

Breeding

It breeds in April–August, solitarily in large platform nests high in tall trees in forests. The nesting of this species was studied for the first time in 2003 by Tshewang Norbu, who observed a nest through the season (RSPN 2003). The nest, found in April, was in a tall Chirpine tree, 20–25 m high. It was well camouflaged, made of leafy branches and twigs. Parents alternated incubation. The literature on egg sizes is unclear, casting doubt on identification. They have been reported as greenish blue (70.2 × 50.8 mm and 69.2 × 49.9 mm) in Arakan but smaller (63 × 49.9 mm and 63.1 × 41.6 mm) in Sikkim (Walters 1976). The nest in Bhutan had a clutch of two, one of which hatched. The young was fed on regurgitated fish. The chick was left unguarded starting in weeks 7–8, fledged at 63 days, leaving the nest completely on day 72 (July 10th).

Population dynamics

The population dynamics is unknown. As a large species in a mature and stable habitat, it is likely a long-lived bird that occupies its territory for many years. As a solitary species with a large range, its population densities were never high.

Conservation

This is an exceptionally rare heron, about which little is known, including its current range, to say nothing of its population status. As a bird of mature forests, its habitat has been under severe threat throughout its range for a century. Although historically nearly inaccessible, its wet forest habitat has been cleared through much of central Asia. As a species that probably has always been scarce, habitat destruction and lack of protection pose the primary threat to its continued existence. Forested habitat is being destroyed by slash and burn agriculture in Arunachal Pradesh. It nests in Namdapha National Park and occurs in other protected areas in the non-breeding season. Lowland wetlands of Brahmaputra are threatened by siltation, plant growth, and fishing. Large-scale habitat alteration of its prime habitat is occurring in Myanmar and Bhutan. The Bhutan nesting area is affected by rock mining operations.

It is a species that apparently is adversely affected by human disturbance, as evidenced by its occurrence primarily in remote forests. Human traffic and movement of domestic animals cause disturbances in the forest. In addition, birds are likely at risk from hunting of all sorts. Local people kill birds at the nest, obtaining them using long bamboo poles. It is likely that hunting has been a principal cause of loss of individuals. The species is endangered (Kushlan and Hafner 2000, IUCN 2003). It is protected in Myanmar and in Bhutan, both legally and culturally. Legal protection for the species is needed in India and Bangladesh and serious protection is needed for parks and reserves in which it is found (Hafner *et al.* 2000). The most important of these is Namdapha National Park in Arunachal Pradesh, Jamjing Reserve Forest in Assam, and the newly discovered sites in Bhutan (near Wangduephodrang and Punaka).

Research needs

Little is known of the biology of this heron. Its current distribution needs to be determined. Surveys need to be conducted in search of this species, and naturalists within its range should be encouraged to report all sightings. Nest sites need to be located (and positively identified), nesting and feeding habitat better defined, and whether the species migrates or disperses needs to be determined in order to locate important areas. A conservation plan is needed to guide conservation action once individuals and especially nests are found.

Overview

This is a very large, solitary heron of Asian forests and marshlands, requiring ever-decreasing mature damp forests and dense marshes. What is known of its biology, such as its dependence on tall trees during nesting, its ability to feed by Standing in fast-flowing rivers, and its occurrence primarily in sites remote from human activity, suggest that it is a highly specialized species. It is however little known biologically, and critically endangered from both habitat loss and human depredation. It is, in fact, one of the most threatened of all the herons.

Sumatran Heron *Ardea sumatrana* Raffles

PLATE 2.6

Other names: Great-billed Heron, Dusky-grey Heron, Giant Heron, Alligator bird, Bull-heron in English; Garza de Sumatra in Spanish; Héron typhon in French; Rußreiher in German; Dangkawan, Dugwak in Pilipino; Cangak laut in Indonesian.

Ardea sumatrana Raffles, 1822. Trans. Linn. Soc. London 13(2), p. 325: Sumatra.

Description

The Sumatran Heron is a very large, uniformly brown-grey heron with a large stocky bill and white throat.

ADULT: The Sumatran Heron has a brown-grey crown with silver-grey head plumes forming a short crest. The very large heavy bill is black above, whitish yellow to greenish below. The irises are yellow to silver-brown and lores yellow to green. The upper throat is light grey, an important field character. Its dorsal colour is grey with a brown-grey wash to head, back, and neck. The lower throat and neck are brown-grey and the rest of the underparts are lighter. Wings are a dark brown-grey and are lighter underneath. Legs are dusky grey to olive-grey.

In the breeding season the grey back and underparts take on lighter highlights. Its throat and centre of the breast become distinctively white, and it develops silver-grey plumes on the lower neck and breast. The bill, while remaining black, develops a yellow-green tip and a yellow patch at its base. Irides are silver-brown to yellow. Lores are olive-yellow to blue-grey. A male has been recorded with blue-grey lores during nesting. Soft-part colour variation needs additional confirmation.

VARIATION: The sexes are alike. Clinal variation is likely. Birds tend to be darker in the humid eastern part of its range and paler in the drier west. Northern Australian birds, described as being browner than northern birds, almost bronze, with a shorter tarsus and longer, have been considered a subspecies, *mathewsae*. But these distinctions have not held up to further study.

JUVENILE: Immature birds are lighter coloured than adults, a rusty brown, with a rufous or sandy-buff wash on the back and underparts, and no crest. The bill and legs are brown.

CHICK: Downy chick is light grey with grey bill and legs.

VOICE: The Sumatran Heron is highly vocal (Dahlberg 1998). Its call is amazing among heron vocalization. It is a deep guttural, resonant 'Roar', emitted especially at night, when disturbed and in defence at the nest. The 'Roar' call is likened to an angry bull or crocodile (likely the origin of two Australian common names). The call lasts several seconds, repeated three to four times, with seconds between calls (Seton 1972), and it may develop into a double roar (Marchant and Higgins 1990). Other calls include croaks, booming, and the sound of a small engine. The vocalizations of this heron clearly deserve additional characterization.

WEIGHTS AND MEASUREMENTS: Weight: 1300–2600 g; length: 110–115 cm.

Field characters

The Sumatran Heron is identified by its size, grey colour, and uniform grey wing colour. It is larger than other herons in its normal range. Its flight is heavy or ponderous with head back and legs dangling. It is generally shy and retiring and usually seen alone. It flies short distances in response to disturbance but is more likely to walk away slowly into the bushes.

It is distinguished from the Goliath Heron by its grey to light grey (not chestnut or white) throat and grey (not chestnut) head and neck. It is distinguished from the White-bellied Heron by its light grey (not white) underparts. It is distinguished from the Grey Heron by its large size, grey-brown (not blue-grey) upper parts, and uniformly sooty grey wings above and below (not bicoloured). In flight, it is distinguished from the Black-necked Stork and Sarus Crane by its uniform grey wings (not contrasting dark and pale wings) and by its coiled (not outstretched) neck.

Systematics

One of the great herons, its appearance suggests that it is probably most closely related to the White-bellied Heron. Additional study of geographic, particularly clinal, variation is needed, especially to re-examine the Australian birds once described as a separate subspecies. It is usually called the Great-billed Heron in Australia, but this common name is not distinctive outside of Australia where it can overlap other great herons, so Sumatran Heron is the preferred English name.

Range and status

The Sumatran Heron is a bird of south East Asia through Australia.

BREEDING RANGE: The Sumatran Heron's range is from coastal Burma (Tenasserim, Mergui, and Arakan coast), Thailand (islands off southwest coast), Malaysia (east coast Peninsula), Singapore, Myanmar (Araken coast), Philippines (Sulu Archipelago), Indonesia in Borneo, Sulawesi (north and southeast, Talaud Islands, Sangihe Islands), Sumatra (coastal including Riau Islands, Beiting Island), Java (offshore islands but not mainland), Sunda Islands, New Guinea (coast), to coastal north Australia and offshore islands (Western Australia, Queensland) (P. Lansdown pers. comm., M. Maddock pers. comm.). It is known as a nesting species in two areas. The northern population

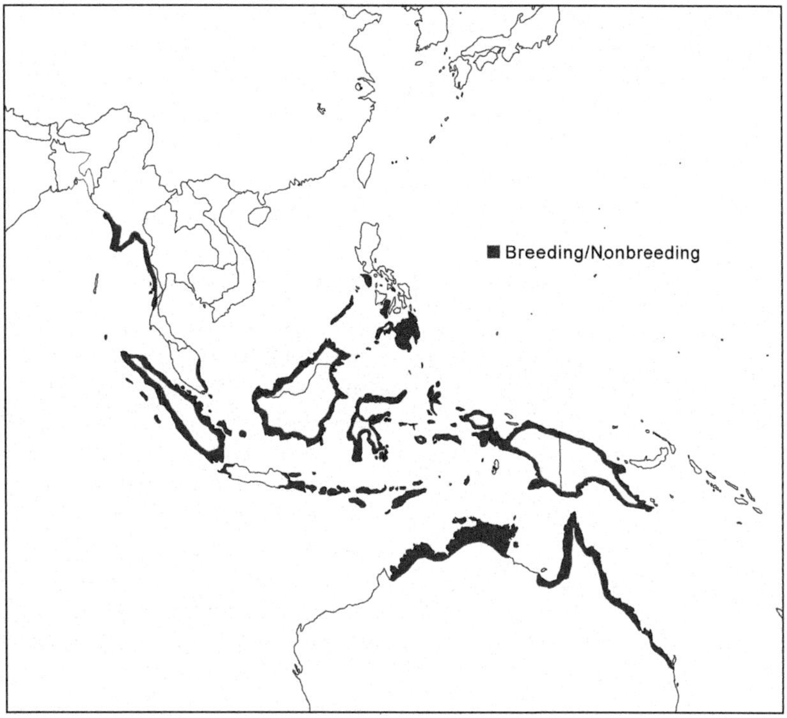

is in coastal Indo-China, islands off the southwest coast of Thailand, and the Andaman Islands. The southern population occurs through Indonesia south to north Australia. In Australia the heron occurs along the coast of north Australia from West Australia to south Queensland, except for the Gulf of Carpentaria (Maddock 2000, Burnham *et al.* 2002). It occurs inland very rarely.

MIGRATION: This is a sedentary species, although postbreeding dispersal may occur, in that there are dispersal records.

STATUS: This species was once distributed widely throughout coastal southeast Asia, but it has been much reduced on the mainland during the past 50 years. Since the 1970s it has been reported in Asia only from coastal islands off Thailand and Malaysia (Lansdown *et al.* 2000). It appears to be extirpated from coastal Vietnam, Cambodia and Gulf of Thailand (J. Tordoff pers. comm.). It is not abundant anywhere in its remnant range and is considered to be scarce in Papua-New Guinea (Beehler *et al.* 1986). It is common in northern Australia, and as far as is known its population is stable there (Maddock 2000).

Habitat

This is typically a coastal heron. It prefers coastal mangrove swamps with tall trees, usually associated with soft muddy intertidal foraging habitat. It frequents shores, mud flats, mangove-lined creeks, and similar shore sites. It also uses forested inland sites, especially forest-lined rivers, billabongs, rocky gorges, swamps, and creeks. It is seldom found in open habitat.

Foraging

This is a solitary hunter, rarely in pairs or in small apparently family groups. It is a shy species, avoiding human disturbance. It feeds typically on outgoing to low tides, roosting at high tide and is principally crepuscular outside the nesting season. Its feeding approach is Standing at the edge of streams or Walking slowly along mangrove-lined creeks with deliberate steps, wading through water up to 30 cm deep. Individuals and pairs permanently occupy feeding locales along rivers and shore, which they probably defend as feeding territories. Forward displays are used against intruders (Lansdown 1989). A strutting terrestrial display was described between two birds (Gill 1979) in which they raise their crests, cock their tails, droop their wings, and fan their tails. This could be a contact display or, more likely, an antagonistic display.

The Sumatran Heron is highly dispersed in mangrove swamps having tall trees and open mudflats nearby. Its density is estimated to be about one pair per 600–700 ha or one pair per 5 km of river (Garnett in Lansdown *et al.* 2000).

It principally eats fish. The massive bill suggests that it must be capable of dealing with very large prey, including mussels and crabs. No detailed studies of foraging ecology or food habits have been conducted.

Breeding

The Sumatron Heron nests in May and June and during breaks in the monsoons. In Australia nesting has been recorded in July, November, February, March, and April. Nesting primarily is in mangrove trees, usually *Avicennia*, large enough to support the large nest. Fundamentally it is a solitary nester, but occasionally nests near mixed colonies (Lansdown *et al.* 2000). There may be multiple nests within what appears to be a single territory, an observation deserving further scrutiny. Few descriptions of nesting are available. Nests are a large (120–130 cm wide), thick (30–40 cm) but flat platform made of large sticks with finer twigs and fresh leaves inside. It is placed low, up to 2.5 m high, usually overhanging water well hidden by dense mangrove branches although more open situations have been reported. As pairs appear to remain together year-round, courtship and pair contact behaviours also probably occur at low intensity over extended periods year-round. Lansdown (1989) described several pair maintenance behaviours away from the nest site. These include Stretch, Fluffed Neck, and Bill Duelling. In one case a bird responded to a Stretch by its companion with a Fluffed Neck. Eggs are pale green-blue, slightly glossy and minutely pitted. The size is 62.5–71 mm × 46.5–49 mm. Clutch appears to be two. Its nesting biology is little known. Both adults tend the nest and the young. Often only one chick survives to fledge. When disturbed, nestlings assume a Bittern Posture and adults give their Roar call, a loud warning call.

Population dynamics

Nothing is known about its population dynamics.

Conservation

The species is rare in continental Asia and not common anywhere in its range. It is considered to be near threatened (Kushlan and Hafner 2000). As is the case with the other great herons, it inherently has small and dispersed populations, is highly specialized, and threatened by habitat changes. The site where five nests were found in Malaysia was subsequently reclaimed (Lansdown 1987a, Lansdown *et al.* 2000). The distribution of nesting birds needs to be better understood and nesting sites protected. Important areas include the Salu–Sudong–Paiwai–Senang group of islands south of Singapore, the Rhio–Lingga Islands, and the Kinabatangan River basin in Borneo (Lansdown *et al.* 2000). The principal habitat for this species is mature coastal mangrove swamp and associated mudflats and coastal islands. These habitats are under particular threat throughout the species' range due to reclamation and stream damming. Protection and management of coastal mangrove habitat are essential for the conservation of this species.

Research needs

So little is known about this species that studies of its basic biology are essential. Its population status throughout its range must be determined and important sites identified and protected. Extensive nesting season surveys for nest sites and feeding habitat are needed.

Overview

The Sumatran Heron is a large, solitary bird of the heavily vegetated coastal zones of east Asia, especially mangroves where it feeds by patiently Standing during outgoing tides. So little is known about it, that even its most basic biological characteristics are guesses. It has always been little understood. It took from its discovery in 1822 until 1906 for its nest to be reported. As a fish eater, it likely specializes in catching a few large fish, which it can dispatch readily. Its food resource is probably generally scarce and the bird has to be as effective as possible in catching what happens by its feeding site.

Goliath Heron *Ardea goliath* Cretzchmar

PLATE 2.5

Other names: Garza Goliat in Spanish; Héron goliath in French; Goliathreiher in German; Goliathreiger in Dutch, Goliathagar in Swedish; Цапля голаф in Russian.

Ardea goliath Cretzchmar, 1826. Ruppells Atlas Reise (Vogel), p. 39, p1. 26: 'Bahhar Abiad' = Bahr el Abiad or White Nile.

Description

The Goliath Heron is a very large grey heron with chestnut head, neck, and belly. It is the largest living heron.

ADULT: Its head is red chestnut, with an elongated, bushy crest. The bill is very large and thick, black above and pale horn below. The irises are yellow rimmed in red, and the lores are green-yellow. The sides of the face and neck and back of the neck are a pale chestnut. The chin and throat are white, with black streaks extending down the neck onto the upper breast. Its back and upper wings are slate grey, with a chestnut 'shoulder' patch at the bend of the closed wing. Lanceolate plumes on the mantle and back are slate grey. The upper breast has white lanceolate plumes that hang over the deep rufous chestnut of the breast and rest of the under parts. Underwings are pale chestnut. The relatively long legs are black, as are the feet.

VARIATION: Females are smaller, with a slightly smaller bill. There is no recognized geographical variation, although if there remains an isolated population in Bangladesh, geographic variation is probable.

JUVENILE: Juveniles are the same as adults, but the chestnut is paler and more rufous. Grey areas are tinged with brown. The undersides are mottled with black and the white striping on neck is less defined than in the adult. It lacks elongated plumes.

CHICK: Chicks are covered with white-grey down. The iris is light green, the bill brown, and the legs and skin are lime green (Cooper and Marshall 1970).

VOICE: Goliath Herons have a distinctive 'Kowoork' call, a loud, repeated, individually distinctive bark audible for up to 2 km. A sharp 'Arrk' call is the disturbance call. The 'Krooo' call is given during the Stretch at the vertical position whereas a 'Huh-huh' call is given at the crouched stage (Salewski and Grafe 1999). 'Organ-like dueting' has been reported at the nest (Brown *et al.* 1982), but more details are needed to confirm this.

WEIGHTS AND MEASUREMENTS: Weight: 4300–4468 g; length: 150 cm.

Field characters

The Goliath Heron is identified by its great size combined with its grey and chestnut coloration. It flies with slow, deliberate wing beats, with the wings markedly bowed. When the bird takes flight its head and neck are at first outstretched, but soon retracted onto the shoulders.

It is distinguished from the Purple Heron by its large size (nearly twice as large), chestnut rather than black cap, and chestnut rather than black-streaked rufous sides of neck. It is distinguished from the White-bellied and Sumatran Herons by its grey and chestnut plumage (not solid grey or grey brown) and the lack the black and white neck pattern.

Systematics

This is one of the great herons, and is probably most closely related to the White-bellied and Sumatran herons.

Range and status

The Goliath Heron is a bird of Africa and the Middle East, and also the Indian subcontinent.

BREEDING RANGE: The Goliath Heron breeds in suitable habitat over much of Africa. It occurs along the east coast from Egypt and Sudan (Goodman and Storer 1987), Yemen on the Red Sea (Porter and Al-Saghier 1998), Somalia (Schels 1993), to Cape Province, South Africa (Pauw 1991), and inland Africa west to the southwest deserts, Gambia, Guinea–Bissau, north to the Sahara. Its specific breeding locations within this larger range are not so well documented.

NON-BREEDING RANGE: Birds occur commonly along both sides of the Red Sea (Egypt including Sinai, Sudan, Saudi Arabia, Yemen) and has been increasing there over the past decades (Bjlsma and de Roder 1982). In this area, breeding is known from northern Sudan, southeast Egypt, and probably Yemen (Brooks *et al.* 1987, Goodman and Storer 1987, Porter and Al-Saghier 1998). Birds also occur, or have occurred in the past, in Syria, northeast Israel, south Jordan, south Iraq, and Iran (Kinzelback 1986, Kasparek *et al.* 1994). The species has nested in the past in Iraq, but there are not recent records. It also has been seen several times recently in Bangladesh (in the Sundarbans) and India (Assam) (Saikia and Bhattacharjee 1990, Thompson *et al.* 1993, Madge 1995). It formerly occurred elsewhere in India and Sri Lanka. Given the distances involved to the core of the known range, it may be that these birds represent a small population breeding in the Sundarbans or Assam, but this needs to be documented. There is also an old record from Madagascar in 1879, but none since, and a misidentification is likely.

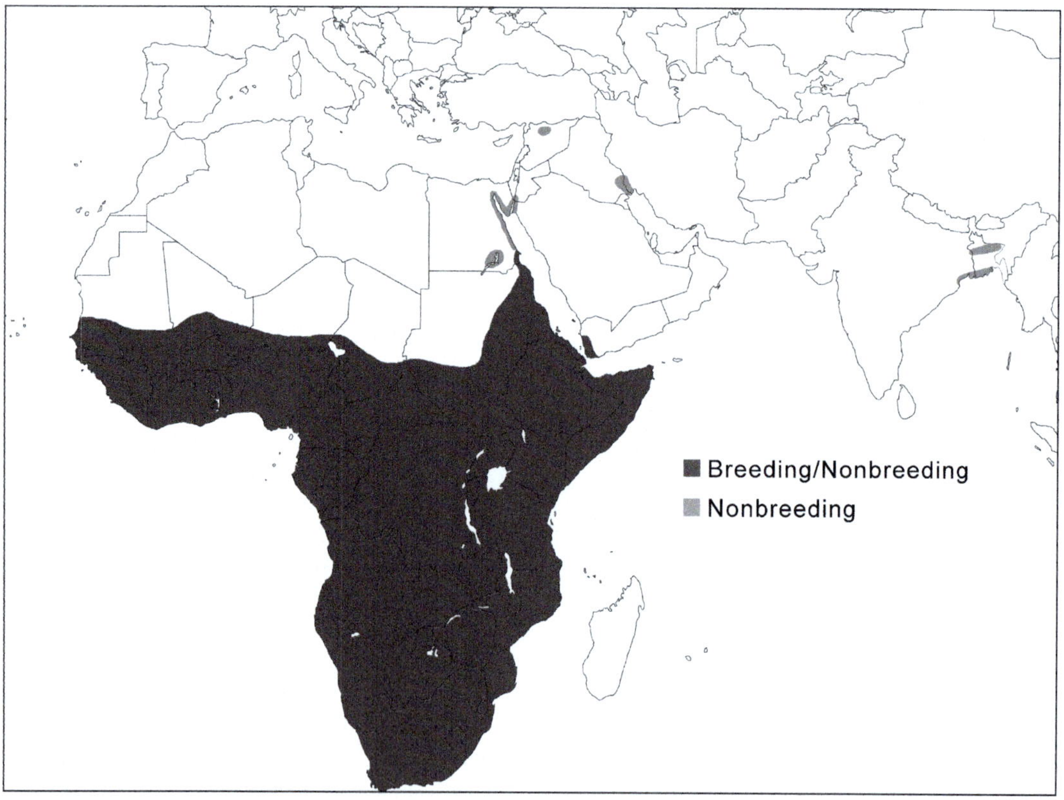

MIGRATION: The species is sedentary and regular migration is not documented. However, local movements occur in response to foraging conditions. The origin of birds outside the known breeding range needs to be assessed to determine if dispersal does occur.

STATUS: No figures are available on population sizes. The species is locally common in east Africa swamps and along some lakes and rivers (Turner 2000). But over most of Africa, it is thinly spaced. For example, it averages only about one per 4 km^2 along a river course (Dowsett and de Vos 1964). An estimate for Tanzania is 2500–4000 birds (N. Baker and E. Baker in prep.). It has increased in South Africa due to dams and reservoirs (Geldenhuys 1984), has expanded its range in Cape Province, South Africa (Pauw 1991), and is increasingly seen along the Red Sea. However, population status of birds in Iraq–Iran and the Indian subcontinent remains unclear but is presumed to have severely declined in both areas.

Habitat

The Goliath Heron is a very aquatic heron of both coastal and inland habitats, rarely wandering far from water. The shallow water margins of large lakes, lagoons, and large river systems are its favoured habitat. It typically feeds out from shore in deep water and near floating plants. Overall, though, it inhabits many types of shallow waters including rivers and lakes, both salt and fresh, marshes and swamps, tidal estuaries, reefs and sometimes mangrove creeks, and even water holes in woodland savanna, from sea level to 2100 m. It is not commonly found around areas of human disturbance.

Foraging

The Goliath Heron is typically a solitary hunter defending large feeding territories from other Goliath Herons (Whitfield and Blaber 1978, Mock and Mock 1980). Very occasionally two may be seen together. It flocks very rarely. It typically feeds diurnally, but also at night (Hustler and Games 1987). The usual hunting method, day or night, is Standing in shallow water or on floating vegetation, grassy shores, or coral. It Stands for long periods, seemingly immobile, waiting for prey to appear. Because of its size, this heron can wade well away from shore. Typically, it begins to hunt by adopting a very Upright posture, but the head is soon lowered and held horizontally and the neck crooked, ready for its thrust. Head Tilting is used infrequently, and it is possible that the bill may be used to attract fish. After long intervals of patient waiting, it may move to a new area by Walking Quickly or Hopping. When it flies to a new site, it often lands on the floating vegetation rather than in the open water. It also Walks slowly, in a very stealthy manner. It fishes along the coast on rising tides.

Fish are caught by a lunging Bill Thrust, catching fish deep in the water, normally spearing them with mandibles held slight apart so both run through the prey. Prey are brought to the surface and shaken off the bill before being killed by restabbing, beating, and poking with the bill. Herons often place their prey on the tops of floating plants while they subdue it. Handling time is high and one-quarter of prey may then be lost, either escaping on their own or through piracy by other fish-predators. Fish Eagles (*Haliaeetus*) are frequent harriers of Goliath Herons (e.g. Abbott 1994), but also pelicans (*Pelecanus*) and Saddle-billed Stork (*Ephippiorynchus*). Herons defend their territories using Forward, Fluffed Neck, Upright, Chasing and Supplanting.

Food consists almost entirely of fish. It is a specialist in large fish, taking a few fish each day (2.3 in one study) and ignoring smaller ones. In Natal the average length of fish caught was 30 cm weighing 500–600 g, but one of 50 cm was too big to swallow. River bream (*Acanthopagnus*), mullet (*Mugil*), tilapia (*Sarotherodon*), and barbel (*Barbus*) are among the fish taken. It also will eat other things that come its way, including prawns, frogs, lizards, snakes, and small mammals.

Breeding

The breeding season generally coincides with the start of the rains. In some places, breeding is year-round with no discernible peak season. In other situations, such as South Africa, birds may not breed every year. The Goliath Heron uses a diversity of nesting habitat. The preferred choice appears to be on islands or on islands of vegetation protected by surrounding waters, as suggested by the observation that birds abandon nest sites when an island becomes attached to the mainland. Colonies are usually located in lakes and other water bodies. Solitary pairs may nest on riverbanks, lakeshores, and small islands. It is a big bird and nests low in the vegetation, generally below 3 m. Nest sites include sedge, reeds (*Typha*), small trees, low bushes, mangroves, and cliffs. On islands any tree, shrub, stone, or bare ground available may be used (Geldenhuys 1984). Nesting dispersion is variable. They nest solitarily, near colonies, within loose single-species colonies, and within mixed-species colonies along with other herons, cormorants, darters, ibis, and gulls (e.g. Dean 1988, Kopij 1996a). Several nests may be placed in one tree, and nests in adjacent tress may be less than 20–30 m apart (Brown *et al.* 1982). In some areas almost 80% of birds nest in colonies (Geldenhuys 1984), and there is evidence that nesting success may be greater for birds nesting in colonies than solitarily (Dean 1988). The former view of this species as fundamentally a solitary nester seems not to be correct.

The nest is a large platform made of sticks or reed stems at least 1–1.5 m in diameter. The nest can be flimsy when materials are limited (Dean 1988). One adult builds the nest alone or both birds of the pair build. The courtship displays of this heron are largely unknown. The rarity of observations suggests that courtship may be subdued, perhaps owing to re-pairing of previously paired birds. The Stretch display has been recorded and possibly a form of Forward (Brown *et al.* 1982).

Eggs are pale blue, 64.8–80 mm × 49–58.9 mm. They weigh about 108 g (92–117 g), fresh. The usual clutch is 3 or 4, ranging from 2 to 5. Incubation begins with the first egg and is by both parents, leading to a size difference among young (Cooper and Marshall 1970). Incubation takes 24–30 days. A replacement clutch has been recorded. Many eggs disappear. Hatching success has been reported to be low, about 25% (Dean 1988). The young are fed by regurgitation on to the nest floor, only twice daily. After hatching, older young can trample younger siblings, leading to brood reduction. Within a few weeks they can scramble out of the nest and defend themselves using Bill Jabbing and size-enhancing postures. At about 5 weeks they leave the nest completely. The parents attend them for a variable time, noted to be from about 40 to 80 days. Survival from hatching to leaving the nest was about 62% (Dean 1988). The White-tailed Eagle is a predator in colonies (Ytreberg 1992). Production is 1 or 2 young per successful nest (Brown *et al.* 1982).

Population dynamics

Nothing is known about the population dynamics and demography of this species. It is likely long lived, with high adult survivorship.

Conservation

In Africa nest sites, particularly colonies, need to be protected. Island colony sites should probably be the focus of protection and management (Dean 1988). Feeding sites also need to be protected and managed. More critical to conservation is determining the status of this species as a nesting bird in south Iraq/Iran and the Indian subcontinent. Birds still occur in these areas, and breeding remains likely. Should birds be found, immediate conservation action needs to be undertaken to protect the birds, their nest site, and feeding habitat.

Research needs

Much is known about the feeding biology and general nesting biology of this species. However, detailed information on courtship behaviour, pair formation, the identity and history of paired birds, nesting success, survival, and dispersal is needed. Long-term banding studies are desirable. The status of the species in the Middle East and Indian subcontinent needs to be determined. Surveys are needed in the Middle East, Bangladesh, and east India to determine the distribution, population status, important sites, and nesting situation of this species.

Overview

The Goliath Heron is a unique heron due to its large size and its dependency on very large fish. It has a suite of adaptations to accommodate this role. Its large bill can handle large fish. It legs, relatively long even for its great size, allow it to access deeper water where the large fish are. Its solitary habits likely are required to access this scarce resource. Catching large fish is neither frequent nor efficient, and handling becomes even more problematical. However, its low-key foraging technique conserves energy. It nests on or near the ground or low vegetation, probably by necessity of its size, and this requires protected sites. It is a very aquatic heron, sparsely scattered in the landscape, selecting shallow, protected waters having large fish.

Purple Heron *Ardea purpurea* Linnaeus

PLATES 3.1, 3.2, 14.2

Other names: Garza Imperial in Spanish; Héron pourpré in French; Purpurreiher in German; Purperreiger in Dutch; Purpurhager in Swedish; РЬґжая цаля in Russian; Cangak merah in Indonesian; Kandangaho (Pilipino), Tagak in Tagalog in (Philippines); Murasaki-sagi in Japanese; Cao lu in Chinese.

Ardea pupurea Linnaeus, 1766. Syst. Nat. ed. 12, p. 236: 'in Oriente', restricted by Stresemann, 1920, to France = 'River Danube'.

Subspecies: *Ardea purpurea manilensis* Meyen, 1834: Philippines; *Ardea purpurea madagascariensis* van Oort, 1910: Madagascar; *Ardea purpurea bournei* Naurois, 1966, Oiseau 36, 89: S. Domingos, Ilha de Sao Tiago, Capre Verde Archipelago.

Description

The Purple Heron a very elongated, narrow-bodied medium-sized heron, with a long thin head and bill and long neck. It is a dark grey heron, with orange-chestnut head and neck, and chestnut breast.

ADULT: The crown is black sporting two black lanceolate plumes up to 15 cm long. The sides of the head and neck are distinctively chestnut to orange-buff to red-buff. A black stripe runs across the ear to the plumes. The long bill is yellow with a horn brown top and tip. The irises are yellow. The lores are yellow-green or dull green. The head and neck together are rather snaky. The chin and foreneck are white. Throat striping consists of elongated black and white spotting. A distinctive black streak runs from the eye down the side of the neck. The back and upper wings are slate grey. Plumes along the lower back are chestnut to buff with feathers, with each feather split into several elongated and plume-like tips. At rest the bird shows a deep red-chestnut 'shoulder' patch. The flight feathers are dark grey to black. In flight the leading edges of the wing are buff-chestnut but the underwings are dark. The chest feathers are elongated into lanceolate tips that are cream-white with heavy black streaky spots that merge into the black striped chestnut breast. The belly is black with chestnut stripes turning black toward the under tail. The legs and feet are brown in font and yellow behind. Toes are elongated.

In the breeding season the long scapular plumes develop chestnut-buff tips. During courtship, soft parts redden, becoming orange to red.

VARIATION: Sexes differ. Males are larger (Boev 1987b), heavier, and darker, tending to slate grey with purple gloss. The scapulars and mantle plume tips are cinnamon-chestnut. Females are olive-grey with buff scapular and mantle plume. Juveniles are smaller than adults. This widespread Old World species shows differentiation in base coloration and throat striping, east–west and on islands (de Naurois 1966, 1988b, Payne 1979, Voisin 1991). *Manilensis* is paler, greyer above, and darker on the underparts, with black throat streaks more broken or absent, chest plumes whiter. *Madagascariensis* is darker with less obvious streaking than *purpurea*. *Bournei* is paler with the orange and purple of the nect being replaced by white and light rufous. The black head and neck streaks remain conspicuous and there are no size differences with the mainland African form.

JUVENILE: Juvenile plumage is lighter, mainly brown above with buff edges to the feathers, more uniform buff underparts, and dark brown streaked breast. Juveniles have a black crown, with a short crest but no long crown feathers or elongated scapular or mantle feathers. Stripes on the side of the head and neck are lacking. The bill and legs are dull brown, with less yellow than the adult. By first autumn, short ornamental scapulars and a few elongated breast feathers develop.

CHICK: Down is sparse and hair-like, rufous brown to dark brown above and white below. The upper down is tipped in white, particularly on the erectile crown feathers. The green skin shows through the down. Considerable areas on the head, neck, back, and belly are bare. The irises are green or yellow. The upper bill is olive green with a tinge of yellow and the lower bill is yellow with a tinge of olive green (Voisin 1991). There appears to be individual and perhaps geographical variation among Purple Heron downy chicks.

VOICE: The 'Frank' call is given in flight, and is higher pitched than that of the Grey Heron. 'Quarr' is the alarm call. A high pitched 'Quawk' call is given with the Forward display. A loud 'Craak' call is used in nest relief and upward part of the Stretch. In the lower part of Stretch a repeated 'Clack' call or soft 'Craak' call is given. Bill Clattering occurs. Food begging is a continuous 'chik, chick, chick' or 'ko, ko, ko, ko'. Young greet parents with a repeated 'Grau-rau'.

WEIGHTS AND MEASUREMENTS: Weight: 525–1345 g; length: 78–90 cm.

Field characters

The Purple Heron is identified by its long bill and neck, narrow body and wings, orange-red head and neck, black plumed cap, and chestnut belly. It is more often seen in flight than on land, flying to and from night roosts, breeding sites and feeding areas. In flight, it appears to be a dark, medium-sized heron, with long feet extending considerably beyond its short tail, and the long neck drooping below the horizontal (Lansdown 1985, Ullman 1985).

It is distinguished from the Grey Heron by being slimmer, smaller, darker, and in flight further distinguished by its dark wings, feet extended, kinked drooping neck, and light wing beats, the body appearing to lift on each down stroke. It is distinguished from the Eurasian Bittern by being larger, longer, slimmer, and thinner necked. It is distinguished from the Goliath Heron by its much smaller size, darker base colour, orange rather than chestnut head and neck, and black crown and crest. It is distinguished from the Black-headed Heron by its orange-red neck and dark upper parts. The juvenile Purple Heron is distinguished from juvenile Grey, Goliath, and Black-necked Herons by being browner, having a dark crown, chestnut neck, and little wing contrast. Melanistic Grey Herons might be most confusing, but the crown, grey wing coverts, and underwing colour can be used to separate them from the Purple Heron.

Systematics

The Purple Heron is clearly an *Ardea*, one specialized for life in reed beds. Natural hybridization has been described between Purple and Grey Herons (Campos 1990, Fenyvesi 1992). Described differences among the subspecies are primarily minor plumage colour distinctions. Given the high degree of plumage and size variation within populations and between sexes and ages, additional study is needed to reassess patterns of geographic variation and assign consistent subspecific ranks, or perhaps specific rank in the case of *bournei* (de Naurois 1988b).

Range and status

The Purple Heron occurs in temperate and tropical Europe, Africa, Asia and its islands.

BREEDING RANGE: The subspecies *pupurea* is the west Palearctic form, breeding from Netherlands and France, through Germany, Austria, Romania, Ukraine, Russia, to Kazakhstan, south through the Mediterranean to Turkey, Israel, Iraq, and Iran. In Africa, it breeds in north Africa in Morocco and

Algeria, in west Africa in Senegal, Mali, Uganda, south Angola, in east and central Africa in Somalia, Kenya, Tanzania (Baker and Baker 2004), north Namibia, Botswana, Zambia, Malawi, Mozambique, and Zimbabwe, and in South Africa as far as Cape Province. *Bournei* breeds on Santiago Island in the Cape Verde Islands. There is a distribution gap between east Europe and Pakistan. *Manilensis* breeds south of the Himalayas from Hindu Kush in Pakistan, India, Sri Lanka, Nepal, Bangladesh to eastern China (Kwamgtung to Shensi, Szechwan, Yunnan, Hainan and Taiwan) and Russia (Ussuriland). It occurs south through Myanmar, Vietnam, Malaysia, Indonesia, and Philippines. *Madagascariensis* is confined to Madagascar and the Seychelle Islands.

NON-BREEDING RANGE: The west Palearctic population winters occasionally within its breeding range, in extreme south Europe and the Middle East (Bahrain), and in north Africa, but most winter in Africa south of the Sahara from west Africa perhaps south to west central Congo to Sudan. The Mesopotanian marshes of Iraq and Iran have long been wintering grounds (Perennou *et al.* 2000). Northernmost populations of the eastern race, *malinensis*, also migrate joining non-migratory populations in southeast China, Korea, Taiwan, Japan (Ryuku Islands), Pakistan, Thailand, Malaysia, and Indonesia to the Lesser Sunda Islands.

MIGRATION: After nesting, northern population (*pupurea*) birds, especially juveniles, disperse from the colony site. The west Palearctic population migrates south as early as July continuing through October (Voisin 1991). In east Europe movement from colonies also begins in July and migration is completed by early September (Knysh and Sypko 1997). Birds migrate along a broad front. Extreme western

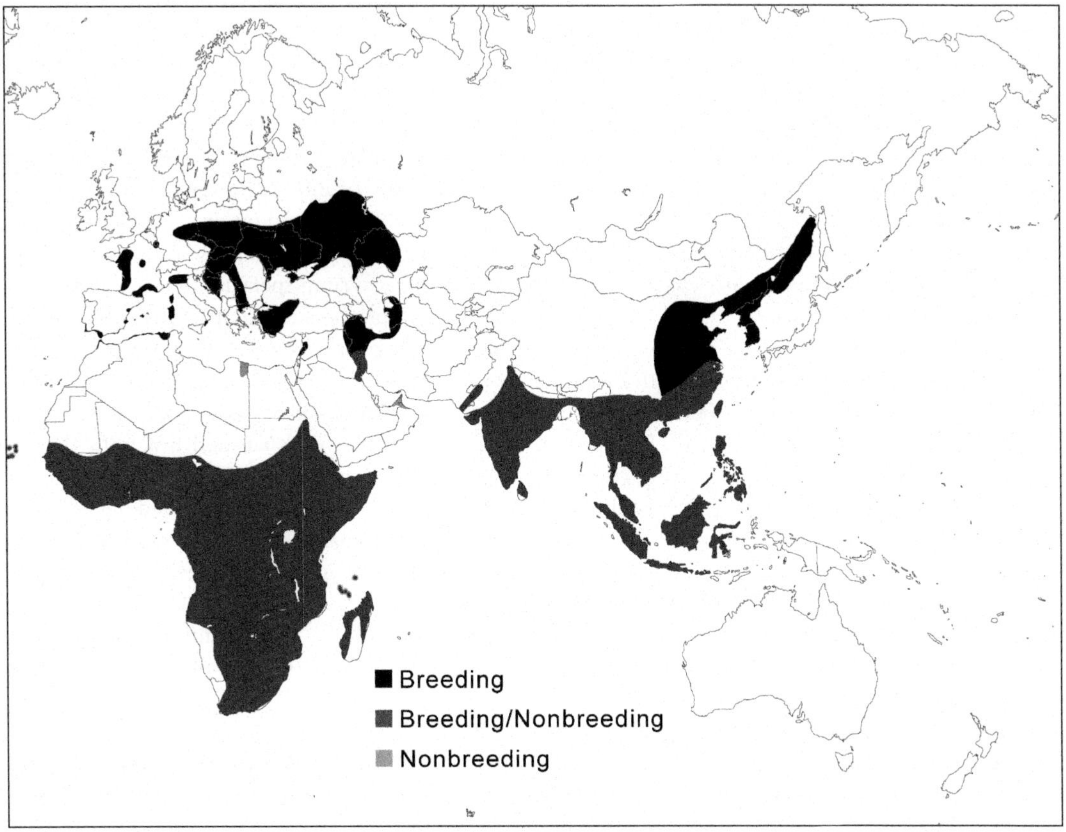

European birds follow the Atlantic coast to Spain while other western birds move to Italy (Voisin 1996). From there, they cross the Mediterranean by the shortest routes. Once in North Africa both groups follow the coast west and south to west Africa. Eastern birds move along Greece and Turkey through Egypt and Eritrea. In the spring, western birds fly to the Niger River and then cross the deserts to reach south Europe directly. Migratory birds from Russia and north China move south to Korea, Thailand, and Malayasia (McClure 1974). Most African and southeast Asia breeders are sedentary. Migration generally occurs by day in small groups, but up to 350–400 birds have been reported in Turkey. Return migration to east Europe is in the first half of April (Knysh and Sypko 1997). Some *malinensis* that breed north of the Yangtze River move across Korea to Japan (Voisin 1991). *Bournei* is not migratory. *Madagascariensis* appears to be sedentary or to undertake local movements within Madagascar; however, the possibility of migration of individuals to Africa should be examined (Voisin 1991). Birds regularly occur outside the breeding range, often in spring overshooting on migration. Birds occur in Britain, south Scandinavia, Atlantic islands from Iceland (Petersen 1985) to Canaries, and across the Atlantic to Fernando de Noronha and Brazil (Teixeira *et al.* 1987, Nacinovic and Teixeira 1989). In Asia, birds occur in central Siberia, Japan, and Korea. In Britain, visitation is an annual phenomenon, in some years exceeding 30 birds (Fraser *et al.* 1999).

STATUS: The population in Europe is relatively large, 49 000 to 105 000 pairs, but only 9000–14 000 pairs are outside Russia (Marion *et al.* 2000). The species increased its population and expanded its range in north Europe during the last century, especially into Germany and the Netherlands. However, a downward numerical trend has occurred since the 1970s across west and east Europe, with a few countries as exceptions (Tucker and Heath 1994, Marion *et al.* 2000). The Dutch population is now isolated. In Russia expansion in the 1970s proved short-term (Knysh and Sypko 1997). The Spanish population in the Ebro Delta dropped from 1000 pairs in the 1970s to about 60 in the early 1980s recovering to 400 by the 1990s (Gonzalez-Martin *et al.* 1992). The recovery in Spain was continuing into the 1990s (Bergerandi *et al.* 1995). In Mediterranean France, a decline from the early 1980s was reversing in the mid 1990s (Deerenberg and Hafner 1999). There are few population data for Asia and Africa. The population in the Cape Verde Islands has decreased to as few as 20 breeding pairs from an estimated 75 pairs (not 200 as early reported) breeding in a single tree (Summers-Smith 1984, Hazevoet 1992, T. Dodman pers. comm.). The population of Niger is estimated to be 1475 (1994–7) (Brouwer and Mullie 2001). A population estimate for Tanzania is 5000–10 000 birds, but it is a difficult bird to count (N. Baker and E. Baker 2004). It is common in scattered locations in Asia, 5000 pairs in Zhalong China in 1980, 1000 at Thale Nol in Thailand, and 4000 at L Tempe in Sulawesi, Indonesia.

Habitat

The Purple Heron is a reed swamp species. It prefers dense, emergent, freshwater, flooded reed or sedge beds. In Europe, low water level in the spring is the most important factor limiting occupation of a reed bed by Purple Herons (Barbraud *et al.* 2001b). It is seldom seen far from well-flooded marshes, and it particularly feeds in shallow water with a sandy or muddy bottom, among or adjacent to emergent reeds, and on or in floating vegetation. In some areas such as Europe, it may be considered a *Phragmites* swamp specialist, but it also uses other emergent plants, such as papyrus. Over its wide range its habitat needs are a bit more generalized. It also uses mangroves, rice fields, man-made ditches, canals, pools, lake shores, river edges, brackish water lagoons, coastal mud flats and hyacinths (Mukherjee and Parasharya 2000). Rice fields are used when the rice is tall. Outside the breeding season it is more often found in open habitats such as river and mud flats. It occurs from sea level to up to 1800 m. *Bournei* is rather exceptional among Purple Herons, using dry hillsides and nesting in trees, especially rubber and mango trees (Voisin 1991).

Foraging

The feeding ecology of the Purple Heron is fairly well known (Tomlinson 1974, Rodriguez and Canavate 1985, Fasola 1986, Moser 1986b,

Gonzalez-Martin *et al.* 1992, Fasola *et al.* 1993, Campos and Lekuona 1997, Knysh and Sypko 1997, Gruell and Ranner 1998, Martinez-Abrain 1998). The Purple Heron feeds during the day, particularly in the morning and evening. It feeds solitarily by Standing in water within dense emergent vegetation. It generally fishes in relatively deep water, up to its legs or even belly feathers. It assumes a horizontal posture and stares into the water for long periods with its bill kept close to the water surface. It also Walks slowly with a methodical gait, carefully placing its very long toes on the emergent or floating vegetation. Birds do on occasion use more active behaviours, e.g. Feet First Diving (Voisin 1991). It more rarely feeds in meadows and fields stalking rodents. It typically defends individual feeding territories, especially when energy demands are high during nesting (Singh and Roy 1995). In Spain where Purple Herons can choose between foraging along rivers or in rice fields, they were more efficient in the former (Campos and Lekuona 2002). Foraging success depends on patterns of prey availability within feeding sites. It strikes at fish, capturing them crossways in a scissor grip. Its feeding appears to depend heavily on cover and it rarely leaves the cover of the reeds. It feeds most often crepuscularly, resting during the day and night, travelling to roosts singly or in small groups. It assumes the Bittern Posture, using its camouflage plumage to good effect.

The Purple Heron feeds principally on small to medium fish (*Esox, Cyprinus, Tinca, Abramis, Scardinius, Perca, Anguilla, Acerina, Lota, Mugil*), although the overall range is 2–55 cm long and larger fish appear to be more profitable (Campos and Lekuona 2000). In the south of France the most important prey and their most frequent sizes were carp (*Cyprinus*) (2–5 cm), mullet (*Mugil*) (4–5 cm), and eels (*Anguilla*) (25–35 cm) (Moser 1984). Its morphology is well suited for a marsh-fish catcher, which accounts for the preponderance of fish in its diet (by biomass) (Campos and Lekuona 1997). It secondarily eats invertebrates, including insects (beetles, dragonflies, bugs), spiders, crustaceans (*Varuna*), and molluscs. It also eats frogs and salamanders (*Pleurodeles, Pelobates, Xenopus*), lizards, snakes (*Natrix*), small and juvenile birds (*Fulica, Tachybaptus, Rallus, Anas*), and mammals. A dietary shift from the predominance of fish (by biomass) to insects has occurred in southern France over a 20-year period (Barbraud *et al.* 2001a). The causes are unclear but may include changes in fish availability due to competition with the increasing Grey Heron population.

Breeding

The Purple Heron is a widespread species, whose breeding has been studied in several areas (e.g. Kral and Figala 1966, Tomlinson 1974, 1975, Moser 1984, 1986a, van der Kooij 1991, Gonzalez-Martin *et al.* 1992, Berthelot and Navizet 1993, Lekuona, J. M. and Campos, F. 1996b, Thomas *et al.* 1999). Its basic breeding biology is known, although much remains to be understood. In Europe and north Africa, breeding is mainly in spring, peaking in April–May, starting a month earlier in the south than the north. In north Asia, breeding is in summer, June–October. Further south breeding depends more on the rainy season cycle. For example, it breeds November–March in south Asia to east Java, but February–August in west Java. In Africa, breeding is mostly in the rainy season but in some places in the dry season (e.g. peaking in April–May in east Africa and September–October in South Africa). In western Europe, the time of nesting has shifted from the third week in April to early May from the late 1970s to late 1990s, suggesting that birds need more time to initiate nesting, perhaps due to habitat degradation (Barbaud *et al.* 2001a).

The nest is typically over water in dense marshes and reed swamps, *Phragmites* swamps in temperate areas but also other species (*Typha, Scirpus, Papyrus*). It nests preferentially in tall reeds, so it uses reed beds that are at least a year old. When reeds are not available and in other areas, it nests more frequently in trees, bushes, or thickets such as in Asia and Africa, and also in Netherlands and France. *Bournei* in Cape Verde nests in tall trees. Nesting is often in small, loose groups, especially in temperate reed swamps. But it also nests solitarily or in medium to large colonies of up to 1000 nests in tropical areas. It nests in both single-species and mixed-species groups, often on the periphery of Grey Heron colonies (Knysh and Sypko 1997). Colony size depends on marsh area, particularly in reed beds of fewer than 30–40 ha (Moser 1984, Broyer *et al.* 1998, Barbraud *et al.* 2001b). Nests are made of reed stems or sticks. The base is constructed by bending over the reeds to form a platform onto which the sticks or other reeds are firmly positioned. Nests are about 36 cm wide (15–76 cm) and 18 cm thick and are added to throughout incubation. In Europe, the Purple Heron nests low, lower than expected on the basis of its size (Fasola and Alieri 1992a), 1–3 m above the water. When they nest in thickets or mangroves, nests are higher, 3–4 m. When nesting in trees, they can be up to 25 m high.

The male chooses the nest site and displays there. Aerial chases and Circle Flights are not uncommon in the early stages of courtship. These can result in fights taking place in the air. Courtship displays are not well known in this species due to the difficulty in observing them. The Fluffed Neck, Upright, and Forward displays with the Quawk call are used as antagonistic displays. The Stretch display, during which the gular region is distinctively puffed out, is a greeting display in this species. In this, the Purple Heron differs from other herons in which the Stretch is primarily a male courtship advertising display. During the upward part of the Stretch the Craak call is given and in the lower part the bird gives a repeated Clack call. Several Greeting Displays are used. In one the returning bird utters the Craak call as it lands away from the nest. The brooding bird turns away from the approaching bird, an unusual response among herons, and gives a Stretch display. The incoming bird stands upright with neck feathers erect. Another Greeting Display used in nest relief consists of the brooding bird bowing its head with tail up, followed by the approaching bird lowering its head. A more complex elaboration of this Greeting Display, not seen in other species, has been described as a Sway and Bob display (Voisin 1991, Tomlinson 1974). Upon being approached, the brooding bird bent forward, head down, tail up, swaying from side to side, and then rapidly bobbed tail and head up and down, before returning to swaying. Allopreening with Bill Clappering and Back Biting are common between the pairs. Upright and Bittern Postures are used upon disturbance.

The eggs are pale blue-green. In Spain they are 55.2 × 40.4 mm; in South Africa 55.4 × 39.5 mm; and in Asia 54.6 × 39.7 mm. Eggs are laid at 1–3 day intervals. The clutch size is 2–8, varying regionally (Moser 1986a, Gonzalez-Martin *et al.* 1992), for example 5.3 in Hungary, 5.1 in Spain and France, 5.7 in Russia (Knysh and Sypko 1997), 3.2 in Zimbabwe, 2.9 in Botswana, and 2.5 in South Africa. Usually a single clutch is produced per year, but replacement clutches can occur (van der Kooij 1997). Eggs are incubated 25–27 days by both parents beginning with the first egg. The chicks hatch asynchronously. Over 95% of eggs in a Russian study hatched (Knysh and Sypko 1997). Nestlings are attended and fed by both parents, at first on food regurgitated onto the floor of the nest, and later on whole items taken from the parent's bill.

Sibling rivalry can be intense and later chicks (5 and usually 4) do not survive. The young begin to clamber from the nest at 7–10 days. The first and second chicks grow more rapidly than subsequent chicks in the brood. In a study in south France, the fourth chick at first grew slowly before speeding up, after its older siblings were no longer in the nest. At 3 weeks they spend most of their time out of the nest. At least another 3 weeks are required before they are fully fledged, and they become independent 2 weeks later. Many predators eat chicks, including otter (*Aonyx*), harriers (*Circus*), crakes (*Limnocorax*), and monitor lizards. Nests can be lost to strong winds (van der Kooij 1997) and rearing also fails if water levels drop rapidly at the colony site, even if it does not dry completely. Nests can be flooded by water levels rising too high. In Africa nesting success was 27% of eggs laid. Survivorship was greater in Spain. Survival of an egg to 16 days was 68%, and survival of a hatched chick to 16 days was 99%. In Russia broods averaged 5.36 young per nest (Knysh and Sypko 1997).

Population dynamics

Survival in migrating populations is affected by conditions on the wintering grounds (Held 1981,

Cavé 1983). Variation in the breeding population size is affected by winter survival of older than first-year birds related to the severity of drought conditions in west Africa. Migration deaths also occur (Nikolaus 1983), and it is possible that most mortality of first-year birds may occur before they reach the wintering grounds (Cavé 1983). In one study, over 60% of birds found dead were in their first year. Recent analyses have shown that both conditions on the breeding ground and conditions on the wintering grounds (rainfall in the Sahel) affect the size of the breeding population (Barbraud and Hafner 2001). The oldest known wild Purple Heron was over 23 years old.

Conservation

Purple Herons are scarce and localized through west Europe because their reed bed habitat is scarce (Hafner 2000). The population decline there has been underway for decades, as reed beds have been lost to development, water management practices, cane harvest (Jongejan 1986, Thomas *et al.* 1999), or elimination for other purposes, e.g. to reduce sedimentation (Broyer *et al.* 1998). This is a bird that is sensitive to disturbance and highly dependent on a specialized habitat and food resource. As a result, it is susceptible to many sorts of feeding and nesting habitat alteration working in combination (Deerenberg and Hafner 1999). Because of this, it is considered to be regionally vulnerable in Europe and north Africa (Hafner *et al.* 2000). Reed bed management for conservation is essential to the well-being of the Purple Heron in west Europe. Unfortunately, commercial reed harvest continues to be detrimental to Purple Herons despite European subsidies aimed at protecting heron colonies (Barbraud and Mathevet 2000). Overall Purple Heron conservation is favoured by maintaining large uncut reed beds with relatively high spring water levels (Barbraud *et al.* 2001b). Purple Herons also have a history of being killed by humans, either by shooting or by fences and other obstructions. The species has been protected since 1975 in France and 1981 in Spain, an essential aspect of its population recovery during that period (Voisin 1991). Population monitoring needs to continue, additional information is needed on habitat requirements and management, and reed beds need to be monitored and protected. This is a population potentially at risk because of conditions on the wintering ground, about which little is known.

The population of Purple Herons on Cape Verde may be down to 20 pairs. This isolated population is globally threatened and requires immediate conservation action (Hazevoet 1992, Hafner *et al.* 2000). This seems to be a Purple Heron of a different sort, apparently independent from wetlands and nesting in a few tall trees. There is no information on habitat use and limiting factors for these birds for most of the year. The reasons for the population decrease and needed conservation measures are not clear. But the current colony sites need to be protected, a survey undertaken of other possible sites, and a conservation plan for this population needs to be developed immediately.

The species is widespread and common over the rest of its range and is not of immediate conservation concern there.

Research needs

The most critical research need is to better understand the ecology of Purple Herons, particularly the relationships of reed bed characteristics to heron use and reproductive success. Information is needed on the distribution, habitat use, winter mortality, and environmental limiting factors in the wintering range on European birds, in sub-Saharan Africa. Study of the geographic variation within the species is desirable, both because of its widespread and disjunct populations and also as a basis for conservation action. Morphological and biochemical study is needed across its range, especially for the threatened population on Cape Verde and the island population of Madagascar. The biology of this species, although well studied, may hold additional surprises. The suggestion that competition for food during nesting with increasing Grey Heron populations has affected population sizes needs to be studied further. The reported high variation in weight, possibly based on food intake (Voisin 1991), suggests profound adaptations for variable food supplies. Recent evidence on the role of spring water levels and further clarification of the relative roles of ecological roles on the breeding ground vs. wintering grounds also deserve further study.

Overview

The Purple Heron is a bird adapted to and dependent on a very specific habitat—dense reed beds. Its long toes, short tarsus, thin body and head, and long bill can be seen as adaptations for living in this habitat (Boev 1988a). It feeds by solitarily waiting patiently for prey or by Walking slowly over and through the reeds. It is also a diet specialist and is among the most morphologically specialized of European herons for fish eating (Boev 1989). Together, these characteristics indicate that it has among the narrowest of ecological niches of the typical herons. Although still an abundant species through much of its range, the Purple Heron's specialization has rendered it vulnerable to habitat alteration, especially drainage and commercial reed harvest. It has responded by population declines, shifts in colony size and locations, changes in prey consumption, and shifts in its annual phenology. It is also affected by hydrological conditions on its wintering grounds, which can affect population size in the following breeding season.

Cattle Egret *Ardea ibis* Linnaeus

PLATES 7.1–7.3, 14.3, 14.4

Other names: Buff-backed Heron, Cow Crane, Cow Bird, Cow Heron, Elephant Bird, Tick Bird in English; Garcilla Bueyera, Garza del Ganado, Garza Ganadera in Spanish; Garça vaqueira in Portuguese; Héron garde-boeufs in French; Kuhreiher in German; Koereiger in Dutch; Kohäger in Swedish; Египетская цапля in Russian; Kuntul kerbau in Indonesian, Talud in Pilipino, Taga kalbao in Tagalog (Philippines) Ama-sagi in Japanese Da Bailu in Chinese.

Ardea ibis Linnaeus 1758. Syst. Nat. ed. 10, 144: Egypt.

Subspecies: *Ardea ibis coromanda* (Boddaert), 1783: Coromandel.

Description

The Cattle Egret is a medium-sized short-legged, white egret with a relatively short, yellow bill, usually seen feeding in flocks in grasslands and pastures following cattle or perched in a haunched posture.

ADULT: The Cattle Egret is a basically white bird. The bill, lores, and irises are yellow. Some nonbreeding birds have buff back or chest plumes, which some individuals tend to keep after nesting is completed. Legs are green to black, light near the leg feathering. In the breeding season they develop long rufous buff plumes on the crown, back, and lower neck (Siegfried 1971c, 1971d). The extent of these plumes differs between subspecies. In courtship, the bill, legs, and irises become bright red and the lores turn purple-pink. At the peak of the courting period, the skin becomes a bright cobalt blue (Lancaster 1970, Maddock 1993). This colour is lost quickly after mating. Postcourtship, breeding birds have yellow-green legs and dark yellow irises. The bill becomes yellow, although often retaining a bit of red to orange colour at the bill base. After nest failure or breakdown of a first pair bond, courting colour is regained after a few days and the mate-attracting process is repeated (M. Maddock pers. comm.).

VARIATION: Males are larger than females in most measurements (Mugica *et al.* 1987) and have more intense breeding colours. The two subspecies differ in their breeding colours and morphology. *Ibis* has rufous breeding colour on it crown, chest, and lower back. *Coromanda* has deep gold to dark cinnamon colour more widespread, covering the entire head, neck, and chest and most of the back. It also has a slightly thicker, longer bill and more extensive feathering on the tibia. Individual variation in the amount of breeding colour is common, and unusual colour forms have also been reported (Willoughby 2001, Herkenrath 2002).

JUVENILE: Juveniles are similar to non-breeding adults. They may show a grey tinge to their plumage and have very dark to black legs. Maddock (1988b, 1989a, 1993) found that for *coromanda*, marked one-year-old birds returning to the nesting colony for the breeding season ranged in colour from all-white, except for a buff patch on the crown, through pale patchy buff colouring to full orange adult plumage. McKilligan (1985) had proposed a two-category colour classification (pale and orange) but Maddock (1989a) found that the colour range could be classified into four categories: white 40%, pale (weak patchy buff) 29%, full pale (full coverage of pale buff) 9%, and full colour (the same strong orange colouration as two-year and older adults) 22%. Birds in all categories were found to be nesting.

CHICK: Hatchlings have white down, with tan to blue-green skin. They have a yellow, slightly down-curved bill. Legs are variable, tan, horn, yellow-green or olive green, but are darker in front than behind. Iris is yellow. The down on the crown is bushy. The skin turns grey with age. The bill turns grey tipped then black by day 10, and diffuse yellow-grey on fledging (Maddock 1993, 1989b). They may have a grey tinge to their plumage and black to very dark legs. Some birds assume adult breeding plumage at 10 months and breed at one year, although others remain all white at one year (Maddock 1993).

VOICE: The 'Rick rack' call is a typical Cattle Egret call. The first syllable is louder and higher than the second. It is given throughout the year in many circumstances, used on alighting and in the nesting Greeting Ceremony. The 'Raa' call is the threat call, given in defending a perch or nest site. The 'Ow-roo' call is a call used by males in the Stretch. The related 'Rooo' call is sometimes given by the female in the Stretch. The 'Thonk' call is also a defensive call, but a soft muffled one given during courtship by the male to the approaching female at the nest site. The 'Ruk, rok' call is a quieter, more hoarse, more muffled version of the Rick Rack call given during courtship. The 'Kraah' call is the defence call, louder, longer, and more aggressive than the 'Raa' call, given while defending the nest or young. The 'Kok' call is the alarm call, given singly or in a series rendered 'kok, kok, kok, kok', often in response to the aggressive chicks. The 'Kaka' call is a soft chattering contact, reassurance call, rendered 'ka, ka, ka, ka', used in several situations by unmated males, a softer version by mated pairs, and a harsh version following the Rick Rack call in the Greeting Ceremony. Young beg with a repeated 'Zit' call, rendered 'zit zit zit'. Bill Snap is used in the Forward, which in the Cattle Egret is a variable advertising display. Wing flaps create a thudding sound during the Circle Flight.

WEIGHTS AND MEASUREMENTS: Weight: 270–510 g; length: 46–56 cm.

Field characters

The Cattle Egret is identified by its short stature (due to its relatively short upper legs), short, stout, yellow bill, relatively large head with a distinctive jowl, thick relatively short neck, dark legs and haunched perching posture, and in breeding by its red bill and legs and buff head, chest, and back plumes. It flies with shallow rapid wing beats usually in flocks, generally low to the ground, often in formation, neck back, and legs extending beyond the body. Upon landing it may spiral from on high, side-slipping to a landing. It is distinguished from the Great, Intermediate, Eastern Great Egrets and white Reddish Egret by its smaller size, shorter and thicker bill, shorter and thicker neck, and during breeding by its buff plumes. It is distinguished from the Little Egret, Snowy Egret and immature Little Blue Heron by its short, thick yellow bill, shorter and thicker neck, shorter upper legs, haunched posture and stocky appearance, dark green to black legs or yellow in breeding (but not black or light green), and during breeding buff (not white) back and head feathers. It is distinguished from the Eastern Reef Heron by its smaller size, yellow (not horn) bill, in breeding buff crown and back, flocking behaviour, and avoidance of marine habitats. In flight, it may be distinguished up close by its buff wash, but at a distance it looks white and its wide wings make it appear a larger bird than it is. It is distinguished from the Squacco Heron by its lighter back, yellow bill, thicker bill, white or buff (not streaked) head, and more vertical posture when feeding.

Systematics

The generic placement of the Cattle Egret has been uncertain, leading to its being assigned to the pond heron genus, *Ardeola*, and to its own genus, *Bubulcus*. The uncertainty was due to difficulty unravelling evolutionary relationships disguised by morphological adaptations to terrestrial foraging. Biochemical studies have revealed it to be an *Ardea* (Sheldon 1987). The relationship of the two currently recognized subspecies deserves additional study as their disjunct ranges and distinctive plumages, and perhaps body proportions, suggest they may be separate species. The taxonomic identity of birds of the Seychelles, recognized as *seychellara*, also needs re-examination based on an adequate sample size. It is possible that this population represents a mixture of an older influx of the Asiatic *coromanda*, followed by a more recent influx of *ibis*.

Range and status

The species now nests across the mid latitudes in North America and South America, Europe, Africa, Asia, and Australia.

Breeding Range: *Ibis* breeds in north Africa (Egypt, Tunisia, Algeria, Morocco (J. Franchimont pers. comm.), Cape Verde Islands). In sub-Saharan Africa it breeds in Mauritania, Senegal, Gambia, Guinea–Bissau, Mali, Ghana, Niger, Nigeria, Chad, Central African Republic, Cameroon, São Tomé and Principe Islands, Congo Brazzaville, DR Congo, Sudan, Ethiopia, Eritrea, Djibouti, and Somalia. Also from Kenya, Uganda, Rwanda, and Burundi south through all central and southern African countries except Zambia. In the western Indian Ocean it breeds on the Comoros, Aldabra, Seychelles, Madagascar, and Mauritius. In Europe and the Middle East it breeds in Spain, south west Portugal, France (Marion *et al.* 1993, Marion 2000), Italy (mainland, Sardinia—Grussau 1987), Romania, Netherlands (Courtens 1996), Greece (Goutner *et al.* 1991), Turkey, Israel, Saudi Arabia, Yemen (Al Safadi and Kasparek 1995), southwest Iran, Russia (Volga Delta), Azerbiajan, and Turkmenistan. In the New World it nests in Canada (Ontario, Saskatchewan), USA (breeding in all but six states), Mexico, West Indies, through Central America and South America except mountains to northwest Chile (Antofagasta), Argentina (Santa Fe, Buenos Aires) (Azevedo 1997, do Nascimento 1999), Hawaiian Islands (introduced, Paton *et al.* 1986, Ohashi and Kimizuka 1988), Galapagos Islands, and Falkland Islands. The Asiatic race *coromanda* is separated from *ibis* by a broad gap across

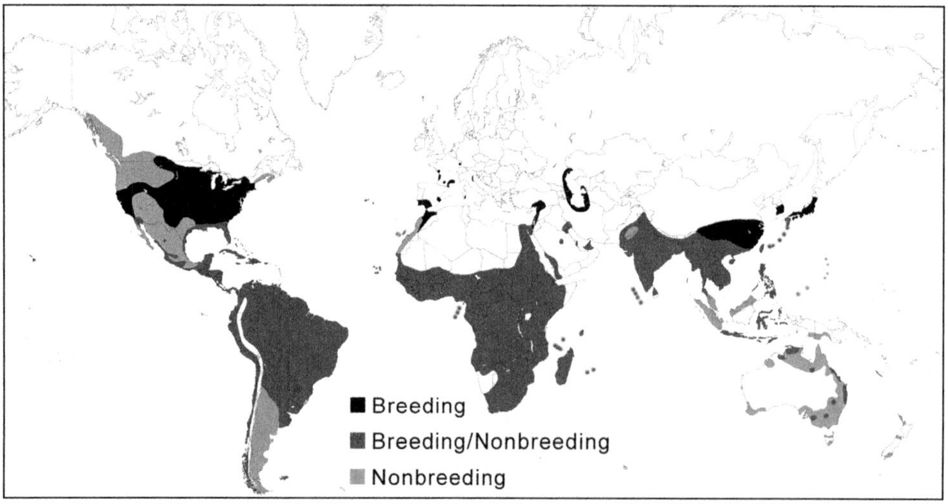

Iran. It breeds south of the Himalayas in Pakistan, India, Sri Lanka, Maldives, Nepal, Bangladesh, southeast China (mainland, Hong Kong, Hainan, Taiwan), South Korea, south Japan (including Ryukyu and Bonin Islands), throughout southeast Asia except peninsular Malaysia (R. Lansdown pers. comm.) in Cambodia, Laos, Vietnam, Myanmar, Thailand (Round 1994), Nicobar and Andaman Islands, Indonesia (Java, Celebes, Moluccas, Sundas, Kalimatan, Sumatra), Australia (Northern Territory and Kimberly, Western Australia) (Jaensch 1994), east Queensland and inland western and coastal New South Wales.

NON-BREEDING RANGE: Cattle Egrets are partially migratory in Europe, some birds remaining in southern France and southern Spain while others move to north Africa. Cattle Egrets occur in the non-breeding season throughout sub-Saharan Africa. Eastern European birds winter in the Middle East (Iraq, Saudi Arabia, Iran), possibly into east and central Africa. Cattle Egrets breeding in Africa shift in the non-breeding season in response to seasonal rainfall patterns, tending to move to the coast in the dry season (N. Baker and E. Baker in prep.). In Asia, Cattle Egrets occur in the non-breeding season in Pakistan, India, south China, Indo-China, Malaysia (peninsular, Sabah—Sheldon *et al.* 2001), Japan (Kyushu south), Philippines, Indonesia (Sumatra, Irian Jaya, Kalimatan), Papua-New Guinea, Micronesia (Carolines), Marinas (USA), and New Caledonia (France). It has been recorded in New Guinea (Papua and Irian Jaya) for 60 years, but no nesting has yet been documented (Beehler *et al.* 1986). In North America, Cattle Egrets winter in the USA south of 32° (with wintering concentrations in Florida, south California, Louisiana) into West Indies, Mexico, throughout the Central American range, into South America. In South America, they occur within the breeding range year-round, but migrate south of their breeding range to Patagonia and the Falklands. Some Australian birds winter outside the breeding range in southeast and northwest Australia, Victoria, Tasmania, New Zealand, Macquarie, Lord Howe, and Norfolk islands (Maddock and Bridgman 1992, Maddock 2000). In both South America and Australia, southward migration to Patagonia and to Tasmania and New Zealand, respectively, results in birds moving from a warmer climate to a cooler one for the winter.

MIGRATION: Many populations of Cattle Egrets are migratory and most show strong postbreeding dispersal, the two processes generally not being clearly separable. Cattle Egrets migrate in small flocks over extended periods mostly during high pressure conditions, avoiding rainy and foggy weather. They use flapping flight without soaring. Siegfried (1978) used an estimate of 50 km/h as the average speed of migratory flight to calculate a potential flight time of 20 hours for a flight range of 1000 km. This estimate was supported by the Maddock and Bridgman (1994) study which would result in a flight to New Zealand taking 42 hours. A satellite study (Bridgman and Maddock 1994) recorded a flight overnight during migration moving at an average ground speed of 43 km/h over land. The optimum altitude of flight has been estimated to be 1 km (Maddock and Bridgman 1992).

Migration in this species has been best studied in Australia. Bridgman and Maddock (1994) found that in eastern Australia southward migrations after the breeding season were carried out in stages during northerly airflow along the western edge of a high, with passage of fronts and shifts to southerly winds halting movement. Movement to New Zealand was possible via southwesterly airflow both from NSW or from Tasmania. Return migration northwards up the coast in spring also took place in stages, but was more direct, with fewer staging points and occurred during southerly airflow. For return from New Zealand, the most likely scenario was found to be around the north of an advancing high in southeast, or easterlies used as tailwinds near the margin of a high (Bridgman *et al.* 1998). At least some migration in Australia has been found to take place at night (Maddock 1990, Bridgman and Maddock 1994). Temperate birds tend to migrate before the local winter, usually southward irrespective of location, whereas more tropical populations of Africa, Asia, and North America are more sedentary but have nomadic and dispersal tendencies.

European Cattle Egrets are partially migratory; birds move south and to the coasts. French Cattle Egrets tend to remain in south France. Birds from

Portugal and Spain move south to the end of the peninsula, some crossing the Mediterranean to Morocco and north Africa. Egrets nesting from Turkey to Russia move south to the Middle East and also to east Africa (Ethiopia).

In Africa, birds breeding in north Africa appear to be mostly sedentary, while west African birds seem to be either migratory or nomadic, moving in response to rainfall. In southern Africa, birds move south to breed during the summer rains (November–February), moving north again to central Africa during the southern winter (April–August) (D. Turner pers. comm.).

In the New World, northern populations migrate, shifting towards a southward movement after postbreeding dispersal, likely in response to the first frosts of the year (Byrd 1978, Telfair 1994). Migration is prolonged, September–November and later. The return is in February–May. Western North American populations migrate to south California and west central Mexico. Central continent populations migrate through Texas, some remaining in coastal Texas, west to Baja, California, central Pacific and Gulf coasts of Mexico, to Panama. Eastern North American Cattle Egrets move south to the Gulf of Mexico Coast and to Florida, passing around the Gulf to join Central Flyway birds, wintering in Florida, passing through Florida to the Greater Antilles and north South America, or passing trans-Gulf from Florida to Mexico, Central America into Colombia. The establishment of a strong migratory tradition by this invader is intriguing. It seems likely that the West Indies–South America to North America route developed by following other migrating herons (Arendt 1988).

In Asia, northern *coromanda* are migratory, while southern populations appear to be more sedentary. As in West Africa, migrations in India correspond to the rain and dry season. Japanese egrets move to the Philippines. South China egrets move to Japan, Philippines, Indonesia (Borneo), and Micronesia (Carolines). Western egrets move to and through Malaysia (Malay Peninsula), Indonesia (Sumatra, Irian Jaya), Papua New Guinea, Micronesia (Carolines), Marinas (USA), and New Caledonia (France).

In Australia, the Cattle Egret is partially migratory, but is likely better considered to be migratory over short distances (Baker 1978, Maddock 1990, McKilligan *et al.* 1993, Maddock and Geering 1993, 1994). Migratory movements may take place in any compass direction but the majority of birds nesting in colonies in central and north coastal Australia move south and, to a lesser degree, north along the coast, although some have been recorded moving inland. Movements vary from short shifts to migrations of 2000 km (Maddock 2000). Egrets move in stages to Victoria, Tasmania, and New Zealand in April–May, returning to colonies in coastal eastern Australia in October–November. Project Egret Watch records show that Cattle Egrets tagged in the Macquarie Marshes in inland western NSW have been seen alive in Victoria in the south and in northern Queensland during winter (M. Maddock pers. comm.). It is possible, but not proven, that northern Australian birds move northwards to New Guinea. Maddock and Geering (1994) have demonstrated faithfulness of individuals to specific wintering sites, of upto 5 years duration.

Among all herons, the Cattle Egret is the most highly accomplished in dispersal. It has occurred as a transient in many places around the world, crossing water gaps to islands, and being seen by ships at sea (Orgeira 1995, 1996, Anonymous 2002d). Individual dispersal is proven to cover long distances, 1900–5000 km in North America (Telfair 1983, 1994). It has been recorded in all Canadian provinces and all but two states in the USA (Telfair 1994). Off South America it occurs on Tierra del Fuego (Chebez and Gomez 1988) and also on South Shetland Islands (Moenke and Bick 1990) and Antarctica (Kaiser *et al.* 1988). In Europe, dispersal records include England, Ireland, Iceland, Belgium, Netherlands, Denmark, Sweden, Greece, Bulgaria, Rumania, Yugoslavia, and Malta. In Africa, South African juveniles have been recovered in Zambia, Zimbabwe, Tanzania, Congo, Uganda, and Central African Republic. Highly nomadic west African birds have reached the Azores, Madeira, Canary, Ascension, St Helena, Tristan da Cunha, and St Peter and St Paul Rocks. It is likely that one-way movements from Africa to South America continue by this route, propelled by trade winds (Orgeira 1996). In the Pacific, it has occurred on Mariana,

Micronesia (Schipper 1985), Guam, and the Aleutians (Alaska, USA—Gibson *et al.* 1987a).

STATUS: This is a widespread and abundant species, undoubtedly the most abundant heron worldwide. Its range expansion during the twentieth century, throughout the Western Hemisphere has been nothing short of sensational. Range change in the Cattle Egret is one of the more intriguing heron stories of the last century (Franchimont 1986a, Arendt 1988, Maddock and Geering 1994, Telfair 1994). *Ibis* occurred historically in tropical and subtropical Africa. But starting in the last century, it began to expand its range. It reached South America in 1877–92, south Europe also by late nineteenth century, nesting proven in 1944, South Africa in the 1920s, the Caribbean in 1933, the USA in 1941 nesting in 1953, Canada in 1952 nesting by 1962, Mexico in 1956 nesting by 1963, and Central America in 1954 nesting by 1958. It was introduced in Hawaii in 1949. *Coromanda* also expanded its range. It occurred in Australia in the late 1800s but was first documented as nesting only in 1954. Once thought to be introduced, it in fact invaded Australia naturally (Maddock and Geering 1994, Maddock 2000). It occurred in New Guinea in 1941. From Australia, it reached New Zealand in 1963, although it is possible that it may have been as early as 1956 (Brown 1980). It has not been shown to nest there.

In North America it achieved a remarkable population expansion and is now the most abundant heron in North American, numbering over 400 000 birds in the east USA alone in the 1970s. Since the 1970s the rate of increase has declined and the population appears to be stabilizing (Telfair 1994, Telfair *et al.* 2000).

In Europe it is slowly expanding and is now the second most abundant heron with a population of 80 000–100 000 pairs, about 66% of which are in Spain and 29% are in Portugal (Marion *et al.* 2000). Winter surveys found 150 000–160 000 individuals in the Iberian Peninsula and expanding (Sarasa *et al.* 1993, Gustin *et al.* 2001). It is increasing in France, recently reaching about 4300 pairs (Marion *et al.* 2000). A sudden increase in nesting numbers and dispersion over most of France in 1992 corresponded to drought conditions in Spain (Marion *et al.* 1993), showing how dispersal events can trigger population expansion. Populations founded after this event persisted (Marion 1997, Rimbert 1997). The nesting population on the Cape Verde Islands appears to have died out (Summers-Smith 1984). There are a few pairs to dozens of pairs in Italy, Russia, Turkey, Romania, Netherlands (since 1998), and Greece. It is increasing in Sardinia.

It is common through most of Africa except in deserts and forests. Populations are increasing and large colonies are being found throughout the continent (Turner 2000), including 20 000 pairs in Morocco, 63 000–65 000 pairs in Mali (inner Niger Delta), 164,000 birds in Niger (1994–7) (Brouwer and Mullie 2001), 1200 pairs in Kenya (Tana River), 10 000 pairs in Tanzania (in the Wenbere Swamp alone) (Baker 1995, Baker and Baker 2004), 2000 pairs in Zimbabwe, and 1700 pairs in South Africa (Hafner 2000, Turner 2000). It is expanding its range in Algeria to inland sites and in Egypt following a decline in the 1970s and 1980s (El Din 1992). In December 1995, a winter roost of 260 000 birds was counted in the Central Niger Delta in Mali. It is common throughout Madagascar, one colony exceeding 1000 pairs. In the Middle East it is increasing in southwest Iran (Perennou *et al.* 1994, Hafner 2000). In Azerbiajan, 10 000– 15 000 pairs were counted (Perennou *et al.* 2000).

Little information is available from west Asia. Farther east, it is common and increasing in Pakistan (Sind, Punjab) (Roberts 1991). Over 9600 pairs were estimated recently in colonies in two areas in China (M. Fasola *et al.* 2004). It is the most abundant heron in India (Subramanya 1996), but it has declined in Japan where it was once abundant.

In Australasia, it is widespread and common wherever suitable pasture with grazing animals is available. The population expanded rapidly in the east of the continent from first recorded breeding in 1954. By 1978–79 there were five nesting colonies with about 2300 pairs in NSW (Morris 1979). By 1991, the number of coastal NSW colonies had expanded to 12, totalling more than 10 000 pairs (Baxter 1994). Project Egret Watch records suggest that there has been no further expansion since the 1989–90 breeding season (Maddock 1999). It is also common in New Zealand as a non-breeding species, peaking at 2200 birds in 1987 before declining.

Habitat

The feeding habitat used by this species is one of its defining characteristics. Throughout its range it forages in native grasslands and in pastures alongside hoofed livestock. In its area of origin, the African subspecies forages in the open grass on savannas and plains subject to annual inundation and slow drying, and associates with native grazing mammals (Siegfried 1978). Such rainfall-dependent grassland has been simulated on a massive scale by man-made pastures managed for livestock through the use of surface irrigation, ponds, tanks, wells, and canals. Nearly worldwide, during the twentieth century, as forest was cleared and improved to provide pasture for livestock, additional habitat was created for Cattle Egrets. It is primarily use of this habitat that allowed the egret's range expansion and consequent near circum-global population explosion. Changes in cattle ranching coinciding with population increases in the Cattle Egret are documented in Guyana, Florida, South Africa, and Australia (Siegfried 1966b, Voisin 1991, Maddock 2000, Turner 2000). Cattle Egrets do use aquatic habitats, and in fact favour damp to recently flooded or poorly drained pastures over dry pastures and surface-irrigated fields over dry fields (Mora 1992, Maddock 1995). They feed at water margins—riverbanks, stream edges, ponds, and shallow marshes—usually in association with other wading birds. They tend to avoid full marine habitats and deep forests. Of critical importance is their use of artificial sites beyond pastureland. These include irrgated crop fields, dumps, parks, athletic fields, golf courses, meadows, rice fields, lawns, and road margins (Franchimont 1986b, Mora 1992). They will form large flocks to follow opportunistically behind ploughs and cultivators at work (M. Maddock pers. comm.). Rice fields are particularly important throughout its range, including India, Sri Lanka, France, Spain, and Italy. In France, Cattle Egrets selectively foraged in rice fields and other agricultural habitats, which afforded them a greater intake than natural habitats (Lombardini 2001).

Cattle Egrets nest in colonies with other wading birds in marshes, reed beds, mangroves, dense thickets, bushes, and trees—usually located on islands over or surrounded by water. Overall the Cattle Egret can be expected to occur where grasslands are present, where temperatures are not limiting to insect activity, where insect abundance is high, and where isolated nesting locations are available with wetlands nearby. It occurs from sea level, to 1500 m in India and 4080 m in Peru (Fjeldsa and Krabbe 1990).

Foraging

The foraging and food of the Cattle Egret has been exhaustively studied throughout its range (Heatwole 1965, Siegfried 1966a, 1966c, 1969, 1971a, 1971c, 1972a, 1973, Fogarty and Hetrick 1973, Thompson *et al.* 1982, Bredin 1984, McKilligan 1984, Ruiz 1985, Newton 1986, Acosta *et al.* 1990, Metz *et al.* 1991, Yen 1991, Telfair 1994, Mathew *et al.* 1997, Kopij 1999a, Lekuona and Artazcoz 2000, Lombardini *et al.* 2001). The Cattle Egret specializes in Walking slowly on land. It characteristically follows moving cattle, native large mammals, birds, or tractors, capturing food that is made obvious by the movement of the beater. In Africa, its primary natural beater was probably the African Buffalo (*Syncerus*), but it also follows other grazing mammals, zebra, elephant, rhinoceros, giraffe, eland, waterbuck, or topi and other large species such as ostrich. Elsewhere, it typically follows cattle, but also uses deer, capybara, camels, horses, feral water buffalo, sheep, goats, chickens, geese (*Anseranas*), and cranes. Cattle Egrets prefer animals that move forward at the most appropriate speed for their own Walking foraging. When following farm, mowing, or land moving equipment, it uses Leap Frog Feeding, in which birds land at the tractor, remain in place foraging, and then fly to the moving head of the flock again. It also uses this behaviour without farm equipment, the flock serving as its own beater (Newton 1986). In commensal feeding, Cattle Egrets primarily hunt for small abundant insects and increase feeding effectiveness by obtaining more food with less effort than do birds feeding alone (Heatwole 1965, Thompson *et al.* 1982, Scott 1984). In commensal foraging, the behaviour used is to Walk slowly adjacent to the animal, usually toward its front quarter, observing prey made obvious by the beater's movements. It walks with its head alternately withdrawn and then pulled forward with each step, a gait characteristic of the species. Cattle Egrets will typically dash a few steps after a prey item, Stabbing, Gleaning off a plant, or Flycatching in flight as it flees the egret or its beater. Cattle Egrets will also stop briefly, Stand and Head Sway, which allows better binocular discernment of stationary prey, or Neck Sway, which probably subtly encourages prey to move. Head Swaying seems to be associated with the hunting hard-to-catch prey. Egrets will often ride on a large mammal's back and Peck, Standing Flycatch, or Hop down to a prey item. In West Africa, Cattle Egrets have been observed Picking flies off Sitatunga antelopes (*Tragelaphus*), also deep probing in the ears (oxpecker-like) (D. Turner pers. comm.).

Cattle Egrets use other supplementary behaviours as called for, including Walking Quickly, Running, Hopping, and Aerial Flycatching. They may use Probing and Pecking at food items and cases of Bill Vibrating have been reported. Egrets feed without beaters, especially in situations where mammals are not available or when stalking high-energy vertebrate prey (Scott 1984). Typically the egret Walks slowly in the grass or along wet margins. Neck Swaying is common during solo feeding, as is Head Swaying. Egrets are attracted to and follow fires, feeding on fleeing insects. They also will Stand, usually peering into water, and wait for a prey to come into view. They can feed arboreally, in trees and hedges where insects are concentrated, and feed on flies attracted to garbage or dead animals.

Cattle Egrets are highly social in feeding. They are seldom totally alone and usually feed in flocks or loose aggregations of dozens to a thousand birds. Flock feeding is so core to the species' behaviour that it is likely that it itself provides some commensal advantage. Birds observe and join other birds foraging, thereby creating aggregations in places where prey is available (Metz *et al.* 1991). In flock feeding, egrets may be aggressive in defence of feeding sites or feeding areas, such as defending its place by the side of a commensal animal or Stealing prey (Garrido *et al.* 2002). They use Supplanting Run with a stiff legged stride to displace a competitor. Raa calls may be given. Runs may become a Supplanting Flight, with Raa and Rick rack calls being given by the alighting bird. Other threat and appeasement behaviours used are the Upright, Forward (in which the aggressor erects plumes, lifts wings, retracts head, rocks forward, and stabs with Bill Snap), and Head Flick. The subordination display is a Withdrawn Crouch. They feed during the day, most actively in the morning and afternoon. During subordination midday and at other times when grazing stock rest to ruminate, foraging flocks often loaf with other birds in trees or on the ground near the resting herd. At night time they roost with other species, sometimes in the thousands. The birds travel from the foraging grounds to the night roost site in late afternoon, often flying low along a watercourse. They form a preroost gathering at a nearby location to loaf, preen, and drink (Blaker 1969a, Siegfried 1971b, Raine 1996, M. Maddock pers. comm.). In the last few minutes before dark, they fly to the roost trees or reeds (M. Maddock pers. comm.). Individual birds tend to take up the same roost perch for the duration of the period the roost is used (Siegfried 1971b, M. Maddock unpublished data). Along the migration routes and at wintering destinations large permanent night roost sites are used each year on both outward and return journeys. Smaller temporary roosts are also established at staging locations (M. Maddock pers. comm.).

DIET: Although having a diverse diet, the Cattle Egret is primarily an insectivore and is, moreover, a grasshopper specialist. Locust, grasshoppers, and crickets are the common element of its diet worldwide, and its specialized feeding behaviour seems designed to find and catch orthopterans. Other insects are eaten including flies (Tabanidae, Calliphoridae), beetles (adults and larvae), caterpillars, dragonflies, mayflies, and cicadas. Ticks are found in the diet but in small numbers; evidence for egrets picking ticks off their commensals is slim. Ticks are probably found on vegetation. It certainly is not a 'tick bird' in the sense of habitually gleaning ticks off cattle. Other prey include spiders, earthworms, molluscs, crayfish, frogs, tadpoles, lizards, snakes, fish, rats, and birds. Vertebrates, especially frogs, are important during late nesting to provide the high-energy packets and calcium needed for development. Chicks are selectively fed vertebrates, the proportion of which increases in the diet with chick age (Kopij 1999a, McConnell and McKilligan 1999). In Australia, advanced chicks were found to have been fed 16 species of reptiles and nine species of amphibians.

Breeding

The breeding biology and behaviour of the Cattle Egret have been well studied (Blaker 1969b, Lancaster 1970, Siegfried 1972b, Hudson *et al.* 1974, Telfair 1983, 1994, Fujioka 1984, 1985a, 1985b, Ploger and Mock 1986, Mock and Ploger 1987, McKilligan 1985, 1987, 1990b, 1996, 1997, Rodgers 1987, Arendt and Arendt 1988, Ranglack and Marion 1988, Maddock 1988a, Voisin 1991, Ranglack *et al.* 1991, McKilligan *et al.* 1993, Wen and Sun 1993, Rao *et al.* 1996, Kopij 1996b, 1999b, Si Bachir *et al.* 2000, Gassett *et al.* 2000, Parejo *et al.* 2001a, Denis *et al.* 2003, Telfar and Bister 2004). The nesting season is variable, depending on food availability. In the temperate north, it nests in spring and summer, April–May in North America, April–July in Europe, and June–August in north India. In southern Brazil nesting begins September–October (Petry and Hoffmann 2002). In the tropics, breeding generally takes place during or immediately following the rainy season as grasslands begin to dry out. Nesting periods are mainly March–November in west, central, and east Africa, September–March over much of southern Africa including Madagascar (D. Turner pers. comm.). The breeding season is November–February in south India, December– March in east Java, February–July in west Java, and September–February in southeast Australia. It nests in reed beds, bushes, shrubs, trees or on the ground in protected locations, usually over or surrounded by water. The nest is made of reeds, twigs, or branches; most are unlined. Characteristically Cattle Egrets often use leafy twigs. The nest is shallow, 40 cm wide (17–61 cm) and 12 cm deep (5.1–27.9 cm), with a pronounced cup. Nests are placed in reed beds, bushes, or low or tall trees up to 20 m high. When nesting in trees; the egrets tend to prefer higher sites, with nests heights decreasing for later nests (Si Bachir *et al.* 2000). Both sexes participate in the building; the male generally brings material, the female doing most of the constructing. Construction appears rather haphazard with many sticks not being successfully integrated into the structure. The nest construction takes 4–5 days, although the pair continues to add to its construction until egg hatching.

Nesting colonies are found on islands in fresh, brackish or salt water, wood lots over dry land, and in swamps and marshes. In the New World, they are attracted to colonies established first by other herons (Belzer and Lombardi 1989). There is not a clear relationship between colony location and environmental conditions around the colony site, probably because most sites are chosen first by aquatic herons. Conversely, Cattle Egret colonies appear to entice other herons to move more inland. Cattle Egrets are highly colonial, breeding in mixed-species colonies of a few hundred pairs to several thousand pairs, including other herons, storks, ibises, darters, and cormorants. They may be anywhere from the first to the last species to nest, depending upon geographical area. Their effect on other herons nesting with them has been a matter of some debate but evidence of meaningful adverse impact is limited. The vast majority of early arrivals back at the breeding colony after migration are two-year-old and older birds (Maddock and Geering 1994), with some one-year-old birds not returning until December, two months after commencement of nesting (M. Maddock pers. comm.). Project Egret Watch records of tagged birds demonstrate faithfulness to natal breeding colony during a bird's lifetime (up to 8 years, M. Maddock pers. comm.) with very few birds found to nest at a different colony (Maddock and Geering 1994).

Small groups of males gather to display together, each male claims its territory, becomes highly vocal and aggressive, and begins displaying. Territory claiming and defence from other males and initially by approaching females is by Standing, Twig Shaking, and Upright behaviour. The primary defensive behaviour, and also a courtship display, is the Forward, in which the plumes are partially raised, bill open, neck extended forward, and the male gives a Raa call, replaced by a Thonk call prior to pair formation. Forwards space out displaying birds. Supplanting Runs displace intruders. If aggressive displays are unheeded, they become the prelude to vicious fights made up of Bill Duels, Bill Jabbing, Face to Face Fighting, and Aerial Fighting.

Males display using several behaviours in addition to defensive ones. Head Shaking is accompanied by the Rick Rack call. Wing Spread is used after a Circle Flight, before a Forward, and while Walking. This is a frequent display, in which the wings are spread downward with rocking movements as if balancing and all plumes partially or fully erect. The Snap is complete with aubible bill snapping. The Stretch is not as obvious a display in Cattle Egrets as in other species. The bird stretches its head and neck vertically, legs partially flexed, body at about a 45° angle. The bird then retracts its neck, flexes legs, raises its scapular plumes, and gives the Ow-roo call. The ow is at the top of the stretch movement; the roo is on the downward movement. Display birds take Circle Flights making loud thudding sounds with their wings and also give Rick Rack calls. The Twig Shake in this species seems to have more of an advertising function than is usually the case for herons.

Females encroach on the male's territory, usually from the rear, being unable to withstand the male's defensive behaviour. Females are attracted to the displaying males, spending time nearby, adopting a characteristic extended necked peering posture with crest partially erected. Once they persist for a few hours, the bond forms quickly. After pairing, birds use several reinforcing displays. Bill Clappering and Back Biting are the primary reinforcing behaviours. Females give a Stretch using the Roo call. Males Twig Shake. In the Greeting Ceremony, the back plumes are fully erected and the crest flattened. The male may present a twig. Pair formation lasts 3–4 days. Different mates are chosen each season, usually in a different tree or location within the colony and different mates are also selected for renesting after nest failure, which is a common occurrence (M. Maddock pers. comm.). Copulation is preceded by a female Stretch, apparently an invitation, or by no obvious display. Unlike other herons, the female crouches and gapes the bill upward during copulation, with the mounted male nibbling her crown feathers. Extra-pair copulation, wanted or unwanted, is common (Fujioka 1987, Fujioka and Yamagishi 1981, McKilligan 1990b, M. Maddock pers. comm.).

The eggs are white with a pale green or blue tinge, broad oval and somewhat pointed, lighter coloured than most other medium-sized egrets. Sizes vary broadly from 40.4–52.1 × 29.8–36.5 mm. The clutch is usually 2–5 eggs, averaging 2.6 in Ghana and East Africa, 2.2 in Zimbabwe, 3.0 and 3.3 at two long-term study sites in USA (Telfair and Bister 2004), 3.0 in South Africa and Madagascar (Kopij 1997), and 3.6 in Australia (McKilligan 1985). Clutch size increased over 28 in Texas, USA (Telfair and Blister 2004). Eggs are laid at two-day intervals. Cattle Egrets will renest after nest failure if not too late in the nesting season. Incubation begins with the first egg, and lasts about 24 days, ranging from 21 to 26 days. Both parents incubate, exchanging one to four times per day. Eggs hatch asynchronously, 1 to 2 days apart. Hatching success varies from 14% in one study in south USA to 97% in north USA, in most areas it is about 50%–90% (Ranglack *et al.* 1991, Kopij 1997, 1999b, Telfair and Bister 2004). Nestlings are altricial. Chicks soon become especially vocal and aggressive. Both parents share care of the chicks in about equal time proportions (M. Maddock pers. comm.). They are guarded until day 10, and shaded in the tropics. Both parents feed them. Chicks initially weakly grab at parent's bill, becoming more aggressive with age. Being the larger, the first chicks usually are fed first, but this preference seems not mediated by the parents. Maximum growth rates are at 9–12 days, as endothermy is attained. They are fully feathered at 13–21 days. Young leave the nest and climb in nearby branches at 2 weeks. They fledge at 30 days, becoming fully independent in 15 more days. Legs change from dark green to black prior to fledging.

Success is usually fairly high; reproductive success was 94% in Texas (Telfair 1983), 37% in South Africa (Kopij 1997), although only 15% in Alabama (Dusi and Dusi 1970). In Australia, the mean number of chicks fledged by successful nests in an 11-year study was 2.42, with 41% of nests fledging 3 chicks and 6% fledging 4 or 5 chicks (M. Maddock pers. comm.). The rate of complete nest failure was high (41%), the majority of which were nests in which one or both partners was one year old. This was not the case in Texas, USA, where only 6.5% of nests lost chicks, averaging 1.16 chicks per brood (Telfair and Bister 2004). An important mortality factor is predation, by lizards (*Varanus*), crows (*Corvus*), ants (*Solinopsis*), owls (*Bubo*), hawks (*Buteo, Parabuteo, Cicus, Falco*), night herons, eagles

(*Aquila, Haliaeetus*), raccoon (*Procyon*), and others. In Australia, unlike other herons there, nesting success is not related to rainfall, except for severe drought (Maddock and Baxter 1991, Maddock and Geering 1994, M. Maddock pers. comm.). In India, drought reduces nesting success (Gopakumar 1991). Strong winds, parasites, and human disturbance also cause mortality (McKilligan 1987, 1996, Kopij 1997). The dominant cause of nestling mortality is starvation. Sibling aggression is frequent and severe in some areas, the extent of which is related to food supplies (Fujioka 1985a, 1985b, Creighton and Schnell 1996). In South Africa starvation of third and fourth chicks is inevitable. When using artificial habitat, resources are in some cases very abundant and nesting is limited by the ability of parents to deliver food rather than food supply itself. In New World studies, 98% of third chicks and 97% of fourth chicks survive to fledge (Telfair 1983), explaining its rapid population increase. In an 11-year study of breeding success in Australia (M. Maddock pers. comm.), very few chick losses of were recorded in 3-, 4-, or 5-chick nests. Forty-two percent of successful nests fledged 3 chicks and 6% fledged 4 to 5. Telfair and Bister (2004) showed that mortality of chicks is related to hatching rank, even when chick loss overall is relatively low.

Population dynamics

Nesting effort is related to rainfall patterns, leading to annual variation in productivity (McKilligan 2001). Birds can breed successfully at one year old (Siegfried 1966a, Maddock 1993, 1989a). In the first year mortality is about 37%–40%, which is a low figure for herons. In older birds, annual mortality is 20%–28% (Siegfried 1970, McKilligan *et al.* 1993, Telfair 1994). Few banded birds are recovered older than 7–8 years (Telfair 1994).), with the oldest bird seen alive in Australia being 11 years (M. Maddock pers. comm.). The Life-span of adults in various populations is 9.1, 9.7 11.1, and 13.7. These are much shorter than is presumed for other herons. Predation and weather appear to be important mortality agents for adult Cattle Egrets. Birds are taken by Peregrines (*Falco*) on migration and exhibit avoidance behaviour when peregrines are hunting (Telfair 1994). The cold winter in 1985 eliminated birds breeding in both the Atlantic and Mediterranean coasts of France, which were later recolonized by birds from Spain (Hafner *et al.* 1992, Marion *et al.* 1993).

Conservation

The Cattle Egret is a highly successful, widely dispersed, abundant species, probably the most abundant heron in the world. Its range expansion is continuing and population gains are being consolidated. While population conservation is not a global concern, any species freshly invading new landscapes, especially one that characteristically uses human-made habitats, can come into conflict with people. The Cattle Egret usually is not persecuted on its feeding ground in that it is perceived to be beneficial or neutral to cattle activities. However, its tendency to develop large new colonies in and near towns and villages creates what is often perceived to be a public nuisance. To alleviate the noise, smell, health issues, water quality concerns, habitat degradation, and inconvenience of these colonies, local pressure arises to remove them or prevent their establishment, or even killing birds that appear to pose a health risk. Such actions have the unintended effects of shifting birds to other inconvenient sites and also can adversely impact other herons nesting with the Cattle Egrets, whose conservation status may be more of a concern (Dusi 1979, 1982, 1983, Fellows and Paton 1988). Control of nuisance colonies needs to be done within a regional planning context including environmental assessment, the use of non-lethal methods, the provision of alternative protected and managed sites, and public education. Beyond clearly nuisance colonies, nesting sites and their associated feeding situations should be protected and managed (Fernandez-Cruz *et al.* 1993). Cattle Egrets can adversely affect the trees and bushes used for nesting, leading to the abandonment of the colony site if it is not managed. Colony sites, and potential colony sites, need to be conserved on a regional basis. There also is a need for attention to the status of foraging habitat along migration routes. For a species that forages in man-managed environments, it could be expected that pesticides and other contaminants would be a conservation issue, but despite many studies none yet indicate severe adverse effects (Heinz *et al.* 1985, Telfair 1994). A likely difference between contaminant burden in Cattle Egrets and other herons is that a insect diet has less potential for bioaccumulation than does a fish diet. Instances of high contaminant burden may reflect local practices, on the breeding or wintering grounds, and so can serve as indicators of environmental conditions (Burger *et al.* 1992).

Research needs

The Cattle Egret is one of the best-studied herons. Research opportunities lie at two scales, the population and the individual. The demography and population biology, such as nesting success, role of brood reduction, and perhaps survival patterns, appear to differ between its historic African range and its newly invaded range. Explicit study comparing the demography of the species in the two areas could unravel fundamental questions of demographic flexibility. There also appear to be fundamental differences in demography between the Cattle Egret and the more long-lived species. Cattle Egrets appear to be rather independent as individuals. Dispersal and fidelity, seem to be a matter of individual discretion. Movement, physiological, and genetic studies at population and individual levels could reveal much about such individual decisions. There is more to be understood about the relationship of the two subspecies of Cattle Egrets.

Overview

The Cattle Egret is an odd heron, being fundamentally a terrestrial insectivore. Although superficially homogeneous, the heron family exhibits a graded series of morphological and behavioural traits. The Cattle Egret is at one end—that of terrestrial adaptation. As a terrestrial walker, it has relatively short tarsus and small feet. It has a short bill, opposite to the long-billed fish specialists (Boev 1989), better for catching small arboreal prey. The bill is also relatively thick to handle hard-shelled prey. Its slow walking feeding is aimed at carefully examining grass and ground for motionless hidden prey, inspiring prey to move, and capturing it before it flies or hops away. Its association with mammals clearly increases its feeding effectiveness. So does its flocking. Although there is no evidence of persistence of membership in any group of Cattle Egrets, flocks develop a functional structure, moving in unison and in fact serving as their own beater.

Its preferred habitat is not grassland *per se*, but grassland that is damp or has been seasonally flooded. Likely this is due to such land being more productive of insects than dry grassland. In Africa, wet and dry periods are highly seasonal, and variable in extent from year to year. As a result Cattle Egrets need to be able to shift their populations in order to find and exploit the variable supply of damp or drying grasslands.

The species has a strong postbreeding dispersal tendency. It is predisposed to take advantage of human-caused habitat alteration, creating improved pasture, complete with buffalo-surrogates in the form of domestic cattle. Its dispersal skills brought it to new lands in sufficient numbers to colonize and to take advantage of such developing agricultural practices.

The importance of an insect diet to such a large bird is problematical. Herons, no matter how successfully redesigned, seem not to be optimal insectivores. That Cattle Egrets can specialize on insects is a wonder, especially when it is calculated that an egret needs to consume 1728 grasshoppers in 12 hours to feed itself and its young, or a grasshopper every 25 seconds (Telfair 1994). This feeding productivity seems unlikely over the long run. In its historic areas in Africa, brood reduction is the rule. Nesting success there depends not on the egret's primary insect prey but on its access to aquatic vertebrates, particularly frogs. So it appears that the Cattle Egret has limits to how far it can stray from its water-dependent ancestry. Brood limitation, while typical of *ibis*'s native situation, seems not the rule in North America. Released from its co-adaptive constraints and nourished by irrigated improved pasture land, the Cattle Egret's enhanced reproductive potential enabled it to explode across the American landscape.

Cattle Egrets need to learn to feed. Juveniles feed in shorter grass, take similar number of steps, make fewer capture attempts, obtain half as much food items per time, have to feed longer, and also feed more alone (Burger and Gochfeld 1989). Some of these results are intriguing. Among what a Cattle Egret needs to know is how to feed in long rather than short grass, how to be efficient in capture, and how to take advantage of flock behaviour. Apparently it learns these things quickly in that it can breed before its first birthday, although not as successfully as adults.

High dispersal tendancy, ability to breed in year one, high nesting success, potentially low chick mortality, low first year mortality but relatively short life span suggest that the Cattle Egret is indeed the "r" strategist among herons, a species well adapted to exploiting changing environments through population increases.

The Cattle Egret is the heron success story of the twentieth century. It has expanded across the world, responding to human habitat change and is becoming an integral part of the human landscape nearly worldwide.

Reddish Egret *Egretta rufescens* (Gmelin)

PLATES 5.6–5.8

Other names: Gaulin (Jamaica) in English; Garzón Rojiza, Garza Melenuda, Garza Piquirrosa, Garza Roja, Garza Rojiza, Garceta Rojiza in Spanish; Aigrette bleue, Aigrette roussâtre in French; Rötelreiher in German.

Ardea rufescens Gmelin, 1789. Syst. Nat. 1, pt 2, p. 628: ex Daubenton p1. Enl. p1. 902: Louisiana.

Subspecies: *Egretta rufescens dickey* (van Rossem 1926): San Luis Island, Gulf of California.

Description

The Reddish Egret is a thick medium-sized dimorphic (dark or white) egret with a shaggy neck, and distinctive bicoloured pink-and-black bill.

ADULT: The dark morph adult has a red-brown to dark cinnamon head and neck and slate grey to blue-grey back, wings, tail, abdomen, and belly. White morph birds are entirely white. In both phases, the irises are straw yellow, set off by dark lores. The bill is dark toward the tip and dusky pink toward the head bright during courtship and duller, but still striking for the rest of the year. Legs and feet are blue-black to black. During breeding, the head and neck become chestnut to pale cinnamon. The outer 45% of the bill becomes shiny black, whereas the basal portion turns striking pink or flesh. The lores are turquoise-blue during courtship changing to flesh during incubation and chick-rearing and then darkening toward the end of nesting. The dark morph is basically dull slate grey. The shaggy lanceolate plumes of the head, neck, and upper breast are cinnamon, while those of the back are grey mixed with some chestnut. In full development the filamentous back plumes extend up to 15 cm beyond the end of the tail, accentuating the bird's characteristically shaggy appearance. These plumes are white on the white morph. The front of the legs and top of the feet remain black, but the sides and back of the legs and bottom of the feet become turquoise-blue. In flight it is graceful and light with head typically retracted.

VARIATION: The sexes are alike, but males are slightly larger. The most important variation in the species is its white–dark dimorphism. The proportion of white birds differs in various populations, ranging from 5% to 12% in Texas, to 15%–20% in south Florida, 50% in Cuba, and almost 90% on Great Inagua in the Bahamas. Additionally, there is considerable individual variation in the plumage shade of dark birds, which may be seasonal (R. Paul pers. comm.). Some dark birds have a number of white feathers, usually flight feathers symetrically on both sides, or patches elsewhere (see photos in Hancock 2000). There are no reports of white birds with dark feathers. Geographic variation has been reported. *Dickeyi* from Baja California is slightly darker than *rufescens* (Payne 1979, Paul 1991).

JUVENILE: The juvenile can be quite variable and rather nondescript. Typically, the dark form is light grey-brown above and paler on the belly. Feathers, such as wing coverts, are tipped cinnamon. Short head and breast plumes, first acquired at 4–6 months, are dull cinnamon. The bill is black to dark horn, usually lighter toward the base. The bill becomes bicoloured with age. The lores are black. Dark individuals range in base colour from medium grey to grey-brown often with a rufous tinge, some being quite rufous. Young white morph birds are all white. By 18 months juveniles resemble adults with worn plumage, and they attain full adult plumage in their third year.

CHICK: The dark morph chick is covered with grey down with a cinnamon grey head crest. The white morph has white down.

VOICE: The Reddish Egret is generally a quiet bird. A soft 'Awh' call is used while feeding and as part of the Greeting Ceremony. The 'Crog' call, rendered 'crog-crog', is used in flight displays and in aggressive encounters away from the nest. The "Raah" call is given in flight and also in attack situations. The 'Arg' is the advertising call. Bill Snaps are used in the Forward and Pursuit Flights while Bill Clappering is used in pair maintenance. Chick beg with a raspy triply-repeated 'Hee' call, rendered 'hee-hee-hee'.

WEIGHTS AND MEASUREMENTS: Weight: 700–850 g; length: 69–81 cm.

Field characters

The Reddish Egret is identified by its thick, shaggy appearance, two-toned bill especially during breeding, leg colour, and its very active feeding behaviour. The dark morph egret is distinguished from the adult Little Blue Heron by being larger and stouter with a thicker straight bill and shaggy lighter neck. The dark morph is distinguished from the Tricoloured Heron by its dark (not white) belly, larger size, and thicker appearance. The white morph bird is distinguished from the immature Little Blue Heron by being larger, thicker, especially thicker neck, a heavier straight bill, dark (not green) legs, and its much more active feeding behaviour. The white bird is distinguished from the Snowy Egret by being larger and stouter with a thicker bill, dark (not yellow) feet, two-toned bill (in breeding), and very active feeding behaviour. It is distinguished from the Great Egret by being smaller, its two-tone (not yellow) bill, and active feeding behaviour.

Systematics

The Reddish Egret is one of the medium-sized egrets. Individual variation among dark birds is little understood, and may be seasonal or geographic. This has lead to some uncertainty is subspecific taxonomy. *Dickyi* is slightly darker than *rufescens*, but weekly differentiated if at all. Other Mexican birds have been assigned to a third race based on plumage differences that more recently have been considered seasonal. Geographic, seasonal and individual variation among Reddish Egrets deserves careful study. The taxonomic history of the species is of some interest. White birds were first identified as a separate species and late, by John James Audubon, to be immature birds, the dimorphism not being worked out until the 1870s.

Range and status

The Reddish Egret is a bird of the New World tropical coast from North America through the West Indies to northern South America.

BREEDING RANGE: This species breeds in east Florida (Merritt Island) and west Florida (Clearwater Harbor), coastal Gulf of Mexico (Alabama, Louisiana, Texas, east Mexico) to Belize (Lopez-Ornato and Ramo 1992), Baja California (Becerril and Carmona 1997), and west Mexico (Sonopa, Sinaloa-Mellink *et al.* 1997), Bahamas (Abaco, Grand Bahama, Andros, Acklins, Inagua), Cuba (Garrido and Kirkconnell 1990, Denis *et al.* 1999) and recently were reported as breeding in the Dominican Republic (E. Ferandez and Rodriguez-Mojica 2002). They formerly nested in Jamaica. They also nest on Bonaire, Los Roques, Venezuela, and likely on mainland Colombia and Isla Margarita (Scott and Carbonell 1986, Paul 1991). It is possible that the species breeds elsewhere than is presently known in Central America, the West Indies, and north South America especially coastal Colombia, but nesting records are non-existent or not recent. Nesting is not documented for Central America, where few records exist (Paul 1991). The race *rufescens* occurs over most of the range. The race *dickeyi* breeds in Baja California.

NON-BREEDING RANGE: Birds occur in winter in breeding areas along the Atlantic and Gulf Coasts from Florida, through Texas, and Mexico. They are increasingly found from north Florida to the Carolinas (R. Paul pers. comm.). On the Pacific Coast, birds winter north of the breeding area in south California (reported as far as Monterey) (Roberson 1980, Small 1994), through west coastal Mexico. Scattered records occur in Central America

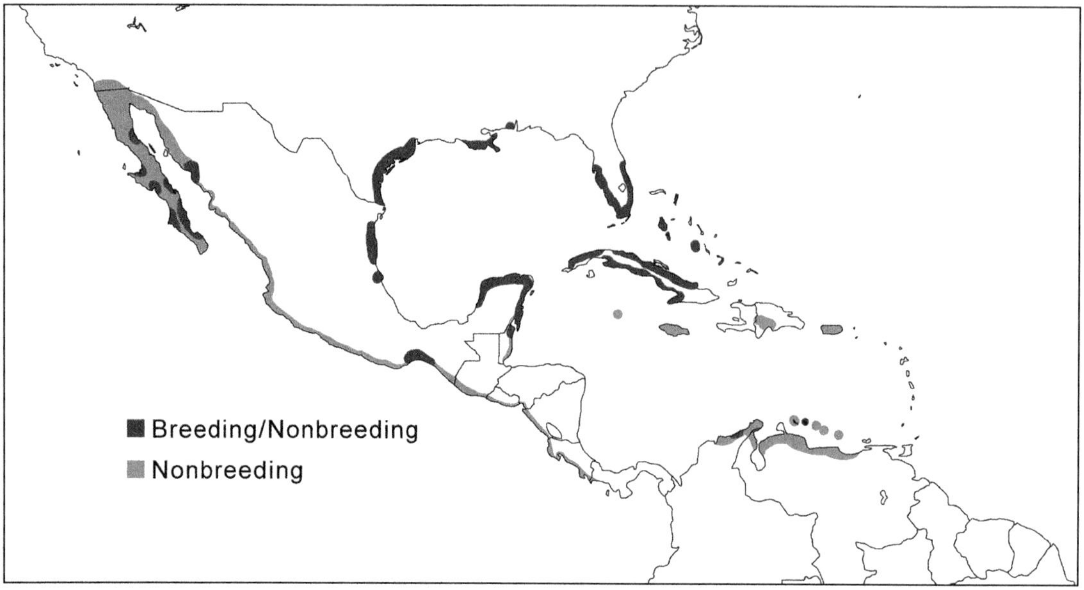

to Belize on the east and Costa Rica on the west. Birds occur year-round on Cuba and the Bahamas, uncommonly on Cayman Island, Jamaica, Hispaniola, and Puerto Rico (Rodriguez-Mojica 2002). They occur with regularity on Venezuela's offshore islands, Colombia, and Netherlands Antilles. There are scattered records for the Lesser Antilles (Norton 1993).

MIGRATION: Florida and Texas populations are best considered to be partially migratory subject to postnesting dispersal and wandering. Distinguishing the two processes is not clear. Texas birds disperse after nesting is completed moving north in June–September then south in September–November (Lother and Paul 2002). Dispersal of Florida-nesting birds begins in March through July, but a southward migration of Florida birds has not yet been documented. Even less is known about movement of Mexican birds. Scattered Reddish Egret records, mostly of immature birds, have occurred in a number of localities in the USA in late summer, autumn and even winter. Dispersal records include Massachusetts (Petersen 1992), New Jersey (Paxton *et al.* 1999), South Carolina (Wooten and Cox 1986), Kentucky, Indiana (Brock 1994), Kansas (Janzen *et al.* 2000), Iowa (Schantz 1996), west Texas, Arizona (Benesh and Rosenberg 1998), and Nevada (Kingery 1991). On the Pacific Coast Reddish Egrets have been reported in Arizona, USA. Individuals ringed in Texas have been found as far south as Campeche, Mexico, Guatemala, and El Salvador, likely crossing Mexico (Paul 1991). Records also occur in the lesser Antilles (Atigua, Dominica, Montserrat, St. Martin) Greater Antilles (Puerto Rico) and Bermuda (Rodriguez-Mojica 2002, Norton *et al.* 2003).

STATUS: The species was adversely affected by plume hunting in the USA and extirpated from Florida by the early 1900s (Powell *et al.* 1989). It was recorded again in Florida in 1937 (Davis 1937, Desmond 1939) and since then has recolonized much of its former range (Paul *et al.* 1975, 1979, Rodgers and Schwikert 1986, Toland 1991). The Texas population decreased from the 1930s to the 1970s but recovered, with birds colonizing Louisiana and Alabama. Both major North American populations have increased and expanded their ranges in the past three decades. However more recently, numbers in Texas may have declined, and there is serious concern for the future status of this population (Lange 1995, R. Paul pers. comm.). Sightings have increased in central and northern South America in recent decades. The documented breeding population in North America is about 2600 pairs. The status of populations elsewhere is poorly known with only fragmentary information available about current nesting numbers. The Belize nesting population is about 100 pairs; the Inagua, Bahamas, population is 200 pairs. Overall, the documented population is about 3000 pairs, suggesting a total global population of no more than 5000 pairs.

Habitat

The Reddish Egret is a bird of open marine flats and shorelines. Few are seen inland, and those are usually juveniles. Typically the species feeds on open, unvegetated tidal flats in very shallow water (less than 15 cm deep) (Paul 1991, 1996, Farmer 1992). They nest on coastal islands, usually low mangroves, bushes, or cactuses.

Foraging

The Reddish Egret feeds typically in very, very shallow, clear, reflective, fish species-poor, marine, tropical habitats. It forages during the day, most actively in morning and afternoon, roosting mid day (Lowther and Paul 2002), depending on tidal cycles, both diurnal and wind driven. It seldom feeds from shore and apparently never on land. Occasionally the egret feeds in flats having scattered mangroves, in somewhat deeper water (to 25 cm), and also use pools where prey becomes concentrated by falling water, either tidal or seasonal. The Reddish Egret is the most consistently active foragers among the herons of the world, and puts on a truly remarkable, if hard to describe, performance. Its behaviour has many components, which it strings together in various ways. Among the identifiable components are Walking, Walking Quickly, Running, Hopping, Wing Flicking, Open Wing Feeding, Peering Over, Foot Stirring, Foot Raking, Underwing Feeding, Double Wing Feeding, Canopy Feeding, and Hovering Stirring. Feeding in exceptionally shallow, clear and reflective water, it moves around vigorously chasing down small fish. Peering Over is a particular compensation for the water's reflective surface. This is sufficiently characteristic to have become a courtship behaviour as well.

How the egret puts all these behaviours together into a foraging bout is fascinating. They Walk Quickly or Runs with long strides usually in semi-crouched posture with head and neck extended forward, continuously Wing Flicking, quickly changing directions, Hopping ahead or to the side, doing Underwing or Double Wing Feeding, and finally stopping when they catch up to the fleeing fish, which they capture by stabbing, often around the feet. They tend to run into the wind perhaps to gain lift using its partially open wings, and tend to use Open Wing Feeding more frequently during cloudy weather, in open water, and with a calm water surface (Rodgers 1983). They also Walk slowly, Foot Raking as they go. They can use Canopy Feeding (Lowther and Paul 2002). Despite the effort put into catching up with their prey, their stab is successful less than a third of the time, less for juveniles (Rodgers 1983). They feed primarily alone, as would be expected, in order to feed in this way. Individuals are well dispersed but not obviously territorial. Even when gathered in loose assemblages, they tend to feed individually. They do aggregate where food becomes highly available, feeding with other species of herons, pelicans (*Pelecanus*), terns (*Sterna*), stilts (*Himantopus*), spoonbills (*Platalea*), and flamingos (*Phoencopterus*). In such situations their usually extravagant feeding behaviour is appropriately restrained. It has been observed, however, that Reddish Egrets sometimes do gather into large, highly compressed flocks, but it is possible that young of the year tend to be the ones assembling into such flocks, perhaps as an aid to their less efficient foraging (Paul 1991), an hypothesis worth testing.

Most prey items are marine and estuarine fish (McMurry 1971, Recher and Recher 1980, Paul 1991). Killifish dominate the diet, particularly minnows of the genus *Cyprinodon*, which has been identified as the main prey of the egret in Texas, Florida, and Bonaire. Other fish include mullet (*Mugil*), other killifish (*Fundulus; Floridichthys*), topminnows (*Poecilia*), insects pinfish (*Lagodon*). Shrimp, crabs, insects are taken less frequently. Small prey are swallowed at once; occasional large prey may be carried to shore for processing by stabbing.

Breeding

Reddish Egrets nest in early spring, February–April on the Florida coasts, Florida Bay, and on the Pacific coast. They nest later in spring, March–June, in Texas and the Gulf coast of Mexico, and in summer in Mexico. However, nesting timing seems to be variable and flexible across the range (Paul 1991). For example, nesting can be year-round in extreme south Florida, where there may be an autumn nesting peak, November–January. The Reddish Egret nests on protected islands near feeding sites, including man-made islands. Specific nesting substrate does not seem very important. They nest on mangroves (*Rhizophora, Avicennia*), bushes (*Schinus, Baccharis*), low herbaceous plants (*Spartina, Sesuvium*), cactus (*Opuntia*), or on the ground, oyster shell, or sand. This is generally a colonial species, although they at times may nest solitarily. The nest is built of twigs lined with grasses and other fine vegetation.

The courtship behaviour is complex and variable (Meyerriecks 1960, Palmer 1962). At the colony, the male claims a display site, which is first large and then contracts. It occupies and vigorosly defends the site by Standing, Crest Raising, Upright (in which the body is uncharacteristically for this species sleeked, with crest raised only slightly and tail flipped), Forward, Pursuit Flights, and Face to Face Fighting from an upright posture. The Forward is quite aggressive-looking; the feathers of head, neck and back fully erect, wings out, and the bird runs at the opponent giving its Raaah call. Fighting is frequent. Pursuit Flights involve Bill Snapping. Circle Flights and Pursuit Flights follow complex paths, the bird flying with its crest erected. In the Stretch display, the bird extends its head fully upward, then brings the head slightly down, points the bill upward, erects the head, back, and breast plumes fully, bulges the eyes, and performs Head Tossing, moving the head up and down. It also does an Aerial Stretch in which it holds its head at a 45° angle while flying from a perch, Head Tossing, and giving the Crog call or Bill Snap. The Stretch and Head Tossing are used in a Greeting Ceremony and in twig passing as is the Awh Call. Pairs use simultaneous peering down displays. Promiscuous mating within colonies has been observed (Lowther and Paul 2002).

There is much interaction among birds away from the nesting site, particularly near the onset of nesting. This dance ground display involves birds Walking, Running or Hopping, with crest raised, Head Tossing, and giving the Crog call. An Aerial Stretch has been described on the feeding site, with the displaying bird flying around the other in a Circle Flight. Another feeding ground display is Peering Down, in which standing birds bend over with bills touching the water for up to a minute or two. Jumping Over is used by both birds on the feeding area. The jumping bird Head Tosses and erects its head, back, and chest plumes. The functions of these interactions are not clear. The most characteristic aspect of several displays is the inclusion of Head Tossing, rapidly raising and lowering its head with crest erected. This features the thick neck plumes and crest to maximum effect. Another common display is Crest Raising, in which the thick crest plumes are raised to make the head appear large.

The eggs are green blue, 46.5–55.6 × 34.6–41.7 mm. The usual clutch size is 2–4, but ranges to 7 (Paul 1991). Incubation is 26 days. Both parents attend young for about 3 weeks. Young leave the nest in 4–5 weeks. They remain nearby and return to the nest to feed. Young fly at 6–7 weeks, but are fed by the adults until 7–9 weeks (Paul 1991). Hatching success is high overall, typically 69%– 85%, and survival from hatching to fledging is variable, averaging about 36% (Paul 1991). Lowered success may be caused by disturbance and food shortage. Reproduction was 1.2 to 1.5 young per pair in Texas and up to 1.8 young per pair in Florida, with lower figures occurring because of disturbance and food situations (Paul 1991). Racoons (*Procyon*), Coyotes (*Canis*), Crows (*Corvis*), grackles (*Quiscalus*), and eagles (*Haliaeetus*) predate eggs and young.

Population dynamics

Some birds breed in their third year, but most in their fourth year (Paul 1991). This age at first breeding is unusually delayed for a medium-sized heron. Adult survival is not known.

Conservation

The Reddish Egret is globally vulnerable (Hafner *et al.* 2000). Its small global population size, small geographic range, and narrow foraging habitat requirements put this species at risk. Continuing urban and aquaculture development and alterations to coastal processes, upland runoff, or hydrography continue to threaten the suitability of these habitats for the species. Because it is a habitat specialist, the Reddish Egret has nowhere else to go when its habitats are lost. The near-impossible task of maintaining natural functioning in the coastal lagoons that it occupies is the primary conservation priority for the species. Sites of particular importance to these birds need to be identified and protected. Given that the maximum global population estimate is no more than 10 000 breeding birds, important areas are those supporting 100 or more birds at some time during the year. In the USA, the most important sites—such as Florida Bay, Tampa Bay, Laguna Madre, Galveston Bay, Aransas Bay, Corpus Christi Bay, Laguna Tamaulipas—need to be the focus of action plans that favour the species, including the management of hydrology and island nesting sites, and including the creation of new nesting sites where appropriate.

Research needs

Much remains to be known about the biology of this species, particularly clarifying its breeding behaviour and the function of display behaviour on the foraging grounds. Geographic variation in morphology, plumage, and biochemistry should be examined throughout the species' range, especially thoroughly re-examining taxonomic status of the birds of Baja California and west Mexico. Surveys need to be conducted throughout the range to locate additional colony sites, and to catalogue locations qualifying as important areas for the species. Feeding habitat for major colonies should be identified and qualified as to potential threats. Nesting seasons throughout the range also need to be clarified. A cooperative multination monitoring programme for important nesting areas should be developed. Production, survivorship, migration mortality, and other demographic parameters should be studied at several sites to assess the basic demography of the species. The concept of energy limitation and risk being a fundamental feature of this species' ecology should be examined. Similarly, there should be additional study of the comparative foraging ecology of juvenile and adult egrets, including the role of dense feeding concentrations, to examine the role of habitat choice, flocking, and feeding effectiveness, particularly in light of this species' delayed maturation. The wintering areas of the several populations of North American birds need to be better understood.

Overview

This is an intriguing bird. Its extravagant foraging technique must use a tremendous amount of energy, its prey are typically very small fish, and its efficiency of capture is low compared with other herons (Rodgers 1983). How its energy requirements are met working hard and inefficiently for small fish is a puzzle. Juveniles forage even more poorly. It is a species that has to learn how to feed well. Reddish Egrets can use their technique only in shallow, open habitats where deep water or vegetation will not impede their movements. But in such situations the predator is highly visible to the prey, thus the need to chase them down as they swim away from the disturbance the bird is causing. These are harsh environments, usually tidal and often periodically hypersaline, which is what restricts plant growth. The attraction of the Reddish Egret to such harsh environments is shown by its use of salt works (Paul 1991). In the places it chooses to forage, diversity of potential prey is not great, consisting of a few highly adaptable fish species. It eats only a few species of fish, and particularly concentrates on cyprinodontid killifish. *Cyprinodon* is found in shallow, coastal lagoons from fresh to hypersaline water throughout the egret's range. The combination of habitat, foraging behaviour, and limited prey base is the essential element of the Reddish Egret's feeding ecology. The species' dependence on coastal tropical lagoon systems is the principal reason for its highly restricted range. It occurs in large numbers in few places. Given that, it seems likely that it occurs now in most of the places it is capable of using effectively.

This is a highly specialized heron. It has a limited habitat scope and depends on a single sort of prey. Given its specialization and energetically costly foraging style, nesting success and subadult survival must be tenuous. That it is able to recover from population crashes, shown twice, suggests that it has resilience. Likely adult mortality is low, especially for non-migratory birds. This is probably a long-lived species trading low reproductive potential within in its difficult environment for a longer adult lifespan. The possibility that resident and migratory birds coexist within a population suggests an even more complex demography. Adding to the complexity is the unknown function of dimorphism in this species. That the ratios of colours differ in different areas may be serendipity or evolutionarily meaningful. Its sexual maturity is delayed, twice as long as other medium herons, a feature likely related to the need to learn how to feed effectively and further suggesting the important role of adult survivorship in the species. The Reddish Egret is an intriguing species indeed.

Pied Heron *Egretta picata* (Gould)

PLATES 5.4, 5.5

Other names: Pied Egret, Allied Egret, White-headed Egret (juvenile) in English; Garceta Pie in Spanish; Aigrette pie in French; Elsterreiher in German.

Ardea (Herodias) picata Gould, 1845. Proc. Zool. Soc. London, p. 62: Port Essington, Northern Territory.

Description

The Pied Heron is a small black and white heron, with black head and head plumes, white neck and black body.

ADULT: The Pied Heron typically has a uniquely black face and crown that forms a crest with elongated black plumes. The bill is straw yellow, dark slate-blue at base. The lower mandible may be grey-yellow. Irises are yellow; the lores dark slate-blue. The throat, lower face, and neck are white. The back, wings, and belly are slate-black to dark blue having a deep slate blue wash. Lanceolate back plumes are slate-blue and long white lanceolate neck plumes hang down over the dark breast. The legs and feet are olive-yellow to orange. During courtship the bill and legs turn very bright yellow and the iris orange-red. The crest plumes become more prominent, and the body plumage takes on a bright slate-blue sheen.

VARIATION: The sexes are alike, and there is no known geographic variation. However, Pied Herons are individually variable in plumage, and depending on age can have variably white heads.

JUVENILE: Juveniles have a white head and neck, lacking a crest. They have a light to white breast, grey-brown to grey-white belly, grey-brown back and wings, which in flight show a white carpal patch. Head colour can vary from white to pale grey, brown-grey, or streaked rufous. Juveniles retain their coloration for over a year.

CHICK: The downy chick is yellow-brown to white with dusky sides. Newly fledged young retain the brown head. The underparts are variably pale slaty, often streaked white, or sometimes entirely white. The black wings are tinged brown, the tips of the secondary feathers are spotted with white.

VOICE: A croaking 'Ohrk' call is the flight and taking off call. A soft cooing is given at the nest. Little is known about vocalizations.

WEIGHTS AND MEASUREMENTS: Weight: males 247–280 g, females 225–242 g; length: 43–55 cm.

Field characters

Its distinctive black head and body and contrasting white neck identify typical adult Pied Herons. It flies with a rapid wing beat, with head drawn back. The adult is distinguished from the White-necked Heron by its smaller size, black cap, yellow legs and bill, and lack of black foreneck spotting. The juvenile is distinguished from the White-necked Heron by its smaller size and yellow bill and legs (not black). In flight the white wing patch of the juvenile Pied Heron is under wing; the similar patch on the White-necked Heron is on the upper wing.

Systematics

This species, along with the White-faced Heron, has sometimes been placed in *Ardea*, but it is better allocated to *Egretta* (Payne and Risley 1976). Juveniles are so distinctive that they were once classified as a separate species.

Range and status

The Pied Heron occurs in northern Australia and the East Indies.

BREEDING RANGE: It breeds in south Sulawesi (White and Bruce 1986) and patchily in Australia (Western Australia, Northern Territory, Queensland) (Chatto 2000, Jaensch 2003). Its breeding range has recently expanded in Australia. In 2000–2001 it bred for the first time in the Macquarie Marshes in the Murray–Darling River system, which may herald an extension further into inland New South Wales (M. Maddock and R. Kingston pers. comm.).

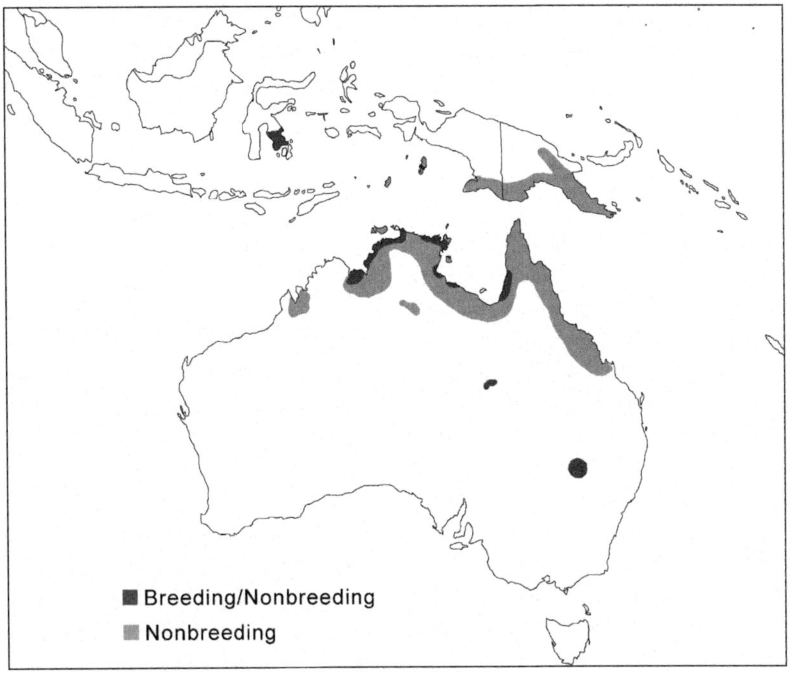

It is possible that it breeds in more areas than are now known.

NON-BREEDING RANGE: In Australia birds occur within a broad band of coastal wetlands inland 100 km, from Wyndham, Western Australia, to Broad Sound, Queensland. They are regularly and increasingly seen in the area of King Sound, Western Australia. A part of the northern Australian population spends the Australian dry season in south Papua New Guinea and Irian Jaya, Torres Straits islands, Tanimbar, and Aru Islands. Migrant birds probably also occur in Sulawesi and perhaps south Borneo.

MIGRATION: Within Australia, Pied Herons undertake regional movements in relation to rainfall and resulting food availability, shifting from coastal breeding grounds to inland swamps and then to tidal mudflats, in June–November. There is a regular migration from north Australia to south New Guinea via the Torres Straits. Dispersal records occur in South Australia and Victoria (Johnson and Eddie 2000).

STATUS: It can be seasonally common within its limited breeding range and in south New Guinea. It is more irregular elsewhere. The nesting population in the Northern Territory is estimated to be 15 000 birds in 15 nesting sites (Maddock 2000). The total population is much larger since flocks of 20 000–30 000 birds have been seen in New Guinea and 22 000 counted along the Middle Fly region (Maddock 2000).

Habitat

The Pied Heron uses a wide variety of coastal and near-coast habitats including coastal swamps, mangroves, mud banks, sand banks, flooded and dry grasslands, tidal river edges, shallow streams, billabongs, and lakes. At times it feeds from floating vegetation or from banks overlooking deep water. It also uses artificial feeding sites such as rice fields, sewage ponds, stock ponds, pig and poultry yards, and garbage dumps and is attracted to burnt grasslands. It generally occurs within 100 km of the coast, from sea level up to 2500 m in Papua New Guinea.

Foraging

The Pied Heron is an active feeder, usually in very shallow water or on land at the water's edge (Recher *et al.* 1983, Gochfeld 1976, Morton *et al.*

1993). It Stands and intermittently Walks slowly in shallow water and channels. It uses Crouched posture to wait for prey. In Walking or Standing, it frequently assumes an Upright posture, often Peering Over. It Stabs at prey with bill angled at about 45°. Most prey are Gleaned from vegetation by Pecking. It chases prey by Walking Quickly, Running, or Walking slowly. As the opportunity allows, it also feeds by Hovering, Plunging, Dipping, Foot Stirring, Wing Flicking, and Probing into mud and grass. It feeds gregariously or alone. When feeding solitarily it may defend feeding territories of up to 50 m^2 with Forward displays, Crest Raising, Supplanting Flight, and Running (Coates 1986). Fights can occur, with both birds Stabbing each other as they fly up together. Feeding in mixed aggregations of over 1000 birds has been reported. When feeding in groups, it is dominated by more aggressive species, such as Little Egrets, but also follows other birds to pirate prey they capture. Immature birds often feed in small loose flocks of a few to 30 birds. On grasslands, it feeds following cattle. When not feeding during the day, especially during high tide, birds roost in mangroves or on exposed banks near the feeding sites to preen or sun bathe, using Droop Wing posture. At night they roost singly or communally in small to large groups. In Papua New Guinea, thousands roost together in bamboo.

Insects predominate in the diet, including grasshoppers, water bugs, diving beetles, and cockroaches. Small fish and frogs also are commonly taken. They eat dragonfly larvae, small molluscs, and crustaceans. They also scavenge, for example by Hovering to take floating debris from the surface of a sewage pond.

Breeding

Nesting takes place in the summer rainy season, usually February–April, although at times as early as January or as late as June, depending on the season. They nest in coastal mangrove swamps, coastal river forests, and tidal creeks, typically in mangrove trees. The Pied Heron is a colonial nester, breeding in large mixed colonies of up to 20 000 nests, including Little Egrets, Eastern Great Egrets, Intermediate Egrets, ibises, darters, and cormorants. Pied Herons tend to clump together within the colonies; their nests can as close as 4 m of each other. The nest is a small, concave platform or bowl of sticks and twigs lined with finer twigs inside (Ely 1976), placed 2–5 m above the water tucked well into the canopy. Nothing is recorded of the courtship of the Pied Heron. The eggs are a deep blue-green and measure 41 × 31 mm. The clutch is 2–4. Incubation, fledging, and juvenile periods are not known, but most young can be expected to be flying by May in Australia.

Population dynamics

The population biology and demography of this species are not known.

Conservation

The species is common and locally abundant along the lowlands of tropical north Australia and south New Guinea. It has been able to take advantage of irrigation and the creation of reservoirs, sewage lagoons, ponds, and stock tanks. Its flood plain habitat is threatened by introduced water buffalo which break levees leading to salt water intrusion. Continued buffalo management and land restoration are required in this area.

Research needs

It is likely that additional breeding sites occur within the core range in north Australia. Surveys should be undertaken to better document breeding sites. Similar surveys in Papua New Guinea should be conducted to determine non-breeding locations and at other seasons to determine if nesting occurs there. This is a little known species. Basic research is needed on the breeding biology, courtship, nesting success, survival, and other aspects of demography.

Overview

The Pied Heron is a species of coastal and near-coastal environments. It is a small, adaptable heron using a variety of feeding situations and a diversity of feeding techniques to catch small prey, mostly insects Gleaned from aquatic plants and also fish caught in very shallow water. It defends these food resources from competitors. In addition to wetlands and pastures, it can use tidal sites as well as human-provided feeding locations. As a result of its large feeding and habitat repertoire its overall diet is robust, taking whatever it encounters from molluscs to floating dead material. It is unusual among typical herons having young that look so different from the adults. It is possible that young also use different social and feeding strategies. When non-breeding, birds move distances to access appropriate feeding sites or avoid drought. Although a species of restricted range, it appears to be a successful one.

Slaty Egret *Egretta vinaceigula* (Sharpe)

PLATE 6.3

Other names: Red-throated Heron, Brown-throated Egret in English; Garceta Giorrgirroja in Spanish; Aigrette vineuse in French; Braunkehlreiher in German.

Melanophoyx vinaceigula Sharpe, 1895. Bull. Brit. Orn. Cl. v, p. xiii: Potchefstroom, Transvaal.

Description

The Slaty Egret is a dark slate grey heron with a red to brown throat, highly localized in the swamps of central south Africa.

ADULT: The base colour of the Slaty Egret is dark blue-grey, including head, side, and back of neck, back, tail, and wings. In the field it can look black (S. Tyler pers. comm.). Rufous patches have been reported on the back and wings (Hustler 1995). The bill is narrow, black with a pale keel to the lower bill. The irises are pale yellow, set off by a fringe of black feathers. It has a red-brown chin, throat, and foreneck. Narrow grey crest plumes extend from the back of the head, lower neck, and back. The chest plumes are mixed with red. The back plumes extend beyond the tail and are conspicuously curved at their tips. Its abdomen and flanks are black mixed with red. The underwing is buff and pale blue-grey. The legs are dull grey-yellow to yellow-green. Toes are small ranging in colour from pale to bright chrome-yellow. The red throat tends to become more buff over time.

VARIATION: Females may be smaller than males. Individuals vary considerably in the amount and distribution of red-brown on the throat as well as in leg and feet colour (Hines 1992, Hustler 1995).

JUVENILE: The immature bird is a pale slate-grey with the upper throat and breast buff-pink to pink-brown sometimes extending down to the belly; the tarsus is grey-green and there is no crest.

CHICK: The chick has black-purple down above, pale-grey below. The skin is pink to yellow turning green-grey. The throat is yellow to orange-yellow. The bill is horn to pink, black-brown toward the tip. The legs are pink to yellow-orange turning green.

VOICE: A triple 'Kraak' is antonistic call. Other calls have not been reported.

WEIGHTS AND MEASUREMENTS: Weight 250–340 g; length: 43 cm.

Field characters

Although the throat coloration is unique, it is variable and not always easy to see (Hustler 1995). The Slaty Egret is distinguished from the Black Heron by its slighter, thinner posture, red throat, and yellow legs (not dark) (Beel 1993). It is distinguished from a dark Little Egret (*sensu lato*) by its throat colour, its slender neck and bill, and yellow legs.

Systematics

One of the least known herons, the Slaty Egret has been a taxonomic mystery (Bock 1956, Benson *et al.* 1971, Vernon 1971, Irwin 1975, Payne and Risely 1976). It was first described from South Africa as a new species resembling the Black Heron but with a vinaceous not black throat. Lacking additional specimens, the consensus grew that these birds were a colour phase of the Black Heron. However, the Slaty Egret's morphology, habitat requirements, and feeding behaviour differ markedly from the Black Heron and the taxonomy of this species was clarified only in the 1970s.

Range and status

The Slaty Egret has an extraordinarily small range, restricted to the floodplain wetlands of south central Africa.

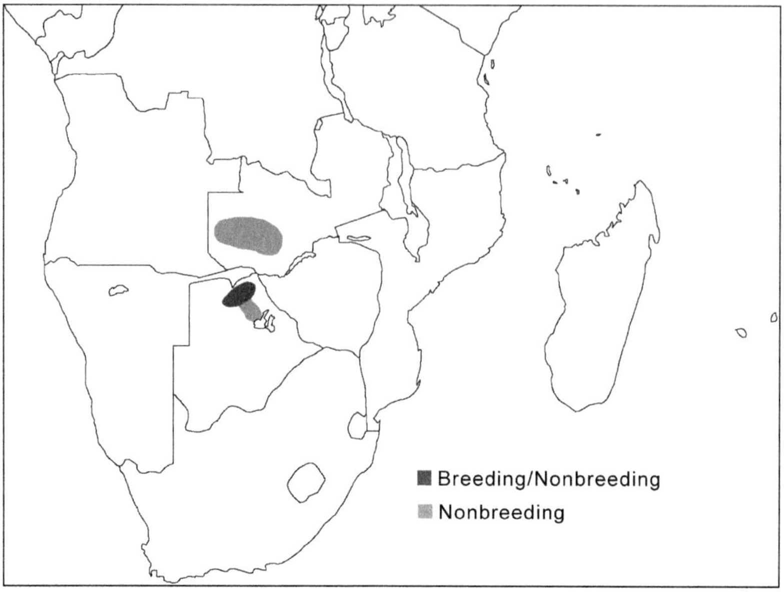

BREEDING RANGE: The core of the year-round breeding range is the inland delta of the Okavango River and along the nearby Chobe River in Botswana (S. Tyler pers. comm.), and perhaps adjacent Captivi Strip of Namibia. It appears to be a sporadic breeder in South Africa, but there is only a single recent record, in the Nye floodplain (Tarboton 1997) (not mapped until proven consistent). Other possible breeding locations are Zambia (Lachinvar), Namibia (Captivi Strip and Bushmanland episodically during floods—Hines 1992), Democratic Republic of Congo (Shaba) and adjacent eastern Angola. But there is no evidence for consistent breeding outside the Okavango/Chobe core, and definitive information is needed from other locations where it is observed.

NON-BREEDING RANGE: It occurs in parts of Botswana (Boteti River), Zambia (Liuwa Plain, Kafue Flats, Lukanga and Bangweula Swamps, Barotse floodplain, Matabele Plain), Zimbabwe (Mashonaland Plateau), Democratic Republic of the Congo (Shaba), eastern Angola, and sporadically in South Africa.

MIGRATION: It is not clear whether this species migrates, in that it is found in some places year round, including locations where it is not known to breed (such as Zambia), but it does move about relative to water conditions and may appear in flooded places to breed (such as Namibia). There are sufficient far-flung records to suggest that dispersal occurs. Records that appear to be dispersals include Mozambique (Zambezi delta but not further confirmed), Botswana (Makgadikgadi Pans, Naxi Pan, Limporo River, Phakalane sewage ponds near Gaborone, South Africa (Marievale outside Johannesburg, Memel, Leeuwdril) (S. Tyler pers. comm.). There are no secure records for the birds' occurrence in Malawi. Clearly more information is needed on the status of the species outside its known core in Botswana.

STATUS: This heron is not uncommon in the core of its range in north Botswana but patchy and uncommon elsewhere. The world population must be small. Its population size is very poorly documented, but likely is fewer than 5000 birds.

Habitat

The Slaty Egret is a bird of seasonally fluctuating wetlands associated with flood plains of large rivers. It feeds generally in small shallow pools, along the margins of small channels, along floating vegetation mats, in temporary shallow pans, and temporary pools flooded by rainfall. It prefers to be in and near vegetated areas and uses open water sparingly. It typically uses an area as water levels decrease from the peak of the wet season.

Foraging

The Slaty Egret is an active forager in shallow water, generally less than 10 cm deep (Milewski 1976, Mathews and McQuaid 1983, Hines 1992). It feeds by Walking slowly, stalking and then Stabbing at prey. It also Stands and waits for prey, typically using foot moving behaviours (Foot Stirring, Foot Probing, Foot Raking, and Foot Paddling) (Mathews and McQuaid 1983). Foot Stirring is used up to a quarter of the feeding time. Standing and Walking slowly are particularly used at dawn and dusk when fish are surfacing. It uses more active behaviours when fish are less available, e.g. mid-day. It Walks Quickly and Runs short distances after prey with partly open wings. It Gleans snails from lily pads, uses Standing Flycatching to catch dragonflies, and Probes. It also Baits, using reed stems to attract fish (Hustler 1995). It feeds solitarily, in pairs, or groups usually of a few birds but up to 60 have been observed (Demey 2003, Tyler 2003). It associates in mixed assemblages with other herons and waders. In groups, feeding sites are defended and aggressive encounters include chasing other birds for many metres. It is a diurnal forager, particularly active at dawn and dusk. It roosts in trees and thick reed beds. It roosts communally with other herons as well as storks and ibiscs, in reeds and water figs (Tyler 2003).

Slaty Herons eat small fish (5–10 cm), and also snails, dragonflies, dragonfly larvae, tadpoles, and frogs. In that no fish occur in temporary wetlands, insects and tadpoles comprise the diet in this situation.

Breeding

Few nesting sites have been reported or studied so the breeding biology is little known. The range of the birds' flexibility in nesting and feeding is only beginning to be understood (Mathews and McQuaid 1983, Fry *et al.* 1986, Hines 1992, Randall and Herremans 1994). It nests in March–June, at or just after the height of flooding in both river flood plains and temporary wetlands (Hines 1992, Randall and Herremans 1994). The Slaty Egret nests in bush thickets of water fig (*Ficus*), *Acacia* and date palm (*Phoenix*) (Atkinson 2003) and in large stands of mature reed beds (*Phragmites*). Nests are in small colonies, perhaps typically with Rufous-bellied Herons. The number of nests reported at sites were 1, 8, fewer than 30–50, and about 50–60. Nests are made of sticks or reed and seem typically to have extensive soft grass and sedge lining, but some reported did not. Vines may be important nest components. Variability may be due to the availability of nesting materials, which are collected very near to the nesting site. Nests are 30–40 cm wide and 10 cm deep, with as shallow central depression. They are placed 1–5 m above the ground, lower in reeds than in shrubs. Nothing is known of the courtship behaviour of this species. Eggs are pale blue-green. They measure 39–45.9 × 29.5–32.6 mm. Clutch is usually 3 (2–4). Incubation is estimated to be around 21–24 days. There is little information on chick-rearing stages. Nesting success appears to be low, about 8% of eggs to fledging in one study, 30% in another. Predation risk appears to be high. A colony in Namibia failed due to a pair of eagles (*Haliaeetus*) and other predators also are implicated in nest failures. In a central Okavango colony, predators killed not only young but also adults. Young leaving the nest die of accidents, such as being impaled by or stuck in the reeds.

Population dynamics

The population biology of this species is unknown.

Conservation

This is a highly restricted species that is at risk to habitat alterations within the wetlands it inhabits. The species uses mature reed and bushy thickets to nest. Fire, reed cutting, drainage, agricultural conversion, and water diversion all threaten these habitats. Birds were eliminated from parts of the flood plain after the Kafue River was dammed. Birds abandoned a reed colony site after it burned. The proposed development of rice farming in Caprivi will alter the Zambezi flood plain. Livestock grazing can impact reed beds. Proposed water diversion and management in the core of its breeding range in the Okavango would be catastrophic for the species. Specific threats include water removal from the Okavango River for urban and agricultural use and the Popa Falls dam in Namibia (S. Tyler pers. comm.). It is considered to be Endangered by the Heron Specialist Group due to it small global population size and restricted habitat that is under multiple threat (Hafner *et al.* 2000). It is considered to be Vulnerable by the IUCN (IUCN 2003) due to habitat loss and degredation. A conservation Action Plan is required for this species. The most important immediate activity is to carry out thorough surveys to locate nesting sites and then to identify and protect important areas. With a likely world population of probably fewer than 5000 birds, sites with 25 nests would be classified as important wetland areas. Important nest sites need to be protected and managed, for disturbance, habitat alteration, and development threats. Development plans in the wetlands it occupies need to be evaluated for potential impact and altered as needed to protect this species.

Research needs

The typical habitats of this species, namely reed beds in extensive shallow inundation zones and temporary rainfall basins, are a not uncommon situation on many African river systems. So why this species, which has some ability to move about, has such a restricted distribution is unclear. Further study of the breeding and feeding biology of this species relative to water fluctuations and specific nesting habitat requirements is needed. The apparent low nesting success and vulnerability to predators need to be better understood in order to assess their true impact.

Overview

The Slaty Egret appears to specialize in some utilizing drying conditions, and as such shows some degree of adaptability as suggested by its diverse foraging behaviour and opportunistic population shifts. Its known colony sites are used when water levels are falling, and nesting is abandoned when drying is delayed. The birds shift seasonally to take advantage of appropriate water conditions, assemble in mixed aggregations to use stranded fish, and even move to breed in places where water levels are temporarily or intermittently suitable. In the river flood plains, drying conditions concentrate fish. In temporary wetlands, invertebrates and tadpoles are taken. But the pattern is consistent—birds utilize a place when food is made available by fluctuating water levels.

Black Heron *Egretta ardesiaca* (Wagler)

PLATES 6.4, 6.5

Other names: Garceta Azabache in Spanish; Aigrette ardoisée in French; Glockenreiher in German.

Ardea ardesiaca Wagler, 1827. Syst. Av. Ardea no. 20: Senegambia.

Description

The Black Heron is a medium, all-black plumaged African heron with yellow feet, usually seen canopy feeding in open shallow water.

ADULT: This is an all slaty-black plumaged heron, having a rather more dumpy appearance than is typical of an *Egretta*. It has a conspicuously luxurient bushy crest composed of lanceolate plumes with a bluish tinge. The bill is black. Irises are dark brown and the eye is ringed by a narrow band of powder blue colour (Zimmerman *et al.* 1996). Lanceolate plumes on the neck and back are black with a blue wash. Individual flight feathers are broad. Toes are yellow to orange. During courtship, the feet briefly turn red. It flies with quick, steady wing beats.

VARIATION: The sexes are alike and no geographic variation has been reported.

JUVENILE: Juveniles are black like the adult, but duller and paler they lack head plumes.

CHICK: Chicks are covered with dark grey down, with a notable downy white crest (Milstein and Hunter 1974). The skin and legs are green-yellow. The feet are yellow.

VOICE: The Black Heron tends to be silent. A low clucking sound has been reported.

Weights and Measurements: Weight: 270–390 g; length: 42–66 cm.

Field characters

The Black Heron is identified by its all-black plumage and yellow feet and also by its unique feeding behaviour and by its distinctive posture for an egret. It often retracts its short neck so that its bushy head is close to its shoulders, making it look rather dumpy as opposed to a typically sleek egret (Beel 1993). In addition to its posture, it is distinguished from the Slaty Egret by being darker, having black rather than distinctively coloured throat, bill, or lores, black legs and yellow feet. It is distinguished from the Rufous-Bellied Heron by its black legs and yellow feet. It is distinguished from the dark phase of the Little Egret by being smaller and darker, with black rather than coloured lores, having a saggy crest and dark (not white) throat. In flight, other medium-sized dark egrets with yellow feet may look very similar.

Systematics

This peculiar heron is one of the medium-sized, yellow-footed *Egretta*, having its flight feathers and plumes modified for feeding. It is likely most closely related to the other endemic African *Egretta*, the Slaty Egret.

Range and status

The Black Egret occurs in Africa and Madagascar.

Range: It is found in Africa south of the Sahara and in Madagascar. It occurs in marshes and lake margins and is not found in desert and forested zones.

Within this large overall range, it is highly localized in suitable habitat and so occurs very patchily. It is absent from many suitable sites. For example, only one breeding site is known in Kenya (Zimmerman *et al.* 1996). The details of this species' range in Africa and Madagascar need to be documented.

Migration: The migratory situation in the species is unclear. It is a consistent local resident in some

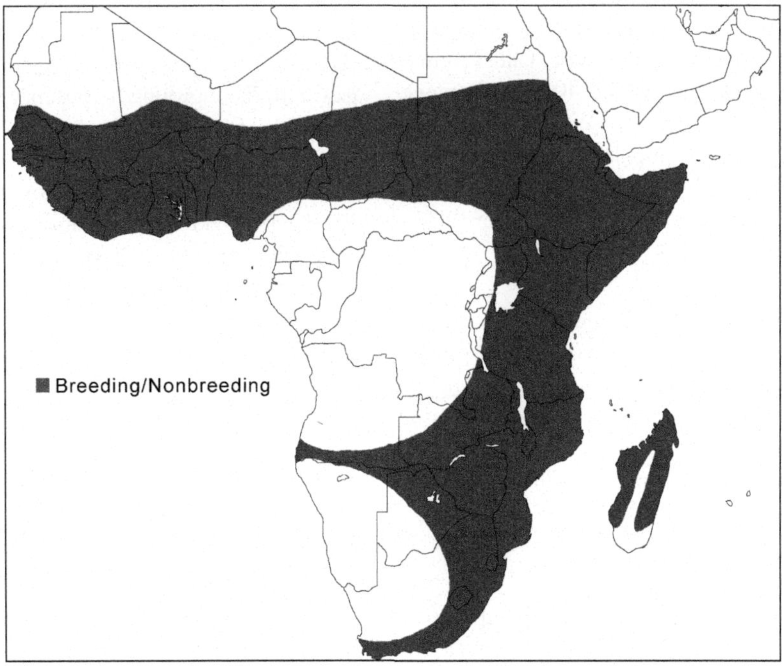

places. It also is a species that rather readily shifts areas in the search for seasonally available shallow-water sites with the right feeding conditions (Hancock 1999). It is, for example, absent from Sierra Leone and the Zimbabwe plateau in some seasons and large numbers migrate in large flocks into Tanzania from Kenya and also through The Gambia (Hancock 1999). Dry conditions desiccate some habitats and seasonal rains create temporary shallow-water feeding areas in other places, apparently inspiring these population shifts. As a result of these movements, dispersion records occur, including in Gabon and Namibia (von Ludwiger and Brand 2000) in the west and Egypt (Aswan) and Yemen in the east (Al-Saghier and Porter 1997).

STATUS: Black Heron populations are widespread but generally small, 10s to 50s, throughout the range (Turner 2000). However it is abundant locally at certain locations and seasons. It is best known in east Africa, where 5000–7000 are estimated in Tanzania (Baker 1995, Baker and Baker 2004). There is no sign of decreases in this area despite expanding human population pressure. The population in west Africa is also large. A remarkable 10 000–20 000 birds were reported from Guinea–Bissau in 1983, but a number of this magnitude requires confirmation (Turner 2000). The population in Madagascar has decreased hugely in the past half century, colonies now numbering in the 40–50 pair range (Langrand 1990). Recent censuses are frightening; the counted Madagascar population decreasing rapidly from 2000 in 1992 to 250 in 1997 (Turner 2000). This population is highly threatened.

Habitat

The Black Heron prefers shallow open waters, especially margins of fresh water lakes and ponds. Inland it also uses marshes, river edges, rice fields, and seasonally flooded grasslands. Along the coast it feeds along tidal rivers and creeks, mangroves, alkaline lakes, and tidal flats. It occurs from sea level, to 1500 m in Madagascar.

Foraging

The most distinctive characteristic of this species is its Canopy Feeding in which it feeds by spreading its wings over its head in a full umbrella with primaries touching the water and the erected nape plumes completing the canopy. The heron Walks slowly or Walks Quickly to a location where it sees a fish or an otherwise appropriate spot. It then takes one or two steps, starting the formation of its Canopy, which is timed to reach completion above the potential prey. If the sequence is interrupted, the partially raised wings are immediately retracted. Otherwise, it completes the canopy and looks under it for 2–3 seconds, often Foot Stirring, and Stabs at fish, presumably those incited to movement by the disturbance caused by the heron (Vandewalle 1985). After the canopy, the heron moves on a few steps to form another one, usually within a seconds of the first. The heron frequently pauses to shake itself off, probably to get feathers back into position. The Black Heron also feeds by Standing and stabbing. The functioning of Canopy behaviour has been argued for some time. While the canopy reduces reflection, provides better visibility, and obscures the silhouette of the heron, there is little evidence that these are primary functions for so complex a behaviour. Whether fish are attracted to or flee the canopy and foot stirring is still debatable. There is no reason why both situations could not occur.

Some resident Black Herons feed solitarily, having habitual, well-defended feeding territories. Others feed in aggregations of a few to about 50 individuals, with a flock of up to 250 being seen in Madagascar. In these groups, herons may tolerate the close approach of other species. A spoonbill (*Platalea alba*) is reported to have stabbed its bill under the canopy, without protest from the heron (Duncanson and Paxton 1994). However, an ibis (*Plegadis*) was driven away when approaching too near (Hartmann 1995). In aggregations, Black Herons use Leap Frog Feeding to keep ahead of the moving canopy-forming group. They feed by day and especially around dusk. On the coast, they move to communal roost on high tides. They also roost communally at night.

Fish are the principal prey of the Black Heron. Aquatic insects and crustaceans are also taken.

Breeding

The Black Heron breeds at the start of the rainy season, when shallow feeding sites develop. The actual timing differs, November–January in South Africa, February–June in Madagascar, March–June in east

Africa, and July–August in west Africa. The Black Heron nests in trees over water. It can also nest on bushes and in reed beds. It nests in single-and mixed-species colonies of herons, ibis, and cormorants. Typically there are 50–100 Black Heron nests in a colony. However, in Africa, a few nests may occur within a larger colony or it may form huge colonies, for example a colony of 10 000 pairs of Black Herons was reported from Madagascar in 1950. The nest is a solid structure of twigs in tree nests, placed 1–6 m above water. They are usually well hidden in foliage of the nesting tree. The courtship behaviour has not been recorded. The eggs are dark blue, averaging 44 × 33 mm. The clutch is 2–4 eggs. Little is known about the breeding biology of this species.

Population dynamics

The population biology of this species is unknown.

Conservation

The greatest threats in Africa are human disturbance and avian predation at nest sites, and threats to the aquatic habitats on which the Black Heron depends. Human disturbance of previously large Black Heron colonies may have led to its nesting in smaller numbers in mixed colonies (Hancock 1999). The situation is complex in that it is found feeding in areas of heavy human use but is absent from areas that would seem to be prime habitat (N. Baker and E. Baker in prep.). Protection of nest sites is essential, especially through involvement with local people. Degradation of feeding habitat may restrict population size or recovery (e.g. Skinner *et al.* 1987). In Madagascar, human interference and habitat change have led to massive population reductions. The Black Heron population on Madagascar should be considered as globally threatened. Immediate development of an action plan for the Black Heron in Madagascar is called for.

Research needs

The feeding biology of this species is well known, but no information is available on its basic breeding biology and/or its demography.

Overview

This is a specialized *Egretta*, adapted to a peculiar feeding behaviour making use of its wings to form a canopy. For example, its flight feathers are particularly broad and shaped to help form the closed canopy. The plumes also help complete the canopy. Its distinctively coloured feet are similar to other egrets, but interestingly, unlike many other species, the Black Heron's legs do not change colour in the breeding season, suggesting that dark legs are important in the functioning of the Canopy behaviour. Feeding alone, in small flocks, and even large aggregations, the ability of the species to adequately catch fish in shallow open water using its unique shadow-casting behaviour is just short of amazing. It may be in more mobile species than previously appreciated.

Tricoloured Heron *Egretta tricolor* (PL Statius-Muller)

PLATES 6.6–6.8

Other names: Louisiana Heron in English; Garceta tricolor, Garza pechiblanca, Garza flaca in Spanish; Garça tricolor in Portugese; Aigrette tricolore in French; Dreufarbenreiher in German.

Ardea tricolor PL Statius-Muller, 1776. Natursyst. Suppl. p. 111: ex Daubenton Pl. Enl. pl. 350: Cayenne.

Subspecies: *Egretta tricolor ruficollis* Gosse, 1847: Burnt Savanna River and Bluefields Creek, Jamaica.

Description

The Tricolored Heron is a medium, thin heron, distinguished by its white underside and dark head, neck, and back, found along the coast in tropical and subtropical Americas.

ADULT: The Tricoloured Heron head and neck are slate-grey to black-grey. The relatively long and thin bill is brown-yellow to orange. The iris is brown. The lores are yellow. The head, bill, and neck are all very long and thin, and together, snake-like. The back and upper wings are blue-grey to black-grey. Lanceolate plumes on the back are purple-maroon. The chin and front of the throat are white or with a thin red-brown stripe continuing onto the chest. The belly and underwing are white. The legs are grey-yellow. During the breeding season the bill turns bright blue with a black tip. The lores become cobalt blue. The iris is scarlet-red. Head plumes are white and the feathers on the neck and mantle are violet-mauve. Filamentous mantle plumes are buff; there is a rufous tinge to the upper scapulars and the back of neck. The legs turn deep pink-maroon.

VARIATION: Males are larger than females (average 415 vs. 324 g) (Frederick 1997). The sexes are alike in plumage, but in breeding the female plumage colour is less intense. Geographic variation occurs. The race *ruficollis* is larger, with slaty upper parts, white chin, and front of neck. *Tricolor* is smaller, with lighter back and back of neck, and a red to chestnut line on the foreneck. Baja California birds are larger than those from east North America. Trinidad birds have the chin and middle of the throat chestnut and front of the neck more rufous. Additional studies of geographic variation are needed.

JUVENILE: Juvenile birds have an olive-brown back and wings and brown neck. The chin, foreneck, and belly are white. The breast is streaked chestnut-grey. The bill is yellow with a dull black tip. The legs are green-yellow.

CHICK: The chicks have dark grey down covering the backs and wings and the white belly. They have a crest of long brown hair-like down. The bill is light green. The iris is yellow-white.

VOICE: The 'Aah' call is the aggression call, used also in Upright and Forward displays. The short guttural 'Aahrr' call is the alarm call. A 'Uunh' call is given as part of the Snap display. The 'Culh', call, rendered 'culh, cuhl' or 'houn, houn', is the flight call and used mutually in the Greeting Ceremony. The 'Scaah' call is used to approach the nest. The Bill Snap is used in Stretch. Wings make a 'Whomp' sound in Circle Flight.

WEIGHTS AND MEASUREMENTS: Weight: 200–415 g; length: 60–70 cm.

Field characters

The Tricoloured Heron is identified by its white belly and slaty dark head, neck, and back. In profile its long bill, head and neck, and slender body are readily identified. It flies with steady wing strokes, head and neck drawn back. It is distinguished from other medium herons in its range by its combination of dark back and white undersides.

Systematics

It is likely most closely related to the other New World medium-sized *Egretta* (Sheldon 1987). Two subspecies are recognized on the basis of size and some colour difference. Birds of Baja California (*occidentalis*) and Trinidad (*rufimentum*) also have been recognized as subspecies, but later synonymized. The taxonomic implications of geographic variation within the species need to be better understood.

Range and status

The Tricoloured Heron occurs in eastern and southern North America, Central America, the West Indies, and South America.

BREEDING RANGE: Its North American nesting range is along the Atlantic and Gulf of Mexico coast from south Maine through Texas. Inland nesting occurs rarely and with varying success in such inland sites as South Carolina (Belser and Post 1987), west Texas, Kansas, and South Dakota (Frederick 1997) (not mapped). It nests along both coasts of Mexico, Central America, the Greater Antilles (Cuba, Dominican Republic, Puerto Rico, Jamaica) and the Virgin Islands, and coastal South America to the mouth of the Amazon River in Brazil on the Atlantic coast, along the Caribbean coast (Venezuela, Colombia, Aruba, Curacao, Bonaire, Trinidad) and to Peru (occasionally as far south as southern Peru) on the Pacific coast. The race *tricolor* occurs in northeast Brazil, to eastern Venezuela, and Trinidad (if only two subspecies are recognized). *Ruficollis* occurs from North America south through Central America, West Indies, northwest Venezuela, Colombia, to south Peru.

NON-BREEDING RANGE: Birds winter through most of the breeding range, including locally in mild winters as far north as New Jersey, New York, and Delaware, USA. Its more common winter range is from South Carolina southward. Birds also winter in small numbers in south California. The core

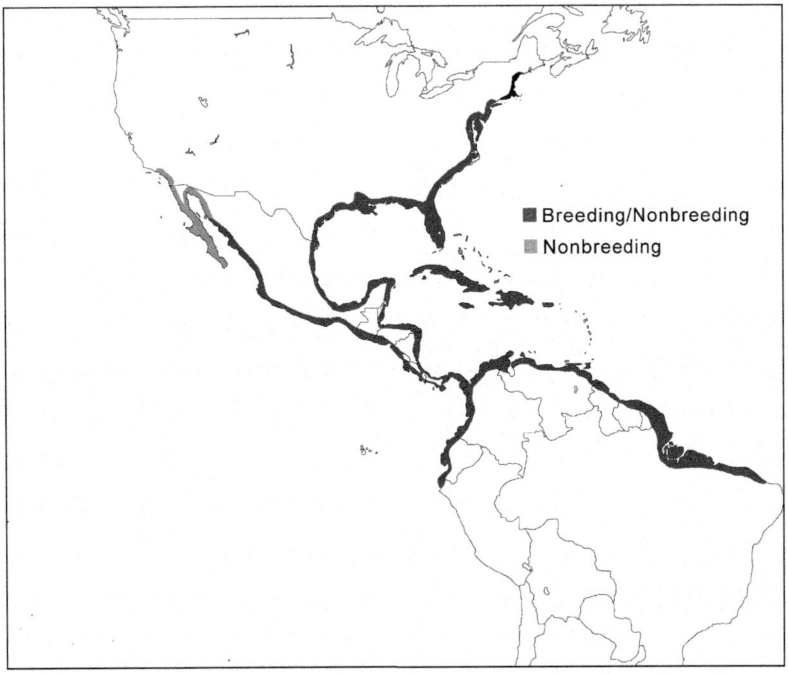

wintering range is probably south of the USA but full documentation of wintering range is needed.

MIGRATION: Birds from east and southeast USA migrate southward following the coast to Florida and then to wintering areas, some through Bahamas, to the West Indies and also to Central America, and Colombia and Ecuador. Birds from Texas migrate south to Mexico and Central America. Return is in February and March. Some birds make a cross-gulf migration. South American birds apparently are sedentary. Wintering and resident birds overlap in unclear ways from southern USA through northern South America. Postbreeding dispersal is characteristic of this species. As the population has expanded, vagrants are often seen throughout much of the continental North America from New Brunswick, Ontario (James 1984), Ohio, Kansas, Colorado, Oregon (Gillson 1994), inland Texas, and Arizona and in South America to Peru. Vagrants also occur in the Lesser Antilles, Fernando de Noronha (Oren 1984), and the Azores (Parrott *et al.* 1987).

STATUS: The Tricolored Heron is common along its coastal range and rare inland. Having not been the target of the plume trade, its numbers were not reduced in the early 1900s to the extent of other species. After the 1940s populations increased range wide, and it substantially expanded its breeding range from east North America, to Virginia through Massachusetts. It also expanded its winter range in east North America and winter populations increased markedly in Louisiana (Fluery and Sherry 1994). In the 1970s the USA population exceeded 200 000 birds. Large scale surveys, such as occurred in the 1970s, have not been repeated. However, information from Florida and Louisiana suggests decreases in nesting birds in areas where they previously were more abundant. Outside the USA information is scarcer. It is especially common and widespread in Mexico and fairly common in parts of Central America. It is especially common in Suriname (10 000 pairs) and Guyana (5000 birds) (Spaans and de Jong 1982, Dejohnge *et al.* 1992, Haverschmidt and Mees 1994).

Habitat

The Tricoloured Heron is a coastal heron but uses many habitats within this zone. It is a bird of shallow marshes, shorelines, and coastal lagoons. Among its specific habitats are mud flats, salt marshes, tidal creeks, mangrove swamps, open shallow bays, fresh water marshes, swamps, and crayfish aquacultural ponds. It particularly uses thick vegetation and shorelines. Although found in salt to fresh water, it nests more frequently in salt or brackish water areas. It has been seen as high as 600 m in Honduras. In Brazil is has been found in dry Ceara habitats (Teixeira *et al.* 1993). It favours natural habitats, unlike some other herons, and flies long distances to feed there (Smith 1995b).

Foraging

The feeding behaviour and ecology of the Tricoloured Heron is fairly well understood through studies over much of its range (Kushlan 1978b, Rodgers 1983, Kent 1986a, 1986b, Acosta *et al.* 1990, Ramo and Busto 1993, Chavez Ramirez and Slack 1995, Miranda and Collazo 1997b, Strong *et al.* 1997, Smith 1997, Frederick 1997). The Tricoloured Heron feeds during the day in shallow to relatively deep water, perhaps deeper than most other medium herons. It often feeds in dense aquatic vegetation, where it Walks or Stands on or among the plants. It tends to feed solitarily, scattered over the marsh or along streams and defends its feeding sites with Alert Posture, Upright, Forward, and Aerial Fighting. It does associate with feeding aggregations, however, when prey is abundant, assuming feeding positions tending the edge of the group. It Stabs prey, often making a nearly horizontal rather than vertical strike. It also Probes and Pecks and may steal food. It will Hop or take a short flight, Lunging at prey on alighting. It feeds by Walking slowly and Walking Quickly. It also feeds by Standing and by Running after prey it sees. It has a characteristic low Crouch while feeding in which the bird squats very deeply, with its neck and belly nearly touching the water. It uses Hopping, Jumping, Diving, Feet First Diving, Wing Flicking, Open Wing Feeding, and Underwing Feeding. In the active pursuit of prey, it may Hop, with flapping wings and lunging strikes, or it may spin its body in the direction of its open wing. It also uses other active behaviours on occasion, including Foot Stirring, Foot Raking, Hovering, Hovering Stirring, Swimming Feeding, Dipping, and Foot Dragging. It mixes up its repertoire of behaviours. It may Walk with slow steps stalking prey then run with wings flapping, turning before stopping to spot another fish. This tactical sequence is called 'Walk Quickly–Run–Open Wing' (Rodgers 1983). It may tend to feed more actively in open coastal habitats than inland (Frederick 1997).

Small fish are the primary food of the Tricoloured Heron, in places making up 90% of the diet. They particularly eat various species of killifish (*Fundulus*), the common small tidal marsh and tidal swamps fish in the region (Hammatt 1981, Kent 1986a, 1986b, Post 1990). They also eat frogs, prawns (*Palaemonetes*), snails, and insects when available.

Breeding

The nesting biology of the Tricolored Heron is fairly well understood owing to studies in several parts of its range (Jenni 1969, Rodgers 1977, 1978a, Alfaro and Russi 1989, Post 1990, Frederick *et al.* 1992). The nesting season is in the spring and summer (March–June) in east North America, although as early as February in Florida. Farther south, nesting depends more on the rainy season, and is variable from February to August in the Guyanas and Trinidad. They nest on islands and vegetation islands within coastal marshes, swamps, lagoons, and rivers, and inland in swamps and marshes. Very occasionally they are recorded on dry ground or other odd places. The nest is typically in trees, bushes, or herbaceous plants usually in thick vegetation and low, less than 3 m above ground in mangroves or willows and even lower when lodged in low reeds, such as *Phragmites*. Tricolored Herons are colonial, in small single-species colonies or in large mixed-species colonies, but generally toward the periphery rather than in the centre of the colony, reflecting their being less social than some of the other herons. The nest is a flat platform of sticks, 25–30 cm wide. It is made of twigs with leaves for lining, gathered nearby. The male stakes out its nesting territory by Alert Posture, Twig Shaking, Preening, Fluffed Neck, and Crest Raising displays and begins building a nest platform (Rodgers 1977, 1978a). It performs its courtship displays from the platform or nearby, especially using the Stretch display with all plumes erect, wings out and down, with and Snap or Twig Grasp at the down part and the Unh call at the peak. The Tricoloured Heron display combines features of usual the Snap and Stretch displays into a single sequence. This behaviour displays the coloured iris, lores, bill, and bill tip. Other behaviours are Twig Shaking, Circle Flight featuring exaggerated wing beats and extended neck, Preening with very rapid strokes, and Bill Clappering near the head and body while making a rattling sound. It defends the nest with Upright and Forward displays, the Forward involving extreme erection of the neck feathers, back plumes, and Crest Raising. Once the pair is established, the Greeting Ceremony includes Crest Raising, the Culh call, and Bill Clappering. After pairing, the male gathers sticks and the female continues to make the nest. The eggs are pale green-blue and measure 44 × 32 mm in North America, and may be smaller in South America, 44 × 30 mm. The clutch is usually 3–4 eggs (2–7), perhaps smaller, 2–3, in South America. The incubation period is about 21–24 days. Both members of the pair incubate. Chicks hatch asynchronously. They are brooded for 12 days and fed by regurgitation, grasping the parental bill crossways. After 2 weeks adults visit chicks only to feed them and by 4 weeks the chicks seek out adults for food. Young are altricial, growing quickly, especially their feet and legs (McVaugh 1972). They can move around the nest in 5 days and can climb from nest in 11 days. By 17 days, they are perching above the nest. They can fly away from the nest locations by 24 days and by 30 days form groups of young in the branches. They moult into juvenal plumage by 30 days. Windstorms, heavy rains, predation, and food limitation cause nestling mortality. Storms can disrupt nesting (Dindo and Marion 1992). Crows (*Corvis*), grackles (*Quiscalus*), and gallinules (*Porphyrula*) take eggs. Hawks (*Buteogallus*), mongoose (*Herpestes*), and Black Vultures (*Coragyps atratus*) take nestlings. Most nestling mortality is due to failure of the food supply. Brood reduction to 1 to 2 chicks is usual (see data in Frederick 1997).

Population dynamics

They probably first breed as second-year birds (21 months). Mortality is 79% in the first year and 32% after that. Longevity is up to 17 years (Clapp *et al.* 1982). Little is known about factors affecting adult mortality.

Conservation

The protection of colony sites from disturbance, the management of colony sites to avoid degradation, the creation of new and replacement colony sites (such as by dredge spoil deposition), the identification, protection, and appropriate management of feeding areas, and the management of crayfish aquaculture in Louisiana are important conservation actions for this species. If a population decrease has occurred, it is likely due to foraging habitat not providing sufficient resources for successful nesting. Feeding habitats along the coast and inland must be protected. In South and Central America and the West Indies, coastal development continues to be a threat. An important portion of the North American population now depends in winter on the availability of crawfish ponds. This resource needs to be continued. Although widespread and still abundant, evidence of population declines in parts of its North American range require further investigation. A near range-wide survey of equal intensity and similar (or improved) methodology to those of the 1970s is needed. This heron nests not only in large, obvious colonies but also in small colonies, so nesting sites are difficult to locate. Furthermore, its dark colour makes it difficult to count accurately by anything other than ground-based censuses. Determining its present population size and trend is its primary conservation need. With this information colony sites and important areas can be prioritized for protection and management.

Research needs

As noted above, a complete survey of the species' North American nesting range is needed to assess population size and trends. Additionally information on the species' biology, distribution and status in Central and South America is needed. Migration pathways, stopover locations, and important wintering locations for various subpopulations of North American birds need to be determined by banding and telemetry studies. Demographic, especially survival, information is needed.

Overview

The Tricolored Heron is a medium egret that feeds in open shallow coastal waters for small fish. Its long bill and neck are a characteristic feature of fish-eating herons. Its wide repertoire of passive and active feeding behaviours allows it to find, run down if necessary, and capture these fish. Mangrove swamps, coastal lagoons, and other open shorelines provide feeding habitat. It is a surprisingly solitary bird, gathering in small groups when food is very abundant. It defends it feeding territory in order to monopolize the scarce fish resource. Food availability depends on hydrology and tidal conditions (Strong *et al.* 1997). This is an example of a heron that is a colonial nester but preferentially a solitary forager. Its unique colour patter seems to support the idea of a solitary (dark back) active (light undersides) feeder. It is likely that the intricate biology of this coastal, small-fish eating, long-billed heron remains to be fully understood and appreciated.

White-faced Heron *Egretta novaehollandiae* (Latham)

PLATES 6.1, 6.2

Other names: Blue Crane, White-fronted Heron in English; Garceta Cariblanca in Spanish; Héron à face blance in French; Weißwangenreiher in German; Canguk australia in Indonesian.

Ardea novaehollandiae Latham, 1790. Ind. Orn. 2, p. 701: New South Wales.

Description

The White-faced Heron is a medium-sized, light grey heron with a distinctive white face.

ADULT: The adult is basically light blue-grey with a distinctive white head including the forehead, crown, chin, and upper throat. The crown pattern is variable with the white sometimes extending down the neck, the variability being sufficient to make recognition of individual birds possible (Maddock 1991). Iris colour also is variable, grey to green to dull yellow to cinnamon. The lores are grey-black. The bill is black with the lower bill tending to pale grey at the base. The feathers have pink-brown tones on shoulders and flanks. The chest has distinctive, long pale chestnut to bronze lanceolate plumes. Underparts are very light pale grey. In flight the wing appears barred, with lighter wing feathers contrasting with darker flight feathers and the leading edge. The flight feathers have black tips. Grey lanceolate plumes occur on the back. The tail is dark grey. The legs are green-yellow to orange-brown. During the breeding season, plumes become brighter, more numerous, and more prominent, grey on the back and chestnut on the chest. Lores turn grey-black and legs pink-orange (Maddock 1991).

VARIATION: The female is smaller than the male. Other variation in size and darkness has been described, which may be geographic. This observation has led to the description of several subspecies, *novaehollandiae* and *parryi* from Australia, *nana* from New Caledonia, and *austera* from Irian Jaya. These are not now recognized taxonomically. Albinistic individuals occur (Heather 1983, Selby 1987). The extent and dispersion of variation within the species requires additional examination.

JUVENILE: The immature White-faced Heron is paler and duller than the adult, with brown-grey shading on the forehead and sides of head, making these features less white than in the adult. There is a narrow pink-buff throat line. The underparts are brown-grey. The chest is buff-pink, without the chestnut of the adult. Immature birds lack lanceolate plumes on either their chest or back.

CHICK: The chick is covered with grey down.

VOICE: The 'Graak' call, also rendered 'graaw', is the species' most common call, given in flight, in interactions, and aggressive encounters such as Supplanting Flights. The 'Gow' call, rendered 'gow, gow, gow', is given on return to the nest. The 'Graak' call is also given upon approaching the nest. A high-pitched 'Wrank' call is an alarm call. An 'Oo' call, rendered 'ooooooh', and an 'Arg' call, rendered 'aaarrrgh', are given as alarm calls including when taking flight. A 'Crock' call, rendered 'crock, crock, crock, crock', is used in the Stretch display. A high-pitched 'Garik' call is a contact call.

Weights and Measurements: Weight: 500–550 g; length: 58–66 cm.

Field characters

The White-faced Heron is identified by its grey body colour and white face. It is a slender and graceful bird with a distinctive flight, slow and deep with bowed wings creating a graceful down beat. The neck is more frequently seen extended when in flight than in most herons, possibly because it tends to move only short distances when disturbed. It is distinguished from the Grey Heron by its smaller size, white face, slenderness, and graceful appearance. It is distinguished from the dark phase of the Eastern Reef-Heron by its more slender bill, white face and darker body, and barred underwing. It is distinguished from the White-necked Heron by its white face (not entire head and neck), white upper throat (without black throat spots), and light green to brown (not black) legs. It is distinguished from juvenile Pied Herons by having a white face (but not head), dark (not light to white) breast, and bicolour wings (without a white patch).

Systematics

This species has always been considered to be closely related to *Ardea*, but for some time it was placed in its own genus due to the absence of the plumes typical of *Ardea*. Its basic appearance and behaviour are more *Egretta*-like. Moreover, biochemical DNA analysis revealed it as a typical *Egretta* (Sheldon 1987). Infra-specific variation is unclear. Although several subspecies have been described in the past, more recently the species has been considered monotypic (Marchant and Higgins 1990). This is an individually variable, eruptive, and fast-spreading species, suggesting that a range-wide study is needed of geographic variation.

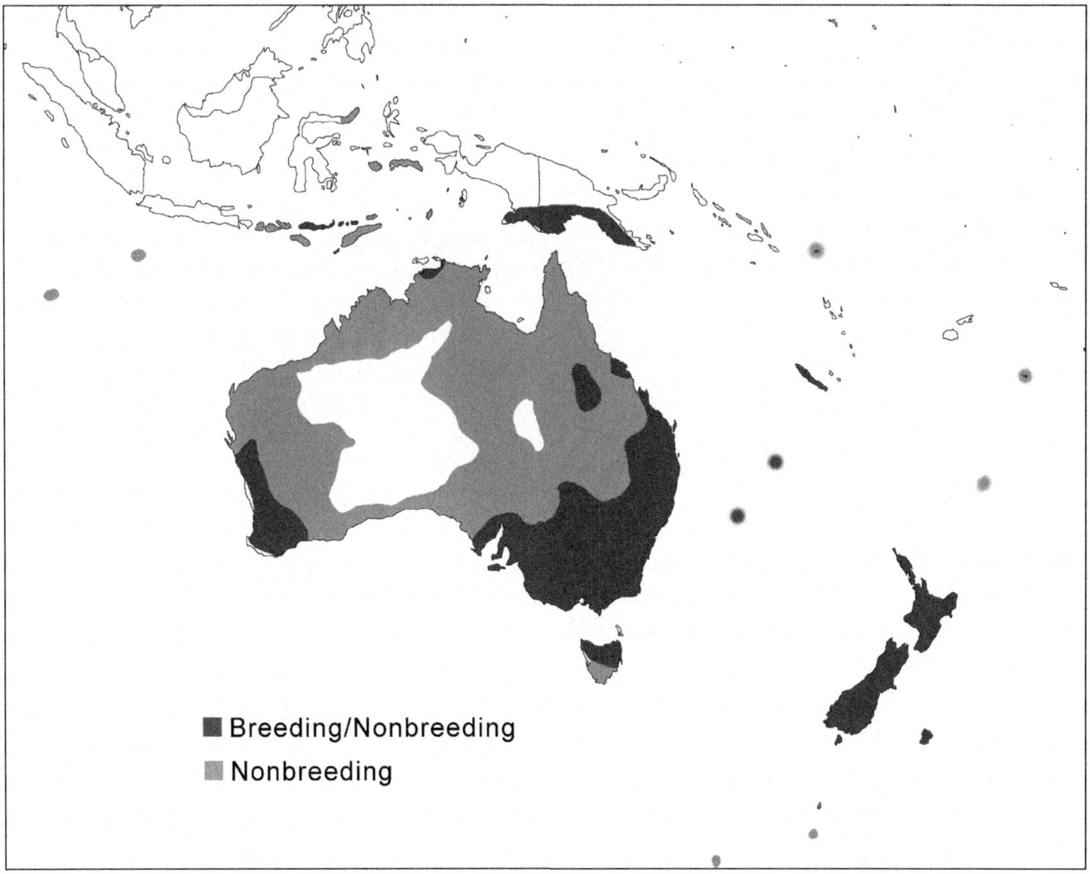

Range and status

The White-faced Heron is a bird of Australia and nearby East Indies.

BREEDING RANGE: The White-faced Heron breeding range includes Lesser Sundas and Moluccas Islands, south New Guinea, Australia except the inland deserts, the near-shore islands, Tasmania, Norfolk and Lord Howe Islands, New Zealand, and Chatham Island (Hemmings and Chappell 1988). Although now resident on Christmas Island, it is not recorded yet as breeding.

MIGRATION: The species is best described as locally nomadic moving about to take advantage of locally-changing hydrologic conditions (Recher and Recher 1980, Recher *et al.* 1983, Davis 1984, Lo 1991). Birds regularly wander widely, sometimes apparently for long distances, given its ability to colonize distant islands. It also has regular seasonal movements. In New Zealand, herons move inland annually in winter. In Australia, they move inland during the nesting season to flooded wetlands and then move back to the coast postbreeding. Herons accumulate in wet areas during annual dry seasons and droughts. Dispersal records occur throughout much of Australasia into the Indian Ocean. It is surprisingly commonly found on Lombok, Flores, and Sumbawa. Other dispersal records include Cocos-Keeling (Stokes *et al.* 1984), Tonga (Gill 1988, 1990), Sulawesi, Kermadec, Aukland, Soloman, Campbell, Snares and Macquarie Islands, Vanuaja (Dutson 2001) and recently as far north as Xiamen Island (China) (Chen *et al.* 2000a, 2000b, Fang *et al.* 2002).

STATUS: It is the most common and widespread heron in Australia and Tasmania. It has been increasing its range in Australia and is now the most abundant heron in New Zealand. The change in status in New Zealand is impressive (Carroll 1970). It was first reported there in 1868; first reported as a breeding species in 1941, and has increased expansively since the 1960s. Although first established along the coast, it is now found throughout both main islands.

Habitat

The White-faced Heron is highly flexible and uses a wide variety of habitats featuring shallow water (Davis 1984). It favours open aquatic habitats with trees or tree stands that flood for extended periods of several months (Briggs *et al.* 1997). Its occurrence in an area often depends on water conditions (Robertson and Heather 1999). Habitats used include tidal mud flats, sea grass beds, mangroves, reefs, saltpans, saline lakes and lagoons, beaches, dunes, rocky shores, lakes, stream margins, natural ponds, bilabongs, farm ponds, ditches, seasonally flooded grassland and pasture, and reservoirs. It also uses drier sites such as pasture, golf courses, urban parks, gardens, orchards, roadsides, and garbage dumps. It occurs as high as 1700 m in west New Guinea.

Foraging

White-faced Herons are active diurnal feeders, although they have been reported as feeding at night as well (Higgins and Smith 1999). Typically, they forage mostly by Walking slowly, Walking Quickly, or Standing, all in Upright or Crouched posture. They use Walking slowly to stalk their prey in shallow water or on dry ground. They Walk Quickly on land, exposed mud, or at the water's edge. They break into Running, chasing after prey, often extending and flapping wings to hop across the water. Their direction of movement is influenced by foraging conditions, such as glare (Davis 1986b). They frequently use Foot Stirring and Foot Raking in shallow water. Other behaviours seen are Head Tilting, Gleaning, and Wing Flicking.

White-faced Herons generally feed solitarily defending well-spaced territories, especially during the nesting season (Lowe 1983, Moore 1984). They also feed rather independently within loose small groups. They use Alert Posture upon being disturbed. Aggressive interactions are the Forward display, Upright display, and Supplanting Flights. In the Forward, birds raise the plumes on their chest and back and walk towards the intruder or walk in parallel, a behaviour that can develop into birds doing parallel strutting runs. In the Upright, the plumes are not raised. Supplanting can begin with one bird running after another or attacking from flight, often with a Graak call. Bill Jabbing has been reported, flying into the air while doing so. The Crouch is used as a submissive posture. Although large birds, they are at times subject to harassment by other species and potential predators (Rowe and Rowe 1998). Although usually solitary, in the non-breeding season and in terrestrial habitats they have been seen to feed in groups of dozens of birds, particularly after heavy rain or flooding (Maddock 1991). They occur in flocks of more than 50, when foraging for insects deep within a tall lucerne crop, and in a similar sized mixed-species flock, following a cultivator (M. Maddock pers. comm.). When feeding outside the nesting season in flocks, aggressive interactions are rare. They also feed commensally following other foraging birds such as ibises, cormorants, and spoonbills (Davis 1984, 1985a, Maddock 1991, 1992), and even defend their 'beaters'.

Adult herons are more efficient than juveniles (Davis 1984, Lo and Fordham 1986). Herons tend to feed on whatever is most available, and feeding intensity and success vary seasonally (Lo and Fordham 1986, Lo 1991). They frequently roost at times during the day on trees, rocky cliffs, or wetlands, particularly between tides. They also may alternate between terrestrial and aquatic feeding sites during the day. They are known to roost at night with other colonial waterbirds such as egrets, ibis, and cormorants (M. Maddock pers. comm.).

White-faced Herons feed on a wide variety of prey especially fish, crustaceans, and terrestrial invertebrates. Prey items appear mostly to be small (Davis 1984). Recorded food includes many species of fish, shrimp, crabs, frogs, worms, snails, amphipods, and insects (mayflies, stoneflies, bugs, lacewings, orthopterans, beetle larvae, lepidopteran larvae, flies). In New Zealand they eat the introduced Australian tree frog (*Hyla*). They consume flies at garbage dumps (Hobbs 1986), but previously accepted reports of this heron eating carrion are now doubted (Garnett 1993). They may eat vegetable material, but this too needs further study. Diet varies between the coast (aquatic invertebrates) and inland (fish and terrestrial invertebrates), and also seasonally (insects in summer to earthworms in autumn) (Lo 1991).

Breeding

This heron breeds during rainy periods when its feeding habitat floods, which occurs variably from year to year and from place to place in Australia. In Australia it has been recorded nesting in nearly all months. The most common starting dates are August–December, with a more constrained season in the southeast. In New Zealand, it nests starting in June and peaking in October. It has been recorded as nesting in many species of trees, from 4 to 40 m high, typically in open country near fresh or salt water and avoiding forests. Nesting trees are solitary or in small groves in open landscapes. It appears to prefer pastures with scattered trees and nearby pools. It flexibly nests also in mangrove swamps, rice fields, along roads, in golf courses, at the edge of wood lots, orchards, salt plains, suburbs and urban parks. On islands, it nests along the shore or shore cliffs on rocks or small crevasses. Sites are reused in successive years. It nests from sea level to over 1000 m. Nesting dispersion is variable. Over most of its range these herons are solitary nesters. They nest within defended areas, measured at 2–3 km apart. They also nest in scattered groups and they have been reported nesting in colonies in the largest wetlands. The nest is a platform of twigs lined with grass. Nests are often flimsy structures, about 40 cm wide and 10 cm thick. But they can also be substantial, particularly when reused in successive years. They are placed in the crotches of trees or out on limbs, sometimes far out from the trunk. Nests are sometimes placed over water, but this is far from required as some nests are many kilometres from water. The adaptability of the species is shown by its willingness to nest on artificial structures (Wall 1986). Both sexes attend the nest site and nest. Building the nest takes 12 days. One bird gathers twigs; the other inserts them with the Tremble Shove motion.

The courtship of this species is hardly ever observed. Perhaps it is truncated in such nomadic herons. Courtship has been observed to take place at the nest site, in the nest tree and nearby trees, and likely (although not proven) on the feeding grounds. Twig Shake, Back Bite, Wing Preen, Bill Clapping, Circle Flight, and mild Forward displays have been reported (Lo 1984, Maddock 1991). The Stretch includes a low Crock call. In the Circle Flight the neck is outstretched and the Grar call is given. The eggs are blue to green-white. They average about 48.51 × 34.3–35 mm. Clutch size ranges from 2 to 7 eggs, more usually 3 to 5. Incubation begins after the second egg. Replacement clutches occur, and a pair has been reported to have raised a brood even after five successive clutches of 5 eggs each had been collected from the nest. Incubation is by both sexes, lasting about 24–26 days. Both sexes attend the nest site and nest. The Greeting Ceremony on nest change-over involves raising the plumes, Bill Clappering, Preening, and giving the Grak call. Young hatch asynchronously. In the early stages of chick rearing, one parent remains with the young but after 9 days the chicks are left unattended while the adults forage for food (Maddock 1991). Both parents feed young and constantly guard them to 3–4 weeks. Food is brought at intervals of 40–70 minutes with frequency decreasing with age. The chicks seize the parent's beak crossways to stimulate regurgitation. Competition occurs among chicks. The young, if alarmed, adopt a Bittern Posture. The breeding season is very extended, lasting up to 8 months. Fledging takes 40 days after hatching. Adults appear to feed fledged young away from the nest site (Moore 1984). The juveniles may stay with their parents as late as the next nesting season, probably learning how to feed (Davis 1984).

There are few studies of breeding success. One study over five breeding seasons recorded a mean success of 1.4 young per nesting attempt, or 1.75 per successful nest (Maddock 1991). Nests and nestlings are lost to storms and predators including Kookaburra (*Dacelo*), magpies (*Gymnorhina*), harriers (*Circus*), and owls (*Ninox*) (Evans 1986, Geering 1988).

Population dynamics

Nothing is known about the population dynamics of this species.

Conservation

This is an abundant and widespread species that has been expanding its range in the past century (Maddock 2000). As a generalist in habitat choice and prey selection, it has been able to take advantage of many of the changes in the Australian and New Zealand landscape brought about by human development. Forest clearing, irrigation of pastures, farm ponds and reservoirs, and introduced prey have been advantageous to this species. Being solitary, it uses and protects its food supply during the nesting season. Being a solitary nester in remnant trees and a species that avoids forests, it is not disadvantaged by land clearing. This is a species for which conservation concerns are local. It is so ubiquitous that it may be an environmental indicator, specifically useful in the continent-wide monitoring of contaminant loads (Davis 1984).

Research needs

The extreme adaptability of this species makes it a candidate for study of its ability to respond to environmental fluctuations. Much is known about its basic foraging and food habits. Nesting success in relation to food supplies, survival in relation to environmental conditions, overall demography, residency and movement patterns of individuals, mate choice and courtship behaviour, and the extended period of juvenile dependence need to be better understood. The parallels between the White-faced Heron and the Cattle Egret, both of which have expanded their ranges and feed in man-altered terrestrial environments, deserve additional comparative study. Its potential role as an environmental indicator might be examined. Understanding interactions and relationships among nearby territorial herons may reveal much about the biology of solitary herons. Studies of marked birds should be used to determine year-round interactions among neighbouring birds and also their long-dependent young. Study is needed of geographic variation to determine if there are bona fide subspecies to be identified. The relation of this species to *Egretta* on the one hand and *Ardea* on the other should be further studied, especially using molecular techniques.

Overview

The White-faced Heron is an opportunistic heron adapted to taking advantage of changeable hydrologic conditions by feeding, nesting, and moving according to wet–dry cycles. It is one of the more generalist of all the herons. It is a solitary species, but one that can also feed or nest in loose aggregations or even large flocks when the situation allows. It is mainly sedentary, but locally nomadic according to conditions. It is a flexible feeder, taking a wide variety of prey from a wide variety of terrestrial and aquatic habitats and changing diets. It appears to depend for nesting on aquatic habitats, where it eats fish and aquatic invertebrates.

However, it just as typically forages on damp grasslands. It defends all-purpose breeding and feeding territories during nesting but is less aggressive outside the nesting season. Much can be learned from this species about flexibility among solitary herons.

Little Blue Heron *Egretta caerulea* (Linnaeus)

PLATES 5.9–5.10, 15.1

Other names: Blue Heron, Blue Gaulin in English; Garceta azul, Garza azul, Garza blanca, Garza pinta, in Spanish; Aigrette bleue, Petit héron bleu, Crabier bleu, Crabier noir, Aigrette bleue, Petit Héron Bleu in French; Blauriher in German; Garça-azul, Garça-morena in Portuguese.

Ardea caerulea Linnaeus, 1758. Syst. Nat. ed. 10, 1, p. 143: North America (=Carolina).

Description

The adult Little Blue Heron is an all-dark medium heron, usually seen stalking with great deliberation in shallow water. Uniquely among dark herons, the juvenile is all white.

ADULT: The Little Blue Heron has a grey-blue body and wings. Its head is purple-maroon. The distal one-third of the slightly downturned bill is black, slate-grey towards the head. The lores are dull green. Irises are yellow. The chin, throat, and upper neck may have some white. The neck has a variable maroon cast, and bears a few slate-blue lanceolate plumes. The legs and feet are grey-green, green or grey (Rodgers 1980a). In breeding, the head and neck turn red-brown. The head develops a tuft of long brown-red lanceolate plumes. The base of the bill becomes cobalt blue, the tip remaining black. The iris is grey-green. Lores are turquoise-blue. Lanceolate plumes on the neck are red-brown. Long lanceolate plumes developing on the back, some extending 10–15 cm beyond the tail, are slate-blue. The legs and feet turn black.

VARIATION: The sexes are alike in colour and plumage, but males are larger than females. No geographic variation is recognized (Parkes 1955), although a Mexican population is characterized as having a proportion of juveniles with chestnut-grey top of head and wings (Dickerman and Parkes 1968). A great deal of individual variation occurs. The red coloration on adults is variable in extent and in colour from chestnut to brown. Birds with slate bodies may have a variable amount of white on the chin, and may have white heads and necks. Birds with white bodies may have grey crowns and wings, and may even have dark plumes. Some birds have plumes tending toward filamentous rather than lanceolate.

JUVENILE: The juvenile Little Blue Heron is entirely white into the second year. The bill is pale green-blue at the base with the tip black. The iris is yellow to white. Lores are pale cobalt. Tips of the white flight feathers are dusky. The legs and feet are green-yellow to light grey. The moult is complicated and probably individually variable but well described (Rodgers and Smith 1995a). During the stage when white feathers are replaced by dark, the white plumage is followed by a 'piebald' or 'calico' intermediate stage.

CHICK: Chicks are covered by white-grey to brown-grey down, with a crest of long hair-like, tawny down. The down turns white within one week. The Bill is pink-grey at the base, pink-blue outward, and turning pinker in the first week. Chicks can be identified up close by their dusky wing tips. The Iris is light grey. Their skin, including the leg and foot, is pink-grey to grey.

VOICE: The Little Blue Heron is generally a quiet bird; and its vocalizations have been well documented (Rodgers and Smith 1995a). The 'Scah' and 'Sken' calls are the threat and aggressive calls, given during alert, defence, or attack. The 'Gwa' call is the flight call. A repeated 'Gerr' call is the disturbance call. The 'Unh' call is used in the Stretch display. The 'Ehoo' call, rendered 'eh, oo, ah, eh, eh', is the Greeting Ceremony call. The nestling begging call is a repeated 'Hee'. The Bill Snap is part of the Stretch.

WEIGHTS AND MEASUREMENTS: Weight 296–410 g; length: 64–74 cm.

Field characters

The Little Blue Heron is identified by its solid blue or white colour, green to green-grey legs (except in breeding), slightly down-curved bill, and its feeding method. It is distinguished from the dark Reddish Egret by being smaller and thinner, having a down-curved bill, and feeding slowly. It is distinguished from the Green Heron by its lack of dark crown, larger size, relatively long legs and neck. It is distinguished from the Tricoloured Heron by its dark belly. The immature Little Blue Heron is distinguished from the Snowy Egret (and white Little Egret) by its grey and black (not all black) down-curved bill, by blue-grey (not yellow in Snowy) lores, by dusky green-grey (not black) legs in non-breeding, by its green (not yellow) feet in all seasons, by dusky tipped primaries (usually not visible), and by its slow feeding behaviour. It is distinguished from the white Reddish Egret by being smaller, thinner, slighter, with green (not blue-black) legs and feet, grey (not flesh) bill base, and by feeding slowly. It is distinguished from the Great Egret and white Great Blue Heron by being much smaller as well as by its soft-part colours.

Systematics

The Little Blue Heron is one of the medium-sized herons, and appears from morphological and biochemical evidence to be closely related to the Snowy Egret. Subspecific variation has not been identified (Parkes 1955).

Range and status

The Little Blue Heron occurs, principally coastally, through much of the Western Hemisphere in North America, Central America, West Indies, and South America.

BREEDING RANGE: The normal breeding range in North America is from south Maine, along the Atlantic coast (more common south of New Jersey), along the Gulf Coast through east Mexico, nesting in a few locations south to Belize. It nests inland regularly in the southeast USA and up the Mississippi River Valley to Illinois, Indiana, west inland to Kansas, Oklahoma, and Texas. In the west it nests patchily in central and south California and Arizona, parts of coastal Baja California, the west coast of Mexico, inland in Jalisco and Guanajuato. It nests in Central America, west coastal South America to Lima, Peru, and north and east coastal South America to south Brazil to about 27°S (Olmos and Silva e Silva 2002, F. Olmos pers. comm.). It also nests sporadically outside this range in central North America—Arizona, New Mexico, Colorado (Ryder *et al.* 1989), South Dakota, North Dakota (Naugle *et al.* 1996), Iowa (Silcock 1984), and Minnesota. Throughout its range, the Little Blue Heron occurs in the breeding season in places where breeding has not been documented, including inland USA, West Indies, and Central America.

NON-BREEDING RANGE: The range during the non-breeding season is along the Atlantic coast from New Jersey (more commonly Virginia), south throughout peninsular Florida, USA, Bermuda, and West Indies to Lesser Antilles, Trinidad, and Netherlands Antilles, along the Gulf of Mexico in the USA and Mexico, in southeast California, Baja California, Mexico, through Central America to Panama, to coastal Peru, Brazil, Uruguay, Paraguay, and Argentina (Contreras 1993, Morales 2000, Casanas *et al.* 2002, F. Olmos pers. comm.).

MIGRATION: The northern populations of the Little Blue Heron are migratory, wintering as far as South America (Dusi 1967, Mikuska *et al.* 1998). In North America, eastern nesting birds move south in September and October along the coast to Florida, to West Indies, and South America. Birds nesting inland migrate south to the Gulf coast and follow the coast to Mexico, Central America, to north South America, some crossing between Yucatan and Cuba, the others Greater Antilles, Bahamas, and north South America (Colombia, Venezuela, Guyanas). Spring return migration is in March–April. Evidence suggests

migrations occur in South America as well (Morales 2000). Prior to migration, postbreeding dispersal is a typical feature of this species, at least in North America. Birds move in all directions, but especially northward before beginning southward migration. The Little Blue Heron is the one of the most dispersive of the North American herons, with a mean dispersal distance of over 1000 km (Melvin *et al.* 1999). Sightings outside the usual range are common. Birds occur sporadicolly across south Canada from Newfoundland west to southeast Saskatchewan, south through central USA, except the Rocky Mountains and Great Basin, and on the Pacific coast north to British Columbia (Innes 1993). In the south of its range, dispersal records occur south of the breeding range in Argentina (Contreras 1993). Sightings have occurred in Hawaii, west Greenland, the Azores, and an exceptional observation in South Africa (Bennett 1992, Brooke and Parkin 1995).

STATUS: It is thought that the Little Blue Heron did not suffer directly from the plume trade as other species did and so its population was not reduced at the beginning of the twentieth century. Populations in southeast USA were estimated to be about 135 000 birds in the 1970s–1980s (Butler *et al.* 2000). This species is hard to census, however, and there are few data on trends. As far as is known, there have been no significant range changes this century. There is some indication of local decreases in the northeast USA. Population and range are increasing in Baja California (Massey and Palacios 1994). The population appears to be stable overall in both North and South America (Butler *et al.* 2000, Morales 2000).

Habitat

The Little Blue Heron feeds in many types of shallowly flooded habitats. Over most of its range it is coastal but uses fresh water and estuarine marshes, tidal flats, salt marshes, mangrove swamps, coastal lagoons, rivers, streams, ponds, lakes reservoirs, canals, ditches, aquacultural facilities, rice fields, and flooded grasslands. It particularly feeds in or on emergent or floating vegetation, such as rush (*Eleocharis*), grass (*Panicum*), or cord grass (*Spartina*). The essential habitat requirements include vegetated areas of shallow water for foraging and isolated forest patches for nesting. It is generally a bird of the lowlands, but does occur to 3000 m, even at 3700 m in the Andes.

Foraging

The foraging ecology and food habits of this species are well known (Meanley 1955, Jenni 1969, Recher and Recher 1969, Mock 1974, Rodgers 1975, 1982, 1983, Kushlan 1976a, 1978c, Willard 1977, Dombe and McFarlane 1978, Caldwell 1981, Scott and Powell 1982, Burchsted 1983, Niethammer and Kaiser 1983, Kent 1983, Bancroft *et al.* 1994, Parsons 1995, Miranda and Collazo 1997a, Olmos *et al.* 2001). The Little Blue Heron typically feeds diurnally, in very shallow water, generally 5 to 15 cm deep, usually among emergent vegetation, in submersed vegetation, on the top of floating plants, or in open water next to vegetation such at the edge of pools, streams, mangrove creeks, and lakes. It typically feeds alone hunting usually by Walking very slowly through the shallow waters while carefully watching for potential prey. In one study, it spent 73% of its time in such slow feeding (Willard 1977). It proceeds in a slow, steady, deliberate, methodical way, Walking then Peering Over and continuing to Walk. It will stop and Stand, but not as frequently as it keeps Walking, but when doing so may focus for some seconds or minutes on a single spot before Stabbing, Probing or Pecking. It also Gleans prey such as crabs from vegetation (Olmos *et al.* 2001) and may use Head Swinging. It frequently uses its feet to stir up prey, usually Foot Raking but also Foot Stirring. Less frequently it uses more active behaviours such as Walking Quickly, Running, Hopping, Swimming Feeding, Wing Flicking, and Open Wing Feeding. In a variation of Open Wing feeding, called Whirling, the heron will spin around, head under its open wing gazing into the water (Burchsted 1983). It occasionally uses aerial feeding techniques, such as Dipping, Diving, Jumping, Hovering Stirring and Hovering over deeper water (Mock 1974, Rodgers 1975). It defends feeding sites, especially when food is not abundant, from all other foragers, including animals that would not be expected to be of concern such as diving terns. It uses Forward, Supplanting Runs and Supplanting Flights. It Steals prey and is in turn robbed by other species when it catches a prey item requiring a high handling time (Clark *et al.* 1989). The existence of territorial defense suggests that it is very advantageous for Little Blue Herons to be alone when stalking hidden or cryptic prey occurring in low abundance. It does join in small groups and less frequently in large groups tending to join mixed flocks rather than its own species and tending to take positions toward the outside of the group. Often it focuses on a single foraging companion, following it as it reveals prey. Feeding commensally it uses beaters as diverse as ibises and manatee, and it very typically follows Snowy Egrets (Kushlan 1978c, Caldwell 1981, Scott and Powell 1982) and can use Leap Frog Feeding. Herons feeding next to Snowy Egrets were relatively more successful than those feeding alone (Caldwell 1981) and it uses less energy feeding in aggregations than alone (Master *et al.* 1993). Even more intriguing, its metabolic rate is only 66% of what would be expected for its weight (Ellis 1980).

Several species of herons have distinct juvenile plumage, but no other starts white and ends dark, two major colour schemes of the typical herons. The selective advantage remains unclear. Juveniles in some circumstances forage less effectively than do adults, but not always. White birds are attacked more by hawks but less by other egrets, are allowed to feed closer to Snowy Egrets, can feed in brighter situations, and have reduced heat stress (Ellis 1980, Mock 1980, Caldwell 1981, 1986).

The Little Blue Heron's diet is globally diverse, including fish, amphibians, crustaceans, and insects. The proportions differ among feeding areas. Fish predominate in coastal east North America; in some areas it appears to be a frog specialist, and in yet other places crabs or shrimp predominate (Miranda and Collazo 1997, Smith 1997, Olmos *et al.* 2001). Invertebrates are often important in the diet, especially crayfish, shrimp, and crabs but also isopods, worms, crabs, beetles, flies, dragonflies, bugs. Prey taken typically are bottom dwelling or otherwise slow moving species.

Breeding

The breeding biology of the Little Blue Heron is fairly well understood (Meanley 1955, Werschkul 1979, 1982, Rodgers 1980a, 1980b, 1980c, 1980d, 1987, Frederick and Collopy 1989a, 1989c, Maxwell and Kale 1977, Smith 1995a, Smith and Collopy 1995, Naugle *et al.* 1996, Miranda and Collazo 1997a, Olmos and Silva e Silva 2002). Time of nesting varies with a peak in May in northeast USA, April in southeast USA, February–August in Trinidad, July–October in Venezuela and September–February in Brazil (Rodgers and Smith 1995a, Olmos and Silva e Silva 2002). In an area, Little Blue Heron breeding chronology tends to be relatively consistent from year to year, despite differing environmental conditions (Rodgers 1987).

The Little Blue Heron nests in trees and bushes, especially on islands. The nest is placed variably from 0.5 to 12 m high. Plants used include cypress, mangrove, pine (*Pinus*), button bush (*Cephalanthus*), willow (*Salix*), cattail (*Typha*), bamboo, cactus, and many more (Rodgers and Smith 1995a). The type of plant, and to some extent the vegetative structure, appears to be relatively unimportant to this species. More important, probably, is the protection offered by the site. It will use artificial islands, such as those created from dredge spoil. The Little Blue Heron is a gregarious, colonial breeder, usually nesting in intra-specfic groups and with other waterbird species. Within a colony they are not particularly synchronous (Olmos and Silva e Silva 2002) and may be affected by other, more aggressive species (Dusi 1968, Burger 1978a, 1978b, Dusi and Dusi 1995). The nest is a platform with a slight depression, built of sticks and other material reflecting local availability, as the male does not wander far from the nest to collect material. The nest is often small, 30–45 cm wide and 21 cm deep. The male starts the nest, collects most of the nest material, and presents it to the sitting female.

The male selects a display site, which it defends using Bill Thrusts and the Forward, which has a Bill Snap component. It advertises from the site, which initially is quite large, contracting as courtship progresses. This leaves space for new birds to set up territories. As a result, Little Blue Heron colonies can be asynchronous within a neighbourhood, very different from the Snowy Egret for example. The courtship behaviour of the Little Blue Heron is relatively simple compared with other egrets. The principal male advertising display is the Stretch, first given soon after the territory is occupied. The bird erects head, back and neck plumes, orients the bill upward to the vertical, lunges upward, gives a Bill Snap, drops down with drooping wings, and then rises up, extends the bill, and gives the soft Unh call, which may not be easily heard. Paired displaying birds also use Standing, Bill Clappering, and Wing Preen displays. Bill Clappering is an appeasement and pair strengthening display. Circle Flights are rare. A Snap display was not seen in the definitive study of this species (Rodgers 1980a). After pairing, birds engage in mutual rubbing, allopreening, and Bill Clappering (called bill-nibbling by Rodgers 1980a), the latter sounding like a rattling chatter. The Greeting Ceremony is used in nest relief. The Ehoo call is given at a distance of 5–10 m as the arriving bird walks to the nest. Copulation usually takes place on the nest, and promiscuous mating behaviour is characteristic of the species. These extra-marital copulations occur when the male intrudes onto a neighbouring female's nest site, usually when her mate is away. A female at the nest may mate with several males within a short period of time, for example 16 times with 3 males in 7 hours. Extra-pair copulations may be forced. Males therefore have to protect their reproductive investment by guarding their nest and mate. From pairing until the final egg is laid, the male does not feed and is seldom away from the nest site for longer than 5 minutes, and then only to collect nest material. The plumage distinction between juvenile and adult Little Blue Herons suggests that adult plumage is required for successful breeding. Rodgers (1978a), however, found that small numbers of subadults, mostly males, do attempt to breed.

The eggs are pale blue, 41–51 × 31–36 mm. The clutch is 3 or 4 eggs, range 1–6 eggs, laid over a period of several days, up to 8 days for large clutches. Average clutch size is smaller in Brazil than in North America (Olmos and Silva e Silva 2002). Replacement clutches can follow within 2.5–4 weeks. Incubation usually begins with the second egg or later and lasts for 21–23 days. Both parents incubate. Rodgers (1978b), however, observed an incubating heron remove four eggs from its nest by carrying each crosswise in its mandibles, presumably to deposit them in a new nest an observation unique for herons. Chicks hatch at 1- to 2-day intervals. Asynchronous hatching of eggs leads to brood reduction. The legs and feet of hatchlings grow rapidly, which allows them to be ambulatory by the time they are 13 days old. At this time both parents can leave the nest to forage. Both parents feed the young. For the first 5 days food is regurgitated onto the nest floor. Thereafter young take the regurgitated food by grasping the adult's bill crosswise. After 21 days the young are independent of parental care except for feeding, which continues until they are 50 days old. The can fly about 30 days. The nesting success of Little Blue Herons is somewhat variable, but overall fairly high. Survivorship is 73% from egg to 2 weeks. Overall survivorship to independence is about 50%, high for medium egrets. One to three young are produced per nest. Nestling mortality is usually the result of younger chicks starving, leading to a modest brood reduction. Predation is relatively low and random in occurrence in North America (Rodgers and Smith 1995a). In Brazil nest collapse, followed by predation, accounted for most nest failures (Olmos and Silva e Silva 2002).

Population dynamics

Given that its maturation is visible, it is known that Little Blue Herons tend to nest early. Most start when they are approaching their second birthday. However, first-year birds do attempt to nest, in one colony site up to 60% of the displaying birds but fewer than 2% of the nesting herons were juveniles. Nothing is known about adult mortality. Longevity is nearly 14 years (Clapp *et al.* 1982).

Conservation

This species, even if declining locally, is not of global conservation concern. Killing of Little Blue Herons at fish farms, legally or illegally, may impact local populations if not managed from a population viability perspective. Contaminant burdens have not proven to be excessive, although there may be heavy metal effects on reproduction (Spahn and Sherry 1999). Probably the most crucial conservation concern is the protection and management of nesting and feeding sites. Degradation and destruction of feeding habitat, particularly wetlands, on both breeding and non-breeding grounds, is occurring throughout the range. Important colonies need to be protected with buffer zones to inhibit disturbance. Dredge material islands should be designed and managed to enhance their use as colony sites.

Research needs

New methods for monitoring nesting Little Blue Herons need to be developed and monitoring programmes established to assess breeding population size and trend, especially in its North American range. Range-wide surveys should be conducted to identify important feeding, nesting, and wintering sites. Studies of the wintering biology of the species should be undertaken. The survival of postfledging juveniles and adults needs to be better studied in order to understand the demography of the species. Populational effects of nesting competition between Little Blue Herons and Cattle Egrets should be re-examined. The adaptive and ecological consequences of age-related plumage dimorphism need definitive examination. Patterns of subspecific population structure should be determined using biochemical methods.

Overview

The Little Blue Heron is a specialist in shallow water stalking, and it benefits from being alone where it can walk very slowly along taking care not to scare its potential prey, or by feeding commensally following another animal capturing the prey it disturbs. Its engagement in feeding aggregations apparently is an extension of its commensal feeding. Although its has a large repertoire of active feeding techniques, it hardly uses them. The fundamental adaptive strategy of this species appears to be minimizing energy use, which may account for its slow movement, its use of commensals, its subsistence on small sized prey, and its surprisingly low metabolic rate. An energy conservation approach to life allows it to catch fewer, smaller, more diverse prey than an active forager like a Snowy Egret. The resulting diet is both globally broad but regionally and temporally distinctive. Fish, crayfish, or frogs can each dominate the diet in different places or at different times. It chooses its diet to match the situation, so that in each area or each ecological situation, it tends to be locally specialized, concentrating on a particular subset of available prey, likely those that are most efficiently caught by its slow, deliberate feeding approach.

Snowy Egret *Egretta thula* (Molina)

PLATES 6.9–6.11

Other names: Garceta Nívea, Chusmita, Garcita Blanca, Garza Chica in Spanish; Aigrette neigeuse in French; Garça branca pequena in Portuguese; Schmuckreiher in German.

Ardea thula Molina, 1782. Sagg. Stor. Nat. Chili, p. 35: Chile.

Subspecies: *Egretta thula brewsteri* (Thayer & Bangs), 1909: San Jose Island, Lower California.

Description

The Snowy Egret is a graceful medium-sized white heron with black bill and legs, and yellow feet, found throughout much of the Americas.

ADULT: As its name implies, the Snowy Egret is all white. It has a long, thin, black bill, grey to yellow at its base. During the non-breeding season, the lores and irises are yellow. The long, thin legs are black with contrasting yellow feet, sometimes with yellow-green extending up the back of the lower legs. In breeding condition, it has luxuriant plumes. The head plumes form a short crest. The aigrette plumes of the neck and back are long and filamentous. The back plumes curve upwards near the tail. In courtship, the lores turn red-pink and the toes orange-red. The lores and feet flush blood red during aggressive encounters (Hancock 1999), which tend to be frequent in this species.

VARIATION: Variation in this species is slight. Males average larger than females but are alike in plumage. Birds breeding in Baja California (*brewsteri*) are larger with more massive bills. Other populations are reported to be larger or smaller than *tula*, but the overall geographic pattern remains unclear (Parsons and Master 2000).

JUVENILE: The juvenile is like the adult, without plumes. Lores are green-yellow to grey-yellow. The bills and legs gradually darken and so may be lighter than in the adult. The bill may be yellow-horn with a dark tip. The legs, although dark, may be lighter than the adult with green-yellow extending up the back of the lower legs from the feet, which also are green-yellow.

CHICK: The chick is covered with white down. The bill is pink at hatching, becoming darker at the tip, either yellow-grey or dark grey at the base. The bill then darkens to black or to yellow with a dark tip. The legs are yellow-grey to grey-black, with toes always being a lighter shade than legs, gradually turning yellow-green. Irises are grey to buff-grey becoming offwhite.

VOICE: This is a highly vocal heron, particularly during interactions among individuals. It has a range of aggressive calls, including the characteristic 'Rah' call, the complexity of which is not adequately captured by our current understandings. These calls become harsher as the intensity of the interaction increases. A high-pitched 'Aarg' call, rendered 'aar-gaarg', is a disturbance call. The 'Aarg' call is also used during supplanting attacks. A harsh high-pitched 'Aarh' is the Advertising Call. The male's Stretch display is accompanied by the distinctive 'Wah' call, a gurgling 'a, wah, wah, wah, wah' sound that characterizes a Snowy Egret colony. Bill Snaps occur during the Snap display. Study of Snowy Egret vocalization could be rewarding.

WEIGHTS AND MEASUREMENTS: Weight 370 g; length: 56–66 cm.

Field characters

The Snowy Egret is identified by its long back plumes, black bill, black legs, and yellow feet. It walks upright with neck slightly arched and flies strongly and buoyantly with deep wing beats, yellow feet trailing visibly behind. It is distinguished from the immature Little Blue Heron and from the white phase of the Reddish Egret by its black legs and bill, its yellow feet and lores, and by its active but not frantic feeding behaviour. These species often have dark or dusky feathers in their white plumage, sometimes having a grey cast. The Snowy Egret is distinguished from the Cattle Egret by its longer, thinner black bill, white not buff plumes, and its slender build. It is much smaller than the Great Egret and white phase of the Great Blue Heron. It is similar to Old World forms, but these are rare in the New World. It is distinguished from the Little Egret (which is rare but increasingly observed in east USA, Lesser Antilles, northeast South America and is now breeding on Barbados) by being smaller, thinner, with longer legs, yellow (rather than green) feet, the colour often extending higher onto the back of the lower leg, and yellow (rather than blue-grey) lores. In breeding, the Snowy Egret is distinguished from the Little Egret also by orange-red feet (rather than yellow-orange) extending up the back of the lower leg, red-pink lores (rather than orange), more elaborate head plumes (crest-like rather than two long lanceolate plumes), and longer, recurved back plumes (Massiah 1996, McLaren 1989). It is distinguished from the Chinese Egret (accidental in Aleutian Islands) by its black (rather than orange-yellow) bill, black (rather than yellow-green legs), and in breeding by its red-pink (rather than blue) lores.

Systematics

The relationships among the medium *Egretta* require additional study, particularly between the Snowy Egret and the Little Egret, which are similar in many ways, including plumage and behaviour (Vanden Berge 1970, Kushlan 1976a, Payne and Risley 1976, Murphy 1992, F. Hayes pers. comm.). Despite their breeding ranges now overlapping in the New Word, in Barbados they appear to interbreed rarely if at all, suggesting that they are reproductively isolated (F. Hayes pers. comm.). Patterns of subspecific geographic variation with the Snowy Egret are unclear (Parsons and Master 2000). While the birds of Baja California differ from other populations, there is much overlap in linear measurements (the only varying characters so far identified) among birds from the rest of the range. Questions include whether all birds west of the Rocky Mountains are appropriately identified as *brewsteri*, differences among the remaining North American birds (sometimes referred to as *candidissima*), and the intra-specific status of Central and South American birds.

Range and status

The Snowy Egret occurs in North America through Central and South America to Patagonia.

BREEDING RANGE: In North America, the principal breeding range is along the Atlantic and Gulf of Mexico coasts. Birds occur in summer in Nova Scotia but have not yet been seen to nest (Erskine 1992). They do nest farther south along the coast from Penobscot Bay Maine (Drennan and Bowman 1993), New Jersey, Delaware and Chesapeake bays, to coastal and inland Florida and Texas. They nest inland in east and central USA in South Carolina (Belser and Post 1987), Pennsylvania, Ohio, south Ontario (Weir 1984, 1986, Curry and Bryant 1987), Wisconsin, South Dakota (Naugle *et al.*

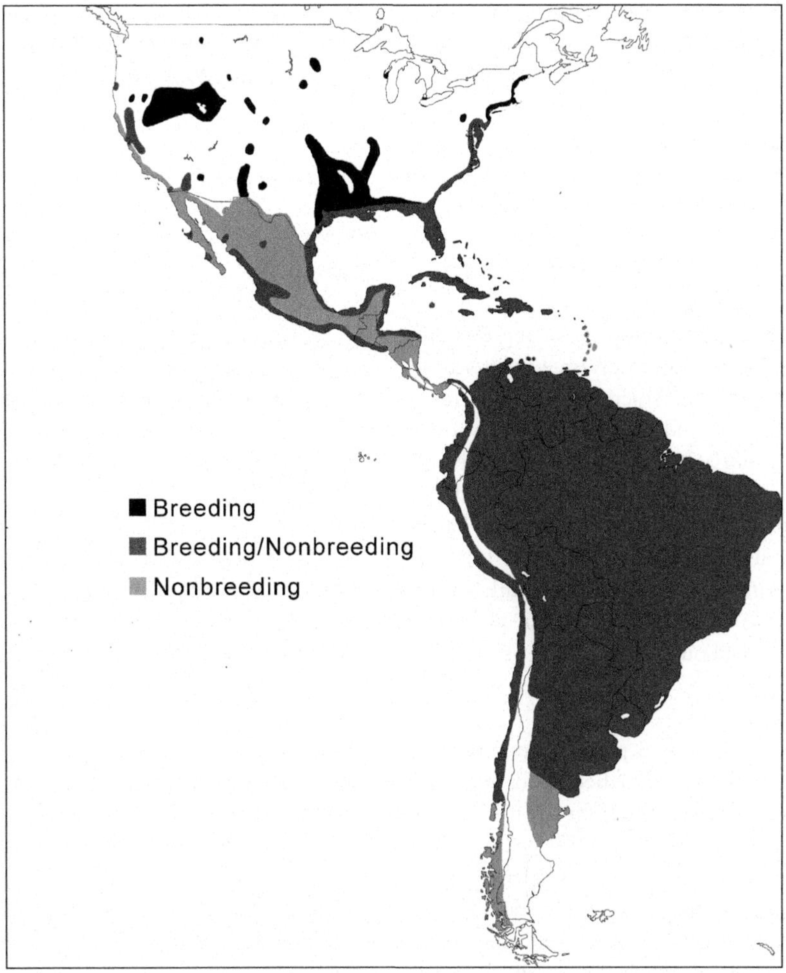

1996), along the Mississippi and Arkansas Rivers, and in a block from Louisiana to east Texas to east Oklahoma to south Kansas. It breeds sporadically elsewhere. In the west USA, it nests in central Oregon, central California, south California at the Salton Sea and Colorado River, and in disjunct areas and pockets in several western states of Idaho, Montana, Wyoming, South Dakota, Nevada, Utah, Arizona, and New Mexico. The breeding range continues down both coasts of Mexico where it breeds patchily in coastal wetlands (Massey and Palacios 1994, Becerril and Carmona 1997), and into coastal Central America, where it is currently documented as breeding in Costa Rica and Panama, but is assumed to breed elsewhere. In the Caribbean islands it breeds in the Bahamas, Greater Antilles, Virgin Islands, Cayman Islands, Guadeloupe, Antigua, Barbados, Bonaire, Aruba, Curaçao, Trinidad and Tobago. In South America, it breeds in extensive and fairly inaccessible wetlands through Colombia, Venezuela, Suriname, Brazil (do Nascimento 1999), as far south as Valdivia, Chile, on the Pacific coast, Cordoba, Argentina, in the interior, and Buenos Aires, Argentina, on the Atlantic coast. Its actual distribution in this vast area is not well documented. And likely it is as patchily distributed in inland South America as it is in North America. A single subspecies, *thula*, occupies most of this range. *Brewsteri* occurs in west North America; its northern limits are unclear.

Non-breeding Range: The usual nonbreeding range extends along the Atlantic and Gulf of Mexico coast from south New Jersey through all of peninsular Florida to Texas (coastal and inland along rivers). Birds are recorded wintering as far north as Maine. On the Pacific coast the non-breeding range includes Oregon, California, Arizona (lower Colorado River valley), throughout Mexico, to coastal Panama and Costa Rica. Birds winter in Bahamas, Greater Antilles, and Lesser Antilles to Trinidad. Birds occur year-round in the rest of the breeding range in South America, but movement patterns are not understood. North American birds occur in Colombia, Venezuela, and Guyana, but likely other areas as well. Important wintering areas for North American breeding birds include southeast USA, Bahamas, Cuba, and parts of the Gulf and Pacific coasts to Central America (Mikuska *et al.* 1997). In south South America, birds winter south of the breeding range to Patagonia, Argentina. They have been reported as far south as the Strait of Magellan (Wheelwright 1978).

Migration: The species is partially migratory in North America, with most northern birds and some that nest farther south migrating for the boreal winter. Snowy Egrets from northeast and east of North America migrate beginning in September and October flying along the coast to Florida, the Gulf of Mexico coast, the east Caribbean, and north South America. One eastern bird was reported from Panama, but this might be unusual for the population. Part of the population breeding in the southeast USA is sedentary. Mid-continent North American Snowy Egrets migrate to the Gulf Coast and through Mexico to Central America. Egrets from western USA are partially migratory, some remaining in central California, others winter along the Pacific Coast to Baja California, central Mexico, to Central America (Ryder 1978, 1998, Parsons and Master 2000). Some North American birds, mostly juveniles, remain in their wintering areas into the next breeding season (Parsons and Master 2000). Postbreeding dispersal and vagrancy lead to birds occurring outside their usual range. In North America, dispersal records occur across south Canada and elsewhere in the interior and, similarly, in South America. It occurs on ocean islands including regularly in Bermuda (Amos 1991), Falkland Islands (Anon. 1989), Tristan da Cunha Islands, Azores (de Heer and Bos 1989, Alström and Colston 1991), Great Britain (Jackson 2001, Rogers 2002), Hawaii, and American Samoa (Scott *et al.* 1983).

Status: As far as is known, in the nineteenth century the Snowy Egret was abundant in the south and southeast USA and common as far north as New York. During the first quarter of the twentieth century, it is probable that plume hunting reduced Snowy Egret populations throughout the Americas, although credible numerical estimates do not exist. It was certainly extirpated from South Carolina and much reduced in Florida. It likely suffered similarly south of the USA. Protection, from about 1910 in the USA but later in Central and South America, allowed recovery. In the USA the population increased and the breeding range expanded beyond the species' known historic limits, including into Oregon, Arizona, Colorado, South Dakota, Wisconsin, Ohio, Pennsylvania, New Jersey, New York, and Maine (Parsons and Master 2000). In the mid 1970s the coastal nesting population in east USA was about 159 000 birds (Butler *et al.* 2000). Populations in the continental interior are relatively small (Findholt 1984b, Findholt and Berner 1988). The Tabasco wetlands of Mexico supported another 60 000 birds (Scott and Carbonell 1986). Given that Central and South America remain unreported, based on these data the range-wide population would be at least 250 000 birds. In recent years local populations have fluctuated, although data are incomplete. Decreases were recorded in Maine, Massachusetts, Delaware, Virginia; stability, in Maryland, South Carolina, Texas; both increases and decreases in Florida; increases in the West Indies, and huge increases in Louisiana (Fleury and Sherry 1995, Raffaele *et al.* 1998, Butler *et al.* 2000, Parsons and Masters 2000). There is not now evidence supporting any overall decrease of this species in North America (Butler *et al.* 2000).

Habitat

The Snowy Egret is primarily a species of coastal wetlands and large river basins and their mouths.

Much of its range is coastal extending inland along river drainages including the California Central Valley, Mississippi River Valley, the Everglades, Orinoco basin, Amazon basin, and Parana basin. This is not universal, however, as it occurs in the continental interior in both North and South America. It has broad habitat preferences. It uses both fresh water and salt water wetlands, especially salt marshes, mangrove swamps, lakes, reservoirs, rivers, coastal lagoons, and bays. Its microhabitat preferences are open pools within denser marshes and swamps, small tidal creeks, open shallows at the edges of rivers, lakes, and lagoons. It requires swamp forest or bushes nearby for roosting and nesting. It also can feed in damp to dry grassland or dry bush, pasture, crayfish farm ponds, catfish farm ponds, rice fields, ditches, and canals.

Foraging

The foraging behaviour and ecology of this species has been well studied and much is known about its approach to feeding (Kushlan 1976b, Willard 1977, Custer and Osborn 1978a, Caldwell 1980, 1981, Hom 1983, Rodgers 1983, Itzkowitz 1984, Kushlan *et al.* 1985, Kent 1986a, 1986b, 1991, Itzkowitz and Makie 1986, Powell 1987, Frederick and Collopy 1989a, Master 1991, 1992, Frederick *et al.* 1992, Master *et al.* 1993, Ramo and Busto 1993, King and Leblanc 1995, Strong *et al.* 1997, Miranda and Collazo 1997a). The Snowy Egret is a diurnal forager in open aquatic habitats. It is an active and aggressive feeder employing a variety of techniques. It forages primarily by Walking slowly, and also by Standing and by Walking Quickly. It specializes in using its yellow feet to startle prey. Foot Stirring, Foot Raking, Foot Probing, and Foot Paddling are employed to capture small fish and crustaceans in shallow water. It frequently uses its wings, Wing Flicking, Openwing Feeding and Underwing Feeding. It uses Head Swaying, while standing. It often forages from the air, using Foot Dragging, Dipping, Hovering Stirring, Jumping, Plunging, Diving, Feet First Diving and Hovering, and even can catch prey in midair (Toland 1999). It uses Bill Vibrating or Tongue Ficking to attract small fish. It can use Standing Flycatching and Gleaning (Toland 1999). It usually stabs at prey, but also Probes and Pecks. It sequences behaviours to fit the circumstance. A typical fast moving sequence is Run–Hop–Open Wing–Foot Stir, similar to but less exaggerated than the Reddish Egret. A slow sequence is Walk–Peer Over–Foot Stir, similar to the Little Blue Heron but less deliberate. These sequences illustrate that the Snowy Egret's hunting is somewhat intermediate between these two species.

It is a highly sociable forager, feeding, traveling, and roosting in single-species and mixed-species flocks with other herons. Despite its social aggregativeness, it is also highly aggressive (Kent 1986b). It defends its individual area within an aggregation and attempts to encroach on other birds' areas, using Upright, Alert Posture, Forward display, Crest Raising, Bill Jabbing, and Aerial Fighting with Aah, Arg, and Raah calls. This bird's combat can appear quite vicious. It attacks the individual area of neighbouring birds by Supplanting Runs and Supplanting Flights. It Robs prey from each other and other species. It feeds with other species as well. On dry land it follows livestock to pick up insects disturbed by their grazing, much like the Cattle Egret and uses Leap Frog Feeding. It roosts with other species when not feeding and at night. Individuals do also hunt alone or in small groups (Itzkowitz 1984).

It is one of the most highly social and consistently flocking herons, and is the core species around which multispecies feeding aggregations develop (Kushlan 1976b, 1977, Caldwell 1981, Master 1992). Central to its sociality must be the symbiotic and commensal advantages of flock feeding, communal food finding, predator protection, and social stimulation. Snowy Egrets, like many other wading birds, are more efficient and have a higher capture rate feeding in aggregations than feeding alone (Kushlan 1977, Willard 1977, Erwin 1983, Master *et al.* 1993), because the active feeding of many birds renders prey more vulnerable by causing movements, deoxygenating water, reducing hiding places, and so forth. Aggregative feeding permits birds to remain productively feeding at a site longer (Master 1992), presumably because the flock prolongs the availability of prey. Feeding, roosting, flying to and discovering feeding sites in groups allow birds to take advantage of ephemeral food supplies and exploit diverse and changeable habitat conditions particularly the search for and discovery of high quality food locations (Gawlik 2002). Within the feeding flock, competition is great, and individual aggressiveness is used to secure resources.

The Snowy Egret has the broadest range of described behaviours of any North American heron (Kushlan 1978a). Different feeding behaviours have different efficiencies under specific conditions (Kent 1987) allowing birds to make optimization choices. Behaviours such as Foot Stirring, using its distinctively coloured feet, and Bill Vibrating or Tongue Flicking are particularly useful for targeting small fishes that are attracted to disturbances (Kushlan 1973a, Willard 1977). Egrets use aerial behaviours to access fish in deep water or scraps of dead fish floating in deep water (Kushlan 1972, Rodgers 1974). Prey robbing is both a threat and alternative feeding style. Snowy Egrets very typically increase their effectiveness by feeding in deliberate association with other species such as other wading birds, grebes, mergansers, or cattle, although the benefit of feeding associations vary with situations (Bennett and Smithson 2001). Standing is energetically efficient, but Walking, Foot Stirring and disturb and chase behaviours are more often used which, while less efficient, can be more productive (Kent 1986a, 1987). Snowy Egrets do not necessarily optimize efficiency or intake by choosing a feeding site (Itzkowitz 1984), but rather appear to adjust behaviour to exploit what prey are available (e.g. Maccarone and Brzorad 2002). Snowy Egrets have to forage for longer during the day than do similar species (Kent 1986b). The species appears to attempt to maximize its food intake as opposed to minimizing its energy expenditure, as contrasted, for example, with large herons.

The daily cycle of feeding and behaviour use reflects the diurnal patterns of prey availability and both diurnal and seasonal patterns of energy demand. In non-breeding season, the usual cycle is for aggregations to form in the early morning (Hom 1983, Erwin 1985, Master 1992). This is when the daily energy demand is high. In some circumstances aquatic prey can be highly vulnerable then, in that in shallow wetlands nocturnal deoxygenation can force them to the surface (Kushlan 1976b, 1978a). This is when simpler feeding repertoires tend to be used. The largest aggregations tend to form in the morning, and as the day goes on the birds tend to disperse to smaller concentrations and begin to use more complex repertoires. They revisit patches during the day, increasing their foraging effort (Erwin 1985, 1989). In tidal situations, feeding timing and flock formation usually depend on falling tides, which expose prey in shallowing water. Seasonally, flocks respond to changes in food supply, for example when spring tides replenish prey supply in salt marshes (Master 1992). During breeding, foraging intensity (such as strike rate) increases (Parsons and Master 2000) and feeding continues longer into the day when the schedule is constrained by nest site attending duties. The interaction of morphology, foraging behaviour, prey captureability, and habitat choice is highly constraining to the Snowy Egret. It has to work harder and longer than similar egrets to meet its energy requirements.

Its diet tends to be relatively restricted to small fish and crustaceans (prawns, shrimp, crayfish). Topminnows (Poecillidae), killifish (Cyprinodontidae), silversides (*Menidia*), and topsmelt (*Atherinops*) are among the fish taken in various locations. Other prey include fiddler crabs, crayfish, snails, aquatic insects, polychaetes, flies, grasshoppers, frogs, lizards, and snakes, and also food from dumps and sewage treatment plants. Thus the diet resulting from the disturb and pursue feeding tactic is specialized (Kushlan 1978a, Parsons and Master 2000), composed mostly of fish and crustaceans that concentrate or otherwise become available under specific circumstances of hydrology, tides, and water oxygenation. There is an inverse relation between the diversity of available prey and feeding behavioural diversity (Kushlan *et al.* 1985), suggesting that complex feeding strategies are relied upon during competition for limited resources. Coastal habitats are favoured throughout the range probably because tidal fluctuations provide diurnally renewable patches of small fish. Inland areas used, such as rivers and tropical swamps, have seasonal water level fluctuations that similarly make food available.

Breeding

Over such a broad range from northern to southern temperate zones, the breeding season is quite variable. In the temperate areas Snowy Egrets nest in the spring, often quite early, March–May in the USA. In the tropics it nests in relation to the rainy season or other conditions that affect prey availability. Other breeding dates are March–May in Puerto Rico, May–August in Panama and Trinidad, and November–January in Ecuador and Chile, September–October in southern Brazil (Petry and Hoffmann 2002). It nests in trees and bushes on islands and on 'islands' of vegetation within wetlands. Mangroves are preferred nesting sites in coastal tropics. Dredge spoil islands in the coastal lagoons of east North America have provided nesting habitat in lieu of natural sites. The Snowy Egret is highly colonial and nests in mixed-species colonies, some very large, typically in the core of the colony. Its nests are close together, often less than 1 m apart. Because courting males are attracted to a displaying group, nearby nests tend to be at the same stage. Snowy Egrets nest with a variety of other species, including other herons, ibises, spoonbills (*Ajaia*), boobies (*Sula*), cormorants (*Phalacrocorax*), darters (*Anhinga*), frigatebirds (*Frigata*), and gulls (*Larus*). Nests are made of sticks with reed, rush, or soft bromeliad (*Tillandsia*) as lining. Nests are shallow, flat, sometimes elliptical platforms, 40 cm wide, 27 cm deep. They are placed 3–10 m from the ground. The nest is built by the female with sticks brought by the male.

Males select display sites and advertise, with a full display of the extravagant plumes. The advertising male principally uses the Stretch. Birds doing low intensity Stretches tend to attract other males to form a displaying group. In the Stretch, the bird points its bill up and then pumps its head up and down from a few to up to 10 times, giving the characteristic repeated Wah call. Males mark and defend display and nesting territory with Upright, Alert Posture, Crest Raising, Raah calls, Forward, Supplanting Flights (of only a few meters), Face to Face Fighting, and Aerial Fighting. As courtship continues, pairing occurs. Males perform Aerial Stretch, Circle Flight, Tumbling Flight, and Jumping Over displays. Eggs are pale blue-green. Size appears to be variable: 42 mm × 30 mm (northeast USA), 33 × 26 mm (Chile), 43 × 32 mm (Trinidad), 40 × 31 mm (Suriname). There is limited evidence of replacement broods, but they will replace eggs lost during the laying period. Clutch varies from 2 to 6, generally 3–5 in northern colonies to 2–3 in the tropics. Both sexes incubate. Incubation is 22–24 days, beginning with the first egg. Asynchronous hatching of young leads to differential mortality, usually through starvation of the youngest chicks. Both parents feed the young, by regurgitation, initially by dropping food onto the floor of the nest and later directly to the young, who grab the parent's bill crossways. Chicks are altricial on hatching. They are brooded by both parents continuously for 10 days. Mass gain reflects feeding conditions during a year. Body and wings are feathered in 2–3 weeks. They are able to clamber from the nest in 20–25 days (McVaugh 1972). Other feathering took 4 weeks. They can fly in less than a month from hatching.

Nesting success is variable from place to place and season to season. The principal cause of nestling death and nesting failure is starvation due to a failure of the food supply, either in the short term or due to the longer-term effects of habitat alteration (Parsons and Masters 2000). Black-crowned Night-Herons, crows, snakes, and owls cause losses of eggs and chicks. Exposure deaths due to rain and storms also occur (Dindo and Marion 1992). For most colonies reported nesting success was 1.6–3 young per nest.

Population dynamics

First breeding is probably when approaching the second birthday, but nesting by first-year birds is suspected. Mortality of young after dispersal from the natal colony is high, 35%–60% (Erwin *et al.* 1996b). Mortality during the first year is 71.6%. Annual adult mortality is 52.4%. Longevity record is 22 years. Mortality risks for adults are predation, food failure, and contaminants. Predation risk is especially prevalent in the tropics (Caldwell 1986). Chlorinated hydrocarbon pesticides have been implicated in reproductive failure in Idaho, probably obtained in Mexico (Findholt 1984a), although range wide in the USA most studies have found below-effect contaminant levels (Mora 1996). Tissue and feather concentrations of mercury in Snowy Egrets in east USA are sufficient to raise concern (Burger and Gochfeld 1997).

Conservation

Uncertainty about recent population trends in the USA suggests the need for a range-wide survey, which should be readily possible given the conspicuousness of the birds. An action plan for this species within the southeast USA is desirable in that it has shown a tendency for population shifts. Provision of a suitable dispersion of nesting and feeding sites is crucial, especially given the tenuous future of crayfish aquaculture. In North America, tri-national conservation efforts need to be organized to monitor pesticide burden and determine any populational impact of contaminants. Important nesting and feeding sites for the species need to be identified and appropriately managed, both in North America and throughout its range. Coastal development is a threat to this species throughout its range, but especially in Mexico and Central America (Massey and Palacios 1994).

Research needs

A range-wide survey of the nesting population is needed over the east and south USA, designed sufficiently to detect changes since the mid 1970s. Surveys should also be conducted through Central America and north South America to better assess population status and dispersion in those areas. The species' distribution should be examined to detect important nesting, foraging, migration, and wintering areas. Colony site dynamics, colony shifts, philopatry, and relationships of food and habitat quality to nesting success require additional study. This is especially the case in declining colonies in the northeast USA, which need to be examined to determine causal factors. The foraging behaviour of the species presents additional opportunities to address fundamental questions of feeding ecology, both inside and outside of the nesting season. Given its simultaneous gregariousness and aggressiveness, it is a species that may show interesting patterns of individual choice of behaviours, if individuals could be identified and followed. Patterns of intra-specific geographic variation over its extensive North to South America range deserve closer scrutiny to determine the genetic populational structure, which could be of importance in designing conservation strategies.

Overview

This is a species characterized by its sociality yet also by its aggressiveness. Its huge repertoire of social and feeding behaviours can be mixed and matched to accommodate the situation at hand. The small prey sought are usually mobile species that require high-energy foraging to obtain. But this egret is small, light, nimble, and quick and it appears to maximize its prey intake, rather than conserve its own energy. Individuals take advantage of ephemerally available food supplies. Populations undertake local movements and the species' range has changed with time, such as its shift from the east USA to Louisiana to take advantage of the development of seasonally drying crayfish farms. This is an adaptable species that exhibits diverse behavioral, social and population approaches to increase energy intake.

Little Egret *Egretta garzetta* (Linnaeus)

PLATES 4.1–4.11, 15.1–15.4

Other names: Lesser Egret, Spotless Egret, Western Reef-Heron, Western Reef Egret, West African Reef Egret, Indian Reef Egret, Arabian Reef Egret, Dimorphic Heron, Madagascar Reef-Heron, Mascarene Reef Egret in English; Garceta Común, Garceta Sombria in Spanish; Aigrette garzette, Aigrette des Récifs in French; Seidenreiher, Riffreiher in German; Kleine Zilverreigre Westelijke Rifreiger in Dutch; Sikeshäger, Revhäger in Swedish; Малая белая цепля in Russian; Kuntul perak kecel in

Indonesian; Tagak (Tagalog) in Philippines; Ko-sagi in Japanese, Bai in Chinese.

Ardea garzetta Linnaeus, 1766. Syst. Nat. ed. 12(1), p. 237: 'in Oriente', restricted to Malalbergo, near Ferrara, NE Italy.

Subspecies: *Egretta garzetta immaculata* (Gould), 1846: Northern portion of Australia (=Port Essington, Northern Territory); *Egretta garzetta nigripes* (Temminck), 1880: L'Archipel des Indes (=Sunda Islands); *Egretta garzetta dimorpha* Hartert, 1914: West Madagascar; *Egretta garzetta schistacea* (Hemprich & Ehrenberg), 1828: Red Sea; *Egretta garzetta gularis* (Bosc), 1792: Senegal River.

Description

The Little Egret is a thin, medium, dimorphic (dark or white) heron, with a long, thin neck and bill, dark legs and yellow feet (in most forms), and active foraging methods. When breeding, the bird acquires distinctive head, chest and back plumes and red lores.

ADULT: The Little Egret is highly variable. Three coastal populations are dimorphic having both white and dark birds. Soft-part colours differ geographically with age, and with season. White forms have predominantly white plumage, although sometimes flecked with dark feathers. Dark forms are blue to grey with variable amounts of white, characteristically under the chin. Individuals with intermediate plumage are common. In non-breeding, short plumes may occur on the back and chest. Bill colours include yellow, brown, or black. The legs are either yellow or dark, dark forms always having dark legs. The feet of most populations are yellow. When breeding, the Little Egret develops characteristic plumes on the back of the head, lower throat, and back. Particularly distinctive are the two or three thin, lanceolate crown plumes, which can be up to 16 cm long. The elongated feathers at the base of the neck are lanceolate distally. Lax back plumes are long, exuberant, and slightly recurved, but do not extend beyond the tail. At the height of courtship, the bill is black and lores turn red, as do the feet in yellow-footed birds. The lores then fade through white to the yellow or blue-grey colour of non-breeding birds. Other soft-part colours may differ among populations or individuals (see below).

VARIATION: Males are larger than females and juveniles are smaller than adults (Nota 2000). Individual variability, the interbreeding of birds representing different racial types, and dispersive tendencies confound patterns of geographic variation, and therefore taxonomic characterization. Individual and populational variability involve morphology (bill shape, bill length, bill thickness, leg length) and soft-part colour (lores, legs, feet). Populations adapted to coastal environments, like some other coastal herons, are dimorphic and tend to have thicker bills (Wassink 1979, Hancock and Kushlan 1984).

Egretta garzetta garzetta is all white. Its bill is straight, sharp, dagger-like, and all black to dark grey, usually shiny and sometimes with a bit of yellow at the base of the lower bill. During non-breeding, the irises are yellow; the lores are blue-grey; the feet are yellow with colour extending up the back of the lower leg. During courtship, the bill remains black, however the lores become pink to red before fading. The feet also turn pink to red before fading back to dull yellow. As far as has been proven, *garzetta* does not have a dark form, although this is debated. Dark birds have been reported periodically within the normal inland range of *garzetta garzetta*, dating back to 1869 (Ashkenazi 1993). A dark phase *garzetta* would likely have a thin black bill, grey to dark grey plumage, blue-grey lores, and bright yellow feet. Whether such a bird would have a white throat is unclear. But some of these characteristics are within the range of variation of *gularis* and *schistacea*. The lack of specimens and molecular data and uncertainty in the underlying subspecific taxonomy have interfered with interpreting these sporadically seen dark, potentially *garzetta garzetta*, birds. Racial hybridization is also a potentially confusing factor. For example, a bird recently appearing in Europe, interpreted by many to be *gularis*, seems more likely the result of the pairing of *garzetta* and *gularis* (photo in Hancock 1999).

Nigripes is white. It is similar to *garzetta* except its feet are black not yellow although occasionally having yellow soles. Non-breeding lores are blue-grey becoming red in courtship. The irises become red in courtship, but the feet remain black.

Immaculata also is white. Non-breeding lores are yellow, an important distinction from *garzetta*. The legs and feet are black, with yellow soles. In courtship, the irises turn red, as do the lores before reverting to yellow. The feet remain black although the soles can turn red. The back plumes are the most spectacularly developed of the races. Interestingly, the long-standing confusion about the identity of yellow-lored Little Egrets may have originated as far back as 1833 when Edward Lear in the *Birds of Europe* seems to have painted an Australian *immaculata* rather than a European *garzetta* (Hancock 1999).

In addition to the two all-white subspecies, three of the named populations are dimorphic. These 'reef herons' are all coastal, as are other dimorphic herons. Each has an all-white and a dark form, as well as intermediate-plumaged birds. The dark forms are blue-grey, usually with a white throat. Some have some scattered white feathers.

Gularis populations are composed predominantly of dark birds, which range in colour from grey-black to pale charcoal-grey, with a distinctive white chin and throat. They often show a white patch on the leading edge of the wing and also a variable amount of white feathering on the upper wing. Light morph birds are rare, occurring mostly in the northern portion of the range and on Sao Tome Island (D. Turner pers. comm.). These birds are white but often have a varying amount of grey feathering scattered about. For both forms, non-breeding lores are yellow to pale green. The relatively long and thick bill ranges from yellow to horn-brown, but is not black. It is heavier and slightly more down curved than the bill of typical *garzetta*. This is a character difference that can be discerned in the wild as illustrated by Barlow and Wacher (1997). The legs are variably yellow-brown to dark brown with dull yellow feet but never black as in *garzetta*. In courtship the bill darkens and the lores become red, fading back to yellow. The legs become dark brown before returning to lighter shades. The feet become red, then return to yellow. Both morphs have two head plumes, but these are somewhat shorter than those of *garzetta*.

In *dimorpha* dark individuals are blackish, slightly lighter than *gularis*, usually with white throats and white wing patches on the carpel joint and sometimes with white speckles. Rare intermediate birds are paler, with mixed dark and light feathers. Light individuals are white occasionally with stray dark feathers. Non-breeding lores are yellow. The bill and legs are jet black, and the feet are yellow. The bill in some populations is thicker than that of *garzetta*, although this is not universally the case, such as in coastal east Africa (N. Baker and E. Baker pers. comm.). In courtship, lores and feet are red returning to yellow.

Schistacea presents a very confusing situation. It has never been properly characterized except geographically. Birds allocated to this subspecies are highly variable. Indian birds were formerly distinguished as *asha*, suggesting the extent of variation in these birds. Light birds are white, often with dark mottling. Dark birds are blue-grey with a white throat, occasionally with white wing patches. Non-breeding lores are yellow, blue-green, or even colourless. The bill too is highly variable, ranging from yellow to pale green to olive-green to light brown to black. It may be somewhat down curved and relatively thick at the base (but again with considerable variation). The bill of some birds is darker toward the tip, sometimes to one-third of its length. Legs vary in the same ways as the bill, but for any bird the bill and leg colours do not necessarily correlate. The feet of most birds are demarcated from the darker legs, being yellow to dark green, but this is not universal. In breeding, lores turn red or orange (Parasharya and Naik 1987). The bill can be yellow with a variably red to orange flush, turning to yellow-brown after egg laying. Legs are black or brown, variably flushed orange. Feet become pink to red returning to dark green. The typical *schistacea* is a somewhat larger, heavier bird than *garzetta*.

Identifying and separating *garzetta, gularis*, and *schistacea* have been a matter of detailed discussion (Hancock 1984b, 1999, Grussu and Poddesu 1989, Ashkenazi 1993, Dubois and Yesou 1995). A typical *garzetta* tends to be slimmer, with a longer, narrower, black (not variable) bill, blue-grey lores, is gregarious, and inland (not coastal). Cramp and Simmons (1977) suggested separating *garzetta* and *gularis* on the basis of tarsus to bill ratio, but due to variability this turns out not to be reliable (Ashkenazi 1993). Such variability in fact, including the intermediacy of many individuals, makes racial identification difficult. Such variation appears to be in part the result of intergradation among populations. The best example of this is in India, where *schistacea* birds have moved inland to nest in the same colonies as *garzetta*. The many examples of intermediate birds, presumably resulting from cross breeding, constitute strong arguments supporting the concept that these populations are not different species. Needless to say, the geographic and polymorphic variation in Little Egrets is not understood in any sufficient way.

JUVENILE: Juvenile plumage is extremely variable (Parasharya and Naik 1984). White juveniles may have varying amounts of mouse brown to grey feathers and can appear intermediate between pure colour phases. Dark colour birds may have white chins. Juvenile *garzetta* have a green bill with black marking, pale green lores, and dull black and green legs. Juveniles lack or have poorly developed plumes.

CHICK: Chicks of white forms have white down chicks of dark forms are dark, both, with longer crown down forming a crest. The bill is yellow-horn, the legs and feet are green. Dimorphic races produce a range of white, grey-blue, grey, and speckled chicks. There is much variation in bill colour of the chicks (Voisin 1991). Bills generally become deeper yellow then brown then usually black. Legs turn dark green then to brown with feet turning from light green to yellow-green. Chicks of *dimorpha* have relatively large legs and feet.

VOICE: The Little Egret is highly vocal. The 'Kark' and 'Kre' calls, the latter rendered 'kre, kre, kre', and other variants are aggression and flight calls. The 'Aaah' call is the taking off flight call and also used on feeding grounds. The landing call is 'Da-wah'. Advertising calls are variable, including a distinctive gargling 'La' call, rendered 'la,la,la,la,ah,h,h,h', in which the syllables are interrupted. The 'Dow' call and 'Po' call, each with 5–9 repeated syllables, are also Advertising Calls. 'Ggrow' is an aggressive call. A gulping 'Ow' call is used during the Stretch. Young call to parents with a 'Ka' call, rendered 'ka, ka, ka, ka, ka'.

WEIGHTS AND MEASUREMENTS: Weight: 300–710 g; length: 55–65 cm.

Field characters

The Little Egret is identified by its combination of thin body form, long neck (half of its total height), medium size, (usually) yellow feet, and when breeding its crest plumes, chest plumes, and long back plumes that do not extend beyond the tail. Flight is slow with intermittent gliding on rounded wings, head and long neck retracted, feet extending beyond the tail. It feeds actively.

It is distinguished from the Cattle Egret by its longer, thinner, usually dark (not yellow) bill, black (not light) legs, slimmer body, longer neck, upright posture, and white (not buff) plumes. It is distinguished from the Squacco Heron by its larger size, upright posture, and white (not tawny) back. It is distinguished from the Intermediate Egret by being smaller, with yellow feet, dark (not yellow bill) and, during the breeding season, by plumes that stop at the tail (not extend beyond). It is distinguished from the Great Egret and Eastern Great Egret by being smaller, yellow feet (in most forms), and in non-breeding plumage black to dark (not yellow) bill. The white form is distinguished from the white Eastern Reef-Heron by its longer, thinner, dark bill (not light with black tip), and dark (not green) legs. It is distinguished from the Snowy Egret in non-breeding by its larger size, longer thicker bill, blue-grey to pale (not bright yellow) lores, flattened head shape, and, in the breeding season, by the two long head plumes and the colour of the feet extending up the lower leg (Murphy 1992, F. Hayes pers. comm.). It is distinguished from the Chinese Egret by its blue-grey (not blue) lores, black (not variably yellow to dark) bill, and, in the breeding season, by fewer, simpler (not bushy) head plumes. The dark form is distinguished from the Black Heron by its lighter grey plumage, thinner bill, two crest plumes (lack of a shaggy crest), longer legs, and white throat. The dark form is distinguished from the dark Eastern Reef-Heron by its thinner bill, white throat (not just chin), longer legs, and upright posture.

Systematics

The Little Egret is the typical egretta heron, and has been demonstrated to be divergent from the larger herons previously thought to be more closely related, the Great Egret and Intermediate Egret (Chang *et al.* 2003). Hancock and Kushlan (1984) first proposed that several medium-sized Eurasian egrets were best considered as representatives of a single polymorphic species. Since then, the concept that these are closely related species has seen common agreement (e.g. Sibley and Monroe 1990) whereas the more radical concept of conspecificity has been accepted by some (e.g. del Hoyo *et al.* 1992) and much discussed by others (Hancock 1984b, Ashkenazi 1993, Yesou 1984, Grussu and Poddesu 1989, McLaren 1989, Newman and Holtshausen 1990, Saleh and Mohamed 1990, Voisin 1991, Murphy 1992, Dubois and Yesou 1995). The issue of the relationships within the little egret group, the classical Little Egrets (*garzetta, nigripes, immaculata*), the Dimorphic (or Mascarene) Reef Egret (*dimorpha*), and the Western Reef-Herons (*gularis, schistacea* including *asha*), is far from settled and remains one of the more intriguing taxonomic and evolutionary problems in the Ardeidae. While the dialogue has been valuable, it has not produced definitive answers. We had hoped that in the interim between our books, additional quantitative studies would have explored this intriguing issue and are disappointed to be unable to report new empirical evidence to either support or negate the taxonomic hypothesis proposed 20 years ago. Morphological studies have not yet examined specimens of all races. Biochemical studies of all the forms have not been conducted. Behavioural observations are not more robust or more helpful than before. What is known is that variability is high, neighbouring populations are hard to distinguish unless differently coloured, and these populations are known to interbreed, such as in Kenya (where birds appearing to be *dimorpha, schistacea*, and *garzetta* all occur) (D. Turner pers. comm.) and in coastal India, the Philippines, north Celebes and Kisite Island (where birds appearing to be *dimorpha, schistacea*, and *garzetta* occur). In retaining the Little Egret as a polytypic species, we again urge definitive, range-wide studies using a combination of biochemical, morphological, and behavioural characteristics. Of particular interest are the areas where the forms appear to interbreed.

Range and status

The Little Egret occurs over much of the Old World, and more recently into the New World. It occurs in Europe, Africa, Madagascar, Asia, East Indies, Australia, Pacific Ocean islands, and West Indies.

BREEDING RANGE: *Garzetta* is the widespread Old World form. In Europe, it nests in Spain including

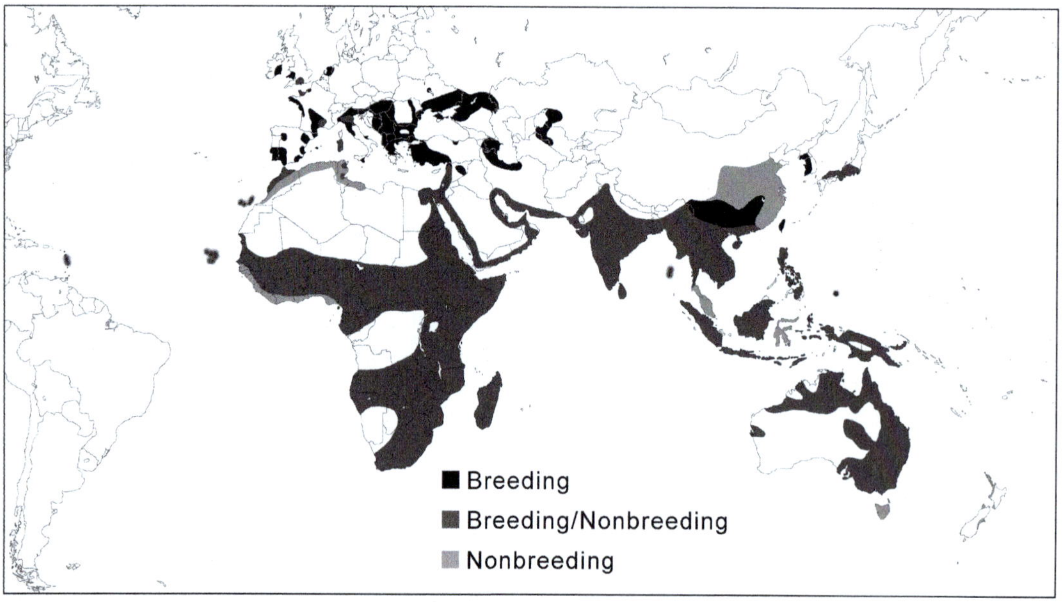

Canary Islands, Balearic Islands, Portugal, France, England, Wales, Ireland, Netherlands, Belgium, Italy (Sardinia), Germany, Czech Republic, Slovakia, Bulgaria, Austria, Hungary, Romania, Croatia, Yugoslavia, Albania, Moldova, Ukraine, Turkmenistan, Russia (to Volga Delta), Greece, and Turkey. In the Middle East, it nests in Turkey, Israel (spilling over into Jordan), Azerbiajan, Kazakhstan, perhaps Iraq, and Iran. In Asia, it nests in Turkmenistan, Uzbekistan, Pakistan, India, Sri Lanka, Nepal, Myanmar, east and south China (Guangdong, Guangxi, Yunnan, Szechwan, Hainan, Taiwan), North Korea, South Korea, Japan (north to central Honshu), Cambodia, Laos, Vietnam, Thailand, Singapore, Indonesia (New Guinea?), Philippines, Palau, New Britain, Australia (West Australia, Queensland, Victoria, Bass Straits Islands, Tasmania). In Africa, it nests in Egypt, Algeria, Tunisia, Morocco, Cape Verde Islands, Mauritania, Senegal, Gambia, Liberia, Mali, Ghana, Chad, Nigeria, and Gabon. Also in Sudan, Kenya, Uganda, Rwanda, Tanzania, Mozambique, Zimbabwe, Zambia, Botswana, South Africa, Namibia, and Angola (Baccetti 1983, Fesenko 1984, Ac and Bohus 1990, El Din 1992, Gustin and Pizzari 1992, Eminov 1993, Thuy 1994, Leibl and Hagemann 1997, Smiddy and Duffy 1997, Rajchard and Novak 1998, Schuster *et al.* 1998, Van den Berg 2000). It has recently become an established breeding species in Barbados in the Lesser Antilles, and is suspected of breeding in east USA, based on the observation of a likely hybrid Little/Snowy Egret (Murphy 1992, Perkins 1995, Massiah and Frost 1998). *Gularis* is the west African coastal form. It is distributed from southwest Mauritania, Senegal, Mali, Gambia, Guinea–Bissau west to Nigeria, Cameroon, Gabon, and Equatorial Guinea, as well as Bioko, Sao Tome, Principe, and Annabon islands, occurring principally along the coasts but also inland in some areas (Walsh 1987). It has been reported as nesting in eastern Spain (Dies *et al.* 2001); *Dimorpha* is the Madagascar and East African form. It occurs on Madagascar, Aldabra, Cosmoledo, Astove Island in the Seychelles, Kisite (southeast Kenya), Pemba, Zanzibar, and Mafia Islands and coast of Tanzania (Zimmerman *et al.* 1996, D. Turner pers. comm., Baker and Baker 2004). *Schistacea* is the northeastern Africa and Indian form. It breeds along the coasts of the Gulf of Suez, Red Sea, Indian Ocean, Persian Gulf, and Arabian Sea in Egypt, Sudan, Eritrea, Dijbouti, northern Somalia, Yemen, Saudi Arabia, Bahrain, Oman, Quatar, Kuwait, Iran, Pakistan, India (mainland west coast), and Sri Lanka. It also has been shown to breed inland at Lake Turkana, records supported by both observations and specimens (Owre 1967, Hancock and Kushlan 1984). Breeding as far south as Somalia, there is a gap in reef heron distribution to the start of the range of *dimorpha* in Kenya. Birds of *schistacea* have been introduced to zoos and parks in Austria and Germany by Pakistani dealers. Birds attributed to this introduction have been sighted in Europe (Wüst 1983). Dark morph birds nest in Israel (Ashkenazi 1993) and the subspecific identity of these and other dark birds in Europe remains unclear. *Nigripes* is the Indonesian form. It nests in Indonesia (Sumatra, Java, Borneo, Lesser Sunda, Mollucas, Palau islands). *Immaculata* is the Australian form. It breeds in northern and eastern Australia.

NON-BREEDING RANGE: *Garzetta* occurs in the non-breeding season widely in south Europe (Spain, Portugal, south and west France, Italy, and recently England) (Barbieri and Fasola 1986, Biondi *et al.* 1994, Hafner *et al.* in press, Smiddy and O'Sullivan 1998, C. Voisin pers. comm.), Middle East (Azerbiajan, Iraq, Iran), sub-Saharan Africa, Southeast Asia, east Asia (Japan, southwest China). Philippines, New Guinea, and Australia. Some Australian birds move for the non-breeding season to Tasmania, New Zealand, New Guinea, New Britain, and Philippines. Other populations spend the non-breeding season in their breeding range. Birds appearing to be *gularis* are reported from Europe (Germany) (Scherner 2001).

MIGRATION: Northern populations are partially migratory, and both northern and coastal populations are highly dispersive. Tropical populations appear to be non-migratory. In northern breeding populations of *garzetta*, dispersal occurs in July–September, migration occurs in late August or September, returning in March–April. Numbers wintering in Europe have increased since the 1950s.

In autumn, European birds move south to and through France and Spain to central and west Africa. Birds from west Spain move south to the south of the Iberian Peninsula to north Africa, whereas birds from east Spain disperse to the Mediterranean and then migrate to west Africa (Bartolome *et al.* 1996). European birds also move south to and through Italy to north Africa, some continuing a trans-Saharan migration to central and west Africa, where they winter in Senegambia, Mali, Guinea, Sierra Leone, Ghana, Nigeria, Cameroon, and Sudan. Birds from farther east in Europe move south and west to and through Israel, Iraq, and Iran, some wintering there others moving down the Red Sea coast to Eritrea, Djibouti, Somalia, and East Africa (Ashkenazai 1993, D. Turner pers. comm.). East Asia birds move to south China, Southeast Asia, and Philippines (Anon. 1987). Japanese birds are partially migratory, with much of the breeding population remaining year-round. Other birds move to the Philippines. Taiwan birds also migrate to the Philippines.

Postbreeding dispersal and spring migration overshoots are common, bringing birds to outside their breeding range. Dispersion records include central Russia, China (Hopeh, Shantung, Knasu), Korea, Cocos-Keeling Island, Norfolk Island, Christmas Island, Hawaii, Germany, Poland, Switzerland, Belgium, Denmark, Norway, Sweden, Netherlands, England, Ireland, Iceland, Faroe Islands, Madeira, Canary Islands, Azores, Newfoundland, Bermuda (Norton *et al.* 2002), Martinique, Puerto Rico, east USA, St Lucia, Trinidad, Tobago, Guyana, Suriname (Haverschmidt 1983, Forster 1989, Holmstrom 1989b, Combridge and Parr 1992, Ólafsson *et al.* 1992, Perkins 1992, Pratt and Bruner 1992, Woods 1994, Ryan 1997, Hayes and White 2001). Dispersion records in the West Indies, e.g. St. Maarten, (Norton *et al.* 2003) are likely now due to birds from Barbados.

Gularis may be partially migratory, although evidence remains sparse. Birds occurring in Gabon, where breeding has not been recorded, may be migrants or wanderers from breeding colonies on nearby Sao Tome and Principe Islands. However, this race certainly does disperse widely, as shown by records from the Cape Verde Islands, Azores, Spain, south France, Sicily, Hungary, Bulgaria, east USA, Puerto Rico, St Lucia, Barbados, Trinidad, and Tobago (Cardillo *et al.* 1983, Smith and Hutt 1984, Murphy and Nanan 1987, Georgiewa 1999, Butler *et al.* 2000, Kenefick and Hayes pers. comm.).

Northern populations of *immaculata* seem to be sedentary. Eastern birds disperse in the non-breeding season to Tasmania, New Zealand, and New Guinea. *Nigripes, schistacea,* and *dimorpha* also appear to be non-migratory. Dispersal records of *dimorpha* exist from Comoros, Seychelles, and Reunion Islands.

STATUS: Populations of *garzetta* were marked by declines and range contractions due to hunting in the nineteenth century. This was followed by a recovery that has now achieved expansion well beyond the historic range. Overall, it is common to abundant within its range in Europe, Africa, and Asia, widespread although not common in Australia, and now has a foothold in the Western Hemisphere.

In Europe, it is expanding its range to north France and the British Isles and populations have increased markedly in France, Italy, Spain, and Portugal (Marion *et al.* 2000, Besancon 2002). The total European population is now about 47 000–70 000 pairs. The largest European concentration of breeding Little Egrets is now in Italy. Also in Europe, its range is increasing with first nesting taking place in Netherlands in 1979, the UK and Ireland in 1998 (Voslamber 1992, Lock and Cook 1998, Anon. 1998, Smiddy and O'Sullivan 1998). In Netherlands, it bred intermittently over several decades, 1979, 1992, 1994–1996, and 2000. In the UK, the breeding population is increasing quickly, 2 pairs in 1996, 7 in 1997, 30 pairs in 1999 (Musgrove 2002). The egret was eliminated from the Volga Delta (Russia) in the early part of the century but is now stable at 3000–4000 pairs.

In Africa, breeding populations along the Mediterranean are few and scattered. About 40 pairs breed in Algeria; it is decreasing in Morocco with sites being abandoned; there are several small colonies in Tunisia and several hundred pairs in Egypt (Hafner *et al.* 2000). A total of 1897 nests of *gularis* were counted on Banc d'Arguin, Mauritania, in May–December 1997 (Hafner *et al.* 1998b). In Senegal several thousand birds breed in the Saloum Delta. In Mali, up to 1000 pairs of *garzetta* and 90–110 pairs of *gularis* breed in mixed-species heronries in the Inner Niger Delta. In Niger, the population was estimated to be 7166 (1994–7) (Brouwer and Mullie 2001). Population estimates for Tanzania are as high as 15 000–20 000 coastal birds (ascribed to *dimorpha*); *garzetta* is surprisingly common around Lake Victoria, where over 25 000 birds have been calculated to occur seasonally (Baker 1995, Baker and Baker 2004). Elsewhere, *dimorpha* is common throughout Madagascar, with up to 400 pairs breeding regularly in a mixed heronry in Antananarivo.

In the Middle East, populations have increased in Israel to 2000 pairs. There are up to 8000 pairs in Azerbaijan, 100 in Bahrain, 920 in Iran, and 110–120 in Oman. In Asia, it is common in India, Thailand, Japan, Java (1000 pairs at Palau Dua), and China (at two sample areas: 2000 nests in 10 heronries near Lake Poyang and 9700 nests in 5 heronries near Lake Tai) (M. Fasola pers. comm.). In Australia, its overall range has expanded during the last century, being recorded first in South Australia in 1952, Tasmania in 1957, Western Australia in 1965, and New Zealand in 1951 (Maddock 2000).

In America, the Barbados population was most recently 40 birds (Butler *et al.* 2000). Little Egret sightings are increasing elsewhere in the West Indies, such as in Trinidad and Tobago (Hayes and White 2001).

Habitat

The Little Egret typically feeds in open or sparsely vegetated shallow to very shallow waters in a wide variety of habitats, both inland and coastal, such as the edges of gently flowing rivers and streams, shallow lakes, pools, lagoons, irrigation canals, flooded meadows, open areas in swamps (*Melaleuca, Eucalyptus*, mangroves) and marshes (*Typha, Phragmites, Scirpus*), coastal mud flats, sandy beaches, rocky shores, coral reefs, mangrove-covered shores, and tidal streams. Use of artificial habitats is very typical, especially rice fields, fish ponds, and irrigation pools. It feeds frequently from floating vegetation. In Lake Victoria, East Africa, it uses floating hyacinth (*Eichhornia*), and may be seen in the hundreds flying about searching for unoccupied rafts to claim (N. Baker and E. Baker in prep.). Terrestrial habitats are not avoided, where it feeds commensally with cattle or other ungulates. It is generally a lowland species but is found up to 1400 m in Nepal, 1740 m in New Guinea, and 2000 m in Armenia. Protected trees, bushes, or islands are needed for nesting. In such sites they may nest in many situations. For feeding, studies contrasting artificial and natural breeding sites have been able to delineate microhabitat preferences (Hafner and Britton 1983, Fasola and Ghidini 1983, Fasola 1986, Hafner *et al.* 1986, 1993). Most typically the Little Egret feeds in shallow (10–15 cm deep), open and unvegetated sites where water levels and dissolved oxygen are fluctuating (tidally, seasonally, or daily) and where fish are concentrated in pools or at the water's surface. Egrets will switch habitats through the year, in one study from fresh water marshes to salt marshes to rice fields (Kazantzidis and Gouter 1996). Natural marshes are preferred to rice fields (Tourenq *et al.* 2001).

Foraging

The foraging behaviour, food, and feeding ecology of the Little Egret has been well studied (Fraser 1974, Clancey 1976, Hafner 1977, Voisin 1978, Fasola *et al.* 1981, Fasola 1982, Hafner *et al.* 1982, Fasola and Ghidini 1983, Hafner and Britton 1983, Davis 1985b, Balanca 1987, Britton 1993, Kersten *et al.* 1991, Berezovikov and Gistov 1994, Zhang *et al.* 1994, Kovacs 2002). The Little Egret feeds almost exclusively by day, when it is able to search for its prey visually. Conditions that disrupt water clarity affect feeding (Cezilly 1992). The bird's usual feeding technique is to Walk slowly through shallow, open water searching for fish or other prey. It periodically stops Walking to Stand and examine a potential prey item. It also may Stand for longer periods in the water, on the shore, or on a floating raft of vegetation. It intersperses Walking with more active behaviours such as Walking Quickly, Running after prey, Foot Stirring, Open Wing Feeding, and Double Wing Feeding (Demey 1986, Voisin 1991, Phalan 1997). Over deeper water it can feed by flying—Dipping, Foot Dragging, Jumping, and Hopping from one spot to another (Hirschfeld 1991). It uses its bill to attract fish (Tsuboshima 1994). In its usual Walking foraging, the neck may be stretched out in any position from vertical to horizontal. It Walks slowly with frequent halts to

Stand and wait for a prey item to show itself. Foot Stirring in the mud, sand or submerged plants is used to inspire prey to move. If no prey is seen after Foot Stirring, the egret may stir again or move on. Egrets pursue swimming fish with Open Wing feeding, Running rapidly back and forth, and Hopping from place to place. At the end of a foraging sequence, the bird will either Stab at a prey or stop to look more carefully. In rice fields, Walking accounted for 62% of foraging time, Walking plus Foot Stirring 32%, Standing 6%; whereas in streams Walking consumed 82% of the time, and Walking plus Foot Stirring accounted for the remaining 18% (Fasola 1994). Types of prey determine much of the variability in the foraging techniques used. Egrets catch large, mobile fish by Standing; medium, mobile prey by Walking and chasing; and small, and non-mobile prey by Walking (Yamada 1994). The catch rate of Little Egrets is high relative to other herons (Voisin 1978, 1991).

Little Egrets feed both solitarily and in aggregations. Solitary feeding occurs along the coast and inland, typically in situations where prey is not concentrated. *Gularis* is considered to be characteristically a solitary feeder (although it does feed in aggregations as well) (Bundy 1985). At sites of high food availability, Little Egrets feed by joining other birds to form single- and mixed-species aggregations. This may occur only in the morning when low oxygen forces fish or prawns to the surface, after which they return to solitary foraging (Kersten *et al.* 1991, Hafner *et al.* 1993). Striking efficiency and capture rates are higher when birds are feeding in aggregations than alone (Hafner *et al.* 1982, Cezilly *et al.* 1990). Juveniles are less efficient then adults when feeding alone, but efficiency becomes similar when they feed in aggregations, showing the energetic advantages of flock foraging for this species (Cezilly and Boy 1988). When feeding in aggregations, interactions with other species are complex. Little Egrets defend feeding sites and individual distances against nearby birds, including other species, notably night herons. Territoriality is strong both when feeding alone, such as along a linear stretch of beach, and even within aggregations (Chu and Wang 1988). In flock feeding they use opportunities to forage commensally, following species as diverse as cormorants (*Phalacrocorax*), ibis (*Threskiornis*), spoonbills (*Platalea*) or even jackals (*Canis*). They also frequently rob other birds of prey and are subject to being robbed by a wide variety of other species (du Plessis 1986, Amat and Aquilera 1989). Other ways of feeding include using floating bread or their bills to attract fish, following cattle, and riding bathing water buffalo (Parasharya and Bhat 1987, Barden 1990, Balasubramanian 1990, Tsuboshima 1994, Wooldridge 1998, Whitehall 1999). Birds roost when not feeding, and in the evening outside of breeding use communal roosts (Itoh 1984a, 1984b).

The Little Egret's food is mainly small fish, generally only 1.2–6 cm long, averaging about 4 cm. Fish include introduced topminnows (*Gambusia*), chub (*Leucisus*), perch (*Perca*), carp (*Cyprinus, Carassius*), stickleback (*Gasterosteus*), eel (*Anguilla*), gobies (*Gobius*), tench (*Tinca*), loach (*Cobitis*), mullet (*Mugil*), sunfish (*Eupomotis),* sand-smelt (*Atherina*), and many more species recorded. The egrets will also eat frogs and tadpoles (*Rana, Hyla, Pelobates*), toads (*Bombina*), and snakes. Insects include beetles (*Dytiscus, Cybister, Hydrobia*), dragonfly larvae, mole cricket (*Gryllotalpa*), and cricket (*Gryllus*). Crustaceans include prawn (*Palaemonetes*), amphipods (*Gammarus*), phyllopod (*Triops*), introduced crayfish (*Procambarus*), and crabs (*Hymenosoma*). Other invertebrates are spiders, worms, snails (*Assiminea*), and bivalves (*Musculus*). They also will eat lizards, snakes, and small birds such as small chickens and warblers (Howe 1989).

Little Egrets are hugely opportunistic, taking advantage of prey that are locally abundant or accessible. It is remarkable in its ability to find sites of temporarily high prey abundance, where if forms aggregations. In such situations interactions are intense, with considerable defence of feeding space, generalized aggression, supplanting, prey robbing, and commensal feeding. Prey capture rates are high, both because of the high prey availability and the commensal effects of the crowd. As feeding efficiency wanes, egrets move off to find another feeding spot with a high density of prey, or they feed solitarily. It is often the case that aggregative feeding occurs in the morning and solitary feeding later in the day. The diet is regionally variable, with fish or invertebrates being relatively more or less common in different places. The diet in natural marshes tends to be dominated by fish, whereas those feeding in rice fields tend to take invertebrates, fish, or frogs depending on availability. Coastal Little Egrets have a high proportion of crustaceans and molluscs, a diet correlated with the relatively thick bills of these populations.

Breeding

The breeding biology of the Little Egret has been well studied throughout much of its range (Blaker 1969b, Voisin 1976, Voisin 1977, Naik *et al.* 1981, Inoue 1985, Parasharya and Naik 1988, Zhang and Liu 1991, Prashant *et al.* 1994, Kazantzidis *et al.* 1996, Rao *et al.* 1996, Thomas *et al.* 1999, Parejo *et al.* 2001b, Wei *et al.* 2002, 2003). As would be expected, the breeding season varies across its large range. In the northern parts, the Little Egret nests in spring and summer, March–July in Europe and north Asia. In tropical areas, breeding generally coincides with periods of high rainfall, most often peaking with or immediately following the height of the rains. The main laying dates vary considerably throughout Africa (D. Turner pers. comm.). They are generally April–August in India, June–August in Vietnam, May–June in Borneo, January–April in north Australia, and November–January in south Australia. Little Egrets nest in many substrates, including trees up to 20 m high, bushes, reeds, on ledges or rocks. It will nest on the ground where protected and also in areas of human habitation (Subramanya 1996, Kobayashi 2000). The Little Egret is a highly gregarious species that nests colonially, sometimes in large mixed-species colonies of other herons (especially Black-crowned Night-Herons), ibises, and cormorants, with some colonies numbering in the thousands. Aggressive behaviour in colonies can expand the distance between nests to 4 m, but in dense colonies nests may be as close as 1 m apart. Coastal birds tend to nest in smaller colonies or alone. Nests are small platforms, 30–35 cm wide, 10–15 cm high, made of sticks. They are built by both sexes usually with sticks brought by the male. Nest material continues to be added through incubation (Zhou *et al.* 1998).

Males display from a succession of temporarily established advertising stations. These are defended vigorously by Upright, Head Flicking, Twig Shake, and Forward displays, first by the male then by both birds of the pair. In the Forward, the egret erects its crest, neck, and back plumes and gives the Ggrow call. Forwards have several intensities. Males also do Supplanting Flights. The Stretch is the advertising display, serving to attract females. The bird crouches, puts its bill vertical, erects its back plumes, and pumps the head straight up and then down, giving the Ow call. It goes up and down in either slow motion or in short, rapid movements, finally returning the bill to horizontal. Also when advertising, birds give the Doo or Po calls, stand Preening, and conduct Circle Flights with neck outstretched producing a loud whomping wing sound. After pairing, Preening, Back Biting with Non Contact Bill Clappering are frequent. The Greeting Ceremony involves the approaching bird giving the Da-wah call and the attending bird giving an Upright display and also calls Da-wah. Both birds give a Forward display with plumes and crest erected, and Bill Clapper.

The eggs are variable green-blue, fading to an off-white. Measurements vary regionally. Eggs from South Africa average 46.5 × 34.2 mm, from Sri Lanka, 42 × 33 mm, and from China 44.3 × 32.3 (Zhang *et al.* 1994). Clutch size varies geographically, within a range of 2–8. Clutches average 4.9 in Japan, 4.8 in Spain, 4.4 in China (Wei *et al.* 2002), 2.4–2.9 in Senegal, 2.6 in East and southern Africa, 2.5 in Madagascar, and 2.2 in Australia. Both sexes incubate the eggs, starting inconsistently with the first egg and providing continuous incubation after the second or third egg (Wei *et al.* 2003). The incubation period is 21–25 days. Chicks hatch asynchronously. The young are semialtricial and nidicolous. Both parents attend them during the guardian period of 10–15 days. However, in some situations one of the parents may abandon the nest after the guarding stage, possibly in order to renest (Fujioka 1989). Chicks are fed by regurgitation, first onto the bottom of the nest and after a few days directly from the parent's bill (Inoue 1985). Growth is rapid. Nestlings compete for food, with the older nestlings being more successful. The result of this competition is usually brood reduction (Inoue 1985). Mortality may be high owing to starvation (Voisin 1976, Voisin 1977). Young leave the nest at 35–50 days, timing that is variable among colonies.

Hatching success is high, 90.2% in Israel (Ashkenazi and Yom-Tov 1997), although egg mortality occurs in large clutches possibly due to limits on the number of eggs that can be effectively incubated (Fasola 1998). Chick survival is influenced by predation, weather, rainfall and hydrology, and prey availability (Hafner *et al.* 1986, Maddock 1986, Baxter 1994, Bellinato and Bogliani 1995). Of these, prey availability is usually the most critical. Birds have the greatest nesting success in years with high fish densities in natural marshes or high prawn densities in rice fields (Hafner *et al.* 1986). A primary cause of brood reduction is nestling mortality due to competitive aggression by siblings. Reproductive success differs among locations and among years at a specific site. In China the number of young per nest exceeded 3.3 with a survival rate for nestlings of nearly 88% (Wei *et al.* 2002). In Australia 1.46 young fledged per nest in dry years and 2.59 for nests in wet years, when food was more available (Maddock 1986). Similarly, a colony site from which egrets fed on prey concentrated in early morning by low oxygen had higher production (3.25 chicks per nest) than colonies without such vulnerable prey (2.69 chicks per nest) (Hafner *et al.* 1993). Predators attack chicks in the nesting colony. These include hawks (*Accipiter, Circus*), owls (*Bubo*), and crows (*Corvus*) (Kayser 1995, 1996). After fledging, young will frequently pursue the adults flying after them through the colony uttering load and raucous Ka calls; the parents will land in a tree to feed them (M. Maddock pers. comm.).

Population dynamics

A substantial proportion of Little Egrets appear to breed in their first year, unusually precocious for herons (Hafner *et al.* 1998a). Adult and juvenile survival, rather than nesting success, is the most critical demographic parameter (Hafner *et al.* 1999, 2000). The average number of eggs in a clutch size appears to be fewer than would be optimal for the production of the most young (Fasola 1998). This finding suggests that it may be more important to reduce stress on the adult than to maximize production of young in any one year. Survival of young varied in the first year from 6.5% to 55.2% (Hafner *et al.* 1998a). Hatch order of juveniles, the effects of mild winter weather, and food supplies immediately postfledging did not influence survival. Hatch order influences subsequent breeding success (Thomas *et al.* 1999). Adult survival is a constant 71.4%. Mortality factors for adults include winter kills, migration risks, and predation (Hafner *et al.* 1992, Combridge 2000). Hard winters in which feeding sites freeze over can kill wintering birds (Hafner *et al.* 1994). It is likely that successive mild winters have contributed to the strong range expansion in Europe (Hafner *et al.* 2000), and by extension its dispersal and colonization of other areas. In France, reproductive parameters, such as clutch size, have decreased over time corresponding to increases in population (Bennetts *et al.* 2000). This finding suggests that reproductive parameters are in part density dependent. Another important aspect of the population biology is the species' ability to shift among nesting sites, even over thousands of kilometers in response to adverse conditions (Hafner *et al.* 2000), although most dispersal among colonies appears unrelated to competitor or other environmental triggers (Fasola *et al.* 2002). Thus local population losses (such as through a hard winter) can be compensated by inflow from other areas.

Conservation

Human populations are large and dense in the range of the Little Egret. Fortunately, the species is tolerant of human presence except where persecuted. In India, 53% of colony sites were in or close to human habitation (Subramanya 1996). Colony sites must be safe from terrestrial predators and secure from direct human disturbance. Consequently, a critical conservation action is to expand surveys range-wide to identify, catalogue, preserve, protect, and manage important colony sites. To protect the species across its range, it is necessary to preserve a network of active and potential colony sites, and given the destruction of natural sites that has occurred, create artificial nesting sites in areas where feeding conditions are adequate but no suitable site exists. Natural wetlands are the traditional foraging habitat of the species. Loss of inland and coastal wetlands has occurred throughout its range, especially in south Europe, north Africa, Middle East, Iran–Iraq, and

India. This has resulted in the large-scale destruction of heron habitat. Threats remain in all areas, especially along the Mediterranean and Black seas, where some of the largest colonies occur. Remaining natural habitats need to be actively managed for prey and protected from further habitat degradation (Kazantzidis and Goutner 1996). Especially important are the few large wetlands that, over the past decades, have been critical to the species (Van Dijk and Ledant 1983, Pyrovetsi and Crivelli 1988, Puglisi *et al.* 1995a, Prosper and Hafner 1996). Some of these, such as the Mesopotamian Marshes and the Black Sea Lowlands, are under immediate threat.

Throughout the Little Egret's range, wetlands already have been drained for agriculture, including for rice farming and fish farming. Agricultural and industrial operations influence contaminant burdens of egrets nesting nearby (Berny *et al.* 2002). Nonetheless, rice fields and fish farms are now important feeding habitats for Little Egrets throughout their range. Rice fields have become exceptionally important feeding areas in some regions such as the Ebro Delta and Albufera de Valencia in Spain, the Po Valley in north Italy, the Axios Delta in Greece, and the Goksu and Cucurrova Deltas in south Turkey (Fasola and Barbieri 1978, Voisin 1978, Fasola and Ghidini 1983, Biondi *et al.* 1993, Fasola and Ruiz 1996, 1997). Rice fields support 39% of the Italian Little Egret population, now the largest in Europe. The Development of rice farming explains a large part of the species' population increase there (Fasola 1983, 1986). An excellent study of these colonies (Fasola and Alieri 1992b) demonstrated the strategy for conserving Little Egrets in developed areas: protection of a network of colony sites from disturbance and ground predators, identification of feeding areas, active management of identified feeding habitat to retain critical characteristics, and establishment of new colony sites in appropriate feeding habitat currently lacking nesting habitat. Changes in rice farming practices around the world raise concern as to the immediate and long-term effects on Little Egrets and other herons (Narusue and Uchida 1993, Lane and Fujiika 1998). Agricultural policy should be developed to protect and continue rice culture techniques of value to egrets. Fish ponds are another artificial resource used by Little Egrets across its range. Interactions of egrets with aquacultural facilities are a conservation challenge. Studies have suggested that there are both costs and benefits to aquaculturalists from egret predation (Ashkenazi and Yom-Tov 1996). In some areas the egrets have become dependent on these ponds (Young and Chan 1997). The loss of fish pond habitat in Hong Kong has caused a decline in the nesting Little Egret population there (Young 1998). Methods need to be found to preserve access to fish ponds where they are essential to population health.

Research needs

The Little Egret, thanks in large part to the work of Heinz Hafner, Mauro Fasola, and their colleagues, is now one of the best-known herons. Its breeding biology and feeding ecology are well understood and provide insight into other species. It also is one of the few species in which demography is at least partially understood, demonstrating the value of long-term banding studies. An expansion of similar studies into other parts of the species' range is highly desirable. A principal research need is to definitively unravel patterns of geographic variation and the taxonomic identity of the populations presently attributed to this species. As discussed above, much still needs to be discovered to fully assess the hypothesis that these populations represent a single polymorphic species. At the core of their ranges Little Egrets in Europe and Reef Herons elsewhere do differ on average in bill shape, body shape, and habitat. However, when light and dark birds come together, such identities become less discernable because individual variability blurs characters that might appear distinctive comparing typical forms. Studies have yet to document whether the Little Egret of the European and Asian mainland is dimorphic, seemingly an important issue. Furthermore, these are birds that disperse widely and shift populations in response to changes in habitat conditions, leading for example to coastal egrets starting to breed inland in India. Thus, some interbreeding may be relatively new, which in itself says little about relationships of the forms other than they are close. It is this variability, uncertainty, and repeated inability to tell specimens apart that suggests the hypothesis of conspecificity. Testing the hypothesis requires range-wide studies of morphological, behavioural, and genetic/biochemical characters. It is as likely that there is one species as it is that there are four or more species in this group.

Overview

The Little Egret is a species characterized by its flexibility, which permits a diversity of ecological strategies across its large range. Among populations currently identified as Little Egrets there is considerable variation in bill, leg, and plumage, some of which have ecological implications. Coastal populations tend to have shorter legs and stockier bills, probably to feed on harder prey. In dimorphic populations, the ratios of colour forms differ geographically, probably as a response to slightly differing selective pressures in different habitat conditions. Populations may be sedentary, migratory or partially migratory, and the proportion of a population migrating changes with time. This behavioural diversity likely balances migration vs. winter risks over time. Individuals move around and populations shift in response to local conditions, for example shifting from Spain to France due to drought. The tendency of individuals to disperse results in their colonizing distant places, including a recent invasion of the New World.

The Little Egret is able to adapt its foraging to the situation at hand. It has a wide repertoire of behaviours, some of which are used in exceptional circumstances, but, the vast majority of time it uses only a few behaviours, Walking slowly with Foot Stirring, Standing and Running, each used to capture different sizes and types of prey. By targeting particular prey types, it has a high prey capture rate, higher than any other European heron. The egrets' opportunistic yet efficient approach to foraging results in a high food intake, despite small prey size, and a resulting high reproductive output. Versatility, high reproductive capacity, early age of first nesting, high adult survival, diverse winter strategies, and dispersive tendancies resulted in the species being able to increase its populations and expand its range so remarkably over the past century.

Chinese Egret *Egretta eulophotes* (Swinhoe)

PLATES 6.12, 6.13

Other names: Swinhoe's Egret, Yellow-billed White Heron in English; Garceta China in Spanish; Aigrette de Chine in French; Schneereiher in German; Kara-shira-sagi in Japanese; Huangzui bailu in Chinese.

Herodias eulophotes Swinhoe, 1860. Ibis 2, p. 64: Amoy, China.

Description

The Chinese Egret is a white, medium-sized east Asian heron, with yellow bill and feet, and dark legs.

ADULT: This egret's plumage is all white. The bill is dusky with inner lower bill yellow to pink. The irises are pale white-yellow. The lores are variably yellow-green, to green, to grey (Long *et al.* 1999). The legs are pale to dark green with green-yellow feet. In the breeding season, the exceptional ornamental plumes develop to their greatest extent. Crest feathers are long and dense, sometimes over 11 cm long. The breast plumes also are long and the loose aigrette plumes on the back extend to beyond the tail. In courtship the bill is yellow-orange; lores are bright blue; legs are black, fading back to green-black; and feet are yellow.

VARIATION: The sexes are alike. There is no known geographic variation.

JUVENILE: Young are like the non-breeding adult, lacking luxuriant plumes.

VOICE: The voice of this egret is not described.

WEIGHTS AND MEASUREMENTS: Weight: 390–540 g; length: 65–68 cm.

Field characters

At some times of year the Chinese Egret is not easily distinguished from other medium white egrets, especially the Eastern Reef-Heron, a situation that caused it to go unrecognized as a species for a long time (Lansdown 1990, Poole *et al.* 1999). In the breeding season it is identified by a combination of black legs, yellow feet, yellow bill, blue lores, luxuriant crest, and also by its feeding behaviour, running back and forth. In the breeding season it is distinguished from the Eastern Reef-Heron by its lack of a jowl, long thin neck, thin (not down-turned) bill, long crest, longer, thinner black legs, and long, loose aigrette dorsal plumes. During the non-breeding season Chinese Egrets and Reef Herons are very similar with yellow-green legs and variable soft-part colours, but the Chinese Egret does not have a dark bill tip, has longer legs with a greater portion of the tibia unfeathered, slightly smaller size, more upright posture, and more active feeding behaviour. In the breeding season it is distinguished from the white form of the Little Egret by its yellow to dusky bill, blue lores, shorter tarsus, and longer shaggier crest. In the non-breeding season it is distinguished from the Little Egret by its smaller size and yellow-green bill and legs. In documenting the occurrence of this species it is critical that good photographs be taken, especially in the nonbreeding season.

Systematics

The Chinese Egret is one of several medium-sized, white, yellow-footed egrets, the taxonomy of which still needs to be fully understood. It was previously considered to be a morph of the Eastern Reef-Heron, from which it differs substantially during the breeding season.

Range and status

The Chinese Egret occurs along the coast from Japan and east Russia, through China, and into Indochina, Malaysia, Indonesia and the Philippines (Murton 1972, Lansdown 1990, Lansdown *et al.* 2000, BirdLife 2003).

BREEDING RANGE: The species currently has a very patchy nesting distribution, at least as far as is currently known. It is likely to nest in more locations than are presently known. Currently its breeding range is far east Russia, North Korea, South Korea, and China. The Russian site is on Furugelm Island (offshore Maritime Province south of Vladovostok) (Litvinenko and Shibaev 2000). Korean sites are on islands off west coast of North Korea and South Korea. Northern Chinese colony sites are in far northeastern China (Liaoning, Shandong). Historically nesting occurred further south along the China coast (Fujian, Taiwan, Guangdong, Hong Kong, Hainan). It is still considered a breeding bird in this area, but documentation is needed as to current breeding status there (X. Chen pers. comm.). It has been found to breed in south Guangxi Province along the north coast of the Gulf of Tonkin (Beibu Bay) (Zhou 2003, pers. comm., Captain Wong pers. comm.). Breeding has also been reported at inland sites in China (Jilan, Henan, Hunan), but given the coastal habits of the species these too require additional documentation as to current status (BirdLife 2003).

NON-BREEDING RANGE: The Chinese Egret occurs in migratory passage in Japan, coastal China including Hong Kong, and Taiwan. The non-breeding range concentrates on the Philippines (Leyte, Palawan, Bohol, eastern Visayas, probably other islands—R. Lansdown pers. comm.), but includes Japan, Indonesia (Sumatra, Java, Sipura, North Natuna, Kalimantan, Sulawesi), Malaysia (west coast peninsular (R. Lansdown pers. comm.), Saba (Sheldon *et al.* 2001)), Singapore (recently shifting from the disturbed Serangoon estuary to the Seletar estuary), Brunei, and Vietnam (Gast and King 1985, Lansdown 1990, Silvius 1991, Round 1992a, 1992b, Mundkur 1993, Lansdown *et al.* 2000, BirdLife 2001). However, records farther west suggest a possibility that the species winters farther west than is now known (BirdLife 2001).

MIGRATION: The Chinese Egret is highly migratory (Lansdown *et al.* 2000). After fledging in July–August, birds disperse to coastal wetlands where they congregate at coastal feeding sites. This is followed by a southward migration in September and October through coastal South Korea, Japan,

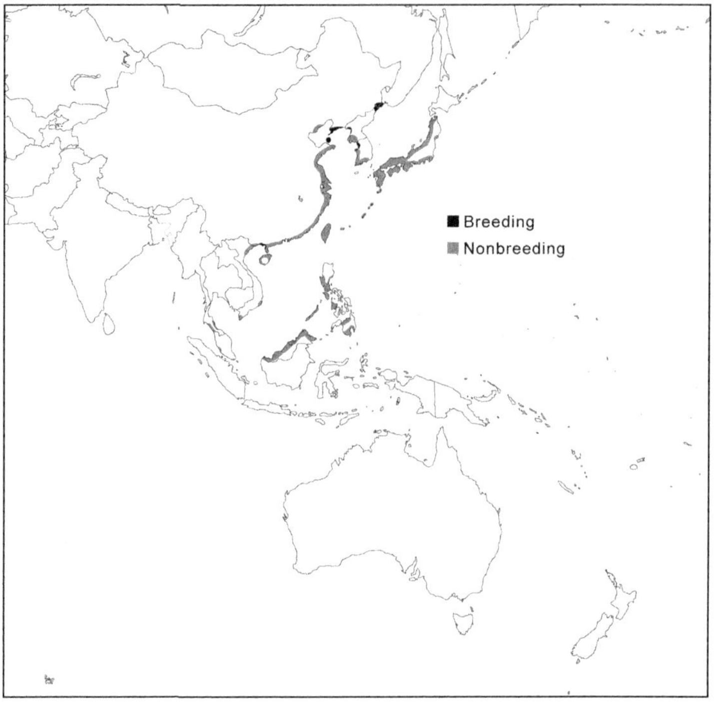

Ryikyu Islands, coastal China, Taiwan, and Batanes Islands. Birds then move to the Philippines and Borneo and others continue down the coast of Vietnam to the Malay Peninsula including Singapore, where a Korean-banded bird was observed (Mundkur 1993). Spring migration is in March and April. Dispersal sightings include Myanmar, Vietnam (Pedersen *et al.* 1998), Sumatra (Silvius 1991), Java (Andrews 1993), Sulawesi, Comoros Islands, and the USA (Aleutian and Pribilof Islands) (Hoyer and Smith 1997). Some of these records may be indicative of more consistent wintering or stopover use than is now understood. Older records from west Sumatra, Christmas Island, India, Myanmar, and Mentawani islands were based on misidentified specimens (BirdLife 2001).

STATUS: This species is rare and vulnerable. The known global population is about 2600–3400 birds (Lansdown *et al.* 2000, BirdLife 2001, 2003). It occurs in small numbers in Russia, the single colony in east Russia having 30–40 pairs (Litvinenko and Shibaev 1999). Estimates include: 100 in Russia, 900–1300 in North Korea, 600–1000 in South Korea, 1,000 in China. The largest breeding concentration in China is on two nearby islets in Liaoning, 210 pairs (Yin and Lei 2002). Its breeding population has fallen in South Korea. It no longer breeds in Hong Kong. In the nonbreeding season, 1600 birds were counted wintering in the Philippines. It is likely that its global population continues to decline. It is considered rare in Japan and also in the Philippines. Determining the species historic population status is difficult, due in part to its similarity to other species. In the nineteenth century, the Chinese Egret likely was common in certain areas of China (Taiwan, Fukien, Kwangtung, Kiangsibut). It may have decreased during the plume-hunting era, but it still remained relatively numerous in Korea and China (BirdLife 2001). It was affected by hunting and habitat alteration, especially the loss of wetlands to agriculture, which led to its gradual decrease in recent decades. Contemporary breeding should be looked for at additional places in mainland China (Jilin, Henen, Hunan).

Habitat

The Chinese Egret is a bird of coastal estuaries and sea embayments. It most often is seen feeding in very shallow water within brackish coastal lagoons and embayments and nearby fresh water sites. Birds may at times use an array of very shallow-water feeding habitats, including shallow tidal estuaries, tidal mud flats, rocky coasts, marshes, lagoons, salt-pans, and river edges. However, they also use rice fields and inland ponds on migration and in winter. In the Philippines in winter, they occur on inter-tidal mudflats and coastal freshwater marshes. They nest on islands near to coastal feeding areas. Most wintering records are also from coastal areas.

Foraging

The Chinese Egret is an active, energetic feeder moving quickly around its feeding site to find and chase fish. It Walks slowly and Stands, but its favourite technique is Running rapidly to and fro, with wings often half-spread and flapped or flicked, and repeatedly stabbing in the water, in a manner like that of the Reddish Egret in North America. It feeds alone or in small groups. Feeding is tidally dependent, the birds following the falling tide and roosting in mangrove trees during high tide. They also feed from perches and also by Feet First Jumping.

The main food is small fish and shrimps, which are susceptible to its active feeding technique. Crabs and insects are also eaten. In one colony in China, it fed only on loaches.

Breeding

The Chinese Egret nests in April–May in the northern summer, with young fledging in July–August. All recent nesting records of this species have been from offshore islands, including on rocky cliffs of sea islands. Nests are situated in trees (e.g. *Quercus*) in Hong Kong, Japan, and China, but also on low bushes (*Sambucus*), herbaceous plants, or even on rocks on the ground. It is a colonial nester, breeding in mixed-species colonies with Grey Herons, Little Egrets, Cattle Egrets, Chinese Pond-Herons, Black-crowned Night-Herons, cormorants (*Phalacrocorax*), gulls (*Larus*), guillemots (*Cepphus*), auklets (*Cerorhynca*), and puffins (*Fratercula*). If disturbed, these egrets will move their nesting attempts from one site to the next (BirdLife 2001). Nothing is known of its courtship. Nests, built of twigs and dry grass, are simple flat structures. The eggs are pale blue-green. Egg sizes reported include 43.5–52.4 mm × 32.6–40.5 mm, 45.56 × 33.57 and 48.4 × 50.6 (Zhang *et al.* 1994, Wen 1997, Yin and Lei 2002). The clutch is usually 3–4 eggs; range is 1–5 eggs. Incubation has been reported to be 21–26 days (BirdLife 2001, Yin and Lei 2002) (correcting previously reported incubation times of 30–35 days that likely were in error). The female undertakes most incubation. Hatching is asynchronous and chicks are attended and fed by the parents about four times per day. Hatching rate was 84%. Birds can move around in the trees at 20 days and are dependent on adults for up to 40 days (Swennen and Won 1993, Zhang *et al.* 1994, Wen 1997). Success rate of eggs was 40% in China (Zhang *et al.* 1994). Much additional information is needed on the breeding biology of this rare species.

Population dynamics

Almost nothing is known about the population biology of this species.

Conservation

The species has been hunted in the past, both for plumes and eggs, which were eaten by women who believed that they improved their skin. Hunting remains a threat in mainland China and Vietnam. Conflicts with aquaculture appear to have led to killing egrets at fish cages, a situation that merits investigation. However, the most important, and continuing, cause of its decline is habitat loss, especially widespread coastal development, including drainage and reclamation of tidal mud flats and estuaries for industry, aquaculture, and other developments. In South Korea, major facilities have been constructed along the coast involving wetland reclamation (BirdLife 2001). Construction of Inchon Airport in the 1990s, led to a decrease and final abandonment of the Sammok Island colony site. Other estuaries and coastal sites face similar threats along the rapidly developing coasts of Korea, Vietnam, and China, as exemplified by the egret's abandonment of its long-term nesting area near Hong Kong. Wintering habitat is also threatened in Philippines and Malaysia (Sabah, Sarawak) by the harvest of mangroves for firewood, development for aquaculture, drainage, and land filling. Disturbance to its few remaining nesting colonies and its intertidal mud flat feeding sites is a continuing and expanding concern. Evidence of such disturbance exists from Russia, North Korea, South Korea, and Vietnam (BirdLife 2001).

The species is vulnerable (IUCN 2003). It was upgraded from Endangered in 1994 owing to the discovery of Korean and Chinese colonies and the Philippine wintering grounds. But with fewer than 3400 birds and its habitat under immediate threat, it truly remains a heron of great concern. It is protected or listed in Russia, South Korea, China including Taiwan. An immediate conservation action plan for this heron is needed. All major colony sites, their intertidal feeding sites, and important wintering sites need to be protected, but especially the sites in North and South Korea, China, and the Philippines. Governments need to fully protect the colony and feeding sites of these birds as the core action in a global conservation strategy for this species. Sites holding more than 30 individuals should be designated as important bird areas and wetlands of international importance. The Russian colonies of Furugelm Island are protected by virtue of being in the Far-Eastern Marine Reserve. However, feeding sites for these colonies, located 15 km away, are not now protected and should be added to the Reserve. The North Korean sites were at threat from egging, but are now protected. They remain threatened by wetland reclamation and coastal development pressures. Feeding areas for the South Korea colony site are threatened by airport construction (Hafner *et al.* 2000). Given the concentration of wintering birds in the Philippines, protecting and managing these sites become critical to the survival of the species. Stopover sites are little understood, but as banding and observations increase, important stopover sites need to be identified and protected. A comprehensive monitoring programme for nesting and non-breeding birds should be developed.

Research needs

Complete and careful surveys of the breeding range are needed to identify additional colony sites and other important areas in summer, winter, and migration. Records of birds in spring and in breeding plumage should be investigated. Banding studies to determine migration, wintering, and survival patterns should be continued and expanded by increasing observer coverage and intercommunication. Additional study is needed on the plumage, moult, and seasonal soft-part colours of this species in order to discover the additional identification keys needed to determine its range and movement patterns in autumn and winter. A monitoring programme should be developed that can assess population trends. A modern taxonomic study, as part of a larger study of the systematics of the medium *Egretta*, is needed to clarify species relationships.

Overview

The Chinese Egret is a coastal species specialized to feed in open tidal lagoons and nest on nearby islands. Its active feeding behaviour is similar to other coastal egrets, most notably the Reddish Egret of North America. Active pursuing methods of foraging are useful when there is little cover, in bright conditions, and in shallow water. It feeds on the falling tide when decreasing water traps and exposes target prey. It nests on sea islands.

Eastern Reef-Heron *Egretta sacra* (Gmelin)

PLATES 6.14–6.15, 16.1, 16.2

Other names: Reef Heron, Pacific Reef Heron, Pacific Reef Egret, Blue Heron, Sacred Heron, White Reef-Heron, White Heron in English; Garceta de Arrecife in Spanish; Aigreet sacrée in French; Kuntil karang in Indonesian; Tagak-laian in Pilipino (Philippines); Kuru-sagi in Japanese, Yan lu in Chinese.

Ardea scra Gmelin, 1789. Syst. Nat. 1, p. 640: Tahiti.

Subspecies: *Egretta sacra albolineata* (GR Gray), 1859: Isle of Pines, New Caledonia.

Description

The Eastern Reef-Heron is a medium-sized dimorphic (dark or white) heron with short yellow-green legs and a thick bill, frequenting ocean shores throughout Australasia.

ADULT: The Eastern Reef-Heron is a variable species, its plumage shade varying among individuals. The dark morph is slate-grey to brown-grey, with differing amounts of white feathers on the chin and sometimes streaking under the throat. The white morph is pure white, although sometimes with dark streaks. Intermediate forms occur. In all forms, the irises are bright yellow. The bill ranges from dirty yellow to dusky to grey-brown, with a dusky horn bill tip and, sometimes, lighter lower bill (see photo in Hancock 1999). The lores are yellow-green. The underwings of the dark form are dark lacking contrast. The stout, relatively short legs are variably coloured, including yellow, grey-yellow, green-yellow to brown-yellow and a combination of dark in front and lighter behind. Most birds have some dark pigment on the front of the tarsus, but this is variable (Edgar 1978). In some, yellow extends up the lower tarsus. The feet are yellow-green with bright yellow soles that are prominent in flight. In breeding the Eastern Reef-Heron develops conspicuous plumes. The small crest is made up of a tuft of short lanceolate plumes. Lanceolate plumes also occur on the foreneck and back. The bill and legs turn yellow. No clear information describes changes in lore colour during nesting.

VARIATION: Variation in the species includes polymorphism, sexual dimorphism, and geographic variation. Males are larger than females. White and dark birds occur together over most of the range, with a greater proportion of dark birds toward the north and south extremes of the range and the white morph being more common in the tropics, perhaps associated with coral reefs (Itoh 1991). Geographic variation occurs in size and in coloration. *Abolineata* has longer wings and its bill may be partly black. The amount of chin striping differs, being most pronounced in Indonesian birds and absent in some birds in the south and east of the range. Birds in Australia are darker than those in Indonesia. In New Zealand, with one exception (Crossland 1992), only grey birds have been reported. Considering this species to be dimorphic does not adequately cover its range of variation. Birds in fact vary from dark grey, to mottled grey to white. Much remains to be deciphered with respect to patterns of plumage variation in this intriguing species.

JUVENILE: Juveniles of the dark morph are pale smoky-grey. White-morph juveniles are white.

CHICK: Chicks have dark grey down, flesh-coloured bills with the upper mandible tipped black, and flesh-coloured legs.

VOICE: The 'Ork' call is the feeding call. The 'Crraw' call is the alarm call. Various harsh croaks are used between paired birds. The Bill Snap is reported during breeding.

WEIGHTS AND MEASUREMENTS: Weight: 330–700 g; length: 58–66 cm.

Field characters

The most reliable features that identify the Eastern Reef-Heron are somewhat subtle: its relatively thick bill, its short legs relative to body size, and a resulting squat, haunched posture. It has a laboured take off extending through short flights. On longer flights, it has more rapid wing beats than most herons, with its neck coiled back, its legs extending beyond the tail showing the yellow footpads, typically flying close to the water and gliding to a landing. It is also characterized by its feeding location on rocky shores and reefs. The white form of the Eastern Reef-Heron is distinguished from the Eastern Great Egret and Intermediate Egret by its smaller size, yellow (not black) legs and feet, shorter lanceolate plumes, and haunched posture. It is distinguished from the Little Egret/Western Reef-Heron by its thicker build, thicker bill, short legs, hunched posture, and yellowish legs and feet. During breeding, the white form is distinguished from the Chinese Egret by its thicker bill, shorter legs, shorter and thicker crest plumes, shorter back plumes not reaching the tail, and yellow legs. During the non-breeding season distinguishing the two species is difficult in that soft-part colors overlap, but the Chinese Egret never has a dark bill tip and only the toe pads, not the entire foot, of the Eastern Reef-Heron are yellow. The two species remain distinctive in relative leg length, posture, and behaviour. The dark form of the Eastern Reef-Heron is identified by the lack of white in the plumage (except for the usually inconspicuous white chin), its uniform grey wings in flight, and yellow foot pads contrasting with less bright feet. It is distinguished from the White-faced Heron by its dark head, thicker and hunched posture. It is distinguished from the great herons by its much smaller size, and other plumage characters.

Systematics

This species is one of the dimorphic medium *Egretta*. Study of the species relationships among birds in this group is needed. This is a widespread, sedentary, variable species, having numerous island populations. Unless it is a fairly recent invader of these islands, geographic variation patterns would be expected. This variation requires study range-wide. Several subspecies may be recognizable, or alternatively, the characteristics of the recognized subspecies of *albolineata* may be well within the range of overall variation of a widespread bird.

Range and status

The species is widespread in coastal east Asia, Australasia, and Oceania.

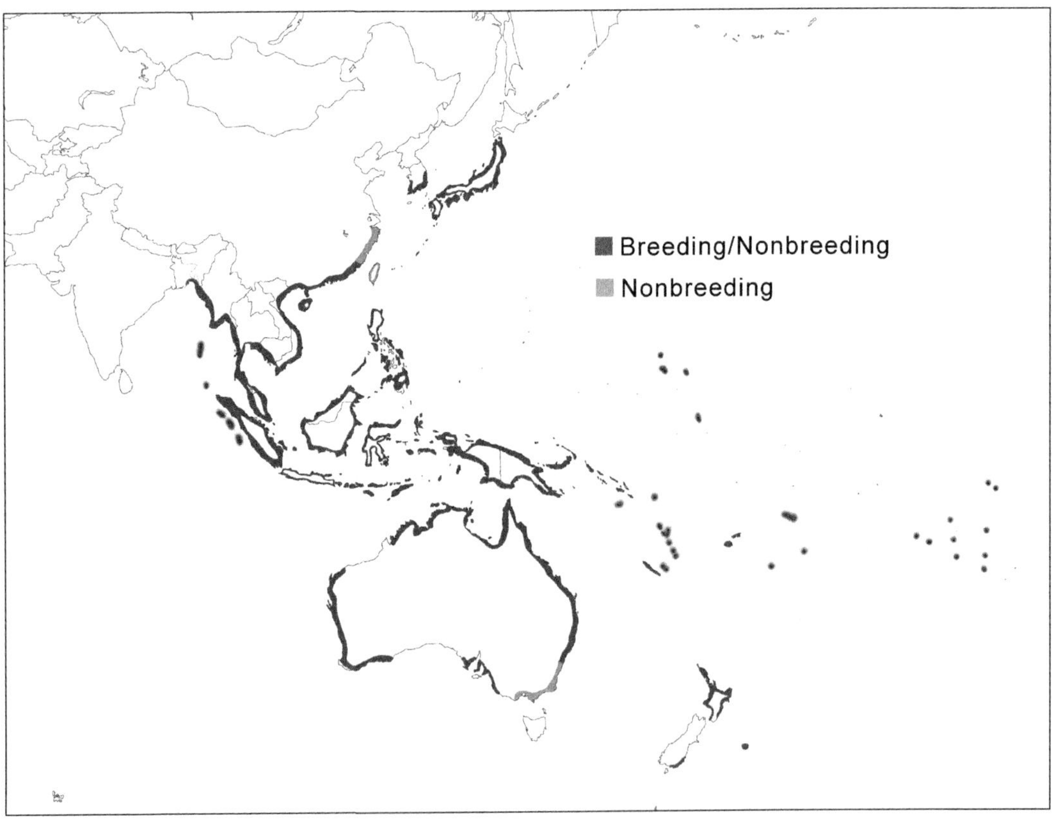

Breeding Range: It ranges from Korea and Japan (Yamagata and Miyagi Prefectures), parts of coastal China, north to islands between Fujian and Taiwan (X. Chen pers. comm.) through southeast Asia to Bangladesh, Andaman and Nicobar Islands (R. Lansdown pers. comm.), south around Australia to New Zealand, east through Polynesia as far east as Marquesas and Tumotus. *Albolineata* is found on New Caledonia and Loyalty Islands. Within this range it is extremely coastal, its inland record being 100 km in New Zealand.

Non-Breeding Range: The Eastern Reef-Heron is sedentary. Some seasonal shifting occurs. Large postbreeding groups assemble in June in New Zealand. Individuals have been found since 1949 in coastal China (Zhejiang, Fujian, Taiwan, Guangdong, Hainan, Hong Kong) (X. Chen pers. comm.).

Status: This species has an extensive linear range, and its population status is not well known. It is most common along the shores of the many islands on which it occurs throughout its range. It is widespread in coastal Australia, but reported breeding is rare. Populations are declining and the range is contracting in the extreme north and south of the range. Populations have declined in south Australia, Tasmania, and New Zealand during the past century. It is now very rare in its breeding range in China, and occurs only as scattered individuals elsewhere in that country (X. Chen pers. comm.).

Habitat

This is a species of the ocean shore. Its usual habitats are islands with rocky shores or shallowly flooded coral reefs. On the mainland and larger islands it prefers sheltered islands with rocky shores with nearby reefs. It also uses mangrove-lined shorelines, intertidal mud flats, beaches, tidal rivers, and creeks. It occurs inland near the coast using lake edges, fields, fresh water marshes, rice fields, and garbage dumps.

Foraging

The Eastern Reef-Heron is overall not gregarious and is a highly territorial species (Recher 1972, Recher *et al.* 1983). It typically feeds alone, in pairs, or small family groups, defending its patch of reef or shore. It uses Alert and Forward displays, attacks and fights to defend its site. On occasion when food, especially small fish, is abundant, reef herons will feed in aggregations. They feed both by day and night, principally depending on tides, feeding on the outgoing and low tides. It typically feeds actively by Walking Slowly in its characteristic deeply Crouched posture. This posture is distinctive, more hunched over than most herons and flattened nearly to the substrate. The posture, achievable because of its short legs, allows it to stand on rocky shores, reefs, or walk in very shallow water while keeping close to the water and having a low, less distinct profile. It also Walks in a less Crouched posture with head and neck retracted, or in Upright posture. It will interrupt its Walking slowly to Walk Quickly, Run, Hop or fly toward a prey item. It periodically assumes an Upright posture to scan for prey. It uses a wide variety of other feeding behaviours, including Wing Flicking, Spread Wing, Gleaning, Foot Stirring, Foot Paddling, Foot Dragging, Hovering, and Diving (Loyn 1985, 1987). It flies short distances, with laboured flight, between reef-top feeding sites, assuming a low crouch on landing. It fishes visually, Stabbing rapidly and repeatedly at fish it sees. It also Lunges at fish, sometimes flapping its wings to travel up to 2 m from its perch to lunge at the prey. On shore it feeds by scavenging, searching under and turning over leaves, by Robbing nesting terns, and by eating tern chicks out of a colony. When not feeding and at night, it roosts alone, in small groups, or in mixed roosts near its feeding sites (Ewins *et al.* 1990). It uses the rocky shore, trees, bushes, docks, breakwaters, piles, and even boats. Many studies of the foraging attributes of morph colour are inconclusive (Murton 1971, Recher 1972, Holyoak 1973). However, differences in foraging have been shown, particularly in habitat choice and in feeding behaviour (Harvey 1975, Rohwer 1990). For example, white-phased birds tended to feed in the surf, whereas dark birds feed by actively running and walking on the reef flat.

The diet primarily is small fish less than 25 mm long, and also small crabs, and molluscs (Recher and Recher 1972). Fish include mudskippers, gobies, blennies, damselfishes, and other coral-or bottom-dwelling species. It eats insects, lizards, and small birds such as tern chicks.

Breeding

The Eastern Reef-Heron is a sporadic asynchronous breeder over most of its range. At the north and south it its range, it nests primarily in the spring, May–July in India and Malaysia and September–December in New Zealand. Its nesting season is year-round in Australia and probably also on various Pacific Islands. It nests on islands, rocks, and coral cays near feeding sites. It uses sheltered sites for nests, on the ground, among rocks, within low vegetation, in small trees, or bushes. It can nest near humans, and has been observed nesting under a pier. The species is highly territorial and typically nests solitarily. It also occurs in small colonies of 20–200 pairs. The nest is a flat platform made of sticks with leaf stalks or grass, 40 cm wide, and a few cm high. Few definitive records exist of the courtship displays, perhaps indicating a tendency for re-bonding by long-standing pairs. The eggs are pale greenish-blue and measure 43–44.8 × 33.3–35 mm. Clutch is 2–6 eggs, usually 2 or 3, but 3–4 in the Andamans, 3–5 in Japan, and 4–5 in Australia. Clutches of 6 were found in Sarawak. Both parents incubate for 25–28 days. Individual incubation stints are exceptionally long, 2 to 6 hours (Mckilligan 2002). Nest attendance corresponds to tide cycle. The young are brooded by adults for a few days and then gradually are left unattended during the day. They are fed by regurgitation. Chicks leave the nest in 3 weeks and hide nearby. Fledging probably takes about 50 days. No information is available on success rates.

Population dynamics

Nothing is known about the population biology of this species.

Conservation

The Eastern Reef-Heron is widely distributed on many islands and along the coasts of New Zealand, Australia, Sumatra, and Thailand. It has very specific habitat requirements, rocky and sandy shores. It is rare and declining in China, where coastal development, siltation and pollution have adversely affected its specialized habitat (X. Chen, pers. comm.). It is adaptable to human presence, using artificial structures for roosting and even nesting. It tolerates non-aggressive human interaction on islands. However, coastal development and aggressive disturbance may adversely affect it (Domm and Recher 1973). Populations in southeast Australia and New Zealand have declined in the past century (Edgar 1978, Peter 1995). This likely is not attributable to disturbance or development, as much as to climatic factors operating at the edge of the range of an otherwise tropical species.

Research needs

The pattern of geographic variation in this species deserves additional and more geographically complete studies. Determining densities over a range of habitats could be the basis of extrapolating a global population estimate. Surveys should be undertaken over its range to define important conservation areas for the bird.

Overview

The Eastern Reef-Heron is a marine bird. Its principal habitat is not mainland shores but islands, where it uses rocky shores and rocky reefs to search for prey exposed as the tide recedes. It has a distinctive feeding posture and concomitant morphology to exploit an environment unique for herons, namely the tops of exposed coral reefs. It also feeds in other shoal habitats especially sandy shores, using a variety of techniques as appropriate. It is a fish and marine crustacean specialist. It likely benefits from knowing its feeding site well, and so is a fierce defender of its territorial prerogatives. It is sedentary in most places, and likely has persistent pairing. It has a small clutch size and long breeding season, likely an adaptation to the uncertainty of food availability along coral reefs, a difficult feeding habitat for a wading bird.

Whistling Heron *Syrigma sibilatrix* (Temminck)

PLATES 8.8, 8.9, 16.3

Other names: Garza Chiflona, Chiflón, Garciolo Real, Pacopaco, Garza Silbadora in Spanish; Maria-faceira in Portuguese; Héron flûte-du-soleil in French; Pfeifreiher in German.

Syrigma Ridgway, 1878. Bull. U.S. Geol. and Geogr. Surv. Terr., 4, pp. 224, 247.

Ardea sibilatrix Temminck, 1824, Planches Color., Livr. 46, p. 271.

Subspecies: *Syrigma sibilatrix sibilatrix* (Temminck) Brazil and Paraguay; *Syrigma sibilatrix fostersmithi* Friedmann, 1949, Smithsonian Misc. Coll., 111, No. 9, p. 1. Caicara, Monagas, Venezuela.

Description

The Whistling Heron is a medium-sized, grey heron with a strikingly coloured yellow-gold chest and neck.

ADULT: The crown is dark to slate-black featuring several distinctive, elongated black plumes, tipped with white or yellow. These rather rigid, lanceolate plumes emerging from the back of the head can be up to 4 cm long. Unlike most heron crests, these feathers actually insert relatively far down the back of the head such that the crest encompasses the upper back of neck (see Hancock 1999). The lores and base of the bill are light blue to violaceous and the long, slender bill is pink to red proximally and black distally to a third of the way from the tip. The iris is pale yellow. A distinctive feature is the yellow to buff neck and chest, which varies in intensity. The yellow cast in the plumage of this species is adventitious, being applied after feather growth, probably from the powder down and/or preen gland. In museum specimens the yellow typically fades toward white. The throat, upper neck, and breast are white. Since these feathers remain white, it is likely that the yellow in other plumage requires a combination of adventitious colour and feather structure. The predominant coloration of this species derives from the bluish-grey of its back, scapulars, and flight feathers. The median and lesser wing coverts appear yellow and cinnamon, streaked black. The abdomen, flanks, tail, and tail coverts are white. The legs (with scutellate tarsi) are greenish-black. The bill and lores brighten as breeding approaches; the bill becomes markedly red.

VARIATION: The sexes are alike as far as is known. The southern race *sibilatrix* has a very black crown, pinkish-cinnamon wing coverts, and a light buff-olive neck and tends to be larger, whereas the northern race *fostersmithi* is smaller in stature, and has a longer bill, a more slate crown, and honey-coloured instead of cinnamon wing coverts with less streaking. A pale grey area is reported between a pinkish lower mandible and a cobalt to azure-blue upper mandible, although a reddish coloration is also reported (Cherrie 1916).

JUVENILE: The plumage of the juveniles resembles that of the adult. The immature plumage is duller, neck feathers streaked, crown more slate coloured, with narrower streaks on lesser wing coverts, and light grey neck and chest. The throat, sides, abdomen, and tail are white as in adult. The yellow in the plumage is acquired after the feathers are fully grown.

CHICK: The chick has not yet been described.

VOICE: The voice, after which this heron is named, is a distinctive, characteristic, far-carrying, melodious whistle, the 'Kee' call. It is given in flight with the neck stretched and then partly retracted. It can be rendered as 'kee, kee, kee'. Calls are well spaced. Also a slow drawn-out whistle may be given as the bird takes flight. There also is a peculiar high-pitched and rapidly repeated call (Friedman and Smith 1950), possibly the 'hard metallic notes' described by Kerr (1950). The young communicate with adults using hisses.

WEIGHTS AND MEASUREMENTS: Weight: 521–546 g; length: 53–64 cm.

Field Characters

The Whistling Heron is identified as a grey heron with golden chest and black cap, which together distinguish it from other medium-sized herons. It is often seen alone or in pairs, in more open and relatively drier situations than most herons. It flies distinctively with rapid, duck-like wing beats, gliding to its landing. Sick (1993) offers that its wings appear attached below the body axis. It seldom retracts its neck completely in flight.

Systematics

The affinities of the Whistling Heron have been long debated as it has been difficult to place within heron phylogeny. However, it seems to be a typical heron, related to the *Egretta* (Sheldon 1987, K. McCracken pers. comm.). It likely is most closely related to the Capped Heron (K. McCracken pers. comm.). Earlier debate centred on whether it was better regarded as a day or night heron, but the affinity of *Syrigma* to the egrets is well supported by behavioural, morphological, and biochemical evidence (Humphrey and Parkes 1963, Short 1969, Kahl 1971, Payne and Risley 1976, Kushlan *et al.* 1982, Sheldon 1987). The only question is how closely it is related to *Egretta*. Interestingly, molecular data suggest it is closer than does evaluation of osteological data (Sheldon and Whittingham 1997), probably due to its secondary adaptation for semi-terrestrial foraging. Because of its apparent distinctiveness and as a result of its specializations, the species has historically been placed in its own genus. Neither recent morphological nor biochemical evaluation provides evidence to undermine such a distinction. Its combination of rigid, lanceolate head plumes, lack of scapular plumes, scutellate tarsi, possession of both day and night heron skeletal features, and measurement proportions is distinctive (Payne and Risley 1976), and it is biochemically separable from its closest relatives in the genus *Egretta* (Sheldon 1987, Sheldon and Whittingham 1997).

Range and status

The Whistling Heron is a South American endemic.

BREEDING RANGE: It has two disjunct populations in the South American lowlands, apparently at least partially separated by the Amazon basin. The northern subspecies, *fostersmithi*, occurs in northern South America east of the Andes in the Llanos and Orinoco basin in Colombia (south to Meta and Arauca—although there are not yet any breeding records from Colombia) and Venezuela (Barinas, Apure east to Monagas and northern Bolívar). The southern race, *sibilatrix*, occurs in Bolivia (Beni, Santa Cruz, Tarija), Paraguay and Brazil (Mato Grasso, Rio de Janeiro, Sao Paulo), Uruguay, Bolivia, Argentina (Tucumá, Sante Fe, Buenos Aires), and Chile (Blake 1977, Antas *et al.* 1986, Sibley and Monroe 1990, Blanco and Canevari 1992, Sick 1993). The species appears to be expanding its range in southern South America. It has recently been documented further north in southern Brazil (southeastern Piaui, eastern Goias) (Morales 2000, F. Olmos pers. comm.).

MIGRATION: It is possible that the species undergoes seasonal migrations, but more information is needed for both populations. Such shifts are likely in response to wet–dry cycles. It is migratory in northeast Venezuela, being absent from November to January (Friedmann and Smith 1950). In the south, it appears to move in response to droughts, but is found in some areas (e.g. Buenos Aires) throughout the year. The dispersal record in east Brazil is from Piaui.

STATUS: The species is locally common to very common in appropriate habitat and common seasonally in the Llanos. It is, however, often patchily distributed in response to habitat conditions. Amerindians used its plumes for barter, but not sufficiently to cause significant pressure on numbers. It is not vulnerable anywhere in its range.

Habitat

This is a heron characteristic of the open seasonally flooded savannah (Kushlan *et al.* 1982). Although very typically found in dry grassland and fields, like other 'dry land' herons, it more typically uses damp and very shallowly flooded situations, such as prairies wet after rains, stream edges, ditches, ponds, pools, lake shores, flooded woodlands, and rice fields. It uses man-altered habitats such as managed pastures and fields, but also road edges and ditches. It uses tree stands for roosting and nesting and is therefore something of an ecotonal species preferring an intermixture both grassland and woodlot.

Foraging

Whistling Herons forage simply, typically by Standing or Walking, usually with the head held high and the arched neck in an erect S (Kushlan *et al.* 1982, Hancock 1999). They hold the bill horizontal or Peer Over or point the head downward, staring at a likely strike spot for many minutes, slowly lowering the head before a Bill Thrust or Pecking on wet

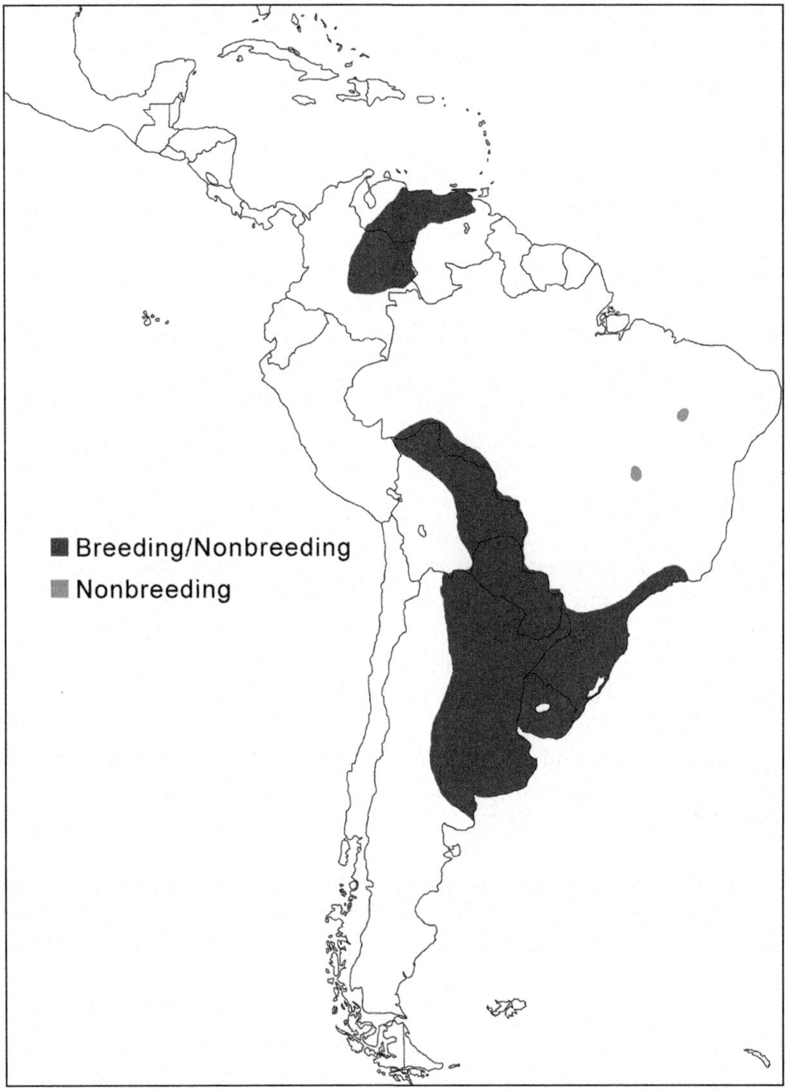

ground. They sometimes Neck Sway before a strike. They also Walk slowly scanning the feeding area and Run after prey in a Crouched posture. Whistling Herons seem ready to catch anything suitable that happens their way. They use Standing Flycatching, Gleaning from plants, and even attempt piracy when the opportunity arises in the course of foraging. Whistling Herons typically feed alone or in pairs in feeding territories (Kushlan *et al.* 1982). Feeding territories are defended by bouts of ground displays, including Crest Raising, Forward displays, Stretch displays, whistles, and jabbing and probing opponents with their bill. A bird appears reluctant to abandon its feeding site, such as a roadway, even when pressed by humans. They do sometimes feed in larger groups, numbering in the dozens, often prior to roosting. They perch on fence posts or in trees within or near their feeding area. They also roost in trees and, despite feeding dispersed, in the evening they may assemble in arboreal roosts in the hundreds. Arriving about an hour or longer before dark, they become highly social, call loudly with raised crests, preen, spar and aggress on each other.

Food taken is varied, depending no doubt on seasonal availability. As is typical of dry land herons, the diet is dominated by arthropods of all sorts, including spiders, beetles and beetle larvae, dragonfly larvae, flying insects, mantis, and also can include earthworms, eels, frogs, tadpoles, and glass snakes (*Ophiodes* lizards).

Breeding

The breeding season appears to be very extended and variable between and even within populations. Breeding is April–September in the north, September–January in Brazil, and January in Uruguay. The Whistling Heron nests in mature trees, such as araucarias or even exotic species, especially in small woodlots and forest remnants (Olmos 1989, Anjos 1990). They nest singly or in small, scattered colonies, traditionally returning to previous nest locations from one year to the next. The nest is a loosely built platform of twigs, usually placed on a thick horizontal branch situated high in the tree, 3–11 m above ground. Very little has been recorded about this heron's display, but it apparently quite often adopts the Bittern Posture (Kerr 1950). There is an aerial display involving flying back and forth and gliding in circles. Payne observed a captive pair raising their long black crest plumes in display (Payne and Risley 1976). The normal clutch is 3 or 4 eggs. Eggs are pale blue and speckled, measuring 46–48 × 35–37 mm. The incubation period is typical of herons, about 28 days (Di Giacomo 1988). Very little is known about chick growth and development. Di Giacomo (1988) reported the nestling period to be 42 days. Adults and young maintain postfledging associations for some months; how they dissolve is not reported. They communicate by whistles and hissing. Young birds are fed by their parents after fledging. Young beg by hissing and drooping their wings (Hancock and Kushlan 1984). In exotic trees, nests were lost in storms; egg survival was 28% and nestling survival was 40% (Di Giacomo 1988) resulting in an average production of only 0.29 fledglings per nest. However, based on observations of young with adults in more natural habitat, it appears that two young normally fledge.

Conservation

Where it occurs, the species is common to very common and expanding its range. As a bird adapted to grassland benefiting from forest conversion, rice growing, and the expansion of cattle ranching, so long as the system is not overly dried and trees and woods for roosting remain. They are frequently seen in human-altered environments, even industrial sites (Olmos 1989). Based on available information, there is not a need for specific conservation planning, other than the protection of flooded grasslands on which they depend. Habitat suitability may depend on maintenance of naturally occurring seasonal fluctuations in water levels and soil moisture and, given the species' dependence on interspersed grassland and trees, complete conversion of forests to pasture would adversely affect it.

Research needs

In many ways, this is a very little known species. Neither its nesting biology nor the variability and seasonality of feeding is well understood. Its population biology is unknown, and could be very instructive given the un-heron-like prolonged dependence of young on adults. It would be intriguing to determine the relatedness among birds occupying the same area and how birds enter the local breeding population. Mortality, survivorship, and dispersal, need to be understood. The study of visibly marked individuals over multiple years could be of significant value in understanding this species. Given the few specimens studied and their tendency to fade with time, a restudy of the racial distinctiveness of the two populations, based on additional fresh specimens, is needed. This is of conservation importance given the recent observations well outside the accepted range (Morales 2000).

Overview

The Whistling Heron remains an enigmatic species. It is a bird of dry to damp grasslands, like several other herons, and so has specialized morphology to accommodate a semi-terrestrial way of life. It Walks and Stands, as would be expected of a grassland heron, and takes what it can capture. However, its ecology is rather distinctive, being sedentary and appearing to maintain family groups. Its calls and exuberant behaviour at the roost are unusual among herons. Furthermore, the prolonged dependence of young on adults is decidedly un-heron-like. Explanations may be the need to teach foraging, but such dependence in other species suggests a saturated habitat and maybe young birds face difficulty in establishing their own territory. Given the difference of this family life to that of most herons, the species is likely to also show unique demography and social structure. It is reported that some local people do not think of the species as a heron, so it likely is more biologically interesting than is now known.

Capped Heron *Pilherodius pileatus* (Boddaert)

PLATE 8.7, 16.4

Other names: Garza Real, Garciolo Real, Garza Capirotada in Spanish; Garça-real, Garcinha, Garça-de-cabeça-preta in Portuguese; Kapppenreiher in German; Bihoreau blanc in French.

Pilherodius Bonaparte, 1855, Consp. Gen. Avium 2 (1857), p. 139. Type by monotypy *Ardea alba* var. β Gmelin = *Ardea pileata* Boddaert.

Ardea pileata Boddaert, 1783, Table des Planches Enluminees, p. 54, based on 'Heron blanc, hupé de Cayenee' Daubenton, Planches Enluminees, pl. 907: Cayenne Guianas.

Description

The Capped Heron is a medium-sized, white heron, sporting with a distinctive black cap set off by long white head plumes.

ADULT: The base colour of the Capped Heron is white, with a distinctive black crown. The bill too is distinctive, predominantly blue-grey, reddish in the centre, tipped in grey or yellow. The forehead is white or grey and lores bright blue. The iris is yellow, green-brown, to brown. The distinctive slender white head plumes may be as long as 20 cm. On the back and wings, it is pearl to pale white with grey shading. The relatively thick (for a heron) neck has a variable amount of cream colour. The legs are blue-grey, with somewhat darker toes. During the nesting season the plumage, especially, the head, neck, breast, and underwings, become tinged with yellow and buff, neck at times achieving an intense cream colour. This coloration is likely to be adventitious, probably derived from powder down. Reports suggest that the blue of the soft parts deepen in the nesting season, often described as cobalt.

VARIATION: As far as is known, the sexes are indistinguishable. There are no races recognized.

JUVENILE: The fledgling's plumage is white like the adult, but it has a streaked crown and is paler grey on the wings. Juveniles retain some white and grey variegation on the head and have greyer wings than adults. Younger juveniles lack head plumes, but do develop them while in juvenile plumage.

CHICK: The chick has white down.

VOICE: The 'Whoop' call, rendered, 'woop, woop, woo', is used in mate communications, and is given as a bird lowers its head in front of its mate and opens its crest (Sick 1998). It also makes gutteral croaks.

WEIGHTS AND MEASUREMENTS: Weight: 518–591 g; length: 51–59 cm.

Field characters

The Capped Heron is identified by its white plumage, dark black cap with white plumes, from which it derives its common English name, its long blue bill, and thick white to cream coloured neck. It is usually found feeding alone, and flies distinctively for a heron, with shallow wing beats never rising above the horizontal, somewhat parrot-like (Hancock and Kushlan 1984). It is distinguished from other white herons in its range, such as the Snowy Egret and immature Little Blue Heron, by its black cap and white body.

Systematics

The Capped Heron is a typical heron, related to the day and night herons. Based on molecular evidence it appears that it is most closely related to the Whistling Heron (K. McCraken pers. comm.). Morphologically it is distinctive enough to be retained in its own genus.

Range and status

The Capped Heron is a tropical South American heron, fundamentally a bird of the Orinoco and Amazon.

BREEDING RANGE: Its breeding range extends northward to the Canal Zone on both coasts in Panama, south through inland Venezuela (Kushlan *et al.* 1982, Morales 2000), east to the Guianas, west to the Andes of Ecuador, Peru, and south to northeast Bolivia, north Paraguay, and south Brazil (Santa Catarina). It is reported west of the Andes only rarely (Ridgely and Gwynne 1993).

MIGRATION: It is non-migratory in most of its range, but has been recorded as seasonal in Darién, Panama, and would probably be expected to shift around in response to the seasonality of habitat conditions, although this remains to be described.

STATUS: The species is widespread, but uncommon in most locations throughout its range. The core of its range seems to be the Orinoco and Amazonia and is less common peripherally in the Guyanas,

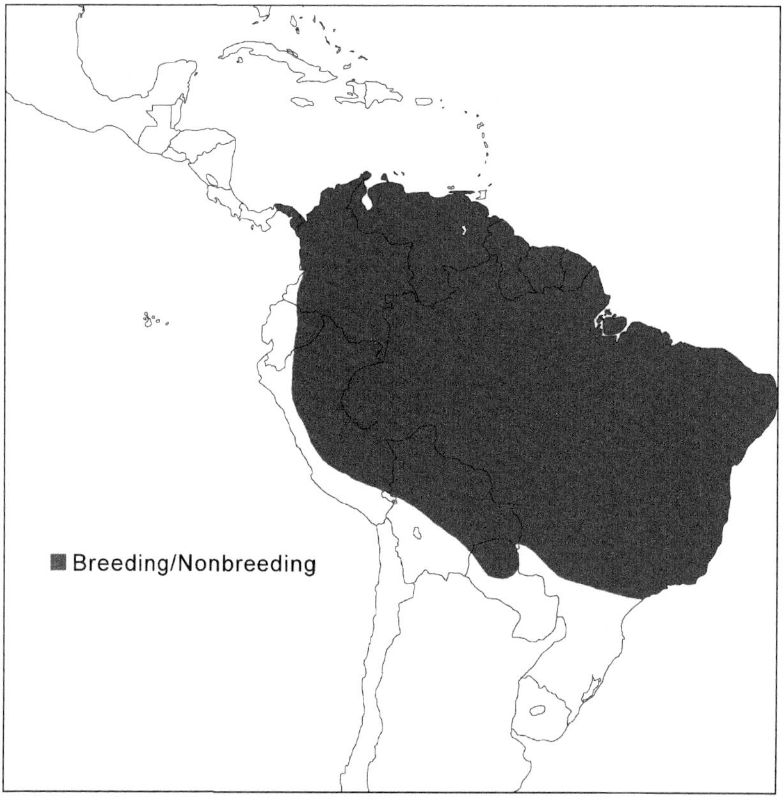

Brazil, Colombia, and Panama. It uses human-altered habitats and perhaps has become more common in Panama (Sick 1998, Butler *et al.* 2000). Even at low density, the total population is probably large and is at no particular risk.

Habitat

The Capped Heron is primarily a bird of forested swamps having rivers and ponds, and of open marshy areas near gallery forests (Blake 1977, Kushlan *et al.* 1982, Hancock and Kushlan 1984, Teixeira and Carvalho1985, Morales 2000). It is found in the lowlands and upland rain forests below 900 m. It can occur more broadly, however, and also is reported from such aquatic habitats as swamps, savannahs, open woodlands, edges of rivers and ponds, wet marshes, road-side pools, sandy, muddy, grassy and rocky river banks, and such man-modified habitats as rice fields, coffee plantations, public gardens, and pools along the Transamazonian Highway. It is dispersed and thinly spread over available habitat. In Colombia, it occurs well dispersed along riverbanks (Hilty and Brown 1986).

Foraging

The Capped Heron is a diurnal, aquatic heron, usually seen feeding streamside or in puddles. It tends to feed alone, although may loosely join larger groups of feeding wading birds. It forages slowly and deliberately, primarily by Standing and slow Walking, remaining in a small area for quite some time reworking it repeatedly. It often stands rather erect, intently looking for potential prey with its neck stretched out or folded back into a compact S shape. When Walking slowly it progresses very slowly, holding a pose for a few seconds while slowly moving a foot to take a new step. It is able to feed by more active methods, such as by Gleaning, Foot Paddling, Dipping, Swimming Feeding, and Bill Vibrating (Hancock and Kushlan 1984). It strikes with deliberation, slowly crouching and extending its neck toward the prey item before making a Bill Thrust, often accompanied by a Body Lunge. It Grasps or Pierces prey and also uses Pecking, sometimes without apparent orientation, or Grabbing. It may Neck Sway or Peer Over turning its head quickly from side to side and holding the neck arched out for minutes.

It eats small fish, primarily, and also adult and larval aquatic insects, tadpoles, and frogs.

Breeding

Specimens in breeding condition have been taken in Venezuela in early March. It nests solitarily in trees in forests, but a great deal more needs to be known. In captivity, birds made a nest of dry twigs placed in trees. Nothing is known of this heron's breeding behaviour. In captivity, clutches consisted of 3–4 dull white eggs averaging 43.2×35.9 mm and incubation was 26–27 days.

Population dynamics

Nothing is known about the population dynamics of this species.

Conservation

The Capped Heron is a widespread species that is expanding its range, and perhaps its use of man made habitats. With a widely dispersed and probably relatively large global population, it is not known to have any present conservation risk. However, it is basically a species of riverine forests, and the loss of this habitat by logging and conversion to pasture would pose long-term threats to population stability.

Research needs

Despite being so widespread in South America, the Capped Heron is one of the least known herons. Most of its basic biology is unknown: feeding habits, nesting biology, courtship, interactions within the pair, individual movements, details of habitat use, survivorship, and even the details of its range. Study of individually marked or telemetred birds over a large enough area to encompass several pairs would be a useful approach to deciphering its fundamental breeding biology. Determining population density and its more precise range are needed to properly assess its conservation status.

Overview

This is a distinctive heron, readily noted in appropriate habitat by not only its plumage but in its erect, solitary, slow foraging, and use of water edges. The Capped Heron is primarily a bird of open edges to water courses. It is a passive feeder, primarily on fish and aquatic insects usually seen Standing still or Walking carefully along its chosen bank. It is most often observed solitarily or in pairs. It is likely that pair bonds are long lasting. Little is known about nesting or parental care. Other herons with a similar biology maintain family groups and care for young after fledging. Much remains to be known about this species.

Green Heron *Butorides virescens* (Linnaeus)

PLATE 7.4, 17.1

Other names: Green-backed Heron, Little Green Heron, Little Crabier, Poor Joe, Gaulin, Water Witch, Pond Bird in English; Garcita Verde, Garzita azulada, Garza Rayada, Chicuaco, Caga Leche, Cuaco, Crá-Crá, Martinete in Spanish; Heron vert, Krabye rivié, Kio in French; Groene reiger in Dutch.

Ardea virescens Linnaeus, 1758, Syst. Nat. 10th ed., p. 144: Based on *Ardea stellata minima* Catesby, Nat. Hist. Carolina, 4:80, Pl. 0. America = coast of South Carolina.

Subspecies: *Butorides virescens anthonyi* (Mearns) 1895: Seven Wells, Salton River, N Baja California; *Butorides virescens frazeri* (Brewster) 1888: La Paz, Baja California; *Butorides virescens bahamensis* (Brewster) 1888: Watling's Island = San Salvador, Bahamas.

Description

The Green Heron is a small, stocky, dark-backed heron with a thick neck, relatively heavy dark bill, thick short legs, and maroon neck.

ADULT: The adult has a glossy green-black cap with erectile plumes forming a shaggy green–black crest. The bill is brown-black with the lower tending to dusky-green and yellow at the base. A cream-buff stripe runs from the bill to under the eye. The irises are orange or yellow. The lores are dull yellow-green, with the lower portion black. The side of the head, neck, breast, and upper back are chestnut-maroon. The throat and the upper breast are streaked and spotted in white, forming a vertical stripe, yellow-brown in the centre and edged in grey. The back is dark grey to black with a green to blue cast. The upper back feathers are elongated to form modest plumes. The upper wings are grey to black with buff edges to feathers. The flight feathers also are black with buff edges. Underparts are brown-grey to grey in most forms. The legs are yellow-orange. In breeding, the head and crest are glossy black, with elongated erectile plumes. The side of the head, neck, breast, and upper back are chestnut. The irises turn deep orange in courtship setting off blue-black lores. The bill becomes glossy black and the legs glossy orange.

VARIATION: Females are smaller (Davis and Kushlan 1994), and also duller and lighter coloured, especially during breeding. There are four recognized geographical forms (Payne 1974). *Virescens* is as described above. *Anthonyi* is distinctly larger and has more birds with paler brown necks. *Frazeri* has a dark neck that appears more purple than brown. *Bahamensis* averages smaller and paler than North American birds. There is additional variation within Central America and the West Indies, birds from this region showing a range of neck coloration from grey through brown to maroon (Payne 1974). Birds from Guatemala through El Salvador and Costa Rica average smaller than North American birds but there is considerable overlap and no consistent coloration differences. Birds from the West Indies, the subspecies *maculatus* of Voous (1986), average smaller than east North American birds but there is considerable overlap with both North American and Mexican birds, so no taxonomic distinction is recognizable (Steadman *et al.* 1997). Birds in Costa Rica show variation and distinctiveness in morphology among populations, although with considerable overlap (G. Alvarado in prep. a).

A red-brown phase occurs in Cuba (Raffaele *et al.* 1998). A partially white heron was reported from California (Garrett 1994). In addition to variation within *Butorides virescens*, intergradation occurs with South American *Butorides striatus* in southern Central America, northern South America, and the southern West Indies which accounts for the range in neck colours in the region (Hayes 2002) Birds from Panama with brown-grey necks were once described as the subspecies *patens* and now are also considered intermediate between western Panamanian *virescens* and South American *striatus*.

JUVENILE: The juvenile plumage differs substantially from that of the adult. The juvenile bird has a green-grey crown with a slight crest. The bill is brown-black above and yellow below. Irises are yellow. The lores are yellow-green. The line under the bill is buff-white bordered with black. The throat and neck are buff-white with dark chestnut streaking. The side of head and neck are yellow-brown broadly streaked in chestnut to grey. The back and wings are dark green-grey. The wings are dark green-grey, with buff-white spots. The flight feathers are dark grey with a green cast, tipped in white. The underparts are brown to buff-white with buff spots. The legs are yellow-green.

CHICK: The hatchlings are grey-brown above, with a downy crest, and light grey below. The skin on the head is green, the lores yellow-green. The eye ring is green-yellow. The bill is yellow with a dark tip. The throat is white. The legs are green, turning yellow-green.

VOICE: The characteristic call of the Green Heron is the 'Skeow' call. Variously rendered as 'skeow, skow, skyow', it is an alarm, flight, and advertising call. The 'Skuk' call is the disturbance call, rendered as 'ku, ku, ku' or 'skuk, skuk, skuk', 'kak, kak, kak', or 'kek, kek, kek'. The 'Showch' call, rendered 'show, ch' or 'ow ch' is an advertising call. The 'Rah' call, rendered 'raaah' or 'raah, raah', is an antagonistic call, used in Forwards and at other times. The 'Arroo' call, rendered 'arroo, arroo', is used with the Stretch display. Adults also use Bill Snaps in courtship. The 'Cuck' call, rendered 'cuck, cuck', is used when approaching young. Young beg with 'tik, tik, tik, tik'.

WEIGHTS AND MEASUREMENTS: Weight: 200–250 g; length 41–46 cm.

Field characters

The Green Heron is identified by its thick maroon neck, its black cap, large bill, and green-tinged, grey-black back. It flies with deep, slow, steady wing beats, 2.8–3.8 per second with its head retracted and legs extended beyond the body. In flight at a distance it appears black. It is often seen flying after being disturbed. Then it flies with head out and feet dangling and gives the characteristic 'Skeow' call. It also usually relieves itself upon taking off, leading to some vulgar common names. It lands after a short glide with head and neck extended. This disturbance flight behaviour, defecation included, is one of its best identifiers. Its chestnut (not grey to rust) neck distinguishes it from the Striated Heron. It is distinguished from all other herons by its combination of dark crown, relatively large bill, uniform maroon neck and side of head, yellow lores and legs, posture, behaviour, and 'Skeow' call.

The immature bird is identified by its brown colour and dark cap, as well as by posture and behaviour. It is distinguished from immature night herons by being smaller and darker, with darker cap, longer neck, and slighter bill. It is distinguished from the juvenile small bitterns by its larger size, dark crown, and darker upper parts. It is distinguished from the American Bittern by its smaller size and lack of a black neck mark. It likely is not distinguishable from juvenile Striated Herons.

Systematics

The systematic position of the *Butorides* has been a matter of uncertainty for some time. These birds are sometimes placed with other small herons in *Ardeola* (Payne 1979), and molecular, behavioural, and morphological evidence confirms their close affinity to these birds (Payne 1974, Payne and Risley 1976, Sheldon 1987, K. McCracken pers. comm.). *Butorides* herons are however distinctive both molecularly and morphologically, and so are best retained in their own genus. *Butorides* and *Ardeola* together are most closely related to the *Ardea*.

Specific and infra-specific systematics have also been a matter of considerable study and discussion, particularly focusing on the relationship between North American birds with their maroon necks and South American birds with their grey necks. In recent decades advocates have suggested combining the forms into a single species on the basis of morphology and intermediacy of the plumage of particular specimens (Payne 1974, AOU 1993). Other advocates have suggested separating them into two species on the basis of lack of intermediacy of particular specimens (Wetmore 1965, Voous 1986, Monroe and Browning 1992, AOU 1998). Birds intermediate between *virescens* and *striatus* are documented from the non-wintering season in the Lesser Antilles, Tobago, Panama, Cocos Islands, Colombia, Venezuela, the Dutch West Indies, and Suriname (Payne 1979, Hayes 2002). But alternative interpretations have been made of these birds too (Voous 1986). Certain specimens have been subject to alternative interpretations as to whether they have or do not have juvenile characteristics or may or may not be migrants. The fact that including or excluding certain specimens is so critical to the findings suggests caution is needed in drawing conclusions based on available specimens. The latest reinterpretation of available specimens shows a range of intermediate birds indicating an extensive zone of hybridization in the south Caribbean islands and also in southern Central America (Hayes 2002).

The North and South American *Butorides* undoubtedly represent forms that have been isolated from each other for some time. Molecular evidence is tending to support differentiation of the two species (K. McCracken pers. comm.). So this book follows AOU (1998) in recognizing *Butorides virescens* as separate from *Butorides striatus*, deviating from the approach in Hancock and Kushlan (1984). However, variability and intermediacy in the zone of contact suggests that the two forms interbreed freely. By traditional concepts, the two *Butorides* might be considered conspecific (Hayes 2000). The final word is not yet in on the species limits of *Butorides*.

Range and status

The Green Heron occurs in North America, Central America, and the West Indies. The status of the species at its southern range edge is uncertain.

BREEDING RANGE: *Virescens* breeds in east, central, and south North America from south Canada (New Brunswick, south Ontario, south Quebec, Manitoba) (Koes and Taylor 1991, Grieef 1991), and north Central USA (east North Dakota, east South Dakota, east Colorado), through central and south USA (Florida, Texas, New Mexico), mainland Mexico, Central America to central and east Panama and Caribbean islands off the coast of Central America and Venezuela (Aruba, Curacao, Bonaire, Los Roques, Margarita Island, Tobago), and the Greater and Lesser Antilles. The situation in Tobago is particularly interesting, in that evidence suggests that there has been a shift in the population from one that was intergrading between *striatus* and *virescens* to one dominated by *virescens* (Hayes 2002). *Anthonyi* breeds in west Canada and northwest USA, in southwest British Columbia (to southeast Vancouver Island) northwest USA (west coastal Washington, Oregon (Jobanek 1988), California, Nevada, south Utah, Arizona), and Mexico (north Baja California, Sonora). *Frazeri* breeds in west Mexico (Baja California Sur). *Bahamensis* breeds in the Bahamas Islands. Green Herons of undetermined subspecies have recently been shown to nest in Bermuda (D. Wingate pers. comm., see also Norton *et al.* 2003).

NON-BREEDING RANGE: *Virescens* occurs in the non-breeding season in the southern part of its breeding range, in south USA (south Carolina, Florida to Texas), Caribbean (Bahamas Islands, Cuba, Cayman Islands, Jamaica), Mexico including the interior, Central America, north South America from south Colombia (Scott and Carbonell 1986), coastal Venezuela and offshore islands (Bonaire). *Anthonyi* occurs in the non-breeding season in the south part of its breeding range, south to west Mexico (Guerrero, Chiapas) more rarely to Central America (El Salvador, Costa Rica). *Frazeri* and *bahamensis* are sedentary, as far as is known, wintering on their breeding range.

MIGRATION: Northern populations have a dispersal or wandering period at the end of nesting, August–September, which then merges with migration in October. Migration is typically, although not always, in flocks at night. They call in flight to maintain contact (Brady 1990–1991). Northern populations tend to be partially migratory, with northernmost birds moving south but some staying rather far north. Western *virescens* migrates south to Mexico and Central America. Birds from the central continent move to Texas, and Florida to Mexico, Central America, and the Greater Antilles. Eastern birds move south to and through Florida, Greater Antilles, and Cayman Islands. Return migration is March–April over most of the range, but May in northwest USA. *Anthonyi* move south to west Mexico and Central America. Dispersal records for *virescens* include Cocos Islands, Guyana, Suriname, Greenland, Iceland (Petursson 2003), England (Ross and Bell 1985, Stanworth 2001, Rogers 2002), Scotland (Gordon 1988), channel islands France (Rault 1995), possibly Azores, Hawaii (Paton and MacIvor 1983), and the Galapagos (Vargas 1996).

STATUS: The Green Heron is widespread and numerous. Populations appear to be increasing in most areas in North America, although reductions appear locally due to habitat modification (Davis and Kushlan 1994). Populations appear to be largest in Florida and Louisiana (Butler *et al.* 2000). The range has been expanding in the middle of the continent along impoundments and managed river

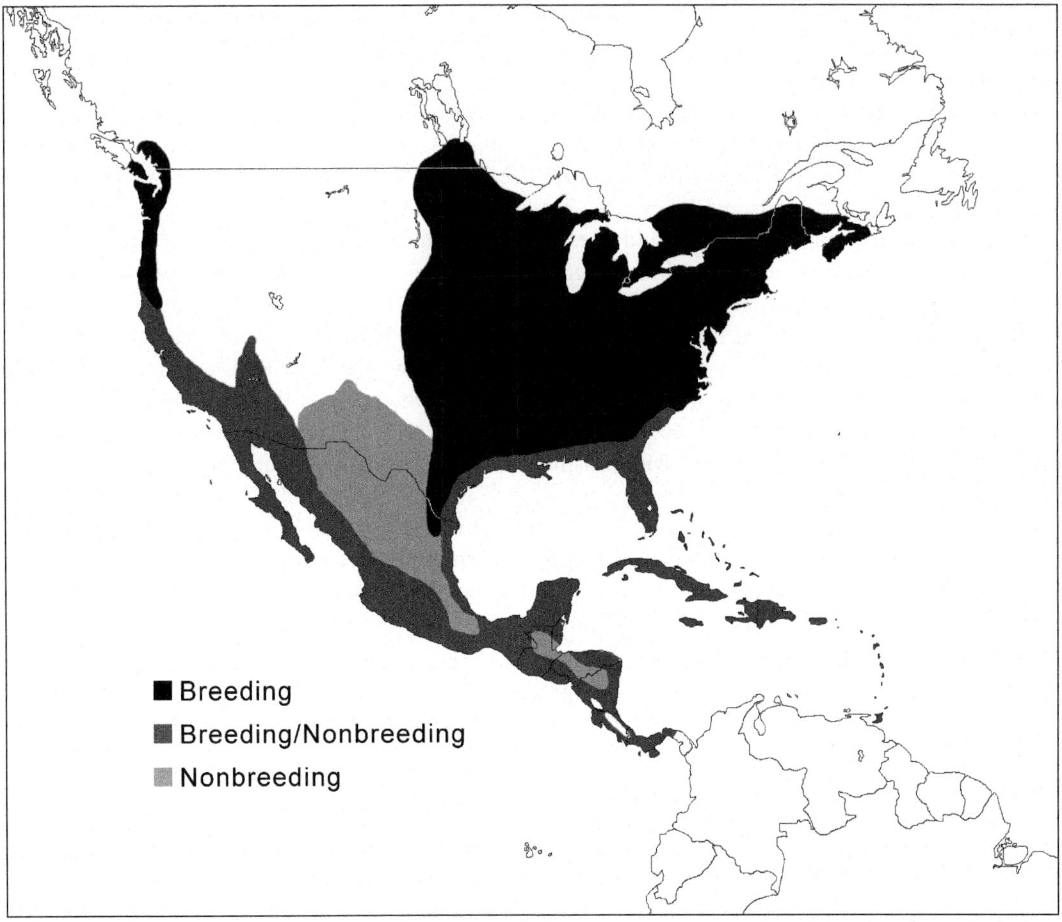

systems and northward along the Pacific Coast. It is abundant in Central America (Butler *et al.* 2000). It is common in West Indies (Raffaele *et al.* 1998).

Habitat

The Green Heron is a bird of swamps and marshes, lined with brush or trees. It occurs in both inland and coastal habitats especially along small and large brushy marshes, river swamps, mangrove swamps, the edges of forested rivers and streams, lake margins, salt flats, woods, sand, muddy or rocky shores, pools, and in human-made habitat such as reservoirs, impoundments, ditches, canals, ponds, and orchards.

Foraging

Green Herons are stand and wait feeders. They feed solitarily, defending feeding sites and sometimes a larger territory. Birds are aggressive in defending their space, and birds that are sedentary may occupy feeding territories long term and may nest within it (Kushlan 1983). Herons defend their sites with Forward displays (see Courtship). They also use Supplanting Flight, Supplanting Runs, and Pursuit Flights. They feed in loose groups and in aggregations if prey is especially available. Feeding occurs at nearly any time, day or night, including at night by artificial light, and along the coast according to the tides. It feeds in very shallow water (<5 cm), and is more successful in shallow than in deep water, presumably because of the avoidance behaviour of the fish (Kramer *et al.* 1983). Its feeding behaviour is primarily Standing, in shallow water or on shore or on a perch overhanging water such as branch, rope, boat, or rock. It uses a very characteristic Crouched posture, legs bent and body parallel to the ground. It Crouches so low that sometimes the lower leg nearly touches the substrate. The head may be withdrawn ready for a Bill Stab or extended out towards the prey ready for a Bill Lunge. It is a patient feeder, standing for many minutes examining the water intently waiting to see a prey item often looking like a part of its perching branch or rock. On its usual feeding grounds, it has favoured perches to which it returns. One common variant of the Standing behaviour is perching above the water. The heron waits patiently, drawing vertically closer to the water as a prey emerges. It then uses its characteristic Forward Throwing behaviour, in which, grasping the branch, it throws its body toward a prey item but keeps its hold on the branch. Sometimes, it Hangs Upside Down to catch fish beneath its perch (Nesbitt 1984). When the prey is too far away, the heron can Dive or Jump from its perch into the water below. It intersperses Standing with Hopping from one feeding station to another or by slow Walking. It also Walks along the edges of the water, or along branches in a slow and methodical manner, continuing in its Crouched posture. It also Walks readily among branches, both from perch to perch and also to transition to an open place from which to fly. While Walking, a heron may use Head Swaying or Neck Swaying to locate prey. It also does things to stir up its prey, such as Walking Quickly, Foot Stirring, and Foot Raking. When feeding actively, it often flips its tail and raises and lowers its crest. It uses Standing Flycatching to catch insects, and the aerial techniques of Feet First Diving and Plunging (Knight 1994), also Swimming, Feeding, and Stealing.

Its most intriguing behaviour is Baiting, which has been reported from a number of places. These birds use a wide variety of baits and lures: bread, maize, popcorn, fish pellets, feathers, twigs, leaves, berries, flies, and plastic (Davis and Kushlan 1994). These observations are clearly examples of tool use (Lovell 1958, Preston *et al.* 1986, Higuchi 1988a, Kurosawa and Higuchi 1993, Harvey 1999, 2000, Hayes and Rooks 2001). Like traditional fishermen, birds have been observed to dig up earthworms and then use them for bait and to break sticks to use for baiting (Davis and Kushlan 1994). Green Herons feed both by day and also at night, especially under artificial light (Hayes and Hardy 2001). They roost when not foraging and spend much of the time preening (Davis and Kushlan 1994). In the evening they roost alone or in small groups. In larger mixed-species groups they tend to segregate themselves. They often give the Skeow call when flying to and from roosts. They respond to disturbance with a Bittern Stance. The Green Heron is primarily a fish eater (e.g. Niethammer *et al.* 1983). Fish include a wide array of species, *Fundulus, Lepomis, Notropis, Aphredoderous, Ictaluris, Esox, Cyprinus, Aplodinotus, Gobius, Dorosoma, Mendia*, and *Anguilla*. This variety suggests that they will eat anything they can catch and handle. It is extremely opportunistic with an overall highly varied diet including earthworms, leeches, adult and larval insects (larval dragonflies, damselflies, water bugs, diving beetles, grasshoppers, crickets, katydids), spiders (*Dolomedes*), crayfish, crabs, prawns (*Palaemonetes*), snails, fish, frogs, toads, tadpoles, salamanders, lizards, snakes, and rodents. Wiley (2001) observed a Green Heron taking nestlings from a weaver's (*Ploceus*) nest, the first record of this species eating a bird. They consume vegetable matter, such as acorns (King 1985), but the frequency and value—if any—of this food remains to be determined.

Breeding

The nesting season varies geographically, both with season and with rainfall. Breeding in the USA is March–July with geographic variation, April–August in West Indies, July–September in San Blas Mexico in the rainy season, and May–June in Jalisco Mexico (Hernández and Fernández 1999). Green Herons nest at the water's edge or overhanging water in shrubs, bushes, and small trees. For nesting, plants used include *Rhizophora, Avicennia, Pinus, Cephalanthus, Salix, Acer,* and *Quercus*. They have been observed to use a bird nest box (Sandilands 1977). Green Herons most typically nest solitarily in defended territories. However, they also nest in colonies (e.g. Maccarone and Gress 1993), usually in small single-species groups of up to 100 nests, and also in mixed-species colonies, where they tend to nest apart from other species. They nest not only with other herons, but cormorants (*Phalacrocorax*), darters (*Anhinga*), starlings (*Sturnus*), and Grackles (*Quiscalus*). The degree of sociality in nesting is likely a function of the local food supply (Kaiser and Reid 1987).

Nests are small structures, about 20 × 30 cm, made of twigs which sometimes are quite long, but usually thin. Nests are generally unlined but with a well-developed cup, about 4.5 cm deep (Dickerman and Gavino 1969). Nests most typically are placed over water in bushes and trees, from ground level to about 2 m. However, higher nests are not uncommon and they averaged over 10 m high in one colony (Kaiser and Reid 1987). Old nests of various species are often reused and reconstructed (Davis and Kushlan 1994). Or they may steal sticks from old nests and start anew. Females build the nest, but both birds make repairs while they incubate. Destroyed nests can be rebuilt quickly.

The Green Heron is an active and aggressive bird in its nesting behaviour. The male chooses an advertising site, often an old nest. In northern populations this occurs upon arrival from spring migration. It has a complex array of nesting behaviours (Meyerriecks 1960). Birds that are sedentary tend to have a shorter and less intensive courtship, probably due to re-pairing of individuals holding adjacent territories (Kushlan 1983). Courtship begins with birds making Circle Flights and Standing at the display site giving Skeow calls. At their most intense, the Circle Flights exhibit intense exaggerated flapping producing a 'whoom, whoom, whoom' sound. In flight, the neck is crooked, legs dangle, crest and back feathers are elevated, and Roo calls may be given on landing. Males give Snap and Stretch displays. During the Snap, the bird moves its head down to its feet and bill snaps, with feathers slightly erect. It also shows bowing and bobbing components. The Stretch starts with the neck, head, and bill straight up, and then bent back until the head almost touches the back. The back plumes are erected but other feathers are sleeked; the bird sways from side to side, the eyes bulge, the iris changes from yellow to deep orange, and the bird gives the Aaroo call. The Green Heron is highly aggressive and defensive. Its Forward display is given with the head and neck fully forward, feathers fully erect, especially the well-developed crest, eyes bulging, bill open showing a red mouth (during breeding), tail flipping, all preceding a Bill Lunge (Davis and Kushlan 1994). As it lunges, it extends its wings and gives the Raah call. It also uses Supplanting Flight, over short distances and Pursuit Flights. As pairing occurs, the displays change. The Stretch becomes more common after pairing. It is part of the Greeting Display, and is responded to by the female who gives a less intense Stretch. The Skeow call, Circle Flights, and Snap (as an advertising display) cease after pairing.

Eggs are pale green-blue, but dull as incubation proceeds. They average 37 × 28 mm. The normal clutch is 3–5, range 1–7. Eggs are laid at 2-day intervals. They can have more than one clutch per year, particularly in the tropics (Dickerman and Gavino 1969, F. Hayes pers. comm.). Incubation is by both sexes, beginning with the second egg, although there is evidence for beginning with the first egg. The incubation period is 19–23 days in Mexico, averaging 19–20 (Dickerman and Gavino 1969, Hernández and Fernández 1999). After the clutch is complete, the male incubates during the day, the female at night. Parents also shade eggs in direct sun and heat. Hatching is asynchronous over 3–4 days. Young are semialtricial, having limited movement. Chicks grow quickly, from 11.5 g at hatching to minimal adult weight in a weeks. They leave the nest and climb about on branches by 1 week, and can jump from branch to branch in 2 weeks. Chick growth is variable due to asynchronous hatching and differential growth, which is not heavily related to sibling competition (Gavino and Dickerman 1972). Chicks call with 'tik, tik, tik, tik', and adults approach the young giving a Cuck call. Young are fed initially into their mouth or on the nest, and then later by 1 week, young grasp the parent's bill. Young are left alone at 9–11 days and leave the nest at 16 days. They fly at 15–22 days. They leave the colony at 25 days. The disappearance of eggs, usually attributed to predation, may be high in some areas. Egg and chick predators include snakes, grackles, crows (*Corvis*), and raccoons (*Procyon*). In Missouri, USA, 78.8% of nests were successful.

Population dynamics

The usual age at first breeding is probably in the second year. There are few reports of mortality by adults. There is little information on survival. Longevity is at least 7 years 11 months.

Conservation

Green Herons are common or even abundant over most of their range. High reproductive potential and broad habitat flexibility appear to compensate for many threats. The heron is subjected to shooting and trapping for food, pesticides, and habitat loss and alteration (Davis and Kushlan 1994). Disturbance to nests by human activity is a continuing problem in many areas. The primary conservation need, range wide, is the protection of habitat, especially wetlands (Kaiser and Fritzell 1984).

Research needs

Nesting success and population biology in various geographic areas need additional study. Further examination of migratory movements of populations and individual birds and the identification of important non-breeding sites in Mexico and the West Indies are needed. The systematics of the *Butorides* herons requires much addition as study, especially at the range overlap of *virescens* and *striatus*.

Overview

The Green Heron's habitat is varied but typically is a thickly overgrown waters edge where it feeds from overhanging rocks, trees, or structures by Standing perched over, or by the water watching for fish to approach. This is a successful technique, but one requiring extraordinary skills at stillness, concentration and patience. This is a successful technique, but one requiring extraordinary skills at stillness, concentration, and patience. It also takes a territory free from disturbance (from other birds and sometimes from people) and sufficiently renewable food supplies. So, like other stand and wait herons, it must keep others from its feeding site and is highly effective in communicating its territory-holding status using its crest and vocalizations. Given the inherent constraints of a Standing-feeding mode of life, it also can make use of a remarkable array of more active and inventive feeding behaviors including using bait. It can feed at all times of day and night, sometimes responding to artificial light. This is a species that is very good at what it does best, but having a remarkable plasticity to deal with novel situations and opportunities.

Striated Heron *Butorides striata* (Linnaeus)

PLATE 7.5, 7.6

Other names: Little Heron, Green-backed Heron, Little Green Heron, Red Heron, Mangrove Heron, Red Mangrove Heron, Red Mangrove Bittern, Johnny Mangrove, Mangrove Jack, Thick-billed Heron, Galapagos Heron, Lava Heron in English; Garcita Verde in Spanish; Socozinho in Portuguese; Heron vert in French; Streifenreiher in German; Mangrovereiger in Dutch; Зеленая кваква in Russian; Kokokan laut in Indonesian; Bakaw in Pilipino, Bakaw Itim in Tagalog (Philippines); Sasagoi in Japanese Lü lu in Chinese.

Ardea striatus Linnaeus, 1758, Syst. Nat. 10th ed., p. 144: Suriname.

Subspecies: *Butorides striata sundevalli* (Reichenow) 1877: Galapagos Archipelago; *Butorides striata atricapilla* (Afzelius) 1804: Sierra Leone; *Butorides striata rutenbergi* (Hartlaud) 1880: Mohambo, northern Madagascar; *Butorides striata brevipes* (Ehrenberg) 1833: banks of Nile and coasts of the Red Sea; *Butorides striata crawfordi* (Nicoll) 1906: Assumption Island; *Butorides striata rhizophorae* Salomonsen 1934: Mayotte, Comoro Islands; *Butorides striata degens* Hartert, 1920: Praslin Island, Seychelles; *Butorides striata albolimbata* Reichenow, 1900: Diego Garcia, Chagos Archipelago; *Butorides striata chloriceps* (Bonaparte) 1855: India, restricted to Hitaura, Chisapani, Garhi district, Nepal; *Butorides striata javanica* (Horsfield) 1821: Java; *Butorides striata amurensis* (Schrenck) 1860: Amurland, Amur Valley, eastern Manchuria; *Butorides striata actophila* (Oberholser) 1912: North Pagai, western Sumatra Islands; *Butorides striata spodiogaster* Sharpe, 1894: Andaman and Nicobar Islands; *Butorides striata carcinophila* Oberholser 1924: Casiguran, Luzon, Philippines; *Butorides striata steini* Mayr, 1943: Dilly–Dìli, Timor; *Butorides striata moluccarum* Hartert, 1920: Buru, Moluccas; *Butorides striata idenburgi* Rand, 1941: Idenburg River, Dutch New Guinea; *Butorides striata stagnatalis* (Gould) 1848: Port Essington, Northern Territory; *Butorides striata flyensis* Salomonsen, 1966: Lake Daviumbu, southern Papua New Guinea; *Butorides striata macrorhyncha* (Gould) 1848: east coast of Australia = Gosford, New South Wales; *Butorides striata solomonensis* Mayr, 1940: Vangunu Island, Soloman Islands; *Butorides striata patruelis* (Peale) 1848: Tahiti.

Description

The Striated Heron is a small, stocky, dark-backed heron with a thick neck, relatively heavy dark bill, thick short legs, and grey neck.

ADULT: The typical adult Striated Heron has a glossy green-black crown with a short erectile crest. The irises are orange-yellow. Lores are dull yellow-green above and black on the lower portion. The bill is dark, the upper being brown-black and the lower dusky green, lighter at the base. A dark line extends from the bill to under the eye. The sides of the head, side of the neck, and breast are light grey to buff, with considerable geographic variation. The back is dark grey with a green cast and elongated feathers especially toward the tail. Upper wings are grey-black with obvious buff edges to feathers. Flight feathers are black, with buff coloured edges. The chin and throat are white, with a row of central grey spots. The throat and the centre of the upper breast are marked vertically in a rufous, rust, or chestnut band, streaked with white and edged in grey. Underparts are brown-grey to grey in most forms. The legs are olive-grey in front and yellow-green behind. In breeding, the head, erectile crest, and back become glossy blue-black. The irises turn deep orange in courtship. The bill becomes glossy black, less so in females. The lores turn blue-black or cobalt blue. The legs turn glossy orange, even in the grey-legged *sundevalli* (Kushlan 1983). Red or pink bills have also been reported.

VARIATION: Females are slightly smaller and duller than males. Variation in this species is widespread, complex, and far from understood. In the Old World, polymorphic variation occurs among grey, rufous and intermediate forms. Rufous forms have the head (other than the crown), neck, and back rufous brown, with feathers fringed slightly darker. The undersides are light rufous brown. Intermediate forms have more white on the foreneck, light grey underparts, and greyer upper parts. In the New World, variation includes birds with brown-grey necks, widely reported throughout South America. There is also variation in the undersides, with some birds being extremely white. While a matter of some controversy, birds intermediate between *striata*-type and *virescens*-type occur where their ranges meet in southern Central America, south Caribbean islands, and north coastal South America (Payne 1974, F. Hayes 2002). Described geographic plumage variation is the largest of any heron, recognized by 23 currently accepted subspecies based mostly on size, and neck and head colour. Many of these variations are found in island populations, but subspecies are also described from within the Asian and Australian mainlands. Some populations are well differentiated, whereas other intergrade into each other (Payne 1974, 1979, Schodde *et al.* 1980).

Butorides striata striata has a dark grey neck and green-grey back plumes. However, birds with brown-grey necks occur throughout South America (Payne 1974). South American birds also differ in the amount of white on the undersides, varying from white, to intermediate, to grey, the whitish birds once being recognized as the subspecies *cyanurus*. *Sundevalli* is dimorphic with herons that are entirely dark and also grey birds similar to those from South America, with perhaps neck streaks being more grey-brown than rufous. The dark birds are dark grey with back plumes of blue-green and grey legs. Intermediate birds also occur. All tend to have large bills and stout legs. The African *atricapilla* has a dark grey neck and unlike South American Striated Herons is lacking the brown-grey morph, has more distinct rufous on the upper breast outside of the lower neck streaks, and also has rufous rather than grey-brown streaks on the lower neck (Payne 1974). *Brevipes* is darker than *atricapilla*, with more bronze and less bright green back and crown. *Solomonensis* is brownish. *Patruelis* has a light buff neck and undersides. *Amurensis* has a light grey neck. The closely related *rutenbergi* from Madagascar is slightly darker. *Chloriceps, javanica*, and *carcinophila* are similar to each other. *Javanica* is less grey and bluer than *chloriceps*. *Carcinophila* has a very dark neck and breast. *Stagnatilis* is a pale dimorphic form, with rufous and grey types, and relatively long bill. *Macrorhyncha* is a dark form with rufous underparts, and a relatively short bill. *Flyensis* is a pale bird with a green gloss to the back plumes and relatively short bill. *Idenburgi* is a small pale bird, very light on its undersides. *Molluccarum* is small and dark, with a relatively short bill.

JUVENILE: The juvenile is an overall brown bird with white spots. It has a brown-black crown with narrow buff to white streaks and a slight crest. The upper bill is brown-black above, the lower bill yellow. Irises are yellow and lores are yellow-green. The sides of the head and upper neck are brown heavily streaked with buff-white. The chin and upper foreneck are white, showing brown tips to the feathers, turning to dark brown with buff streaks on the lower foreneck. The back and wings are olive-brown spotted in buff-white. The underwing is light grey. Flight feathers are dark grey with a green cast, tipped in white. The underparts are brown to dark brown streaked or spotted in buff. The legs are yellow-green, greyer on the front. Juveniles gradually lose the streaking on their head, neck, and underparts and the white on the flight feathers.

CHICK: The hatchlings are covered with grey down on the upper sides. Down is longer and darker on the head forming a crest. Irises are green-yellow, becoming yellower or white. The skin on the head is green; the lores are yellow-green. The eye ring is green-yellow. The bill is yellow with a dark tip. The throat is white. The down is light grey on the undersides. The legs are green becoming yellow-green in fledging.

VOICE: The voice is not well characterized, especially compared with that of the Green Heron. It is reported to be a relatively silent bird and so appears to be far less vocal than the Green Heron. The primary vocalization is the 'Keeuuk' call, used as an alarm, flight, and advertising call. It is also rendered as 'k-yek', 'baaek', 'tyong', 'tyah', 'kyah', 'skuk', or 'tchack'. 'Skow' calls are given by displaying males (at least in the Galapagos—Kushlan 1983). A sneezing 'Tech-aah' call and a 'Hoo' call are reported in courtship. In *sundevalli*, the female answered the male's stretch display with a soft 'Coo' call. 'Tcheeunk' is used on arrival at the nest. Other calls, especially squawks and croaks, have been reported but their context is not documented.

WEIGHTS AND MEASUREMENTS: Weight: 193–235 g (Africa); length 35–45 cm in South America, 40–41 cm in Africa, and 43 cm in Australia.

Field characters

The Striated Heron is identified by its thick grey to rufous neck, black cap, relatively large bill, and dark grey back. It flies with slow wing beats and legs extended beyond the body making it appear short tailed. It is distinguished from the Green Heron by its grey to rust (not chestnut) neck. It is distinguished from other herons by its combination of dark crown, relatively large bill, uniform colour pattern (although geographically variable) of the sides of the head and neck, yellow lores, yellow to orange legs, crouched feeding posture, and behaviour including tail flicking. It is distinguished from the dark little bitterns (Dwarf and Black) by being larger and lighter coloured (and lacking the white or yellow throat stripes) and from other little bitterns by its larger size, dark cap, and grey back. The rufous form is distinguished from the adult Rufous Night-Heron by being smaller, having dark (not white) underparts, and green (not yellow) legs. The immature Striated Heron is identified by its dark cap and dark, spotted back. It is probably not distinguishable from the juvenile Green Heron, although it may tend to be greyer on the neck (F. Hayes pers. comm.). The Striated Heron is distinguished from immature night herons and the Rufous-bellied Heron by being darker with less spotting, with a darker cap, and being much smaller with a longer neck and slighter bill. It is distinguished from immature pond herons by being darker, lacking head and breast streaking, and having dark wings. It is distinguished from the juvenile small bitterns by its larger size, dark crown, and darker upper parts.

Systematics

As noted in the Green Heron account, the systematic position of the *Butorides* herons has not been clear. *Butorides* appears to be closely related to *Ardeola*, and together these species are most closely related to the *Ardea* herons. Species-level systematics of the *Butorides* has been a matter of considerable study. In this book *striatus* is recognized as distinctive from the North American *virescens* based principally upon several studies, the conclusions of the North American Check List Committee (AOU 1998), and the trend of molecular evidence suggesting distinctiveness of the order generally attributable to species differences (K. McCracken and F. Sheldon pers. comm.). But these findings are not yet conclusive and cannot be until the breeding birds are thoroughly studied in the wild in their zones of overlap. Fortunately, new information is emerging (G. Alvarado pers. comm., F. Hayes pers. comm.). Although controversial, further examination of available specimens suggests an extensive hybrid zone in the south Caribbean islands, north South America, and south Central America. The variability and intermediacy in the zone of contact suggests that *striata* and *virescens* interbreed freely (Hayes 2002). Although not definitive, and requiring additional molecular, behavioral, and perhaps morphological study, these finding suggest that by traditional concepts, the two *Butorides* might be considered conspecific. So it is likely that the question of specific distinctiveness of the two forms is not yet settled. Additional questions remain also as to the relationship between the South American and Old World forms, and similarly within the Old World, as to the relationships among the mainland and many island forms. It seems likely that at least Old World and New World forms might be different species.

Individual variation also occurs. In the New World, *striata* morphs occur in various places with neck coloration ranging from grey to brown; such variability is not shown in African or Asian birds (Payne 1974). There appears to be clinal variation among the Old World populations that may not be captured adequately by the present array of recognized subspecies.

At the subspecific level, the considerable range of variation in colour and morphometrics is recognized in 23 subspecies, many of which are island populations. In the New World, two subspecies are recognized, the nominate form and *sundevalli*, a particularly distinctive, dimorphic island population with large feet, sturdy legs, and a majority of birds with an all-dark grey plumage, unique in the group. *Sundevalli* has often been considered to be a separate species. However, South American colour forms are common on the Galapagos. Despite the morphological differences, intriguingly molecular differentiation has not yet been found between the Galapagos birds and mainland birds (K. McCracken pers. comm.). In the Old World, 21 named forms are recognized. In this book, we diverge from the Old World subspecies listed in Hancock and Kushlan (1984) in following the findings of Schodde *et al.* (1980) who recognized five Australasian subspecies: *stagnatalis* including *rogersi* and *cinerea* (of Hancock and Kushlan 1984), *macrohyncha* including *littleri* (of Hancock and Kushlan 1984), *flyensis* (not recognized in Hancock and Kushlan 1984), *idenburgi*, and *moluccarum* including *papuensis* (of Hancock and Kushlan 1984). As noted above, variation over the Old World range of the species deserves additional attention. Much of the plumage variation seen may be individual and developmental, and may not have a geographic component. Study of geographic variation is hampered by the lack of specimens from important places and seasons. Molecular evidence may hold the best clues for disciphering distant relationships among forms. It may well be that some currently recognized subspecies may better be considered full species. The systematics and evolutionary relationships among *Butorides* herons are clearly interesting research issues within the Ardeidae.

Range and status

The Striated Heron is one of the more cosmopolitan herons, occurring in South America, Africa and Madagascar, Indian Ocean islands, south and east Asia, the East Indies, Australia, and Pacific islands. Information on range relative to subspecies is from Payne (1974, 1979) and Schodde *et al.* (1980).

Breeding Range: *Striata* is the South American form. It breeds in east Panama, Colombia, Venezuela, Trinidad, Guyanas, Ecuador, Peru, Bolivia, Brazil, Paraguay, Uruguay, and Argentina (to Buenos Aires, La Pampa). *Sundevalli* breeds in the Galapagos Islands. *Atricapilla* breeds in Mauritania, Senegal, Gambia, Guinea–Bissau, Guinea, Sierra Leone, Liberia, Ivory Coast, Mali, Burkina Faso, Ghana, Niger, Nigeria, Chad, Cameroon, Gabon, Principe, and São Tomé; also in Sudan, Ethiopia, Somalia, Kenya, Tanzania,

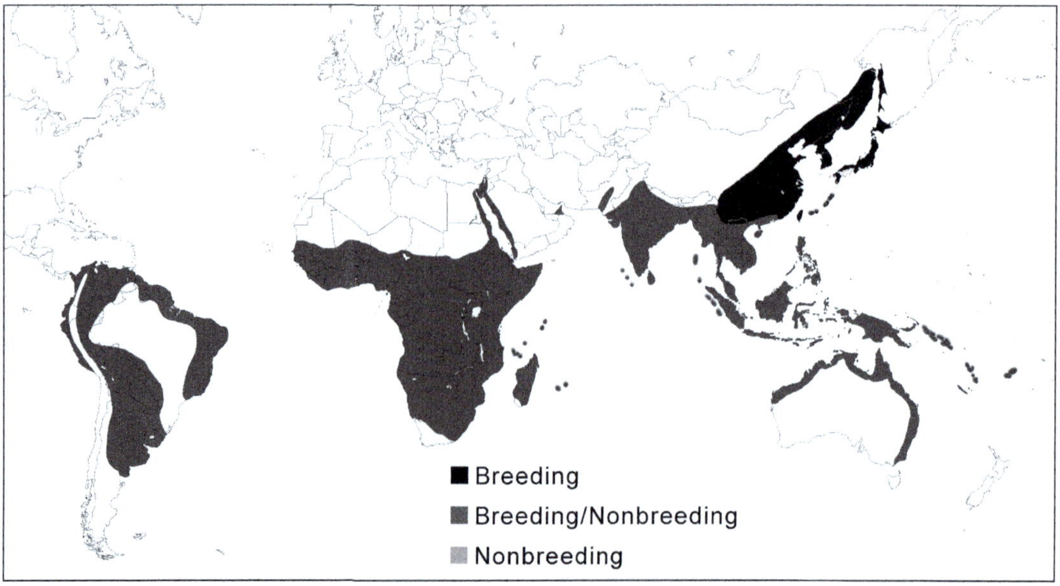

Uganda, Rwanda, DR Congo, Angola, Botswana, South Africa, Zimbabwe, Zambia, Malawi, and Mozambique (van Niekerk 1992, D. Turner pers. comm.). It is probably this subspecies that is reported to breed in Aswan, Egypt, as opposed to the Red Sea form, *brevipes* (Dijkstra 1997). *Brevipes* breeds on the Red Sea coasts of Egypt, Sudan, Eritrea, Djibouti, Somalia, Yeman, United Arab Emirates, Oman and Saudi Arabia (Baccetti 1983, Stagg 1984, Goodman and Storer 1987, Porter and Al-Saghier 1998, C. Tourenq pers. comm.). The species has recently been reported breeding in Israel (Yosef *et al.* 2002). In the Indian Ocean, *rutenbergi* breeds on Madagascar; *crawfordi* breeds on Aldabra, Assumption, the Amirantes, Cosmoledo, Astove, and Farquhar; *rhizophorae* breeds on the Comoro Islands; *degens* breeds on the granitic islands of the Seychelles; and *albolimbata* breeds on Diego Garcia, the Chagos archipelago, and the Maldives. *Chloriceps* breeds in India (Sind, Punjab, Gujarat (Mukherjee and Parasharya 1998), Kasmir, east to Assam, Manipur, Laccadive Islands), and Sri Lanka. *Javanicus* breeds in Myanmar, Thailand, Malaysia (peninsular), and Indonesia (Sumatra, Java, Kalimatan, Bali). Birds from Mauritius and Rodgrigues islands (Rowlands 1987) are assigned to this subspecies but are difficult to distinguish from *chloriceps*. *Amurensis* breeds in Russia (Amurland, Ussuriland, Sakhalin Island), North Korea, South Korea, northern China (Heilongjiang to Shandong), and Japan (Honshu south, Ryukyu and Bonin islands). *Actophila* breeds in south China (Sichuan, Yunnan, to south Shaanxi, Hainan), north Vietnam, north Laos, north Thailand, north Cambodia, and north Myanmar. *Carcinophila* breeds in Taiwan, Philippines (Luzon, Negros, Cebu, Samar, Mindanao), and Indonesia (Sulawesi and nearby islands). The birds from Taiwan and Sulawesi are assigned to this subspecies on the basis of their darkness. *Steini* breeds in Indonesia and the Lesser Sunda Islands (Sumba, Flores, Alor, Timor). *Moluccarum* breeds in west New Guinea and nearby islands including the Mollucas (Schodde *et al.* 1980). *Idenburgi* breeds in Irian Jaya (Schodde *et al.* 1980). It has not been confirmed if this is the subspecies of birds from north Papua New Guinea. *Spodiogaster* breeds on west Sumatran islands (Indonesia), and the Andaman and Nicobar Islands (India). *Stagnatalis* breeds in coastal mid and northwest Australia and north Australia (Schodde *et al.* 1980). *Macrorhyncha* breeds in coastal east Australia (Schodde *et al.* 1980). *Flyensis* breeds on the south coast of New Guinea (Schodde *et al.* 1980). *Solomonensis* breeds in New Ireland, Solomon Islands (Schodde *et al.* 1980), Vanuatu, west Fiji Islands. *Patruelis* breeds on Tahiti, Society Islands.

NON-BREEDING RANGE: Most subspecies are sedentary. Northern populations are migratory or partially migratory in the non-breeding season. *Amurensis* occurs in the non-breeding season in south China, north Indo-China, Taiwan, Philippines, Malaysia (north Borneo), and Palau Islands. *Actophila* is partially migratory. Some remain in the breeding range (Desholm and Wegeberg 1997) but also occur in the non-breeding season as far south as the Nicobar Islands (India), and Indonesia (west Sumatra islands, west Kalimatan). Some *atricapilla* populations show local rain-dependent movements (e.g. Gambia—Barlow and Wacher 1997), while in South Africa erratic seasonal movements have been noted during flood years (Brown *et al.* 1982, Becker 1984, de Swardt 1992, Esterhuizen 1994). In the New World, *striata* occasionally occurs north of its zone of overlap with the Green Heron, such as St. Vincent and Virgin Islands (St. John) (Norton *et al.* 2003).

MIGRATION: Migration patterns for *amurensis* and *actophilus* are poorly known. Similarly, local movement patterns are poorly documented, other than the coming and going of birds from a specific area. Although regular dispersal is not as dramatic as for some other herons, it has colonized many islands. Island dispersal records include Christmas Island (Indian Ocean) and Tonga (Gill 1990).

STATUS: Overall this is a widespread species and common where it occurs. It is common and widespread in Africa, although populations are not estimated and nests are seldom documented (Turner 2000). It is abundant in some west African mangrove swamps, often nesting in the hundreds (D. Turner pers. comm.). A population estimate for Tanzania is 10 000–15 000 birds (Baker 1995, Baker and Baker 2004). It occurs throughout Madagascar, but is rare on the High Plateau. It is widespread on the

Comoros and is relatively common to abundant on the Seychelles, where mangroves remain intact (J. Gerlack pers. comm.). It may be expanding in Melanesia (Dutson 2001). While present on the Mascarene Islands it is now extirpated from Reunion. It is widespread and abundant in east Asia (Lansdown *et al.* 2000). It has a limited range in Australia, occurring in mangroves. Its status is unclear there, but local reductions have occurred (Maddock 2000). It occurs sparsely in coastal areas of New Guinea. It has declined in Tahiti over the past century, now numbering about 100–120 individuals (Monnet and Varney 1998). It is widespread in South America, although its status is little known (Morales 2000). The subspecies on the Galapagos is common.

Habitat

The Striated Heron uses a variety of inland and coastal habitats, and ocean islands. It especially uses edges—brushy river edges, river swamps, of forested streams, lakes, salt flats, woods, shores, pools, rice fields, canals, and artificial ponds. It forages typically at the edge of pools, streams, and channels, often from perches overhanging the water. Over much of its eastern range it is a mangrove heron seldom found far from the coast and particularly using mangrove swamps. In Australia it is so restricted to mangroves that inland occurrences are notable (Gynther 1995, Britton and Britton 1996, Mitchell 1996). It occurs from sea level to considerable altitude, up to 4000 m in the Andes (Fjeldsa and Krabbe 1990).

Foraging

The Striated Heron is a skulking predator feeding primarily in and under bushes adjacent to pools and watercourses venturing also onto adjacent mud flats and similar open areas. It characteristically feeds solitarily, defending its feeding site and larger territories. In southern Africa it occurs along a river at a density of only 1.9 birds per 13 km transect (Allan and Davies 1999). It uses its crest to communicate its aggressive and territorial intentions, especially in response to other herons. On occasion, it will feed in loose groups and among aggregations. It feeds by day, at night, or in the evening, and is attracted to lights. It feeds primarily by Standing in shallow water or next to the water in a characteristic Crouched posture, with its body parallel to the water. The head is withdrawn or extended as it waits for long periods between strikes, once measured at 29 and 42 minutes between successful strike (Oliver 1995). It also Stands on branches overhanging the water and Stabs into the water, sometimes submerging itself. It Walks slowly in very shallow water, along the shore, or along branches. It is so slow that its foot may be kept raised for 30 seconds between steps. Its rate of progress has been measured at 1 metre per minute (Recher *et al.* 1983). In one study the number and frequency of strikes during Walking far exceeded that in Standing (Oliver 1995). It also uses Standing Flycatching. It feeds by Walking Quickly and Running in such situations where active feeding is more successful than standing (Oliver 1995). It also uses Feet First Diving and Plunging (Davis 1983). Characteristically it flicks its tail up and down, at almost any time. It also assumes a Bittern Posture to hide itself. Reports of Striated Herons using Baiting are frequent and widespread (Walsh *et al.* 1985, Higuchi 1986, 1988b, English 1987, Foxall and Drury 1987, Crous 1990, Richards 1992, Kurosawa and Higuchi 1993, Robinson 1994, Pyle 2000, Hayes and Rooks 2001). Japanese herons particularly are noted to have used a wide variety of baits (Higuchi 1988b).

The Striated Heron is primarily a fish eater (e.g. Niethammer *et al.* 1983), but it is also opportunistic. Fish include a wide array of species-*Belone, Periophthalmus, Boleophthalmus, Puntius, Mystus, Gobius, Mugil, Anbassis*, and *Gambusia*. It also eats insects (water beetles, dragonflies, grasshoppers), earthworms, polychaetes (*Nereis*), crustaceans (*Metapenneus, Scylla, Portunus, Varuna, Macrobrachium*), frogs (*Rana*), reptiles, and birds (Stocker 1994, Hull 1999). Striated Herons also may eat vegetation (Beltzer 1983).

Breeding

The nesting season tends to be in the rains. It is March–June in Java, November–April in north Australia, September–January in west Australia, and September–December in south Australia. In Africa the breeding season is diffuse, with nesting recorded in almost every month of the year but with well-defined peaks during the rains (D. Turner pers. comm.). In West Africa, while there are some dry-season breeding records, egg-laying peaks during the July–September rains. In east and southern Africa, breeding has been recorded year-round, with peaks in April–June, September–October, and again in December–January following periods of prolonged rainfall. On islands in the western Indian Ocean, breeding is generally September–January. In Tahiti 22 clutches were laid in nine different months (Monnet and Varney 1998). The situation in the Galapagos is typical of the coastal tropics (Kushlan 1983). Nesting begins with the onset of the rainy season, but herons may nest three times annually and so the nesting timing of individuals can be variable throughout the season (Kushlan 1983). They nest in shrubs, bushes, and trees usually overhanging water, in well-concealed locations. However, they occasionally nest over dry land. Plants used especially include mangroves (*Rhizophora, Avicennia*) and also *Allocasuarina, Myoporum, Callistemon, Hibiscus, Casuarina, Syzygium*, and *Inga*. Unusual sites are also used, such as a TV antenna in Japan (Kobayashi and Hironaka 1992). Most Striated Herons nest solitarily, typically spaced along the shoreline of a river, linear swamp, or sea coast, where they nest within their defended territories (e.g. Kushlan 1983). However, they also nest at various degrees of proximity from loosely spaced nests in single species groups of 5–15 nests to breeding aggregations in the hundreds, such as have been reported from West Africa. They tend not to nest with other species, but when they do they tend to nest apart from the main colony core. Nests are small, shallow, loosely constructed, often flimsy structures, about 20–40 cm in diameter. Nests are typically low, about 1 m above the surface. They are made of twigs and, as far as is reported, unlined. Building takes up to 14 days, but the parents also continue to improve the nest during nesting.

Courtship behaviour has been surprisingly little documented. Courtship begins when the male chooses an advertising site. Territories are defended by the bird assuming an aggressive presence at the nest site, emphasized by Crest Raising as well as Pursuit Flights. Early displays include Circle Flights and Pursuit Flights. Males give the Skow call from the display site. Pairing birds give simultaneous Snap and Stretch displays, which lack a side-to-side swaying component. The female gives a Coo call as the male does the Stretch display. Observations of the Stretch display in *sundevalli* (Kushlan 1983) differ from those of the Green Heron by lacking the swaying motion and by the Stretch being performed mutually by both males and females prior to pairing. Mock Preening and Allopreening are common between members of the pair. Crest Raising occurs upon alighting and is undoubtedly an important part of the landing display.

Eggs are pale green-blue, becoming duller as incubation proceeds. They average 41 × 31 mm in Australia, 37 × 28 mm in Africa, and 37.7 × 29 mm in Madagascar. The normal clutch is 3–5, range 1–7. Average clutch size is 4 in France, 2.37 in east Africa, 2.7 in Zimbabwe, 2.6–2.8 in South Africa, 3 in Madagascar, and only 1.05 in Tahiti (Monnet and Varney 1998, Delord *et al.* 2003). Eggs are laid at 2-day intervals. Both sexes incubate, beginning with the second egg. Incubation is recorded as 21–22 days in Galapagos, 21 days in South America (Mosso and Beltzer 1992), 21–25 days in Africa and Australia, and 25–29 days in Tahiti. Hatching is asynchronous. Young hatch over period of 3–4 days. At hatching chicks are semialtricial, having limited movement. Chicks grow quickly and climb about on branches by 1 week. Young grasp the parent's bill and are fed by regurgitation. Both parents attend the young, which leave the nest in 14 days (Mosso and Beltzer 1992). Adult care appears to continue in a family group for some time after the young have left the nest. The disappearance of eggs, usually attributed to predation, may be high in some areas. Nest predators include water dragons (*Physignathus*), shrikes (*Colluricincla*), crows (*Corvus*), and eagles (*Haliaeetus*). Nests are also lost due to high tides and storms. In France, nest survival was high through the nestling period (Delord *et al.* 2003). The survival rate of young in France and East Africa is reported at less than 2 per nest (Brown *et al.* 1982, Delord *et al.* 2003).

Population dynamics

The usual age at first breeding is probably in the second year, but birds in immature plumage appear at nesting colonies and so may be breeding or attempting to breed in the first year. Mortality is poorly documented. The demography and population biology are not understood.

Conservation

Loss of mangroves in Australia and Seychelles and elsewhere in its range is creating cause for concern. This is a polymorphic species with many island forms. Each of these forms has its own set of conservation issues that need to be addressed. Conservation of each genetic population is crucial.

Research needs

There is surprising little known about the biology, especially the behaviour of this species. Its courtship behaviour, vocalizations, and aggressive behaviour need especially to be documented, given the taxonomic uncertainty in the group and the substantial knowledge base existing on the behaviour of the Green Heron. Basic comparative information is needed on populations in South America, Africa, Asia, and Australia. Foraging, habitat selection, and prey taken all need to be better documented throughout its extensive range. Documentation of population sizes and trends also are needed throughout the range. The present status of island populations especially needs to be documented. Perhaps the most intriguing and important research need is for further systematic study of the various geographic populations. Determining the evolutionary relationships among the populations will require much additional molecular and behavioural study range-wide. Determining the genetic structure of the species is not only of biological interest but need to underlie a conservation strategy that protects distinctive populations and preserves the inherent variability in this widespread group of birds.

Overview

The Striated Heron is a solitary species, of mangroves, streams, ponds and lakes, and inland wetlands. It particularly uses the edges of watercourses, either wading in very shallow water or feeding from overhanging branches. It feeds quietly, with great patience, and skulks from place to place, often returning to favoured sites. It nests solitarily, or in very loose groups. Most populations appear sedentary, although some shift as the wet and dry seasons make habitats alternatively available and unavailable. This is a highly successful species, especially on islands and in mangroves swamps throughout the Old World tropics.

Squacco Heron *Ardeola ralloides* (Scopoli)

PLATES 7.13–7.16

Other names: Garcilla Cangrejera in Spanish; Crabier chevelu, Héron crabier in French; Rallenreiher in German; Ralreiger in Dutch; Желтая цапля in Russian.

Ardea ralloides Scopoli, 1769. Annus I Historico-Nat., p. 88: Carniola (formerly Austro-Hungarian Empire now Kranjska in Yugoslavia).

Description

The Squacco Heron is a tawny buff-brown heron with a streaked head and back, and in breeding a black and white mane.

ADULT: The adult non-breeding Squacco Heron has a head that is finely streaked in black, brown, and grey, forming a modest crown but no elongated plumes in the non-breeding season. The relatively large and powerful bill is pale green-yellow with a black tip and top. The lores are dull yellow-green. The irises are yellow. The hind neck, like the head, is finely streaked in black, brown, and grey. The upper parts are buff brown with a slight tawny tinge. The white wings are mostly concealed at rest by the back plumes, are shorter than in the breeding season. The rump and tail are white. The foreneck and breast are bright buff coarsely streaked in dark brown. The remaining underparts are white. The relatively short legs and the feet are dull yellow-green. In breeding plumage, the upper parts become brighter and deeper. The crown is a mane of yellow-buff or straw-coloured feathers. The crown feathers are slightly elongated (1–5 cm) and are bordered with black. Several very elongated feathers (13–14 cm long) occur on the back of the crown. These are white bordered with black, and extend over the upper back. The lores are green or blue. The lower neck and back plumes are golden to cinnamon-buff. The foreneck and breast are red-gold. During courtship, the bill becomes bright blue except for the dark tip. The lores turn briefly blue before reverting via emerald to yellow-green. The irises in courtship become rich yellow. The back is pink-brown, with the longest back feathers being golden and drooping over the wings. The legs turn bright red in courtship, fading to pink after pairing. The other soft-part colours return to normal after the eggs are laid.

VARIATION: The sexes are alike. Geographic variation is not recognized taxonomically. South and central African birds were once considered to be the subspecies *paludivaga*.

JUVENILE: The immature bird is similar to adults in non-breeding, but drabber. It lacks the crest and back plumes. The bill is more uniform with a dark tip. The breast is more strongly streaked in dark brown. The underparts are grey rather than white. The flight feathers have a brown tinge and shafts, so that the wings appear to be mottled with brown. The tail also is washed with a brown tint.

CHICK: The chick has dark grey down, with long brown-grey down on the head forming a crest. The irises are yellow and lores green-yellow. The rest of the skin is olive-green. The bill is yellow with a dark tip. The belly is naked for the first two weeks. The legs are olive-green in the front and pale yellow behind.

VOICE: The Squacco Heron is generally a quiet bird. The characteristic 'Squawk' call, rendered variously as 'caw', 'kak', 'rrra', or 'kak', is given often at dusk, especially during breeding, as part of the Forward, and in flight. The call becomes a 'kek, kek, kek, kek' in an aggressive Forward. The 'Grr' call, rendered 'grrrr, grrrr, grrrr, grrrr', is the alarm call. The young beg with 'kri, kri, kri, kri'.

WEIGHTS AND MEASUREMENTS: Weight: 230–370 g; length: 42–48 cm.

Field characters

The Squacco Heron is identified by its tawny buff-brown colour, streaked head in non-breeding season, and its crest during breeding. It flies fast with rapid wing beats, its white wings contrasting with its dark body to form an important field character, especially upon take off. When feeding, it appears as a small solitary totally nondescript brown-buff heron. Juvenile and non-breeding birds are difficult or impossible to tell from other pond herons. It is distinguished from the Indian Pond-Heron by its head streaking (vs. no streaking). In breeding it is distinguished from the Malagasy Pond-Heron by its tawny (not white) plumage. It is distinguished from the juvenile Malagasy Pond-Heron by being more slight, its tawny (not dark) plumage, lighter streaking, slighter bill, and narrower wing tips. It is distinguished from the Cattle Egret by its slender build, darker back, white wings, darker (not yellow, orange or red) bill, and streaked head.

Systematics

The Squacco Heron is an *Ardeola*, closely related to the other pond herons. Four of the pond herons (Indian, Javan, Chinese and Squacco) are considered to form a superspecies.

Range and status

The Squacco Heron occurs in Europe, Africa, Madagascar, and the Middle East to Iran.

BREEDING RANGE: It breeds in scattered localities in Europe from south Spain, Portugal (Dias 1991), France (Loire Atlantique, Rhone Delta) (Caupenne 1993, (Fernandez *et al.* 1992), Hafner *et al.* 2001), Italy (Fasola *et al.* 1981, Grussu 1987, Ciaccio and Siracusa 1989), Germany (Kolbe and Neumann

1990), Latvia, Hungary, Croatia, Albania, Greece, south Romania, Bulgaria (Delov 1993), central and south Ukraine, Moldova, Azerbaijan, south Russia, Kazakhstan (Gordienko 1987), Uzbekistan, Tajikistan, Turkmenistan, Iran, and Saudi Arabia (James 1991). In North Africa, it breeds in small numbers in Morocco, Algeria, and Tunisia and also in Egypt on the Nile Delta and Answan (El Din 1992). In sub-Saharan Africa, it breeds over a wide area in Mauritania, Senegal, Gambia, Sierra Leone, Mali, Ghana, Nigeria, Chad, DR Congo, Sudan and Ethiopia, and from Kenya, Uganda, and Tanzania south to Mozambique, Zimbabwe, Zambia, Angola, north Namibia, Botswana, and South Africa (to Cape Province) (Euston-Brown 1993), and over much of Madagascar.

NON-BREEDING RANGE: Northern populations winter in very small numbers in Spain, France, and Italy (Kayser 1994, Hafner and Didner 1997) and in North Africa (Morocco, Algeria, Tunisia, Egypt). Most migrate further south into the Middle East, Iraq, and Iran (Mesopotamian marshes), Oman, Saudi Arabia, United Arab Emirates. The main wintering areas are in tropical subSaharan Africa, from Mauritania, Senegambia, Guinea–Bissau, Sierra Leone, Ghana, Nigeria, and Cameroon east to Rwanda, Uganda, Kenya, and Tanzania. However, the southern limits of wintering northern birds are not clear because of overlap with local African populations.

MIGRATION: Northern populations in Europe and west Asia are migratory. Southward migration is in August–November, and appears to occur at a rather leisurely pace. Birds move in a broad front across Europe, the Middle East, and the Sahara. Return migration is in February and March to as late as May. Migration appears to be along a broad front in that migrating birds are seen widely throughout the Mediterranean and its islands. Migration in both seasons crosses the Sahara. Widespread postbreeding

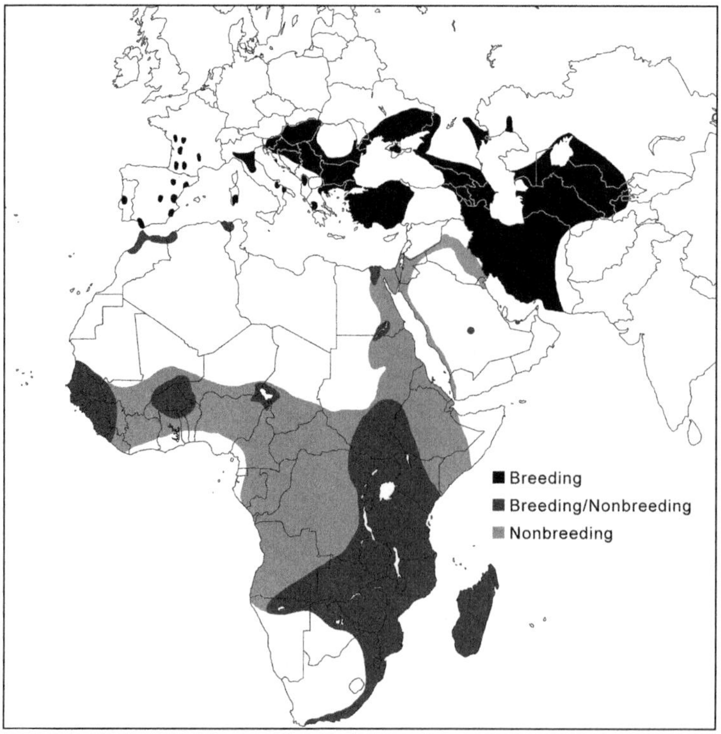

dispersal of juveniles begins in July continuing until migration commences. Spring migration leads to frequent overshoots. So, outlying dispersal records are common place. Dispersal records of note include Iceland, Britain (Moon 1997, Grear-Grearson 1998), Belgium, Luxembourg, Netherlands, Denmark, Germany, Sweden, Finland, Austria, Poland, west central Asia, Azores, Madeira, the Canary and Cape Verde Islands, and Brazil (Fernando de Noronha Island) (Nacinovic and Teixeira 1989). In the Indian Ocean, there are dispersal records from the Comores and Mauritius, undoubtedly from Madagascar.

STATUS: Populations in Europe have fluctuated in interesting ways (Marion *et al.* 2000). By the end of the nineteenth century populations had decreased due to plume hunting. They recovered but then declined again, attributed largely, although indirectly, to habitat destruction. Between 1940 and 1960 the Western Palearctic population increased by 30%; but, between 1970 and 1990, two-thirds of European national breeding populations decreased by 20% or more (Marion *et al.* 2000). A more specific example is the history in the Volga Delta, where 7000 pairs bred in 1970 but then decreased to an annual maximum of 300 in the 1990s. In recent years the species has again increased in south Europe—Spain, Italy, and south France. In the latter area, the increase has been extraordinary, from 105 nests in 1999, which was a number typical of the 1990s, to 200 in 2001 (H. Hafner and M. Kaiser pers. comm.). The total European population is presently estimated to be 14 300–26 800 pairs, with the main concentrations being in Russia (35% of the European population), Turkey (32%), Bulgaria, and Romania (Marion *et al.* 2000). Nesting sites are widely scattered in Europe and around the Mediterranean. Nesting numbers are in the low thousands at the most important sites, such as the Danube Delta, but nesting in west Europe is in the low to mid hundreds (Hafner 2000). The species is rare in north Africa. It has recently been recorded as breeding in the Nile Delta (El Din 1992). Otherwise there were a few pairs nesting in Algeria and Tunisia and 15–85 in Morocco in the early 1980s (Hafner and Didner 1997). Elsewhere in Africa, few data are available. It is a common and widespread species, although large colonies are well scattered (Hafner and Didner 1997, Turner 2000). Indi-cations are that the species is increasing in range and abundance in Africa (Turner 2000), with the known breeding range increasing markedly in West Africa. In East Africa, the population in Tanzania may number up to 20 000 birds (Baker and Baker 2004). Numbers in various breeding areas range from the hundreds to the low thousands. In Madagascar, the breeding population around Antananarivo has increased from 350+ pairs in 1969 to 1000+ in 1990 (Langrand 1990, Turner 2000). In addition to breeding birds, much of Africa supports nonbreeding birds from Eurasia. Drought in African wintering grounds may increase mortality (Hafner and Didner 1997). Thousands winter in areas of West Africa, such as Guinea–Bissau.

Habitat

This is a species that prefers dense, shallow, fresh marshes having cover of tall reeds or dense bushes near by. It occurs usually in lowlands, river valleys, marshes, and estuaries. It can however nest to 2000 m. The Squacco Heron particularly uses reed beds, marshes, rice fields, ponds, canals, ditches, irrigated land, and similar shallowly flooded areas. Sea coasts, reefs, and islands are used on migration. Its principal habitat throughout its range is now rice fields, the extent of which is a likely cause for its recent range increases. The species avoids both dry habitats and those with very high rainfall. Climatic changes can affect population status (Hafner and Didner 1997). For nesting, it prefers dense trees and shrubs convenient to its feeding areas.

Foraging

The Squacco Heron typically feeds by Standing very patiently and perfectly still, waiting for prey to approach. It also Walks, very slowly, searching for prey in a highly Crouched posture, either in the open or among the reeds. It uses Gleaning to take insects and snails off grass stems, Standing

Flycatching, and Baiting (Prytherch 1980, Crous 1994). It usually feeds solitarily defending its feeding territory against other Squacco Herons using Forward displays and Supplanting Flights. More rarely it feeds in small groups of 2–5 birds, especially during the nesting season. It feeds in large flocks in winter and on migration. Feeding success is higher for solitary birds than for those feeding in flocks (Hafner *et al.* 1982). In mixed species groups it defends its territory but is also subjected to interspecies interactions, such as prey robbing (Stival 1987). It feeds during the day, but especially concentrates on crepuscular activity; feeding bouts start an hour before sunrise. It has been observed feeding at night. It appears to prefer to feed under cover, from the stems and branches of overhanging plants or wading in shallow water. However, it does feed in the open, especially in rice fields. It also feeds with cattle, in the same way as the Cattle Egret. It roosts in groups, using sheltered woods and, less frequently, reed beds. Roosting may continue at the colony site after nesting is completed. Roosts in the non-breeding season tend to be large, drawing birds from up to 80 km (Hafner and Didner 1997). Upon disturbance it assumes an Upright posture and may fly off.

The main diet consists of relatively small prey, particularly fish, frogs and tadpoles, insects and insect larvae, dominance depending on place. Insects are most important in Russia, fish in Italy, frogs in France, and a balance of insects, fish, and frogs in Hungary (Molnar 1990). In rice fields it eats more insects than fish. Insects include water bugs (*Naucoris, Notonecta*), water beetles (*Dytiscus, Hydrophilus, Cybister, Hydrous*), grasshoppers, crickets, mayfly larvae, dragonfly larvae (*Libellula, Aeshna*), and fly larvae. Amphibians include frogs and their tadpoles (*Rana, Hyla, Bombina*). Fish include *Scardinus, Cyprinus, Carassius, Rhodeus, Alburnus, Lipomis, Tinca, Leiciscus, Abramis, Anguilla*, and the introduced *Gambusa* and *Lepomis*. Other prey include amphipods (*Gammarus, Apus*), crustaceans, spiders, snails, earth worms (*Lumbricus*), lizards (*Lacerta*), snakes (*Natrix*), and mammals (*Sorex*). It can eat rather large prey, including fish to 10 cm and small birds (*Sylvia, Anthus*) (Gordon 1986), and will attempt to eat foolishly larger prey, such as a Wood Sandpiper (*Tringa*) (van der Ham 1984).

Breeding

Squacco Herons in Europe nest from late April to late July, the season stretching from mid April to August in Spain. In North Africa, egg laying is April–June. In Africa, breeding generally takes place during or at the end of the rainy season, the latter particularly the case in inundation areas, though there are many West African records of nesting in the dry season. In southern Africa, breeding has been recorded throughout the year, particularly in years when summer (November–December) rains are far above average. In Madagascar, breeding takes place from October to March, peaking in November–December. This species nests in dense bushes or small trees (although sometimes to 20 m), near or overhanging water. It also nests less frequently in reed beds and papyrus swamps, using either the reed or small trees. The Squacco Heron typically nests colonially in small or sometimes large mixed colonies, including with cormorants (*Phalacrocorax*), ibises (*Pleagadis*), and other herons, especially with Little Egrets, Black-crowned Night-Herons, and notably the Malagasy Pond-Heron in Madagascar. In colonies, nests tend to be distant, 5–10 m apart, but may be as close as 0.5 m. Nesting colonies occur within proximity to feeding areas, generally within 5 km. It also nests solitarily. Nests are small, bulky, and compact, 17–2 cm in diameter, either flimsy or substantial, up to 20 cm deep, made of reeds, grass, and sometimes twigs. They are built low to the water from almost on the ground to 1.5 m in reeds, papyrus, or low trees. In some colonies they are at intermediate heights, above ground nesting species but below the canopy nesting species (Fasola and Alieri 1992b). The nests are built by both sexes, usually with the male bringing the sticks, though building by each sex alone has been reported. The nest structure can be completed in 6–8 days, but material is added later as well.

The male builds a platform from which to advertise. It defends its site and a larger area with Forward displays. The Squacco Heron's Forward is given with legs bent, bill aiming at the opponent, folded wings slightly aside, in Crouched posture, neck bent back, all plumes fully erected nearly doubling the bird's apparent size. The Squawk call is commonly given. The heron also uses an Upright display with all feathers sleeked. It is often a prelude to a Forward. When alarmed, it gives its Grrr call, which is a sound characteristically heard in a Squacco Heron colony. The Stretch display (also called Foot Lifting in this species) is the principal courtship display (Voisin 1991). The heron perches on a branch, with feet apart, bends slightly, grasping the branch tightly with one foot while loosening the grip on the other foot, which it lifts and extends, then alternates with the other foot. At the same time as it sways from side to side, it stretches its neck downward.

Copulation occurs on or near the nest site. Upon pairing, contact Bill Clappering and Back Biting are common between the pair. In the Greeting Display, the approaching bird fully erects its plumes. The receiving bird does the same. As they meet they mutually do a Forward display, which is soon dissipated (Voisin 1991). The male often brings a stick. Eggs are green-blue. They average 39 × 28 mm in Europe, 38 × 29 in southern Africa, and 37.3 × 27.5 in Madagascar. The clutch is 4–6 eggs in Europe but only 2–4 in Africa and Madagascar. Clutch sizes have decreased in south Europe (Camargue) over several decades (Hafner *et al.* 2001). Eggs are laid at 1 to 2-day intervals. Incubation begins before completion of the clutch, likely after the second egg. Incubation is 22–24 days in Europe but is reported to be only a surprising 18 days in Madagascar (Milon *et al.* 1973). Incubation appears to be mostly by the female. Eggs hatch at 1–2 day intervals. Chicks are fed by both parents, at first by regurgitation and then by food deposited at the nest. Later the chicks rundown the parent and grab the adult's bill. Young begin to clamber from the nest into branches at 14 days. They can fly at 30 days, and are fledged at 45 days. Young form groups at the colony site. They become independent soon after 45 days. In the Camargue, 79.2% of eggs produced young (Hafner 1978). The average brood size is 3.2–3.3 (Hafner and Didner 1997). The loss of eggs and nestlings include predation by hawks (*Accipiter*) (Kayser 1995).

Population dynamics

Generally Squacco Herons breed at two years, but a banded bird was observed to breed at one year. Nothing is published about the long-term demography of the species.

Conservation

The greatest conservation issue is the protection and restoration of habitat for this species (van Dijk and Ledant 1983, Pyrovetsi and Crivelli 1988, Fernandez-Cruz *et al.* 1993, Hafner 2000). It is likely that habitat degradation has been the cause of various population declines. The species' adoption of rice fields has been responsible for range and population increases, where they are occurring. Both natural and artificial habitats are under threat. Rice field suitability depends on cultivation practices, which affect prey availability (Hafner and Didner 1997). The species also depends on wooded sites near feeding areas for nesting, and these too are under threat. Conservation should be focused on protection and management of appropriate feeding and breeding habitats.

Research needs

The Squacco Heron is a well-studied species. Perhaps the most important further work is that linking populations and nesting success to habitat features. This information is needed in order to design habitat conservation strategies and to better understand the role of habitat in population fluctuations. The comparative biology of temperate and tropical populations could provide interesting insights into the relative roles of breeding and non-breeding ecological conditions in the life history and demography of the species. The evolutionary relationships of the several pond herons require additional study to determine their most appropriate classification. The four species making up the *ralloides* group of pond herons are closely related, generally allopatric forms, differing mostly in plumage characteristics. The situation is analogous to *Butorides*. Study using molecular techniques should be conducted to determine the species limits of this group of birds.

Overview

The Squacco Heron is a brownish-buff bird that blends completely into its surroundings. It typically is a lone feeder, standing, crouched low and still, watching for prey in the water below. It is bird of dense marshes and rice fields. The alarm and flight call is highly recognizable, giving the bird its name. Populations fluctuate, the causes not being clear; hunting, habitat change, and perhaps climate have all been advocated. The increase in rice fields are a likely cause of its recent population growth. Northern populations are highly migratory, and so may be affected by conditions in sub-Sahara Africa. Population fluctuations and shifts seem an integral aspect of the biology of this widespread and successful species.

Indian Pond-Heron *Ardeola grayii* Sykes

PLATES 7.7, 7.8, 17.2

Other names: Paddybird, Indian Paddybird, Pondbird, Indian Pond Bird, Pond Heron in English; Garcilla India in Spanish; Crabbier de Gray in French; Paddyreiher in German.

Ardea grayii Sykes, 1832. Proc. Zool. Soc. London, p. 158: Dukhuin (=Deccan).

Description

The Indian Pond-Heron is a small, drab, brown, or in breeding pink-buff, heron with pure white wings and rump, visible in flight.

ADULT: The non-breeding adult Indian Pond-Heron has a head and neck streaked dark brown and buff. The bill is yellow-horn with a long black tip. The lores are green and the irises are yellow. It has drab brown back feathers streaked with buff covering much of its upper wing. The upper wing and flight feathers are white, except that the outermost flight feather is dusky, with a brown-grey shaft; the two or three adjacent flight feathers have dusky tips. Its rump and tail also are pure white. Its underparts are white with heavy buff streaking on the breast. The legs and feet are dull green. In the breeding season the head and neck are unstreaked brown-buff. Three or four cream-white lanceolate plumes extend from the back of the head over the back. The lores and base of the bill are reported to be blue or orange, with the rest of the bill remaining yellow and black. The upper parts are pink-buff. Long feathers of the upper back and wings are deep maroon. With time these feathers become duller and mottled dark brown when worn. The breast has elongated brown-buff feathers. These plumed areas together form a loose coat. The remaining underparts are white, elongated buff feathers growing from the buff-tinged breast. Seasonal changes in leg coloration in the Indian Pond Heron remains a matter for additional study range wide (Relton 1996, Wesley 1993, 1996, Sundar in prep.). What is known is that whereas most individuals have bright yellow legs during breeding, a small percentage (1% in Uttar Pradesh, India) have red legs (Sundar in prep.). This colour develops during prenuptial moult and deepens and is brightest in concert with full breeding plumage.

VARIATION: The sexes are alike. Individuals vary, perhaps with a geographic component. In the Maldive Islands males have a longer, deeper bill, the females have a shorter tarsus, and both sexes have wholly white outermost three or four primaries. This variation has been recognized taxonomically (*phillipsi*) but measurements and feather colour appear to be within the overall range of variation.

JUVENILE: Immature herons are streaked brown, generally similar to non-breeding adults. Their head, neck, and breast feathers are shorter, and their back feathers do not form the loose coat. Their breast is spotted pale brown rather than streaked brown; their upper back feathers may have some dull cream shaft streaks; and the tail may have some faint brown mottling. The bill is darker, the upper mandible ranging to wholly dark and the lower one having yellow only at the base.

CHICK: The downy chick is grey, buff, and white.

VOICE: The voice has not been thoroughly documented. The high-pitched 'Squawk' call is a flight call. The 'Wakoo' call, rendered 'wa, koo', is a breeding call. It is described as very human-like.

WEIGHTS AND MEASUREMENTS: Weight: 230–276 g; length: 42–46 cm.

Field characters

The Indian Pond-Heron is identified by its brown coloration in non-breeding and combination of subtle colorations. It is a drab-looking bird that merges inconspicuously into its habitat but becomes noticeable when its contrasting white wings and tail are revealed in flight. Its flight is fast and agile. It is difficult to distinguish from other pond herons. In fact, in most situations the non-breeding adults and the juveniles may not be separable. The adult is distinguished from the adult Squacco Heron by its brown-buff (not tawny brown) back and the lack of any hint of black streaking on the head and crest. It is distinguished from the Chinese Pond-Heron by its brown-buff (not dark chestnut) head and neck, maroon (not slate) back, and brown-buff (not maroon) breast. It is distinguished from the Javan Pond-Heron by its lighter back (maroon not slate), and brown-buff (not cinnamon) breast and lower neck, and white underparts without any hint of cinnamon and its bright yellow legs during breeding.

Systematics

The Indian Pond-Heron is an *Ardeola* heron, closely related to the other typical pond herons. Four of the pond herons (Indian, Javan, Chinese, and Squacco) are sometimes considered to be subspecies of *Ardeola ralloides*.

Range and status

The Indian Pond-Heron is a bird of south and east Asia.

BREEDING RANGE: It breeds from south Iran, Pakistan, India, Myanmar, Sri Lanka, Nepal, the Laccadive, Maldives, Andaman, and Nicobar Islands.

NON-BREEDING RANGE: It occurs but has not been documented to breed in Kuwait and the United Arab Emirates (Richardson and Bannon 1991).

MIGRATION: This is a sedentary species. Movements may occur due to local flood or drought events. Dispersal records include Thailand (Boesman 1990, Round 1995b) Saudi Arabia, Oman, and Belgium (Claeys *et al.* 1989).

STATUS: It is common in its range in India, Nepal, and Myanmar, and is one of the most readily seen herons in those countries. Little information is available on population numbers.

Habitat

The Indian Pond-Heron uses many types of shallow aquatic habitats, fresh or salt, natural or human made. It appears to somewhat prefer very shallow, non-flowing, or slowly flowing situations especially the muddy margins of ponds, jheels, ditches, marshes, rice fields, and irrigated pasture. But it also occurs along rivers, streams, lakes, tidal flats, mangrove swamps, and reservoirs. It is frequent in human-dominated areas, nesting in towns and cities, and takes advantage of feeding opportunities made by human activities. It has been observed to feed in clusters of floating water hyacinth (*Eichhornia*) (Parasharya and Bhat 1987). It ranges from sea level to 2150 m in India.

Foraging

This species feeds primarily by Standing, either in or out of water or mud. It Stands for long periods

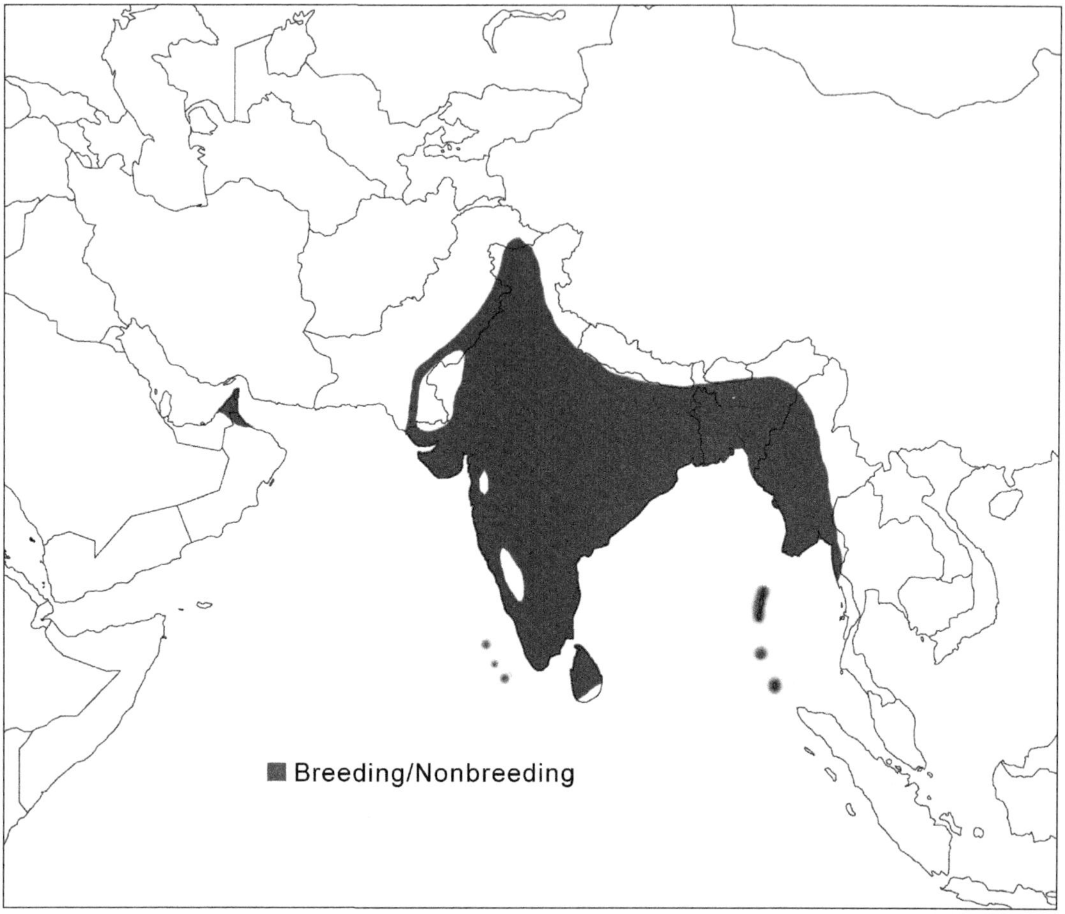

waiting for fish to come into its strike area (Sodhi 1986). It also feeds by Walking, usually with extremely slow and stealthy steps. When Walking, it often uses Head Bobbing (Sodhi 1986). It tends to use different behaviours in different habitats. It feeds by Standing in marshes and other shallow water and Walking along the shallow muddy edges of water bodies and in flooded pasture. It also uses more active behaviours. It Runs after insects and uses Bill Vibrating, Diving and Jumping from branches and rocks, Dipping, and Hover Feeding (Grimwood and Brocklehurst 1984, Sodhi 1986, Sivasubramanian 1988). It feeds solitarily or in small loose groups, as would be suggested by its cryptic plumage (Sodhi 1986). It can gather in large aggregations when food becomes highly available such as in drying situations. Fifty birds were observed at a rice field being ploughed and 15–25 birds at a site of shoaling fish (Grimwood and Brocklehurst 1984, Andrews and Mathew 1997). It follows ploughs, eating prey that are exposed and feeds commensally with other species, such as cormorants (*Phalacrocorax*). In aggregative situations, territory guarding and prey robbing occur (Goenka and Pandit 1986). It feeds in the morning and late afternoon and roosts at midday in the shade (Andrews and Mathew 1997). During midday roosting, it devotes about a quarter of its time to preening (Sodhi and Khera 1984).

Its diet is rather broad, including fish, insects, and frogs (Mathew *et al.* 1980, Sodhi 1986). Fish include *Barilius, Nemacheilus*, and *Ophiocephalus*. It takes mudskippers (*Periophthalmus*) off tidal flats. Insects include dragonfly larvae, grasshoppers, crickets, ants, bees (Prasad and Hemanth 1992), flies, caterpillars, water bugs, water beetles and other beetles, and caterpillars. It also eats earthworms (Raza 1993), crabs (*Ocypoda*), spiders, tadpoles and adult frogs (*Rana*), and baby turtles. The fishes it catches are relatively large. One analysis found plant material in high enough proportion to suggest deliberate consumption. This observation should be further investigated.

Breeding

The breeding season varies with latitude and with the timing of the monsoons. It nests in May–September in north India, and December–May in south India and Sri Lanka. Breeding may begin earlier and end later at both ends of the range. It nests in a variety of trees, bushes, and other plants. Nesting sites may be isolated trees or a clump of tress. The site may be over water or even some distance from the water. A great variety of trees are used, including bamboo, willows, eucalyptus (Pandey 1991), and mangroves. It often nests close to human habitation and urban activity and occurs in the same colony site for many years. It generally nests in small single-species colonies (e.g. Begum 2003). But it also nests alone and in colonies with other species of herons, where it assumes peripheral positions. The nest is a very small platform made of twigs. It is placed in a fork of a tree or bush, 2–16 m high. The male collects material and the female does most of the building. Very little is reported on the courtship displays of this heron. Because the birds are sedentary and nesting sites are used year after year, it has been suggested that the birds form long-term pair bonds, but this has not been demonstrated. Eggs are dark blue-green, but fade quickly. They average 38 × 28.5 mm, tending to be a bit larger in the south of the range. The normal clutch is 4 eggs, range 3–6 eggs. Both parents incubate, lasting 24 days. Very little information is recorded on the latter stages of nesting. The chicks are fed by regurgitation. Chicks can clamber out of the nest when about two-thirds grown. The fledging period of the young is not known. Nothing is known about nesting success.

Population dynamics

The population biology and demography of this species has not been studied.

Conservation

The species is widely distributed and common through much of its range. It is one of the most common herons in its region and has acclimatized well to the human-dominated environment, nesting and feeding in artificial habitats and near rural and urban settings.

Research needs

For such a widespread bird, its biology has been surprisingly little studied. Its foraging repertoire is large and interesting. Studies are needed of its foraging and breeding behaviour and biology over its variety of habitats and geography.

Overview

The Indian Pond-Heron is an abundant and successful species. Using stealthy feeding techniques, primarily Standing alone, waiting for prey to approach. Its cryptic colours during most of the year allow it to feed undisturbed enhancing the effectiveness of its Standing hunting. By concentrating on large dispersed fish but eating whatever it finds and can catch by various means, the heron has a significant dietary scope. It also is prepared to take advantage of various kinds of feeding opportunities, including human activities and food concentrations, and in those situations uses a wider variety of feeding behaviours and engages in interspecific and intra-specific interactions. So this is a species that is specialized yet has a substantial behavioural scope to deal with novel situations, especially those presented by man.

Chinese Pond-Heron *Ardeola bacchus* (Bonaparte)

PLATE 7.9

Other names: Paddybird in English; Garcilla China in Spanish; Crabier chinois in French; Bachursreiher in German; Kara-shira-sagi in Japanese Chi lu in Chinese.

Dupus bacchus Bonaparte, 1855. Consp. A. 2(1), p. 127: Malay peninsula.

Description

The Chinese Pond-Heron is a small white bird with golden-buff streaks on its head and back; in the breeding season it turns dark with wine-chestnut head and neck and purple-black back with white wings.

ADULT: The non-breeding adult has a golden brown to buff, streaked head. The bill is yellow with a dusky upper bill and dark tip. The irises are gold-yellow, and lores are green-yellow. The chin is white, whereas the neck and throat are streaked white, brown, and golden-buff. The back is olive-brown, with the feathers tending to cover the wings. Wings are white with the outer flight feathers having an olive-brown tinge. The legs are yellow and feet green-grey. In breeding, the head is wine-chestnut, with a crest that includes long rufous lanceolate plumes. The bill is yellow with a dark tip and blue base. Lores are green-yellow. The throat is white and the sides of the head, hind neck, and upper breast are rufous. The upper breast has long maroon plumes and the lower breast, dark grey plumes. The back is dark slate with a purple-bronze cast. Long black back plumes extend beyond the tail and cover much of the white wing. The belly is white. The legs during breeding become pink-red.

VARIATION: The female in breeding has less developed plumes, a paler foreneck, and lacks the black component of the breast plumes. There is no geographic variation described.

JUVENILE: The juvenile is like the non-breeding adult, with buff head heavily streaked in dark brown. The underparts are white with heavy brown streaks.

CHICK: The downy chicks are brown and white.

VOICE: This is generally a quiet bird outside the nesting season. Calls have not been documented.

WEIGHTS AND MEASUREMENTS: Length 45 cm.

Field characters

The Chinese Pond-Heron is identified in the breeding season by its dark back and red-chestnut head and neck. It flies with rapid wing beats, its head well tucked back. Despite its overall darkness, at a distance the bird appears white in flight due to its white wings and, closer, it appears black and white. It is distinguished from the Cattle Egret by being darker. It differs from the Javan Pond-Heron by its darker head (chestnut not golden), chestnut (not cinnamon) breast, and chestnut (not white-buff) crest plumes. Non-breeding birds and juveniles are very difficult or impossible to distinguish from other pond herons. Immature and non-breeding Chinese Pond-Herons are distinguishable from the equivalent Javan Pond-Heron only by being larger, with a larger bill, and having darker tips to the flight feathers. These features are hard to distinguish in the field. The immature and non-breeding birds are distinguished from the adult Squacco Heron by its dark (not tawny brown) back and the lack of any hint of black streaking on the head and crest.

Systematics

The Chinese Pond-Heron is an *Ardeola* heron, closely related to the other typical pond herons, so much so that four of the pond herons (Indian, Javan, Chinese, and Squacco) are sometimes considered to be subspecies of *Ardeola ralloides*. In the extreme west of its range, it overlaps and nests in the same colonies as the Indian Pond-Heron. The appropriate relationship among the several forms requires additional, molecular study as well as further observations in zones of apparent overlap. Whether four species or one is still to be definitively determined.

Range and status

The Chinese Pond-Heron occurs in east Asia through Indonesia.

Breeding Range: It breeds from east India (Assam, Manipur), Bangladesh, central and east China (Shanxi, northwest Kansu, Neimongol, Liaoning, Jilin, Yunnan, Guangxi, Guangdong, Hainan, Taiwan), Andaman Islands, north Myanmar, and Indo-China. There is a single record from north Japan (Niikura and Nakamura 1987, Mitsunaga and Imamura 1996). Breeding status in north Thailand is uncertain.

Non-breeding Range: The non-breeding range is south China (to Hainan and Taiwan), Japan (rarely Kanagawa, Ryukyu Islands) (Niikura and Takehiro 1986), Cambodia, Laos, Vietnam, Myanmar, Andaman Islands, Malaysia (peninsular, Sabah, Sarawak), Brunei, Indonesia (lowlands of north Sumatra, Kalimantan), and Philippines (Luzon, Mindanao, Negros) (Kennedy *et al.* 2000). It is difficult to ascertain the exact wintering range given that it overlaps other pond herons, which cannot be distinguished in winter.

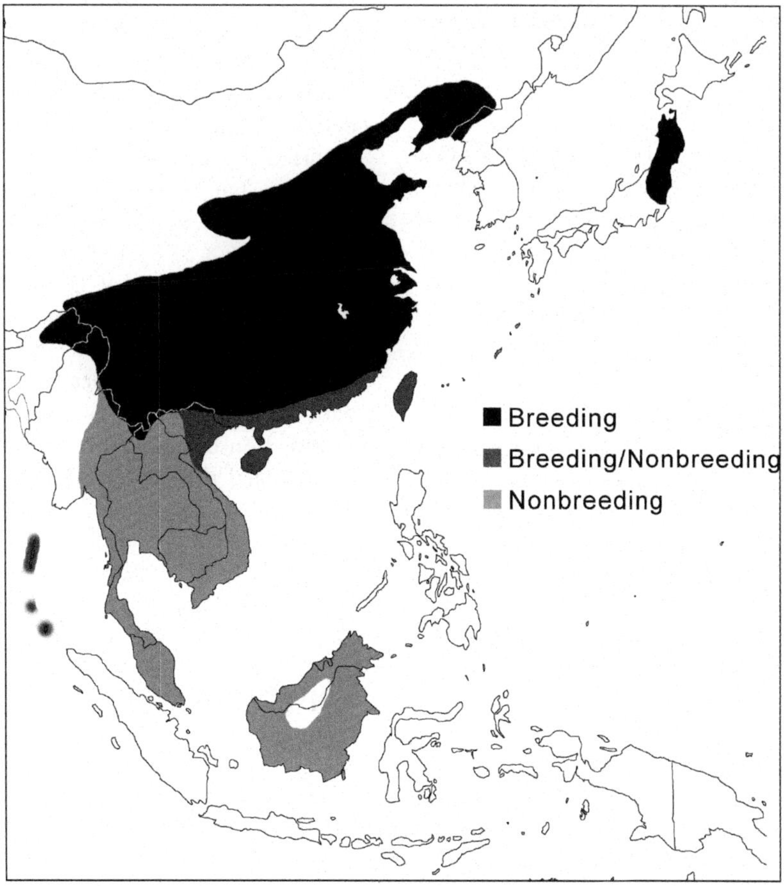

MIGRATION: Northern birds are migratory, moving south to and through south China in September– October. Indian and more southern birds are sedentary, with some local movement. Return migration is May–April. Dispersal records include Russia (Kolomiitsev 1986), northeast China (Heilongjiang, Xinjiang—Ma *et al.* 2001), east India (Parasharya 1983), Sri Lanka (Hoffmann 1996), South Korea and Japan. They have also been reported in very distant places such as Norway, Hungary, and Alaska (Hoyer and Smith 1997, Ecsedi *et al.* 2000).

STATUS: It is abundant through most of its core range. It is common and expanding in east China. Large colonies are reported from some areas, such as Yunnan. Over 6000 nests were found in two areas of China (Fasola *et al.* 2004). It is common in winter in peninsular Malaysia, Vietnam, Thailand, and Borneo. It is rare in winter in Philippines and Sabah (Sheldon *et al.* 2001).

Habitat

It is observed primarily in rice fields, but uses many less obvious, shallow aquatic habitats including marshes, swamps, riverbanks, mangroves, tidal pools, streams, and dry grassland. Fish ponds are a very important habitat locally, especially for young (Young 1998). It occurs from sea level to a relatively high altitude in Thailand.

Foraging

These birds feed in the normal pond heron manner of Standing and Walking very slowly, in a hunched over posture, from which they lunge or stab at prey. They usually feed solitarily or in pairs, but sometimes in small flocks up to 60 birds. Upon seeing a prey they lunge or stab at it. This pond heron also feeds actively in diverse ways. They feed within trees, for flies (Melville 1990). They dive from overhanging branches into the water, as many as 10 m high. They also feed on the wing. They feed by Feet First Diving and Hovering and used Dipping when fish were surfacing in distress, (Syler in prep.). It has been observed feeding commensally with otter (*Lutra*) (Kruuk et al. 1993). It feeds throughout the day but peaks in activity just after dawn and before dusk. It roosts in mixed-species roosts, flying to roost in evening in small groups of 2 or 3 birds. It feeds on insects, including flies (Melville 1986, 1990), aquatic insects, molluscs, worms, fish, frogs, and small birds (Melville 1995, Zhou *et al.* 1998).

Breeding

The Chinese Pond-Heron breeds in May–August in India and April–July in China. It nests in trees and bushes next to or near marshes, swamps, rivers, and ponds, in mixed-species colonies with night herons and Little Egrets. Breeding density in China was 1401 individuals per km^2 (Zhu *et al.* 1994). Nests are made of small sticks, sometimes lined with leaves and grass. They are placed high in trees, often near the top (Shen *et al.* 1987, Zhu *et al.* 1994). In one study over 50% of nests were above 10 m. Trees and bushes used in nesting include *Populus, Sophora, Quercus, Ulmus, Ficus, Cunninghamia, Tea*, and *Sassafras.* They also nest in bamboo. The courtship behaviour has not been described. Eggs are green-blue. Their size has been reported as averaging 37.7 × 28.4 mm and 39 × 29.7 mm, but eggs in this species appear to be highly variable, in one study ranging from 32.6 × 27.2 to 43.6 × 31.8 mm (Shen *et al.* 1987). Clutch is 4–6 with a range of 2–8 eggs. The Chinese Pond-Heron is typically double brooded. A second clutch is laid soon after the first young have become independent. Incubation is 18–22 days, with both parents brooding the eggs. Not much is recorded on hatching and rearing of young. Both parents feed the nestlings. Chicks can have their food stolen by mynas (*Acridotheras*) (Leader 1995a). Growth is relatively fast, until fledging at 30 days (Zhu *et al.* 1994). They are independent by 36 days. There is no information on nesting success.

Population dynamics

There is no information about the population biology and demography of this species.

Conservation

This is a common bird in its range, and one that has acclimatized to the human environment by using rice fields and other artificial habitats. The use of these are now likely critical to population stability. The loss of fish pond habitat led to a decrease of herons in Hong Kong (Young 1998). The management of occupied wetlands, including artificial environments occupied by the birds, is a critical conservation measure for the long-term stability of the populations. Eggs and birds are taken for food in China (Parish and Melville 1985). However, range expansion continues despite these losses. Hunting and other human use need to be evaluated.

Research needs

For such a wide ranging bird, little has been documented regarding its biology, although much can be assumed from that of other pond herons. The feeding biology of the species especially needs to be documented quantitatively. Several studies of breeding biology exist, published in Chinese. It would be most worthwhile to have a widely available, thorough review of what is known about the breeding biology of the species in China.

Overview

This is a cryptic, solitary, slow-moving hunter, usually within and at the edge of dense vegetation. It frequents both natural and man-made habitats. As a result, it is a common to abundant bird over much of its range.

Javan Pond-Heron *Ardeola speciosa* (Horsfield)

PLATE 7.10

Other names: Garcilla Indonesia in Spanish; Crabier malais in French; Prachtreiher in German; Blekok sawah in Indonesian.

Ardea speciosa Horsfield, 1821. Trans. Linn. Soc. London 13, p. 189: Java.

Subspecies: *Ardeola speciosa continentalis* Salomonsen, 1933: Bangkok, Thailand.

Description

The Javan Pond-Heron is a small, brown-streaked heron with a golden brown head and black back in the breeding season.

ADULT: In non-breeding plumage, the head is streaked olive-brown and buff. The bill is yellow with dusky on the upper bill and a blue base. The lores are green-yellow and the irises are yellow. The neck is streaked brown and the back and rest of the upper parts pale olive-brown. The tail and wings are white. The breast is streaked brown, with the rest of the underparts white. The legs are pale green-yellow. In the breeding season it has a pale gold-yellow head, crest, and neck with two long buff-white crest plumes. The bill becomes bright yellow with a black ring at or very near the tip and blue at the base. The lores are dull greenish-yellow but the irises turn orange. Elongated slate-black back feathers extend to the tail tip. The wings are white and are covered by the back skirt when standing. The plumes of the lower neck and breast and a collar at the lower neck are rufous to cinnamon, normally lacking black feather tips. These plumes extend to the upper belly. The rest of the underparts are white, tinged in orange. The legs are pale yellow, becoming green-yellow soon after the eggs are laid, and later dark green.

VARIATION: The sexes apparently are alike. Geographic variation in the mainland population is recognized by the subspecies *continentalis*, which has longer wings and bill.

JUVENILE: Juveniles are streaky brown, like the non-breeding adult.

CHICK: The downy chick is brown, white, and grey.

VOICE: The typical 'Squawk' call is its flight call. The 'Krak' call is the disturbance call.

WEIGHTS AND MEASUREMENTS: Length: 45 cm.

Field characters

The Javan Pond-Heron is identified by its pale golden-yellow-brown head, crest feathers and neck, cinnamon breast, and black back. It flies with slow, short wing beats. It is difficult to distinguish from other pond herons. In fact, in most situations the non-breeding adults and the juveniles are unseparable. It is distinguished in the breeding season from the Indian Pond-Heron by its black (not maroon) back, gold-brown (not brown) head, cinnamon (not brown-buff) breast, and lack of grey-brown tips to its flight feathers. It is distinguished from the Chinese Pond-Heron by its golden (not chestnut) head and cinnamon (not maroon) breast. The adult is distinguished from the adult Squacco Heron by its black (not tawny-brown) back and the lack of any hint of black streaking on the head and crest.

Systematics

The Indian Pond-Heron is an *Ardeola*, closely related to the other typical pond herons. Four of the pond herons (Indian, Javan, Chinese, and Squacco) are sometimes considered to be subspecies of *Ardeola ralloides*. Infra-specifically, the continental population differs modestly in size, and specimens collected outside the breeding season may have been confounded with wintering birds. Details of superspecific and infraspecific variation should be re-examined. It is likely that recognition of two subspecies is not justified.

Range and status

The Javan Pond-Heron occurs in southeast Asia and the East Indies.

BREEDING RANGE: *Continentalis* occurs in central Thailand, southeast Myanmar, south Vietnam, and Cambodia. *Speciosa* occurs in Indonesia (Sumatra, Java, Sulawesi, Kalimatan, and Lesser Sundas including Bali, Flores, Sumba, Sumbawa).

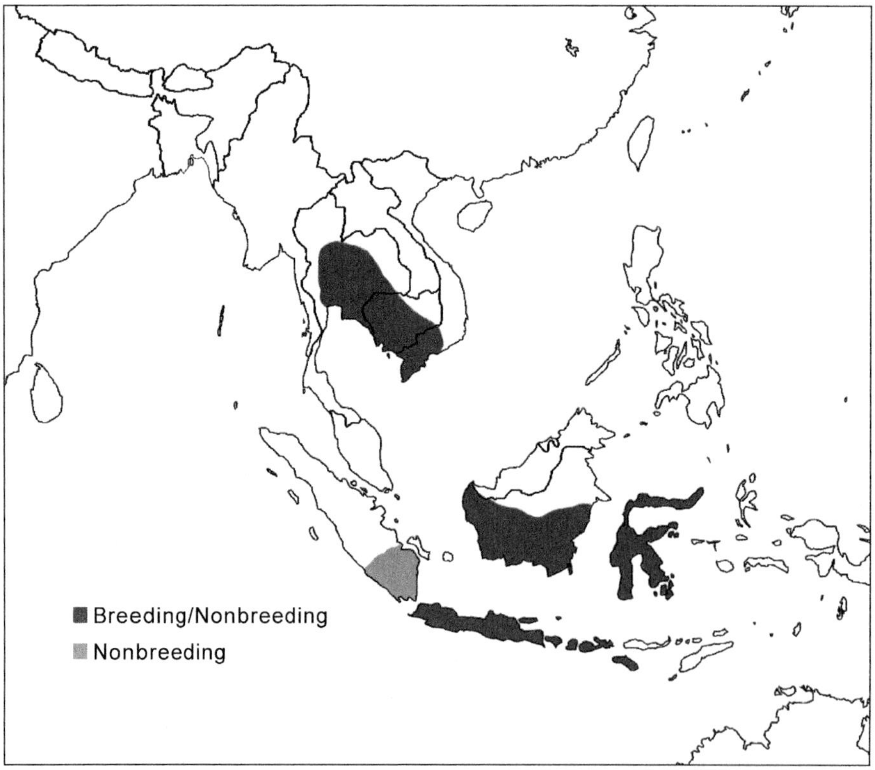

NON-BREEDING RANGE: It recently extended its range into south Philippines (Mindanao), where breeding is suspected but not yet documented (Kennedy *et al.* 1984, 2000) and rarely in Saba and Sarawak (Sheldon *et al.* 2001) and in south Malaysia, in that vagrant birds moulting into breeding plumage have been recorded in west peninsular Malaysia (Howes 1986, Helbig 1987, Lansdown *et al.* 2000) but breeding is yet not documented.

MIGRATION: The Javan Pond-Heron is sedentary, but it also occurs where breeding has not been determined, such as peninsular Malaysia (Helbig 1987) and south Sumatra (Verheugt *et al.* 1993). Whether such records are migrants or breeders is uncertain.

STATUS: It is common throughout its range, notably in Thailand, Vietnam, Cambodia, Borneo, and Java. Roost counts number in the hundreds to as many as 10 000 birds. Colonies also number in the hundreds to thousands. It recently colonized the south Philippines and may be continuing to expand its range northward. It is rare on the outskirts of its range, in Indonesia.

Habitat

The Javan Pond-Heron is probably most commonly seen and associated with rice fields. It also occurs on a variety of flooded natural grasslands, marshes, ponds, and the margins of lakes. Along the coast it uses mangroves, coastal flats, and reefs. It is primarily a lowland bird, occurring from sea level to up to 1500 m, in Java.

Foraging

The Javan Pond-Heron feeds primarily by Standing motionless, in low Crouched posture with head retracted. It also feeds by Walking with slow steps. It feeds not only under cover but even more typically hunts in the open. It is diurnal but likely is most active during the morning and evening. It primarily feeds solitarily, but it also occurs in small dispersed groups and occasionally in large groups. It flies to roost in the evening in groups of 2 or 3 birds. Its diet includes fish, frogs, and insects. Insects include grasshoppers, beetles, ants, and termites. It also eats earthworms and crustaceans. Both tadpoles and frogs are eaten.

Breeding

It breeds in December–May in east Java and January–August in west Java. The breeding season on the Asian continent is not clear. It nests in trees and bushes near marshes and ponds, usually overhanging the water. It nests solitarily, in small colonies, and also in large colonies in association with other species including Cattle Egrets and Little Egrets. The nest is built of twigs, not reeds. It is placed at varying heights within the colony, not just low. The nests are in the leafy branches of trees. Apart from an alleged, but undescribed special call, no observations have been made of the courtship behaviour of this species. Eggs are dark green-blue, 37.7 × 28.7 mm, about the same size as those of are the Indian and Chinese Pond-Herons. Clutch is 3–5 in mainland colonies, 2–3 in Java and Bali (MacKinnon and Phillipps 1993). The incubation period is not known and very little is known about the nesting biology of the species, chick development, or nesting success.

Population dynamics

Nothing is known about the population biology and demography of the species.

Conservation

Conservation requires additional fundamental information on the biology of the species and surveys to identify important areas.

Research needs

Despite its abundance, the biology of the species is little known. Even its food habits are barely documented. Basic studies are needed to develop an understanding of its distribution, important areas, foraging ecology, nesting biology, and demography. The breeding status of the species in the Philippines and peninsular Malaysia needs to be determined.

Overview

So much has been documented about the ecology of this bird. As a typical pond heron it is a lonely, cryptic species that feeds by Standing and by walking slowly in shallow water or on the edge of the water. It feeds in the open, and is typically a bird of rice fields over much of its range. It is seen frequently and easily, and can be very numerous. It likely consumes a wider array of prey than is now known, and probably occupies a more diverse habitat. It is a successful species, benefitting from its association with agriculture.

Malagasy Pond-Heron *Ardeola idae* (Hartlaub)

PLATES 7.11–7.12

Other names: Madagascar Pond Heron, Malagasy Squacco Heron, Madagascar Squacco Heron in English; Garcilla Malgache in Spanish; Heron crabier blanc in French; Dickschnabelreiher in German, Mpiandry-voditatatra in Malagasy.

Ardea idae Hartlaub, 1860. Journ. f. Orn. 8, p. 167: east coast of Madagascar.

Description

The Malagasy Pond-Heron is a small buff-brown streaked, heron, which when breeding, is snow white with a blue bill and red legs.

ADULT: In non-breeding plumage, the crown and back of the head are buff, broadly streaked black. The bill is green-grey, black towards the tip. The irises are yellow and lores green. The sides of the head and throat are yellow-buff streaked narrowly with blackish-brown. The upper and lower back is dark brown with narrow white or buff streaks. The rump and tail are white. The upper wings are brown with white or buff streaks. The flight feathers, in contrast, are white, conspicuous in flight. The upper underparts are heavily streaked black and buff-brown, contrasting with the lower belly and under tail coverts. The legs are green-yellow or green-grey. In the breeding season, the Malagasy Pond-Heron becomes entirely snow white. The crown has a full crest. Dense plumes also occur on the back of the neck, back, foreneck, and breast. These give the bird a characteristic fluffy appearance. The bill is a deep azure-blue with a black tip. The irises are yellow and the lores are green with some red on the orbital skin. The legs become rose-pink, with green toes. The intermediate plumage is a mixture of new brown feathers, starting on the back.

VARIATION: The sexes are alike. There is no known geographic variation over its small range.

JUVENILE: Juveniles are similar to non-breeding adults, but are sooty-brown on the outer flight feathers and tail. The iris is pale green. The bill is dull orange, with a dark tip and culmen.

CHICK: The chick has thick buff-yellow down.

VOICE: The 'Squawk' call rendered 'Krruk', is louder, more raucous and less shrill than the equivalent call of the Squacco Heron. The 'Burr' call is the threat call.

WEIGHTS AND MEASUREMENTS: Length: 45–48 cm.

Field characters

In breeding plumage, the Malagasy Pond-Heron is unmistakable, identified as a small all-white heron with bushy plumes, blue bill, and green lores. In non-breeding, it is identified by its brown plumage and stocky build. Non-breeding and juvenile birds of the Malagasy Pond-Heron and Squacco Heron are not easy to separate in the field. It is distinguished by being stockier with darker browner (not tawny) plumage, darker heavier streaking, heavier bill, and in flight having a squarer wing tip. There are also some behavioural differences. Unlike the Squacco Heron, it is often found near trees and will fly into trees when disturbed.

Systematics

The Malagasy Pond-Heron is an *Ardeola* pond heron, confined to the Malagasy region. It is clearly closely related to the other pond herons. Pond herons are sometimes considered to be subspecies of *Ardeola ralloides*. But *idae* is a different species, its range overlapping *ralloides* without interbreeding. The relationships among all the pond herons clearly need to be restudied using molecular techniques.

Range and status

The Malagasy Pond-Heron occurs in Madagascar and the Aldabra Islands and in the non-breeding season in central and east Africa.

BREEDING RANGE: It breeds only in Madagascar and Aldabra. In Madagascar, it breeds on the west of the island, especially around Antananarivo. In Aldabra, it has been reported breeding on Ile aux Aigrettes and Ile aux Cedres since 1967.

NON-BREEDING RANGE: Although breeding in Madagascar, the Malagasy Pond-Heron spends the winter in Africa. It's winter range is concentrated in Tanzania and Kenya (G. Young pers. comm., Zimmerman *et al.* 1996). It also has been reported from Uganda, Rwand, Burundi, Zambia, D.R. Congo, northeast Mozambique, northeast Zimbabwe, Mali and south Somalia the Comores and Seychelles (Medland 1989, D. Turner pers. comm.). Status as a persistent winter resident outside the core range requires additional documentation.

MIGRATION: This pond heron migrates between Madagascar and Africa. It moves westward during May, returning in October. Some birds remain in Madagascar during non-breeding and some young individuals may remain in Africa during their first year. Unsurprisingly for a migratory species, dispersal records occur, such as the Seychelles (Feare 1975) and southern Somalia (May and August 1979). Dispersal has no doubt led to colonization of Aldabra.

STATUS: This species has undergone a decades-long decline in breeding numbers. The species once nested throughout the island and was common during the 1940s. By the 1970s populations had been severely reduced and its rapid decline continues. It remains fairly widespread, but apart from around Antananarivo, numbers are especially declining on the high central plateau (D. Turner pers. comm.). Colony size has decreased from over 700 to dozens, and it now breeds only in small numbers in a few colony sites (Langrand 1990, Lyles 2000). In Aldabra, two colony sites have been reported and the population was estimated to be less than 100 individuals in 1967–68 (Benson and Penny 1971). Recent information is lacking. It is generally uncommon outside of the nesting season. But little is known of status in Africa, outside of Tanzania.

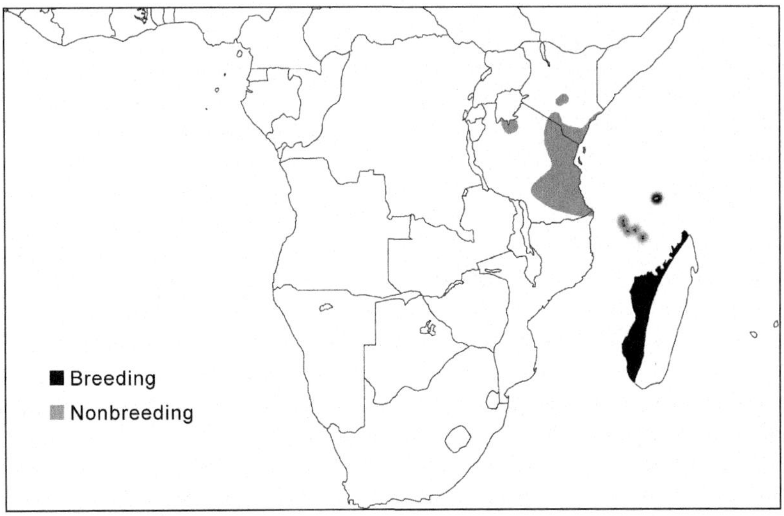

Habitat

In Madagascar, the bird's major habitat includes small grassy marshes, small lakes, ponds, streams, ditches, and rice fields usually near trees and bushes. It appears to be more dependent on trees than other pond herons. In Aldabra, it uses inland pools, lagoon shores, and also areas away from water. In East Africa, it occurs in fresh water marshes and along the banks of small streams. It ranges from sea level to 1800 m.

Foraging

The Malagasy Pond Heron is a solitary feeder. It Stands and Walks slowly along the banks and edges of muddy lakes, ponds, shorelines, and even reefs or on floating plants. When disturbed it will often fly into nearby trees or bushes. In Aldabra, adults feed in mixed groups of egrets. Its diet includes fish, frogs, small reptiles (skinks, geckos), and insects (grasshoppers, beetles).

Breeding

In Madagascar the breeding season coincides with the start of the rainy season, from late October to December, but will continue through March if two clutches are laid. On Aldabra, eggs are laid in November–December. It nests in trees and bushes in and near marshes, lakes, or ponds. In Aldabra it uses mangroves (*Rhizophora*) and small bushes. The species is a colonial nester, historically joining the Squacco Heron in colonies exceeding 1000 pairs. In a typical mixed-species colony, it places its nests in a higher and more secure position than does the Squacco Heron. The nest is a bulky structure made of twigs. It is placed fairly high in trees or bushes. In the mangroves in Aldabra, nests were placed 1.5–2.5 m above dry ground under the canopy of small bushes. Courtship behaviour has not been described. Its threat display involves erecting its plumes, flapping its wings, and jabbing its bill forward. The eggs are pale green-white, as in the Squacco Heron. They average 40 × 29.9 mm in Madagascar, and in the range 40–42 × 32–33 mm in Aldabra. The usual clutch is 3–4 eggs, with a range of 2–4. Eggs are laid at intervals of up to 3 days. Although usually single brooded, a second clutch in February–March has been recorded in Madagascar (Milon *et al.* 1973). Incubation begins with the last egg and lasts 20 days. Both parents incubate, the male from mid-morning to mid-afternoon. Chicks hatch asynchronously. They are fed by regurgitation and development is rapid. They can begin to clamber from the nest at about 15 days. At less than 4 weeks they leave the nest and are independent. Nothing is known about nesting success.

Population dynamics

Nothing is known about the population biology or demography of this species.

Conservation

This species is one of great conservation concern. The decrease of its population in Madagascar has been substantial to the point that conservation action is required. It is considered to be Vulnerable by IUCN (IUCN 2003) and near Threatened by HSG (Hafner *et al.* 2000). Habitat loss through clearing, drainage, and conversion to rice fields has adversely affected this species. This species is an excellent candidate for establishing a captive breeding population. All herons on Madagascar appear to be suffering from population distresses owing to habitat loss and hunting. This species has not adjusted to rice growing as well as has the Squacco Heron. Conservation action involves continued surveys of the nesting areas in Madagascar and the identification of important wintering areas, which then need to be the focus of preservation and conservation action.

Research needs

In Madagascar, surveys need to be conducted to provide quantitative information on population size and trend and to identify additional important breeding areas. Surveys also should be conducted on Aldabra to reaffirm its breeding status there. So little is known about the biology of the species that fundamental studies need to be conducted on both the breeding and non-breeding ranges.

Overview

The Malagasy Pond-Heron is typically a pond heron in depending on marshes, small lakes and ponds, streams and ditches. Having trees and bushes nearby appears to be important for escape. Its distinctive seasonal plumage change and migration pattern between Africa and Madagascar are unique to the species. The African part of its annual cycle is particularly mysterious.

Rufous-bellied Heron *Ardeola rufiventris* (Sundevall)

PLATE 7.17

Other names: Garcilla Ventrirroja in Spanish; Héron à ventre roux in French; Rotbauchreiher in German.

Ardea rufiventris Sundevall, 1851. Ofv. K. Vet. Akad. Forh., Stockholm, for 1850, p. 110: 'Caffaria superior' = Mooi River, Potchefstroom, Transvaal, South Africa.

Description

The Rufous-bellied Heron is a medium-sized, dark grey and chestnut heron.

ADULT: The adult has blue-black crown and head. The bill is basically yellow; most of the lower bill is pale yellow with a black tip. The lores, orbital skin, and irises are yellow. The chin, throat, neck, and upper back are dark slate-grey. The upper wing is chestnut with black flight feathers. The lower back is dark brown-grey, whereas the rump and tail are chestnut. The undersides are chestnut. The legs are bright orange-yellow and the toes may have a red tinge. In breeding, it develops long back plumes, up to 10 cm long. It is likely that the irises turn orange; the bill and lores, coral-red; and the legs, pink (Beasley and Beasley 1985, Richards 1992). More information is needed.

VARIATION: The species is sexually dimorphic. The female is duller and the grey of the plumage is sootier or browner. The chin and foreneck have a buff-white streak. The lores are paler, almost white, whereas the bill is horn-brown, with the basal two-thirds of the lower bill green or orange-yellow. The neck and back are sooty-brown with the back plumes being shorter, 7–8 cm long. The chestnut parts are the same as the male but the underparts are lighter.

JUVENILE: The immature resembles the female, but has buff-brown streaking on the sides of the head, neck, and upper breast.

CHICK: The young have grey down on the back and white to grey on the underparts; the bill is pale pink-yellow to brown-yellow at the tip; the legs are green-yellow, more yellow to the toes (Fry *et al.* 1986).

VOICE: The 'Kar' call is the typical call of the species, rendered as 'caw' or 'kraak'. The 'Cherr' call, rendered 'cherr, cherr', is used in pursuit flight. A 'Twaa' call is used when bringing food to the nest. The nestling call is a soft 'po, po, po'.

WEIGHTS AND MEASUREMENTS: Length: 38–39 cm.

Field characters

The Rufous-bellied Heron is identified by its slate-black upper parts and chestnut underparts, yellow bill and face skin, and yellow legs and feet. It is distinguished from the Black Heron by being smaller, its chestnut plumage and yellow (not black) legs. It is distinguished from the Striated Heron by being larger, with a thicker neck, heavier bill and longer wings. It is distinguished from the Squacco Heron by being smaller, with dark slaty-black (not buff) upper parts and chestnut wing and underparts. The adult is distinguished from the Malagasy Pond-Heron by its dark-slaty black upper parts and chestnut underparts. Its immature form is distinguished from the immature Squacco Heron and Malagasy Pond-Heron by its dark colour and limited streaking.

Systematics

The Rufous-bellied Heron appears to be similar to both *Ardeola* and *Butorides*. It appears to fit better with the pond herons, given its sexual dimorphism, overall appearance, behaviour, and habitat. However, some morphological characteristics resemble the Striated Heron (Payne and Risley 1976). For many years it was placed in its own genus, *Erythrocnus*. Resolution of the relationship of this species to *Butorides* and *Ardeola* is a candidate for molecular study.

Range and status

The Rufous-bellied Heron occurs year-round in central and southern Africa.

RANGE: It occurs locally in southwest Kenya, south Uganda, Rwanda, southeast DR Congo, Zambia, north Namibia, southwest and northeast Angola, north Botswana (Okavango Delta), Tanzania (Elliott 1983), Zimbabwe (Rockingham 1998), Malawi, and South Africa (Natal, Transvaal—Robertson 1987, Vincent 1990). Its exact breeding range outside its strongholds remains poorly delineated. Its breeding status needs to be clarified for Burundi, Mozambique, and South Africa.

MIGRATION: It may undertake limited seasonal movements associated with the flooding of river flood plains. Additional documentation of the seasonal occurrence of this species relative to hydrological conditions is needed. Dispersal records are reported from Nigeria and elsewhere near its breeding range, such as eastern Zimbabwe and the Natal and Transvaal regions of South Africa (Lockhart 1996, James 1988).

STATUS: The species overall is local and generally uncommon. It is more common in the larger swamps, river systems, and flood plains including northern Congo, the Okavango Delta of Botswana, Kagera River swamps, the Usangu Plain of Tanzania and Rwanda, Lake Bangweulu in Zambia, and coastal Mozambique. It is patchy and rare outside the large wetland areas over its mapped range. It appears to be expanding its range in Tanzania and Kenya, numbering about 3000–5000 birds in Tanzania (Baker 1996, Baker and Baker 2004).

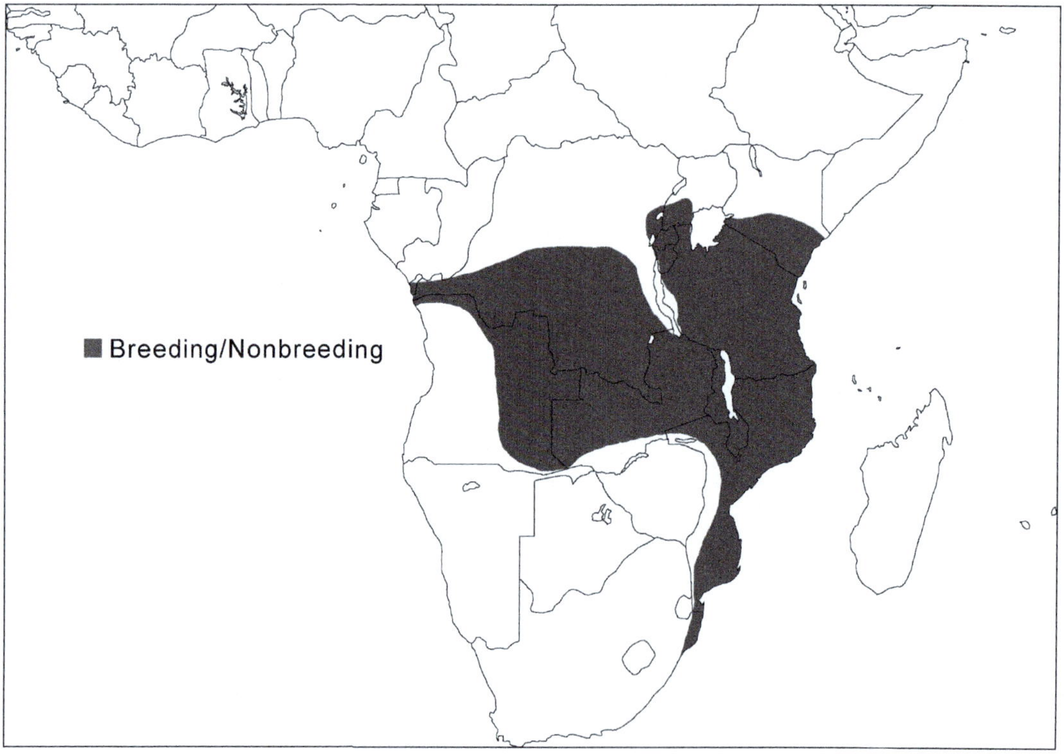

Habitat

The Rufous-bellied Heron occurs primarily in seasonally flooded grassy flood plains, reed swamps, and the vegetated edges of water bodies. Among the habitats from which it has been reported are dense reed beds, papyrus swamps, ambatch swamps, lake edges, and river edges. It appears to benefit from rice fields in the Usanga Plain of Tanzania (N. Baker and E. Baker in prep.).

Foraging

This heron is usually a solitary feeder. But it does sometimes occur in small groups of a dozen or more and at times with other species. It typically is diurnal, but also appears sometimes to feed at night. It feeds by Walking slowly in a Crouched posture. When disturbed, the species assumes a tall Upright posture. It roosts in trees when not foraging. The diet includes worms, insects, crustaceans, frogs, and fish (*Tilapia, Barbus*).

Breeding

The breeding season depends on water conditions, taking place during the rainy season. Where water levels reflect river flooding, it nests near peak water levels, even if this extends into the local dry season. It also is variable in its nesting chronology from year to year, depending on conditions. This variability can be seen in available data on laying dates (Brown *et al.* 1982): January–April and July–December in Zimbabwe, January–June and August in Zambia, December–January in Congo and Malawi, August in Botswana, April–May in Uganda, September in Namibia, and February–March, May and December in South Africa. It nests in reeds (*Phragmites, Cyperus*) or in trees and shrubs (*Aeschynomene, Ficus, Acacia*) usually but not always standing in water. The species is a colonial nester. It typically nests in small groups of 6–30 pairs, but 60–80 nests were reported at Lake Bangweulu in Zambia. It nest frequently in mixed colonies including Purple Herons, Marabou Storks (*Leptoptilos crumeniferus*) and Yellow-billed Storks (*Mycteria ibis*), typically at the edge of the colony. The nest is a small platform 25–35 cm in diameter, 10–12 cm thick. It is made primarily of reeds or twigs lined with reed stems (Fry *et al.* 1986), depending on the vegetation. Nests are placed within the reeds or lodged in shrubs or trees. They may be situated close to upright stems or over the water and are placed low, generally from 0.5 to 4 m above water or ground level. The courtship behaviour of the Rufous-bellied Heron has not been described. Eggs vary from pale blue to uniform deep turquoise. They average 37.7 × 28.6 mm. Clutch size is 2–3, with a range of 1–4. Incubation begins with the first egg but the incubation period is not documented. The young hatch asynchronously, within intervals of 2 days. The young mature rapidly and are active. They move around the nest using wings as props by 5 days, stand at 7 days, leave the nest at 17 days, fledge at 24 days, and can fly strongly at 32 days. At 5–7 days, the bill is yellow becoming orange distally with a dark tip (Fry *et al.* 1986). Feathers begin to replace down at 14 days. The irises turn brown and the legs turn yellow at 20 days. When alarmed young crouch in the nest or jump into adjacent vegetation. Little is known about breeding success. Eggs are eaten by the Black Crake (*Limnocorax flavirostra*). Brood reduction occurs through starvation of smaller chicks and ejection from the nest by larger ones. Chicks are eaten by lizards (*Varanus*) and hawks (*Circus*).

Population dynamics

Nothing is known about the population biology and demography of this species.

Conservation

The principal conservation needs of this species are to obtain better information on its basic biology and to better understand its true distribution. Full protection of the Nyl River flood plain is essential for its continued survival as a breeding species in South Africa. Indications of expansion in Tanzania in response to increases in rice production (N. Baker and E. Baker in prep.) suggest the importance of rice fields in its conservation strategy.

Research needs

The biology of the Rufous-bellied Heron is little known. There is a need for studies to determine its habitat choice, feeding behaviour, breeding biology and success, and seasonal movements. Its breeding distribution needs to be understood in more precise detail.

Overview

The Rufous-bellied Heron, like other *Ardeola,* appears to be a tropical reed bed heron. Many observations suggest that its habitat use is rather more complex though. It is a colonial nester, joining other herons in mixed colonies during the local high water period. It relies on quiet stalking techniques, as do the other pond herons. However, much more must be known to understand this species.

Yellow-crowned Night-Heron *Nyctinassa violaceus* (Linnaeus)

PLATES 9.1, 9.2, 18.2

Other names: Crabcracker, Crab-catcher in English; Martinete Coronado, Pederete Enmascarado, Chicuaco Enmascarado Rey Congo, Yaboa Común, Guanabá Real, Crabier Bois, Crabbier, Gaulin in Spanish; Martirão, Savacu-de-coroa in Portuguese; Bihoreau violacé, Crabbier gris, Coq de nuit in French; Geelkruinnachtreiger in Dutch; Krabbenreiher in German.

Ardea violacea Linnaeus, 1758, Syst. Nat. ed. 10(1), p. 143: North America = Carolina ex Catesby.

Subspecies: *Nyctinassa violacea cayennensis* (Gmelin), 1789: Cayenne; *Nyctinassa violacea pauper* (Sclater & Salvin), 1870: Indefatigable I., Galapagos archipelago; *Nyctinassa violacea bancrofti* (Huey), 1927: Scamion Lagoon, Lower California; *Nyctinassa violacea caliginis* (Wetmore), 1946: Panama.

Description

The Yellow-crowned Night-Heron is a medium blue-grey heron with a light cap and cheek, and a thick bill.

ADULT: The adult has a white crown with a yellow or rusty tinge. The side and back of the head are black. The bill is black with a yellow patch of skin at its base. The lores are yellow-grey and the irises are orange. A broad white streak runs along the side of the head from the bill, below and beyond the eye. The neck is grey. The back and upper wings are grey, patterned by the feathers each having a dark centre and light margins which form distinctive back stripes. The legs are dull orange-yellow. In breeding, the cap is cream-white and has several long (upto 150 mm), thin creamy white plumes. The bill becomes glossy black and the lores dark green. The irises turn scarlet and the legs turn pink to scarlet or bright orange.

VARIATION: Males average larger than females (Watts 1995a). There is geographic variation in the base colour of the birds, and both geographic and individual variation in bill shape, which is thought to vary in part in response to the size of local crabs (Watts 1995a). Five subspecies are recognized. *Violacea* is the dark form, described above, with broad dorsal streaks and variable bill (Payne 1979). *Bancrofti* has paler plumage (both adult and juvenile) with narrower dorsal stripes and larger bill. *Caliginis* has a large and thick bill and darker grey plumage, especially juveniles who have a streaked head. *Cayennensis* is darker, with narrow dark back streaks and slender bill. *Pauper* is smaller and darker.

JUVENILE: The juvenile is brown streaked with white and cream. The crown is dull black streaked with buffy white. The bill is black to brown. The iris is yellow-orange. The back and wings are brown spotted in buff-white. The underparts are streaked brown and white-cream. The legs are dull yellow-green. The plumage becomes duller through the first year.

CHICK: The downy chick is white-grey, with longer head tufts forming a crown. The eyes are dark yellow. The skin is pink. The upper bill is pink to grey. The lower bill is pale pink to yellow-pink with an ivory tip. The legs are flesh pink.

VOICE: The Yellow-crowned Night-Heron is highly vocal (Bagley and Grau 1979). The 'Scaup' call is the often heard, characteristic call of this species. It is similar to the Squawk call of the Black-crowned Night-Heron but at a higher pitch. It is given throughout the day and night and in all seasons. It is used variously as a contact call, disturbance call, flight call, taking flight call, and territorial call. The 'Squawk' call is the aggression call, used in the Forward and by the young defending themselves at the nest, as early as 3 weeks of age. The 'Ahhh' call, rendered 'ahhh, ahhh', is a guttural aggressive call. The 'Whoop' call is a high-pitched call used in the Stretch. The 'Yup' call, rendered 'yup, yup', is a communication between paired birds and is used in the Greeting Ceremony. The 'Huh' call is given during the courtship period when several birds are roosting together and also just prior to the Stretch. Birds use Bill Clappering within the pair bond. Young beg with 'chu, chu, chu, chu' or 'chee chee, chee, chee'.

WEIGHTS AND MEASUREMENTS: Weight: males average 716 g, females average 649 g; length: 55–70 cm.

Field characters

The Yellow-crowned Night-Heron is identified by its grey body and white head and cheeks. The head and neck appear disproportionally large compared with the slim neck. It flies with slow wing beats, its legs and feet extending beyond the tail. It is distinguished from the Black-crowned Night-Heron by its white (not black) crown and all grey (not black and white) body, and in flight by extended legs. The immature bird is distinguished from the immature Black-crowned Night-Heron by its shorter, deeper bill, longer legs, heavier appearance, erect posture with longer neck, darker brown base colour, and fewer and smaller white spots on the back and wings. In flight its legs and feet (not just feet) extend beyond the tail. The immature bird is distinguished from the immature Green Heron and Striated Heron by its larger size, streaked (not dark) crown, thicker bill, and extensive spotting on the back and wings.

Systematics

Once thought to be related to the *Nycticorax* night herons, this species is distinctive. It is a member of the typical heron group and is likely no more closely related to *Nycticorax* than it is to the other typical herons, but additional molecular study is desirable.

Range and status

The Yellow-crowned Night-Heron occurs in North America, the West Indies, along the coastlines of Central America and South America and on Pacific Islands including the Galapagos.

BREEDING RANGE: In the USA, *violacea* breeds from Massachusetts and Connecticut, south along the Atlantic coastal plain to Florida, inland to north Alabama, west Tennessee, west Kentucky, southwest Ohio, south Illinois, south Indiana, Missouri, Michigan (Wolinski 1988), southwest Virginia to southeast Tennessee (Williams 1978, Henderson 1985, Robinson 1990), Illinois, Wisconsin, and Minnesota, and west to central Oklahoma. It has bred in Michigan and Ontario. It ranges south along the east coast of Mexico, Belize, Honduras, the West Indies from the Bahamas, through the Greater and Lesser Antilles, and Tobago (Payne 1979), and on Bermuda, where it was reintroduced from the Florida population (Wingate 1982). *Bancrofti* occurs in Baja California (Becerril and Carmona 1997, Whitmore *et al.* 1999), Socorro Island, Tres Marías Islands, and the Pacific coast of Central America from Mexico (Mazatlán) to Guatemala, El Salvador, and Nicaragua (Payne 1979). This night heron occurs along both coasts of Costa Rica but it is not clear what subspecies (G. Alvarado pers. comm.). *Caliginis* occurs on the Pacific coast of Panama, Pearl Islands, Cocos Islands, and the Caribbean coast of Panama (Almirante, Bocas del Toro, Puerto Olbadia, San Blas), west Colombia, Ecuador, to Peru. *Cayennensis* occurs in east Panama, east Colombia, coastal and inland Venezuela (Scott and Carbonell 1986), through the Guyanas, Trinidad, north Brazil (extreme north west Amazonas), to central Brazil (Rio Grande do Sul). *Pauper* occurs on the Galapagos Islands.

NON-BREEDING RANGE: The northern population winters in southern parts of the breeding range including south Florida, the Gulf of Mexico coasts of Alabama, Mississippi, Louisiana, and Texas, Gulf coastal Mexico, Baja California, the Pacific coast of Mexico, the central volcanic belt in Mexico, Nicaragua, Costa Rica, and the Caribbean coast of Panama, Colombia, Venezuela, Ecuador and Peru. More tropical and island populations appear to be sedentary, but there actually is little information.

MIGRATION: In north North America herons undergo an extensive postbreeding dispersal, generally to the north and west. This turns into a southward migration in September. The migratory portion of the population has expanded as the range expanded. The southern limits of their wintering grounds are poorly known, but extend at least to Costa Rica, Cayman Islands, Barbados, and Grenadines. Return migration is in March. Dispersal records include Canada (south New Brunswick, Newfoundland, Nova Scotia, Clipperton Island, south Quebec, south Ontario, south Saskatchewan, south Manitoba), and north and west USA (Maine, New Hampshire (Petersen 1992), north Wisconsin (Wolinski and Pike 1987), Vermont (Kibbe and Boise 1984), central California, south Arizona (Rosenberg and Stejskal 1992), south New Mexico, Colorado (Kingery 1983b), Montana, North Dakota, Wyoming (Kingery 1983a, Findholt 1984b, Findholt and Berner 1988), and Nevada (Kingery 1993).

STATUS: In the USA the breeding range expanded northward in the 1940s and 1950s along both the coast and inland, in part reoccupying its historic range from the 1800s (Watts 1995a). Range expan-

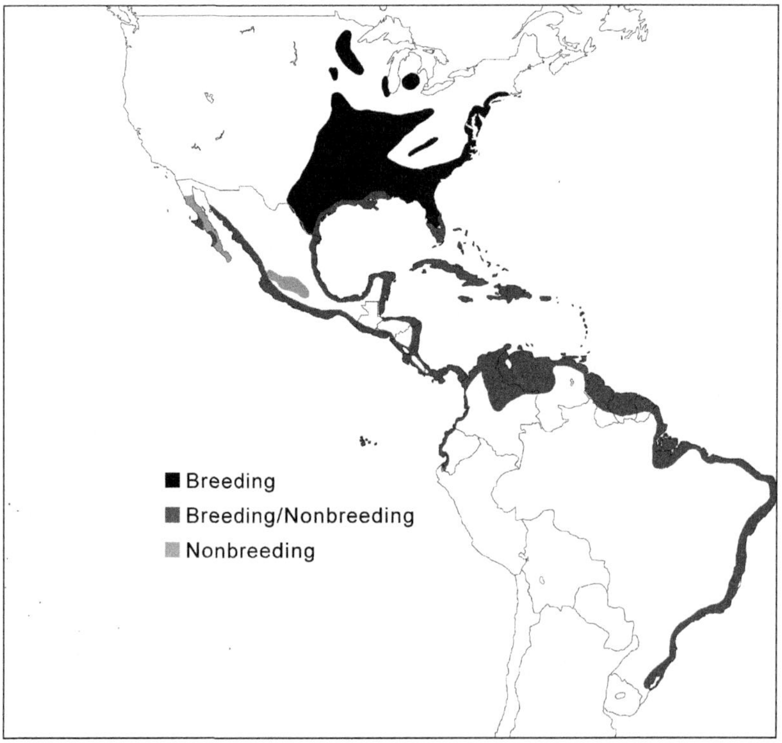

sion is also occurring in South America (Morales 2000). It is usually a common bird in its restricted coastal habitat, such as in the Bahamas, Greater Antilles, Virgin and Cayman Islands, and north Lesser Antilles (Raffaele *et al.* 1998). It occurs in low numbers in South America (Morales 2000). It is common in the Galapagos.

Habitat

This is fundamentally a coastal heron, occurring inland only where crustaceans are readily available. It prefers sites with significant edges along which to feed. Its principal habitats are coastal marshes, swamps, lagoons and shores, and inland river basin swamps and marshes. Its specific feeding habitats are the edges of tidal mangrove swamps, tidal mudflats, rocky shores including tide pools, sandy shores including feeding in the surf, banks of slow moving rivers, swamps, ponds, and reservoirs. It feeds in upland sites along arid island shores, pasture, lawns, parks, and ploughed fields. The nesting habitat is usually forests and wooded areas near suitable foraging sites, usually within 1.5 km.

Foraging

The foraging ecology and food consumption of the Yellow-crowned Night-Heron has been well studied (Riegner 1982a, b, Rodgers 1983, Niethammer, and Kaiser 1983, Clayton 1985, Laubhan *et al.* 1991, Austin 1995, Watts 1995a, King and Leblanc 1995, Martinez 2004). It feeds by Standing and waiting for prey in either the Upright or Crouched posture. This takes a preponderance of its feeding time, 77% in Florida and 64% in Missouri. Standing, the bird waits patiently for a prey item to appear, and if it does not the bird moves by Walking or Hopping to a new site. In feeding for crayfish, a bird targets a burrow and waits for the crayfish to emerge. It also feeds by Walking, usually along the shore or in shallow water with its head either extended or retracted. It feeds carefully and deliberately. The balance of time spent in Walking and Standing differs with habitat; Walking is used more on vegetated than on unvegetated shores (Rodgers 1983). It very typically uses Head Swaying and Neck Swaying to observe prey especially on land. It catches prey by a Bill Lunge. It Pecks at small prey, merely Grasping them. It will also Run between places to feed. It is primarily a solitary forager, and defends its feeding space. It feeds in loose aggregations at sites where prey is abundant. It also defends its feeding sites in these aggregations, leading to the birds being well spaced. It is characteristic for Yellow-crowned Night-Herons when feeding together to be uniformly spaced at distances on the order of 5–10 m. Given the difficulty of capture and the hard shells of crustaceans, learning appears to be an important aspect of its feeding. Adults spend less time handling prey than do juveniles. In one study juveniles took more easily captured tadpoles, whereas adults caught larger, burrowing crayfish (Laubhan *et al.* 1991), again suggesting the importance of learning. Young Walk while feeding more than do adults (Rodgers 1983). Handling the hard-shelled, pincer-equipped prey is a fundamental issue for this species. Small prey are swallowed whole, but a large prey is repeatedly grabbed and shaken. The heron will jab its bill through the carapace, which appears to stun the crab (Watts 1995a). In shaking, the claws and legs are broken off. The body is further broken up or eaten whole. The legs are then eaten. The size and defensive behaviour, and also morphology of prey influence handling time (Riegner 1982a). Larger prey are often carried away from the feeding site for handling. Undigested shells are cast as pellets.

Although called a night heron, it is not dependently nocturnal. It does feed at night, but also crepuscularly and during the day. Daytime feeding occurs especially on islands and to take advantage of tidal cycles, the birds feeding for several hours around low tides (Austin 1995, Watts 1995a). It is a visual forager by both day and night (Clayton 1985). Night-time feeding is often done around artificial light.

This heron is a crustacean specialist. It feeds principally on various types of crabs and crayfish. Its crustacean diet is varied, including mud crabs (*Uca, Rithropanopeus, Panopeus, Ucides*), shore crabs (*Pachygrapsis, Grapsis*), land crabs (*Gecarcinus*), mangrove and marsh crabs (*Goniopsis, Sesarma*), swimming and shallow water crabs (*Callinectes, Ovalipes, Carcinus, Cancer Hyas*), beach crabs (*Ocypode*), mole crabs (*Emerita*), and crayfish (*Procambarus, Cambarus*). It can eat relatively large crabs but consumes smaller crabs more efficiently than large ones. Size and species taken differ geographically. As a percent of the diet, crustaceans were 79% in south USA, 74% in Louisiana, 97% and 91% in northeast USA, 76% in Missouri, and 97% in Bermuda (Watts 1995a). Its efficiency as a crab predator is demonstrated by the Bermuda experiment (D. Wingate, pers. comm.), in which the introduced population succeeded in controlling land crab numbers. It also takes other food (Riegner 1982b, Niethammer and Kaiser 1983, Watts 1988). These prey include leeches, earthworms, marine worms, centipedes, snails, amphipods (*Ampithoe*), insects (grasshoppers, crickets, katydid, water beetles, water bugs, soldier fly), scorpions, tadpoles (*Rana*), frogs, lizards, snakes, fish (*Anguilla, Micropterus, Lepomis, Notropis, Fundulus*), and small mammals (*Mus, Peromyscus, Sylvilagus*). The use of alternative prey is important for less efficient juveniles. Also, seasonal reduction in crab abundance leads birds to switch to insects (Clayton 1985).

Breeding

The breeding season in the USA, Trinidad, and Suriname is March–July. In the northern parts of the range, the breeding season is constrained by the date of crab emergence, which depends on temperature (Watts 1995a). In tropical areas, nesting is associated with the rainy season but has been reported over much of the year. It nests in various situations, always near water given its prey dependence. On the coast, nesting is generally on islands, including barrier islands, artificial dredge material islands, and in coastal mangroves. Inland, it nests in swamps and uplands near lakes, rivers, and creeks. It also nests in human-dominated environments, such as residential areas. It uses a variety of trees, often nesting near or over the water. Trees used include *Rhizophora, Pinus, Quercus*, and *Liquidambar*, and bushes include *Baccharis, Morus, Myrica*, and *Celtis*. On dry islands it can nest on the ground, rock ledges, and cactus. The most typical situation is for the species to nest in scattered pairs or in small loose colonies. The average colony size in Virginia and in Missouri was about 4 nests (Laubhan and Reid 1991, Watts 1995a). In Virginia, a quarter of all pairs nested singly. In larger colonies it tends to occur at low density, mixed among the other species. Total nests may number in the low hundreds. Seldom do more than one pair nest in a tree. However, larger colony sites may be persistently used for decades.

The nest is a platform with a slight depression, made of twigs and variably lined with grass, leaves, or moss. Dimensions appear to be variable, but surprisingly little documented. Reported nests range from 1.25 to 0.5 m wide. Size may depend on the availability of sticks and the reuse of old nests. In northern latitudes nests are begun about 10 days after arrival at the nesting site. Several nests may be started before the final one is chosen. Nests are constructed by both birds, but it is more the case that the male gathers sticks and the female works to put them together. Nests are repaired throughout nesting by both birds. Rebuilding after nest failure may be part of the renesting cycle (Watts 1995a). Nest height varies, from the ground to 35 m (Laubhan and Reid 1991) depending on the situation. High nests on branches out from the trunk may be the most typical situation (Laubhan and Reid 1991). On islands, nests may be on shorter plants, on the ground, or even in caves.

Males claim and defend advertising sites and display in them (Bagley and Grau 1979). Circle Flights and other kinds of flying around the colony are common at this time. In the Circle Flight, the wing beats often are flapped to produce a loud 'womp, womp' sound. They aggressively defend their sites using Forward, Pursuit Flights, and Supplanting Flights. The main display is the Stretch, given by the male and sometimes by the female in response. Just prior to the Stretch the birds may give a Huh call. In the Stretch, the bird extends its head up in a slow, steady movement; the bill is horizontal; the head is lowered to touch the back; the back plumes are erected to form a fan behind the head; the legs are flexed; the bill is then tilted up and a Whoop call is given. Bill Clappering is common between the paired birds at the nest site, as is Preening and Standing. Circle Flights also are common at this time. Copulation occurs on the nest site. After pairing, the Greeting Ceremony is used upon the approach of one of the pair. It consists of Crest Raising and the greeting birds giving the Yup call.

Eggs are pale green-blue. Size averages are 51 × 37.4 mm in the USA, 47.5 × 35 mm in Trinidad, 50 × 36 in Tobago, and 49.1 × 34.3 mm in Suriname. Clutch size is 4–5 in northern populations (averages are 2.9, 3.5, 4.3), 2–4 in the tropics, and the overall range is 2–8. Birds renest after egg loss but in the north they are not inherently double brooded as was long thought (Watts 1995a). Both birds incubate, beginning with the first egg. Birds attend the nest continually after laying. The non-incubating parent spends much time staying near the nest early in the incubation period, presumably for guarding. They sit on eggs up to 88% of the time (Bagley and Grau 1979). The incubation period is 21–25 days (Weeks 1976). Hatchlings are semi-altricial and nidicolous. Eyes open on day 1. Both parents attend and care for the chicks. They feed the young by regurgitation onto the centre of the nest. Grasping of the parent's bill does not seem to be important in this species (Bagley and Grau 1979). Parents recognize their young. After fledging, the young return to the nest to be fed. Growth is rapid over 5 weeks (Watts 1995a). Pinfeathers emerge on day 5. Flight feathers appear on day 12 (McVaugh 1975). Over the development period, the irises become orange-yellow, legs yellow-green, and skin grey-green. Young fledge in about 30 days and can fly independently at 50 days. At independence the young are below adult size. Nesting success varies: 3.6 young per nest in Alabama, 2.25 in Florida, 2 in Bermuda, and 1.7 in Virginia. Nesting success in Virginia varied from 0.6 to 3.1 among years. Success depends on the extent of losses, but also the timing in that later losses are not replaced by renesting. Egg and nest predators include crows (*Corvis*), raccoon (*Procyon*), opossum (*Didelphis*), and domestic dogs (*Canis*) and cats (*Felis*).

Population dynamics

Typically Yellow-crowned Night-Herons first nest at 2 years when they acquire full adult plumage. However, birds in juvenal plumage do nest successfully. Immature birds accounted for less than 0.5% of the breeding population (Watts 1995a, 1995b). No information exists on life-span or survivorship. Adults may be killed by owls (*Bubo*) and hawks (*Buteo*); the latter has been observed drowning a heron (Richter 1985). Because they use human-dominated areas, young are susceptible to being killed by dogs and cats.

Conservation

The Yellow-crowned Night-Heron is a widespread, even if not a locally very abundant species. It has expanded its range in the last century. It is considered a species of concern in some states in the USA, due its peripheral status and perhaps a lack of information. But overall the population appears stable in both North and South America, to the extent that is known. Quantitative trend data are for the most part lacking. Certainly the most important conservation issue is habitat protection. The dependence of the species on a small array of prey types and very selective habitat choice makes it a candidate for special consideration within habitat management plans. For example, the timing and speed of water level draw down of wetlands may affect nesting viability (Laubhan and Reid 1991). Loss and alteration of mangrove forests and other coastal habitats, especially in the southern portion of its range, is a long-term threat. In the USA nesting near and even over houses, roads, and driveways leads to conflict (Watts 1995a). Disturbance in colony sites is an important issue, and needs to be managed in sites near people.

Research needs

Studies are needed on the demography, movements, survival, and distribution of juvenile Yellow-crowned Night-Herons (Watts 1995a). Feeding habits, maturation of feeding efficiency, potential residency outside the breeding range, survival, and site fidelity are all issues that deserve detailed scrutiny. Additional study is needed on breeding phenology, particularly on the relationships among weather, prey emergence, and breeding cycle in more northern latitudes and other factors influencing prey availability elsewhere.

Overview

The Yellow-crowned Night-Heron is an intriguing species, in that it shows the most specialized development among crab-eating herons. Its stout, strong bill is used for catching and handling these hard-bodied and somewhat threatening prey. That the bill is geographically variable is thought to be an evolutionary response to the size of prey. Island populations eat relatively large prey, such as land crabs. Other populations eat smaller and softer prey, such as inland crayfish. Its dark coloration is cryptic, but it is not solely a night heron. In fact, it feeds whenever is best for the prey available. Although not thoroughly studied, its nesting season appears to be geared toward the timing of availability of crabs, whether based on rainy season or spring temperatures. Developing skills to capture and handle crustaceans appears to take some time. Very few birds attempt to nest before year 2. Younger birds are less efficient, more affected by having other birds nearby, and take easier to catch (and probably not as profitable) prey. Much is yet unknown about what may be the distinctive ecology of juvenile birds. This is a successful species despite its narrow niche. It is widespread and not declining seriously anywhere in its range. It is likely that its flexibility and adaptability are the keys to its success.

Black-crowned Night-Heron *Nycticorax nycticorax* (Linnaeus)

PLATES 9.3, 9.4, 17.3

Other names: Night Heron, Black Capped Night Heron, Quawk, Quok, Crabcracker in English; Martinete, Martinete Común Perro de Agua, Guaco, Garza Bruja, Huairava, Boca d'agua Rey Congo, Yaboa Real, Guanabá, Guanabá de la Florida in Spanish; Savacu in Portuguese; Bilhoreau gris, Héron bilhoreau, Coq de nuit, Coq d'eu, Crabier bois, Crabier grosse tête Bilhoreau a couronne noire in French; Nachtreiher in German; Kwak in Dutch; Кваква in Russian; Natthäger in Swedish; Goi-sagi in Japanese; Lapay in Pilipino (Philippines); Kowak in Indonesian; Ye lu in Chinese.

Ardea nycticorax Linnaeus, 1758. Syst. Nat. ed. 10(1), p. 143: southern Europe.

Subspecies: *Nycticorax nycticorax obscurus* Bonaparte, 1857: Chile and Patagonia.

Description

The Black-crowned Night-Heron is a stocky, black and white heron with a glossy black crown and back.

ADULT: The adult has a black cap that goes forward to a white line above the bill. The sides of the head and thick neck are white. The thick, down-curved bill is black. The lores are green-blue; the irises are crimson-red. The back is black, and upper wings, rump, and tail are dark grey. The belly is white to pale grey. The relatively short legs and feet are pale yellow. During courtship, the lores are black and the legs and feet are red to pink. The black plumage of the head and back takes on a blue-green gloss and white head plumes develop that may reach a length of 25 cm.

VARIATION: Females average smaller than males in most measurements and have shorter head plumes during the breeding season. Black-crowned Night-Herons vary in size and colour geographically and up to four races have been recognized on these bases. However, variation among individuals is high, light, dark, and intermediate colour birds occur in South America, and very dark and cream-coloured birds have been reported (Pitelka 1938, Gochfeld *et al.* 1982, Davis 1993). *Obscurus* is larger with a slate-grey back and brown-grey face and chest. Other subspecies, *hoactli* and *falklandicus*, have been described but are doubtful and not recognized here.

JUVENILE: Juveniles have brown plumage, very different from that of the adults (McVaugh 1972). The head and upper parts are grey-brown with buff, white, or rufous spots. Lores are green and the irises are orange-yellow to brown-red, changing to red at 2–3 years. The stout bill is dark and horn. The upper bill is black with yellow or green sides, becoming black with green sides at 1 year. The lower bill is horn, turning yellow with a horn tip or yellow-green with black tip about 1 year, and black by 2 years. The upper wing is grey-brown with lighter spots; the flight feathers are grey-brown with white tips. The upper tail coverts are brown. The rump is grey-brown streaked with white. The tail is grey. The underparts are grey with dark brown streaks. The legs are yellow-green to olive-green, turning yellow by 2 years. At the end of its first year, the juvenile still has a brown wash, brown cap and back, with some spots remaining, and is striped below (Davis 1993). Older juveniles (2–3 years) gradually take on adult characteristics, becoming more solidly dark above, with the spots disappearing, and lighter below, with some remnant brown feathers persisting on the head, back, or wings.

CHICK: The chick has buff-brown down on its back, and light down below. The outer portions of the crown down are white, creating a crest. The bill is pink with a dark tip. Iris is gray olive, changing to light yellow, and then bright yellow by 30 days. Lores are grey at hatching becoming grey-green. The legs and feet are brown at hatching, turning yellow-olive, then olive-grey, and finally light green at 30 days.

VOICE: The Black-crowned Night-Heron is a noisy, quarrelsome bird. Its typical call gives it similar colloquial names around the word—Quawk, Quok, Coq, Kwak, and Kowak. Its vocalizations are the most completely studied of any heron (Bari and Piazza 1990, Bari *et al.* 1990, Venchi *et al.* 1994). This 'Quok' call is a hoarse raucous flight call. The 'Plup' call is the advertising call. This is a unique call in which the sound of a rubber band being plunked is followed by a buzz or hiss, rendered 'snap-hss' or 'plup-bzz'. In addition to advertising, it may also serve a group coherence function (Albert and Wollesmann 1986). The 'Rok' call, rendered 'rok, rok', is the threat call. The 'Wok' call, rendered 'wok, wok' is used for disturbance, in the Greeting Ceremony, and during precopulation. The 'Kak' call, rendered 'kak, kak', is the landing call. Young beg with 'yip, yip, yip' becoming 'yak, yak, yak', and finally 'chuck, chuck, chuck' (Davis 1993). Birds may use Bill Clapping in copulation and greeting.

WEIGHTS AND MEASUREMENTS: Weight: 727–1014 g; length: 58–65 cm.

Field characters

The Black-crowned Night Heron is a black, white, and grey bird, identified by its black cap and back, white neck and grey wings, stocky build, and when breeding, white head plumes. In flight, it has a distinctive compressed outline with broad, round wings. The wing beats are faster than most herons; its toes only barely extend beyond its tail, and it often gives its typical Quok calls while flying. It is distinguished from the Rufous Night-Heron by its black (not chestnut) upper parts. It is distinguished from the Yellow-crowned Night-Heron by its black (not yellow-white) crown and three-toned (not grey) body. It is distinguished from the large *Ardea* by its smaller size, shorter, thicker neck, relatively short legs, haunched posture, black back, and all-grey wings in flight. The juvenile looks like other herons with cryptic plumage. It is distinguished from the juvenile Yellow-crowned Night-Heron by having buff-white spots, darker ventral streaking, narrower, longer bill, thicker neck, more crouched posture, foot tips (not feet) projecting beyond the tail in flight, and lower-pitched call. It is distinguished from the several large bitterns, whose ranges it overlaps, by its prominent white spotting on the back, smaller, brown (not gold-brown) colour, spotted pale brown (not dark brown) wings, lack of black face markings, taller stance, in flight having toes (but not feet) projecting beyond the tail, and rapid (not slow) wing beats. It is distinguished only with difficulty from the immature Rufous Night-Heron by its grey-brown (not rufous-brown) base colour, grey-brown, white-tipped flight feathers (not rufous-brown with dark brown inner edges), brown (not dark brown) upper tail coverts, grey-brown (not white) rump, and grey (not rufous-brown) tail.

Systematics

The Black-crowned Night-Heron is closely related to the Rufous Night-Heron, its Australasian counterpart. They have been considered conspecific

(Dickerman *et al.* 1982). They have hybridized where their ranges have come into contact in Java and Borneo (Hubbard 1976, Sheldon and Marin 1985). They also now overlap in the Philippines and in the past in Celebes (White 1973). Field observations there did not indicate inter-nesting. Additional observations in the field are needed. Patterns of geographic variation among *Nycticorax nycticorax* populations are not clear. Particularly unclear are the status and ranges of the two subspecies occurring in South America. Intra-specific geographic variation deserves additional study, as does the relationships among the Black-crowned and Rufous Night-Herons and the several now extinct island night herons from Reunion (*dubosi*), Mauritius (*megacephalus*), and Rodrigues (*megacephalus*) (Hilton-Taylor 2000).

Range and status

The Black-crowned Night-Heron occurs across the temperate and tropical world from North and South America, Europe, Africa, and Asia to the East Indies.

Breeding Range: The subspecies *obscurus* occurs south of the Atacama Desert in Chile and Rio Negro in Argentina, and is found as far south and east as Tierra del Fuego and the Falklands. Its northern and western range boundaries are not clear and require additional study. The subspecies *nycticorax* breeds elsewhere in the species' range. It ranges in North America from Canada (British Columbia, Alberta, Saskatchewan, Manitoba), west USA (Washington, south Idaho, Montana), mid USA (south Minnesota, south Wisconsin, Illinois, Ohio,) and east Canada (Ontario, Quebec, New Brunswick), south through the rest of the USA except mountains, the Greater Antilles, larger Bahama Islands, Cayman Island, to the Virgin Islands, and coastal and inland Mexico (to Jalisco and Campeche and also Chihuahua) (Gomez de Silva 2003). It nests in Panama and into South America. It is now known to breed in the Amazon and high Andes of Peru through Chile (Morales 2000), to Argentina. It is also found in Hawaii. In Europe it nests in Spain, Portugal, France, Belgium, Netherlands (Verbruggen 1989, Erhart and Kurstjens 2000), England (as escapees), Germany, Austria, Czech Republic, Slovakia, Poland, Ukraine, Belarus (Samusenko and Pinchuk 1999), Moldova, Russia, Romania, Bulgaria, Hungary, Italy, Slovenia, Croatia, Albania, Greece, and Turkey (Marion *et al.* 2000). In Africa, it breeds in Nigeria, Central African Republic, DR Congo, northwest Angola,

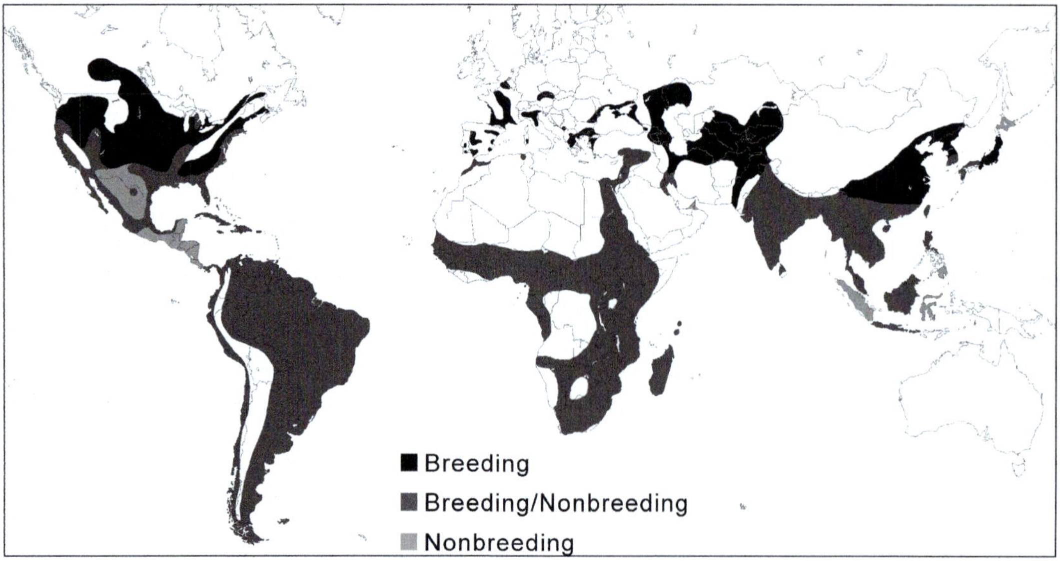

Tanzania, Malawi, Zimbabwe, Botswana, Namibia, South Africa, Madagascar, Seychelles and possibly the Amirantes (J. Gerlach pers. comm.). In Asia, it occurs in places in the Middle East and extends more continuously across the continent, nesting in Israel, Saudi Arabia (James 1991), United Arab Emirates, Iran, Azerbaijan, Turkmenistan (Eminov 1993), Uzbekistan, Kazakhstan, Kyrgystan, Tajikistan, Russia (Komarov and Lipkovich 1986), Afghanistan, Pakistan, India, Nepal, Sri Lanka, Bangladesh, Myanmar, Cambodia, Laos, Vietnam, China (south east Sichuan, Henan (Zhang *et al.* 1994), Hunan, Guangxi, Guangdong, Zhejiang, Hong Kong, Taiwan), south Korea (Yu and Hahm 1997), Japan, Philippines (Luzon, probably also Mindanao and other islands—R. Lansdown pers. comm.), Malaysia (single colony on west coast of peninsula—R. Lansdown pers. comm.), Singapore, Brunei, and Indonesia (to Borneo, Java).

NON-BREEDING RANGE: Temperate populations tend to be partially migratory, and tropical and subtropical populations tend to be more sedentary, although dispersal occurs in both. Throughout its northern hemisphere range, the Black-crowned Night-Heron winters fairly far north. North American populations are partially migratory. Some remain as far north as south Oregon, Great Lakes, and south Maine, but most winter in the southern USA (south California, south Arizona, south Texas, to Florida), the Mississippi Valley, West Indies through the Lesser Antilles, especially Cuba, west and east Mexican coast, central Mexico, and Central America. European populations also are partially migratory. Some remain in Europe along the Mediterranean coast and even inland from Spain to Greece north to Ukraine (Rusev 1984), and also Israel and Morocco (Tosi and Toso 1979, Barbieri and Fasola 1986, Ham 1986, Brugiere 1999, Hafner 2000). Most European birds winter in tropical Africa, ranging at least to Mozambique (Fasola *et al.* 2000). Herons breeding in Africa are mostly sedentary, but some South African herons disperse north to DR Congo and Mozambique. They winter in Arabian Gulf region (United Arab Emirates, Oman-C. Tourenq pers. comm.), Iran, Iraq, Pakistan (especially Sind), India (especially Tamil Nadu), Maldives, southeast China, south Korea (Kim *et al.* 1997), Japan, Philippines, and Indonesia. Southern Asia birds are sedentary.

MIGRATION: In the Black-crowned Night-Heron, an extensive postbreeding dispersal merges into migration. Birds migrate at night alone or in small groups. In North America eastern birds move southward in late September and October along the Atlantic coast, down the Mississippi Valley, and along the Pacific coast. Birds overlap broadly in destination: Florida, Gulf Coast, coastal Mexico, central Mexico, Central America, and the West Indies. Return is early, mid March–mid May (Byrd 1978, L'Arrivee and Blokpoel 1990). Similar movements may take place in South America, but information is not complete. In Europe southward migration is in August. Broad front movements cross the Mediterranean and Sahara to tropical Africa. They return to the breeding grounds in March–May. East Europe and west Asia birds move south to Iran, Iraq, west and central tropical Africa, and to Pakistan and India. In West Africa, herons are sedentary; in southern Africa some migrate north. In east Asia, northern populations are migratory; southern populations are sedentary. West Asian birds move south to southeast China and Japan and then to the Philippines and Indonesia. Black-crowned Night-Herons, especially juveniles, overshoot their return to the breeding grounds. Dispersal bring them to casual or even regular occurrences beyond the breeding range, such as Alaska (Gibson *et al.* 1987b), Newfoundland, Greenland, Iceland, England, Ireland, Micronesia, Wake Island, the Galapagos, the Azores, Sumatra, Cocos-Neeling, North Mariana, and Guam. Individuals wander widely from their natal colony. French-bred birds were found later in Romania and in Russia. This dispersal likely is an important factor inhibiting geographic isolation among populations.

STATUS: The Black-crowned Night-Heron is widespread and abundant locally throughout most of its range. In Europe, populations are increasing following both population and range decreases in the late nineteenth century due to habitat destruction and hunting (Fasola and Huduc 1997, Marion

1997). The European population now is 50 000–75 000 pairs, concentrated in Italy, which has 14 000–24 000 pairs, and in Russia with 10 000–15 000 pairs (Marion *et al.* 2000). About 1000 pairs nest along the Mediterranean coast in France, 2200 pairs in Spain, and 100–200 pairs in Portugal (Hafner 2000). In some areas, such as Netherlands, Germany, and Belgium, colonies established artificially aided in the recovery. In other areas (Italy) the development of rice field cultivation supported the population increase. Populations fluctuate but are stable or increasing overall in Europe. Regionally they appear to be decreasing in Spain, Mediterranean France, Croatia (500 pairs), and Romania (Danube Delta decreased 47% to 3100 pairs in the mid 1980s) (Hafner 2000). The wintering population in tropical Africa is estimated to be 70 000–100 000 birds, including 10 000 at the Senegal Delta alone. Conditions in Africa are critical in that 95% of European herons winter south of the Sahara and a high proportion of European juveniles may remain there for 2 to 3 years. The African breeding population is common and widespread in sub-Saharan Africa with colonies in the hundreds to low thousands (Turner 2000). In 1962 a single colony in Tanzania was counted to have 1000 pairs and it remains common in that country (Baker 1995, Baker and Baker 2004). It is rarer across north Africa, with small numbers breeding (and expanding their range inland) in Algeria, and 500–1500 pairs in Morocco. It is also fairly common on Madagascar. Overall the resident population in tropical Africa (including Madagascar and Seychelles) is estimated to be between 25 000 and 100 000 birds (Rose and Scott 1994). In the Middle East, Israel has 2000 pairs nesting (Paz 1987) and the species appears to be increasing its range including in the Arabian Peninsula using artificial wetlands (Perennou *et al.* 2000). In central and east Asia, the species is common and widespread, although difficult to census. It is increasing in Malaysia (one area increasing from 4000 to 12 000 birds in 20 years to the mid 1980s). Colonies numbering in the thousands have been reported from India, Pakistan, Malaysia, Indonesia (Java), China, and Japan. In central China, partial censuses in 2000–2001 located 10 heronries with a total of 22 500 Black-crowned Night-Heron nests in a 2000 km^2 area near Lake Poyang, and 5 heronries with 45 700 nests along a 200 km transect near Lake Tai, accounting for over 130 000 nesting birds in the two areas (Fasola *et al.* 2004). In North America the species' range has been stable, although regional population changes may be occurring—increasing in the northeast and decreasing in mid-continent and mid-Atlantic states (Butler *et al.* 2000). In the 1970s over 48 000 birds were censused in part of the USA (Spendelow and Patton 1988). The population is increasing in Hawaii (Engilis and Pratt 1993). It is common and abundant in South America, with 50 000–250 000 pairs estimated in Venezuela (Morales 2000).

Habitat

The Black-crowned Night-Heron has wide habitat preferences. It occurs most frequently on the vegetated margins of shallow fresh water or brackish rivers, streams, ponds, lakes, marshes, swamps, mangroves, and mudflats. It often feeds from on top of aquatic vegetation, for example on kelp beds 500 m offshore in the Falklands. It also uses grasslands and coastal habitats, especially on migration, and high mountains, nesting to 4800 m in Chile—exceptional for a heron. The species frequently, perhaps characteristically, uses human-made habitats. These include pastures, ponds, reservoirs, canals, ditches, fish ponds, rice fields, wet crop fields, and dry grasslands. Rice fields are especially important habitats in much of its range (Fasola *et al.* 1996). In areas where rice fields are available, up to 96% of the food resources of a colony were taken from that habitat (Fasola *et al.* 1993, 1996, Fasola and Ruiz 1996). Despite the wide range of habitat use, large nesting colonies seem to be associated with protected sites in large wetlands (Alieri *et al.* 1988, Hafner and Fasola 1992, Davis 1993). Colonies are on islands protected from predators and often—although not necessarily—away from human disturbance. Since night herons rest during the day, roosts become particularly important habitat features. The heron is flexible in its choice of roosting substrates. It uses trees, bamboo, pines, tamarisks, mangroves, oaks, jacarandas, drowned trees, and mixed woodlots. Roosts need not be over water, and often are in parks and other human environments. Habitat essentials appear to be dense roosting cover, not necessarily free from human disturbance, and fresh, brackish or salt water feeding grounds nearby.

Foraging

The foraging behaviour and food habits of this species are well studied (Drinkwater 1958, Wolford and Boag 1971, Kushlan 1973b, Watmough 1978, Parasharya 1982, Fasola and Ghidini 1983, Fasola 1983, 1994, Fasola *et al.* 1993, Prigioni *et al.* 1985, Sodhi 1985, Voisin 1991, Yen 1991, Davis 1993, Fasola and Hafner 1997, Mugica *et al.* 1998, Boukhemza *et al.* 2000). Black-crowned Night-Herons typically feed by Standing intermittently mixed with Walking slowly. They Stand in a Crouched posture and carefully stare at the water until seeing or hearing something and then make a lunging strike. Herons tend to switch from Standing to Walking when the feeding success on smaller prey increases (Fasola 1984). This is a nocturnal species and most often and typically feeds at night, locating prey by sight and by sound. But they do readily feed during the day and often are required to do so to meet the energy demands of raising young (Fasola 1984, Endo and Sawara 2000). It has relatively large and more widely separated eyes than do the more diurnal herons, which is likely associated with maintaining visual ability over a range of light levels (Katzir and Martin 1998). The wide placement of the eyes does not, however, provide any better binocular vision. Feeding behaviour is the same day or night, but during the day they catch fewer, although larger prey (Watmough 1978). Daylight also allows use of more active behaviours such as Running, Bill Vibrating, Hovering, Feet First Diving, Swimming Feeding, Plunging, and Underwing Feeding (Tyler 1979). Along with very few other herons, this species is known to use Baiting (Davis and Zickefoose 1988, Riehl 2001). Daylight feeding brings it into conflict with typically day-feeding herons. During the day it is are displaced and robbed by other herons, and appears to be definitely subordinate to the day-feeding herons (Kushlan 1973b, 1978a, Watmough 1978).

The Black-crowned Night-Heron is mostly a solitary forager. It usually feeds alone, maintaining exclusive feeding territories. In defence it uses Forward displays, Supplanting Runs, and Supplanting Flight. It more often attacks other night herons than other species (Fasola 1986). It feeds in loose, dispersed aggregations and also in denser aggregations where prey is highly concentrated (Fasola 1982). These sites are found by following birds flying to feed and also by joining birds already feeding at a site (Fasola 1982). It roosts during the day, (except during breeding season) either alone or communally in groups of 10–100 birds (Perlmutter 1992, Igarashi 1996, 1997). It prefers to roost in dense trees, presumably for protection. It flies out of the roosts at dusk, 10–20 minutes after sunset, heading to feeding grounds, and giving its characteristic Quok call. It returns to roost the following morning, averaging 17 minutes before sunrise (Fasola 1984). Time budgets are constrained by season (Kim *et al.* 1997).

The food habits of this heron have been very well documented. Insects, frogs, and fish dominate the diet. A diet might include all three types of prey, having different relative importance in different areas or at different times. For example, in one study, in rice fields amphibians predominated, whereas in rivers and lagoons, fish predominated. The overall food list is quite large, given that the night herons are opportunistic foragers and will eat what they encounter. Insects eaten include beetles, bugs, grasshoppers, crickets, flies, and dragonflies. Other invertebrates include spiders, leeches (*Hirudo*), earthworms, prawns, crabs (*Uca*), crayfish (*Cambarus*), mussels (*Mytilus*), clams (*Venus*), and squid (*Loligo*). It also sometimes eats carrion (Parejo 1997). Many types of fish are taken including carp (*Cyprinus, Carassius*), cichlid (*Tilapia*), mosquitofish (*Gambusia*), perch (*Perca*), roch (*Rutilus*), tench (*Tinca*), herrings (Clupeidae), suckers (Catostomidae), minnows (Cyprinidae), killifish (Cyprinodontidae), eels (*Anguilla*), barbels (*Barbus*), catfish (*Ictaluris*), sticklebacks (Gasterosteidae), pickerels

(*Esox*), sunfish (*Lepomis*), mullet (*Mugil, Liza*), flounder, sleepers (*Dormitator*), and many more (Davis 1993, Fasola and Hafner 1997). Amphibians include frogs and toads (*Rana, Bufo, Pseudacris, Hyla*) and newts (*Ambystoma, Triton*) (dead or alive). Other vertebrates include lizards (*Lacerta*), snakes (*Thamnophis, Natrix*), rodents (*Rattus, Apodemis, Microtus*), screws (*Sorex*), bats, birds and their eggs. The propensity of the Black-crowned Night-Heron to eat birds is unusual among herons, and these may constitute an important part of the diet of some individuals and their chicks. This species often take the young of other colonial nesting waterbirds such as terns, herons and ibises, which it obtains by walking through the colony. In eastern North America, reports include taking nestling White Ibis (*Eudocimus*), Cattle Egrets, and Great Egrets. It also eats non-colonial marsh birds such as stilts (*Himantopus*) and duck (*Anas*). In one example, a heron perched on nest boxes and ate swallows (*Tachycineta*) as they emerged (Miller 1988–89). In a captive situation a juvenile devoured a hatch of baby Guinea fowl (*Numerius*).

Breeding

The courtship behaviour and breeding biology have been well studied, including the distinction of being the object of some of the classical studies in behaviour (Gross 1923, Lorenz 1938, Noble *et al.* 1938, Allen and Mangels 1940), other studies are Palmer 1962, and Voisin 1970, 1991, Chapman *et al.* 1981, Custer *et al.* 1983a, 1983b, 1984, Prigioni *et al.* 1985, Alvarado 1986, Custer and Frederick 1990, Custer 1991, Custer and Peterson 1991, Erwin *et al.* 1996b, Blus *et al.* 1997, Yu and Hahm 1997, Voskamp and Zoetebier 1999, Zhu *et al.* 2000, Petry and Hoffmann 2002, Hothem and Hatch 2004). The nesting season differs geographically and sometimes varies in one region (Mackrill 1987). In temperate areas it is in the local spring, often early, but in tropical and subtropical areas nesting is variable and generally coincides with the rainy season. Breeding takes place in April–June in Canada, March–July in east USA, July–August in Costa Rica, September–October in southern Brazil, April–June in southern Europe and North Africa, April and August–September in Ethiopia, December–January and again April–June in East Africa, November–May in DR Congo, January–April in Zambia, August–March peaking September–January in southern Africa, August–January peaking November–December in Madagascar, December–April in East Java, and February–July in West Java. The nesting habitat is highly variable. It nests in bushes and trees of many species, up to 50 m tall, but also in reeds, sedge, grass tussocks, and on the ground (Alvarado 1986, Davis 1993, Kobayashi 2000). Colony sites are reused in most places; however, in other places, sites change from year to year (Subramanya 1996). Colony site movement can be a response to birds destroying their trees through their nesting activities. The species often nests in rural, suburban, and urban settings. Across its range, it seems to be particularly attracted to nesting in zoos. Some populations began as captive colonies and continued as birds became feral. Nesting is usually colonial, in single-species colonies or in mixed-species colonies with other herons, cormorants (*Phalacrocorax*), or ibises (*Eudocimus, Plegadis*). Nests may total in the thousands, but the night heron usually remains in discrete single-species groups within the larger colony. They tend to nest very close to each other, with as many as 20–30 nests in one tree. Colonies are dispersed on the landscape in relation to distance from feeding areas. Nests are a platform of sticks and reeds, 30–45 cm wide and 20–30 cm deep. The nests start as fragile structures, which, however, are added to through incubation. Nests that survive are reused, enlarged, and become quite bulky. In one study, 86% of birds used old nests (Davis 1986a). Males collect the twigs and presents them to the females for insertion into the nest. Sticks are often stolen from other nests.

The male chooses a site from which to display, often an old nest. During courtship the display plumes are in full development, and the white head plumes are particularly featured. The male claims its site and defends it using Standing, Preening, Upright, and Forward. In the Upright threat, the head and neck are up and forward, the crown, neck, and back feathers are partially erect, and gives the Rok call. In full display the Forward may involve maximum erection of feathers and head plumes, the eyes bulge out, and the Rok call is given. Night herons will attack with minimal previous display, launching a Supplanting Flight with strong blows of the bill. Similarly, they will engage in Face to Face Fighting, with one bird striking or grabbing the bill or wing of the opponent. The Snap display (also called the Bowing Display) in this species is performed in a crouch, the head is lowered and extended, the feathers of the head, back, and neck are partially erected, and it holds a twig, or it snaps the bill. Unmated males use it as an advertisement display. The most common advertising display is the Stretch–Snap, a seemingly unique display derived from the Stretch (Voisin 1991) that has also been called Reverse Stretch, Foot-lifting, Snap–hiss, Song-and-dance, and Danse (in French). The bird stands erect, treading from one foot to the other (or lifts one foot repeatedly), extends its head and neck forward and down, feathers are not raised, and the eyes bulge. With its bill down, it gives Plop calls, which can go on for many minutes. Between songs it Preens its belly and Twig Shakes. After the Stretch–Snap, a bird may lift its head upwards with the bill pointing to the sky, in a more typical Stretch display. The male also does a Twig Shake display as part of pair formation. The female manages to enter and maintain herself in the male's territory to form the pair. After pair formation, Wing Touch, Bill Clappering, Billing, and Back Biting are all performed to establish and maintain the pair. In the Greeting Ceremony both sexes bend horizontally with outstretched neck, head plumes raised, and touch bills, which they may rattle together, and also give the Wok call. Analysis of this display originates with Conrad Lorenz (1938), but it is still unclear if individual recognition is possible from the call (Venchi *et al.* 1994). The twig-holding Stretch display is also used in greeting and in twig passing, and Billing and Back Biting also commonly attend nest relief. Both birds actively defend the nest site. Copulation is not preceded by a set display, but may follow a Greeting Ceremony.

The eggs are green to pale blue-green (Custer 1991). They average 50 × 36 mm in Europe, 52.3 × 37.37 mm in North America (Davis 1993), 51.35 × 37.43 in Central America (Alvarado 1986), 48.0 × 34.3 in China (Zhang *et al.* 1994), 49 × 35 in Africa (Brown *et al.* 1982), and 50.9 × 35.8 in Madagascar. The clutch is usually about 2–5 eggs (Custer *et al.* 1983a), and the range is 1–7 eggs. Averages include: 3.2–3.8 in northeast North America (Parsons *et al.* 2001), 3.9 in west North America, 3.6 in inland North America, 2 in the Falklands, 3.3 in East Africa, 2.8 in Zimbabwe, 2.2–2.4 in South Africa, 2.5 in southern France, 2.6 in northern Italy, 4.1 in Korea, and 3.55 and 3.49 in China. Eggs are laid at 2-day intervals. They generally will have only one nest per season, but will readily produce additional clutches if the first nest fails (Blus *et al.* 1997). Incubation is continuous, starting with the first egg. Both parents incubate and attend the nest. Incubation averages about 23 days, with a range of 21–26 days. Eggs hatch asynchronously. At hatching the young are semialtricial and nidicolous. Hatching success is fairly high, 75% in China (Zhang *et al.* 2000). Both parents attend the nest and brood the young for 10 days after hatching. They are then left increasingly alone, and are not guarded at all after 20 days. Adults feed the young by regurgitation; the young grab the adult's bill and force the regurgitation. Later the adults put the food on the nest floor. They feed about seven times per day. The young develop rapidly (Chapman *et al.* 1981, Galeotti 1982). Temperature is regulated by day 10 and the young begin clambering out of the nest by 3 weeks. At 35 days they are fully feathered. They fledge in 6–7 weeks. For much of the time they are extremely noisy, constantly enlarging their repertoire from the high-pitched cackle of their first two weeks. Brood reduction is common, especially for the last-hatched chick. Growth is slower for the third chick but there is little evidence of chicks killing siblings (Custer and Peterson 1991, Emond and Medeiros 1999, Medeiros *et al.* 2000). Chick death appears to occur predominantly in the first few days after hatching (Prigioni *et al.* 1985). Nesting success (young per egg) is often high, 75%–85% in North America (Henny 1972, Parsons *et al.* 2001) and 70% in France (Hafner 1978). Production is variable, 0.47–1.94 young per successful nest in studies in North America (Blus *et al.* 1997), 2.08 from 58 successful broods in East Africa (Brown *et al.* 1982), and up to 2.9 in Central China (M. Fasola pers. comm.). Food supply, drought, storms, and predators reduce production. Predators of eggs and young include raccoons (*Procyon*), muskrats (*Ondatra*), owls (*Bubo*), gulls (*Larus*), caracaras (*Polyoborus*), crows (*Corvus*), jays (*Cyanocitta*), lizards (*Iguana*), and snakes (*Boa*). Low productivity in some places may be linked to contamination (Parsons *et al.* 2001).

Population dynamics

Most individuals nest at 2–3 years old, but yearlings sometimes breed (Custer and Davis 1982). In central China, 41% of the breeders are birds in juvenile plumage, probably in their second year. Most of the pairs are strictly assorted by age, and the reproductive success of pairs of adults is higher than the success of pairs of juveniles (Fasola *et al.* 2001). The mortality rate in the first year may be high, up to 69% (Henny 1972), but decreases later. One longevity record of 16.5 years has been reported (Houston 1974). Juveniles from Europe depend on African habitats for up to several years, so it would be expected that conditions there could affect nesting populations; 10%–20% of the population variability is explained by the African climate (Fasola *et al.* 2000).

Conservation

Although widespread across one of the most cosmopolitan ranges among herons, the Black-crowned Night-Heron's nesting is limited to a few areas in most regions. As a result conservation of these sites becomes crucial. The protection of nesting sites and the management of feeding sites are both required to ensure regional population stability. One successful conservation scheme, in Italy, involved the protection of existing colony sites, the active management of habitat, and the creation of a network of new breeding sites spaced appropriately (Fasola and Alieri 1992b, Fasola *et al.* 1992). Colony and roost site management takes a landscape perspective. Colony sites are destroyed by the bird's activities, and so alternative sites need to be available. Colony impacts are only important if there are no other sites to move to (Marion *et al.* 2000). They readily nest on artificial sites, such as spoil islands, prepared for them. The attraction and accommodation of Black-crowned Night-Herons to zoos and other urban settings provide many opportunities for environmental education. Zoos within the range of the species should be encouraged to establish nesting sites for wild populations (of the local genetic stock) that can be managed for their education value. The flexibility of this species allows options management of colony sites at massive scale (Crouch *et al.* 2002).

The dependence of regional populations on large wetlands for both breeding and non-breeding, places the conservation of the species in the full context of wetland conservation. Wetland drainage historically undermined their population stability in Europe. Wintering sites on the Mesopotamian marshes in Iraq are threatened by massive development schemes. Large wetland ecosystems will need to be protected and managed for night herons to maintain stable regional populations. Wetland drainage and the destruction of vegetation has led to a decline on Black-crowned Night-Herons in many parts of Africa (D. Turner pers. comm.). It is only in Zimbabwe and South Africa that through the provision of artificial water bodies and the planting of trees are there signs of an increase in both the range and numbers of this species.

Other human interactions must also be considered in developing conservation strategies for this species. In North America, populations declined due to pesticides, particularly up to the 1960s. Since then populations have recovered, contaminant loading has decreased, and present contaminant burdens tend to be minor or not proven to be of significant populational consequence (Henny *et al.* 1985, Hothem *et al.* 1995, Movalli *et al.* 1995, Hoffman *et al.* 1986, Blus *et al.* 1997, Fasola *et al.* 1998, Rattner *et al.* 2000). Chronic contamination can still occur where persistent pesticides are still being applied (Henny and Blus 1986). It has been thought for some time that high contaminant levels in herons from the western USA were due to accumulation in winter in Mexico, but careful re-examination suggests that data supporting this view are slim (Mora 1997). Herons also are at risk from poisoning by lethal materials (Hunt *et al.* 1995). Monitoring of contaminant burdens in this species is a good environmental indication of contaminant availability in breeding or wintering areas (Custer *et al.* 1991, Custer and Custer 1995). Interactions with aquaculture and agriculture also need to be considered. Herons traditionally have been shot at fish hatcheries, although they are less seen and obvious when they feed at night. The true impact of the species at aquaculture facilities needs to be determined and conservation-friendly measures employed (Ashkenazi and Yom-Tov 1996). Beyond antidepredation killing, hunting mortality continues. In some areas, like Madagascar, herons are still taken for food. Rice fields support some of the largest populations, and the culture techniques conducive to these herons need to be continued. Black-crowned Night-Herons are fairly tolerant of non-aggressive human activities, and often nest and roost near humans. This can bring them into conflict with towns that would prefer not to have colonies and roosts nearby. The management of sites and public education are required. Local disturbance, especially at egg laying, can provoke abandonment of the nests (Tremblay and Ellison 1979, Parsons and Burger 1982, Marion *et al.* 2000).

Research needs

A primary research and conservation need is to unravel the patterns and taxonomic consequences of geographic and species variation in the *Nycticorax* herons. The proper allocation of populations to species and to subspecies requires a re-examination of *Nycticorax nycticorax, N. calidonicus*, and *N. leuconotus* as well as population variation within *N. nycticorax* using plumage, morphometric, and genetic characters. Also, the now extinct island populations of *Nycticorax* should be examined from the larger perspective of species and subspecies limits in the genus. The breeding biology and ecology of the species are well known. Intriguing questions remain however. One is its apparent subservient position among herons during daytime foraging, which seems unexpected for such an aggressive heron. The question might be cast within the relative costs and benefits of aggressive behaviour during day and night. Prey capture at night is not completely understood, and studies of the physiology and behaviour of night foraging may prove interesting. Conservation research is needed to determine the best approaches to landscape-based population protection and management. Successes in Italy suggest ways that colony site protection, creation, and habitat preservation can be used to create a regional conservation strategy. This model can be tested elsewhere. The status of the species throughout its range needs to be better understood. The recent discovery of large colonies in China is an example of how little is actually understood about the status and trends of this widespread species.

Overview

The core of the Black-crowned Night-Heron's ecology is its adaptation to nocturnality. Night-time foraging limits the options of suitable foraging behaviour. As a result the heron is primarily a stand and wait predator. It intermixes slow Walking to shift locations, to investigate a potential prey, or to cover more ground. Its two behaviours produce different results, with Walking being better for small abundant prey and Standing better for larger prey. The heron defends it feeding site and only forages communally when prey abundance is high. It is morphologically designed for its slow foraging. It has relatively short legs and neck for standing in a crouched posture close to the water. It has a plumage that is cryptic at night, light below from the fishes' view and dark above from a predator or other heron's view. It has huge eyes to see better under a variety of conditions. Its daily schedule is to feed at night and sleep during the day. The nocturnal habit of the species must, however, break down during nesting. As it is unable to secure sufficient food for itself and its young foraging only at night, so it must feed during the day, on larger, scarcer prey in competition with the day-feeding herons, to which it is subordinate. It sometimes resorts to active behaviours under these conditions. Dense food supplies or high availability are essential for the heron to meet the energy demands of nesting. One place of high availability is the colony itself where it eats other birds. Its habitat choice is broad, but it clearly depends on large natural wetlands and on artificially provided situations such as rice fields and fish ponds, on which it has become dependent. The loss of habitat in Europe led to a population collapse, and the creation of fish farms, rice fields, and polders brought it back. It similarly is able to use human-provided nesting sites. So, it has adopted living in close proximity to humans. It has three separate lines of food: fish, insects, and frogs, which can vary in importance depending on their availability. It is a species that takes advantage of a varied and changeable array of food—a food generalist. The Black-crowned Night-Heron is adaptable not only in food and nesting, but in its travels. A strong dispersal tendency brings it to places outside the nesting range, which can ensure colonization of likely locations. The species, and its close relative *N. calidonicus*, together cover much of the world.

Rufous Night-Heron *Nycticorax caledonicus* (Gmelin)

PLATES 9.5, 9.6, 17.4

Other names: Nankeen Night Heron in English; Martinete Caneto in Spanish; Bihoreau cannelle in French; Rotrückenreiher in German; Hashibuto-goi in Japanese; Lapay degar in Tagalog (Phillippines); Baku gabi in Pilipiro; Koward merah in Indonesian.

Ardea caledonica Gmelin, 1789. Syst. Nat. 1(2), p. 626: New Caledonia.

Subspecies: *Nycticorax caledonicus manillensis* Vigors, 1831: Manila (Philippines); *Nycticorax caledonicus mandibularis* Ogilvie-Grant, 1888: Aola, Guadalcanal (Solomons); *Nycticorax caledonicus hilli* Mathews, 1912: Parry's Creek, NW Australia; *Nycticorax caledonicus pelewensis* Matthews, 1926: Palau Islands.

Description

The Rufous Night-Heron is a stocky rufous and white heron with a black crown and white head plumes when breeding.

ADULT: The crown and nape are grey-black. The face is white with rufous wash. The short, heavy bill is black. A line surrounding the eye may be white or rufous. A facial mask consisting of the lores and skin at the base of the lower bill is green-yellow. The irises also are yellow. The neck is short and thick. The hind neck, back, tail, and upper wing vary from light to dark—pale cinnamon, cinnamon-rufous, chestnut-brown, maroon. The underparts are white, with a rufous or similar wash. The underwing is white, with cinnamon-rufous flight feathers. The neck and upper breast often have a pale chestnut wash, with the white of the undersides extending over the front of the wing. The short, thick legs and the feet are green, cream-yellow, to orange-yellow. In breeding plumage, two or three long white plumes emerge from the crown down the back. When fresh, the plumes are tipped with black. The lores turn blue; irises are red turning orange; and legs turn bright pink or red.

VARIATION: Sexes are alike but females average smaller in most measurements. Five subspecies are differentiated primarily on size and back plumage colour. In addition there is much colour variation that is not geographically based, so that even in a small area one can see various shades of colorations. *Caledonicus* is dull rufous-chestnut above with a light rufous belly. *Manillensis* is relatively large and dark, having dark maroon upper parts and chestnut-washed underparts; the eye circle is lacking or poorly developed; and crown plumes are smoky black or white with dusky base and tips. Within the Philippines, southern birds are larger than northern birds and may have a paler base to the bill; they have been described as a different subspecies, *major* (Hubbard 1976). *Hilli* is chestnut-brown above white on the belly. *Pelewensis* is dull maroon above. *Mandibularis* is relatively smaller, with rich chestnut upper parts, the colour extending around the neck and chest; it lacks the eye circle; and it can have crown plumes with a black wash. Birds of New Britain and New Hanover Islands have been described as intermediate between *hilli* and *mandibularis* (Amadon 1942). Birds intermediate between *hilli* and *manillensis* are described from Java and Borneo (Amadon 1942, Hoogerwerf 1966).

JUVENILE: The juvenile is rufous-brown with much buff and white spotting and streaking. The crown is black-brown with buff streaking. The iris is yellow to orange-yellow and the facial mask of the lores and skin next to lower bill is yellow-olive, developing a blue tint. The upper bill is grey-black with a yellow edge, whereas the lower bill is yellow-olive with a grey black edge and tip. The chin and upper throat are white. The neck is buff and brown streaked. The back is rufous-brown with buff spots. The rump is white but the upper tail is brown with large buff spots at the tips of the feathers. The tail is rufous-brown with dark brown barring and narrow white feather tips. The upper wing is brown-rufous with large buff spots at the end of each feather. The underwing is white with grey-brown streaks and flight feathers are rufous-brown with dark brown inner edges. The underparts are brown and cream streaked. The legs and feet are lime-green to olive-grey. With age the crown becomes blacker and the upper parts more rufous and less spotted.

CHICK: The chick is covered with dark brown down on the back and white on the undersides. The bill is cream with a grey-black edge. The iris is orange-yellow to yellow and the lores are white to orange-pink. The legs and feet are olive.

VOICE: The characteristic 'Quok' call is the flight call. The 'Rok' call is the threat call, used in feeding aggregations.

WEIGHTS AND MEASUREMENTS: Weight: 810–1014 g; length: 55–65 cm.

Field characters

The Rufous Night-Heron is identified by its back cap and chestnut to brown back. The adult is distinguished from the closely related Black-crowned Night-Heron by its rufous (not black) upper parts. The juvenile is similar to other cryptic herons. It is distinguished only with difficulty from the immature Black-crowned Night-Heron, by its rufous-brown (not grey-brown) appearance, flight feathers rufous-brown with dark brown inner edges (not grey-brown), dark brown-buff spotted upper tail covers (not brown), white (not grey-brown) rump, tail rufous-brown with dark brown barring (not grey), and slightly smaller size, on average. It is distinguished from the Australasian Bittern by its shorter neck, hunched posture, rufous (not brown) upper parts, spotted (not barred) wings, and occurring in less dense habitat. It is distinguished from the juvenile Striated Heron by its larger size, light spotting on the back and wings, and pale underparts.

Systematics

The Rufous Night-Heron is closely related to the Black-crowned Night-Heron and these species have hybridized in Java and Borneo where their ranges have come into contact. They overlap in a minor way in the Philippines and in the past in Sulawesi (Hoogewerf 1966, Hubbard 1976, Erftemeijer 1989). There is no evidence of persistent present-day hybridization (Sheldon and Marin 1985). It is likely that the ranges have come into overlap due to relatively recent invasions of the Indonesian islands by the Rufous Night-Heron and of the Philippines by Black-crowned Night-Herons. The two birds are none the less mostly allopatric. The Rufous Night-Heron has been in the Philippines and islands to the southeast long enough to have diverged in size and colour. As far as is known, the Black-Crowned Night-Heron has not racially differentiated in the area, suggesting that it is the more recent invader. It would appear that attempts by the Rufous Night-Heron to colonize other islands in the past century have had limited success. Given the intermittent history of range overlap, and non-persistence of hybrids, it seems likely that these are two good species.

Range and status

The Rufous Night-Heron occurs in Indonesia (Java, Borneo, Celebes, Lesser Sundas), New Guinea, the Philippines, Palau, Micronesia, New Caledonia, Bismarck Archipelago, Solomon Islands, Australia, and New Zealand.

BREEDING RANGE: *Caledonicus* occurs in New Caledonia. *Manillensis* occurs throughout the Philippines (Luzon, Mindanao, Marinduque, Negros, Panay, Palawan, Sulu Archipelago), west Indonesia, and north Borneo (Hubbard 1976). *Pelewensis* occurs on Palau and Micronesia. *Mandibularis* is found in the east Bismarck Archipelago, New Britain, and the Solomons. *Cassirostris* previously occurred on the Bonin Islands (Japan). *Hilli* breeds in Australia and in very small numbers in recent years in Java (Erftemeijer 1989, MacKinnon 1993), Borneo (Sheldon and Marin 1985), and New Zealand (Marsh and Lövel 1997). In Australia it occurs in north, east, and southwest Australia. Nesting populations are concentrated in the southeast in Queensland, New South Wales, and Victoria, but records indicate widespread nesting throughout its Australian range including the Great Barrier Reef. Introduction into New Zealand in the mid 1800s failed (Heather and Robertson 1996). It may have bred in 1957–9 and was discovered nesting in 1995 (Marsh 1995, Marsh and Lövel 1997). It also has nested in the past on Celebes and Cocos-Keeling, but recent records of nesting are not available.

NON-BREEDING RANGE: *Hilli* occurs in the non-breeding season in New Guinea, west Bismarck Archipelago, Lesser Sundas Islands, New Zealand, and Cocos-Keeling.

MIGRATION: The Rufous Night-Heron is partially migratory in the south of its range. Birds move north in the Australian winter as far as Papua New Guinea. Depending on the persistence of suitable feeding and nesting conditions, some populations are sedentary, others more nomadic, moving around in response to rainfall and flooding events to take advantage of newly opened foraging conditions. In Australia, birds move into and out of the Darling–Murray River system depending on water conditions, large colonies forming in wet years. Substantial postbreeding dispersal of young takes place, which gives rise to quite long-distance wanderings. Dispersal records occur in Lord Howe Island and Christmas Island.

STATUS: *Hilli* is widespread, common, and stable in Australia in the north, east, and southwest, but rare or absent from the interior and northwest Western Australia. It is common to abundant in

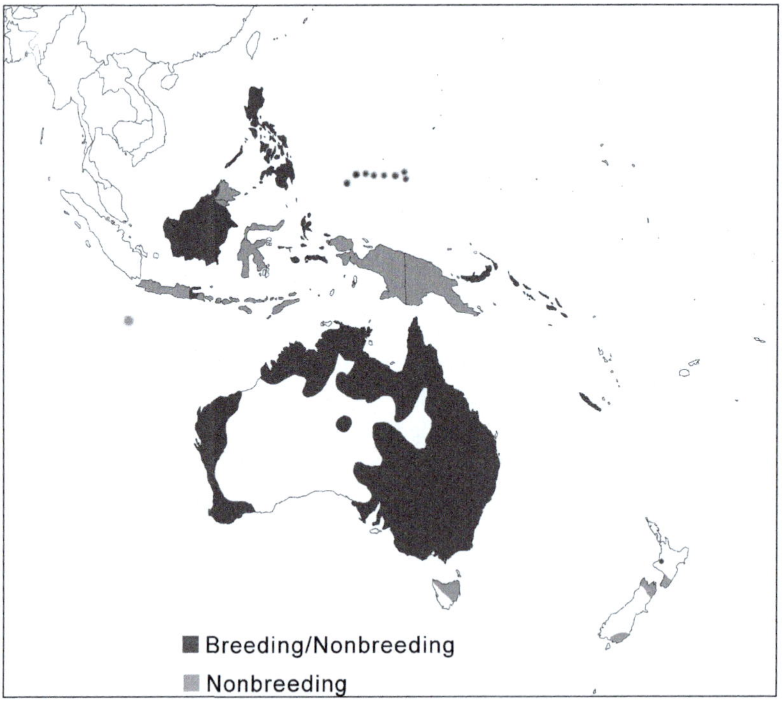

the non-breeding season in New Guinea, numbers fluctuating from year to year. The breeding population is very small on Java. It breeds regularly throughout north Borneo (Mann 1989, Lansdown *et al.* 2000). It is rare in Celebes and Lesser Sundas (Mason 1994). *Caledonicus* is common in New Caledonia. It is one of the most common birds on Cocos-Keeling (Stokes *et al.* 1984). *Cassirostris*, discovered on the Bonin Islands in 1820, became extinct by 1889.

Habitat

The Rufous Night-Heron feeds in a variety of habitats, mostly shallow, still or slow-moving water and tall but not overly dense vegetation that offers some degree of cover. It is usually on the margins of deeper water or in flats and open areas within marshes, swamps, and forests. Its habitat ranges from the coast to inland fresh water, from permanent water bodies to temporary pools, and from swamps to dry grasslands. It most commonly uses the varied habitats of river floodplains. Among specific habitats used by night herons are wooded swamps, creeks, pools, lakes and river margins, reed swamps, paper bark forest, eucalyptus forest, damp meadows, and dry grassland. Birds occur less often in saline areas but are found on offshore islands, exposed rocky shores, reefs, beaches, estuaries, mangrove swamps, salt marshes, and shallow lagoons. They readily use human situations including flooded pasture, dry fields, urban wetlands, ponds, parks, gardens, airports, road edges, grain bins, and garbage dumps. Most of its habitat is near sea level, but it is found to 1600 m in New Guinea. Being nocturnal, it requires day resting spots. It roosts in dense vegetation, such as leafy trees, dense bushes, reeds, and other waterside situations, such as a log jam (Marchant and Higgins 1990). It has been seen to use dead trees, but such situations are generally too exposed. In coastal areas it uses dense low shrubs where available. It chooses roosts in developed or even urban areas using introduced trees such as cypress and pines.

Foraging

The Rufous Night-Heron is a nocturnal bird, feeding primarily at night and in the morning and evening. Its typical nocturnal foraging behaviour is Standing and Walking slowly in shallow water (Recher and Holmes 1982, Recher *et al.* 1983). It Stands in Crouched or Erect posture peering into the water looking for prey. It also will Walk slowly through the water searching for prey. Night herons often are solitary foragers, maintaining exclusive well-defended feeding territories. They also are gregarious when prey is concentrated or habitat limiting, assembling into feeding groups of a few to several hundred birds. Although a nocturnal species, the Rufous Night-Heron feeds during the day during breeding in order to increase the food available to feed young. During the day it can use more active behaviours, especially for feeding in deep water such as Feet First Diving and Swimming Feeding (Loyn 1985). It also is an opportunistic feeder taking advantage of food resources such as garbage dumps, mouse plagues, and sea turtle nesting. Except during breeding it usually roosts during the day, in thick foliage. Roosts are usually communal, with a few to hundreds of birds using the same place. They roost singly but more often in large groups, flying from and to roosts after dusk and before dawn giving the Quok call. Roosts are located convenient to feeding grounds and use varies with feeding conditions, some being used for many years.

The Rufous Night-Heron typically eats crustaceans, insects, fish, and frogs. Crayfish (*Cherax, Euastacus*) appear to be a dominant prey in many situations. It also frequently eats shrimps and crabs (*Holthuisiana*). Fish taken include introduced mosquitofish (*Gambusia*) and carp (*Carassius*). Insects include water beetles and larvae, crickets, water bugs, caterpillars, wasps, ants (*Camponotus*), dragonfly larvae, and termites. Frogs and tadpoles, mice (the introduced *Mus*), and hatchling sea turtles are important in the diet in some situations. It is a common predator on small and hatchling birds especially colonial birds, cormorants (*Phalacrocorax*), ibises (*Threskiornis*), and spoonbills (*Platalea*), and also non-colonial species such as starlings (*Sturnus*), ducks (*Anas*), and house sparrows (*Passer*). It feeds on refuse at garbage dumps.

Breeding

In Australia the breeding season is variable depending on rainfall and water conditions. The usual breeding season is October–March in Australia, February–June in the Philippines, and February–June in Java. Nesting on the Great Barrier reef in December–July coincides with the hatching of sea turtles. Given the development of appropriate feeding conditions, nesting seems to be able to occur anytime in the year. In colonies nesting is not synchronized (Braithwaite and Clayton 1976). It typically nests in swamps and marshes on islands or islands of vegetation in the water. It most commonly places nests in dense cover in trees, up to 20 m high, or in large bushes. Along the coast, it nests in mangroves. On offshore islands, it nests on the ground, within short scrubby vegetation, on rocky shores, or even under rock ledges and in caves. The night heron nests in many urban and suburban settings and is especially attracted to urban parks and zoos (Melbourne, Adelaide). In developed areas, it nests in tall introduced trees, particularly pines and cypress. The Rufous Night-Heron is a highly gregarious colonial nester in single- or mixed-species colonies of 200–3000 birds, including other herons, spoonbills (*Platalea*), cormorants (*Phalacrocorax*), and ibises (*Threskiornis*). Within large colonies, it typically nests in separate groups, with many nests per tree, typically high in the densest part of the tree. It occasionally nests alone.

The nest platform is a small, loose construction of sticks that may be only 20–30 cm in diameter and 3–4 cm thick. Nests on the ground on islands may be only a few sticks that keep the eggs from rolling away. Larger nests also occur, 35–50 cm wide, 8–15 cm thick. Nests are lined with leaves. Males collect the twigs, and both parents build the nest. Sticks may be stolen from other nests. Males claim and defend territory in the nesting colony. The Forward is given with the bird standing erect and then crouches, feathers and plumes erect, points and snaps the bill. The courtship display has not been described, but is likely to be similar to that of the Black-crowned Night-Heron. However, this should not be assumed. Both adults defend the nest site. The eggs are pale green-blue. They average 50 × 37 mm. The clutch is usually 2–3 eggs; the range is 2–5. It is normally single brooded but can lay replacement clutches and nest more than once in a year, given proper water conditions. Incubation starts with the first egg and lasts 21 days. Both parents incubate. Hatchlings are semialtricial and nidicolous. Both parents attend and care for young. Parents feed young by regurgitation, initially by putting food into their mouth and later into the nest. Young leave the nest at 14 days, returning to be fed. After 21 days, parents will feed them away from the nest. They fledge in 6–7 weeks. No information is available on breeding success.

Population dynamics

Birds generally nest in the third year after adult plumage develops, but birds in juvenile plumage are able to breed within in their first year, especially under particularly favourable conditions (Braithwaite and Clayton 1976). No information is available on demography.

Conservation

The species is widespread, common, and stable in the core of its range in Australia. Colonies can be modest (hundreds of birds) and sometimes very large (several thousand birds) under exceptional water conditions. Smaller persistent colony sites are repeatedly used, and these can be subject to local conservation and management. Larger colonies, probably representing the bulk of the generational productivity, need to be protected through conservation of the large, variable wetlands in which they occur. Throughout Australia, drainage, clearing, burning, grazing, increased salinity, groundwater extraction, and flood control have modified habitat in natural fresh water wetlands. Much of the landscape used by the heron is on private lands, and these owners must be take responsibility for conservation (Corrick 1995).

Research needs

For such a widespread species, surprisingly few studies have been conducted. Its basic breeding biology, courtship behaviour, feeding behaviour, and food choice under differing conditions need to be better determined. It should not be assumed a priori that these will be the same as the Black-crowned Night-Heron. Rather it might be hypothesized that such a flexible and nomadic species requires individual and populational responses that differ from those of the Black-crowned Night-Heron. It is intriguing that despite several invasive attempts, the species has not prospered where it overlaps with the Black-Crowned Night-Heron. The study of the comparative biology of these species in their zone of overlap, such as north Borneo, is needed. The range and pattern of geographic variation within the Rufous Night-Heron deserves more exhaustive study. Given both individual and geographic variation in the species and the existence of putative hybrid specimens, a range-wide study of geographic variation is called for.

Overview

The Rufous Night-Heron is a preferentially nocturnal forager that depends on Standing and Walking. Its diet is very broad as is the case for the Black-crowned Night-Heron. However, its published diet seems to have a smaller fish component. There have not been detailed studies of the feeding or breeding biology of the species, and its similarity to the Black-Crowned Night-Heron should not be assumed. One aspect of its ecology that is clear is its flexibility in nesting. While some populations appear to be relatively sedentary around permanent feeding areas, large colonies develop to take advantage of exceptional water conditions. It is intriguing that, for a species characterized by delayed maturity and prolonged juvenile plumage, under these exceptional conditions juveniles nest in great numbers. This suggests that the prolonged maturation is a reflection of normal conditions, but is superseded when foraging conditions are such that immature birds can nest successfully.

White-backed Night-Heron *Nycticorax leuconotus* (Wagler)

PLATE 9.7

Other names: Martinete Encapuchado in Spanish; Bilhoueau à dos blanc in French; Weißrückenreiher in German.

Ardea leuconotus Wagler, 1827. Syst. Av., Ardea, No. 3: Senegambia.

Description

The White-backed Night-Heron is a shy, nocturnal, medium-sized dark heron with a large white eye ring on a dark head and white plumes on an otherwise dark back.

ADULT: The White-backed Night-Heron has a black head and a short, thick crest. Most strikingly is a large white patch and an eye ring of yellow skin that around a relatively huge eye. The bill is black, with a yellow lower base. The irises are yellow and the lores are green-yellow. The chin is white and the neck is rufous-brown. The back is darker, black-brown with long white lanceolate plumes forming the dorsal patch, which is conspicuous at times and inconspicuous at others, probably related to breeding. The upper wings are brown; the flight feathers are slate. The short tail is black-brown. The underparts are pale buff and brown; the belly is white-brown. The legs and feet are green to orange-yellow. In breeding, the irises turn red, chestnut, amber, yellow, probably in a sequence not yet sufficiently described. The lores become lemon-yellow.

VARIATION: The sexes are alike. Geographic variation is unclear, but South Africa birds have been described as *natalensis*.

JUVENILE: Immature birds are lighter than adults and have a dark brown streaked crown and forehead. The body is buff-brown with heavy white streaking and speckling.

CHICK: The chick is covered in olive-brown down with buff feather tips. The bill is green-grey. The irises are brown, the legs are green-grey, and the feet are yellow-green.

VOICE: The species is generally quiet. The growling 'Taash' call, rendered 'taash, taash, taash', is the disturbance call heard at night. The 'Kroak' call is the alarm and flight call.

WEIGHTS AND MEASUREMENTS: Length: 50–56 cm.

Field characters

The White-backed Night-Heron is identified by its relatively large, dark head and huge eyes accentuated by the white eye ring and its white back patch which is conspicuous in flight (Zimmerman *et al.* 1996). It is distinguished from immature Black-crowned Night-Herons by its smaller size, large head, white back patch, and more spotting on the back and wings.

Systematics

The generic identity of the species is not well settled. It has been variously assigned to *Nycticorax* and *Gorsachius*. The taxonomy of these night herons needs to be further examined to determine both the genus and species limits.

Range and status

The White-backed Night-Heron occurs in Africa, south of the Sahara.

BREEDING RANGE: It breeds widely from Senegal, Gambia, Guinea–Bissau, Cote d'Ivoire, Nigeria, Gabon, Sudan, Ethiopia, Congo, Tanzania (Baker and Baker 2004), Uganda, Botswana, Zambia, Zimbabwe (Jonasi 1994), Malawi (Nyirenda 1992), Mozambique, north Angola, and east South Africa (Raymakers 1992).

MIGRATION: The species appears to be sedentary, although it is likely that they undertake movements

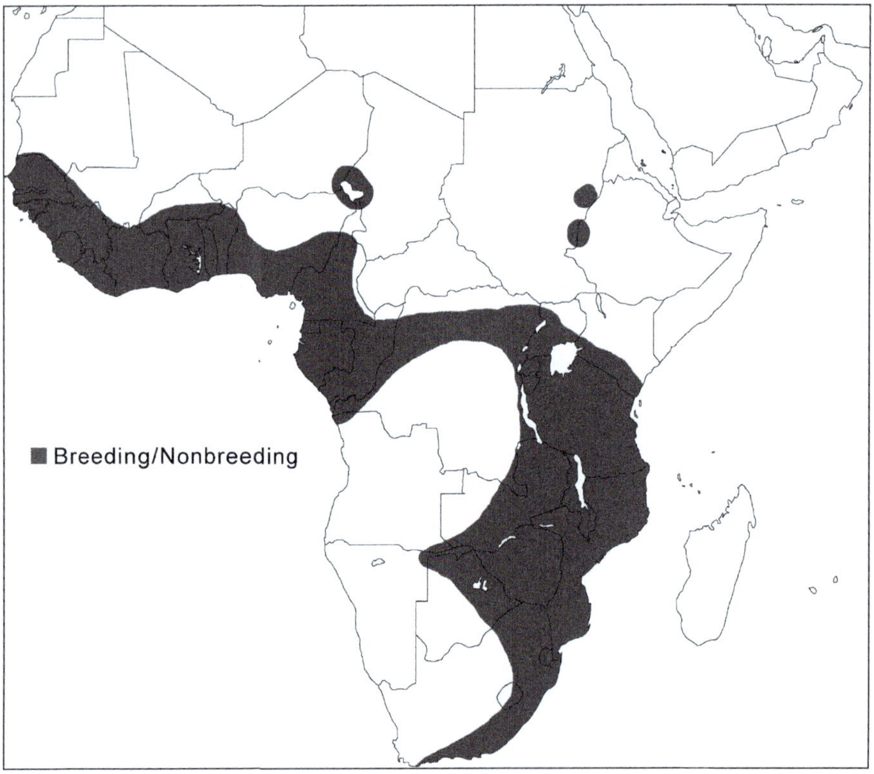

correlated with the onset of rains, but more information is needed on this secretive species.

STATUS: It is widely distributed through central and southern Africa but rare to uncommon. It may be most abundant in the Congo swamp forests and upper Zambezi in Zambia (Turner 2000).

Habitat

The White-backed Night-Heron is a solitary forest heron. It occurs in densely vegetated forests near water, especially along with tree-fringed streams, on islands in large rivers and lakes (Jonasi 1994), mangrove swamps along both African coasts, and forested lakes. It does also uses reed beds and dry forests. In places (Gabon) it is found close to human habitations, but this is not general. As a nocturnal species it requires secure day resting spots. It roosts in very dense marsh or forest vegetation, often high in trees.

Foraging

This appears to be very much a strictly nocturnal species. It has been observed feeding primarily at the water's edge, along muddy banks, and in the reeds along water courses. It feeds by Walking slowly along and in shallow water and by Standing on vegetation such as rafts of floating reeds. It feeds alone or in pairs in and near dense vegetation. Both adults and young assume a Bittern Posture on disturbance (Bock and Medland 1988, James 1988). Its diet is little known but it does consume fish (Blasdale 1984), amphibians, molluscs, and insects.

Breeding

The breeding season is variable, usually at the height of the local floods, which changes from place to place, and time to time. It nests in July–January and also March–April in South Africa, August–December and also January–June in Tanzania, May–June in Kenya, and September–October in Nigeria. The White-backed Night-Heron nests solitarily low in trees, bushes, reed beds, mangroves, and on rocky shores in caves (Edwards 1988). They are large and bulky platforms made of sticks or reeds lined with smaller pieces of the same material, 25–30 cm across. Nests are generally close to the water, usually within 1 m but sometimes 3 m. Nothing is known about courtship or nesting behaviour. The eggs are green-white, 45.7 × 35.1 mm. They are probably laid at intervals of 48 hours or more. The clutch is 2–3 eggs; the range is 2–5. The herons will sometimes double clutch. Incubation is 24–26 days. The young often grow to considerable size before leaving the nest at 6–7 weeks old.

Population dynamics

Nothing is known about the population dynamics of this species.

Conservation

The White-backed Night-Heron is a widespread but scarce bird, and is secretive, making surveys and monitoring difficult. A first step in conservation is to determine the species' distribution, habitat needs, and population status. This will require surveys throughout its range. The dependence of the species on swamp and riparian forests makes it vulnerable to the effects of harvesting of such forests throughout its range.

Research needs

Owing to its secretiveness, dispersed range, and difficult habitat this is one of the least understood herons. Studies are needed of its basic biology, including nesting, courtship (including the function of the dorsal plumes), foraging, food selection, dispersal, and habitat requirements. The evolutionary relationships of the night herons are not clear and a comparative study of the species is needed.

Overview

This night heron is a nocturnal, forest dweller that feeds secretively on the edges of cover tending to nest in the local flood season. Little is known about its feeding behaviours, food taken, or environmental correlates of foraging or nesting success.

White-eared Night-Heron *Gorsachius magnificus* (Ogilvie-Grant)

PLATE 9.8

Other names: Magnificent Night-Heron in English; Martinete Magnifico in Spanish; Bihoreau superbe in French; Hainanreiher in German; Hainan yan in Chinese.

Nycticorax magnifica Ogilvie-Grant, 1899. Ibis, p. 586: Five Finger Mountain, Hainan.

Description

The White-eared Night-Heron is a medium brown heron with a brown-streaked breast and a white patch on the side of the head.

ADULT: The adult male has a black crown with a brown-black crest. The rest of the head is distinctly patterned in white and brown patches. A brown-black cheek patch extending from the crown to the bill is separated from the cap by a white wedge extending back from the eye—the white ear of the species' common English name. A smaller white line descends diagonally from the eye. The chin is white and the upper throat also is white, but dusky centres to the feathers results in a mottled appearance. The irises are yellow, and the lores and skin around the eye are yellow-green forming a conspicuous pale patch in front of the eye. The bill is basically dark, the upper bill being black and the lower bill having a green-yellow tinge at the base. The front of the throat is vertically streaked down the middle, blacker towards the head and browner onto the breast. Three broad vertically oriented patches cover the area of the side to the back of the neck: lateral to the dark-mottled central throat stripe is a white line descending from the white of the lower cheek, lateral to that is a broad black-brown line ascending from the dark shoulder, and lateral to that is a broad buff-yellow patch merging to chestnut-orange at the back. The upper parts are dark grey-brown with a purple tinge, sometimes with a few white spots on the lower back. The flight feathers are slate. The brown stripe on the foreneck merges on the breast onto the mottled brown and white undersides. The thighs are dark reddish-brown. The legs are green.

VARIATION: In the female, the colour of the head and neck is less prominent; the back and wings are more mottled, streaked, and spotted with white, especially on the upper wing. The crest feathers are shorter than the male's.

JUVENILE: The juvenile has brown instead of brown-black feathering spotted with buff or white. The back and upper wings are brown with heavy buff or white spotting.

CHICK: The chick is undescribed.

VOICE: Nothing is known of the voice of this species.

WEIGHTS AND MEASUREMENTS: Length: about 54 cm.

Field characters

The White-eared Night-Heron is identified by its dark base colour contrasting with white ears and white throat, and by its short bill. The contrasting pattern on the throat and sometimes head is visible in flight (Fellows *et al.* 2001). It is distinguished from the Malayan Night-Heron by its tricoloured (not rufous) neck, dark (not chestnut) sides, and slate (not black) flight feathers. It is distinguished from the Black Bittern by its shorter, thicker bill, stouter appearance, and shorter neck.

Systematics

The White-eared Night-Heron is related to the Japanese Night-Heron, Malayan Night-Heron, and the larger night herons. The relationships among the night herons remain unclear.

Range and status

The present known range of this species is confined to southeast China and Vietnam.

RANGE: This is a species with the most restricted known breeding range of any heron. Observations in 1999–2004 have clarified the current status.

These have been sponsored by the Heron Specialist Group, Kadoorie Farm and Botanic Garden (KFBG), and BirdLife International (KFBG, H. Hafner, O. Pineau and A. Tordoff pers. comm.). Based on these surveys at present, there are two known breeding sites in the world, both located in south Guangxi Province, China, and two other areas where breeding is suspected, in Guangdong Province, China, and Bac Can Province, Vietnam. Its presence in Hainan has not been documented in recent years.

MIGRATION: It is most likely that this species is sedentary (BirdLife 2001). However, information is very uncertain. There are both winter and spring records from its most recent breeding areas in south China (Hubei, Fujian, Guangdong, Guangxi). But there also is a suggestion that dispersal occurred to the north prior to southern movement and that Hainan is, or was once, a wintering area for birds from farther north. More information on time of occurrance is needed.

STATUS: Very few individuals have been reported over the past 30 years. Of the two known breeding sites, the one near Fusui is highly degraded and is unlikely to support the heron for much longer. In the early 1990s the species was present in Hubei Province, in the Shennongjia Nature Reserve, but a re-examination of the site in 2001 revealed that what was prime habitat has been converted into a reservoir, and both the habitat and species have disappeared (KFBG, O. Pineau, and H. Hafner pers. comm.). As a result this area is no longer shown as part of the known range. In the early 2000's the bird was seen in Guangdong Province in the Chebaling Nature Reserve suggesting another breeding site exists there. In 2001, single birds were observed in Vietnam, the first record there in over 30 years (A. Tordoff, pers. comm.). It has not been

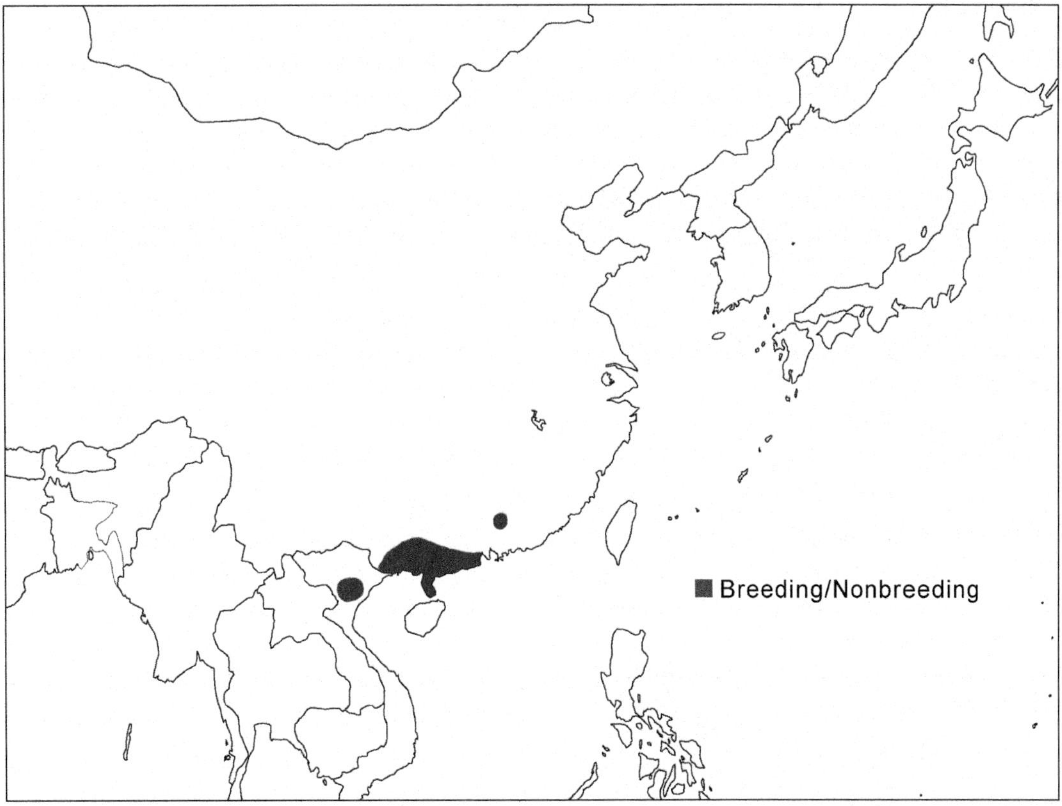

observed in recent years in other parts of its historic range (Zhou 1996, Gao *et al.* 2000, Fellowes *et al.* 2001, BirdLife International 2001, 2003, H. Hafner pers. comm.). There are no exact population figures, but the world population likely fewer than 1000 birds. Certain remote areas in south China, not yet surveyed, need to be visited in the near future. In these forested mountain habitats, difficult to access, isolated breeding sites may have gone unnoticed so far. This is the most endangered of all the herons (Kushlan and Hafner 2000).

Habitat

The White-eared Night-Heron is a forest bird. Optimal habitat appears to be extensive, dense, primary forests with streams and adjacent marshes. It currently is observed in mid-altitude mountains and has not been found higher than 1300 m. In Vietnam, it was seen in trees next to a stream in limestone forests. (A. Tordoff pers. comm.). However, it likely was originally also a lowland species, and its current restriction to hilly areas is probably because there are no remaining lowland forests (Fellowes *et al.* 2001). Given that very little pristine habitat remains within the historic range of the species, it is now observed primarily in degraded habitats in and near forest reserves. Recent observations have been near streams, a reservoir, rice fields, irrigation facilities. All feeding sites were located within 40 km of subtropical forests. These observations do not necessarily suggest that it can survive long term in altered environments, but rather that it has been forced into these sites because more appropriate habitat is now lacking (BirdLife 2001).

Foraging

This is a very poorly known species. Observations of the species' foraging are few (Caldwell and Caldwell 1931). Herons have been reported as feeding singly or in isolated pairs on the ground in ways likened to a bittern. They roost high in trees. All evidence suggests that they feed primarily at night, as they have been observed flying from roosts in the evening (Fellowes *et al.* 2001). However, daytime activity has also been reported (Zhou 1996). The diet includes fish, shrimp, and insects, but is very little documented (Zhou 1996).

Breeding

Nearly nothing is known about the breeding biology of this species. Various records of nesting activity indicate that it nests in trees including oak (*Quercus*), pine (*Pinus*), bamboo (although a nest was not found) and, likely, broad leafed trees (Caldwell and Caldwell 1931, BirdLife 2001). Information on nesting by this species is confounded by indications that it nested with Malayan Night-Herons. Presumed nests were very near the tops of the trees, 4–10 m above the ground near the trunks. Presumed nests were simple flat platforms (Fellowes *et al.* 2001).

Population dynamics

With nothing known about nesting, there is also no information on population biology.

Conservation

The White-eared Night-Heron is the world's most endangered heron; immediate action is required for its conservation. It is clearly endangered (Kushlan and Hafner 2000 IUCN 2003). Fortunately, Guangxi and Guangdong Forestry Departments and county governments are strongly committed to conserving the species (Fellowes *et al.* 2001). In Guangxi, the Fusui White-eared Night-Heron reserve was declared in 1999. In that only two breeding sites are known thorough surveys need to continue and expand. The recent surveys in southeast China were important first steps in identifying locations where birds remain (Fellowes *et al.* 2001). The surveys began by canvassing the local population, gaining public involvement, providing publicity, and following this with ground surveys. This model could be followed to great benefit in other areas where the species is expected. Where sightings occur, immediate protection of the area should be envisaged. Specific conservation issues include hunting, deforestation, reintroduction, and education. Hunting is a threat to this species, apparently even in nature reserves. Young herons are captured for the market. In fact, discovering an animal for sale in a Nanning market sparked the current conservation effort (Lee 1998). Hunting pressure of all kinds is intense in the rural agricultural areas that constitute the species' current range. Night herons (including both this and the Malayan Night-Heron) are trapped, shot, and noosed (Fellowes *et al.* 2001). In Fusaui, where empty night heron nests were found, villagers have reported that they killed adult birds at the nesting site the year before (BirdLife 2001). An immediate conservation effort should be undertaken to eliminate the use of this and other rare species to meet the strong local market demand for birds. This practice threatens all the herons in the area, especially those that nest colonially. The optimal approach would be to attempt to divert hunting from all the colonial species. The night heron is a nationally protected species in China.

The principal habitat threat to the species, in fact to other *Gorsachius* herons as well, is habitat loss, including deforestation, reforestation with pine monoculture, reservoir construction, and gold mining. Deforestation has been underway for centuries, and the remaining diverse forests are very confined, mostly in reserves. Most of the existing forest in the area is monoculture pine plantation. This species occurs in areas of south China recognized for their importance to a collection of endemic species, many of which are now endangered. Even in reserves, degradation and disturbance continue. Tourism is increasing in remaining reserved areas, also posing threats for this heron's well-being. Another very important threat, especially to this nocturnal species, is the numerous electric power lines concentrated usually along streams because the streams provide openings in the forest facilitating the erection of the lines. To provide habitat, truly protected forest areas are essential. At the nature reserves where the species still occurs, specific management plans are needed, including control of tourism effects, forest management and restoration, management of river bottoms and marshes as feeding habitat, realignment of reserve borders, creation of reserve corridors, and protection from hunting. When new breeding sites are found, they and the feeding habitat will require absolute protection. These qualify as Important Bird Areas for the species and need to serve as the basis of a network of reserves, managed sites and other conservation efforts aimed to protect the remaining birds. To the extent possible, a captive breeding programme should be considered. Fellowes *et al.* (2001) recommended that no more birds be taken from the wild. However, birds that are rescued from market could form the core of a captive population which could be used to re-establish or enhance the wild stock as sites are protected. Educational awareness of the existence and plight of the species among local populations is essential. The campaign that accompanied recent field surveys leading to the establishment of a reserve is an excellent example of meeting such needs. The species will not likely be saved except by local people.

Research needs

Clearly, with nothing known about the biology of the species, observational studies should be undertaken at every opportunity, particularly focusing on establishing the habitat requirements of the species. Surveys of known and suspected areas used by the species need to be undertaken to identify additional important areas. In order to render these surveys more efficient in the future, any opportunity should be taken to record vocalizations, for instance by captive birds. Once vocalizations are known, the periodicity of calls during the year, the day, and the night should be established in the known breeding areas. Surveys of this secretive bird in other forest habitat could then focus on the periods and hours with the highest probability of hearing a call. It is essential that the migratory status and year-round habitat needs of the species be understood. The potential feeding sites within flying distance of possible colony sites need to be determined.

Overview

The White-eared Night-Heron is the most endangered of all the herons, and biologically nearly unknown. Only a few have been seen and reported in recent years, and the world population likely numbers not many more than are presently known. This is fundamentally a lowland forest heron, which occurred in dense, wet, primary lowland forests with steams and adjacent wetlands and rice paddies. This habitat hardly exists at present. Herons that persist due so at the edge of their ecological tolerances in secondary forest in hills and mountains near residual forests. This is a species that seems to not tolerate disturbance well, and so requires reserves. Where found now it feeds in rice fields and irrigation systems. When rediscovered, the birds were in demand for the market in the area they are found, and their rarity did not provided much protection. The only hope for the species is for its welfare to become a concern of the local human population, its nesting sites be completely protected from disturbance, and feeding sites secured. The only successful conservation for this species will be that provided by and in cooperation with the local people.

Japanese Night-Heron *Gorsachius goisagi* (Temminck)

PLATE 9.9

Other names: Japanese Bittern in English; Martinete Japonés in Spanish; Bilhoreau goisagi in French; Rotscheitelreher in German; Mizo-goi in Japanese, Bakaw-gabi in Pilipino (Philippines). Litou Yan in Chinese

Nycticorax goisagi Temminck, 1835. Pl. Col. Livr. 98, pl. 582: Japan.

Description

The Japanese Night-Heron is a small, small-billed chestnut and buff heron.

ADULT: The Japanese Night-Heron has a chestnut cap with a short chestnut crest. The bill is relatively short and stout for a heron. The upper bill is dark brown; the lower bill is yellow. The irises are yellow and the lores are green-yellow, forming a notable patch in front of the eye. The sides of the neck are buff-brown. The upper parts are pale chestnut. The darker wings have blackish primaries with prominent broad tawny tips, giving the appearance of a black wing bar on an otherwise tawny wing. A pale patch on the forewing can be seen in good light. The undersides are brown with darker streaking and the flanks are slightly mottled. The legs are black-green. In breeding, the lores turn blue, and the bill turns green-black.

VARIATION: The sexes are alike.

JUVENILE: The juvenile is similar to the adult with a black-brown crown and less chestnut head. Its neck is more spotted and streaked. The upper wings are paler.

CHICK: The appearance of the chick has not been recorded.

VOICE: A croaking 'Buo' call, rendered 'buo, buo', is the principal call in feeding and breeding. It is probably a contact and identification call.

WEIGHTS AND MEASUREMENTS: Length: 49 cm.

Field characters

The Japanese Night-Heron is identified by its small size, tawny cap, short stout bill, and red-brown upper parts. It is distinguished from the Malayan Night-Heron by its chestnut (not black) cap, shorter crest feathers, shorter and stouter bill, and tawny (not white) tips on primaries. It is distinguished from the Cinnamon Bittern by its bicoloured back, chestnut wings, and thicker bill.

Systematics

The Japanese Night-Heron is closely related to the Malayan Night-Heron. The relationships among the night herons are unclear.

Range and status

This night heron occurs in Japan, south China, Indonesia, and the Philippines.

BREEDING RANGE: The Japanese Night-Heron breeds only in Japan (Honshu, Shikoku, Kyusgu, Izu Islands). It has also bred on Taiwan (Severinghaus 1989).

NON-BREEDING RANGE: Its principal wintering range is the Philippines (Luzon, Semirara, Leyte, Negros, Panay, Siquijor, Mindanao, Palawan). In the northern winter, the heron also occurs in south Japan (south Kyushu, Ryukyu Islands) and more rarely southeast China (Guangxi, Fujian, Zhejiang, Taiwan).

MIGRATION: The Japanese Night-Heron is highly migratory. It moves south at the end of October and occurs widely through Japan on passage. Some may remain in Kyushu but most move south through the Ryukas to and through coastal China And Taiwan to the Philippines. Return is in March–May. The species has dispersal tendencies, possibly overshooting on southward or northward migration. Birds occur in spring and summer in east Russia (Primorye, Sakhalin) and South Korea (Gore and Won 1971). Some birds remain in the Philippines in summer (Rand and Rabor 1960). It has occurred on Palau, Indonesia (Sulawesi, Halmahera), and Brunei (Borneo) (Elkin 1993). Given the paucity of records and difficulty of observing it, it should be sought after in these areas to determine if it might not be more regular than thought.

STATUS: The species is rare and extremely local in Japan. There is no exact population estimate available

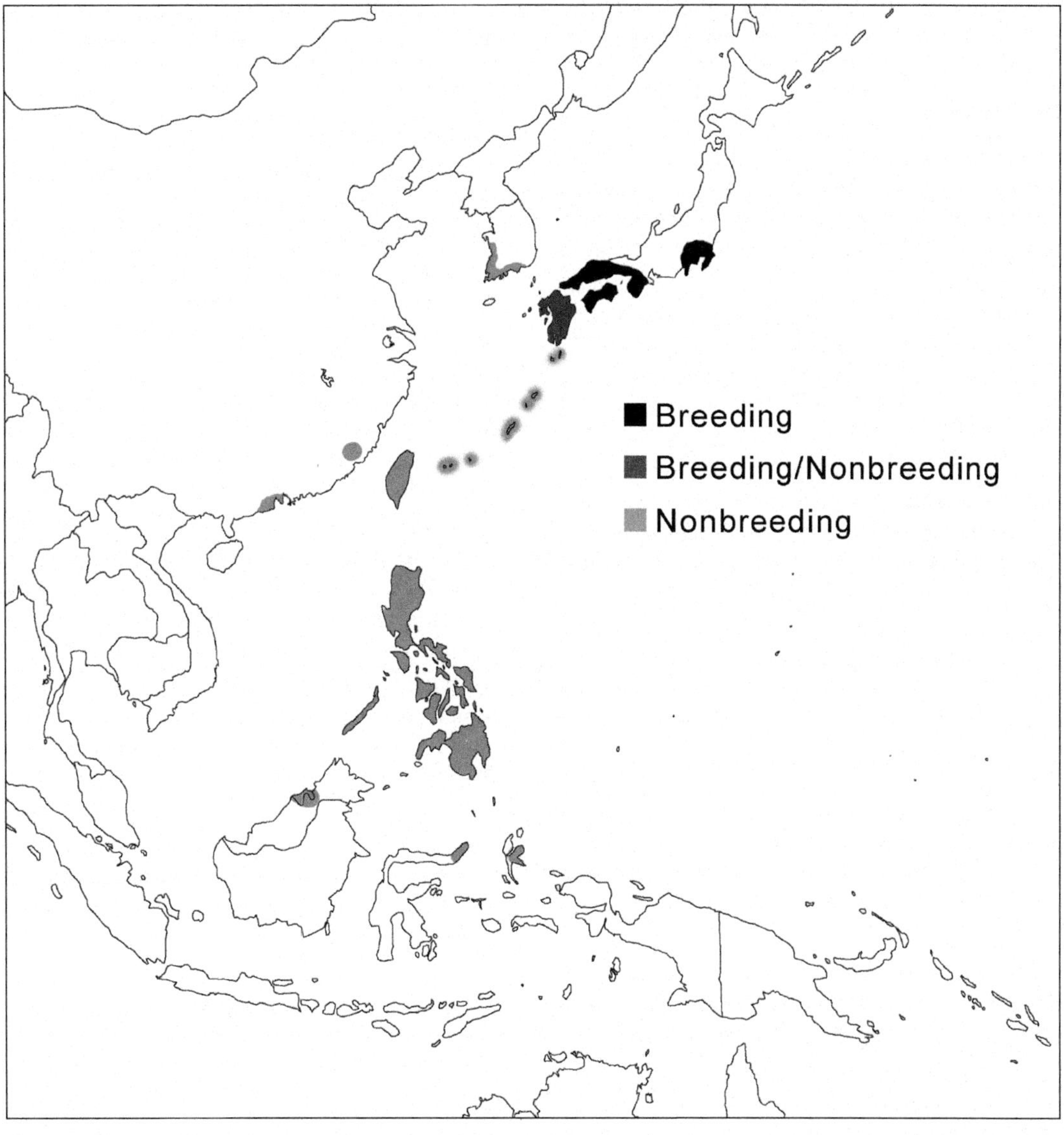

and it is not counted on surveys due to its secretive nature. Only six nesting sites are known (Lansdown *et al.* 2000). It is likely that the population numbers less than a thousand birds (BirdLife 2001). At the beginning of twentieth century it was more common and remained relatively so into the 1940s and 1950s. It disappeared from many traditional areas in Japan in the 1980s and 1990s. Although it was reported to have bred in Taiwan, it is now rare as a winter visitor and passage migrant there and in south China (Sykes 1996). It is rare but regular in South Korea. It is uncommon in the Philippines, its main wintering area (Dickinson *et al.* 1991). This is a species facing severe population decrease.

Habitat

This is a nocturnal forest heron. It occurs in heavily forested rivers, streams, and swamps in the low mountains. This habitat is now extremely scarce in Japan. Similar habitats appear to be used in the non-breeding range (Sykes 1996). It occurs generally from 50 to 240 m in elevation, but has been reported as high as 1500 m in Japan and 1350 m in the Philippines (with two records from as high as 2400 m).

Foraging

The Japanese Night-Heron is primarily crepuscular, feeding solitarily or in small groups. It also feeds during the day in cloudy weather. It feeds by Walking Slowly within the dense vegetation presumably along watercourses and pools but also on dry ground. It forages more rarely in the open. Habitats used include swamps, forest, farms, and rice paddies. The diet appears to consist mainly of crabs and possibly other crustaceans, insects, earthworms, and small fish.

Breeding

The Japanese Night-Heron breeds in May–July. It nests in tall trees, 7–20 m tall. Species include cedar (*Cryptomeria*), cypress (*Chamaecyparis*), oak (*Quercus*), zelkovi (*Zelkova*), and chestnut (*Castanea*). On migration in China it occurs in tall trees and bamboo forests along rivers and in the hills. The species nests solitarily. However, in favourable situations these nests can be gathered within relatively close proximity (250–500 m), although not truly a colony. Nests are flimsy, crude platforms of sticks, placed on more or less horizontal and usually densely foliaged boughs, 7–20 m above ground. Nothing is known of courtship display in this species, and little of its nesting habits. The incubating bird will adopt the Bittern Posture when alarmed. Eggs are dull white and measure on average 47.5 × 37mm. The clutch is 3–4. The incubation period was estimated to be 17–20 days. Nothing is known about the hatching, growth, or breeding success of this species other than fledging occurs in 35–37 days. Late eggs found in July suggest renesting.

Population dynamics

Nothing is known of the population dynamics of the species.

Conservation

The Japanese Night-Heron is endangered (IUCN 2003). With only six confirmed and 15 possible breeding sites (Lansdown *et al.* 2000), the species is certainly under threat of extinction. If the 21 sites in Japan each supported 10 pairs, which is unrealistically high, the total known breeding population could be as few as 200 birds. It is certainly less than a thousand birds. Conservation of this species faces two great challenges: lack of information and destruction of the species' habitat. The lack of information on the species constrains conservation action. The present status and distribution of the species must be determined as rapidly as possible. The 21 identified active and possible breeding sites should be re-examined for nesting, and those supporting nesting birds should be protected in a reserve system (Hafner *et al.* 2000). Threats to the known habitat of the species, both during breeding and non-breeding, are the principal conservation issues facing this species. In Japan, hill forests have been and continue to be deforested and converted to plantations. Remaining forests have developed a scrubby undergrowth (BirdLife 2001). Very little pristine forest habitat remains in Japan. Remnant stands need to be protected. Although the species is now found in degraded habitat, thoughts that this is an adequate habitat are probably incorrect. In the Philippines, lowland primary forest is similarly threatened by deforestation, rendering some areas totally unsuitable for the species. Protection of habitat on the wintering grounds is essential to the species' conservation. Hunting may in the past have been a factor in its conservation. They have been found for sale in Taiwan in the 1970s and *Gorsachius* herons have been confiscated in trade in the 1980s. Introduced predators have led to its decline on islands.

A species action plan covering both breeding and non-breeding areas is desperately needed for the species.

Research needs

Research is difficult given the secretive nature and tenuous status of the species. None the less, much could be learned from observational study. So little is known about the biology of this species, which represents an adaptive arm of the herons—a small nocturnal upland forest heron, that its basic biology needs to be understood and related to both conservation needs and heron biology. Surveys need to be conducted to locate breeding and wintering areas. Given the apparently small population size, any site with a few birds constitutes an important area for this species. These sites need to be identified and placed in a range-wide reserve network to be monitored, protected, and managed. The taxonomy of the night herons is not well understood, and a thorough analysis is needed.

Overview

Nearly nothing is known about the ecology of this rare and declining species. In fact, it is not understood what constitutes the basic ecological characteristics of a nocturnal forest heron. It is certainly a very rare species that has depended on a forested habitat that has been under severe threat on both the breeding and wintering areas.

Malayan Night-Heron *Gorsachius melanolophus* (Raffles)

PLATE 9.10

Other names: Malay Tiger Bittern in English; Martinete Malayo in Spanish; Bihoreau malais in French; Wellenreiher in German; Zuguro-mizo-goi in Japanese; Bakaw-gabi in Pilipino (Philippines); Kowak melayu in Indonesian; Helguan yan in Chinese.

Ardea melanocephala Raffles, 1822. Trans. Linn. Soc. London 13, p. 326: western Sumatra (probably near Bengkulu).

Description

The Malayan Night-Heron is a small chestnut and brown nocturnal heron of low- to mid-elevation wet forests.

ADULT: The crown and nape are black, ending in a long black crest. The bill is stocky, down-curved, dark brown above and green below. The lores and orbital skin are blue-green. It is possible that in some seasons the facial skin is greenish-yellow (sykes 1966). The irises are green-yellow. The sides of the face and neck are rufous. The chin is white with a central row of black streaks. The back and upper wings are chestnut, finely barred with black. The flight feathers are brownish-black with white tips, visible in flight. The tail is black. The underparts are dark brown finely speckled with black. The flanks and undertail coverts are mottled brown and white. The legs and feet are olive. During breeding the lores become blue. The males change well in advance of nesting, whereas the females change during courtship (Chang 2000). After courtship the lores fade to blue-green, to green, to grey-green.

VARIATION: The sexes are alike, but the male may have a longer crest and bluer lores in courtship (Tsao 1998). The species shows a high degree of individual variability in plumage. Previously three subspecies, differing in plumage colour, were recognized, but their variability appears to be within the overall range of individual variation of the species. Geographic variation does occur, however. Immature birds of the herons breeding on Palawan are buff, heavily overlaid with black. Herons from the Nicobar Islands are small.

JUVENILE: The juvenile is cryptically coloured, strikingly different from the adult. The upper parts are dull grey to brown spotted and barred with white and buff. The underparts are white, spotted, and barred brown. The crown is blackish, barred and spotted with white; facial skin is bluish-green; bill is horn.

CHICK: The chick has not been described.

VOICE: The 'Arh' call, rendered 'arh, arh, arh', is the flight call. The 'Kwok' call is a single low pitched call similar to that of the black-crowned Night-Heron (Sykes 1966). The 'Wu' call, rendered 'wu, wu, wu' is given during courtship.

WEIGHTS AND MEASUREMENTS: Weight: 417–450 g; length: 49 cm.

Field characters

The Malayan Night-Heron is identified by its rufous neck, barred chestnut back, black cap with crest, and white-tipped primaries. The Malaysian Night-Heron is distinguished from the Japanese Night-Heron by its chestnut brown (not darker) body colour, black (not tawny) crown and longer crest, slightly longer thinner bill, and white (not tawny) tips on primaries (Sykes 1966). It is distinguished from the Cinnamon Bittern by its short stubby (not long and pointed) bill, black crown, blue-green lores, larger size, barred cinnamon back (not plain cinnamon or brown with buff spots), and black and cinnamon wings with white flight feather tips (not uniformly cinnamon). It is distinguished from all bitterns by its white wing tips and stout bill. The immature Malayan Night-Heron is distinguished from the immature Rufous and Black-crowned Night-Herons by its short, stubby, and downward-arched bill, smaller size, finer spotting on the back and wings, and mottled (not striped) underparts. It is distinguished from the immature Striated Heron by its slightly smaller size, spotted (not plain) back, shorter, stubbier bill, and green (not yellow) legs and feet.

Systematics

The Malayan Night-Heron is closely related to the Japanese Night-Heron. The relationships among the *Gorsachius* night herons are unsettled. Patterns of individual and geographic variation, and the appropriate subspecific taxonomy, need to be clarified.

Range and status

The Malayan Night-Heron occurs in India, south-east Asia, the Philippines, and the east Indies.

BREEDING RANGE: The Malayan Night-Heron breeds in India (west India, Assam and Manipur, Nicobar Islands), Nepal, Thailand (north), Laos, Cambodia, Vietnam (Cambodia not certain), south China (Hainan, Guangzhou, Taiwan—Sykes 1996, Yunan, Gaunxi, Kwangsi, Guangdong (R. Lansdown pers. comm.), the Philippines (including Palawan), and Japan (Ryukyu Islands). Recent information reporting immature birds in Malaysia (Peninsular, Taman Negra), Java, and Borneo suggests nesting, but this needs to be confirmed (Perennou *et al.* 2000, R. Lansdown pers. comm.).

NON-BREEDING RANGE: It occurs during non-breeding in west India (Perennou *et al.* 2000), Sri Lanka, Thailand, Laos, Cambodia, Vietnam, south China (Hainan, Guangzhou, Taiwan), Malaysia (Malay Peninsula, Sarawak), Indonesia (Sumatra, Java, Kalimatan, Halmahara, Talaud) (Coates *et al.* 1997), the Philippines (to Palawan), and Japan (Ryukyu Islands).

MIGRATION: Because it is a solitary species in difficult habitat away from surveys and bird watchers, its migration is not very well understood. Evidence exists for both migratory and sedentary habit in most populations, including west India, Nicobar Islands, Indo-China, south China, Taiwan, the Philippines (including Palawan), and south Japan. It is likely that only partial migration occurs in northern populations such as Taiwan and India, but this needs to be confirmed. Some west India birds migrate south in November–December through Sri Lanka (which is not the important wintering ground it once was thought to be), to Nicobar, Malaysia, and across the Straits of Malacca to Indonesia. East India, north Myanmar, and Thailand birds migrate south in August–October, presumably through Myanmar also to and through Malaysia to Sumatra, Java, and Borneo. Return migration through the Malay Peninsula is in April. Whether postbreeding dispersal occurs in unclear. Birds have been seen outside the usual range, but not very far

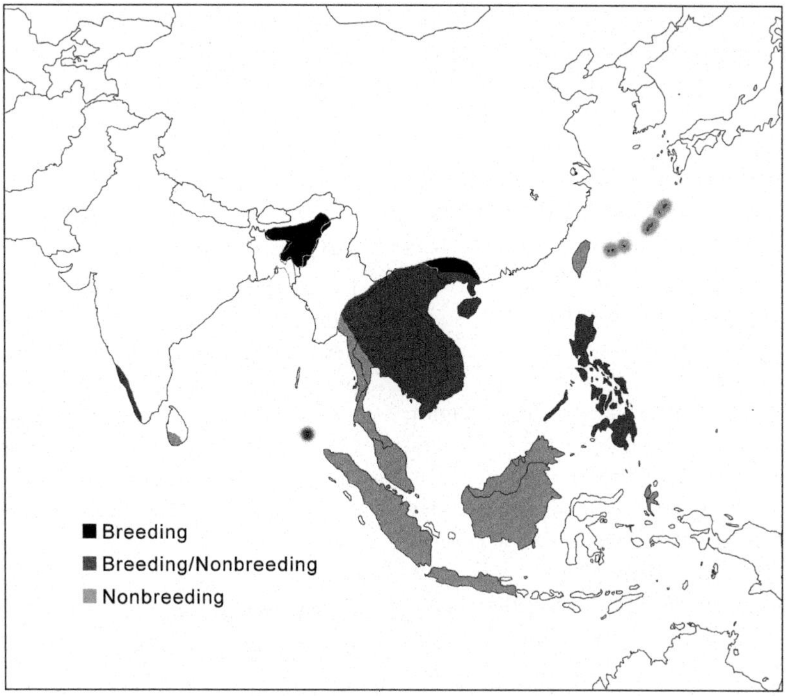

(Shikoku Island), Palau Island, Indonesian islands (Bandai) and Christmas Island.

STATUS: No data are available on population sizes. Recent information suggests that the stronghold of the eastern population is in North Thailand, Vietnam, and Laos (Perennou *et al.* 2000). It is uncommon throughout much of its range. It is local and uncommon in both seasons in Japan (Ryukyu Islands) (Wild Bird Society 1982), uncommon in the non-breeding season in Sri Lanka (Perennou *et al.* 2000), uncommon in both seasons in China (mainland, Taiwan—Sykes 1996), uncommon in the Philippines in both seasons, and rare in the non-breeding season in Java (McKinnon and Phillips 1993).

Habitat

This is a heron of mature, high-rainfall, subtropical forests. But it also uses streams, marshes, swamps evergreen forests, secondary scrub, and reservoirs. It occurs as high as 800 m in west India, 1200 m in Thailand, 1800 m in Sri Lanka, and 2300 m in east India. It uses reed beds in migration and is often reported in human environments, particularly flooded rice fields, pastures, and vegetable gardens, and nesting near houses.

Foraging

The Malayan Night-Heron feeds by both day and at night. It appears to be preferentially a bird of deeply shaded forest floor. It may be more often observed feeding in the open, but this is likely due to its visibility in such situations. It feeds by Standing in a characteristic posture with its neck and bill pointing forward, and also by Walking slowly along the forest floor, and along the edges of water bodies and the edges of fields. In the forest, it alternates Walking and Standing (Sykes 1996). It catches fish in aquatic situations, but in the forest it seems to specialize on capturing frogs and very large (10 × 300 mm) earthworms. It catches the worms by probing them from the soil (Chang 2000), whereupon it nips them apart with a headshake and swallows them in pieces.

It defends itself with Forward displays, in which it raises its crest, opens its wings, and Stabs at its opponent. When disturbed while foraging, it runs away a little distance, stops, and flies to a hiding place nearby. Based on the observation that it doubles back when so disturbed, it appears to hold territories. When not feeding, it roosts well hidden in reed, bamboo, and other dense vegetation.

Terrestrial prey tend to dominate the known diet, especially earthworms and frogs, but also beetles, mollusks, crab, snakes, lizards, birds and small fish (Shen and Chen 1966, Yao *et al.* 2000, Yao 2002).

Breeding

The Malayan Night Heron nests in May–August, during the big rains, in India, May–June in Assam, and April–September in Taiwan. In the latter location, courtship calling occurred in February–May, nesting in April–July. It is sometimes, a colonial nester, alone or with other species, with nests tightly packed–as many as 20 close together. It also nests solitarily, near but not in the same nest as in previous years. The nest is a small, fragile, flat platform of sticks sometimes lined with leaves and grass, well placed in a fork of a tree, 5–10 M high under the canopy. In Taiwan, the mean nest height was 6.8 M, the mean diameter 0.7 M × 0.59 M, and mean thickness 0.105 M. There does not seem to be much preference for tree species. It occasionally is reported to nest in reed beds. Both birds of a pair build the nest.

Courtship behaviour is undescribed, but it gives its monotonal guttural honk call, the Wu call at dawn and disk and during the night prior to nesting, and ceases calling when nesting begins (Yao *et al.* 2000, Yao 2002). It appears to be typical for the species to nest while still in juvenile plumage (Chang 2000). Nest building takes nine days. Eggs are laid 2–3 days apart. The eggs are chalky white with green tinge. A single measurement was 46 × 37 mm. Clutch is usually 3–4, range 3–5. Egg laying takes 10 days and incubation 32–34 days (Yao *et al.* 2000). Parents brooded for 30–32 days; Young fledge at 43 days. Brood reduction is reported in Taiwan, with the youngest chicks being ejected from the nest by older ones (Chang 2000). Both parents attend the young, which are fed mostly during the day but also at night. Food was regurgitated into the chick's mouth. Fledging success ranged from 0 to 100% with a mean of 40.3% in Taiwan, with a mean of 1.3 young fledged per clutch (Yao *et al.* 2000). Juveniles remain with parents after fledging. Young birds have also been reported to help at the nest in feeding young (Chang 2000), an unusual observation in herons.

Population dynamics

It is an intriguing observation that birds in juvenile plumage nest (Chang 2000). The observed birds nested months prior to moulting into adult plumage, so they were probably two years old; however, the sequence of juvenile moults is not known. Nothing is known about the population biology of this species but it appears young return to their hatching area (Yao *et al.* 2000), and mature in 2–3 years.

Conservation

Populations, except that of Nicobar, are vulnerable due to habitat loss (Heinz *et al.* 2000). The deficiency of data on this species is a critical impediment to conservation action. No doubt its habitat, like that of other *Gorsachius* herons, has been reduced due to deforestation and degradation. But the habitat needs of the species are not well understood, given that it is often reported to be using human-made habitats, and on Taiwan it nests in parks and near human residences (Chang 2000). It has been proposed that the species might be an indicator of environmental quality in wet tropical and subtropical forests, but this needs to be demonstrated (Perennou *et al.* 2000). Given its inaccessible habitat and secretive habits, it may be that the species is much more widespread and abundant than thought, particularly if it is accustomed to human-managed habitat as now is becoming clear in Taiwan. Intensive year-round surveys need to be conducted in suitable habitat in order to determine its status in various places.

Research needs

Much of the basic biology of the species remains to be determined. Information coming out of recent observations hints that the species may be of considerable biological interest. Nesting in juvenile plumage, helpers at the nest, and multiple clutches within a season are all very unusual among herons. This is a species that could reveal new elements of heron behaviour. Range-wide, year-round surveys of its annual occupancy of potentially suitable habitat need to be carried out to provide the basis for a conservation strategy.

Overview

The situation is much like that in the Japanese Night-Heron. So little is known of this heron that much of its basic biology is speculative and it is not certain what constitutes the essential characteristics of a nocturnal low- to mid-elevation wet forest heron.

South American Bittern *Botaurus pinnatus* (Wagler)

PLATE 12.3

Other names: Pinnated Bittern in English; Avetoro Mirasol, Garza Tamboruda, Mirasol, Mirasol Grande in Spanish; Socó-boi-marron, Socó-boi-baio in Portuguese; Butor mirasol in French; Südamerikanische Rohrdommel in German.

Ardea pinnata Wagler, 1829. Isis, col. 662: Bahia State, Brazil.

Subspecies: *Botaurus pinnatus caribeaus* Dickerman, 1961: Cantemo, Tabasco, Mexico.

Description

The South American Bittern is a stocky, medium brown and buff, cryptic heron with a sooty-brown forehead and barred crown, white chin, and broadly brown-streaked neck and breast.

ADULT: The adult male South American Bittern has a sooty-brown forehead with dark vermiculated bars over buff on the crown and neck. The sides of the head are ochraceous-buff. The bill is pale yellow with a dusky edge and tip. The irises also are pale yellow and a bright yellow streak at its base merges with the yellow-green loral skin and yellow line over the eye. The chin and upper throat are white. The back and upper wings are buff barred and spotted with sooty-black giving a variegated look. The flight feathers are sooty-black. The foreneck and sides are white, vertically barred with rufous-brown. The sides of the neck create an expansible ruff, which while typical of the large bitterns, is particularly prominent in this species. The undersides are white to yellow-buff streaked with dark rufous-brown. The feathered upper legs are buff, barred in light brown. The legs and feet, with very long toes, are yellow-green. Courtship-related changes have not been described; however, reports of orange-brown and cinnamon-rufous irises seem likely related to courtship (Wood 1986).

VARIATION: The sexes are similar but females average smaller. Geographic variation in the apparently disjunct population in south Mexico is recognized as the subspecies *caribaeus*. Compared with *pinnatus*, this population is generally paler, less ochraceous, has sand buff-sides of the face, less buff, almost white undersides, less streaking on the lower throat, and a bill averaging shorter but with overlap (Dickerman 1972). Considerable individual variation also exists throughout the range of the species.

JUVENILE: Immature birds resemble adults, but are paler, and have less bold barring on the upper parts.

CHICK: The chick is not described.

VOICE: The 'Boom' call is the booming roar characteristic of the large bitterns, rendered 'oong, kachoonk', used in territorial defence and advertising. The South American Bittern's Boom call is higher pitched and less hollow than the pumping call of the North American Bittern (Birkenholz and Jenni 1964). The 'Kwak' call is the alarm call, rendered 'kwaak, kwaak'. The 'Ro' call, rendered 'ro, ro, ro', is the call used when taking flight.

WEIGHTS AND MEASUREMENTS: Weight: 584 g; length: 65–76 cm.

Field characters

The South American Bittern is identified by its buff and black-barred appearance. In full flight the wing beats are noticeably quick. It is distinguished from the North American Bittern by its black-barred (not plain buff) crown, rufous-buff (not grey-buff) base colour, barred and vermiculated (not finely spotted) upper parts, larger size, lack of a black mustache, and higher pitched 'Boom' call. It is distinguished from immature night herons by being larger, its barred (not light spotted) back, buff (not brown) base colour, barred (not spotted) wings, stocky stature, and even more rapid wing beats when in full flight. It is distinguished from the Rufescent Tiger-Heron by its buff, black-barred neck, and stockier stature. It is distinguished from the Bare-throated Tiger-Heron by its sooty-brown forehead and buff barred (not black) cap, ochraceous-buff (not grey) cheeks, smaller size, and stockier build. It is distinguished from the Fasciated Tiger-Heron by its buff (not slate) colour, courser barring, sooty-brown and buff barred (not black) cap, and stockier build. It is distinguished from the immature tiger herons by its stockier build and appearance, its fine black barring becoming streaked on the upper parts (not bold, broad black and buff banding in early juvenal plumages), darker sooty-brown crown, buff base colour, longer toes, and marsh habitat.

Systematics

The South American Bittern is one of the four large *Botaurus* bitterns, which all have streaked brown plumage, scutellate tarsi, 10 tail feathers, and a booming call. It differs morphologically from the North American Bittern and from the two Old World bitterns (Payne and Risely 1976). It is likely more closely related to the North American Bitterns than to the Old World species (Payne and Risely 1976, Sibley and Monroe 1990). As currently understood, the species has three disjunct populations, one of which has been identified as a distinctive subspecies.

Range and status

This bittern occurs in central and north South America and Central America to Mexico.

Breeding Range: *Pinnatus* occurs in El Salvador (San Miguel), southeast Nicaragua (Managua, Río San Juan), east Costa Rica (Río Frío, Guanacaste, Turrialba), Colombia (Cauca, Patia, Boyacá, Cundinamarca, Arauca, Meta), west Ecuador, east of the Andes in south Venezuela (Mérida, Aragua, Guárico, Delta Amacuro, southeast Bolivar), Trinidad (Belcher and Smooker 1934), the Guyanas, east and south Brazil (Pernambuco, Mato Grosso, Rio de Janeiro (Teixeira and Nacinovic 1985), São Paulo, Uruguay (Blanco and Canevari 1992), Paraguay (Itapua) (Barnett *et al.* 2002) north Argentina (Formosa) (Di Giacomo 1997), Córdoba, north Buenos Aires). *Caribeaus* occurs in south Mexico (Dickerman 1961, 1972) (Tamaulipas, Tabasco, Veracruz, Oaxaca (Schaldach *et al.* 1997), Chiapas, Yucatan, Quintana Roo, Campeche), and Belize (Wood 1986). There are a few records from northwest Guatemala. The species has not been reported in El Salvador, Honduras, or Panama, creating gaps in its known distribution (Ridgely and Gwynne 1993).

Migration: There is no evidence of migratory movements in this species in the north of its range. However, in the southern portions of north Argentina, Uruguay, Paraguay, and south Brazil, it has not yet been reported outside the breeding season. Migration would be expected in more temperate areas based on the biology of the other large bitterns.

Status: Throughout its range, its known status is rather localized. It is common in some localities in Mexico. Its status in Brazil is unclear but is reported throughout the country (Sick 1993), but its occurrence in the Amazon and Mato Grosso has not been recently confirmed (Morales 2000). It is rare to uncommon in north Argentina, its first confirmed nesting having been only recently reported (Di Giacomo 1997).

Habitat

The typical habitat is dense, flooded fresh water marshes and swamps, thickly overgrown with tall reeds or rushes. They also occur in wetlands bordering rivers, rice fields, and sugar cane fields. Altitude is not restrictive in that it occurs up to 2480 m in the Andes.

Foraging

The South American Bittern feeds by Walking slowly, creeping forward in a very deliberate

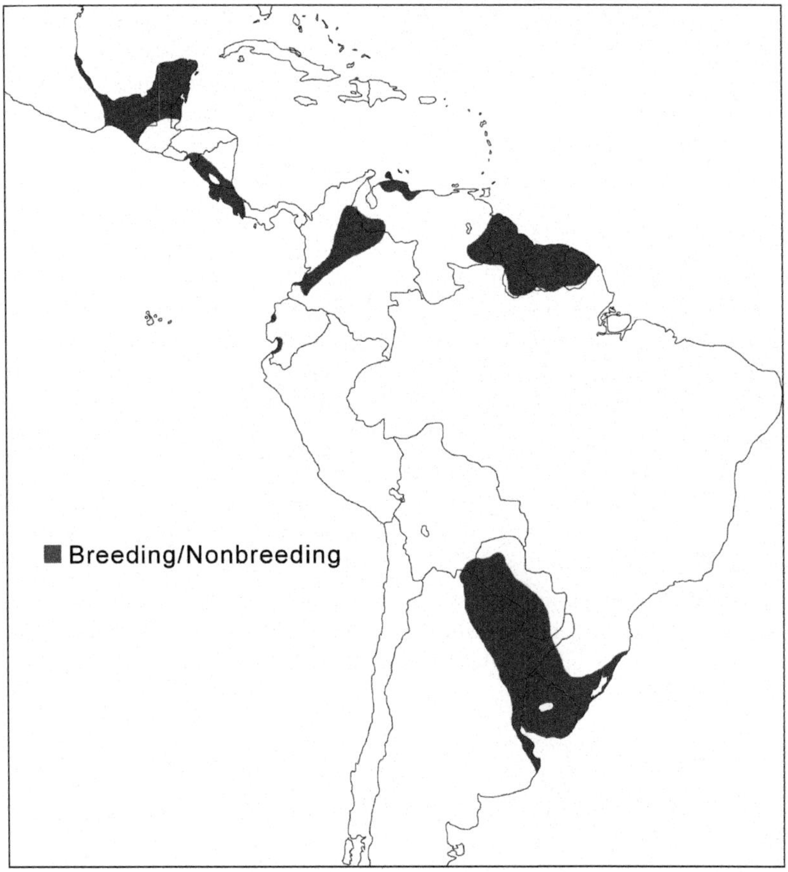

manner in Crouched posture, with its body almost parallel with the ground. It seems that it mostly feeds at night, and feeds alone or in loose groups. The bittern's diet consists primarily of fish (*Myrophis*), frogs (*Pseudis*), snakes (*Ophis*), and insects such as dragonflies (Odonata) and bugs (Hemiptera).

Breeding

The nesting season is variable across its range and not well documented. It is July–October in Trinidad, May–July in Mexico and Central America, February–April in Brazil, and October in north Argentina. They nest in herbaceous marshes with tall, dense vegetation over standing water or waterlogged grassland. Solitary nests are situated within dense plant cover including *Typha, Scirpus, Phragmites, Paspalium*, and *Eryngium*. The nest is a platform built of dried reeds and other plants. In Argentina, a nest was 35 cm wide and only 3 cm deep (Di Giacomo 1997). It is built into the herbaceous vegetation just above water level. Detailed descriptions of nests or range of placements are scarce. Breeding displays have not been well reported. The Boom call is given in the nesting territory as both a territorial and advertising call. Up close males defend their territory with displays that include erecting their neck ruff (Teixeira and Nacinovic 1985, Nacinovic *et al.* 1986). Eggs are reported to be olive and straw-yellow, measuring 49.5 × 36.3 mm in Trinidad and 51.2 × 40.3 mm and 52.4 × 40.7 mm in Argentina. The clutch is 2 or 3. Incubation may be only by the female, but the details and incubation times are unrecorded. Very little is known about the nesting biology of the species.

Population dynamics

Nothing is known about population biology or demography.

Conservation

The status and actual dispersion of the South American Bittern are very poorly known. It is widespread over a large range so there is no evidence that it is threatened. Because it is seldom seen, it is likely more common than presently documented. The first step of conservation action must be to better define its dispersion, breeding population status, seasonal movements, and habitat requirements.

Research needs

Little is known about the behaviour and ecology of this species, including such basic information as precise habitat requirements, breeding behaviour, degree of polygamy if any, nesting season, food and foraging, and demography. Study of the basic biology of the species is needed. The actual range of the species in Central and South America requires clarification, particularly in the apparent range gaps in north Central America, Panama, the Amazon, and also in west Brazil. As more specimens become available, the subspecific taxonomy, especially of the apparently disjunct populations, should be re-examined. It is not clear whether the species, particularly southern populations, are migratory. Seasonal movement patterns need to be better understood.

Overview

Like the other large bitterns, this is a species of deep, dense reed swamps and marshes. They feed alone or in small groups by Standing and Walking on the vegetation waiting for fish and other prey to appear. Much remains to be understood about this species.

North American Bittern *Botaurus lentiginosus* (Montagu)

PLATE 12.2, 18.1

Other names: American Bittern in English; Avetoro Lentiginoso, Mirasol, Ave Torro, Yaboda Americana, Guanabó Rojo, Tocomón in Spanish; Butor d'Amérique in French; Nordamerikanische Rohrdommel in German.

Ardea lentiginosa Montagu, 1813. Suppl. Orn. Orn. Dict., unnumbered text and plate: Piddletown, Dorsetshire, England.

Description

The North American Bittern is a stocky, medium brown heron marked in grey and black with a plain rusty-brown crown, white chin, and black moustache.

ADULT: The crown is plain rusty-brown. The bill is dull yellow, with a broad dark top line and tip. The irises are yellow. The loral area is dark with a green-yellow streak with a cream-buff line extending above the eye. The sides of the face are brown and a black moustache streak runs from the gape down the throat and becomes a more triangular cheek patch on either side of the neck. The throat is white with rufous streaks outlined in black shading to buff on the breast. Feather tufts, on either side of the breast, are white. The hind neck, back, and upper wings are rich brown finely freckled, spotted and vermiculated with grey and black. The underwing is dark brown and the flight feathers are unbarred black to dark brown and contrast with the rest of the wing. The feathered thighs are buff. Its relatively short legs are yellow-green.

VARIATION: The female is similar to the male in plumage, averaging smaller (Azure *et al.* 2000a). The species shows a high degree of individual variation that is not geographically based (Parkes 1955).

JUVENILE: The young are similar to the adults but lack the cheek patches.

CHICK: The chicks have yellow-olive down, and pink bills with black-tips. The irises are olive.

VOICE: The 'Boom' call is the loud booming or pumping call characteristic of the large bitterns, rendered for this species as 'oong, kach, oonk', or 'pump, er, lunk' with the accent on the last note. It may be preceded or followed by a series of soft gulping sounds. The Boom call of this species is invariably described as resembling the noise of an old-fashioned hand plump, or of hammering a stake into muddy ground, leading to two of the bird's nicknames—'stake-driver' and 'thunder pumper'. It is lower pitched than the similar call of the South American Bittern (Birkenholz and Jenni 1964). The 'Kok' call is the alarm and flight call, rendered as a hoarse 'kok, kok, kok' or a nasal 'haink'. The 'Chu-peep' call is the copulatory call (Gibbs *et al.* 1992b).

WEIGHTS AND MEASUREMENTS: Weight: 550–1072 g; length: 60–80 cm.

Field characters

The North American Bittern is identified by its stocky build, medium size, brown base colour with grey and black markings, plain grey-buff crown, black moustache, broadly brown streaked neck and breast, and finely vermiculated and spotted upper sides. It flies rather heavily and hurriedly, using rapid wing beats. It is distinguished from Old World large bitterns by its finely spotted and vermiculated upper parts, smaller size, longer, thinner bill, longer neck and less stocky build, darker upper bill, browner base colour, and brighter green-yellow soft parts (Lansdown 2000). It is distinguished from the South American Bittern by its plain buff (not black barred) crown, brown (not rufous buff) base colour, fine spotted, freckled, and vermiculated upper parts (not barred), smaller size, black moustache, and lower-pitched Boom call. It is distinguished from immature night herons by being larger, lighter brown, having a finely freckled and vermiculated (not light spotted) back, freckled (not spotted) upper wings, contrasting wings with dark flight feathers and paler body, and faster, steadier wing beat. It is distinguished from the Rufescent Tiger-Heron by its brown (not red) neck, black moustache, and stockier stature. It is distinguished from the Bare-throated Tiger-Heron by its grey-brown (not black) cap, black moustache along the cheeks, and stockier build. It is distinguished from the Fasciated Tiger-Heron by its brown (not slate-black-brown) colour, black moustache, grey-brown (not black) cap, and stockier build. It is distinguished from the immature tiger herons by its stockier build and appearance, its fine black and grey freckling and vermiculation on brown upper parts (not bold, broad black and buff banding in early juvenile plumages), general brown base colour, longer toes, and marsh habitat.

Systematics

The North American Bittern is one of the four large *Botaurus* bitterns, which all have streaked brown plumage, scutellate tarsi, 10 tail feathers, and a booming call. It differs morphologically from the South American Bittern and from the two Old World large bitterns (Payne and Risely 1976). It is likely more closely related to the South American Bittern than to the Old World species (Payne and Risely 1976, Sibley and Monroe 1990).

Range and status

The North American Bittern occurs in North America to south Mexico, and the northern West Indies.

BREEDING RANGE: The North American Bittern breeds well north in southeast Alaska, Washington, across Canada (British Columbia (55°N), Northwest Territories (at 61°N), Manitoba, Ontario, Quebec, Newfoundland—Etcheberry 1990), south through all but interior USA to California, Nevada, New Mexico, north Texas, central Kansas, central Missouri, west Tennessee, west Kentucky, Ohio, south Pennsylvania, northeast West Virginia, Maryland, and east Virginia. It also has been recorded in the past as breeding in Florida, Arizona, Louisiana, and central Mexico (Banks and Dickerman 1978, Howell and Webb 1995).

NON-BREEDING RANGE: During the non-breeding season, the North American Bittern occurs where waters do not freeze over, from the coastal and south

USA to Central America and Greater Antilles. It occurs in British Columbia, in west coastal USA (west Washington, west Oregon, California), south and east USA (Arizona, New Mexico, Texas through Florida north to New Jersey and infrequently farther north) through Mexico and Guatemala (Land 1970), also Bahamas and the Greater Antilles (Cuba, Cayman Islands, Jamaica, Hispaniola, Puerto Rico). Its non-breeding status in south Mexico and farther into Central America requires additional documentation. It appears to be irregular in Guatemala (Land 1970). Although its winter range is usually noted to extend to Costa Rica, where it may have once been infrequent (Stiles and Skutch 1989), and Panama, this is based on a few very old records (G. Alvarado pers. comm., Ridgely and Gwynne 1989). Additional information on status is needed from south Mexico through Central America in order to determine if its usual wintering range extends into these areas.

MIGRATION: North American Bitterns migrate south in the northern winter to below the freeze line. Southward movement is September–November; return is February–mid May. Migration routes are not known. Migration is nocturnal, alone or in small groups.

Extensive postbreeding dispersal occurs starting in July. Birds occur north and south of the usual breeding and wintering ranges in Labrador, Greenland, Virgin Islands, Montserrat, Guadeloupe, Martinique, and Barbados but also occurs at great distances, especially on Atlantic islands and in Europe-Azores, Greenland, Canaries, Iceland, Clipperton Island, Faeroes, England, Scotland, Ireland, Norway, Spain (Brazier 1983, Madge 1983, Holt 1991, Skipnes and Folvik 1997,

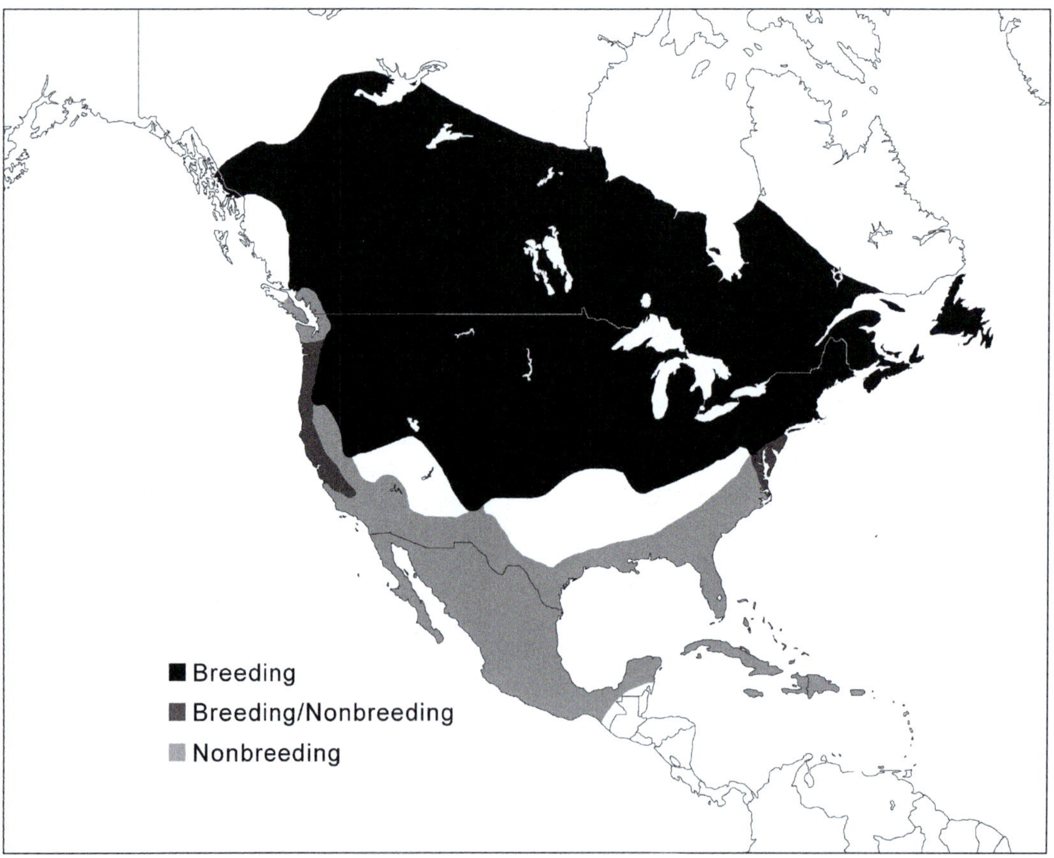

Ortvad *et al.* 2002). The bird's occurrence east of the Atlantic is nearly routine. It has occurred over 60 times in Britain from 1957 to 1998 (Lansdown 2000), and in fact, the species was first described in 1813 from a vagrant specimen captured in southern England.

STATUS: The North American Bittern is most abundant in Canada and less so in the more southerly portions of the range during breeding (Gibbs and Melvin 1992). In winter it is most abundant in the southeast USA, Florida, Louisiana, and the southwest (Salton Sea and San Joaquin Valley California). It is uncommon and local in Cuba and larger islands of Bahamas, rare in Cayman Islands, Greater Antilles, Costa Rica, and Panama. Data suggest that a significant population decline is occurring in the USA, especially in the north central states (Hands *et al.* 1989b). Canadian data show no downward trend. Wintering populations may be declining in Louisiana (Fluery and Sherry 1995). This trend is probably part of a long-term change of range and status, fundamentally a result of climatic warming. Records from the early 1900s of breeding in the southern tier of states and central Mexico and reduction in the records from Britain may all be part of this range shift northward.

Habitat

The North American Bittern, like the other large bitterns, is a bird of the herbaceous marshes. Although the marshland habitats of this bittern differ in detail over its extensive breeding range, they have common features of tall vegetation, shallow (5–20 cm) fresh to brackish water. Marsh plants include *Typha, Scirpus, Mariscus, Sparganium*, and *Polygonum*. It actually has a wider than expected range of habitat preferences in fresh or brackish marshes, including coastal marshes, mangrove swamps, open meadows, and even relatively dry grasslands.

Foraging

North American Bitterns feed by Standing or by Walking extremely slowly. When Standing, the bird holds its bill horizontal as it watches beneath its head, moving the head very slowly, with bill held low to the water. In Walking, each foot is raised slowly as the bird advances. A similar extremely slow movement is made in changing positions slightly while Standing. From its low position, the bill strikes out rapidly Stabbing at the prey. It can also Walk Quickly or Run, even in tall grass. It uses Gleaning to take insects from grass stems and Standing Flycatching to catch them in the air. It frequently uses Neck Swaying, often synchronized with the wind. This bittern is a very solitary feeder and has an extremely fierce-looking Forward display. It crouches with wing half spread and bill forward. When another bird or other animal approaches, it quickly assumes a Bittern Posture with up-pointed bill and neck stretched out. Often it gently sways in this posture like a reed, but perhaps also to help vision. It also uses Supplanting Flights and Supplanting Runs. It defends its territory not only from bitterns but from other fish-eating species as well (Phillips 1983). It feeds in shallow water, usually along the fringes of marshes and from banks along the shoreline, tending to avoid deep marsh interiors and older and overly dense vegetation. It feeds crepuscularly but also during the day and at night. The daily cycle needs to be better determined.

Although this species' food is diverse, fish dominate most diets. These include *Perca, Esox, Lepomis, Ictalurus, Anguilla*, and *Fundulus*. They also eat crayfish, snakes (*Thamnophis*), tadpoles and frogs (*Rana, Hyla*), salamanders (*Ambystoma*), insects (dragonflies, water bugs, water scorpions, water beetles, grasshoppers), small mammals such as ground squirrels, pocket gophers, voles and shrews, and birds such as a Sora (*Porzana*) and sparrows (*Passer*) (McBride 1993, Austin and Slivinski 2000).

Breeding

The breeding season is the spring, April–June, earlier in the south and later in the north. Eggs are laid early May in Michigan and New York, June in Quebec, and May in Minnesota (Gibbs *et al.* 1992b). Bitterns nest in dense cover of herbaceous marshes. They also can nest upland just adjacent to the wetland. They use both small and large marshes, although there is evidence in some areas of preferring larger tracts (Brown and Dinsmore 1986). They occupy various sorts of marshes such as beaver ponds, artificial impoundments, lake edges with floating and emergent plants, and especially shallow cattail marshes. The American Bittern nests solitarily, in defended territories. Nests may on occasion be found within 40 m from one another. Calling-males density figures range from 2.6 to 40 per 100 ha (Gibbs *et al.* 1992b). The nest platform is 25–40 cm. It is composed of the adjacent marsh plants and is built up from the water or lodged within the vegetation just above the water (8– 20 cm). It also nests on firm ground, but nearly always in dense cover. The comings and goings of the bittern make flattened approach paths through the reeds or rushes. It is thought that the female builds the nest, but this needs to be confirmed. The male advertises from its territory with the Boom call. It defends its sites against intruders with Forward displays, either Standing or Walking. In the latter display the males approach each other in a deep crouch, erecting the white chest plumes over their backs so as to frame the head, dark crown, and moustache. Supplanting Flights and Aerial Fights occur, with the birds stabbing at each other. Birds also display with Circle Flights, legs dangling. An interesting observation of aerial behaviour was a bird's response to a tape play back of a call; it flew from the marsh in an upwards spiral before diving back, clearly a territorial display (Devine *et al.* 2000). The Boom call is usually given from a perch and increases in frequency as courtship proceeds. The male probably chooses the nest site. Copulation takes place on the nesting territory, but not only on the nest, and is preceded by active pursuits. Unlike the Eurasian Bittern, evidence of polygamous nesting is slim, inferred from a few observations of closely spaced nests. The mating system should be assumed to be monogamous unless shown otherwise.

Eggs are unmarked brown or olive-buff, averaging about 48.6 × 36.6 mm. The usual clutch is 4–5 eggs, range 2–7. Mean clutches are 3.8 in North and South Dakota, 4.1 in Illinois, 4.4 in Minnesota and 3.7 range wide (Duebbert and Lokemoen 1977, Graber *et al.* 1978, Hands *et al.* 1989b, Svedarsky 1992). Renesting can occur (Azure *et al.* 2000b). Incubation begins before the clutch is completed. It lasts 24–29 days. Incubation is by the female only, based on presence of incubation patches. Care and feeding of young are believed to be confined to the female, but more information is needed. Chicks are fed by regurgitation. They begin to leave the nest after about two weeks. The length of time they remain dependent on the parents is unknown. Information on nesting success is very limited. Eggs hatched in only 41 of 72 upland nests (Duebert and Lokemoen 1977).

Population dynamics

There is very little information on population biology or demography. Longevity is at least 8 years (Clapp *et al.* 1982). It is likely that adult winter mortality, migration losses and long-term reproductive success are the key demographic features.

Conservation

Canadian populations appear to be secure. But clear population declines in the central USA indicate an overall downward population trend for the species and perhaps an overall long-term reduction in breeding range. Because of its decline, it is considered to be near Threatened by HSG (Hafner *et al.* 2000). More and better information on status and trends are needed (Hafner *et al.* 2000). It is critical to identify important nesting areas and to concentrate management and protection efforts on them. Habitat management is the key to successful conservation of this species (Hands *et al.* 1989b), focusing on wetlands greater than 10 ha and adjacent habitat. It is probably necessary to actively manage important areas within a region to maintain mid-successional habitat for the species.

Research needs

More information is needed on the basic biology of the North American Bittern, including space use, courtship, nesting biology, success, demography, and population regulation. This information is needed to understand the causes and circumstances of long-term population and range reduction in North America. There is a need to develop more precise standard survey and monitoring methods for populations and habitats, and to use these and other techniques to assess distribution and habitat use in winter. Research is needed on case histories to develop management strategies for regionally based habitat management.

Overview

The North American Bittern is a solitary bird of dense emergent marshes. It feeds and breeds in isolation, in defended territories deep in the marsh thicket. It feeds by Walking carefully and very slowly, working its way through the vegetation, particularly near the edges of tall vegetation. It Stands for long periods and transitions to Walking extremely slowly. It hunts its prey, primarily fish and insects, with great patience. Its Boom call claims its territory and advertises its personal presence. It appears to be a very northern species, not breeding over much of the usual heron range in North America. Nesting very far north, populations migrate southward in the autumn. Although somewhat flexible, it none the less depends on a few sorts of marsh habitats. Its range appears to be retreating perhaps as a response to post-Pleistocene climatic amelioration.

Eurasian Bittern *Botaurus stellaris* (Linnaeus)

PLATE 12.1, 12.4

Other names: The Bittern, Great Bittern, Common Bittern, European Bittern, African Bittern, Boomer in English; Avetoro Común in Spanish; Butor étoilé in French; Rohrdommel in German; Boerdomp in Dutch; Выпь in Russian; Rördrom in Swedish; Sankano-goi in Japanese; Da mayan in Chinese.

Ardea stellaris Linnaeus, 1758. Syst. Nat. ed. 10, p. 114: Europe; restricted type locality, Sweden.

Subspecies: *Botaurus stellaris capensis* (Schlegel), 1863: South Africa.

Description

The Eurasian Bittern is a stocky, thick-necked, medium-sized, mottled, black crowned, golden brown heron.

ADULT: The Eurasian Bittern is the largest of the large bitterns. It has a black cap. The long bill is yellow with the top being brown and becoming almost black at the tip. The iris is yellow and a line above the eye is pale buff. The loral area is green, extending on to the lower bill. The sides of the head are uniform buff-brown. The neck is buff-brown narrowly barred brown. The chin and throat are cream-white with a rufous-brown median streak. A weak black to dark-brown moustache extends from the bill gape onto the neck. The hind neck and back are golden-brown, consisting of black and buff mottling and barring. The 'shoulder' tufts consist of elongated feathers with a brown centre and large white border, usually concealed by the folded wings. The upper wings are pale rufous-buff, darker on the leading edge, and blotched with black. The flight feathers are pale red-brown to black, with black spots. The foreneck and breast are yellow-buff with longitudinal brown streaks and finely spotted black. The streaks are widest on the chest and narrowing on the belly. The feathers of the foreneck are thick and elongated into a bib. The belly is yellow-buff. The underwing is pale buff with grey mottling. The legs and feet are pale green. The soles of the feet and back of lower legs are yellow. In breeding the loral area is light blue in the male and light brown in the female. It is likely that reports of orange-red irises are indicative of a colour change in courtship (Wood 1986).

VARIATION: Males are significantly larger than females. The females may tend to be less well marked than the males. There is much individual variation in the dark plumage markings, some individuals appearing quite pale. There also is geographical variation. Birds from east Asia tend to be more boldly barred, and were formerly considered a subspecies. The disjunct birds of southern Africa, *capensis*, are slightly smaller and darker, with more narrow and irregular barring on the flight feathers than *stellaris*.

JUVENILE: Juveniles are very similar to adults. They are paler with browner stripes, have a browner crown and less distinct brown (not black), moustache streak.

CHICK: The chick has cinnamon-tawny to rufous-brown down on the upper parts. The chin and throat are white. The underparts are rufous-buff. The bill and legs are blue-green.

VOICE: The 'Boom' call consists of two to four deep, resonant booms repeated at 1–2 second intervals, preceded by a few short grunts or pumps and sometimes accompanied by Bill Clappering. The sequence of pumps and booms in the Boom call is variable among individuals (McGregor and Byle 1992, Gilbert *et al.* 1994, Puglisi *et al.* 1997). The call has been rendered as 'up, up, up, rumb', 'umb, uh, uh, uh, ub', 'up, up, up, rumb', and 'hu, hu, umb, ub'. The 'Boom' call is used during the breeding season for advertisement and territorial defence (Puglisi *et al.* 1997) and is audible for up to 5 km. The female in answering the 'Boom' call uses the 'Wumph' call. The 'Kau' call is a flight call, contact call, and disturbance call; perhaps the multiple functionality of a resonant call is not unexpected for a bird from dense habitat. It has been rendered as 'kau', 'aargh', 'aark, aark', and 'awk, awk'. The 'Kro' call is a disturbance or warning call, rendered as a rapid, short 'ko, ko, ko', and 'kro, kro, kro'. The 'Kra' call is a threat or warning call. Young beg with a crackling 'rah, rah, rah'.

WEIGHTS AND MEASUREMENTS: Weight: females 867–1150 g; males 966–1940 g; length: 70–80 cm.

Field characters

The Eurasian Bittern is identified by its stocky build, thick neck, medium size, golden-brown mottled plumage, black crown, and dark moustache. If disturbed it flies up with neck extended and legs dangling, and quickly drops back into the reeds. In long flight it has steady slow wing beats, on broad rounded wings. Described as owl-like, it flies low over the marsh, with head retracted and legs trailing. It is distinguished from the immature Black-crowned Night-Heron by being larger with a heavier, stockier build and less hunched appearance, having the back and upper wings mottled with black (not densely spotted white), black crown, and moustache. It does not flock or perch in trees. It is distinguished from the North American Bittern by being larger, with a black (not rusty-brown) cap, indistinct black (rather than prominent large) moustache stripe on the head, more boldly marked plumage, and broader wings without dark flight feathers (Lansdown 2000).

Systematics

The Eurasian Bittern is one of the four large *Botaurus* bitterns, all of which have streaked brown plumage, scutellate tarsi, 10 tail feathers, and a booming call. It is most closely related to the Australasian Bittern (Payne and Risely 1976, Sibley and Monroe 1990).

Range and status

The Eurasian Bittern is a bird of the Old World temperate and tropical zones of Europe, Asia, and Africa.

BREEDING RANGE: The Eurasian Bittern breeds from England, Wales, Sweden, Finland, Germany, Lithuania (Matiukas 1990), Italy, Belarus, European Russia, south through Europe, north Africa (north Morocco, Algeria), Albania, Bulgaria, Turkey (Kasparek 1986), north Iran, Kazakstan, Azerbaijan, Turkmenistan, Uzbekistan, Tajikistan, Kirghizstan, Afghanistan, Mongolia, north China (Sinkiang to Heilongjiang), Russia (Siberia to 64°N, Yakutia, Amurland, Ussuriland, Sakhalin Island), North Korea, South Korea, and Japan (Hokkaido). *Capensis* is presently known to nest in Natal and Transvaal in South Africa and probably Zambia.

NON-BREEDING RANGE: *Stellaris* winters from central Europe, east Europe to Ukraine, east Saudi Arabia, Iraq, Iran, Turkmenistan, Pakistan, north India, Nepal, Bangladesh, southeast China (including Hong Kong—Leader 1999), Korea, Japan (Honshu), Myanmar, Cambodia, Vietnam, south Thailand, north Malaysia (peninsular), also north Africa in Libya, Egypt, sub-Saharan Africa in Mauritania, Senegal, Ghana, Gambia, Gabon, Nigeria, Cameroon (Robertson 1993), Kenya, north Zaire, Sudan, Ethiopia, and Eritrea.

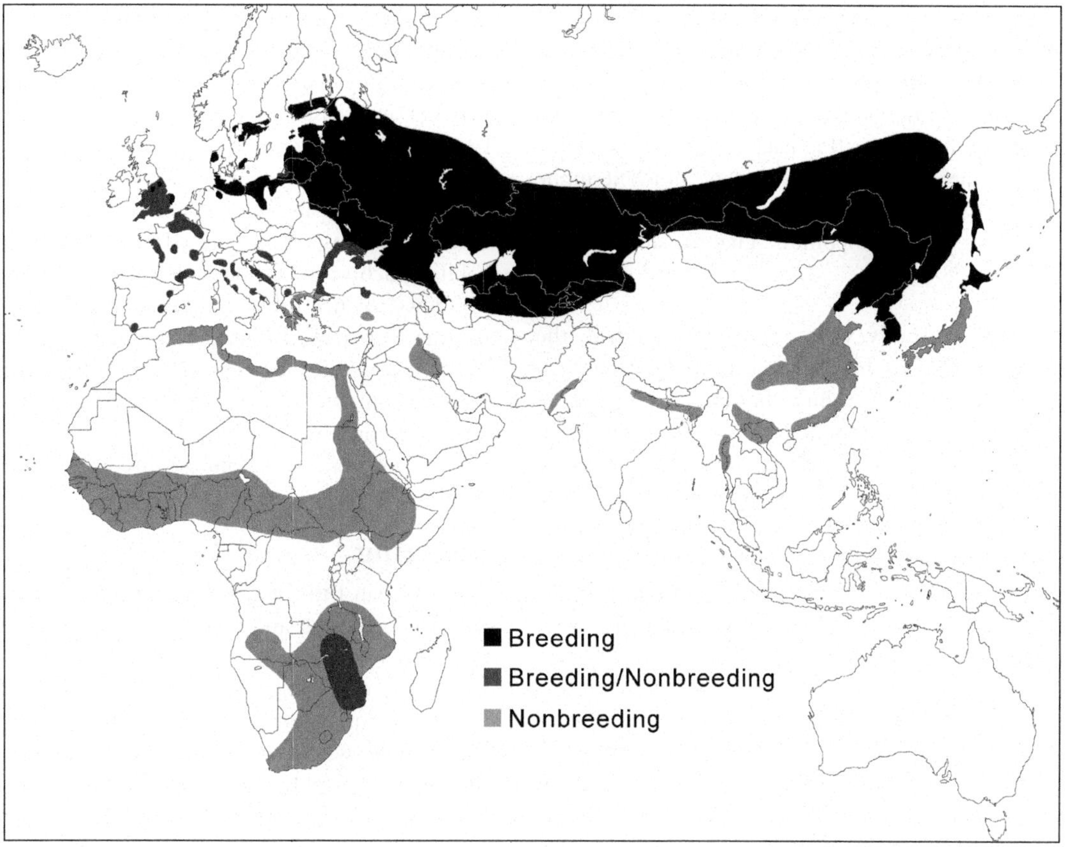

The known breeding sites for *capensis* are few. However, it occurs over a much wide range in central and south Africa. Until proven otherwise, this is considered part of the non-breeding range. It occurs in central Angola, Namibia (Hines 1985–7), north Zambia, Zimbabwe, south Tanzania, Malawi (Johnston 1989), central Mozambique, northwest Botswana, north Seychelles, and South Africa (Vernon 1993). It has been reported but not confirmed from Kenya (Zimmerman 1996).

MIGRATION: The north Eurasian populations are partially migratory. Some birds remain in breeding areas where water does not freeze, or stay farther north in mild than in harsh winters (Bibby 1981, Andersen 1989, Nankinov 1991, Kolland 1996, Mundt 1996, Bernasconi *et al.* 1999, Gauggel 1999). Some populations, such as in England and Netherlands, are nearly residential. Both autumn and spring migrations are primarily at night (Puglisi and Baldaccini 2000). Birds move singly or in small flocks. Southward migration is variable, seriously getting underway after the first hard frost, generally September to as late as December. West European birds move south through France and Spain; others move through Greece and Italy (Voisin 1991). They cross the Sahara where they occur in oases. Return migration is February–April. The origin of birds wintering in Iran and Iraq is not known, but is probably Russia (Voisin 1991). Asian populations move south to the northern half of the Indian subcontinent, Myanmar, and throughout central China to the Chinese coast. Migration here is little understood (Leader 1999). The African race *capensis* is mainly sedentary. Some movements take place, mainly to the southwest, and these are governed principally by rainfall. Postbreeding dispersal is extensive, especially of immature birds. Dispersal records include Iceland, Faeroes, Madeira, Azores, Canary Islands, Norway (Jabekk 1996), Syria, Jordan, north Russia (Pleshak 1999), south India, Myanmar, Sri Lanka (Gunawardene and Wijesinghe 1986), Ryukyu Islands, Taiwan, Malaysia (peninsular), and the Philippines (Luzon), and Brunei (Mann 1987).

STATUS: Formerly widespread and abundant, the Eurasian Bittern has suffered a steady decline in numbers in Europe, since the nineteenth century, notably from 1970 to 1990. It was extirpated from England in 1868. It began a comeback through Europe in the early 1900s, returning to England in 1911. It increased in population and range in the 1960s and then began a second decline. At present, the bittern remains widespread in Europe overall, but with relatively few remaining in the west. Declines in some areas have been rapid and alarming, such as in Spain, but is reversed at other sites (Duhautois 1984, Smith and Tyler 1993, Puglisi *et al.* 1995a, Vicens 1997, Kayser *et al.* 1998, Marion *et al.* 2000). The total European population is 20 000–44 000 pairs (Marion *et al.* 2000). Russia supports 63% of the European population, 10 000–30 000 pairs, but its status there is little known. Ukraine, Poland, Belarus, and Romania account for 24% of the European population. It is locally abundant in Turkey (Kazilirmak Delta) and Romania (Danube Delta) (Hafner 2000). In North Africa, it is very rare, occurring only in a few pockets of fresh water. In Africa records of *capensis* are sparse and it has clearly has undergone a considerable decline, especially in recent years. It is known to breed in South Africa and is rare in Zambia. It is common in Iran but considered to be uncommon through its breeding and non-breeding range in Asia.

Habitat

The Eurasian Bittern is a bird of densely vegetated wetlands having extensive beds of tall grass or sedge. Usually these marshes are shallow, fresh or brackish, with a stable water level, low acidity, low elevation (less than 200 m), and having reed stands of varying ages interspersed with open water. Marshes dominated by cane, *Phragmites*, are favoured in Europe. It also uses *Scirpus* marshes, papyrus swamp, brackish marshes, and coastal dune wetlands. Rice fields may be serving as a potentially important habitat (Alessandria *et al.* 1992). Outside the nesting season its requirements are less restrictive. Beyond typical reed beds, it uses more varied and open aquatic habitats such as small ponds, gravel pits, wet grassy meadows, ditches, fish ponds, watercress beds, other floating leafed plant beds, and sewage lagoons (le Mair *et al.* 1995). During winter in the north, it tends to come out of the reeds to find open or running water in which to feed. In very cold weather the species may occur almost anywhere there is open water.

Foraging

The Eurasian Bittern is a solitary feeder. It fiercely defends its feeding and nesting area during breeding, even attacking in the air (Soderholm 2001). It may occur in loose groups outside of nesting. It is mainly crepuscular, but also hunts during the day and at night, within and at the edges or reed beds. The main method of hunting is Walking slowly, with great stealth. It walks in an extreme Crouched posture, the bill pointed forward, the feet lifted high with each step. It also feeds by Standing, often as an intermission in Walking. It Walks Quickly, especially across open spaces and Swims across open water between patches of cover (Lindblad 1995). It has been observed to Run across open habitat picking up prey items. It Foot Stirs to inspire fish into movement. The bittern tends to feed at the edge between emergent reeds and open water, such as at a pool, channel, or ditch. It tends to avoid unflooded ground and walks by climbing over the emergent stems, grasping them with long toes. It also climbs up the stems for observation, and to sunbathe. It avoids shrub, which has consequences when marsh succession occurs. Large areas are covered in the search for food. It captures prey by Stabbing using its entire body. Fish are shaken, beaten, and stabbed before consumption. It often uses the concealing Bittern Posture in response to disturbance, but also for surveillance as it can peer directly forward when its bill is raised; it has been observed to hold this posture for hours. The bittern generally is found on the ground or in herbaceous growth, but on occasion uses trees (Lunn 1992).

A wide variety of foods is taken, varying with locality and season. Fish, amphibians, and insects usually dominate the diet. Fish taken are diverse depending on availability. Species include *Esox, Eupomotis, Cyprinus, Salmo, Perca, Thymallus, Tinca, Anguilla, Cottus, Leuciscus*, and *Rutilus*. The diet also includes frogs, tadpoles, and salamanders (*Rana, Triton, Bombina, Leuciscus*), worms (*Lumbricus*), crustaceans, insects (water beetles, water bugs, grasshoppers, dragonflies, leaf miners), small mammals (*Avicola, Mus*), birds (*Anas*) (Loison 1991), and snakes (*Natrix*). In some areas eels, frogs, or tadpoles dominate the diet. The diet is particularly challenging for birds that remain north in the non-breeding season, where they use atypical habitat and prey (Windisch 1999).

Breeding

The Eurasian Bittern nests in March–July in Europe, sometimes earlier; September–January in South Africa. Generally, nesting is in the rains in the tropics, but there is little information. During the breeding season the bittern is nearly restricted to dense reed beds. The ideal habitat is restrictive: dense plants, with both old and young stems, stably flooded, shallow, with clearings or channels. It may use either fresh or brackish water, but not saline. In addition to *Phragmites*, they nest in *Scirpus*, rice, *Papyrus, Bolboschoenus, Cladium*, and *Juncus*. Breeding is essentially solitary. Density varies from 2 per 100 ha to 100 per 100 ha, depending on the habitat quality. However, under suitable habitat conditions, nests may be rather close together, as many as 10 were counted in 0.5 ha (200 per 100 ha). This is highly unusual. Nests are also close in situations where there are more than a single female associated with a male. The nest is a pad of matted reeds and other marsh vegetation, lined with finer material. It appears carelessly put together and becomes flattened by use. Nests are 30–40 cm wide and 10–15 cm thick, placed on and within the reeds at water level. It is built by the female alone. The bittern's movements soon create radiating tracks through the nearby marsh. Material is added through the nesting period. In spring, the male claims its territory and advertises by giving its 'Boom' call (Fontanelli *et al.* 1995). Non-migrating birds begin to call in late winter. The male will occupy and defend its territory through July. The Boom call is produced with the bill closed and pointing downwards and neck feathers fluffed out. The Boom call is most given at dawn and dusk, but also throughout all hours of the day and night and decreases under cloudy and rainy conditions (Puglisi *et al.* 1997, Poulin and Lefebvre 2003). Males defend their home range territories (Puglisi *et al.* 1999). They are extremely aggressive in defence. Supplanting Flights occur, as does Aerial Fighting (e.g. Soderholm 2001), which may end in the death of one of the bitterns. Forward displays are used to defend sites at close proximity. The bird has its legs bent, neck bent back, bill open, wings bent forward, crest, neck feathers, scapulars, and shoulder tufts erected (Gorman 1995, Alessandria *et al.* 2003). Attacks are made by a strong Jab, perhaps Jumping forward. Chest pecking may be a ritualized outcome of low intensity attacks. Subservient birds respond with a Withdrawn Crouch. Bitterns sometimes kill other bitterns. It defends its area against other species, particularly the Marsh Harrier (*Circus*) and Carrion Crows (*Corvus*), and is known to lose prey to pirating birds, such as Grey Herons (Kington 1991). Interactions between pairing individuals are seldom described. In close proximity, the shoulder plumes are displayed (Gruell *et al.* 1988). In addition to the Boom call, displays include individuals or several birds indulging in extensive Circle Flights over the reed beds. The flying is erratic, at heights of up to 60 m and for as long as 10 minutes, ranging over large areas of the habitat. When flying together the bitterns circle around each other or shoot upwards and downwards again.

This is the heron species in which polygamous mating has best been shown. It is not unusual, and it appears a male may have up to five mates. The eggs are olive-brown, occasionally with some fine spotting at the large end. They average 53 × 39 mm. The normal clutch is 4–5 eggs; the range is 3–7. Eggs are laid at 2–3 day intervals. The species has long been known to be single brooded; however, double brooding has recently been documented

(Mallord *et al.* 2000). Only the female incubates. Incubation begins with the first egg and lasts 25–26 days. The male continues to advertise and defend the nesting territory for several months, through June–July in Europe. Eggs hatch very asynchronously. In a large clutch the last egg may hatch 12 or 13 days after the first; due to interruption of incubation to tend to first-hatched young. These last eggs often fail to hatch. Young are semialtricial and nidicolous. They are cared for by the female. Polygamous males apparently play no part in the rearing of the young. They are driven away by the female should they approach the nest. In monogamous pairs, the male may bring food near to the nest for the female to use. Vary rarely, the male joins the female on the feeding grounds. Food is regurgitated into the nest. Large prey not consumed by the young is often reconsumed by the parent. After 15–20 days, the young are fed off the nest site. The young soon become very active, exhibiting various self-defence attitudes and actions. They clamber from the nest and walk through the reeds after the first week. They fledge at 50–55 days. Breeding success in one study in Germany was 56%, to 14 days.

Population dynamics

Bitterns first breed as early as one year. Populations in which some birds remain north in winter can suffer significant mortality in hard winters, up to 40%–80% (Marion *et al.* 2000). After a winter kill in the early 1960s, nesting sites were never reoccupied. In Sweden the population declined from 450 to 270–300 pairs after a hard winter and needed 3–7 years to recover (Marion *et al.* 2000). Adult survival (local) in England ranged from 65.2–82.8% (Gilbert *et al.* 2002). Longevity is 9 years.

Conservation

The population in Europe has been in decline for many years. An original contributing factor was habitat alteration, mainly through drainage and human persecution. In some countries protective measures brought back the population by the middle of the last century, but in the past 30 years the decline has resumed. Although factors are unclear, it is likely that core to the decline is the continued loss of *Phragmites* marshes. Conservation activities in Europe need to concentrate on these habitats. In evaluating the decline of European bitterns, it is important to consider that similar declines are occurring in other large bitterns. Whereas habitat alteration is clearly involved in some areas, such in west Australia, causes remain puzzling in other areas such as New Zealand, and large-scale climatic events appear to be implicated elsewhere, such as in North America. The long-term and short-term impact of winter mortality on populations is clear (Bibby 1981). Information on the trend of the Eurasian Bittern in east Europe and Asia is important to begin distinguishing among causal possibilities.

The HSG considers the European and North African population to be Regionally Vulnerable (Hafner *et al.* 2000). Habitat loss, management practices that fail to maintain the habitat characteristics needed by bitterns, disturbance by humans, and pollution have been implicated in the declining status of the bittern in Europe (Marion *et al.* 2000). Human disturbance is pervasive in Europe. Cane cutting, recreation, motor vehicles, hunters, and boats all disturb nesting bitterns (Day and Wilson 1978, Newton *et al.* 1994, Tyler *et al.* 1998, Whiteside 1989, Piskorski 1999). Protection and management of important bittern habitats are needed across its European range. Fortunately research is demonstrating appropriate approaches (Bibby and Lunn 1982, Tyler 1994, Hobbs 1995, Welch 1995, Vicens 1997, Puglisi *et al.* 1995b, Dvorak *et al.* 1997, Tyler *et al.* 1998, Soto-Largo *et al.* 1998, Schlumprecht 1999, Smith *et al.* 2000). Reed cutting, for example, if properly managed, can provide multiage stands desirable to the bittern, but if badly managed—such as permitting large-scale cutting in late winter—habitat quality is destroyed. Regional management schemes, monitoring, education, and local action are needed, perhaps linked to conservation of the Purple Heron in these same habitats (Hafner *et al.* 2000). Wintering birds remain common in Iraq and Iran. There drainage of the Mesopotamian marches is a threat to their welfare and that of other water birds that use this critical area. Information on wintering status in Africa is nearly non-existent. The southern Africa population is even more at threat, given its rapid decline over the past several decades. Conservation action in its remaining known locations in South Africa and Zambia is essential.

Research needs

There have been important advances in our understanding of the biology and conservation needs of this species. Studies should continue on the breeding, habitat needs, and demography of this species especially outside western Europe, where most information is from. Surveys are needed to determine the status and distribution of nesting of Eurasian Bitterns in southern Africa and wintering birds in northern Africa. The status and trends of the species in Russia need to be understood, not only to develop appropriate conservation strategies but also to compare with the situation in west Europe.

Overview

The Eurasian Bittern is a skulking species seldom seen far from its reed-bed habitat. It spends most of its time Walking very slowly, but covering large distances around its feeding territory. It catches what it can, with considerable specialization in different locations. The Eurasian Bittern has polygamous tendencies and exhibits lack of a parental role by the male. This means that the feeding of the chicks is left mostly to the female, the male's role being to defend the territory and protect food supplies for the female. The bittern has the largest span for asynchronous hatching among herons, the span of ages within the brood perhaps serving to separate periods of high energy demand among the siblings. Much remains to be understood about the biology of polygamy in bitterns.

Australasian Bittern *Botaurus poiciloptilus* (Wagler)

PLATE 12.5

Other names: Australian Bittern, Boomer, Brown Bittern, Bull-bird in English; Avetoro Australiano in Spanish; Butor d'Australie in French; Australische rohrdommel in German.

Ardea poiciloptila Wagler, 1827. Syst. Av., Ardea, sp. 28: New South Wales.

Description

The Australasian Bittern is a stocky, thick-necked, medium-sized, mottled dark brown and buff heron with a black moustache.

ADULT: The Australasian Bittern's crown is brown. The upper bill is yellow to buff with the top of the bill dark brown to grey-black, and is somewhat small relative to the New World large bitterns. The loral area is variable, green-grey to blue-grey, extending onto the lower bill. The irises are yellow. A narrow stripe of skin from the nostrils to the bill is dark olive-brown. A light line above the eye is buff-white. The side of the face is buff. A dark moustache-like streak runs from the gape to the sides of the neck. It is dark brown near the bill becoming paler brown as it blends into the neck. The hind neck and back are dark brown with buff streaks and freckling giving a general mottled and streaked appearance. The upper wings are finely barred in buff and brown with dark brown flight feathers irregularly barred or spotted with buff. The tail is brown with buff fringes. The chin and upper throat are white with a mottled brown central stripe. The lower throat and foreneck are buff-white with dark brown longitudinal stripes and freckling. The belly and feathered thighs are white. The underwing is buff-white with freckled brown. The legs and feet are green-yellow to dark-olive. There is no information on changes during courtship or breeding, but it is likely that reported blue-green lores and orange-brown irises are breeding colours.

VARIATION: Females are much smaller than males but have the same plumage. However, individual plumage is highly variable. Dark, light, in-between, and red birds occur. Darker birds have a darker moustache-like stripe. The upper parts are dark brown with little of the pale streaking or freckling. The dark feathers make the bird appear uniformly dark to black at a distance. Birds are also reported that are reddish, with a cinnamon-buff side of face and rufous upper parts and upper wing.

JUVENILE: Immature birds are similar to adults but are paler, with a yellow tinge to the basic colour, and underparts streaked rufous. The iris is yellow. The bill is yellow-green to olive-grey with a grey-brown top of bill. The facial skin is light green to green-yellow, with the narrow stripe between the eye and bill being black-green. The legs are dark olive.

CHICK: The hatchlings have long, sparse, dark brown down. As they get older, they become buff to yellow-brown above and yellow-brown below with a white chin and throat.

VOICE: The 'Boom" call is the loud, resonant call given during the breeding season by territorial males. It starts with four short quiet gasps followed by 'woomph, gasp, woom'. Each call typically includes two or three booms. Males call back and forth to each other. The 'Craak' call is a short harsh alarm call and also the flight call when disturbed. A bubbling sound is reported from a female returning to the nest. This and other vocalizations need more study.

WEIGHTS AND MEASUREMENTS: Weight: females 571–1135 g; males 875–2085 g; length 66–76 cm.

Field characters

The Australasian Bittern is identified by its stocky build, thick neck, medium size, brown and buff mottled plumage, plain brown crown, dark brown moustache, broadly brown streaked neck and breast, and finely vermiculated and spotted upper sides. When disturbed it flies up heavily on broad bowed wings with legs dangling, and quickly plunges back down into the vegetation without circling. In full flight, this bittern flies with steady slow wing beats, reported to be rather owl-like (Marchant and Higgins 1990). It is distinguished from the immature Rufous Night-Heron by being larger with a heavier, stockier build and less hunched appearance, having mottled (not densely spotted) back and upper wings. It also is not usually found in flocks nor perched in trees. Its bill is relatively shorter (shorter than the length of the head vs. slightly longer in the night heron) (Marchant and Higgins 1990). It is distinguished from the Black Bittern by being larger, stockier, and brown (not black) upper parts. It is distinguished from the Eurasian Bittern by having darker upper parts, much darker in some, especially on the neck and back, which also are less strongly marked in paler shades.

Systematics

The Australasian Bittern is one of the four large *Botaurus* bitterns, which all have streaked brown plumage, scutellate tarsi, 10 tail feathers, and a booming call. It is most closely related to the Eurasian Bittern (Payne and Risely 1976, Sibley and Monroe 1990). The birds of southwest Australia and New Zealand in the past have been recognized as separate subspecies.

Range and status

The Australasian Bittern occurs in Australia, New Zealand, and nearby islands.

BREEDING RANGE: The range of the Australasian Bittern includes southeast Australia (south Queensland, New South Wales, Victoria, east South Australia), west Australia (south Western Australia), Bass Strait Islands, Tasmania, New Zealand (North Island, South Island, Stewart Island, Great Barrier Island, Mayor Island), New Caledonia, and Loyalty Islands. There are in fact few records of actual breeding within this range, which is inferred from sightings year-round.

MIGRATION: The species is generally sedentary. It likely undertakes some seasonal population shifts associated with the wet and dry seasons, and there is some suggestion of winter influxes occurring along the coasts of Australia and New Zealand (Whiteside 1989, Marchant and Higgins 1990). There also appears to be movement in exceptionally dry or wet situations. Dispersal occasionally occurs, perhaps more so in the past when populations were higher. Dispersal birds are observed in Western Australia, and on islands including Chatham Island (more frequently in the past), Kapiti Island, Great Mercury Island, and Lord Howe Island, and recently in the Torres Strait (Stokes 1983), suggesting that the species should be looked for in New Guinea.

STATUS: It is widespread New South Wales, Victoria, and South Australia, being most common in the Murray–Darling basin and coastal areas. In Tasmania, it is most common in the east. Populations in Western Australia have decreased in the last century to about 100 pairs, and it is now confined to high-rainfall areas along the west coastal plain. It was once common on the largest of the Loyalty Islands, but its current status is unclear and is likely extirpated. It is not clear if the Chatham Island population was resident or dispersive, but it has not been reported there since 1910. It is widespread but thought to be declining in New Zealand. The population there is 580–725 birds. New Zealand distribution maps for the species (Marchant and Higgins 1990, Heather and Robertson 1996) show it as occurring throughout the whole of the North and South Islands, which can be somewhat misleading (M. Maddock pers. comm.). There are mountains in the North Island and along the whole spine of the South Island that are up to more than 2000 m in altitude, capped with permanent snow and glaciers and heavily forested on the slopes below the snow line. However, deep valleys with lakeshore wetlands suitable for the bitterns, within the habitable altitude range for the species, penetrate well into the mountains. For example, Chambers (1989) recommends Lakes Manapouri and Te Anau, surrounded by ranges of 1000 m or more in altitude, at the eastern margin of Fiordland, as suitable locations for observing the bird.

Habitat

The Australasian Bittern typically uses permanent fresh water marshes with dense reed and rush beds situated in shallow (less than 30 cm) water. It occurs along rivers, pools, lakes, and swamps. It particularly

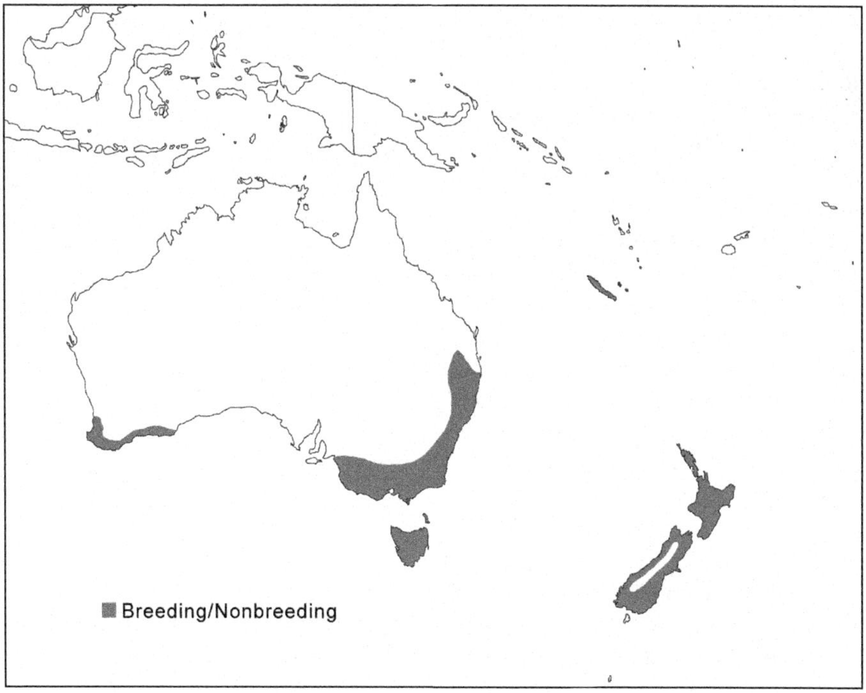

uses places near the edges of pools and waterways through the marsh. Typical vegetation includes *Typha, Phragmites, Juncus, Baumea, Eleocharis, Gahnia, Bulboschoenus, Ludwigia, Eragrostis*, and small bushes. It usually feeds and nests within the marsh, near the interface of the marsh and open water. It sometimes occurs in more open situations, on mud banks and in open water. It also uses temporary pools, tidal marsh near the mouths of fresh water creeks and seeps, mangroves (Miller and Miller 1991), rice fields, tall wet pasture, and drainage ditches (Grant and Bennett 1986). It is primarily a bird of lowland swamps but occurs to 300 m in New Zealand.

Foraging

The Australian Bittern is generally a solitary hunter, although it also feeds in pairs and sometimes in loose groups of up to 12 birds. It feeds primarily at night, but is also observed to feed crepuscularly and by day especially in winter. More information is needed on the daily cycle. It feeds by Standing motionless, waiting for prey from the edge of thick vegetation, from the bank, or from overhanging branches. It establishes feeding spots by flattening areas in the reeds, which often are littered with discarded remains of crustaceans and frogs. It also Walks slowly in a Crouched almost horizontal posture, moving forward with great deliberation, often knee-deep in water or along a bank. Each foot in turn is lifted high as it progresses. It frequently Neck Sways. It also has been recorded as Baiting using pieces of grass to lure fish. On locating prey it makes a rapid Bill Strike by pivoting forward on its legs with neck and back straight (Whiteside 1989). The upper legs are short relative to the lower legs, providing the pivot. It also Lunges from a crouched position, leaving the ground. It rapidly assumes a Bittern Posture when encountering an intrusion but also uses it for surveillance (Whiteside 1989). When disturbed it flies up and returns into the marsh quickly.

The diet is variable, but fundamentally fish (eels, trout, *Crassius*), frogs (*Hyla*), and crayfish (*Cherax*). The overall diet includes insects (cutworms, weevils, crickets, grasshoppers), spiders, molluscs, lizards, rats, mice, and small birds, including white-eye (*Zosterops*). In New Zealand the introduced frog (*Hyla*) appears to constitute its main prey. But in all areas, the diet has not been studied in any detail.

Breeding

Breeding is October–February, but more information is needed on details and geographic variation. It nests in reed beds up to 2.5 m tall, usually near openings such as pools or streams. It uses small to large wetlands, 5–300 ha, reportedly usually only one bird per wetland. Four and five birds were reported in 200 ha swamps, or 2 birds per 100 ha (Marchant and Higgins 1990). The bittern is a solitary nester. However several (up to 7) nests can be found in close proximity in the same reed bed. These are assumed to belong to one male and may represent polygamous breeding. More details are needed. The nest is a platform 30–40 cm across, 20–22 cm thick. It is made of reeds and occasionally grass placed within the marsh vegetation seldom more than 10–30 cm above the normal water level. Females appear to build the nest alone. The male announces and advertises its territory in the spring with its Boom call. This far-carrying call is given mostly at dawn and dusk. Circle Flights have been recorded with feet dangling. A flight display has also been described involving two birds alternating slow flapping and gliding. It is likely the Bittern Posture is used as a display, in which the brown moustache-like streak is expanded into a ruff on either side of the head and neck. The eggs are brown-olive or green-cream, average 52 × 38 mm. The clutch is 4–5; the range is 3–6 eggs. Incubation starts with the first egg and is 25 days. The female alone incubates. Renesting occurs when eggs are lost. The chicks are fed by regurgitation into the nest, apparently by the female. The chicks can leave the nest in 2–3 weeks, and begin to wander into the surrounding reeds. They fledge in about 7 weeks. Nothing is known about nesting success.

Population dynamics

Nothing is known of the population biology or demographics of the species.

Conservation

Information on status is slim because of its occurrence in dense, inaccessible habitats. Although locally common in the Murray–Darling river system, long-term population declines elsewhere in Australia and New Zealand are matters of grave concern. In Australia, declines are caused by habitat alteration—drainage, salinization, damming, grazing, clearing for agriculture, and peat extraction. In New Zealand, swamp drainage and, previously, hunting for feathers sought-after by trout fishermen caused the decline. That these two presumably isolated, declining populations have been thought distinctive enough to merit subspecific status suggests the importance of focusing conservation measures on these disjunct areas. In many ways, however, the viability of the species as a whole depends on the health of the Murray–Darling system. In west Australia and New Zealand habitat restoration may be required. Like other large bitterns, this species favors *Typha* marshes, and could use disturbed wetlands. Properly managed disturbed and artificial wetlands can be an integral part of a conservation strategy. In addition to protecting habitat in parks and reserves, the importance of habitat management on private land cannot be overstated (Corrick 1995). Conservation of this species needs to become a higher priority (Hafner *et al.* 2000). It is categorized as Critically Endangered by HSG. It is considered to be Vulnerable by IUCN (IUCN 2003). It also is a Vulnerable species in Schedule 2 (updated 1999) under the NSW Scientific Committee Threatened Species Conservation Act 1995 and in Victoria, its conservation status is classified as Endangered (NRE 2000). One reason why its conservation status is inconsistent is that its actual status is surprisingly poorly known. Although its estimated world population is well under 10 000 birds (Hafner *et al.* 2000), it may be far smaller in that fewer than 100 pairs were surveyed in west Australia, and fewer than 800 birds estimated in New Zealand. The nesting population in southeast Australia has not been documented. Actions required include a complete survey of the breeding population in east Australia, especially in the Murray–Darling system, and New Zealand, identification of important areas (those that support 100 or more birds), development of habitat management recommendations including enhancing the use of artificial and disturbed wetlands, implementation of a range-wise strategy for habitat protection and management, and establishing a monitoring programme. Conservation should focus independently on west Australia, east Australia, and New Zealand.

Research needs

So little is known about the biology, breeding range and dispersion, behaviour, and habitat requirements of this species, that comprehensive studies of its biology are needed. Actual records of nesting are few. Surveys should be undertaken to determine the actual breeding range and dispersion of nesting across the range. Especially important are the areas where declines are occurring and also the Murray–Darling basin, which supports the majority of the population. The species' use of *Typha* marshes should be carefully studied in order to determine required habitat characteristics within this marsh type. Such information can be used to develop habitat management recommendations for enhancing the bird's use of artificial and disturbed *Typha* marshes. The species is highly variable in plumage characteristics. It is unclear what part of this variation is individual and what has geographic components. Intra-specific variation in plumage and morphometrics should be studied to re-examine the distinctiveness of west Australia and New Zealand populations. The results may have important conservation implications.

Overview

This is a reed-bed species. It characteristically uses patches of emergent marshes of grass, reed, or sedge, located in swamps, along rivers, or as isolated wetlands. Its habits and morphology are designed to make maximal use of concealment within the thick marsh vegetation. Its long toes allow it to support itself on the grass; its short legs, particularly the upper legs, allow it to crouch low to the water while Standing or Walking and still be able to lunge forward to strike. It eats fish, frogs, and insects, but overall has a broad diet of whatever it can encounter in the marsh. It has the potential for being adaptable in that its favoured habitat is also one that responds well to human hydrological alteration and it is able to use more open and even human-created habitats. It none the less occurs primarily in inaccessible situations, and its biology remains very much unknown.

Streaked Bittern *Ixobrychus involucris* (Vieillot)

Plates 11.3–11.5

Other names: Stripe-backed Bittern, Azara's Bittern, Pygmy Bittern, Little Red Heron, Variegated Heron in English; Avetorillo Listado, Garza Enana Amarilla, Mirasol Común in Spanish; Socoí-Amarelo in Portuguese; Blomgios varié in French; Streifendommel in German.

Ardea involucris Vieillot, 1823. Tabl. Enc. Meth. 3, p. 1127—based on Azara No. 361: Paraguay.

Description

The Streaked Bittern is a small, sandy heron with a black crown stripe and heavily buff and black streaked back.

Adult: The Streaked Bittern has a prominent black stripe along the centre of its crown from the forehead down the back of the head. The very slender bill is yellow-brown, dusky at the tip and upper surface and pinker at the lower base. The irises are pale yellow. The hind neck is sandy-buff. The back and upper wings are sandy-buff heavily streaked with black. The flight feathers are dark buff with tips rufous or cinnamon-buff. The throat and undersides are white streaked with buff. The legs and feet are green-yellow.

Variation: The sexes are alike in plumage. Northern populations are smaller than southern populations.

Juvenile: The juvenile is not described but is presumably similar to the adult.

Chick: The down of the chick is buff above, greyer below.

Voice: The "Auk" call is a soft gulping or croaking. The "Gargle" call is a rapid stuttering. Both are used together as territorial calls (Behrstock 1996).

Weights and Measurements: Weight: 73.4–104.0 g; length: 28–33 cm.

Field characters

The Streaked Bittern is identified by its black streaks on a yellow-buff background. It flies weakly, and quickly flies into cover when disturbed. It is distinguished from the Least Bittern by its streaked yellow-buff (not green-black) back, light wings, and narrow dark crown (not full dark cap) and by voice.

Systematics

The Streaked Bittern is one of the small bitterns, *Ixobrychus*, which share similar plumage, white eggs, scutellate tarsi, and 10 tail feathers. It is the only one with streaking on the back. The similarity or distinctiveness of the northern and southern populations is unclear. The northern birds tend to be smaller, than southern birds, but intermediate-sized birds are also known.

Range and status

The Streaked Bittern occurs in northern and southern South America.

Breeding Range: The species has a wide but apparently disjunct distributions in South America. It has been thought that the northern, southern, and western populations were separated, but recent records may be beginning to fill in distribution gaps within the Amazon basin (Terborgh *et al.* 1984, Olmos and Souza 1988). It breeds in Colombia (Atlántico, Magdalena, Meta), north Venezuela

(Carabobo, Aragua, Portuguesa), Guyana, Suriname, Trinidad), southeast Bolivia (Tarija) (Scott and Carbonell 1986), Paraguay, Brazil (Rio Grande do Sul to Minas Gerais—Parrini and Pacheco 1997), Uruguay, north and central Argentina (Rio Negro, Corrientes—Darrieu and Camperi 1997), and central west Chile (Aconcagua, Llanguihue).

NON-BREEDING RANGE: The status of the species in the Amazon is not clear. It has now been reported outside its breeding range in Brazil (Piaui, Maranhão, and Sergipe) (Olmos and Souza 1988, Parrinl and Pacheco 1997, F. Olmos pers. comm.) and perhaps Ecuador (northern Napo) (Behrstock 1996). Similarly, a sighting in southeast Peru (Madre de Dios—Terborgh *et al.* 1984) covered a few months without evidence of breeding. Determining whether breeding or non-breeding birds occur in these areas is crucial to understanding the range and status of the species.

MIGRATION: The migratory situation in this species is similarly unresolved. In some places birds occur all year (Chile, Brazil, Argentina) but others it is reported seasonally (Suriname, Argentina, the Amazon) suggesting migration or other types of populations shifts.

STATUS: There is very little known about the status of this species as no quantitative or trend information

is available. It is widespread in South America and probably occurs over a much wider area than is now known. It is reported to be common in some parts of its range, such as Argentina and Colombia, and is likely more common than is presently appreciated (Morales 2000).

Habitat

The Streaked Bittern usually uses dense fresh-water marshes and reed beds with tall emergent grasses or sedges (*Scirpus, Typha*). However, it now appears that it is not restricted to this habitat in that it also has been recorded in rice fields, along fast mountain streams, and in the high Andes surrounded by agriculture and pasture (Scott and Carbonell 1986). In Brazil it was recently encountered and reported on the shore of an artificial lake surrounded by dry land vegetation (Olmos and Barbosa 1988). Although occurring primarily in the lowlands, a breeding population occurs at 3500 m in the Argentinean Andes (Scott and Carbonell 1986).

Foraging

Like other small bitterns, this species probably feeds by Standing and by Walking. It has been reported to move quickly through the reeds, running up and down with incredible agility. It feeds alone or in pairs at night. When disturbed it quickly adopts the Bittern Posture, freezing and elongating itself and keeping its underside facing the intruder. This bittern is seldom flushed and flies clumsily before diving back into cover. It eats small fish (*Lebestes*), crustaceans, insects, including dragonflies, dragonfly larvae, and water beetles (*Dytiscus*).

Breeding

Nesting is reported from July to October in the north (Trinidad) and October–December in the south (Argentina). It nests solitarily in marshes in thick patches of sedge or rush. Nests are small inverted cones or platforms made of stems of herbaceous plants (rushes, etc.). Nests are attached to the emergent herbaceous plants and placed 30–60 cm above the water. No observations appear to have been made of courtship rituals. Eggs are yellow, green, and green-yellow. Egg size is 32.7–33.7 mm × 25.7–26.1 mm in Chile and 32–33.6 mm × 25.3–25.8 mm in Argentina, with an average of 31 × 24.5 mm in Trinidad. Most of the few clutches found have been 3 eggs. Nothing is known of the incubation, nestling, fledging, or postfledging periods of the reproductive cycle.

Population dynamics

The population biology and demography of this species is unknown.

Conservation

This is a little known species, and the first step in its conservation is to increase basic knowledge of its distribution, breeding locations, and habitat requirements. It is considered Data Deficient by the HSG (Hafner *et al.* 2000).

Research needs

The Streaked Bittern remains the most enigmatic and least-known representative of the small bitterns. Patterns of geographic variation, if any, among the northern, western and southern populations need to be examined. There is a need to understand its habitat use, especially its use of man-altered habitats. It is necessary to determine if the species is migratory and where it nests or spends non-breeding periods and its biological overlap with the Least Bittern. Determining the range and breeding range of the species in the Amazon is a top priority. The species should be looked for from the Andes slopes to the Atlantic and also in the Orinoco.

Overview

Very little is known about the basic biology of this species. It occurs in extremely dense vegetation where observations are difficult. Consequently, we know only that it feeds within dense reed beds or marshes probably on fish and insects. Recent information showing its use of atypical and human-altered habitat broadens the perspective on its habitat use. Similarly, recent information on its occurrence over a much larger range, may lead to revising our understanding of its distribution, movements, and annual cycle. Much more needs to be understood about this species.

Least Bittern *Ixobrychus exilis* (Gmelin)

PLATES 10.6, 10.7

Other names: Nitlin, Gaulin in English; Avetorillo Panamericano, Garcilla, Garcita, Martinetito, Mirasol Menudo, Gallito, Alcaravancito, Garza Enana, Ardeola in Spanish; Socoí-vermelho in Portuguese; Petit blongios, Blongios minute, Kio jaune, Crabier, Mang lanmé, Martinet in French; Amerikanische Zwergdommel in German.

Ardea exilis Gmelin, 1789. Syst. Nat. 1(2), p. 645: Jamaica.

Subspecies: *Ixobrychus exilis erythromelas* Vieillot, 1817: Paraguay; *Ixobrychus exilis bogotensis* Chapman, 1914: Suba marshes, Bogota, Colombia; *Ixobrychus exilis pullus* van Rossem, 1930: Tobari Bay, Sonora, Mexico; *Ixobrychus exilis peruvianus* Bond, 1955: Vegueta, Lima, Peru.

Description

The Least Bittern is the smallest heron, a pale buff bird with a dark crown and back and buff-coloured wing patches.

ADULT: The adult male has a green-black cap. The irises are bright yellow, sometimes almost orange, and the lores are lime-green. The bill is yellow with brown-horn along the top ridge. The sides of the face and neck are buff to tawny-chestnut, and the hind neck and shoulders are bright buff to chestnut. The chin is white, as are the throat and underparts, with buff-brown throat streaks. Conspicuous tufts of feathers on either side of the upper breast, characteristic of the small bitterns, are black-brown with pale edges. The back, rump, and tail are glossy green-black. The flight feathers are black, whereas the secondary feathers are chestnut, and the upper wing coverts are buff, forming a contrasting wing patch. White lines bordering the upper wing are conspicuous both at rest and in flight. The legs are green in front and yellow behind. The soles of the feet are yellow. The legs and feet may be suffused with brown or black. In courtship, the Greeting Ceremony, and at other times the base of the lower bill flushes red (Kent 1986c).

VARIATION: The female averages larger than the male. It has chestnut, rather than black upper parts, a less prominent crown, darker neck stripes, dark brown chest streaks, and paler wing patch. The Least Bittern varies geographically and individually (Dickerman 1973b). Compared with *exilis* described above, *erythromelas* has the sides of its face and neck rufous; the female has less streaky underparts. *Peruvianus* is larger, with the longest bill, and duller colour. *Bogotensis* has a richly coloured ochraceous side of the head and buff underparts; the front of its legs and feet are almost entirely black; the female lacks the rufous tinge to the blackish-brown back found in *erythromelas*. *Pullus* is notably dark above, its coverts scarcely contrasting with its flight feathers. Other subspecies have been proposed such as *limoncochae* in Ecuador. Some variation appears to have both geographic and individual components. A variant once known as 'Cory's Bittern' has the pale areas, especially on the neck, undersides and wing coverts, replaced by dark chestnut. This form has been reported from Ontario, Florida, and Brazil (Teixeira and Alvarenga 1985). Additional study in needed of plumage variation in this species.

JUVENILE: Immature birds are similar to adult females but have a paler, browner crown and back, and dusky streaks on the wing patch. The streaking on their throat and breast is browner than in the female, which gives a striped effect.

CHICK: The young's head, back, and neck are covered with pale buff down, with the longest down on the head forming a crest. The underparts are white. Chicks vary overall from pale to deep buff. The skin of the head, neck, and back is pink with the abdominal area grey. The bill is pink and black toward the tip, and the irises are white. The legs are pink.

VOICE: The Least Bittern is very vocal. The 'Coo' call is the low-pitched, quiet, dove-like advertising call of the male. It is rendered 'uh, uh, uh, oo, oo, oo, oo, ooah', 'poo, poo, poo'or 'woh,woh,woh'. The 'Tut' call is the year-round disturbance and alarm call, rendered 'tut, tut, tut', 'hah, a, tut, tut, tut', or 'rick, rick, rick'. The 'Gack' call, rendered 'gack, gack', is the contact call, also given from the nest. The 'Ank' call, rendered 'ank, ank' or 'kuk, kuk' or 'gik, gik' is the disturbance and flight call. The 'Gra' call, rendered 'gra, a, a', is an aggressive call.

WEIGHTS AND MEASUREMENTS: Weight: 73–95 g; length: 28–36 cm.

Field characters

The Least Bittern is identified by its small size, buff and black colour scheme, with black back and light lines on the side of the back, and buff wing patches in flight. It tends to fly rapidly with fast wing beats, but weakly and usually for short distances. When startled into flight, it appears clumsy and flies with its legs dangling and neck extended. It is distinguished from the immature Green Heron by its smaller size, wing patch, and deep buff to chestnut colours. It is distinguished from the Streaked Bittern by its green-black back (not streaked yellow), full black crown (not narrow black crown streak), and dark (not buff) wings.

Systematics

The Least Bittern is one of the small bitterns, *Ixobrychus*, that share similar plumage, white eggs, scutellate legs scales, and 10 tail feathers. It is closely related to the Little and Yellow Bitterns, with which it shares a slender bill, uniform dorsal coloration, and moderate plumage sexual dimorphism. The dark colour morph was once considered a separate species (Pittaway and Burke 1996). Geographic variation is classified into five subspecies.

Range and status

The Least Bittern has an extensive breeding range in North America, Central America, West Indies, and north, west and east South America.

BREEDING RANGE: *Exilis* breeds from southeast Canada (south Quebec, south Ontario, south New Brunswick), east USA (Maine south to Florida and the Gulf Coast), central USA and Canada (north Michigan, central Wisconsin, northwest Minnesota, northeast North Dakota, south Manitoba, south to Texas), west USA (south Oregon, central and south California (Kirk 1995), Wyoming (Grove and Henny 1990), Arizona, New Mexico), Mexico (Baja California, Baja California Sul, Tamaulipas south on the Gulf Coast to Tabasco), the north West Indies including Cuba, Jamaica, Bahamas, Cayman Islands, Puerto Rico (Norton *et al.* 2003), Hispaniola, Guadeloupe (Benito Espinal and Portecop 1984), and Central America including Belize although the status is unclear (Jones 2001a), Guatemala, El Salvador, Honduras, Nicaragua, Costa Rica (Lago Caño Negro, Poloverde) (Alvarado 1998, Jones 2001b), to Panama. *Pullus* occurs in northwest Mexico (Sonora and Sinaloa coastlands). *Erythromelas* occurs in Central and South America in south Panama (south of Chagres River), north Colombia (Cauca, Antioquia, Santa Maria, Boyacá, Cundinamarca), north Venezuela (Guárico, Miranda), Trinidad, Guianas, east Ecuador (Napo), east and south east Brazil (east Amazonas to Gois, Mato Grosso (Cintra and Yamashita 1990), and Santa Catarina), north Bolivia (Beni), Paraguay (Presidente Hayes, Cordillera—Lowen *et al.* 1997), and northeast Argentina (Misiones, Corrientes). *Bogotenisis* occurs in the savanna east of the Bogotá, Colombia. *Peruvianus* occurs in west coast Peru (Libertad and Lima province).

NON-BREEDING RANGE: Although some birds remain far north, the northern populations in North America winter in the USA Gulf Coast (Louisiana, Texas, Florida), south California, Mexico (Baja California), Central America, Greater Antilles, Lesser Antilles (Dominica—Norton 1993) to north South America.

MIGRATION: In the north of its range, the Least Bittern migrates from areas where frost prevails in winter. Southward migration occurs in October. Western birds migrate through western Mexico to about Costa Rica. Eastern birds migrate along the Gulf of Mexico coast and the West Indies, reaching northern Colombia. Return migration is the end

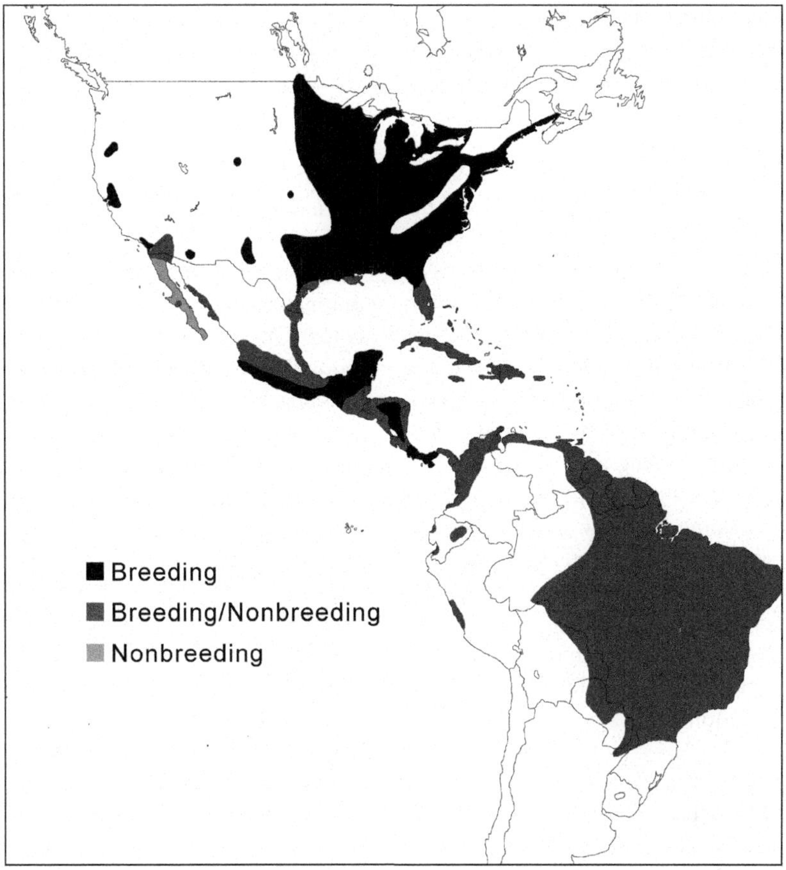

of February to early April, or later in the far north. Postbreeding dispersal occurs as well. Dispersal records in the north include Newfoundland, northeast Manitoba, British Columbia, Iceland, Azores, and Barbados. The southern subspecies appear to be sedentary.

STATUS: The Least Bittern is widespread in North America, but little is known of its population status because of the difficulty of censusing it. Calling male density ranges from 0.4 to 12 nests per hetare (Gibbs *et al.* 1992a). Although decreasing in some areas and increasing in others, populations in North America appear stable overall, but information is poor (Morales 2000). It is common on Cuba, Cayman Islands, and Puerto Rico; uncommon on the Bahamas, Hispaniola, and Guadeloupe; rare on Dominica and Virgin Islands. It is uncommon in Costa Rica (G. Alvarado pers. comm.) and uncommon to rare in the rest of Central America. It is common in Trinidad and Colombia. As far as is known, populations are stable over most of South America and its known range is being extended by new observations there (Morales 2000). *Bogotensis* is still common in its restricted range but its wetlands are threatened (Scott and Carbonell 1986). The status of *peruvianus* is not clear, but it occurs in a small range and uses disturbed areas (Scott and Carbonell 1986).

Habitat

The habitats used by the Least Bittern differ considerably from place to place but usually include very dense vegetation in water, some woody growth, and open water patches. In addition to fresh water marshes, they use lake edge marshes

(Rodgers and Schwikert 1999), salt marshes in temperate areas, and mangroves in the tropics. It occurs from the coast to 3100 m in South America.

Foraging

Characteristically, the Least Bittern feeds by Walking slowly, stalking through the reeds or along the edge of dense reed stands, or on branches over the water. It walks in Crouched posture, with its neck extended and its bill nearly touching the water. From this position it launches a rapid strike. It also feeds by Standing in place, and will build feeding platforms to use while Standing. It uses Neck Swaying and catches insects by Gleaning and Standing Flycatching. Among active techniques is Wing Flicking. It tends to change foraging spots after a strike by Walking to a new spot.

The principal prey of Least Bitterns is small fish (*Fundulus, Umbra, Lepomis, Perca*). Its overall diet also contains crabs, crayfish, leeches, frogs, tadpoles, salamanders, small mammals (shrews, mice), and insects (grasshoppers, dragonflies). It is likely that it eats small birds on occasion.

Breeding

In the north, the Least Bittern nests in the spring and summer, May–August. Nesting in the tropics is more varied and more information is needed. It nests in February–March in Colombia. The nest primarily is in thick herbaceous marshes, with woody vegetation, near open water. The most commonly used habitat is cattail marsh (*Typha*) (Frederick *et al.* 1990, Gibbs *et al.* 1992a). Other herbaceous plants used include *Mariscus, Carex, Scirpus, Sagittaria, Phagmites*, and mixtures of these. Woody plants include *Cephalanthus* and mangroves (*Rhizophora*). It nests solitarily and in small groups. Nests in groups can be as close as 1 m a part resulting in densities of 3–4 nests per 100 square meters (Kushlan 1973b, Rodgers and Schwikert 1999). Home ranges during nesting averaged 9.7 ha (Bogner and Baldassarre 2002b). The nest is a small platform of herbaceous stalks or sticks. Nests are 15–20 cm wide and 5–12 cm thick placed over the water usually in herbaceous vegetation. A canopy may be made by pulling reed stalks down over the nest. Nests are also placed in bushes next to or over the water. The nests are less than 1 m above the water and near open channels or pools within the march (Gibbs *et al.* 1992a). Primarily the male constructs the nests, which continues into incubation (Weller 1961). Male bitterns advertise with the Coo call, but little else is known about courtship. Calling rates are highest during the nest initiation period as contrasted to later in nesting (Bogner and Baldassarre 2002). After pairing Non-contact and Contact Bill Clappering occur. The Greeting Ceremony is used in nest relief. The bird on the nest gives a low Gra call. The arriving bird erects its crest and body feathers and both birds Bill Clapper. Territorial defence is well documented. The Bittern Posture seems to be used as a defensive display. It also uses a more aggressive Forward display, with head withdrawn and wings spread.

The eggs are white and measure about 31 × 24 mm. Clutches in the northern part of the breeding range tend to be of 4–5 eggs. Mean clutch sizes were 3.8 in South Carolina and 4.08 in Florida (Post 1998, Rodgers and Schwikert 1999). In the tropics clutches of 2 or 3 are more usual. The range is 2–7. Incubation begins with the first or second egg and takes 17–20 days. Both sexes incubate and rarely leave the nest unattended. The female incubates more than the male, which eventually abandons the nest if a mate is lost (Aniskowicz 1981). Double broods have been reported (Bogner and Baldassarre 2002b). Hatching is asynchronous occurring over 3–4 days. The young are semialtricial and are fed by regurgitation, the male apparently doing about twice as much of the feeding of young as the female (Weller 1961). Chicks develop quickly, particularly their legs and feet. They can grasp by day 3, assume the Bittern Posture by day 4, and leave the nest temporarily by day 5. They begin to wander from the nest after 6 days and leave the nest completely in 13–15 days. They return to be fed in the vicinity of the nest until they fledge in 3.5 weeks. Hatching success is 70%–84% and the success of a nest to fledge young was 20%–73% (Gibbs *et al.* 1992a, Post 1998, Rodgers and Schwikert 1999, Bogner and Baldassarre 2002b). In Florida, only 47% of the nests produced 2-week-old young (Rodgers and Schwikert 1999). Nesting success in mixed-species colonies was greater than for bitterns nesting alone (Post and Seals 1993). Losses are due to flooding, wave action, abandonment, and predation, including mink (*Mustela*), wrens (Cistothorus), snakes, raccoons (*Procyon*), crows (*Corvis*), turtles, and raptors.

Population dynamics

Nothing is known about the population biology or demography of this species.

Conservation

Conservation of this species requires protection and management of its marsh habitat. Water impoundments and wetland construction for various purposes increase the potential habitat for the species as it often nests in cattail marshes created by human activities (Post 1998, Gibbs *et al.* 1992a, Horstman *et al.* 1998). In North America, many wetlands are managed by the government for waterfowl and similar purposes, and these provide important opportunities to also provide habitat for bitterns (Gibbs *et al.* 1992a). Wetland management should take the needs of Least Bitterns into consideration when designing management actions. Standardized monitoring needs to be conducted to evaluate population trends. In Central and South America, information is needed on the breeding distribution and habitat needs of the species.

Research needs

Nothing is known of the courtship and social systems of the species, and these topics need to be studied further. In North American, standardized techniques need to be developed and employed regionally to assess population sizes and trends. Studies are needed of habitat use and the distribution of bitterns during migration and non-breeding to identify major stopover and wintering sites for North American birds. Management techniques need to be developed to create optimal combinations of marsh and open habitat conducive to Least Bittern nesting. In South America, surveys need to be conducted to determine the breeding range of the species through the Amazon and Andes. The status of bitterns in west Peru, and their habitat requirements there, need to be addressed.

Overview

The Least Bittern is a widespread species that characteristically depends on thick emergent marshes. Its small size, thin body, and long toes and nails allow it to move through dense thicket with ease. It feeds quietly by Walking and Standing and often changes locations. Where food is abundant it makes feeding platforms to use within its feeding territories. By feeding from the vegetation it uses and seems to prefer deepwater situations. Its most common habitat is cattail marshes, which tend to increase in disturbed areas. Its diet seems to be mostly fish. This cryplic, inconspicuous small heron can be locally abundant, and much remains to be learned about its sociality.

Little Bittern *Ixobrychus minutus* (Linnaeus)

PLATES 10.8, 10.9

Other names: Minute Bittern, Leech Bittern, New Zealand Bittern, Australian Bittern, Black-backed Least Bittern in English; Vetorillo Común in Spanish; Blongios nain in French; Zwergdommel in German; Woudaapje in Dutch; Волчок in Russian; Xiao weijan in Chinese.

Ardea minuta Linnaeus, 1766, Syst. Nat. ed. 12, p. 240: 'Helvetia, Aleppo', restricted type locality, Switzerland.

Subspecies: *Ixobrychus minutus podiceps* (Bonaparte), 1855: Madagascar; *Ixobrychus minutus payesii* (Hartlaub), 1858: Casamanse, Senegal; *Ixobrychus minutus novaezelandiae* (Potts), 1871: Westland, South Island, New Zealand; *Ixobrychus minutus dubius* Mathews, 1912: Hermann's Lake, South West Australia.

Description

The Little Bittern is a small heron with a dark back and cap and buff-white neck and wing patches.

ADULT: The male has a green-black crown with elongated feathers forming a modest crest. The bill is yellow or yellow-green with a dark brown upper edge. The irises are yellow, and the lores are yellow or green. The side of the face is grey washed with a vinaceous tinge. The chin and throat are white with a buff centre. The back and tail are green-black. The flight feathers are green-black, which contrast on the upper wing with buff-white wing patches. The sides of the upper breast have small tufts of elongated black feathers. The undersides are buff-white with minimal brown streaking that is variable among individuals with the underwings white. The legs also vary among individuals from green, green-grey, yellow, green in front, and yellow behind. The toes are long. In breeding the plumage is brighter and the upper breast feathers are longer and looser. In courtship the lower bill (of both sexes) flashes red briefly during copulation, nest relief, and other excitement. The lores and orbital skin flush dull red.

VARIATION: The species is sexually dimorphic. The female is smaller and a duller colour. Its crest is black and less glossy than the male. It has a brown or rufous tinge to the dark colours, which also show some streaking. Wing patches are pale brown-buff, slightly streaked and smaller than in the male. The underparts are striped in brown. There are no known differences between sexes in soft-part colours. Geographic variation has been recognized in five subspecies. *Payesii* is smaller with shorter wings than *minutus*; the neck and wing patches are more red-brown to chestnut rather than the buff of *minutus*; the irises become red-brown in courtship; the legs in breeding are olive-green in front and yellow behind. *Podiceps* is smaller than *minutus* or *payesii*; the adult male has deep rufous on the neck extending over the whole underparts and underwing and becoming chestnut on the upper parts; the immature bird is darker than *minutus*. *Dubious* has a shorter, thicker bill; the neck and wing are chestnut to rufous; the flight feathers are dull black or dark grey-brown contrasting with the buff wing patch; the immature has pale primaries with fulvous tips; the female is not well marked. *Novaezelandiae* was larger and darker, the back and scapulars are dark red-brown with rufous lining to the feathers; the hind neck is chestnut; the wing patch is spotted chestnut; the female is more heavily streaked below.

JUVENILE: The juvenile is more boldly streaked below and above with a red-brown tinge. The crown is streaked brown. The wing patches are heavily mottled with brown and buff.

CHICK: The downy chick is pink-buff above (*minutus*), reddish-buff (*payesii*), pink-buff to brown (*dubious*) and white below. The irises are black-brown. The bill is dull pink turning to grey. The facial skin is blue-grey becoming olive-yellow. The legs and feet are olive-grey with pink toes.

VOICE: The 'Kohr' call is the distinctive and characteristic grunting or barking advertising call used during breeding. It is variously rendered as 'kohr, kohr, kohr, kohr', 'hork, hork, hork', 'cor, orr, orr, orr', or 'gogh, gogh, gogh, gogh' and also 'hogh', 'rru', and 'woof'. The 'Kwer' call is a flight call. It is rendered as 'kuk-kuk, kuk-kak', 'cuck, cuck,cuck cuck', 'cra, a, a, a, k', 'quer', or 'ker-ack'. It is low pitched and abrupt, and sometimes proceeded by a higher pitched 'quee'. The 'Koh' call is the disturbance call. The 'Gek' call is a repeated call given frequently at the nest site, rendered as 'gek, gek, gek, gek' or 'ek, ek, ek, ek'. A similar 'Gak' call is the alarm and threat call. It can be rendered as 'kuk', 'gat', 'gack', or 'yick'. The 'Aark' call is an anxiety call. The 'Goo' call, rendered 'goo, goo', is used with the Greeting Ceremony. Young beg with 'tu, tu, tu, tu, tu'.

WEIGHTS AND MEASUREMENTS: Weight: 140–150 g; length: 25–35 cm.

Field characters

The Little Bittern is identified by its small size, dark cap and back, and buff-grey wing patches offsetting dark flight feathers are conspicuous in flight. Its flight is rapid for a heron, flying with rapid, shallow, clipped wing beats, legs dangling, often dropping into cover. It is distinguished from the Yellow Bittern by being slightly larger, having a shorter bill, its black (not brown) back, and white to grey-buff (not yellow buff) wing patch. It is distinguished from the Cinnamon Bittern by its dark (not cinnamon) back and cap. It is distinguished from the Dwarf Bittern by being lighter (not dark grey) and having wing patches.

Systematics

The Little Bittern is one of the small bitterns, *Ixobrychus*, that share similar plumage, white eggs, scutellate tarsi, and 10 tail feathers. It is closely related to the Least and Yellow Bitterns, with which it shares a slender bill, uniform dorsal coloration, and moderate plumage sexual dimorphism. The Little Bittern covers a large discontinuous range, with other small bitterns filling in the range gaps. *Novaezelandia* was sometimes considered a different species, due to its larger size (e.g., IUCN 2003). *Payesii* and *podiceps* are also sometimes considered to be separate species. The intercontinentally disjunct range suggests that the Little Bittern, as currently recognized, may indeed represent more than one species, a question that deserves additional investigation using molecular techniques.

Range and status

The Little Bittern occurs in Europe, west Asia, Africa, Madagascar, north India, Australia, and New Guinea.

BREEDING RANGE: The north boundary of the breeding range of *minutus* includes England (Allport and Carroll 1989), Netherlands (Bekhuis 1990), Belgium, north Germany, to Estonia, Russia (west Siberia), Azerbaijan, Kazakhstan (Lopatin *et al.* 1992), Turkmenistan, Uzbekistan, Tajikistan, Kyrgystan, and west China (Sinkiang). It breeds in North Africa (Morocco to Tunisia, north Egypt—El Din 1992), Turkey, Israel, Saudi Arabia, Bahrain (possibly breeding), Iraq (possibly breeding), Iran, Pakistan (Sind), India (Kashmir (Holmes and Hatchwell 1991), Uttar Pradesh, Assam), and Nepal. *Payesii* occurs in Africa south of the Sahara in Mauritania, Senegal (Morel and Morel 1989), Gambia, Guinea–Bissau, Mali, Sierra Leone, Ghana, Gabon, Principe, Nigeria, south Sudan, south Somalia, Uganda, Rwanda, Kenya, Tanzania (Baker and Baker 2004), southeast Congo, Zambia, Malawi, Mozambique, Zimbabwe, Namibia, east and south South Africa (Transvaal, Natal, Cape Province). *Podiceps* is confined to Madagascar. *Novaezelandiae* occurred only on South Island, New Zealand. *Dubius* breeds in Australia (Queensland, New South Wales, south Western Australia, north Western Australia) (Jaensch *et al.* 1987, Jaensch 1988).

NON-BREEDING RANGE: *Minutus* from Europe and west Asia move south in winter. A few birds remain in west and south Europe (Mediterranean, Ireland, Britain, Germany, Netherlands, and north Africa) (Cramp and Simmons 1977, Nankinov and Kantarzhiev 1988). Most birds winter in Africa south of the Sahara, mostly in east Africa but also west to Senegal and south to as far as South Africa. *Minutus* from north India appear to remain there during non-breeding but the species migrates, rarely, to west Xinjiang China (MacKinnon and Phillips 2000). *Podiceps* is probably sedentary; there is some evidence of its moving to Africa in the dry season (Brown *et al.* 1982) but this is refuted (N. Baker and E. Baker in prep.). *Dubius* also appears to be mostly sedentary but it also occurs in winter in New Guinea, mostly in the southern lowlands but with a single record in the north (Jaensch 1995, 1996). As far as is known these are wintering birds from Australia, although there remains a possibility of its breeding in New Guinea (Beehler *et al.* 1986).

MIGRATION: *Minutus* is migratory across most of its breeding range and has a significant postbreeding dispersal. Birds in Europe move south in August–October. They fly singly and in small groups at night. Western birds move through Italy, Spain, and France and along the Atlantic coast, (Nankinov 1999). They cross the Mediterranean and Sahara in a broad wave. Birds from the east cross Israel, Iraq, Arabia, and Egypt, also in a broad front. Movement in Africa is less clear. Return migration is in March–April. Birds regularly overshoot and land north of the breeding range. *Minutus* in the Middle East are partially sedentary. *Minutus* from north India, *payesii*, and *podiceps* are at least partially sedentary, with local movements that

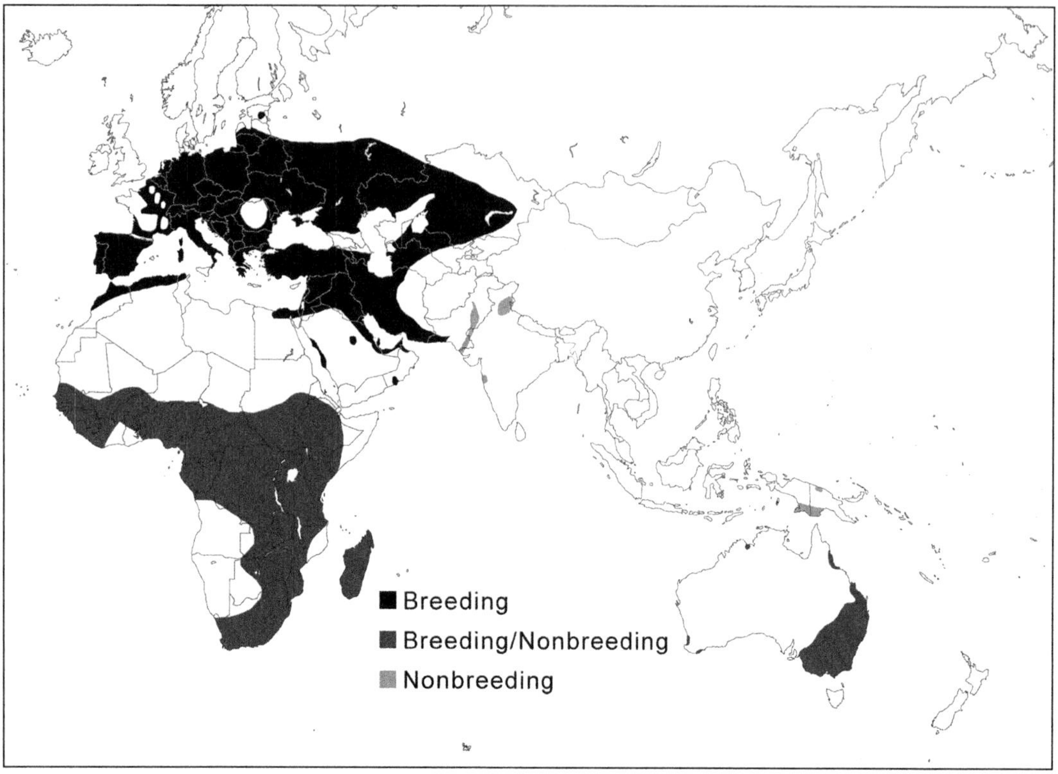

are not clearly understood. *Payesii* shifts in response to rainfall and drought. *Podiceps* is now understood not to migrate occasionally to Zanzibar as suggested by Brown *et al.* (1982). *Dubius* is probably migratory, shifting after wetlands dry out (March–April) from south to north and inland to the coast (e.g. Carpenter and de Jong 2001), and also to south Papua New Guinea. Return migration to the southern breeding areas in Australia is in August–September. The Little Bittern ranges widely in postbreeding dispersal, moving in all directions. Dispersal records include Iceland, Faeroes, Azores, Madeira, Canary Islands, and Scandinavia. Dispersal records in the east include Lord Howell Island and New Zealand (O'Donnell and Dilks 1988).

STATUS: The species is widespread and common in many areas within its range. It has been decreasing in Europe, especially from the 1970s to the 1990s, due to habitat loss (Nankinov 1999). Its nesting distribution is now fragmented, and the species appears to be in rapid decline in west Europe (Marion *et al.* 2000). Its overall population is 37 000–107 000 pairs, the range reflecting uncertainties in status in eastern Europe—Romania, Ukraine, and Russia—which together support the greater portion of the European population (Marion *et al.* 2000). The Little Bittern is common in north Africa, is increasing in Egypt (El Din 1992), and is more common in Arabia than previously appreciated. It has been under-represented on surveys in Tanzania; a guess at its population there puts it under 10 000 adults (N. Baker and E. Baker in prep.). It is rare in South Africa, under 100 pairs. It is uncommon in Madagascar and known from only a few places. It is abundant in parts of India (1000–2000 pairs in Kashmir). The population in New Zealand went extinct for unknown reasons—it is one of a few contemporary herons that has suffered extinction. The Little Bittern is rare and very localized in Australia its core being in the Murray-Darling Basin. It has declined in west Australia due to habitat loss but may be more common in other areas than presently appreciated (Jaensch 1989). Its status (breeding, wintering) in New Guinea is uncertain.

Habitats

The habitat used by the species is varied across its huge range. Most typically it uses fresh water wetlands having thick herbaceous vegetation with trees or bushes interspersed nearby. These habitats include peat bogs, reed swamps, edges of lakes, pools, reservoirs, oases, swamps, wooded and marshy edges of streams and rivers, wet grasslands, mangroves, salt marshes, and lagoons. In east Africa it prefers smaller, well-vegetated swamps, marshes, and drainage ditches (N. Baker and E. Baker in prep.). It also can be found in forests. It occurs in lowlands and up to 1500 m in Madagascar and 1800 m in the Himalayas. Typical herbaceous plants used in these habitats include *Scirpus, Typha, Phragmites, Baumea*, and *Juncus*. Shrubs and trees used include *Muehlenbeckia* and *Melaleuca*. It uses human habitats including rice fields, ponds, crop fields, vegetable gardens, and sugar cane fields. Little Bitterns can be very tolerant of humans and nest in places regularly visited by people (Cempulik 1994).

Foraging

The Little Bittern feeds by Walking slowly at the water edge stalking prey, either from the ground or more characteristically from a perch. It also Walks Quickly using a Crouched posture, with head forward, in rapid steps. It Stands at the edge of cover on a perch. It feeds with its head and neck withdrawn. As it sees a prey item, it slowly extends its neck and then Stabs. It sometimes feeds by Pecking, jabbing its bill in the water, and using an insect for bait (Baumann 2000). It is a solitary feeder generally within territories held long term. Its activity periods appear to vary. It is primarily crepuscular over much of its range, but feeds at night and also at times during the day. African birds are primarily diurnal (Langley 1983). When alarmed it assumes the Bittern Posture.

The diet is varied, comprising fish (*Perca, Esox, Alburnus, Blicca, Cyprinus, Gambusia, Gobio, Eupomotis, Leuciscus*), frogs and tadpoles (*Rana*), reptiles, eggs and young birds (Olioso 1991), shrimp, crayfish, worms, insects such as crickets (*Gryllotalpa*), grasshoppers, caterpillars, water bugs, beetles (*Notonecta, Naucoris*), beetle larvae, dragonflies (*Libellula, Aeshna*), and spiders. The diet differs in various places. In some places it has primarily a fish diet (Langley 1983, Holmes and Hatchwell 1991) and in other places, such as Italy, insects predominate.

Breeding

The nesting biology of the Little Bittern has been well studied (Langley 1983, Darakchiev *et al.* 1984, Gerard 1986, Holmes and Hatchwell 1991, Hoyer 1991, Bozic 1992, Lopatin *et al.* 1992, Cempulik 1994, Martinez-Abrain 1994, Caballero 1997). As expected over such a large range, its nesting season is variable. Nesting occurs in the spring in the north of the range, May–July in Europe and India. It nests in the rainy season or just after the rainy season in the tropics. Nesting is May–July in north Africa; July–October in west Africa; June–September in Nigeria; May–September in Congo; July, November–December in Uganda; March–April; June in Zambia; April–May in Malawi; February, September, November–December in Zimbabwe; March in Namibia; June–February in South Africa; and October–January in Australia. The species nests in thick herbaceous vegetation, especially near open water pools, but also in trees or bushes usually over water, and has also been found nesting in trees over dry land. The Little Bittern nests solitarily, but also and perhaps more typically in loose colonies with nests as close as 5 m but usually 30–100 m apart. It likely is extremely residential, in that nests may be reused in consecutive years (Barbieri and Boileaum 2000). The nest is a platform with a conical base, 15–20 cm across, and 10 cm thick. In South Africa more substantial nests were 20–35 cm across. The nest is made of stems of herbaceous vegetation, lined with finer material. The nest is typically inserted in reeds, rushes, grass, or papyrus. However, in some areas and situations they nest in trees and bushes and make stick nests.

It is built by the male, who starts during the display period. Early in the breeding season, males establish breeding territories and give the Kohr call, staking out the territory and advertising. When calling, the lower bill flushes red. Territories are defended by an Upright display, Ground and Aerial Supplanting Attacks, and a threat display in which the bird places its side to the opponent, spreading the wings, lifting one and lowering the other. Males choose a nest site and begin building while continuing to advertise with the Kohr call. The males also use Circle Flights as part of the display. A flight also has been described in which the neck is extended and the head held below the body. Upon formation of the pair bond, birds participate in Contact and Non-contact Bill Clappering, during which the pair cross their necks. The Greeting Ceremony includes the arriving bird approaching the nest, with Bill Clappering, feathers raised, Crest Raising, and gives the Goo call. The bills flush red during the Greeting Ceremony. Upon completion, birds will Bill Clapper. Paired birds will remain together through the nesting season.

Eggs are chalky white. They are laid at intervals of 1–3 days. Size averages 36 × 26 mm in Europe and 34.6 × 26.6 mm in South Africa. Clutch size varies geographically, 5–6 in Europe and 3–4 in the tropics and South Africa (Langley 1983). The overall range is 2–7 eggs. Replacement clutches occur if eggs are destroyed but also after the young fledge. In some case three broods are raised per year (such as in South Africa). Clutch size decreases later in the season (Cempulik 1994). Incubation, by both parents, begins with the first egg and lasts 16–20 days (Langley 1983, Holmes and Hatchwell 1991). Hatching is asynchronous and chicks have their eyes open and legs are fairly developed after hatching. Young are fed in the first 2 days by food deposited on the nest floor. The parents guide the bills of the nestlings to the food. Thereafter, young grasp the parent's bill and are fed directly. Chicks are brooded 8–10 days.

Chicks grow relatively fast. By 3 days they beg by grasping the parent's bill. Chicks assume the Bittern Posture when disturbed. Pinfeathers develop at 4 days. Sibling rivalry is low, despite asynchronous hatching. And there was not found to be a difference in growth rates relative to hatch order or brood size (Holmes and Hatchwell 1991). Chicks grow quickly, climb out of the nest in 1 week and can leave the nest entirely by 14–16 days. Maximum growth takes place at 15 days (Langley 1983). The birds fledge in 27 days flying strongly. Success was 56.6% of eggs hatching to nest departure in South Africa and 70%–71% in India (Langley 1983, Holmes and Hatchwell 1991).

Population dynamics

Females can nest before their second birthday (Langley 1983). Nothing is known about the demography of this species.

Conservation

The conservation of this species is made difficult by its large range and multiple forms. Given the disjunct distribution of the presently recognized subspecies each with its own situation, conservation should be focused on each population. The European population has been deceasing for some time in some areas but is increasing or even expanding in other areas. It is more widespread in Arabia and the Middle East than previously appreciated (Perennou *et al.* 2000). The situation in west Europe is clearly critical, however. The population there is continuing to decrease. In Belgium, Netherlands, and France decreases are associated with habitat alteration. However, the general European decrease continues despite this being a species that is tolerant of human alterations and despite the continued availability of apparently suitable habitat. The European population is migratory, and it may be that juvenile or adult mortality in migration and wintering may be a one cause for the widespread decrease. Its populations in east Europe are probably high, but the data are insufficient to detect trends. The Indian population appears to be stable. It is crucial to locate the important nesting areas for this population and undertake protective measures.

The African breeding population is widespread but uncommon in all regions (Turner 2000). Although the two subspecies can be told apart, the existence of widespread wintering by European and west Asian birds makes interpretation of the African breeding population difficult.

The Australian population is considered to be uncommon, but rather appears to be concentrated seasonally in a few highly suitable breeding sites. It has declined and had to shift its distribution due to past hydrological changes in its rivers (Maddock 2000). It is classified as Endangered in Victoria (NRE 2000), where it represents only 0.7% of bird observations in the state since 1970 (Scientific Advisory Committee 1997). Management and conservation of these river systems are critical to the species in Australia. It has been found breeding outside its known range, and evidence of further range expansions should be sought.

The Madagascar population may be down to 100 pairs and is known from a few widely scattered sights. This is a population that is at risk due to habitat alteration and should be considered Vulnerable. The New Zealand population is extinct, having last been seen over 100 years ago. The extinction of this recognized and distinctive subspecies (Hilton-Taylor 2000) suggests the potential precariousness of populations elsewhere. With dispersal arrivals from Australia it is possible that the species will re-establish itself there. The birds in New Guinea are probably migrants from Australia.

Research needs

The causes of the population decrease in west Europe need to be investigated through demographic study of the roles of migration and nonbreeding adult and juvenile mortality. This may require additional banding studies on the breeding ground and also related study in the wintering area. Comparative simultaneous study of east European populations would allow comparisons to be made. Drought, habitat change, loss of staging wetlands in the Sahara and Sahel, and conditions of wintering wetlands could be key environmental factors to be investigated. Surveys should be undertaken of the Madagascar population of the Little Bittern to determine important areas for remaining birds. Surveys should also be undertaken to determine the breeding status in New Guinea. The taxonomy of the species and subspecies of the small bitterns, particularly of the Little/Least/Yellow Bittern group, should be re-examined. At the same time, patterns of geographic variation and the subspecific taxonomy of the several disjunct populations of the Little Bittern should be undertaken, including using molecular techniques.

Overview

The Little Bittern is a bird of dense marsh vegetation, in which it feeds and nests. It forages in the typical bittern manner of Walking and Standing on marsh plants, old nests, or branches. It catches a diversity of prey, but primarily fish or insects, depending on the locality. It is a solitary feeder returning to its territory daily. By maintaining its territory, it has knowledge of its feeding site and avoids competition or disturbance of its prey. It remains paired through the nesting season and typically renests one or more times if the breeding season is long enough. Although young hatch asynchronously, in the situations so far studied, there is no difference in growth or survival of late-hatched young.

Yellow Bittern *Ixobrychus sinensis* (Gmelin)

PLATES 11.6, 11.7

Other names: Long-nosed Bittern, Chinese Little Bittern, Little Yellow Bittern, Least Bittern in English; Avetorillo Chino in Spanish; Blongios de Chine in French; Chinadommel in German; Yoshigoi in Japanese; Bakaw-tumana in Philipino (in Philippines); Bambangan kuning in Indonesian; Huang Weiyan in Chinese.

Ardea sinensis Gmelin, 1789, Syst. Nat. 1(2), p. 642: China.

Description

The Yellow Bittern is a small, slender, light buff to light brown heron with a black cap and tail, and long thin bill

ADULT: The Yellow Bittern has a dull black to blue-grey crown and crest. The forehead is brown. Its long, thin bill is pale ivory, darker (to black) on the top with a pink tinge, contrasting with yellow-green lores and yellow irises. The throat is sandy-brown with browner stripes, with the sides of the neck buff, often with a pink blush. The back is rufous to sandy-brown with a pink to maroon wash. The wing is characteristically patterned, with a light inner patch and black outer feathers. On the upper wings the major feathers (primaries, secondaries, and coverts) are black, whereas a patch of wing coverts is light sandy-buff. The underwing patch is white, contrasting with the grey of the undersides of the flight feathers. The tail and its coverts are black. The underparts are sandy-brown. The legs are yellow-green. In courtship the base of the bill flushes red.

VARIATION: The species is sexually dimorphic. The female's crown is indistinctly streaked instead of dark. The hind neck and upper back are rufous. The underparts are streaked dark red-brown and buff, with black on the sides of the breast. The throat has chestnut-buff streaks. The wing patch is darker than in the male. There is considerable individual and geographic variation in the shade of the upper and underparts.

JUVENILE: Immature birds lack the dark crown and have a dark streaked head, back, and undersides. The upper bill is dark brown, whereas the lower bill is yellow, as are the irises. The wings and tail are the distinctive black of the species. The legs are olive-yellow.

CHICK: The chick is covered by a pinkish down.

VOICE: The 'Kakak' call, rendered 'kakak, kakak', is the flight call. The 'Crrw' call, rendered 'crrw, crrw', is given in the Stretch and also by near-fledging juveniles (Lansdown and Rajanathan 1993). The 'Ohr' call is a loud, repeated, guttural grunt. A gentler call is given when the bird brings food to the nest. Deep wing flaps are used in Circle Flights.

WEIGHTS AND MEASUREMENTS: Weight: 92–104 g; length: 38 cm.

Field characters

The Yellow Bittern is identified by its tawny-brown colour, black cap in the adult male, black and tawny wings, and black tail (Leader 1995b). It flies erratically, flashing its black wing edges. It is distinguished from the Little Bittern by being slightly smaller and having a longer bill, brown (not black) back, and yellow-buff (not grey-buff to white) wing patch. It is distinguished from Schrenck's Bittern by being slightly smaller, its lighter upper parts, and more contrast of black and buff (not chestnut and grey) on the upper wing. It is distinguished from the immature Schrenck's Bittern by lacking white spots on the upper wing. It is distinguished from the Cinnamon Bittern by being slightly smaller and tawny-brown (not cinnamon).

Systematics

The Yellow Bittern is one of the small bitterns, *Ixobrychus*, that share similar plumage, white eggs, scutellate tarsi, and 10 tail feathers. It is closely related to the Least and Little Bitterns, with which it shares a slender bill, uniform dorsal coloration, and plumage sexual dimorphism. Several subspecies have been proposed on the basis of differences in colour, but the variation appears to be within the range of individual variation, which is substantial within the small bitterns.

Range and status

The Yellow Bittern ranges through southeast Asia through Indonesia, the Philippines, and western Pacific islands.

BREEDING RANGE: The Yellow Bittern occurs in Pakistan, India (Mundkur 1985), Nepal, Sri Lanka, north and east Myanmar, southwest Thailand, China (south Shanxi, Yunnan, Sichuan, Heilongjiang, Guangxi, Guangdong, Hainan, Taiwan), Russia (Ussuriland, Sakhalin Island, Kurel Islands), Malaysia (peninsular, Sabah) (Lansdown 1987b), Singapore (Lansdown 1986, 1987b), Indonesia (Sumatra, perhaps Kalimantan-Borneo, perhaps Sulawesi) (Smythies 1986, Holmes and Burton 1987, Lansdown 1987b, Coates *et al.* 1997), Japan, the Philippines, New Britain (Bishop 1983), Palau, Micronesia (Yap, Woleai), and Guam (USA). Breeding in some parts of this range has been documented only rarely. Disjunct breeding occurs in the Seychelles and in Oman (Eriksen and Eriksen 1999, Eriksen 1999, Gerlach and Skerrett 2002).

NON-BREEDING RANGE: During the non-breeding season it occurs from India, Sri Lanka, south China, Indo-China, through Malaysia, Indonesia (Java, Sumatra, Timor, Kai archipelago), New Guinea, and the Philippines.

MIGRATION: North and central China populations move south in October. Southern populations are sedentary although there is evidence of local movements in India. Some birds remain in south China, Japan, north Thailand, and Burma, but other northern birds move south to Indonesia. Birds migrate at night. Northward migration is in mid April to late May. Dispersal records include Australia (Queensland, Western Australia), and Christmas Island (Stokes 1990).

STATUS: Little is known about population sizes. Numbers counted in various places are in the hundreds to low thousands. The species is common to abundant throughout its range and is undergoing range expansion (Lansdown *et al.* 2000). For example, it is the commonest of the three small Bit-terns in Hong Kong (Leader 1996). It is seriously declining on the Seychelles (J. Gerlach pers. comm.).

Habitat

The Yellow Bittern occurs in fresh water marshes and swamps and rice fields. It uses reed beds, shrubs, or other dense vegetation especially around paddy fields, as well as lakes, and sluggish rivers, and in mangrove swamps. In the Seychelles, it occurs in fresh water marshes behind the coast. It is primarily

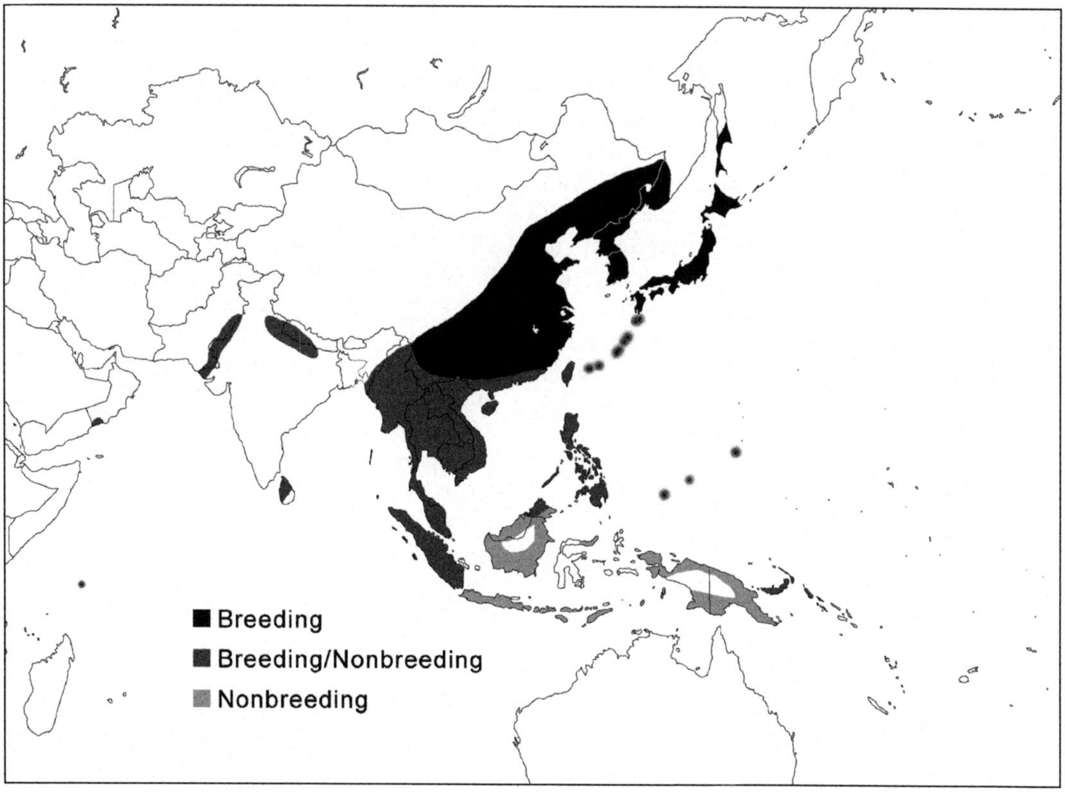

a lowland species but occurs up to 1500 m in Sumatra. It may roost in the hundreds in mixed-species aggregations (Galbraith *et al.* 1999).

Foraging

The Yellow Bittern feeds by Standing and by Walking slowly, climbing along reeds but also uses Standing Flycatching and Gleaning to catch small insects. It readily adopts the Bittern Posture when disturbed. It uses more active foraging techniques such as Wing Flicking, and also Running and Hopping to pursue active prey (Lansdown and Rajanathan 1993). It tends to be crepuscular, roosting at night and during midday on rice stalks (Lansdown and Rajanathan 1993). It is less shy than most of the small bitterns and upon landing perches for a moment instead of plunging straight into cover as do many other small herons.

Its diet appears to be principally insects, and also fish (such as loaches), frogs, crayfish, fresh water prawns, and molluscs (Ueda 1992, Lansdown and Rajanathan 1993). Terrestrial insects are taken in rice fields (Lansdown and Rajanathan 1993). The species is welcomed by rice growers for eating stem-boring crickets (Smythies 1986).

Breeding

The onset of breeding activity coincides with spring or, in the subtropics, with the rains. It is June–September in India, July–October in Malaysia, May–August in Japan and China (Shi-quan 1990), and September–April in the Soloman Islands. They nest primarily in bushes and herbaceous plants adjacent to extensive rice cultivation systems (Lansdown *et al.* 2000). Nests are especially placed in bushes surrounding or adjacent to water, for example on levees, next to ditches, around rice fields (Lansdown and Rajanathan 1993), and in dwarf mangroves along the coast. They also nest in trees including tall banyan (*Ficus*). In herbaceous vegetation, nests are placed in reed beds, grass, marsh vegetation over standing water, including cattail (Ma 1990), bulrushes, rice, wild sugarcane, bamboo, and sword grass. They tend to be less dependent on less dense vegetation than the other small bitterns, but water under the nests sites, probably as protection, appears to be a critical point in nest site selection (Uchida and Matsuda 1992). Nests are solitary, clumped in small dispersed groups, or in small colonies (Lansdown 1987b, Shi-quan 1990, Uchida and Matsuda 1992, Ueda 1996). As many as six nests can occur in a single tree (Lansdown and Rajanathan 1993). Nest density in China averaged 11.4 pairs per hectare reaching 31.18 pairs per hectare in irrigated fields (Shi-quan1990). The nest is a pad, 15–25 cm in diameter. It is made of fronds and stalks of marsh vegetation, usually canopied by foliage bent over and woven together. The female builds the nest. Nests typically are low, about 1 m above the water or ground. The male advertises with a Stretch display, in which it assumes a hunched position, puffs out its throat, gives the Crrw call, and flushes the basal area of the bill (Lansdown and Rajanathan 1993). There does not seem to be an actual stretch, as there is in other small bitterns. The male also performs Circle Flight, with a slow wing flap, flying from, then around, and then back to the advertising site. It frequently Crest Raises, stretches one or other wing, and the jerks the short black tail from side to side at the end of which it depresses the tail to a right angle from the body (Henry 1971). Sites and nests are defended with Snap displays.

The eggs are milky green-blue, 32–33 × 24.5 mm. Eggs are laid 3 days after the nest is complete. They are laid at 1-day intervals. The clutch size is usually 4–6, with a range of 3–7. Egg laying within a colony may be prolonged, for example June–August in Japan (Uchida and Matsuda 1992). It is believed that two or even three broods may be raised in a season. If so, that is a large number of annual nesting attempts for a heron. Moorhens (*Gallinula*) are reported to dump eggs in Yellow Bittern nests (Ueda 1993). The incubation period is 19 days, range 17–20 (Uchida and Matsuda 1992). Feeding is by regurgitation, with the chick grasping the parent's bill. The nestling period is remarkably short (Lu 1987, Uchida and Matsuda 1992). Nestlings can leave the nest as early as 5–8 days. Within 15 days the young begin to clamber out of the nest with the agility characteristic of this species, and abandon the nest completely by 20 days.

Population dynamics

Nothing is known of the population biology and dynamics of this species.

Conservation

As a small bittern that is fairly common and not as shy as other species, it is fairly well known, especially by rice farmers, who consider it to be beneficial. Rice fields are a critically important habitat for the species and may be responsible in part for its recent range expansion (Lansdown *et al.* 2000). It is probably a species that has benefited from the clearing of lowland forests for rice. As a result, its future in many ways depends on the distribution and continuation of rice farming practices conducive to the species. Integrating bittern conservation into rice field management is an important conservation action. Important nesting areas need to be identified so as to emphasize the importance of some rice systems.

Research needs

As a possibly expanding species, its breeding and non-breeding ranges need to be clarified so that future expansion can be documented. Range-wide surveys should be conducted to determine the most important nesting areas, and in each such area the importance of rice cultivation should be evaluated. Individual, sexual, developmental, seasonal, and geographic variability within the species should be further studied to determine if there are populations that merit taxonomic distinctiveness. As an expanding species, additional study of its courtship, foraging, nesting success, and demography are desirable.

Overview

The Yellow Bittern is a bird of fresh water marshes feeding on fish, insects, and other invertebrates. It particularly favours dense water edge vegetation, either reeds or woody plants and rice fields. It feeds in typical small bittern style, Walking and Standing, but also uses surprisingly active behaviours to startle or run down prey. This active behaviour is not as common among egrets. Perhaps the most intriguing aspect of its little known biology is the rapid nesting cycle. Chicks grow exceptionally fast and are quickly independent, and as a result it is possible that the species has multiple clutches during the season. As this is an unusual situation in herons, it suggests an interesting demography. Patterns of survivorship of both young and adults may similarly differ from the heron norm. It is a successful species, one that is in the process of expanding its range, especially in response to rice field development. Its dependence on rice fields becomes a critical aspect of its biology, as it nests more densely in flooded fields. Therefore agricultural practices affect the species.

Schrenck's Bittern *Ixobrychus eurhythmus* (Swinhoe)

PLATES 11.8, 11.9

Other names: Von Schrenck's Bittern, Little Bittern, Schrenck's Little Bittern in English; Avetorillo Manchui in Spanish; Blongios de Schrenck in French; Mandschurendommmel in German; O-yoshi-goi in Japanese; Bammangan coklat in Indonesian; Zibei Weiyan in Chinese.

Ardetta eurhythma Swinhoe, 1878, Ibis, p. 74, pl. 2: Amoy and Shanghai.

Description

Schrenck's Bittern is a small, dark heron with a chestnut (or spotted) back and lighter wings, a black stripe down the chest, a black cap, black tail, and in flight bicoloured buff and black wings.

ADULT: In the adult male the crown and back of the head are black-brown and the sides of the face are chestnut. The bill is relatively short and thick at the base (compared with the other small bitterns). The upper bill is dark horn-brown, whereas the lower bill is green-yellow. The irises are yellow and the lores are olive-green to purple-pink. The back of the neck, upper back, upper wing scapulars, and rump are bright chestnut-brown. The tail is black-brown. The upper wing has a pale grey-buff patch consisting of the wing coverts contrasting with dark grey flight feathers, paler at the tips. The throat and foreneck are buff-white, with a distinctive vertical dark rufous centreline. The upper breast is pale red-brown. The lower breast and undersides are pale buff, clearly demarked from the darker back. The underwing is silver-grey. The Legs are dull green-brown, yellow at the joint, and the undersides of the feet are yellow. The bill is black during the breeding season.

VARIATION: Schrenck's Bittern is sexually dimorphic. The female is smaller, with crown, face, back of neck, back, and upper wing dark chestnut, and the back and upper wings have large white to buff spots. On the upper wing, the flight feathers are dark brown and so show less contrast with the upper wing patch than in the male. The tail is black. The throat is cream with black vertical stripes. The undersides are cream broadly streaked in brown. The underwing is silver-grey. Geographic variation is unreported.

JUVENILE: The juvenile is dark chestnut spotted buff on the back and upper wings. The undersides are broadly streaked in brown and white. The underwing is grey.

CHICK: The chick has not been reported.

VOICE: The 'Squak' call is the flight call. The 'Gup' call, rendered 'gup, gup, gup' is a breeding call, probably a territorial and advertisement call.

WEIGHTS AND MEASUREMENTS: Length: 33–36 cm.

Field characters

Schrenck's Bittern is identified by size, relatively deeper, shorter bill, its chestnut-brown upper colour demarked from paler undersides, dark throat stripe, black cap, black and buff upper wings, and grey underwings (Leader 1995b). It is distinguished from the Yellow Bittern by being slightly larger, having a bill that is deeper at the base and more dagger-like, its darker upper parts, less contrast of black and buff on the upper wing, and grey underwing. It is distinguished from the Cinnamon Bittern by being smaller (Payne and Risley 1976; but contrast with Leader 1995b), its chestnut-brown (not cinnamon) back, grey underwings, and dark (not rufous) flight feathers. Females and juveniles are distinguished from other small bitterns by the white and buff spots on the back and upper wing.

Systematics

Schrenck's Bittern is one of the small bitterns, *Ixobrychus*, which share similar plumage, white eggs, scutellate tarsi, and 10 tail feathers. Its range overlaps that of the Cinnamon Bittern, but it differs in size and colour. The species is distinctive among the small bitterns in the spotted colour pattern of the female.

Range and status

Schrenck's Bittern occurs in China, east Russia, and Japan and in winter, Indonesia and the Philippines.

BREEDING RANGE: This is a northern species, breeding in east Russia (Amurland, Ussuriland, Sakhalin), Korea, east China (Heilungkiang, Guangdong), and north Japan (Honshu northward).

NON-BREEDING RANGE: It occurs in the non-breeding season regularly in the Philippines, and also scattered in southeast China (Taiwan, Hainan), Japan (to Ryukyu and Bonin Islands), Vietnam, Indonesia (primarily north Sulawesi (Coates *et al.* 1997) with few records from Sumatra and west Java (R. Lansdown pers. comm.), and Kalimantan (Van Balen *et al.* 1994). There are a few records in summer from Thailand and peninsular Malaysia (R. Lansdown pers. comm.).

MIGRATION: Schrenck's Bittern is migratory. Southward migration begins in August peaking in October. Return migration is April–May in the south and June in the north of its breeding range. Dispersal records occur from Myanmar in the west to Palau in the east. Extraordinary dispersion records include Europe.

STATUS: Little is understood of the population status of this species (Lansdown *et al.* 2000). It is

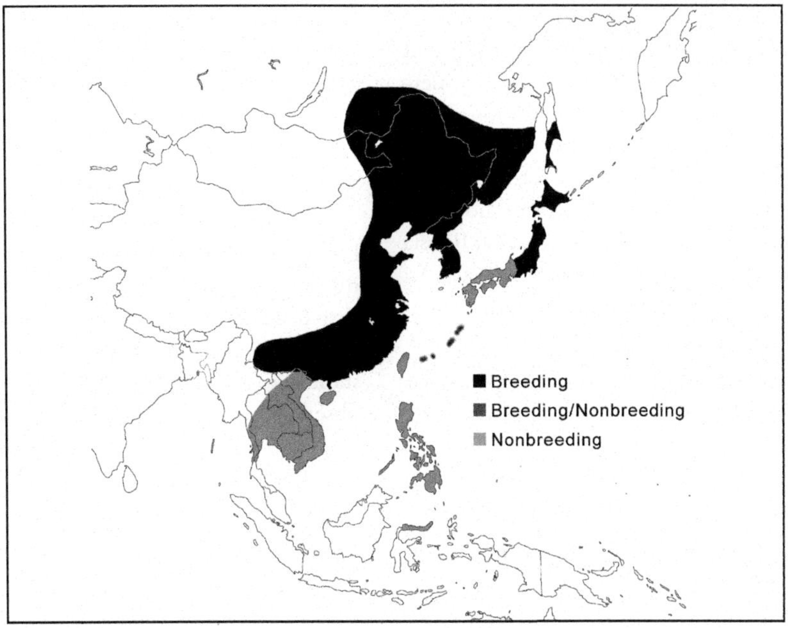

uncommon in Russia. In Japan, it is known to breed at only three sites. It is more common in south China. Migrants are recorded regularly through southeast Asia and Taiwan. There are very few records, mostly of individuals, from wintering areas and so it is rare in the Philippines and Indonesia.

Habitat

Schrenck's Bittern occurs in low-lying areas in broad river valleys. It uses open grassy meadows and open marshes along river valleys, reed beds, and also rice fields. It seems to use relatively drier areas such as dry pastures with brush. It has been recorded as occurring in village ponds in east Russia.

Foraging

Nothing is known about the feeding behaviour of this species. It hides in the grass or assumes a Bittern Posture when alarmed. It feeds crepuscularly and at times during the day. It is known to eat small fish, shrimp, isopods, dragonfly larvae, and ground crickets. But overall there is a marked paucity of information on its ecology.

Breeding

The nesting season is in the summer, May–August in Russia, China, and Japan. It nests solitarily in reeds and shrubs over or next to the water. The nest is a small cone, made of dry grass stalks with a lining of dry leaves and grass. It is placed within 1 m of the ground, well concealed in tall grass or shrub. Nothing is known of the courtship behaviour of this species. Eggs are white and measure 32 × 20 mm. The usual clutch is 3–5 eggs. Both parents incubate starting with the first egg. Incubation is 16–18 days. The male is reported to do most of the chick feeding.

Population dynamics

Nothing is known about the population biology or demography of this species.

Conservation

This species is relatively easy to see, frequenting open marshes and rice fields. None the less, it remains uncommon to rare over most of its breeding and wintering areas, with the possible exception of south

China. Mostly there is little information available. It is considered to be in the Category Data Deficient by HSG (Hafner *et al.* 2000). The principal conservation need for this species is for clarification of the breeding distribution and the identification of potentially important areas for their protection. Surveys of known and potential breeding sites should be undertaken and reported, especially in east Russia and east and south China (Hafner *et al.* 2000).

Research needs

So little is known about the biology of this species that studies of its basic biology are urgently needed. Its status needs to be determined in the breeding range and its distribution in the non-breeding range.

Overview

The ecology of this species is essentially unknown. As one of the small bitterns, some aspects of its biology can be assumed, such as it is likely a marsh-dwelling, Standing predator. But there are few first-hand reports or observations of its biology or behaviour. Contributions can be made to understanding this species by general biological observations.

Cinnamon Bittern *Ixobrychus cinnamomeus* (Gmelin)

PLATES 11.10, 11.11

Other names: Chestnut Bittern in English; Avetorillo Canelo in Spanish; Blongios cannele in French; Zimtdommel in German; Ryukyu-yoshi-goi in Japanese; Bakaw-kanela, Lapay in Pilipino; Kekondangan, Kokokan in Indonesian; Li weiyan in Chinese.

Ardea cinnamomea Gmelin, 1789. Syst. Nat. 1(2), p. 643: China.

Description

The Cinnamon Bittern is a small, predominantly rufous heron with paler underparts.

ADULT: The crown, sides of head, nape, back, wings, and tail of the adult male are cinnamon-rufous. The bill is yellow to green-orange with the top edge horn. The lores are yellow. The irises are yellow to orange. The chin and throat are buff-white, and a broad dark brown rather ragged median streak runs down the front of the neck. A tuft of loose black and dark brown feathers on the side and at the base of the neck is expansible for display. The upper wing is rufous. The underwing is pink-cinnamon to chestnut-grey. The underparts are cinnamon-buff, paler than the upper parts. The tail is chestnut-brown. The legs are yellow-green. With the onset of breeding, soft-part colours brighten, more markedly in the male. The iris has been described as yellow, orange, or pinkish-red, the last two probably being associated with courtship. The bill becomes more orange, and the skin of the lores turns to a dark or purple-red in the male. The legs also are said to brighten. The dark head takes on a silvery sheen from the powder down (Lansdown 1988).

VARIATION: The female is smaller and has a sooty crown. The lores are yellow. The back has a generally brown-chestnut coloration with a grey wash and large buff spots on wings. The neck has several vertical dark stripes running the length of the neck. The neck plume cluster is smaller than the male's, dark brown edged in buff. The upper wing is cinnamon-rufous. The underwing is rufous-buff.

JUVENILE: The juvenile's crown is a drab chestnut-brown with sandy-brown streaking. It has large sandy spots on the wings. The throat is white with dark brown stripes going its entire length. The flight feathers are bright chestnut-brown contrasting with the darker upper wing. The tail is chestnut-brown. The underwing is chestnut-grey.

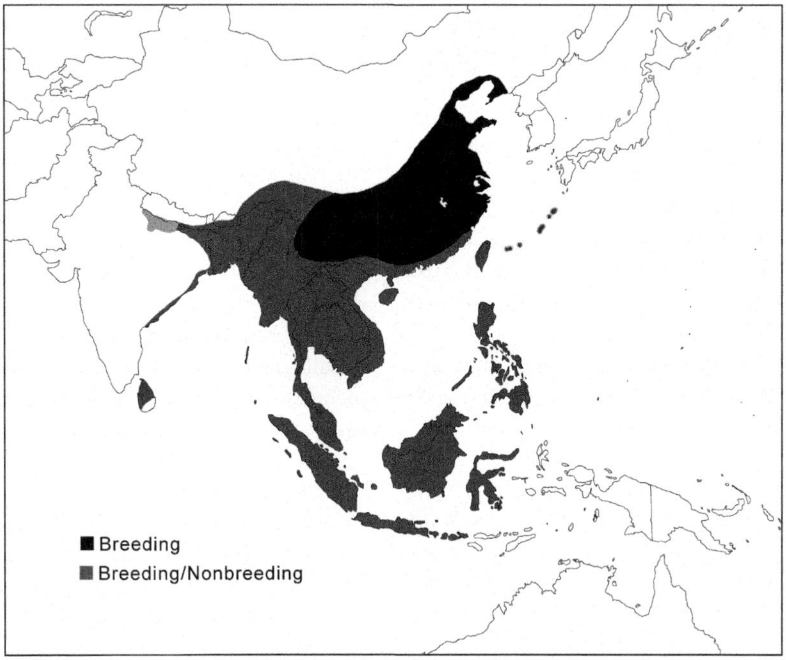

CHICK: The chick has pink-chestnut down.

VOICE: The throaty 'Keuch' call is the alarm, flight, and disturbance call, also rendered 'kk, kk, kk, kk' by Lansdown (1988). The 'Kwok' call, rendered 'kwok, kwok, kwok, kwok', is a territorial, flight, and contact call. The 'Kok' call, a quiet 'kok' or 'geg' repeated 10–14 times, is an advertisement call.

WEIGHTS AND MEASUREMENTS: Weight: 106 g; length: 40–41 cm.

Field characters

The Cinnamon Bittern is identified by its uniform cinnamon back and upper wings, which lack the contrast of other small herons (Leader 1996b). It is slightly smaller than Schrenck's Bittern (Leader 1996). It is distinguished from the Yellow and Schrenck's Bitterns by its cinnamon wings and back and less spotting in the female. It is distinguished from pond herons by its cinnamon (not white) wings. It is distinguished from the Malayan Night-Heron by its smaller size, thinner bill, olive-yellow (not green) legs, and lack of large black crown.

Systematics

The Cinnamon Bittern is one of the small bitterns, *Ixobrychus*, that share similar plumage, white eggs, scutellate arsi, and 10 tail feathers.

Range and status

The Cinnamon Bittern occurs in Pakistan and India, through southeast Asia, Indonesia, and the Philippines.

BREEDING RANGE: It breeds in Pakistan, India (Assam, Kerala, Andaman, Nicobar Islands), Nepal (R. Lansdown pers. comm.), Sri Lanka, Bangladesh (R. Lansdown pers. comm.), Maldives, Myanmar (mainland, Andaman), China (southeast Sichuan, west Yunnan, Guangdong, Hainan, to Liaoning, Taiwan), Japan (Ryukyu Islands), Thailand (Round 1995c), Malaysia (peninsular, Saba—Marin and Sheldon 1987), Indonesia (Sumatra—Voous and van Marle 1988), Java, Kalimatan, Sulawesi (north), Bali, Lombok, Timor (Noske and Ueda 1996), and the Philippines.

NON-BREEDING RANGE: Northern populations winter from coastal China and Japan (Shimanc, Ryukyu Islands) (Wakisaka 2002), south through Indonesia (to Timor).

MIGRATION: Southern Cinnamon Bitterns are sedentary; northern birds are migratory. They move south by September and October. In India, movements may be governed by rainy seasons. Return migration is in April–May. Dispersal records include the Seychelles (Lucking 1995), Christmas Island (Carter 2003), east Russia (Labzyuk and Glushchenko 1984), and an extraordinary record from Morocco (Robertson 1995).

STATUS: This is a common to abundant bird through most of its range, including Thailand, Malaysia, Sumatra, Java, Bali, and the Philippines.

Habitat

This is a rice paddy heron. It occurs especially in rice fields, but also drainage ditches in coconut plantations, reed beds, rank grass and scrub bordering marshes and lakes, grassy areas, and coastal mangroves. It occurs along the coast but also up to 1800 m in the wetlands of the Himalayan foothills. It occurs in grassy and marshy areas near human habitation.

Foraging

The Cinnamon Bittern feeds by Walking slowly. It also takes insects by Standing Flycatching, Head Tilting, Peering Over, and Head Swaying (Lansdown and Rajanathan 1993). It is solitary forager, primarily crepuscular (Lansdown 1988) but also feeds at times during the day. On alert it assumes a Bittern Posture. It is not overly shy, and will feed in rice paddies with human workers. It often chooses to roost during the day. Its diet includes insects, molluscs, frogs, and fish. Its food is relatively large.

Breeding

In India and southeast Asia nesting is July–August, during the monsoon season. It also nests in September–December and February–July in the Malay Peninsula, January–March in lowland Sumatra, October–December in upland Sumatra, and October–June in Java. It nests primarily in reed beds and rice fields. It also nests in bamboo, bushes, or flooded trees. Nests have been found on the ground in southeast China, and at the base of grass tussocks. The nest solitarily, in small loose associations, and near mixed colonies, including Cormorants (*Phalacrocorax*), Cattle Egrets, and Black Bitterns. The nest is small, made of small sticks, sedge, reed, or *Pandanus* leaves, lined with a few dead leaves or fine grass. Nests are situated low, about 1 m or so above the water.

The male advertises with a Stretch display. The bird positions itself on rice plants, a bush, or a tree. It stands straight, the base of the bill flashes bright orange, it crouches low, bill tilted upward, exposing the neck tuft of plumes. The head is lowered and it gives a Kok call. Between active displays, the bird Preens and rubs its head against powder down tracts. It also does Circle Flights with stiff beats and also a display in which it flies up to females with a deep flapping flight and giving the Kok or Keuch call (Lansdown 1988). It also gives the Kwok call when it alights at the end of Circle Flight. It defends its nest site with a Forward display. Eggs are white, 33–36.5 × 25.5–26.4 mm. The clutch is 3–4, with a range of 2–5 eggs. Eggs are laid at daily intervals. There is no evidence of double brooding. Incubation is apparently by both sexes and starts with the first egg. The incubation period is unreported. The young hatch asynchronously and growth is fast. They can clamber from the nest at 10 days. Fledging periods are not known.

Population dynamics

Nothing is known of the population biology or demography of this species.

Conservation

This is a species that is widespread, locally abundant, and well adapted to human-altered environments. It is hunted in peninsular Malaysia (Lansdown *et al.* 2000).

Research needs

Understanding the breeding biology of the Cinnamon Bittern is at an early stage. Additional information is needed on the biology, behaviour, and population biology of this species.

Overview

This is not a well-understood heron, despite being relatively easy to observe. It feeds in rice fields and other open herbaceous wetlands by Walking slowly stalking prey, taking invertebrates and relatively large fish. It has a substantial repertoire of behaviours to use in locating and capturing prey and can also make use of a variety of habitats, especially agricultural situations. Much remains to be learned.

Dwarf Bittern *Ixobrychus sturmii* (Wagler)

PLATES 11.12, 11.13

Other names: Rail Heron, African Dwarf Bittern in English; Avetorillo Plomizo in Spanish; Blongios de Sturm in French; Graurückendommel in German.

Ardea sturmii Wagler, 1827. Syst. Av., Ardea, No. 37: Senegambia.

Description

The Dwarf Bittern is a very small, dark heron with black and buff striped throat and undersides.

ADULT: The male has a dark slate-grey head and neck, with elongated head and neck feathers. The bill is dark. The upper bill is black to dark green with a black tip. The lower bill is yellow. The lores and orbital skin are blue to yellow-green. The irises are red-brown to dark red. The back, upper wings, and other upper parts are dark slate-grey. Its throat and upper breast are pale buff, darkening to tawny on the abdomen, heavily streaked black. The legs and feet are green-yellow in front and yellow behind. In courtship the legs and feet turn bright orange.

VARIATION: The female is paler, with a more rufous tinge on its belly and yellow irises.

JUVENILE: The immature bird is more buff and pale than the adult. It has buff tips to the wing covert feathers. Its underparts are more russet. Its foreneck has narrower streaking. The bill is dull yellow grey. The irises are orange-red and the legs and feet are yellow.

CHICK: The down of the chick is ginger-brown, slightly longer on the crown, creating a crest. The irises are dark brown. The skin, bill, legs, and feet are pink.

VOICE: The 'Hoo' call, rendered 'hoo, hoo, hoo' is the advertising call. A loud croak is uttered when disturbed.

WEIGHTS AND MEASUREMENTS: Weight: 142 g; length: 27–30 cm.

Field characters

The Dwarf Bittern is identified by is uniformly dark upper sides, streaked undersides, and yellow feet. Its flight is slow and steady, often keeping its neck outstretched. The bird is most often encountered flying into the bushes when disturbed. It is distinguished from the Black Bittern of Asia by its striped belly, dark neck tufts, shorter, darker bill, and much smaller size. It is distinguished from the Striated Heron by its smaller size and slow flight, its pale and heavily streaked underparts, and its uniform (not barred) upper parts. It is distinguished from the Little Bittern by being darker, smaller and lacking wing patches.

Systematics

The Dwarf Bittern is one of the small bitterns, *Ixobrychus*, which share similar plumage, white eggs, scutellate tarsi, and 10 tail feathers. The species is unusual among the small bitterns in having only a slight sexual dimorphism. It was once placed in a genus of its own (Payne and Risley 1976).

Range and status

The species ranges throughout much of Africa south of the Sahara, except for deserts.

BREEDING RANGE: It breeds from Senegal, Ghana, to Sudan, southwest Ethiopia, south Somalia, Kenyan, Tanzania (one record), southeast Zaire, Zimbabwe, south Mozambique, Namibia, and northeast South Africa.

NON-BREEDING RANGE: Birds occur in tropical Africa during the northern and southern non-breeding dry season.

MIGRATION: This is a species that is migratory, or even nomadic, over the northern and southern portions of its African range. Movements of equatorial birds are less clear. Birds occur in South Africa in the November–April wet season. In the dry season, birds from both the north and south migrate toward the equator (Brown *et al.* 1982). They appear to move about to find flooded pans in which to nest. The species does wander outside its usual range, including farther into South Africa (Ingram 1998), Canary Islands (Velmala *et al.* 2002), and France.

STATUS: The species is widespread but rare to uncommon in its range. It becomes more obvious and common after the rains (Zimmerman *et al.* 1996) at breeding sites, such as in Namibia and south Angola.

Habitat

The Dwarf Bittern uses a variety of habitats, but generally thickets and forests characterized by fluctuating water levels. Habitats include the margins of

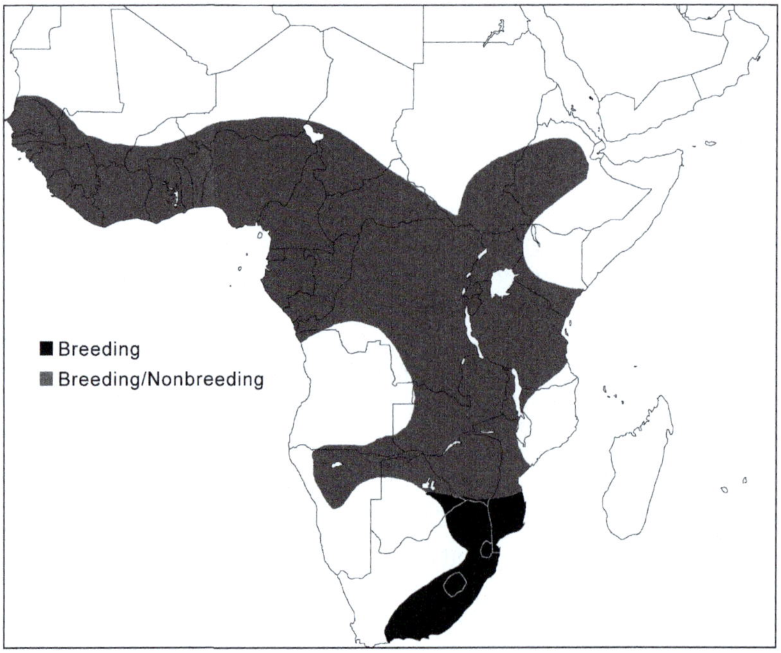

flood plain rivers, edges of creeks, rivers and streams overhung by dense foliage, rice fields, roadside pools, open grassy swamps, lake islands, cattail marshes, seasonal pans, and papyrus. It nests in thickets and trees, such as *Combretum, Ziziphus, Terminalia*, and *Diospiros* (Hustler and Williamson 1985) near its feeding area.

Foraging

The Dwarf Bittern feeds solitarily or in pairs, hunting by night and, especially if it is cloudy, by day. This it does silently and unobtrusively by Standing and by Walking slowly. This daily schedule overlaps the availability of frogs (Tarboton 1980, Hustler and Williamson 1985). The Dwarf Bittern sometimes prefers to sneak through vegetation and reeds rather than climb or perch on them. When disturbed it stays put at the water's edge for long periods in an exaggerated version of the Bittern Posture or it often flies up into trees moving only short distances. It is likely that individuals defend their feeding territory, and its use of a Forward display has been described (Riddell 1987). Insects, such as grasshoppers and water bugs, spiders, small fish, crabs, snails, and frogs form the main items of the diet. Local populations specialize in grasshoppers, and can take advantage of seasonal prey availability such as frogs in fishless seasonal pans (Hustler and Williamson 1985). It may move around to take advantage of eruptive food supplies.

Breeding

The breeding season is highly variable geographically. The Dwarf Bittern nests primarily in the rains, but in some places in the early dry season. Nesting begins in September in Senegal, July in Sierra Leone and Ghana, August–October in Mali, December– January in Zaire, May–June in Kenya, January– March and September–October in Zimbabwe, February–March in Namibia, and January–April and September–December in South Africa. It nests in forests along the margins of flood plains, usually when the water inundates the forests. It appears to prefer thicket vegetation in which it nests solitarily or in small groups. Although not usual, it has been seen to nest near Squacco Herons and Rufous-bellied Herons. The nest is a platform of twigs lined with coarse grass, 23–38 cm across and 7 cm deep. It often uses thorn bushes for nesting and thorny twigs in the nest. It builds its nest quickly (Tarboton 1980). The nest is placed near the central trunk of the tree, on a branch, or in bushes overhanging the water. The nest is 0.5–4 m above the water.

Courtship behaviour is not known. The eggs are white. Some fresh eggs are faint blue, but this fades within days (Hustler and Williamson 1985). Eggs are about 34 × 28 mm. The clutch is 3–4 eggs, and the range is 2–5 (Hustler and Williamson 1985). Incubation is 18 days and longer. Nests have not been reported with asynchronously hatched young, so it is possible that incubation begins with the last egg (Hustler and Williamson 1985). The young are fed by regurgitation in the evening. The chicks develop quickly, and at 7 days they leave the nest if disturbed, but return. When fully feathered they wander widely. When sitting on a branch, they adopt a Bittern Posture and are hard to see. In one study only 69% of eggs hatched (Hustler and Williamson 1985).

Population dynamics

There is no information on the population biology or demography of this species.

Conservation

The species is uncommon but also is widespread. Information is limited, and the most important conservation measure is to increase our knowledge of the species.

Research needs

The biology and behaviour of the species is relatively unknown. There is a particular need for a better understanding of its seasonal movements and its nesting and non-breeding distribution. The suggestion that this is a species adapted to take advantage of ephemeral food resources made available by rain and fluctuating water levels should be studied further. This is a species for which a satellite transmitter study of individual birds could provide important information to clarify its continent-wide movements. Seasonal, individual, sexual, and developmental changes in soft-part colours are poorly understood. Additional study is needed to assess these individual differences.

Overview

The Dwarf Bittern is a forest bittern, and one that appears to be to take advantage of ephemeral food supplies. One such habitat is seasonally flooded pans, which dry out in the dry season. When the rains come, birds find the pans. The bitterns build their nest and lay the clutch quickly, have a short incubation period, perhaps synchronous hatching, and chicks develop quickly and soon leave the nest. This nomadic tendency allows it to take advantage of both the predictable and unpredictable weather patterns that characterize Africa.

Black Bittern *Ixobrychus flavicollis* (Latham)

PLATES 11.1, 11.2

Other names: Yellow-necked Bittern, Mangrove Bittern in English; Avetorillo Negro in Spanish; Blongios à cou jaune in French; Schwarzdommel in German; Takasago-kuro-sagi in Japanese; Mangansile in Pilipino (Philippines); Kokokan sungai in Indonesian Hei Yan in Chinese.

Ardea flavicollis Latham, 1790. Index Orn., 2, p. 701: India.

Subspecies: *Ixobrychus flavicollis australis* (Lesson), 1831: Timor; *Ixobrychus flavicollis woodfordi* (Ogilvie-Grant), 1888: Solomon Isles, Aloa.

Description

The Black Bittern is a medium-sized, black or dark brown heron with a buff-yellow.

ADULT: In the male, the head and hind neck are dark grey-black with a blue sheen. The relatively long bill is dark but variable. The upper bill is dark brown to red-brown. The lower bill is pale green or purple below, but sometimes suffused with a reddish tinge with a yellowish tip. The irises are yellow to red-brown and the skin around the eye is variable, dusky to green or pink. The chin, throat, neck, and upper breast are deep buff streaked with black and white and bordered on each side by a paler gold-yellow band from the chin to the base of the neck. A loose tuft of elongated feathers on the foreneck overhangs the upper breast. The back, tail, and upper wing are grey-black, with the wing coverts tipped in buff. The underparts are dark grey-brown, often with a yellow wash. The legs are variable black, dark brown, or olive-green with yellow-green soles.

VARIATION: The female is slightly paler, generally brown-grey instead of dark grey-black, with a more yellow wash to the underparts. The female also has the throat, foreneck, and upper breast streaked with yellow, white and black and has yellow breast plumes. Birds vary individually and geographically. In Australasia for example, six plumage morphs have been recognized (Mayr 1945). All-black and mostly white individuals have also been reported (Payne and Risley 1976). Described geographic variation is ascribed to three subspecies. *Flavicollis*, as described above, has the least amount of plumage variation. Compared with *flavicollis, australis* tends to be larger and more variable. Males from Bismarck archipelago are mostly black with brown streaking on the throat, and some females have brown-grey underparts. Other morphs have been reported within this subspecies' range. Compared with *australis, woodfordi* is slightly smaller, has a relatively longer bill, and males are darker below with narrow white streaks on the breast and foreneck. Patterns of geographic variation are not clear given the vast amount of individual variation in the species.

JUVENILE: The immature bird has a grey-black crown, but appears duller with a rufous wash. The breast is brown streaked in yellow. The back feathers are edged in buff. The lores are green-brown. The irises are yellow. The legs and feet are brown-grey.

CHICK: The downy chick is white with a light brown tinge to the neck and rump. The irises are black-brown becoming dark brown. The bill, lores, legs, and feet are light pink.

VOICE: The 'Whoo' is a low booming drum-like call, a peculiarity of this species among the *Ixobrychus* bitterns. It is long, drawn-out call, rendered 'wh, o, o, o, o, o, o', given at 15 second intervals and answered by neighbouring birds. It undoubtedly is a territorial and advertising call. A soft 'Coo-oorh' call is given approaching the nest. Other calls have been reported but need to be clarified.

WEIGHTS AND MEASUREMENTS: Weight: 275–358 g; length: very variable, 55–65 cm.

Field characters

The Black Bittern is identified by its size, dark blue-black back, striped neck bordered by a striking gold-yellow streak, and its long bill. It is usually seen flying away into thick cover. It is distinguished from the other *Ixobrychus* bitterns by its dark back colour with yellow neck and its larger size. It is distinguished from the *Gorsachius* night herons and the New Guinea Tiger-Heron by its uniformly dark plumage and disproportionately long dagger-like bill. It is distinguished from the Striated Heron by its uniformly dark upper parts, yellow patch, and breast stripes.

Systematics

The Black Bittern is one of the *Ixobrychus* bitterns, but is the largest of the group. It is similar morphologically but uniquely has a booming call (Payne and Risley 1976). It also has a distinctive skull shape, related to its very long bill and so was previously assigned its own genus but is clearly an *Ixobrychus* bittern based on molecular genetics (Chang *et al.* 2003). Its skull and colour pattern are most similar to the Dwarf Bittern.

Range and status

The Black Bittern occurs in India, Southeast Asia, Indonesia, the Philippines, Australia, New Guinea to the Solomon Islands.

BREEDING RANGE: The breeding range of *flavicollis* is Pakistan (Sindh), India (Jamdar and Srivastava 1990), Maldives, Nepal, Sri Lanka, Bangladesh, Myanmar, Thailand (Round 1992b, 1995c, Lansdown *et al.* 2000), Cambodia, Vietnam, to central and south China (Yunnan, Hainan, Guangdong, Fujian, Taiwan, Hubei, Anhui, Shanxi), Malaysia (Sabah, Sarawak, Indonesia (Sumatra, Timo)), and the Philippines. *Australis* occurs in Indonesia (Moluccas, Irian Jaya), Papua New Guinea (including Bismarck archipelago). It breeds in coastal far south-west of Australia (Jaensch pers. comm.) and occurrances elsewhere in its historic range in Australia are now rare (Morris 1971, Pollock and Noden 2002). *Woodfordi* is confined to the Solomon Islands.

NON-BREEDING RANGE: Northern populations in China are migratory, wintering in western Malaysia (peninsular, Sarawak), through Indonesia (Lesser Sunda Islands), and the Philippines. There has been no recorded breeding in southwest Western Australia since the 1950s (Marchant and Higgins 1990).

MIGRATION: Chinese birds move south in October to peninsular Malaysia, Greater Sundas, and the Philippines. The return is April–May. Postbreeding dispersal occurs and is likely the origin of island populations of this species. In India, postbreeding dispersal and drought-based movements occur. A Malaysian bird has been recovered in India (Manipur) and an Indian bird was recovered in Maldives. Most Australian birds appear to be sedentary, but some regional movements occur and those in the south appear to be migratory (Pollock and Noden 2002). Dispersal records include Japan (Fukumoto *et al.* 1992), Christmas Island (Indian Ocean), Mollucas, and Guam.

STATUS: The status of the species is unclear. It is common in Pakistan, parts of India, and Sri Lanka.

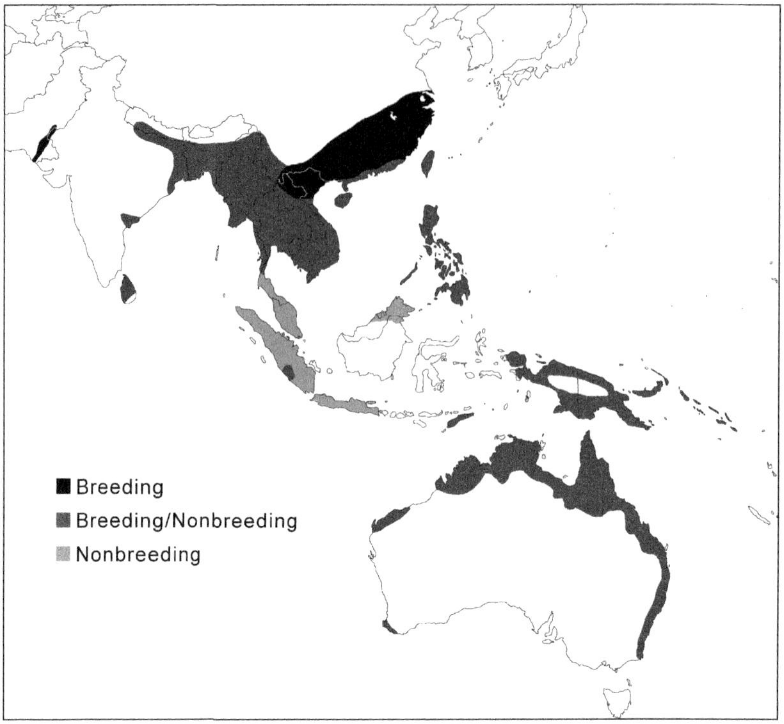

It is common and abundant in Myanmar and uncommon to common in southeast China (Mackinnon and Phillipps 2000, Lansdown *et al.* 2000). In Australia it has declined over the past 50 years due to the destruction of river-edge habitat for farming, pasture, and also because of salinization (Marchant and Higgins 1990, Maddox 2000, Pollock and Noden 2002) persisting in the southwest only along the coast (R. Jaensch pers. comm.). It is common in winter in Sarawak (Malaysia). It is uncommon in winter in Sumatra (Indonesia) and the Philippines. It is widespread in New Guinea.

Habitat

This species is a bittern of forests, mangroves and other thick scrubby vegetation particularly near streams and pools with permanent water. It also uses swamps, mud flats, reed beds, wet forests, flooded bushy areas, marshes, forested streams, *Pandanus* fringed channels in swamps, *Malaleuca* swamps and and coastal mangroves in Australia. Although primarily a lowland species, it occurs up to 1200 m in India. It winters in reed beds around lagoons, pools, and reservoirs in peninsular Malaysia and Borneo. It uses trees and reeds to rest and roost.

Foraging

The Black Bittern forages singly or as a pair in thick vegetation. It feeds by Standing with its neck and head pulled in and by Walking slowly in Crouched or Upright postures (Recher *et al.* 1983). It also can scuttle agilely along mangrove roots and creep stealthily in a distinctive crouch. It feeds crepuscularly and also during the day when it is overcast or rainy. It is seldom seen because it hunts in densely vegetated marshes and creek edges. When disturbed, it runs into cover or flies away. It roosts, when not foraging, in reeds, trees, or on the ground.

Food items recorded include relatively large fish, 12–115 cm long (*Perca, Amniataba, Gobiomorphus*), frogs, lizards, crabs, crayfish, shrimp, mussels, and insects, including dragonflies, beetles, and flies.

Breeding

Nesting timing differs geographically. In the north, breeding is triggered by the advent of the monsoon, at any time between May and September. In Sri Lanka nests have been found in April, May in Java, February in New Guinea, and September–January in Australia. Overall it breeds in densely vegetated, secluded wetlands. But its nesting habitat is surprisingly diverse, requiring mostly dense cover near water. In India and Sri Lanka, it nests in reeds, thorn bushes or cane breaks low over water. In China, it nests in bamboo and trees as high as 6 m. In Australia nests are located in gums standing in water and in mangroves. It occasionally nests away from water and is it not unusual to nest close to human habitation. The Black Bittern nests solitarily or in loose nesting groups. At places nests are interspersed in mixed colonies with other herons. The nest is a relatively large, untidy platform, 23–35 cm wide and 7.5 cm thick, with a shallow central depression. It is made of twigs or reeds often with thorns. The nest is built by both birds. Its placement is variable, from 1 to 15 m above the water or ground.

Little is known about the courtship of this species. It and the chicks adopt the Bittern Posture when disturbed, and when alarmed, raises and lowers its crest. It is likely that the plumes at the base of the neck are used during courtship display, but there is no specific information. The eggs are white, although there have been reports of a blue tinge to some. As recorded, they vary in size averaging 41.6 × 31.4 mm in India and Myanmar, 45 × 35 mm in Australia, and 43 × 34 mm in New Guinea. The clutch size is 3–6 eggs. Although it is probably usually single brooded, double broods do occur. Incubation is by both sexes. The incubating bird sits very tightly. Both sexes feed the chicks with regurgitated food. The chicks soon leave the nest if disturbed. Nothing is known about late nesting.

Population dynamics

Nothing is known of the population biology and demography of this species.

Conservation

In Australia the species has declined owing to clearing for agriculture and the increasing salinity of the rivers. It was once common in southwest Western Australia but is not known to have bred there in 50 years. It is widespread elsewhere in Australia, but Garnett (1992) listed it under Taxa of Special Concern. In NSW it is listed (updated 1999) as Vulnerable under the NSW Scientific Committee Threatened Species Conservation Act 1995 and in Victoria its conservation status is classified as Critically Endangered (NRE 2000). The actual breeding distribution is little understood and not quantified. In many cases the range is based on a few observations. A basic conservation need is to clarify the breeding locations and distribution of the species and to locate the most important breeding areas.

Research needs

Patterns of geographic and individual variation are not clear. The variation in plumage and size in this species needs to be re-examined. There also is a need to clarify the status of the species throughout its range.

Overview

The Black Bittern is a forest bittern, a bird of thick cover, generally scrubby vegetation along mangrove shores and other watercourses. It is not a river heron in the sense of using running water, but thick vegetation along water courses appears to be a habitat requirement. It feeds in normal bittern fashion, but is relatively large for a 'small' bittern. With its long bill and neck, it appears to be a fishing heron, working along the banks of watercourses from vegetation. It appears to take relatively large prey, and so may not eat large numbers of prey items per day. It likely feeds from territories, which it knows well and undoubtedly defends. Much remains to be learned about this heron.

Zigzag Heron *Zebrilus undulatus* (Gmelin)

PLATES 8.1–8.3

Other names: Vermiculated Heron in English; Avetigre Enana, Garza Cebra in Spanish; Socof-ziguezague in Portuguese; Onoré zigzag in French; Zickzackreiher in German.

Ardea undulata Gmelin, 1789, Syst. Nat. 1(2), p. 637: Cayenne.

Description

The Zigzag Heron is a small, short-tailed, black-crested, cryptically coloured dark heron of dense South American forests.

ADULT: The Zigzag Heron's head and neck are mottled grey with a black crest forming a prominent mane, giving the bird a thick-headed appearance (Davis *et al.* 1980). The bill is relatively short and dark. It may be all black, but it may also have a black or grey upper bill and horn or yellow lower bill. The irises are yellow and the lores are grey to black. The brown and black back and upper wings are barred and speckled and with buff-white vermiculation. Underparts are buff-cream, heavily barred, speckled, and streaked with black. The legs are grey-horn with yellow toes. The upper part of the leg is nearly fully feathered.

VARIATION: The sexes are alike. There appears to be considerable individual variation that requires further study.

JUVENILE: The juvenile plumage differs from that of the adult. The immature bird has a rufous forehead, sides of head, and neck. The bill is yellow-green. It is paler on the back with more rufous zigzag barring and grey wings. The undersides are tawny, with sparse indistinct black stripes. The legs are yellow-green. The immature plumage was once thought to be a rufous phase.

CHICK: The chick has yellow-cream or white down. Chicks darken to rufous on the face and back; the bill and legs are yellow-green and the irises are yellow.

VOICE: The 'Kreeik' call is the alarm call. It is rendered 'kreeik' becoming a 'krik' (Mathews and Brooke (1988). The 'Oooop' call is a hollow hooting used by adults for advertising, to which birds respond aggressively on playback (English and Bodenhorst 1991). The 'Ahnnn' call also is a distance communication call used in chorusing between individuals (Ascanio-Echeverria and Davis 2000) given most frequently by juveniles. It too is a call also responded to by both adults and juveniles. The 'Oooop' and the 'Ahnnn' calls are also combined and given in a series. These are loud, long-carrying calls, such as would be expected in a forest heron, and appear to be recognition, distancing, and, very likely, territorial calls.

WEIGHTS AND MEASUREMENTS: Weight: 123 g; length: 28–30 cm.

Field characters

The Zigzag Heron is identified by its small size, dark vermiculated colour scheme, and cryptic habits. It flies in an indirect weaving manner (A. Lyles pers. comm.). It sits uncharacteristically for a heron in a perched posture like a songbind, rather horizontal with the neck pulled in as would be expected of a short necked heron (Davis *et al.* 1980). It is distinguished from an immature Striated Heron and small bitterns by the white barring on its black back.

Systematics

The systematic position of this species has been a matter of great debate, its being moved about taxonomically to reflect relatively closer affinities to the tiger herons or to the bitterns. Molecular studies revealed the Zigzag Heron to be a bittern (Sheldon *et al.* 1995). In addition to the biochemical evidence, it is similar to bitterns in having 10 tail feathers (unlike the 12 of typical herons), scutellate tarsus, a long inner toe, and white eggs (not splotched like tiger herons). The Zigzag Heron's habit of using thorny twigs in the nest is similar to the habits of two forest bitterns, the Dwarf Bittern and the Black

Bittern. It also is similar to some of the tiger herons in having three powder down patches, rather than two as in bitterns. It is encouraging to have the lineage of this long time problem-heron resolved.

Range and status

The Zigzag Heron occurs in northern South America.

BREEDING RANGE: The breeding range has been poorly known, with reliable sight records and museum records providing limited information. However, the known range has increased substantially since Hancock and Kushlan (1984), such that it is now clear that it encompasses the Amazon and Orinoco river drainages. It occurs in east Colombia (Arauca, Vaupés), Venezuela (Montagas, Cojedos, Apure, Delta Amacuro, north Bolivar, south Amazonas), the Guianas (Ingels *et al.* 2003), Brazil (Amazona, Mato Grosso), north Paraguay, Bolivia (Beni, Santa Cruz), east Peru (Loreto, Madre de Dios), east Ecuador (Napo, Pastaza) (Davis *et al.* 1980, Scott and Carbonell 1986, Mathews and Brooke 1988, English and Bodenhorst 1991, Robbins *et al.* 1991, Biazis pers. comm., A. Lyles pers. comm., Ascanio-Echeverria and Davis 2000, Morales 2000).

MIGRATION: As far as is known, the Zigzag Heron is sedentary.

STATUS: Nothing is known about population sizes of this species, but it is now understood to be widely distributed in the forests and wetlands of north South America, and it is increasingly observed as it is now being sought by ornithologists and ecotourist guides.

Habitat

The Zigzag Heron's primary habitat is tropical forest near and along the edges of rivers and oxbow

lakes (Davis *et al.* 1980, Mathews and Brooke 1988, English and Bodenhorst 1991). It specifically occurs along slow-moving rivers and streams lined with thick overhanging trees, exposed roots, and fallen logs, from which it can fish. It uses this habitat within undisturbed tropical forest and also the narrow gallery forests of the llanos (Ascanio-Echeverria and Davis 2000). A second habitat is the edge of depression ponds, river pools and oxbow lakes, surrounded by either trees or grass. Although observations are few, it appears that a third principal feeding habitat may be the forest floor. It occurs from the lowlands to about 200–500 m.

Foraging

The Zigzag Heron has a diversity of feeding methods (Davis *et al.* 1980, Mathews and Brooke 1988). Most typically it feeds by Standing on a perch above the water on branches, logs, or roots. It Stands with its neck retracted. When it sees a fish, it wags its tail back and forth, bends more horizontally, extends its neck and then lunges, while holding onto the branch. When it lunges vertically it depresses its tail, which catches the branch to provide additional support. It also uses Standing while wading in the water. It Walks slowly in the water, along the shore, or on the forest floor. While Walking or wading, it captures prey by Pecking at the surface rather than Stabbing. It has been described as Hopping along the ground on the forest floor (Scott in Sick 1993). It uses Standing Flycatching from a Bittern Posture to catch flying insects. It feeds during the day, particularly at dawn and dusk, when it also calls. When not feeding, particularly midday, it perches 5–10 m high in trees, where it is most often seen Standing, Preening and Tail Flicking. When perched it has been observed to jump into the air and turn 180° on its perch (Davis *et al.* 1980). It eats small fish and insects. Probably its diet is varied, depending on what is available.

Breeding

The Zig Zag Heron breeds April–July in the west of its range. It nests solitarily in trees near and overhanging water. There may be other herons nesting nearby, particularly the Rufescent Tiger-Heron. The nests are shallow round platforms, averaging 17 × 22 cm (English and Bodenhorst 1991). They are made of twigs and often incorporate thorny branches along the periphery. Nests are placed low in the trees or bushes next to the water, 1.5–3 m above the ground. They may be low enough to be washed away in unexpected flood events. Nothing has been reported on courtship. The egg is white and unspotted. Evidence so far suggests that the species lays only one egg, which would be very intriguing if it turns out to be so. There also is suggestions of multiple clutches in a year (English and Bodenhorst 1991). Incubation time has not been recorded. Both adults attend the young. One adult is usually at the nest while the young are small. After that, neither remain in attendance. When disturbed at the nest the adult assumes the Bittern Posture. There is no information on breeding success other than on individual nests.

Population dynamics

Nothing is known of the population biology or demography of this species.

Conservation

Once considered to be near threatened by IUCN and HSG (Hafner *et al.* 2000), the species is now known to occur over much of north South America and is readily seen in some areas. Likely the total global population is substantial. The species now should be considered to be widespread although in low density and not threatened. None the less threats do remain. Morales (2000) points out that critical to the conservation of this species is the preservation of the great river landscapes of South America, the Amazon, and the Orinoco. This bird may be an indicator and symbol of undisturbed ecological conditions within these watersheds. The conservation of this species may be tied also to the conservation of the Rufescent Tiger-Heron that occupies much the same habitat.

Research needs

Much has been learned about this species over the past 20 years, enough to confirm that it is, as expected, an intriguing species. Its breeding biology, particularly clutch size, success, and nesting cycles, need to be further examined. It is likely to reveal interesting twists in heron biology which may in turn reveal basic understanding of the biology of forest herons. Observations and surveys need to continue to locate concentrations that can be identified as important areas for conservation. Perhaps touring associations and lodges can adopt this species as one to be sought after, recorded, monitored, and reported upon. Some of the morphological peculiarities of the heron, particularly the use made of its tail in grasping the branch during a vertical fishing lunge are intriguing. A study of its functional morphology might reveal an interesting suite of adaptations.

Overview

Our understanding of the ecology of this species, while imperfect, has improved markedly through several observational papers. It is clearly a forest heron that uses the wooded banks of slow-moving streams and pools, which it fishes by perching and lunging at prey below. This is similar in some ways to the Green Heron. However, it does this with some unique twists, such as grasping with its tail, not a typical heron ability. Moreover, it has a much wider feeding repertoire than ever was suspected and also a wider range of habitat preferences. An extreme dependence on deep forests is not the case. Rather, it uses wooded stream margins, the width of which seems to be not so critical. Additionally, it can use grass-lined pools and the forest floor, on which it feeds Pita style. Its behaviour is so much more versatile than ever thought. Even more intriguing are hints about its society. That young respond aggressively to play back suggests they may be defending familial territories. That the species may only lay one egg, suggesting limits to successful reproduction, further strengthened by the hint of multiclutching. Intriguing suggestions from the anecdotal literature so far assembled are that this may be a species with limited annual reproductive potential implying a long life-span perhaps in secured long-standing territories, a classical K-selected scenario. There is much here to generate testable hypotheses.

Agami Heron *Agamia agami* (Gmelin)

PLATES 8.10, 8.11, 18.5

Other names: Chestnut-bellied Heron in English; Garza Agamí, Garza Pechicastaña in Spanish; Garça-da-mata, Garça-beija-flor, Socó-beija-flor, Garça-da-Guiana in Portuguese; Héron agami in French; Speerreiher in German.

Ardea agami Gmelin, 1789. Syst. Nat. 1(2), 629: Cayenne.

Description

The Agami Heron is a medium, elongated, extremely long-billed and long-necked heron of wooded tropical streams.

ADULT: This is an exceptionally striking bird. The face and crown are bluesh black. The lores are pale yellow to yellow-green. The irises are variably light red-brown, orange, or yellow. The bill is extraordinary; it is long and rapier-like, averaging 140 mm but sometimes reaching 163 mm, about one-fifth of the bird's total length. The bill too is variably coloured from yellow with a dusky top edge to pale blue with a black tip. The base of the bill is yellow. The chin is white. The neck is very long, snake-like in appearance. The upper neck is chestnut with a central white stripe bordered by black. The lower neck is slate-grey and features a characteristic mat of shaggy light blue feathers that vary from normal to uniquely sickle-shaped plumes. The back, wings and tail are glossy green, sometimes appearing powder blue. The breast is pale grey to slate-grey-green. The belly is chestnut. The legs and toes are relatively short for the size of the bird, yellow to olive-green with black claws. In the breeding season it has ribbon-like light blue crest feathers up to 125 mm long and also broad powder-blue plumes on the lower back. During courtship and at other times during the breeding season, the lores can turn from yellow or light green to red, with intermediate pink-cream stages as well (G. Alvarado in prep.). This occurs rapidly and temporarily (Alvarado 1992). These colour changes are part of the interactions within pairs, occurring when the birds are together, touching, or displaying (G. Alvarado pers. comm.). During breeding the legs and feet turn gradually from black to grey (G. Alvarado pers. comm.).

VARIATION: The female is smaller than the male. No other variation has been reported.

JUVENILE: Immature birds are drab brown above, blacker on the crown and back, with whitish underparts streaked with black and buff on the lower breast. The throat has narrow white and brown vertical streaks. The irises are grey white; the upper bill is black; the lower bill is green-yellow; the facial skin is dull green-yellow except for a dusky band across the lores; the legs and feet are green-yellow and grey-green anteriorly (Marin 1989).

CHICK: The downy chick is uniformly sooty black with pink skin and dark blue around the eye (Marin 1989). The irises are bright yellow. The upper bill is black with green at the base. The lower bill is flesh coloured with a black tip. The legs and feet are blue-grey (Marin 1989).

VOICE: Several calls were described by Alvarado (1992). The 'Gr' call, rendered 'gr, gr, grr', or 'krr, krr, krr', is the characteristic, low-pitched rattling call. The 'Chup' call is rendered 'chup, chup, chup, chup', but really is much more complex (G. Alvarado pers. comm.), is the greeting call at the nest. The 'Chuc' call, rendered 'chucuchuc, chucuchuc, chucuchuc', is the contact call. A similar sound, 'Cchucuchucú, chucuchucú, chucuchucú', is the call of the chick.

WEIGHTS AND MEASUREMENTS: Weight: 475–535 g; length: 60–75 cm.

Field characters

The Agami Heron is identified by its long neck and bill, its chestnut belly, and its powder blue head and back plumes. It is distinguished from all other herons by these features.

Systematics

The Agami Heron has long been thought to be related to *Egretta* and *Ardea* herons, distinguished from them by its proportionately long neck and bill and distinctive plumes. However, its plumage is quite distinct, as apparently is its courtship behaviour. Recent molecular evidence suggests that it is not a typical heron, but rather a surviving representative of a different lineage of herons, basal to the main evolutionary stem. Like the Boat-billed Heron and tiger herons, it appears to be sufficiently distinctive to afford it its own subfamily. Although this radical proposal is subject to future findings, it serves to suggest how different this species might be.

Range and status

The Agami Heron occurs in coastal Central and north South America and in the Orinoco and Amazon basins.

BREEDING RANGE: In the north, it occurs in east Mexico (Veracruz, Chiapas, Quintana Roo, Oaxaca, Yucatan—Michner *et al.* 1964, Schaldach *et al.* 1997), east Guatemala (Petén), Belize, Honduras (La Ceiba), Nicaragua (I. Davidson pers. comm.), Costa Rica (Caribbean and Pacific coasts) (Campos 1988, Marin 1989, Alvarado 1995), Panama, Colombia, Venezuela (Michner *et al.* 1964, Ramo and Busto 1982), Trinidad (Herklots 1961), Guianas, southeast Peru (Blanco in Morales 2000), north and central Brazil (Amazona, Mato Grosso) (do Nascimento 1990), north and east Bolivia, and west of Andes to northwest Ecuador.

MIGRATION: There is no indication of migratory movements or movements of individual birds, but dispersal records include east Brazil, the Andes of Colombia at 2600 m (Fjeldsa and Krabbe 1990), and Nuevo Leone, Mexico (Contreras-Balderas 1988).

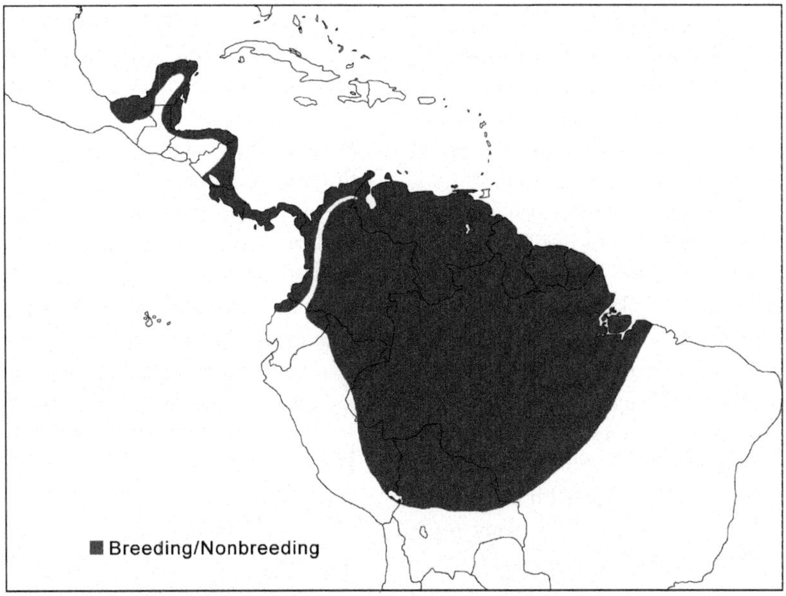

STATUS: The Agami Heron is seldom observed and so its status is poorly known. In Panama it is considered common, at least between January and June. It is poorly reported in South America, but is locally common in varzea forests. In appropriate habitat 20–30 may be seen in a day (R. Williams pers. comm.). Individuals nest together, suggesting that reasonable regional populations exist within the range. Large colonies are reported in Belize, over 80 nests, (Jones 2002) and French Guiana, over 900 nests (Reynaud and Kushlan in prep.), indicating substantial regional populations in these areas.

Habitat

The preferred habitat is dense, tropical, lowland forest where it is concentrated along margins of streams, small rivers, and swamps. It is also found, although less commonly, along the margins of pools, oxbow lakes, and other small water bodies. It is very rarely observed in dry forests along streams. It occurs from coastal mangrove swamps to 300 m elevation. It is a bird of the understory, and is nearly always seen in shaded places.

Foraging

The Agami Heron is a specialized bank fisher. It Stands in Crouched posture on banks, dykes, bushes, or branches overhanging the water. Its long neck and short lower legs permit a long lunging Strike. It also Walks slowly in shallow water at the edge of a stream or pond. With its long neck and bill, it is primarily a fish-eating heron. It is known to eat cichlids (*Aequidens*) and various characins (*Triportheus, Astyanax*).

Breeding

Nesting is during the wet season, June-September in Venezuela and Trinidad and March–July in Costa Rica. Young were in the nest in May in Brazil (do Nascimento 1990). In French Guiana, during May nests were in all stages from construction to having fledged young. It nests in dense vegetation, isolated clumps of mangroves, dead branches of drowned trees in an artificial lake, trees standing in water, and bushes within marshes. The Agami Heron is a colonial breeder usually nesting in very small numbers (6–11 nests) in colonies, and at times near other species, including other herons (such as the Striated Heron, Boat-billed Heron, Cattle Egrets–although these sometimes nest later), Anhinga (*Anhinga*) and ibis (*Phimosus*). In Venezuela, Agami Heron nests were near but separate from a mixed colony of 2000 nests (Ramo and Busto 1982). Colonies of over 80 nests (Belize) and 900 nests (French

Guiana) have been reported (Reynaud and Kushlan in prep.). The nest is a loose, thick platform of sticks or twigs. In French Guiana they averaged 15 cm in diameter and 8 cm thick. They are always located well within canopy cover low to the water averaging about 2 m high.

Alvarado (1992, in prep.) observed a courtship display that appears to be derived from the Stretch, but distinctively so. She calls it a 'Dance and Bow' display. With legs fixed in place, the bird puts its breast out and rocks its body smoothly from side to side. At the central position, it shakes its head, bill, and crest while producing a snap sound. Then with wings semi-open and crest somewhat erect, the heron, moving gradually, lowers its head toward one foot, then towards the other with a trembling motion. This behaviour was observed many times, and seems to be the principal pairing display for the species. It appears to be unique among herons. Members of a pair also engage in a Bill Duel display, consisting of two birds threatening by jabbing their bills simultaneously toward each other and withdrawing immediately. Bill Grabbing or the female lowering her head were not described as a part of the display in this species. The eggs are characteristic of typical herons, pale blue-green or a pale dull blue, unmarked. In Trinidad measurements averaged 49.6 × 35.75 mm. In Costa Rica the average was 49.9 × 36.4 mm (Marin 1989). In Brazil the average was 48 × 35 mm (do Nascimento 1990). Most nests found contained 2 eggs or 2 young. The clutch range overall is 2–4 eggs. Nothing is reported on incubation. The young gain weight quickly, more than doubling in the first week (do Nascimento 1990). They are nearly motionless in the nest, but will readily climb out on branches when disturbed. Chicks often interact and fight bill duels. It appears one to two chicks fledge. Young are tended by adults off the nest, for example, an adult may use its wings to shade a fledged young (Reynaud and Kushlan in prep.).

Population dynamics

The population biology is unknown.

Conservation

There is very little information on the conservation status of this species. Once considered to be near Threatened by IUCN and HSG (Heinz *et al.* 2000), this classification was mostly due to deficient knowledge. It is infact widespread, locally common and probably having a large population size over its extensive range. Given its known habits, it would be threatened by significant deforestation and damming of rivers. Nesting site protection is critical as the birds are easily disturbed. In Costa Rica, villagers cut trees adjacent to the colony site, and the birds then moved to a more secluded site (Marin 1989). However, birds allowed nestlings to be moved in the face of rising waters, suggesting active colony site management is possible (do Nascimento 1990).

Research needs

Surveys, inventories, and recording of observations are needed to better assess the distribution, status, and, eventually, trends for the species. Pursuing the hypothesis that the Agami Heron represents a lineage of herons separate from the typical herons and bitterns puts new emphasis on increasing our knowledge of its basic biology, courtship behaviour, nesting, and feeding habitat requirements. Its unusual morphology should be further examined for additional insights into what may be the ultimate development of the long-billed fishing heron.

Overview

The Agami Heron is one of the more intriguing herons. It is an extraordinarily elongated heron. The details of its ecology remain little known, but it is certainly fundamentally a branch or bank-feeding heron. Its long neck and bill may be the maximum evolutionary development of fish-eating heron morphology. In hunting, it crouches low on its shortened legs and feet, which creates a pivot from which to lunge at its prey. It nests in colonies, which accumulate individuals from the region. Aspects of the relationships between colonial nesting and solitary feeding in this species deserve close investigation. Its actual breeding distribution and therefore its habitat requirements are only slowly becoming understood. Of particular interest are recent findings that suggest that the Agami Heron may be evolutionarily distinctive from other herons, perhaps the last remnant of an ancient lineage of South American herons.

Bare-throated Tiger-Heron *Tigrisoma mexicanum* (Swainson)

PLATE 10.1

Other names: Cabanis Tiger Heron, Mexican Tiger Heron, Mexican Tiger Bittern in English; Avetigre Mejicana, Garza Tigre, Garza-tigre Gorjinuda, Garza-tigre Cuellinuda in Spanish; Onoré du Mexique in French; Nacktkehlreiher in German.

Tigrisoma mexicanum Swainson, 1834. In Murray's Encyclopedia of Geography, p. 1388: Read del Monte (Hidalgo), Mexico.

Description

The Bare-throated Tiger-Heron is a medium, cryptically coloured heron, with a green face and throat.

ADULT: The crown and nape are black. The side of the head is grey. The upper bill is black with a blue-horn cutting edge and the lower bill dull yellow. The irises are yellow to silver. The lores, skin around the eye, and featherless throat are yellow-green, this colour running onto the lower bill. A black line runs from the eye to the throat. The front of the neck is tawny and white striped. The hind neck is finely barred in brown and buff. The back is dark olive-brown with very fine buff vermiculation. The flight feathers are black. The underparts are cinnamon and the thighs are grey. The legs are dark grey-olive to slate-green. During nesting the bare throat may become bright yellow to orange.

VARIATION: The sexes are alike. Birds in Sonora and Sinaloa are larger with broader striping, and had been recognized as a distinct subspecies.

JUVENILE: In the immature bird, the head, neck, and chest are barred and spotted with cinnamon-buff and brown. The throat is yellow. The upper parts are dark brown boldly barred in cinnamon. The underparts are pale buff with broad brown barring. Its tail and wings are blackish, with 4–5 narrow white bars.

CHICK: The nestling has pale grey-white down with white, erectile, filamentous plumes on the crown.

VOICE: The 'Woh' call, rendered 'who, woh, woh, woh' or 'howk, howk, howk', is the disturbance call. The 'Ohrr' call, rendered 'ohrr, ohrr, orrr', is a roaring or booming call used in the evening and at night, especially in the breeding season. It is probably an identification, contact, and distancing call. The voice also has been described as croaking, snoring, barking, and bellowing, so more study is needed.

WEIGHTS AND MEASUREMENTS: Weight: 1046–1274 g; length: 70–80 cm.

Field characters

The Bare-throated Tiger-Heron is identified by its black cap, grey head, yellow face, and white feathered throat. It is a solitary, secretive species that is not often observed. Its flight is heavy. It is distinguished from the Rufescent Tiger-Heron and Fasciated Tiger-Heron by its black cap, yellow face and throat, and vermiculated brown (chestnut) neck. It is distinguished from the *Botaurus* bitterns by its slightly larger size, black cap extending further down the back, yellow face and throat, dark bill, and grey head.

Systematics

The bare throat of this species was recognized for years as being highly distinctive. It is now placed in the same genus as the other two South American tiger herons. The tiger herons are a distinctive lineage of the Ardeidae, adapted to a forest existence. The five species constitute a sister group to all the other herons and bitterns and to the Boat-billed Heron (Sheldon *et al.* 1995).

Range and status

The Bare-throated Tiger-Heron occurs in coastal Mexico, Central America, and northern South America.

RANGE: In Mexico, it occurs, along both coasts as far north as south Tamaulipas and south Sonora,

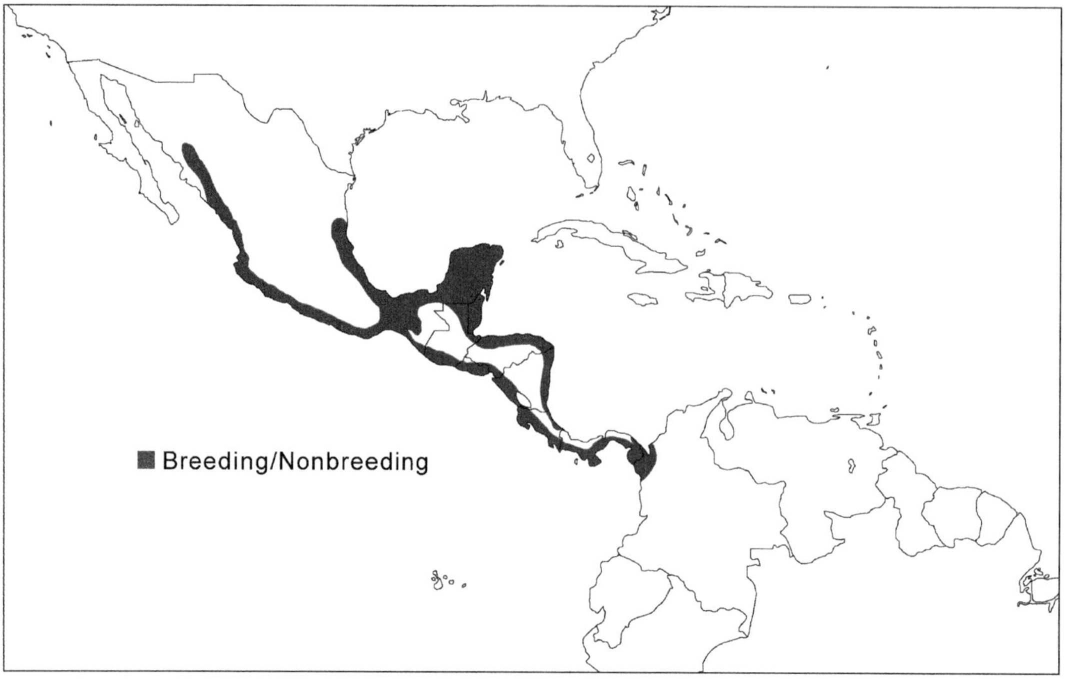

across the Isthmus of Tehuantepec, in all of the Yucatan, coastal in Central America through Panama, including Coida Island and Perles Archipelago, but of limited distribution on the Caribbean slope, into extreme northwest Colombia from the Gulf of Uraba' to the lower Atrato River valley.

MIGRATION: It apparently is resident throughout it range as there is no report of migratory movements.

STATUS: Nothing is known of the population size or trends of this species. It is widespread throughout its range. It is reported to be or have been common in Honduras, Guatemala, Belize, Nicaragua, and Costa Rica; rare to uncommon and perhaps declining in Panama (Ridgely and Gwynne 1989); and possibly at risk in north Colombia (Kushlan and Hafner 2000).

Habitat

The Bare-throated Tiger-Heron is a bird of tropical swamps. Along the coast, it is typically a mangrove species. It is found, however, in a variety of coastal and fresh water situations, typically characterized as forest-edged aquatic habitats. It feeds in coastal lagoons, mangrove swamps, fresh water marshes, swamps, gallery forests along rivers and streams, and wet meadows. It feeds more in the open and in larger habitats than other tiger herons. It is primarily a coastal species occurring below 1000 m. It avoids human habitats.

Foraging

The primary foraging behaviour of this cryptic heron is Standing in shallow water or on the edge of the water, with neck stretched out diagonally, waiting motionless for long periods. It also Walks slowly, often in situations away from cover. It is a solitary species, usually foraging alone. However, a dozen have been seen in a loose group when food is readily available. It is primarily crepuscular but also partially nocturnal. When disturbed it takes flight and perches in a nearby tree. It eats fish, frogs, crabs, and flies.

Breeding

The breeding season is variable. In Costa Rica nesting may be year-round but peaks at the early rainy season (Stiles and Skutch 1989). It is May–August in most of its range but February–April in Panama. The tiger heron nests solitarily in trees above the water, especially in mangroves (*Rhizophora*). Nests are small to large platforms made of sticks, usually lined with leaves. They are placed on branches 4–15 m above the ground. Courtship is little understood in this species, and the displays reported have little context to them (Dickey and van Rossem 1938, Sutton *et al.* 1950, Wetmore 1965). The primary display appears to be a version of the Stretch—a pair faced each other, necks and bodies crouched horizontally, feathers fluffed up; the birds raise their bills and neck vertically, depressing the neck and body feathers, and emitting two or three hoarse booms or roars. They may sway the neck at the vertical. This may be repeated, and one bird may move toward the other. A mutual Bittern Posture has also been described. The eggs are dull white with a green tinge, rarely flecked with brown or buff. The usually unspotted eggs are not typical of the tiger herons (Dawn 1964). Two measured 56.6 × 43.5 mm and 58.1 × 45.3 mm. The clutch is 1–3 eggs. There is no information on incubation, chick development, or fledging and little is known about the nesting success of this species. Nestlings are at risk to hawk (*Buteogallus*) predation (Lewis and Timm 1991).

Population dynamics

Nothing is known about the population biology of this species.

Conservation

A principal conservation concern for this species is the lack of information available on its distribution, populations, and trends. As a cryptic, solitary, forest-edge species it is difficult to census. None the less there is sufficient reason to be concerned, especially in the south of its range. It is considered to be declining in Panama due to persecution, drainage, and development for agriculture (Ridgely and Gwynne 1989). The population in north Colombia is considered to be near Threatened due to habitat alteration in the Gulf of Uraba' and Rio Atrato (Hafner *et al.* 2000).

Research needs

The ecology of the Bare-throated Tiger-Heron is clearly poorly understood. This is the case with most of the tiger herons; the accessibility of the habitats of this species should make it a prime candidate for detailed studies. The population status and distribution of the Bare-throated Night-Heron need to be assessed throughout its range, but particularly in the south. Its foraging biology and behaviour are poorly known and its family and territorial structure, courtship behaviour, and nesting biology are similarly little known. Study of its basic biology is needed to understand this species within the context of the other tiger herons, which differ by tending to occupy denser habitat, and also to provide a basis for conservation action. An intriguing study would be to examine variations in the egg colour of Tiger-Herons. Except for a single record, the species is reported to have plain, unmarked eggs unlike other tiger herons.

Overview

The Bare-throated Tiger-Heron is bird of tropical swamp forests, typically of coastal mangroves and gallery forests. Its feeding strategy is solitarily Standing patiently and Walking slowly looking for fish and other prey. It is a cryptic, but accessible species.

Fasciated Tiger-Heron *Tigrisoma fasciatum* (Such)

PLATE 10.3

Other names: Salmon's Tiger Heron in English; Avetigre Oscura, Garza-tigre Barreteada, Pájaro Vaco Oscuro, Pájaro Vaco, Martín Peña in Spanish; Socó in Portuguese, Ororé fascié in French; Streifenreiher in German.

Ardeola fasciata Such, 1835, Zool. Journ. 2, p. 117: Brazil.

Subspecies: *Tigrisoma fasciatum salmoni* (Sclater & Salvin), 1875: Medellin, Cauca Valley, Colombia; *Tigrisoma fasciatum pallescens* Olrog,

1951: Quebrada El Pilón, Rio Alisos, Salta, Argentina.

Description

The Fasciated Tiger-Heron is a medium-sized, cryptic, slate-coloured heron with fine buff barring on its neck and body.

ADULT: This tiger-heron has a black crown and the side of its head is slate coloured barred with black. The bill is short, heavy, slightly down-curved, and dark. The upper bill is black with a horn tip. The lower bill is horn at the tip, dark near the end, and yellow-green toward the face. The colour of the facial skin (the lores and skin at the base of the bill) is variable. The lores are yellow to black, perhaps changing seasonally. The skin at the lower gape is yellow to yellow-green. The irises are yellow. The neck and back are slate-black to black-brown with narrow, widely spaced pale buff barring, with the buff of the lower neck more distinct than the back. The foreneck and breast are white streaked with brown. The belly is tawny or rufous. The flanks are grey uniformly. The underwings are indistinctly barred slate and white. The legs are brown.

VARIATION: The sexes are alike. Geographic variation is notable. *Fasciatum* has long wings and tail and darker rufous abdomen, and its throat feathering extends a short way over the base of the lower mandible, as in the Rufescent Tiger-Heron. *Salmoni* is smaller and lacks mandibular feathering. *Pallescens* is larger and paler, especially on the abdomen.

JUVENILE: Juvenile birds go through several plumage stages over the four years of immaturity. The juvenile initially is broadly banded black and buff. The striping becomes narrower with age with darker black and lighter buff bands.

CHICK: The chick is not described.

VOICE: The voice has not been described.

WEIGHTS AND MEASUREMENTS: Weight: 850 g; length: 60–70 cm.

Field characters

The Fasciated Tiger-Heron is identified by its small size, dark head and neck, its robust bill, darker crown, and white throat. It is distinguished from the Rufescent Tiger-Heron by being slightly smaller, with shorter tarsi, its short, slightly down-curved and robust (not straight and narrow) bill, its barred (not chestnut) head and neck, its uniformly grey (not banded) flanks, its tawny (not grey and ochre abdomen), and by its higher (not lowland) habitat choice. It is distinguished from the Bare-throated Tiger-Heron by it smaller size, feathered white throat, and dark sides and underparts. In the early broad-barred plumage stages, juveniles are very difficult to distinguish from juvenile Rufescent Tiger-Herons. It is distinguished with difficulty from the juvenile Rufescent Tiger-Heron by its smaller size, arched and thicker (not thin and straight) bill, thicker appearance, generally less strong patterning, and in hand by the distribution of powder down patches.

Systematics

The tiger herons are a distinctive lineage of the Ardeidae, adapted to a forest existence. The five species constitute a sister group to the other herons and bitterns, and to the Boat-billed Heron (Sheldon *et al.* 1995). Subspecific variation is well documented in the Fasciated Tiger-Heron. The northern and southern races have been considered separate species.

Range and status

The Fasciated Tiger-Heron occurs in Central and South America, south to Argentina.

BREEDING RANGE: *Salmoni* occurs in east Costa Rica, east Panama and Darien, Colombia (both coasts), east of the Andes and north and west of the Amazon in Venezuela (to Sucre), east Colombia, east Ecuador, east Bolivia (La Paz, Cochabamba, Santa Cruz), and northwest Argentina (Jujay). *Pallescens* occurs in northwest Argentina (Salta, Tucumán) (Vides-Almonacid 1988). It has recently been reported from Surinam, although more information is needed to understand the extent of its range along the Atlantic Coast of South America (Ottema 2002). *Fasciatum* occurs in southern Brazil (Rio de Janeiro, southern Sao Paulo, and northern Parana) (Yamashita and de Paula Valle 1990, Straube 1991, F. Olmos pers. comm.) and northeast Argentina (Missiones).

MIGRATION: Like all *Tigrisoma* tiger herons, this species is basically non-migratory but juveniles undertake dispersal. In Peru they move to the temperate zone up to 3300 m into semi-arid valleys. In Costa Rica they have been reported from the Pacific slope (Jones 2001a).

STATUS: This is a species of naturally low density and perhaps patchy distribution. It is rare in Central America through Colombia. It is expanding its range into the mountains in Costa Rica due to the increase in artificial fishponds (G. Alvarado pers. comm.). It is very rare in southeast Brazil and listed as endangered in Brazil (F. Olmos pers. comm.). In the past 15 years it has been documented in only six locations in this overall southern range (F. Olmos pers. comm.). There is no information on numbers, and information on its distribution is not complete (e.g. Yamashita and de Paula Valle 1990). Its apparent rarity and lack of data makes knowledge of its status uncertain over most of its range (Morales 2000).

Habitat

This is a bird of clear, fast-flowing streams of mountains and foothills, in contrast to the Rufescent Tiger-Heron, which is its equivalent in lowland streams. It occurs the eastern length of the northern Andes, up to 3300 m in Peru (Fjeldsa and Krabbe 1990), along rapidly flowing, forested streams. In Central America, it occurs on the

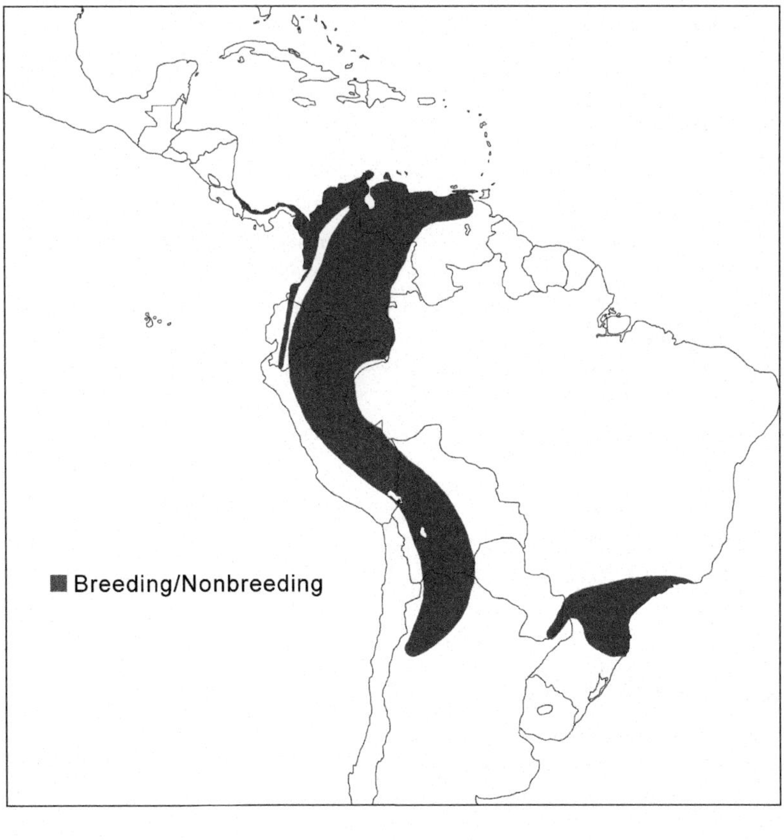

well-watered Caribbean slope in forests up to 750 m. In Colombia, it occurs along mountain streams up to 1539 m. The southern race is found in forested hill country, and in more open forest (Yamashita and de Paula Valle 1990).

Foraging

The Fasciated Tiger-Heron is a solitary species. It feeds by Walking slowly along the banks of fast-flowing rivers or along mid river gravel beds. It also Stands very still waiting for prey on river rocks, from the bank, or on gravel or sand bars. It feeds in the open, not necessarily under tree cover, but none the less is not very conspicuous (Ridgely and Gwynne 1993). It is not clear if the species feeds by night or by day. Virtually nothing is known of its diet, other than it eats fish and insects. It is likely that in such fast-flowing streams fish are the principal prey.

Breeding

Virtually nothing is known of the breeding biology of this species.

Population dynamics

Similarly, nothing is known of its population biology or demography.

Conservation

The rarity of this species and the lack of knowledge of its distribution, numbers, trends, and basic biology make conservation extremely difficult. Stream-edge forests are a vulnerable habitat, and it is likely that the heavy deforestation and urbanization of its range are threats to its population stability and likely have led to decreases (Morales 2000). It is thought to require pristine, low nutrient streams (Yamashita and de Paula Valle 1990). If so, the situation is of particular concern because these are in severe threat from siltation and other results of watershed disruptions. It is considered to be near Threatened (Hafner *et al.* 2000). Conservation action first requires better information on its distribution and habitat requirements, and then implementation of plans for conservation of its upland forested stream habitat.

Research needs

There is almost no information on food, details of foraging behaviour, or territorial structure, and nothing on its breeding biology. Basic information is needed on the distribution, habitat requirements, and conservation status of the Fasciated Tiger-Heron throughout its range. Study of the biology, distribution, and taxonomy of the apparently disjunct populations is desperately needed to understand its basic biology, habitat requirements, and to clarify specific and the subspecific relationships of these herons.

Overview

This is a solitary heron of fast-flowing rivers of the forested hills and mountains sharing habitat with torrent ducks (*Merganetta*) and mergansers (*Mergus*). It feeds by quietly Standing or Walking awaiting what are probably rare feeding opportunities. It appears to be uncommon, and likely well dispersed along its habitat.

Rufescent Tiger-Heron *Tigrisoma lineatum* (Boddaert)

PLATE 10.2, 18.3–18.4

Other names: Lined Tiger Heron, Lineated Tiger Heron, Banded Tiger Heron in English; Avetiger Colorada, Garza-tigre, Castaña, Garza Serpiente, Pájaro Vico, Martín Peña in Spanish; Socó-boi in Portuguese; Onoré rayé in French; Narmorreiher in German.

Ardea lineata Boddaert, 1783. Table Pl. enlum., p. 52: Cayenne, ex Daubenton Pl. enlum., No. 860

Subspecies: *Tigrisoma lineatum marmoratum* (Vieillot), 1817: Paraguay.

Description

The Rufescent Tiger-Heron is a medium-sized, cryptic, brown heron with a rufous neck and streaked throat.

ADULT: The Rufescent Tiger-Heron's head, sides of the head, and long, thick neck are rich chestnut-brown to rufous-cinnamon. The bill is relatively long, up to 10 cm, appearing slightly up-tilted, and varies in colour according to age and season in ways that are not well understood. The upper bill is dark (olive-black, yellow-brown, or black) and the lower bill is light (yellow, green, or pale olive-buff). The skin at the base of the bill is yellow. The lores appear to have three stripes running from the bill, yellow above and below, grey in the centre. The skin around the eye is also yellow. The feathered chin is white. The irises are yellow, but sometimes pale brown. The neck is relatively long and can be snake-like as it is held out while the bird is feeding. The foreneck has a wide, rufous vertical stripe brown in its centre, bordered by white stripes on either side, which are particularly effective when the bird raises its head or assumes a Bittern Posture. The back and tail are green-black and wings dark grey, all having multiple delicate brown lines, speckles, and flecks that extend variably onto the head (Hancock 1999). Not visible, but a character distinguishing this species from the other tiger herons, is a lack of interscapular powder down patches. The flanks are slate, horizontally striped with thin white lines, also variable among individuals. The abdomen is grey with an ochre tinge. The underwing is slate with sharp white banding. The legs, with relatively long tarsi, and are dark olive or black in front, and green behind. In courtship the soft parts brighten and the irises turn bright red.

VARIATION: The sexes are alike, although there is an indication of sexual difference in plumage details (Hancock 1999). Geographic variation is recognized by two subspecies. *Lineatum*, as described above, is the smaller of the two and has less extensive throat feathering. *Marmoratum* is somewhat larger. Its throat feathering extends a short way over the base of the lower mandible, as in the Fasciated Tiger-Heron. Birds from Argentina have a dark crown, sometimes with bold streaking, and were formerly treated as a separate species, *Tigrisoma bolivianium*. However, dark crowns, sometimes with additional black streaks occur in individuals in other populations, and the Argentinian birds are therefore considered to be *marmoratum* (Eisenmann 1965).

JUVENILE: Juveniles are typically bright cinnamon buff, broadly barred and spotted in brown and black. Immature birds are variable, undergoing gradual plumage changes through the fifth year (Rossi 1958). The head and neck are thinly barred in brown. The irises are yellow, and the pale horn bill is shorter than in the adult, starting as short as 8 cm. The lores and base of the bill are white. The neck, back, sides, and upper wing are boldly barred with brown. The undersides are pale to white. The tail is narrowly banded with white. The thighs are lightly banded in white and light brown. The legs are yellow. The first juvenile plumage of pale buff barred brown and black is gradually replaced at 8–12 months by bolder black and tawny barring, black spots on the breast, and an indication of chestnut on the nape. From about 16 months the chestnut becomes more pronounced, and over the next two years the black barring on the lower neck and crown gradually disappears.

CHICK: The young is covered with white down with a short white crest. It has a bare yellow throat, yellow irises and bill, and dull black legs and feet.

VOICE: The 'Wok' call, rendered 'wok, wok', is the alarm. The 'Quoh' call, rendered 'quoh, h, h', is the night-time contact call. In courtship, a groaning or nasal humming call has been reported, as has the Bill Snap.

WEIGHTS AND MEASUREMENTS: Length: 66–76 cm.

Field characters

The adult Rufescent Tiger-Heron is identified by it rich chestnut to rufous brown neck and head, dark and white lined throat, dark flanks with narrow white bands, white banded underwings, and relatively long neck and lower legs. The adult is distinguished from the Bare-throated Tiger-Heron by its rufous-chestnut (not black) head, and chestnut (not vermiculated brown) neck. It is distinguished from the Fasciated Tiger-Heron by being slightly larger with longer lower legs, by its slender, straight bill, its chestnut (not barred) head and neck, white banded flanks (not uniformly grey), grey abdomen with ochre tinge (not tawny), and having its slaty underwing coverts much more sharply banded in white (Eisenmann 1965). In the early broad-barred plumage stages, juveniles are very difficult to tell from juvenile Fasciated Tiger-Herons; but it is distinguished by its straight, slender (not arched and thicker) bill, slender appearance, and in hand by the distribution of powder down patches. It is distinguished from the *Botaurus* bitterns by its slightly larger size, cinnamon-buff with brown and black barring on the upper sides, pale to white undersides, and barred black and white sides.

Systematics

The tiger herons are a distinctive lineage of the Ardeidae, adapted to a forest existence. The five species appear to constitute a basal, or primitive, group as do the Agami Heron and the Boat-billed Heron (Sheldon *et al.* 1995, K. McCracken pers. comm.). Variation is high, with individual, developmental, and geographical components, so subspecific differentiation may deserve reanalysis.

Range and status

The Rufescent Tiger-Heron is a bird of Central and South America.

Breeding Range: *Lineatum* is the northern form breeding through Central America from northeast Guatemala, southeast Honduras, east Nicaragua, east Costa Rica (Jones 2001a), east Panama (Dyrcz 1999) and the Pacific coast at Darian (Dyrcz 1999), Colombia, along the Pacific coast to southwest Ecuador, east of the Andes in Venezuela, the Guyanas, Trinidad, and east, south central and south Brazil (Amazon basin). *Marmoratum* occurs in Brazil (Mato Grosso, Rio Grande do Sul), central and south Bolivia, Paraguay, north Uruguay, and northeast Argentina (Cordoba, Mendoza, Santiago del Estero, Santa Fe, Entre Rios) (Michelutti 1992).

Migration: As far as is known, the species is sedentary.

Status: Very little information is available on status throughout its range, and it is not often reported nor seen frequently on surveys. For example, only 85 birds were reported in the winter Neotropical Waterbird Survey in southern South America (Blanco and Canevari 1992). However, it is widespread, probably with good overall numbers, and locally numerous in some areas, such as in Nicaragua, Costa Rica, and the llanos of Venezuela and Argentina (Hancock 1999). It is likely slowly increasing in Panama (Ridgely and Gwynne 1989, Dyrcz 1999).

Habitat

The Rufescent Tiger-Heron is a bird of low altitude wooded tropical swamps. It occurs especially along slow-moving rivers in swamp forests, gallery forests, and mangrove swamps and also in other extensive forested wetlands. It is a bird typical of the great wetlands of South America, the Amazon, the Paraguayan and Argentinean Chaco, and the Brazilian Mato Grosso, and is also found in hilly swampy forests such as the foothills of the Andes.

Foraging

This species feeds primarily by Standing, typically along the edge of a stream or on floating plants. It freezes in its Standing position for long periods, with its neck withdrawn or, often, with its head and neck thrust forward on its elongated snake-like neck. It has been reported to Walk slowly on floating vegetation and also to use bread as bait to catch fish (Teixeira 1991). It feeds by night but also during the day, alone or in pairs. It vigorously defends its feeding territory, including from other herons such as Whistling Herons (Hancock 1999). It on occasion feeds in small groups. It appears to readily habituate to human presence and activities (Hancock 1999).

Its diet is little known but includes fish (*Hoplosternum, Synbranchus*), grasshoppers, water beetles, dragonfly larvae, and snakes (Colubridae) (Beltzer 1990). Its long tarsus, bill, and neck suggest a primary adaptation for fishing.

Breeding

The breeding season is not well documented. On the basis of a very few observations, it is April in Panama, May–June in Suriname, and January–February in Argentina. Much more information is needed. It is a solitary nester, in tall trees in forests and small woods, generally near wetlands. The nest is a large platform of sticks, placed fairly high in the tree, many up to 7 m high (Haverschmidt 1968, Hancock 1999). Most reports are that the nest is a large untidy pile of sticks (Dyrcz 1999, Hancock 1999). The nest is about 50 cm in diameter.

The courtship has not been described in detail. In captivity, a mutual display included elements of the Snap and Stretch. The pair bowed with entwined necks, Bill Snapping. Then one stretched its neck from side to side, while the other adopted a Bittern Posture (Rossi 1958). The Courtship display of captives was accompanied by a 'nasal humming'. Eggs are blue-white to dirty white blotched in pale violet, dull red, and brown. In Suriname, they averaged 56.8 × 43.2 mm (Haverschmidt 1968). The clutch is 2 eggs. In captivity only the female incubated, taking 31–34 days (Rossi 1958). In one nest observed in the wild, changeover did not occur during the day and one of two eggs hatched (Hancock 1999). In captivity both adults of *marmoratum* fed the young by regurgitation of fish, at first well digested (Rossi 1958). The chick is fed only at night, apparently only once per night. Adults shade the young from the sun, both adults and young using gular fluttering to stay cool (Dyrcz 1999). Nothing is documented on chick development and nothing is known about breeding success.

Population dynamics

Nothing is known of the population biology of the species.

Conservation

The species is very widespread, large overall, and probably stable (Morales 2000). However, as a species of forested wetlands and streams, it is not often reported, although not particularly shy (Hancock 1999). It can be expected to be affected by deforestation, reforestation, damming of lowland forests, its principal habitat, and human disturbance of nesting.

Research needs

This is a species that is little known despite its wide range. Given the range of variation in the species and the difficulty of distinguishing juvenile Rufous from Fasciated Tiger-Herons, individual and geographic variation deserves closer scrutiny. The two (or three) identified subspecies appear to overlap geographically (Morales 2000). Individual, developmental, and geographical components of plumage, size, and habitat variation need to be re-examined. Movements, or lack of them, also need to be described, as this might confound observable patterns of variation. Tiger heron courtship behaviour deserves careful comparative study, given their evolutionary distance from the other herons and the distinctive nuances of behaviour so far reported. It may be that courtship differs substantially from the typical herons. The small clutch size and prolonged juvenile period suggests an interesting, long-lived demography. Study of its demography and ecology could provide considerable insight into the species and its less common relatives. This is a heron deserving much additional study.

Overview

The Rufescent Tiger-Heron is a forest heron that uses the edges of tree-lined streams and other water courses. It hunts by standing and waiting for prey, presumably fish. It is likely to feed nocturnally, but its daily schedule is not at all clear. Most intriguing is the small clutch size and extremely delayed maturation. These suggest that it is a long-lived species with low annual reproductive potential and likely is sedentary. The ecological correlates of this way of life could be intriguing.

New Guinea Tiger-Heron *Zonerodius heliosylus* (Lesson)

Plate 10.4

Other names: New Guinea Bittern, Forest Bittern, Zebra Heron in English; Avitigre Papúa in Spanish; Onoré phaeton in French; Bindenreiher in German.

Ardea heliosylus Lesson, 1828: 'Coquille', Zool. 1, Livr. 7, pl. 44; libr. 16, 1830, p. 722: New Guinea.

Description

The New Guinea Tiger-Heron is a medium-sized, brown and buff striped heron with a black crown.

Adult: The crown is black. The bill is relatively long, 90 mm, longest of all the tiger herons. The upper bill is black; the lower bill, ashy grey. The chin is white-buff. The irises are yellow and the lores and orbital skin are sulphur-yellow. The sides of the head and neck are barred in brown and light buff. The back and the upper wing are dark brown with fine buff lines. The rump and lower back are white with dusky tipped feathers. The underparts are buff to white with the feathers of the upper breast tending to fall over the belly like a bib or pinafore

(Iredale 1956). The legs are described as lemon-yellow, with the back of the tarsus washed with dusky brown.

VARIATION: The female is slightly smaller than the male. It is reported that the bill of the female has a pearl-green tinge below, the lores and orbital skin have a greener tinge, and the legs are more uniform greenish-yellow. More information on this and other attributed sexual differences are needed.

JUVENILE: The juvenile has not been described.

CHICK: The single chick described was covered in yellow-white down.

VOICE: The usual call of the New Guinea Tiger-Heron is said to be a low grunting. The species has also been credited with a bittern-like boom, but more information is needed.

WEIGHTS AND MEASUREMENTS: Length: about 70 cm.

Field characters

The New Guinea Tiger-Heron is identified by its barred plumage, black cap, light underparts, and long bill. It is distinguished from the Australasian Bittern by its longer bill and its barred plumage, including heavy horizontal (not vertical) barring on the throat. It is distinguished from the Rufous Night-Heron by its black crown and transverse barring. It is distinguished from the Black Bitten by its barring.

Systematics

The tiger herons are a distinctive lineage of the Ardeidae, adapted to a forest existence. The five species constitute a sister group to all the other herons and bitterns, and to the Boat-billed Heron (Sheldon *et al.* 1995). The New Guinea Tiger-Heron is considered to be more closely related to the African *Tigriornis* than to the American tiger herons (Payne and Risley 1976). The two Old World tiger herons have been regarded as belonging to the same genus. Additional material is needed.

Range and status

The New Guinea Tiger-Heron is a species of the East Indies.

BREEDING RANGE: This species occurs in New Guinea (Indonesia, Papua New Guinea) and on a few of the larger islands off its west coast (Aru, Salawati).

MIGRATION: There is no indication of migration in the species.

STATUS: No information is available on population sizes, either present or past. It is rarely seen or reported (Beehler *et al.* 1986). There have been only 30 confirmed sightings in 15 years and only a single nest has ever been reported (Hafner *et al.* 2000).

Habitat

The New Guinea Tiger-Heron occurs in rain forest streams. It is primarily a lowland bird but has been recorded up to 1450 m in the Central Highlands (Beehler *et al.* 1986).

Foraging

The New Guinea Tiger-Heron feeds solitarily along forest waterways. There are no observations reported of its feeding behaviour, other than being seen feeding in this habitat. It is recorded to eat crayfish, crabs, water beetles and other insects, small fish, lizards, and snakes.

Breeding

The breeding season is timed to the local rainy season. Few observations exist, but a nest was discovered in a tree in fairly open forest bordering a stream. This may be typical for the species. Undoubtedly it is a solitary nester, but this needs to be confirmed. Only one nest of this species has ever been discovered (Rand and Gillard 1967). This was a substantial structure of sticks and twigs, 12 m up in a tree in fairly open forest bordering a stream. Almost nothing is known of the breeding behaviour of this heron. The courtship

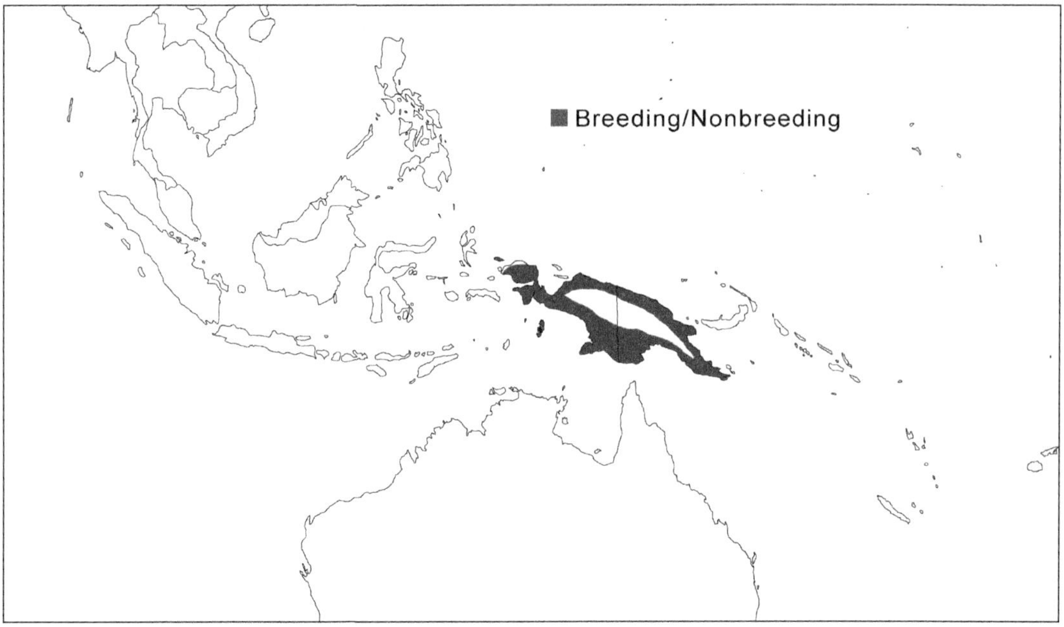

has never been observed. Nothing is known of the eggs or the clutch size. The one observation is of a single chick. Similarly, nothing is known about the basic reproductive biology of this species.

Population dynamics

The population biology of the species is unknown.

Conservation

It is likely that the species has declined in New Guinea (Rose and Scott 1997). It is considered to be near Threatened (Hafner *et al.* 2000, IUCN 2003). The lack of systematic, long-term knowledge is a huge impediment to being able to specify conservation action. With only about 30 sightings in 15 years, the species already likely is more than 'near' threatened. Conservation action on behalf of the species should focus on preserving forest-edge habitat in New Guinea.

Research needs

Almost nothing is known about the basic biology of this species. Research is needed into the basic biology of the New Guinea Tiger-Heron. The details of its distribution, status, and trends are virtually non-existent, and so there is a need to conduct range-wide surveys for the species.

Overview

As a tiger heron it feeds along the edge of forest streams, generally in the lowlands. This is a species that has been observed and reported very infrequently. Much needs to be understood about its basic biology and conservation needs.

White-crested Tiger-Heron *Tigriornis leucolophus* (Jardine)

PLATE 10.5

Other names: White-crested Tiger Bittern, African Bittern, White-crested Bittern in English; Avetigre Africana in Spanish; Onoré à huppe blache in French; Weißschopfreiher in German.

Tigrisoma leucolopha Jardine, 1846. Ann. Mag. Nat. Hist. p. xvii: 'Western Africa' = Old Caladara or Bonny Rivers region (Nigeria).

Description

The White-crested Tiger-Heron is a slender, medium, cryptically coloured, black and buff barred heron with a white crest, visible when erected.

ADULT: The adult male has a black crown and nape, and a white crest, which is typically concealed by the darker feathers except when raised. The bill is long and slender. The upper bill is black-brown, whereas the lower bill is yellow-green. The lores and orbital skin are green and the irises are yellow. The sides of the head, neck, back, and upper wings are transversely barred black and buff. The tips of the flight feathers are white. The underparts are streaked buff and spotted with black. The tail is barred black and white. The legs are brown, yellow behind. The undersides of the toes also are yellow.

VARIATION: The female tends to be smaller and has narrower buff barring on the upper sides and pink-cinnamon on the underside.

JUVENILE: The immature bird has broader barring. The lighter colour goes from rufous to sandy with age, probably in successive plumages, as in other tiger herons.

CHICK: The chick's down is initially is yellow and then white.

VOICE: The voice is a boom, like a bittern. It is a single or double note repeated slowly for several minutes.

WEIGHTS AND MEASUREMENTS: Length: 66–80 cm.

Field characters

The White-crested Tiger-Heron is identified by its transversely barred black and buff plumage, black cap, and white crest. It is distinguished from the European Bittern by its darker plumage (narrower light barring), and slender (not stout) appearance, and its longer neck (Rodwell 1996).

Systematics

The tiger herons are a distinctive lineage of the Ardeidae, adapted to a forest existence. The five species constitute a sister group to all the other herons and bitterns, and to the Boat-billed Heron (Sheldon *et al.* 1995). The African *Tigriornis* is a sister taxon to the American *Tigrisoma*. Morphologically it appears to be more closely related to the Asian New Guinea Tiger-Heron than to the American tiger herons (Payne and Risley 1976).

Range and status

RANGE: The White-crested Tiger-Heron occurs in the equatorial rain forest of west Africa in Senegal (Morel and Morel 1990), Guinea–Bissau (Rodwell 1996), Guinea, Sierra Leone, Liberia, Ivory Coast, Cameroon, Gabon, and the Central Africa Republic to Zaire.

MIGRATION: The species is sedentary but there may be dispersal.

STATUS: This species is seldom seen and very seldom reported on. It is locally common in parts of Gabon and the forests of the lower Congo swamp forests (Turner 2000).

Habitat

The White-crested Tiger-Heron is a bird of the tropical rain forest, occurring along small streams located in marshes or swamps within the forest, gallery forests along rivers, and mangrove-lined streams on the coast.

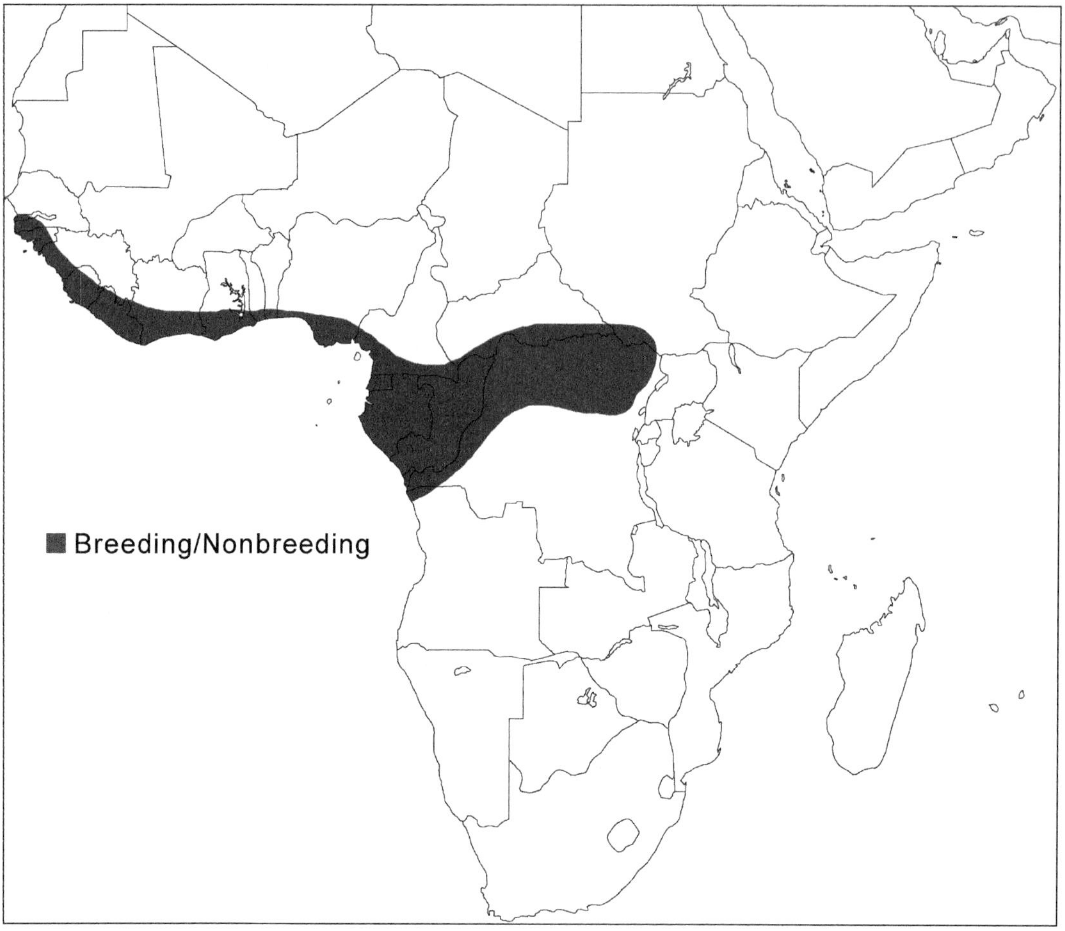

Foraging

The White-crested Tiger-Heron is a solitary feeder, and probably nocturnal. Its feeding behaviour has not been described. It assumes a Bittern Posture when disturbed. The details of the diet are not understood. It eats small fish, crabs, crayfish, frogs, snakes, and insects. It likely has a broad diet of whatever it can catch along the river shore.

Breeding

Its breeding biology is virtually unknown. Nesting appears to be in the rainy season. Egg-laying dates have been documented as: Sierra Leone, September–October; Liberia, October; Cameroon, May; and Gabon, January. It nests solitarily as far as is known, in trees. One display has been described, apparently having a large threat component. A bird in captivity spread its tail and primaries, retracted its neck, spread its white crest, and moved rhythmically from side to side while hissing. The single nest described was a loose, flimsy structure of twigs. It was placed 6 m high in a tree. One egg described was beige-yellow with purple and blue-grey blotches and smaller red-brown spots. It was 55 × 45 mm. The estimated incubation period is less than 29 days.

Population dynamics

There is no information on the breeding biology of the species.

Conservation

The lack of data on the distribution and status, combined with almost no information on its biology, poses a significant conservation risk. There is nearly nothing known about this heron and so it is not possible to assign a degree of conservation concern or develop a conservation strategy. It is listed as Data Deficient by IUCN (IUCN 2003).

Research needs

Any information on its biology would be valuable. Surveys need to be undertaken to determine the distribution and density of the species within its habitat and its habitat requirements. Studies are needed on its status and distribution, its basic biology and ecology in order to understand its conservation needs.

Overview

The White-crested Tiger-Heron is a stream-side heron of the African tropical rain forest. It uses various aquatic habitats, including small streams or swamps within the forest, gallery forests along rivers, and mangrove-lined streams on the coast. It is a solitary, primarily nocturnal feeder. Its feeding behaviour has not been described. It is hard to find and harder to observe.

Boat-billed Heron *Cochlearius cochlearius* (Linnaeus)

PLATES 8.4–8.6

Other names: Martinete Cucharón, Pato Pico de Barco, Garza Bruja, Huairavo, Boca d'agua in Spanish; Coá, Arapapá in Portuguese; Savacou huppé in French; Kahnschnabel in German.

Cancroma cochlearia Linnaeus, 1766, Syst. Nat. ed. 12, 1, p. 233: Cayenne (Guiana).

Cochlearius cochlearius zeledoni (Ridgway), 1885: Mazatlan (Sinaloa), Mexico; *Cochlearius cochlearius panamensis* Griscon, 1926: Corozal, Canal Zone; *Cochlearius cochlearius phillipsi* Dickerman, 1973: Atasta, Campeche, Mexico; *Cochlearius cochlearius ridwayi* Dickerman, 1973: Coyoles, Yoro, Honduras.

Description

The Boat-billed Heron is a stocky, medium-sized, mostly black and white and sometimes buff heron, with a huge black bill.

ADULT: The Boat-billed Heron's head and its crest of long, lanceolate occipital plumes are black. The broad, thick, flat—entirely outlandish—bill is black with a yellow wash at the lower base. The forehead and cheeks are white. The eyes are huge and bulge out. The irises are dusky or brown. The lores and orbital skin are grey with a green-yellow spot on the eyelid. The gular pouch, visible only when distended, is variably dusky sulphur-yellow, yellow and grey, and pink. The upper back is black, the rest of the back and upper wings are grey. The underparts are a rich rufous, and the flanks and leg feathering are black. The mantle is buff-brown merging into vinaceous on the sides of the neck. The legs are green. During breeding the mouth lining, lores, and gular area become black. The peak of the crest plumes is achieved just prior to nesting.

VARIATION: The sexes are alike, but males have longer occipital plumes. The five recognized subspecies differ in size and plumage colour (Dickerman 1973a, Dickerman *et al.* 1982). *Cochlearius* is the palest form, with light grey upperparts, and a pure white breast and neck. *Panamensis* is the darkest form, olive-grey above and darker lavender below; the face, throat, and upper breast are suffused with olive-brown. The remaining subspecies are paler, the upper parts are lavender grey, and the throat and breast are buff-brown. They differ slightly from one another in size and colour intensity on the underside. *Phillipsi* is the largest. *Ridgwayi* is a little bigger than *zeledoni* and also darker below, though not as dark as *panamensis*. Specimens from western Costa Rica are considered to intergrade between *ridgwayi* and *panamensis*. Specimens from northern Guatemala intergrade between *ridgwayi* and *phillipsi*.

JUVENILE: Juveniles have a dull rufous back and wings, and are dingy white below with a pinkish or buff wash. They have a shorter crest than the adult. They move through six plumages from hatching to adult (Dickerman *et al.* 1982). The juvenile *cochlear-*

ius has a tawny-brown back. Juveniles of the other races are browner than adults, but lighter than *cochlearius*.

CHICK: The downy chick is pale grey above, dull white to pale grey below, and has a blackish crown lacking erectile tufts. In its colour pattern and lack of a crest, it differs from other nestling herons (Dickerman and Juarez 1971). It has greenish-yellow skin. The chick has a triangular short beak, broad at the base and tapering to a blunt point anticipating its adult form.

VOICE: This is a vociferous heron. Its raucous, jeering, laughing call is commonly heard along the mangroves and forests in which it occurs and nests. The 'An' call and its variations are characteristic of the species. But they are varied in style and probably meaning. The 'An' call is a repeated vocalization, rendered 'an, an, an, an' or 'oh, oh, an-an'. The 'Ha call' rendered 'Ha, go, go, go, go', is the disturbance, threat, and general presence call used in the colony and on the feeding grounds. This single call is highly variable and used on many occasions. The An call ('an, an, an, an') has been called the Short Chant. When the 'an, an, an, an' notes rises, it indicates enhanced agitation. A more complex 'An' call is called the Long Chant, starting with several notes, then a series of paired notes. This can elicit a chorus from birds in the colony. It is the antagonistic, territorial call used in the Greeting Display, Supplanting Flight, when fleeing an attack, during disturbance, and in chorus. A variation is also the flight call, used in flying back and forth to roost. A faint 'Ump' call, made in the throat with the bill closed, is a contact call. The 'Pop' (also called 'Clap-clap'), is the sound made by the snapping of their substantial bills together. The 'Pop' call is usually doubled and may be joined by an 'An' call. They give a 'High Cry' call when fleeing other herons.

WEIGHTS AND MEASUREMENTS: Weight: males 680–770 g; females 503–726 g; length: 45–51 cm.

Field characters

The Boat-billed Heron is identified by its broad, thick bill, its moderate size, black cap, white forehead and cheeks, and grey back and wings. It flies with slow wing beats, which may be deep or shallow. It is distinguished from any other bird by its bill. Enough said.

Systematics

The systematic relationship of the Boat-billed Heron is one of the best-studied questions of heron biology (Cracraft 1967, Dickerman and Juarez 1971, Dickerman 1973a, Payne and Risley 1976, Sheldon 1987, Sheldon *et al.* 1995, McCracken and Sheldon 1998). The conclusions have ranged from being placed in its own family to being classified as a night heron. Molecular studies have now brought clarity to the situation. The Boat-billed Heron is a heron but constitutes a sister group to other herons (Sheldon *et al.* 1995, K. McCracken pers. comm.). Thus, it is properly placed in its own subfamily, Cochleariinae. This taxonomy is coherent with available morphological and behavioural evidence. Recognizing the ancient lineage and distinctiveness of the Boat-billed Heron suggests that the relationships among the several recognizable populations of Boat-billed Herons deserve additional scrutiny. The northern and South American birds have sometimes been separated as different species, and in fact the several forms may be more distantly related than is now appreciated.

Range and status

The Boat-billed Heron is a bird of South and Central America.

BREEDING RANGE: The Boat-billed Heron occurs in coastal north central Mexico south to Peru, Bolivia, and northeast Argentina. The race *zeledoni* (formerly including all boat bills from Mexico to the Nicaraguan border) is restricted to the Mexican west coast from Sinaloa to Guerrero. *Ridgwayi* occurs in Chiapas (Mexico), coastal Guatemala, El Salvador, and west Honduras. In northern Guatemala *ridgwayi* intergrades with *phillipsi*. *Phillipsi* occurs on the Caribbean slopes of Mexico from the latitude of Cuidad Victoria to the tip of the Yucatan peninsula and south to Belize, and likely, but not proven, in Nicaragua. *Panamensis* and

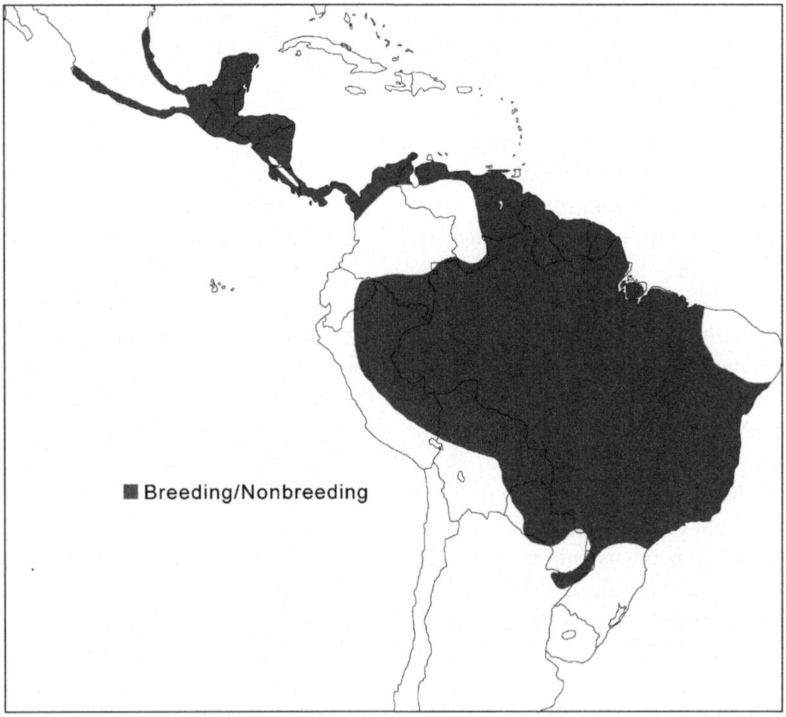

ridgwayi intergrade in west Costa Rica. *Panamensis* occurs through coastal Costa Rica through most of lowland Panama. *Cochlearius* occurs throughout most of the South American range except the northwest coast and reaches east Panama.

MIGRATION: The Boat-billed Heron is rather sedentary. Non-breeding records occur outside the breeding range. *Panamensis* occurs along the Pacific coast of South America down to Lima, Peru, suggesting migration or regular dispersal. It occurs occasionally in northern Argentina and south Brazil (Rio Grande do Sul). Local movements to tidal areas in the non-breeding season are reported.

STATUS: The species is widely distributed and not uncommon where it occurs.

Habitat

The Boat-billed Heron uses wooded or mangrove fringes of fresh water creeks, lakes and marshes with thick bushes or trees overhanging water, which it requires for daytime roosting and nesting. It is a lowland species, occurring to 650 m.

Foraging

Despite its great bill, this species for the most part feeds like a typical heron. It feeds by Standing, sometimes for many minutes, usually in the Crouched posture. It Walks slowly, with very deliberate steps on the ground or along branches and roots. It proceeds in a very Crouched posture, described as hunch-backed (Mock 1975b). It sometimes uses Neck Swaying, a rhythmic movement of the body but not the head, and Peering Over when observing food items. It also feeds by Walking Quickly and by Running, frantically dashing about, stopping, and Stabbing. It moves short distances by Hopping up and uniquely flapping once or twice (Mock 1975b). Using these behaviours it appears to feed visually (Biderman and Dickerman 1978). It captures prey both by Stabbing and by full-bodied

Lunging. It also Probes and uniquely uses its large bill in Ploughing (Wetmore 1965, Willard 1979). In the latter, the individual wades along, with the bill partially submerged, thrusting forward and scooping with each step, thus using the wide bill to catch sedentary prey in mud and leaf litter. Probing and Ploughing are non-visual techniques.

It feeds nocturnally and crepuscularly, as suggested by its huge eyes. It is easily disturbed while foraging, and readily flies off, calling. During the day it perches in dense trees and bushes, and also retreats there when disturbed. Its resting posture is characteristic, and being used during the day, is more often observed than its foraging. At roost, its head appears to rest on its shoulders, bill forward and slightly down. When sleeping, the bill falls further and slumps on its breast. Or it may lean its head to one side and rest it under the wing (Sick 1993). It spends its time at roost, sleeping, resting, and preening. It preens by both nibbling and stropping in a typical heron fashion, but also uses the hook on the end of its bill. It occasionally feeds during the day. It feeds alone, flying from the communal roost to independent, probably defended, feeding areas or in tidal pools and other situations, it will feed in aggregations.

The diet is broad and includes insects, shrimp, fish, amphibians, and small mammals. Crustaceans are particularly important to the diet. Shrimp include *Panaeus* and *Macrobrachium*. Fish include mullet (*Mugil*), sleeper (*Dormitador*), snook (*Centropomis*), and catfish (Ariidae).

Breeding

The breeding biology of the Boat-billed Heron has been the subject of several studies (Dickerman and Juarez 1971, Juarez and Dickerman 1972, Mock 1975a, Alvarado 1992, Hernández and Fernández 1999). The breeding season is variable and may be dependent on a combination of factors, such as the rains and food availability. Breeding is in the rainy season in most places but can be in the dry season in coastal habitats. It nests in April–August in Mexico, June–June in Honduras, February–April in coastal east Mexico and Panama, August in Suriname, June–October in Trinidad, and November in Brazil. Nesting seasons need to be further clarified throughout the range of the species.

The Boat-billed Heron nests in trees and bushes, especially in mangroves. It uses flooded forests, swamps, and trees along and overhanging water, such as streams and pools. The heron nests solitarily or in small groups of about a few to a dozen pairs. It also forms part of mixed heronries, and seems particularly to nest with Striated or Green Herons. In mixed colonies Boat-bills often nest near each other but away from the main colony. Nests are a flat platform made of sticks, with leafy branches, 30–35 cm wide, and 10–15 cm thick (Hernández and Fernández 1999). They are often small, although old ones are typically reoccupied or enlarged or bigger nests of other species taken over. In one site 55% were reused nests (Hernández and Fernández 1999). They are usually placed 0.5–10 m above the ground on thick supportive interior branches. Contrasted with other herons nesting with it, the Boat-billed Heron's nest is found in denser, more closed vegetation.

The courtship displays of the Boat-billed Heron differ from those of other herons, taking advantage of the display opportunities of the great bill, and perhaps constrained by mechanical limitations of that bill. Mock (1975a) found few that he considered to be homologous with other known heron displays. It is a highly vocal heron in its courtship behaviour. Boat-billed Herons appear to arrive at the nesting colony already paired. As a result, males do not perform the typical advertising displays from nesting territories before pair formation. But they do show a series of behaviours in and near the nest site (Mock 1975a), most of which are threats of various sorts. Crest Raising is particularly common, and is used when the bird is disturbed and also in other displays. The degree of crest raising is variable but at full expansion it forms a black fan, contrasting sharply with the white forehead. As a display, the heron may also extend its neck up and forward in a slight arch. Crest Raising is a disturbance and Greeting Display. Tall Rocking consists of a heron extending its neck up and forwards, raising its crest, and rocking slowly from side to side. It is a mild threat. Twig Shaking is also a threat. In a visually striking Wings Out display, given usually after landing, the heron extends both wings out to the side. The Pop Display features the Pop call. The male flexes its feet slightly, snaps the head up with the bill pointing down, lowers the rump making the body vertical, erects its crest half way, flicks the wings out, and gives the loud resonating Pop call by snapping the bill, sometimes with an An call. Bill Snaps are also used in territorial defence. In the Forward display (also called the Single Clap display) the bird walks towards another, extends its neck up and forward, erects its crest and snaps its bill once. The display also sometimes includes a shrieking An call (also called an An–Snap), which is more aggressive. Between members of a pair, both Contact and Non-contact Bill Clappering are the most common displays and often follow a Pop Display. The paired birds spend time together in and around their display territory. They walk around together, appearing to examine possible nest sites. Bill Clappering occurs as the birds remeet during the walk. A most intriguing display is the Bill Duel (Alvarado 1992, Sick 1993). The birds face each other or the male is behind the female, and they touch bills at the tips. When the female's bill is within the males, he closes it and they begin a series of rhythmic movements for several minutes before separating.

The eggs are pale blue to green. They are usually spotted with cinnamon on the larger end, sometimes forming a ring. The spotting is usual but not invariably reported, likely because the eggs fade to white or blue-white with a calcareous wash during incubation or in egg collections more like an ibis or tiger heron than a typical heron (Sick 1993). Eggs average 49.5 × 33.0 mm in Mexico, 49.7 × 35.9 mm in Trinidad, 50.3 × 35.3 mm in Suriname (see Dickerman and Juarez 1971). The clutch is usually 3 eggs, the range is 1–4 eggs. Incubation is 26 days, with a range of 25–27 days. Incubation is shared by both parents and begins with the first egg. The young are at first fed entirely at night. The adult is aggressive in defending young from all intruders, a behaviour not typical of herons. The defence is with a Forward, in which the bird stretches its neck forward, crest raised, wings half open, calling, and Bill Snapping. Chicks develop quickly and can take to branches with great agility (Sick 1993). Nesting success is little studied. Hernández and Fernández (1999) determined that it was relatively low. They inferred predation by grackles (*Quiscalus*). The winds of tropical storms also cause chick mortality.

Population dynamics

Nothing is known about the population biology of the species.

Conservation

The species is widespread and found in suitable habitat throughout its range. There is little information available on population sizes and status. In areas counts are 200–300 pairs. Overall it is common in places but rarely abundant (Morales 2000). It is particularly common in parts of Mexico, Costa Rica, Belize, Nicaragua, and Suriname. It is uncommon in Panama.

Research needs

Despite the interest and the studies, we still do not really understand much about the biology and ecology of this unique species. The foraging ecology of this species is not yet understood in its entirety. The role of its bill in feeding needs to be more thoroughly examined to determine its function within an annual cycle. It seems likely that there is more to this heron's use of its large bill than is presently known. The calls and songs of this species should be explicated and functions determined, again with the involvement of the bill in mind. The demography of this species is unknown and should be examined through banding studies. An interesting result of Boat-billed Heron being understood as occupying an old and distinctive line in heron evolution is the possibility of there being unrecognized differentiation among Boat-billed Heron populations. It may be that these populations are more distinct than is now appreciated. So it would be worthwhile to further examine not only their systematic relationships but whether there are differences in their behaviour and ecology as well.

Overview

The Boat-billed Heron is a heron of a different sort, evolutionarily, behaviourly and anatomically. At least one feeding behaviour, Scoop feeding, makes use of its grandest feature. Its courtship behaviour is very different from typical herons, as is its vocal repertoire of songs and calls. It is basically a solitary to modestly aggregative, noturnal, shallow water wading species. Its broad range, and local commoness shows it continues to be a successful species, one that suggests the potential for diversity within the family of herons.

Bibliography

Abbott, T. (1994). A Goliath Heron at Ndumo. *Albatross (Durban)*, **317**, 32–3.

Abdulali, H. and Alexander, H. G. (1952). Ardeidae with red legs. *Ibis*, **94**, 363.

Ac, P. and Bohus, M. (1990). [First record of nesting of *Egretta garzetta* in Podunajsk.] *Ziva*, **38**, 183–5.

Acosta, M., Mugica, L., and Martinez, P. (1990). Segregación del subnicho trófico en seis especies de ciconiformes cubanos. *Ciencias Biológicas*, **23**, 68–81.

Acosta, M., Mugica, L., Torres, O., and Abad, Y. (1990). Alimentación de *Bubulcus ibis ibis* (Linneo) (Aves: Ardeidae) en la provincia de Pinar del Río. *Ciencias Biológicas*, **23**, 82–91.

Adams, C. E. and Mitchell, J. (1995). The response of a Grey Heron *Ardea cinerea* breeding colony to rapid change in prey species. *Bird Study*, **42**, 44–9.

Al Safadi, M. M. and Kasparek, M. (1995). Breeding observations on the birds of the Tihamah, Yemen. *Zoology In the Middle East*, **11**, 15–20.

Al-Saghier, O. and Porter, R. F. (1997). The first Black Heron *Egretta ardesiaca* in Yemen. *Sandgrouse*, **19**, 140–1.

Albert, A. and Wollesmann, M. (1986). Spectrogram and oscillogram comparative analysis of Night Heron (*Nycticorax nycticorax*) advertising call. *Tiscia (Szeged)*, **21**, 123–7.

Alessandria, G., Carpegna, F., and Della Toffola, M. (2003). Vocalizations and courtship dispays of the Bittern (*Botaurus stellaris*). *Bird Study*, **50**, 182–4.

Alessandria, G., Della Toffola, M., and Carpegna, F. (1992). Nuovo sito di riproduzione di tarabuso, *Botaurus stellaris*, e di airone rosso, *Ardea purpurea*, in Piemonte. (Aves, Ardeidae). *Rivista Piemontese di Storia Naturale*, **13**, 97–102.

Alfaro, M. and Russi, D. (1989). Estimacion del exito de anidamiento de la Garza Morena *Egretta tricolor ruficollis* (Gosse, 1847) en la Laguna de San Lorenzo, Cartagena (Colombia). *Boletin Ecotropica*, **19**, 3–25.

Alieri, R., Gariboldi, A., and Fasola, M. (1988). [Breeding habitat requirements of herons in the central Po Valley.] *Naturalista Siciliano*, **12(Supplement)**, 29–32.

Alan, D. G. and Davies, G. B. (1999). The birds (Aves) of the middle Komati River Valley, Swaziland. *Durban Museum Novitates*, **24**, 22–42.

Allen, R. P. and Mangels, F. P. (1940). Studies of the nesting behavior of the Black-crowned Night Heron. *Proceedings of the Linnean Society of New York*, **50–51**, 1–28.

Allport, A. M. and Carroll, D. (1989). Little Bitterns breeding in South Yorkshire. *British Birds*, **82**, 442–6.

Alström, P. and Colston, P. (1991). *A field guide to the rare birds of Britain and Europe*. HarperCollins, New York.

Alvarado, G. M. (1986). Estudio de la amidacion del martinete coronegro o chocuaco *Nycticorax nycticorax* (Ardeidae) en Isla Pajaros, Rio Tempisque, Costa Rica. Thesis. Universidad de Costa Rica, Costa Rica.

Alvarado, G. M. (1992). Comportmento reproductivo in tres especies de garzas: *Cochlearius conchlearius* (Linnaeus), *Agamia agami* (Gmelin), y *Butorides striatus* (Linnaeus) (Ardeidae) en Westfalia-Limon, Costa Rica, 1988–1989. MS Thesis. Ciudad Universirtaria 'Rodrigo Facio', Costa Rica.

Alvarado, G. (1995). Nueva distribucion de la especie *Agamia agami* en Costa Rica. *Brenesia*, **43–44**, 93–4.

Alvarado, G. M. (1998). Extension del ambito reproductive del avetorrilo pantanero, Ixobrychus exilis en Costa Rica. *Brenesia*, **49–50**, 109–10.

Alvarado Q., G. M. (in prep a). Estudio de la variacion morfometrica de la Garcilla Verde *Butorides virescens* en diferentes poblaciones de Costa Rica.

Alvarado, G. (in prep b). Observations on the breeding behavior of the Chestnut-bellied Heron (*Agamia agami*).

Amadon, D. (1942). Birds collected during the Whitney South Sea Expedition. XLIX. Notes on some non-passerine genera, I. *American Museum Novitates*, **1175**, 1–11.

Amadon, D. and Woolfenden, G. (1952). Notes on the Matthew's collection of Australian birds: The order Ciconiiformes. *American Museum Novitates*, **1564**, 1–16.

Amat, J. A. and Aguilera, E. (1989). Some behavioural responses of Little Egret and Black-tailed Godwit to reduce prey losses from kleptoparasites. *Ornis Scandinavica*, **20**, 234–6.

Amat, J. A. and Aguilera, E. (1990). Tactics of Black-headed Gulls robbing egrets and waders. *Animal Behaviour*, **39**, 70–7.

American Ornithologists' Union (1998). The AOU Check-list of North American Birds, 7th ed. AOU, Washington.

Amies, P. (1990). Grey Herons hunting by artificial light. *British Birds*, **83**, 425.

Amos, E. J. R. (1991). *The birds of Bermuda*. Corncrake, Warwick, Bermuda.

Andersen, L. N. (1989). [Bitterns in winter.] *Fugle*, **9**, 26–7.

Ando, Y. (1993). [Nest site selection of six heron species in a breeding colony.] *Japanese Journal of Ornithology*, **41**, 29–38.

Andrew, D. G. (2004). Medieval Little Egrets and others. *British Birds*, **97**, 44–5.

Andrews, M. I. and Mathew, T. K. (1997). Some observations on the feeding behaviour of the Cattle Egret *Bubulcus ibis coromandus* (Boddaert) and the Pond Heron *Ardeola grayii grayii* (Sykes). *Pavo*, **35**, 67–73.

Andrews, T. (1993). The first record of the Chinese Egret on Java. *Kukila*, **6**, 133.

Aniskowicz, B. T. (1981). Behavior of a male Least Bittern incubating after loss of mate. *Wilson Bulletin*, **93**, 395–7.

Anjos, L. D. (1990). [Avian distribution in an araucaria forest of Curitiba City (South Brazil).] *Acta Biologica Paranaense*, **19**, 51–64.

Anonymous (1987). Ringing recoveries. *Stilt*, **11**, 50.

Anonymous (1989). Significant sightings. *Falkland Islands Birding News*, **1**, 8–10.

Anonymous (1994). From the field. *Oriental Bird Club Bulletin*, **19**, 65–7.

Anonymous (1998). European news. Little Egret *Egretta garzetta*. Ireland. First breeding record. *British Birds*, **91**, 242.

Anonymous (2000). Great White Heron. *North American Birds*, **54**, 431.

Anonymous (2002a), Bird news. November 2002. Birding World, **15**, 443–57.

Anonymous (2002b). Bird news, December 2002. Birding World, **15**, 487–500.

Anonymous (2002c). Great White Heron. *North American Birds*, **56**, 123.

Anonymous (2002d). Cattle Egret. *North American Birds*, **56**, 254.

Antas, P. T. Z., Silva, F., Alves, M. S., and Lara-Resende, S. M. (1986). Brazil. In *A directory of neotropical wetlands* (eds. D. Scott and M. Carbonell), pp. 60–104. IUCN and IWRB, Slimbridge, UK.

Arendt, W. J. (1988). Range expansion of the Cattle Egret (*Bubulcus ibis*) in the greater Caribbean basin. *Colonial Waterbirds*, **11**, 252–62.

Arendt, W. J. and Arendt, A. I. (1988). Aspects of the breeding biology of the Cattle Egret (*Bubulcus ibis*) in Montserrat, West Indies, and its impact on nest vegetation. *Colonial Waterbirds*, **11**, 72–84.

Ascanio-Echeverria, D. and Davis, W. E., Jr. (2000). Notes on the habitat, behavior and vocalizations of Zigzag Herons in Venezuela. *Waterbirds*, **23**, 521–3.

Ashkenazi, S. (1993). Dark-morph individuals of *Egretta* spp. in Israel. *Colonial Waterbirds*, **16**, 202–7.

Ashkenazi, S. and Yom-Tov, Y. (1996). Herons and fish farming in the Huleh Valley, Israel: conflict or mutual benefit? *Colonial Waterbirds*, **19**, **Special Publication 1**, 143–51.

Ashkenazi, S. and Yom-Tov, Y. (1997). The breeding biology of the Black-crowned Night-Heron (*Nycticorax nycticorax*) and the Little Egret (*Egretta garzetta*) at the Huleh Nature Reserve, Israel. *Journal of Zoology (London)*, **242**, 623–41.

Atkinson, G. (2003). A new Slaty Egret breeding colony. *Babbler* **43**, 49.

Austin, H. M. (1996). Observations on daylight feeding periodicites of wading birds (Ardeidae) in Sarah's Creek, York River, Virginia. *Journal of the Elisha Mitchell Scientific Society*, **111**, 137–45.

Austin, J. E. and Slivinski, M.V. (2000). American Bittern depredates Sora. *Prairie Naturalist*, **32**, 59–60.

Azevedo Junior, S. M. D. (1997). Colonizacao da Garca-boieira *Bubulcus ibis* em Pernambuco, Brasil. *Airo*, **8**, 48–50.

Azure, D. A., Brininger, W. L., Toepfer, J. E., Huschle, G., and Crawford, R. D. (2000a). First described renesting attempt by an American Bittern. *Wilson Bulletin*, **112**, 271–3.

Azure, D. A., Naugle, D. E., Toepfer, J. E., Huschle, G., and Crawford, R. D. (2000b). Sexing American Bitterns, *Botaurus lentiginosus*, using morphometric characteristics. *Canadian Field-Naturalist*, **114**, 307–10.

Baccetti, N. (1983). Heronries of Somalia, a preliminary report. *Monitore Zoologico Italiano Supplemento*, **18**, 173–85.

Bagley, F. M. and Grau, G. A. (1979). Aspects of the Yellow-crowned Night Heron reproductive behavior. *Proceedings of the Colonial Waterbird Group*, **3**, 165–75.

Baillon, F. (1989). Decouverte d'une colonie nicheuse de heron cendre (*Ardea cinerea L.*) au Senegal. *Alauda*, **57**, 233.

Baker, E. C. S. (1935). *Nidification of the birds of the Indian Empire*. Taylor & Francis, London.

Baker, N. E. (1995). Tanzania waterbird count: The first coordinated count on the major wetlands of Tanzania, The Wildlife Conservation Society of Tanzania and BirdLife International, Dar es Salam.

Baker, N. E. and Baker, E. M. (2002). *Important bird areas in Tanzania: A first inventory*. Wildlife Conservation Society of Tanzania, Dar es Salam.

Baker, N. E. and Baker, E. M. (2004). *Tanzania: bird atlas. http://tanzaniabirdatlas.com*

Baker, R. R. (1978). *The evolutionary ecology of animal migration*. Hodder & Stoughton, London.

Bakka, S.V. (1989). [Colonies of the Heron (*Ardea cinerea*) in Gorki region.] *Ornitologiya*, **21**, 127–8.

Balanca, G. (1987). Etude des strategies alimentaires de l'aigrette garzette, *Egretta garzetta*, sur la cote Atlantique du Maroc. *Gerfaut*, **77**, 443–62.

Bancroft, G. T., Strong, A. M., Sawicki, R. J., Hoffman, W., and Jewell, S. D. (1994). Relationships among wading bird foraging patterns, colony locations, and hydrology in the Everglades. In *Everglades: the ecosystem and its restoration* (eds. S. M. Davis and J. C. Ogden). pp. 615–57. St. Lucie Press, Delray Beach, Florida.

Banks, R. C. and Dickerman, R.W. (1978). Mexican nesting records for the American Bittern. *Western Birds*, **9**, 130.

Bara, T. (1988). Expansion rapide d'une colonie de herons cendres *Ardea cinerea* dans la basse vallee de la Durance. *Faune de Provence*, **9**, 89–90.

Barbieri, F. and Fasola, M. (1986). Wintering Little Egrets and Night Herons in Italy. *Supplemento alle Ricerche di Biologia della Selvaggina*, **10**, 377.

Barbieri, L. and Boileaum, N. (2000). Breeding of the Little Bittern *Ixobrychus minutus* in the same nest in three consecutive years in Romelaere (France). *Nos Oiseaux* **47**, 77–80.

Barbour, D. and Barbour, M. (2001). Aerial fishing by Grey Herons. *Promerops*, **245**, 16.

Barbraud, C. and Hafner, H. (2001). Variations des effectifs nicheuride Hérons Pourprés *Ardea purpurea* sur le littoral Méditerranéen française relation avec la pluviometrie sur les quartiers d'hivemage. *Alauda*, **69**, 29–36.

Barbraud, C., Lepley, M., Lemoines., and Hafner, H. (2001). Recent changes in the diet and breeding parameters of the Purple Heron (*Ardea purpurea*) in southern France. *Bird Study*, **48**, 2–10.

Barbraud, C., Lepley, M. Mathevet, R., and Mauchamp, A. (2002). Reedbed selection and colony size of breeding Purple Herons *Ardea pupurea* in southern France. *Ibis.*, **144**, 227–35

Barbraud, C. and Mathevet, R. (2000). Is commercial reed harvest compatible with breeding Purple Herons, *Ardea purpurea*, in the Camargue, southern France? *Environmental Conservation*, **27**, 334–40.

Barden, P. J. (1990). Little Egret feeding in association with fishing Cormorant. *British Birds*, **83**, 279–80.

Bari, A., Gonella, P, and Piazza, R. (1990). Vocalization study of two species of herons (*Nycticorax nycticorax* and *Egretta garzetta*). *Ethology, Ecology and Evolution*, **2**, 296–7.

Bari, A. and Piazza, R. (1990). Preliminary spectrographic analysis of the vocalizations of the Night Heron *Nycticorax n. nycticorax* (L.). *Atti della*

Accademia delle Scienze di Torino I Classe di Scienze Fisiche Matematiche e Naturali, **124**, 62–72.

Barlow, C. and Wacher, T. (1997). *A field guide to the birds of The Gambia and Senegal*. Yale University Press, New Haven.

Barnett, J. M. Coconier, E. G., Veazquez, M. and Clay. R. P. (2002). Primer registro de nidificacion actualizacion sobre la presencia de *Botaurus pinnatus* en Paraguay. *Hornero*, **17**, 49–51.

Bartolome, J., Fernandez, C. M., and Campos, F. (1996). Band recoveries of Spanish Little Egrets, *Egretta garzetta*. *Colonial Waterbirds*, **19**, 220–5.

Bauer, K. M. and Glutz von Blotzheim, U. N. (1966). *Handbuch der Vogel Mitteleuropas*. Vol. 1. Akademische Verlagsgesellschaft, Frankfurt.

Baumanis, J. (1998). Hybridisation zwischen Silberreiher *Egretta alba* und Graureiher *Ardea cinerea* in Lettland. *Limicola*, **12**, 142–5.

Baumann, W. (2000). Zwergdommel fischt mit Koeder! *Falke*, **47**, 346.

Baxter, G. S. (1994a). The influence of synchronous breeding, natal tree position and rainfall on egret nesting success. *Colonial Waterbirds*, **17**, 120–9.

Baxter, G. A. (1994b). The location and status of egret colonies in coastal New South Wales. *Emu*, **94**, 255–62.

Baxter, G. S. (1996). Provision of supplementary nest material to colonial egrets. *Emu*, **96**, 145–50.

Baxter, G. S. and Fairweather, P. G. (1989). Comparison of the diets of nestling Cattle Egrets and Intermediate Egrets in the Hunter Valley, New South Wales. *Australian Wildlife Research*, **16**, 395–404.

Baxter, G. S. and Fairweather, P. G. (1994). Phosphorus and nitrogen in wetlands with and without egret colonies. *Australian Journal of Ecology*, **19**, 409–16.

Baxter, G. S. and Fairweather, P. G. (1998). Does available foraging area, location or colony character control the size of multispecies egret colonies? *Wildlife Research*, **25**, 23–32.

Bayer, R. D. (1981). Regional variation of Great Blue Heron longevity. *Journal of Field Ornithology*, **52**, 210–13.

Bayer, R. D. (1985a). Interactions between Great Blue Herons and gulls. *Wilson Bulletin*, **97**, 538–41.

Bayer, R. D. (1985b). Vocalizations of Great Blue Herons at Yaquina Estuary, Oregon. *Colonial Waterbirds*, **7**, 35–44.

Bayer, R. D. and McMahon, E. (1981). Colony size and hatching synchrony of Great Blue Herons in coastal Oregon. *Murrelet*, **62**, 73–9.

Beasley, A. and Beasley, J. (1985). An enigmatic heron at Mana Pools. *Honeyguide*, **31**, 212–13.

Becerril, M. F. and Carmona, R. (1997). [Nesting of water birds in Ensenada de La Paz, Baja California Sur, Mexico (1992–1994).] *Ciencias Marinas*, **23**, 265–71.

Becker, J. J. (1985). Fossil herons (Aves: Ardeidae) of the late Miocene and early Pliocene of Florida. *Journal of Vertebrate Paleontology*, **5**, 24–31.

Becker, J. J. (1986). An early heron (Aves, Ardeidae, *Ardea*) from the Middle Miocene of Nebraska. *Journal of Paleontology*, **60**, 968–70.

Becker, P. (1984). Mangrovereiher (*Butorides striatus*) in Swakopmund, S.W.A. *Mitteilungen der Ornithologischen Arbeitsgruppe SWA (Sued West Afrika) Wissenschaftlichen Gesellschaft*, **19**, 2–3.

Beehler, B. M., Pratt, T. K., and Zimmerman, D. A. (1986). *Birds of New Guinea*. Princeton University Press, Princeton.

Beel, C. (1993). Slaty Egret field characters. *Honeyguide*, **39**, 193.

Begum, S. (2003). Colonial nesting behaviour in Indian Pond Heron (*Ardeola grayii grayii*) of Bangladesh. *Zoos' Print Journal*, **18**, 1113–6.

Behrstock, R. A. (1996). Voices of Stripe-backed Bittern, Least Bittern and Zigzag Herons with notes on distribution. *Cotinga*, **5**, 55–61.

Bekhuis, J. (1990). Hoe long nog broedende Woudaapjes *Ixobrychus minutus* in Nederland? *Limosa*, **63**, 47–50.

Belant, J. L., Tyson, L. A., and Mastrangelo, P. A. (2000). Effects of lethal control at aquaculture facilities on populations of piscivorous birds. *Wildlife Society Bulletin*, **28**, 379–84.

Belcher, C. and Smooker, G. D. (1934). Birds of the colony of Trinidad and Tobago. *Ibis*, **4**, **13th Series**, 572–95.

Bellinato, F. and Bogliani, G. (1995). Colonial breeding imposes increased predation: experimental studies with herons. *Ethology, Ecology & Evolution*, **7**, 347–53.

Belser, C. G. and Post, W. (1987). Tricolored Herons and Snowy Egrets breeding in the interior of South Carolina. *Chat*, **51**, 68–71.

Beltzer, A. H. (1983). Alimentacion de la garcita azulada (*Butorides striatus*) en el valle aluvial del rio Parana medio (Ciconiiformes: Ardeidae). *Revue d'Hydrobiologie Tropicale*, **16**, 203–6.

Beltzer, A. H. (1990). Notes on the food of the Rufescent Tiger Heron *Tigrisoma lineatum* (Aves: Ardeidae) in the Middle Paraná River Floodplain, Argentina. *Studies on Neotropical Fauna and Environment*, **25**, 93–6.

Beltzer, A. H. (1991). Aspects of the breeding biology and the death rate of the *Butorides striatus* (Aves: Ardeidae). *Revista de la Asociacion de Ciencias Naturales del Litoral*, **22**, 35–40.

Belzer, W. R. and Lombardi, J. R. (1989). Cattle Egret symbiosis and heronry abandonment. *Colonial Waterbirds*, **12**, 115–17.

Benesh, C. D. and Rosenberg, G. H. (1998). The migration season: August 1–November 30, 1997. Arizona Region. *Field Notes*, **52**, 100–4.

Benito Espinal, E. and Portecop, J. (1984). Installation d'*Ixobrychus e. exilis* (Gmelin) en Guadeloupe et a Marie-Galante (Petites Antilles). *Oiseau et la Revue Francaise d'Ornithologie*, **54**, 149–52.

Bennett, D. C., Whitehead, P. E., and Hart, L. E. (1995). Growth and energy requirements of hand-reared Great Blue Heron (*Ardea herodias*) chicks. *Auk*, **112**, 201–9.

Bennett, G. (1992). First record of Little Blue Heron in southern Africa. *Birding In Southern Africa*, **44**, 71.

Bennett, J. and Smithson, W. S. (2001). Feeding associations between Snowy Egrets and Red-breasted Mergansers. *Waterbirds*, **24**, 125–8.

Bennetts, R. E., Fasola, M., Hafner, H., and Kayser, Y. (2000). Influence of environmental and density-dependent factors on reproduction of Little Egrets. *Auk*, **117**, 634–9.

Benson, C. W. (1960). The birds of the Comoro Islands, results of the British Ornithologists' Union centenary expedition. *Ibis*, **103b**, 1–106.

Benson, C. W., Brooke, R. K., and Irwin, M. P. S. (1971). The Slatey Egret *Egretta vinaceigula* is a good species. *Bulletin of the British Ornithologists' Club*, **91**, 131–3.

Benson, C. W. and Penny, M. J. (1971). The land birds of Aldabra. *Philosophical Transactions of the Royal Society of London, Series B.*, **260**, 417–527.

Bentzien, D. (1991). Graureiher (*Ardea cinerea*) erbeutet Bisamratte. *Ornithologische Mitteilungen*, **43**, 16.

Berezovikov, N. N. and Gistov, A. P. (1994). Fodder behaviour of Little Egret (*Egretta garzetta* L.) and Large Egret (*E. alba* L.) in the Ural delta. *Selevinia*, **2**, 73–6.

Bergerandi, A., Arzoz, M. J., Campos, F., and Lekuona, J. M. (1995). Aumento de la poblacion reproductora de Garza Imperial, *Ardea purpurea*, en Navarra, norte de Espana. *Ardeola*, **42**, 201–3.

Bernasconi, R., Maroni, V., Vigano, A., and Zanetti, G. (1999). Il tarabuso *Botaurus stellaris* sul lago di Varese. *Avocetta*, **23**, 85.

Berny, P., Sadoul, N., Dol, S., Videman, B., Kayser, Y., and Hafner, H. (2002). Impact of local agricultural and industrial practices on organic contamination of Little Egret (*Egretta garzetta*) eggs in the Rhone Delta, southern France. *Environmental Toxicology and Chemistry*, **21**, 520–6.

Berthelot, J. Y. and Navizet, G. (1987). Aigle criard(?) predateur dans une colonie de herons cendres en Dombes. *Nos Oiseaux*, **39**, 67–9.

Berthelot, J. Y. and Navizet, G. (1993). De la parade nuptiale du heron pourpre . . . aux mecanismes d'apparition d'une espece. *Nos Oiseaux*, **42**, 79–95.

Besancon, T. (2002). L'aigrette garzette (*Egretta garzetta*) en Lorraine. *Ciconia*, **26**, 55–64.

Besson, J. M. (1994). Heron cendre (*Ardea cinerea*) predateur de jeunes grebes huppes (*Podiceps cristatus*). *Nos Oiseaux*, **42**, 480.

Bharos, A. M. K. (1998). Use of Cattle Egret *Bubulcus ibis* organs as medicine. *Journal of the Bombay Natural History Society*, **95**, 503.

Bibby, C. J. (1981). Wintering bitterns in Britain. *British Birds*, **74**, 110.

Bibby, C. J. and Lunn, J. (1982). Conservation of reed beds and their avifauna in England and Wales. *Biological Conservation*, **23**, 167–86.

Biderman, J. O. and Dickerman, R. W. (1978). Feeding behavior and food habits of the Boat-billed Heron (*Cochlearius cochlearius*). *Biotropica*, **10**, 33–7.

Bildstein, K. L. (1980). Adult Brown Pelican robs Great Blue Heron of fish. *Wilson Bulletin*, **92**, 122–3.

Biondi, M., Guerrieri, G., and Pietrelli, L. (1993). Stress antropogenici ed evoluzione de una comunità ornitica in una zona umida artificiale dell'Italia centrale. *Avocetta*, **17**, 23–7.

Biondi, M., Pietrelli, L., and Guerrieri, G. (1994). Presenze invernali e status degli Ardeidae lungo la costa del Lazio (1986/1992). *Rivista Italiana di Ornitologia*, **63**, 137–44.

BirdLife International (2001). *Threatened birds of Asia: The BirdLife International red data book. BirdLife International*, Cambridge.

BirdLife International (2003). *Red data book: Threatened birds of Asia*. BirdLife International, www.rdb.or.id.

Birkenholtz, D. E. and Jenni, D. A. (1964). Observations on the Spotted Rail and Pinnated Bittern in Costa Rica. *Auk*, **81**, 558–9.

Bishop, K. D. (1983). Some notes on non-passerine birds of West New Britain. *Emu*, **83**, 235–41.

Biswas, B. (1960). Birds of Nepal. *Journal of the Bombay Natural History Society*, **11**, 126–7.

Bjlsma, R. G. and de Roder, F. E. (1982). Goliath Herons in Egypt in 1981. *Dutch Birding*, **4**, 82–4.

Blake, E. R. (1977). *Manual of neotropical birds*. University of Chicago Press, Chicago.

Blaker, D. (1969a). Behaviour of the Cattle Egret *Ardeola ibis*. *Ostrich*, **40**, 74–129.

Blaker, D. (1969b). The behaviour of *Egretta garzetta* and *Egretta intermedia*. *Ostrich*, **40**, 150–5.

Blanco, D. and Canevari, P. (1992). *Censo neotropical de aves acuáticas*. Programa de Ambientes Acuático Neotropicales, Buenos Aires, Argentina.

Blasdale, P. (1984). Some observations on Black-crowned and White-backed Night Herons. *Malimbus*, **6**, 85–9.

Blus, L. J. and Henny, C. J. (1981). Suspected Great Blue Heron population decline after a severe winter in the Columbia Basin. *Murrelet*, **62**, 16–8.

Blus, L. J., Henny, C. J., Anderson, A., and Fitzner, R. E. (1985). Reproduction, mortality, and heavy metal concentrations in Great Blue Herons (*Ardea herodias*) from three colonies in Washington and Idaho. *Colonial Waterbirds*, **8**, 110–16.

Blus, L. J., Henny, C. J., and Kaiser, T. E. (1980). Pollution ecology of breeding Great Blue Herons in Columbia Basin, Oregon and Washington. *Murrelet*, **61**, 63–71.

Blus, L. J., Rattner, B. A., Melancon, M. J., and Henny, C. J. (1997). Reproduction of Black-crowned Night-Herons related to predation and contaminants in Oregon and Washington, USA. *Colonial Waterbirds*, **20**, 185–97.

Bochenski, Z. (1995). History of herons of the Western Palaearctic. *Acta Zoologica Cracoviensia*, **38**, 343–62.

Bock, K. R. and Medland, R. D. (1988). White-backed Night Heron, *Nycticorax leuconotus*, adopting 'bittern' posture. *Nyala*, **12**, 76.

Bock, W. J. (1956). A generic view of the family Ardeidae (Aves). *American Museum Novitates*, **1779**, 1–49.

Boesman, P. (1990). Indian Pond Heron in Thailand in March 1989. *Dutch Birding*, **12**, 20–1.

Boev, Z. N. (1986). [A comparative morphological characterization of the flying apparatus of *Ixobrychus minutus* and *Ardea cinerea* (Aves, Ardeidae).] *Acta Zoologica Bulgarica*, **30**, 15–24.

Boev, Z. N. (1987a). [Morphometric characteristics of sexual dimorphism and individual variation of herons (Aves, Ardeidae) from Bulgaria I. Somatometric indices.] *Acta Zoologica Bulgarica*, **33**, 44–59.

Boev, Z. N. (1987b). [Morphometric characteristics of the sexual dimorphism and the individual variability of the herons (Aves, Ardeidae) in Bulgaria. III. Indices.] *Acta Zoologica Bulgarica*, **35**, 53–64.

Boev, Z. N. (1987c). [Morphometric features of the sexual dimorphism and individual variability of herons (Aves, Ardeidae) from Bulgaria II. Osteometric features.] *Acta Zoologica Bulgarica*, **34**, 53–67.

Boev, Z. N. (1988a). [Morphometrical characteristics of the adaptations of herons (Aves, Ardeidae) in relation to their movement along the substrate.] *Acta Zoologica Bulgarica*, **36**, 63–71.

Boev, Z. N. (1988b). Osteological features for the identification of herons (Aves, Ardeidae). *Acta Zoologica Bulgarica*, **36**, 56–62.

Boev, Z. N. (1989). [Morphometrical characteristics of the adaptations in relation to the getting of food of the herons (Aves, Ardeidae).] *Acta Zoologica Bulgarica*, **37**, 49–61.

Bogner, H. E. and Baldassarre, G. A. (2002a). The effectiveness of call-response surveys for detecting Least Bitterns. *Journal of Wildlife Management*, **66**, 976–84.

Bogner, H. E. and Baldassarre, G. A. (2002b). Home range, movement, and Least Bitterns in western New York. *Wilson Bulletin*, **114**, 297–308.

Boldreghini, P., Tinarelli, R., and Rizzoli, M. (1995). Strategie trofiche di airone cenerino (*Ardea cinerea*) e di nitticora (*Nycticorax nycticorax*) in un'area di pescicoltura della Pianura Bolognese. *Supplemento Alle Ricerche di Biologia Della Selvaggina*, **22**, 197–205.

Boonman, M. and de Vrieslaan, H. (1992). Buzzard *Buteo buteo* examines a nest of a Grey Heron *Ardea cinerea*. *Limosa*, **65**, 29–30 in Dutch.

Borrero, J. I. and Cruz-Millán, C. A. (1982). Notas sobre la historia de la garza morena *Ardea cocoi* (Aves) en Colombia. *Acta Biologica Colombiana*, **1**, 51–75.

Boukhemza, M., Doumandji, S., Voisin, C., and Voisin, J. F. (2000). Disponibilites des ressources alimentaires et leur utilisation par le heron garde-boeufs *Bubulcus ibis* en Kabylie, Algerie. *Revue d'Ecologie la Terre et la Vie*, **55**, 361–81.

Bourne, W. R. P. (1999). The past status of the herons in Britain. *Bulletin of the British Ornithological Club*, **199**, 192–196.

Bousquet, J. F. (1998). Statut de la grande aigrette *Egretta alba* en Midi-Pyrenees. *Pistrac*, **17**, 83–5.

Bowey, K. (1997). Grey Heron catching common Starling in flight. *British Birds*, **90**, 112–13.

Boozic, I. A. (1992). Breeding biology of Little Bittern *Ixobrychus minutus* at Draga Ponds near Ig. *Acrocephalus (Ljubljana)*, **13**, 76–84.

Bourne, W. R. P., Ashmole, N. P., and Simmons, K. E. L. (2003). A new subfossil night heron and a new genus for the extinct rail from Ascension Island, central tropical Atlantic Ocean. *Ardea*, **91**, 45–51.

Bracko, F. (1985). [A threat to the Grey Heron *Ardea cinerea* colony at Bis.] *Acrocephalus (Ljubljana)*, **6**, 38–9.

Brady, A. (1990–1991). Major offshore nocturnal migration of Blackpoll Warblers, herons, and shorebirds. *Cassinia*, **64**, 28–9.

Braithwaite, L. W. and Clayton, M. (1976). Breeding of the Nankeen Night Heron *Nycticorax calidonicus* while in juvenile plumage. *Ibis*, **118**, 584–6.

Brazier, H. (1983). American Bittern—a correction. *IWC News*, **35**, 3.

Bredin, D. (1984). Regime alimentaire du heron garde-boeuf a la limite de son expansion geographique recente. *Terre et la Vie*, **39**, 431–45.

Bridgman, H. A. (1994). Meteorological conditions and Cattle Egret migration: An update. *Notornis*, **41**, 189–204.

Bridgman, H. A. and Maddock, M. (1994). Meterological conditions and Cattle Egret migration. An update. *Notornis* **41**, 189–205.

Bridgman, H. A., Maddock, M., and Geering, D. J. (1998). Relationship between Cattle Egret migration and meteorology in the southwest Pacific: A review. *International Journal of Biometeorology*, **41**, 143–54.

Briggs, S. V., Thornton, S. A., and Lawler, W. G. (1997). Relationships between hydrological control of river red gum wetlands and waterbird breeding. *Emu*, **97**, 31–42.

Britton, A. R. C. (1993). Feeding behaviour of the Little Egret at Mai Po, Hong Kong. *Hong Kong Bird Report*, **1992**, 176–84.

Britton, P. L. and Britton, H. A. (1996). Additional sightings of the Striated Heron on inland freshwaters. *Australian Bird Watcher*, **16**, 349.

Brock, K. J. (1994). The Reddish Egret: an addition to the Indiana avifauna. *Indiana Audubon Quarterly*, **72**, 131–2, 169.

Brooke, R. K. and Parkin, D. T. (1995). Little Blue Heron *Egretta caerulea* in South Africa. *Bulletin of the African Bird Club*, **2**, 40–1.

Brooks, D. J., Evans, M. I., Martins, R. P., and Porter, R. F. (1987). The status of birds in north Yemen and the records of the OSME expedition in the autumn. *Sandgrouse*, **9**, 466.

Brouwer, J. and Mullie, W. C. (2001). A method for making whole country waterbird population estimates, applied to annual waterbird census data from Niger. *Ostrich*, **Supplement No. 15**, 73–82.

Brown, A. W. (1985). BTO Grey Heron census 1985. *Lothian Bird Report*, **7**, 83–90.

Brown, B. (1980). Possible early record of Cattle Egrets in New Zealand. *Notornis*, **24**, 285–6.

Brown, L. H., Urban, E. K., and Newman, K. (1982). *Birds of Africa*, Vol. 1. Academic Press, London.

Brown, M. and Dinsmore, J. J. (1986). Implications of marsh size and isolation for marsh bird management. *Journal of Wildlife Management*, **50**, 392–7.

Broyer, J., Varagnat, P., Constant, G., and Caron, P. (1998). Habitat du Héron Pourpré (*Ardea purpurea*) sur les étangs de pisciculture en France. *Alauda*, **66**, 221–8.

Brugiere, D. (1999). Hivernage de bihoreaux gris *Nycticorax nycticorax* dans le val d'Allier et le moyen val de Loire. *Alauda*, **67**, 73–4.

Bundy, G. (1985). Communal feeding by Western Reef Herons. *British Birds*, **78**, 107–8.

Burchsted, A. E. (1983). Whirling and territorial behaviors of feeding Little Blue Herons. *Florida Field Naturalist*, **11**, 18–19.

Burger, J. (1978a). Competition between Cattle Egrets and native North American herons, egrets, and ibises. *Condor*, **80**, 15–23.

Burger, J. (1978b). The pattern and mechanism of nesting in mixed-species heronies. In *Wading Birds* (eds. A. Sprunt, IV, J. C. Ogden, and S. Winckler) pp. 45–58. National Audubon Society Research Report 7. National Audubon Society, New York.

Burger, J. (1981). A model for the evolution of mixed species colonies of Ciconiiformes. *Quarterly Review of Biology* **56**, 143–167.

Burger, J. and Gochfeld, M. (1989). Age differences in Cattle Egrets *Bubulcus ibis*, foraging with wild ungulates in Kenya. *Ardea*, **77**, 201–4.

Burger, J. and Gochfeld, M. (1990). Vertical nest stratification in a heronry in Madagascar. *Colonial Waterbirds*, **13**, 143–6.

Burger, J. and Gochfeld, M. (1997). Risk, mercury levels, and birds: Relating adverse laboratory effects to field biomonitoring. *Environmental Research*, **75**, 160–72.

Burger, J., Parsons, K., Benson, T., Shukla, T., Rothstein, D., and Gochfeld, M. (1992). Heavy metal and selenium levels in young Cattle Egrets from nesting colonies in the northeastern United States, Puerto Rico, and Egypt. *Archives of Environmental Contamination and Toxicology*, **23**, 435–9.

Burkholder, G. and Smith, D. G. (1988). Great Horned Owls (*Bubo virginianus*) nesting in a Great Blue Heron (*Ardea herodias*) heronry. *Journal of Raptor Research*, **22**, 62.

Burnham, M., Gynther, I., and Hobson, R. (2002). Records of the Great-billed Heron *Ardea sumatrana* Raffles, 1822, from the Tin Can Bay area, southern Queensland. *Memoirs of the Queensland Museum*. **41**, 45–8.

Burton, K. M. and Smith, S. D. (2001). First report of the Grey Heron in the United states. *Western Birds*, **32**, 88–90.

Butler, R. W. (1992). Great Blue Heron, *Ardea herodias*. In *The birds of North America*. No. 25 (eds. A. Poole, P. Stettenheim, and F. Gill), pp. 1–20. The Academy of Natural Sciences, Philadelphia and The American Ornithologists' Union, Washington.

Butler, R. W. (1997). *The Great Blue Heron: A natural history and ecology of a seashore sentinel*. UBC Press, Vancouver.

Butler, R. W., Kushlan, J. A., and Davidson, I. J. (2000). Herons in North America, Central America and the West Indies. In *Heron conservation* (eds. J. A. Kushlan and H. Hafner), pp. 151–75. Academic Press, San Diego.

Butler, R. W., Whitehead, P. E., Breault, A. M., and Moul, I. E. (1995). Colony effects on fledging success of Great Blue Herons (*Ardea herodias*) in British Columbia. *Colonial Waterbirds*, **18**, 159–65.

Byrd, M. A. (1978). Dispersal and movements of six North American ciconiiforms. In *Wading birds*. (eds. A. Sprunt, IV, J. C. Ogden, and S. Winckler), pp. 161–85. National Audubon Society Research Report No. 7. National Audubon Society, New York.

Caballero S., J. (1997). Fenología reproductiva, status y distribución del avetorillo común *Ixobrychus minutus* (Linnaeus 1766), en la región de Murcia. *Oxyura*, **9**, 53–83.

Caldwell, G. S. (1980). Underlying benefits of foraging aggression in egrets. *Ecology*, **61**, 996–7.

Caldwell, G. S. (1981). Attraction to tropical mixed-species heron flocks: proximate mechanisms and consequences. *Behavioural Ecology and Sociobiology*, **8**, 99–103.

Caldwell, G. S. (1986). Predation as a selective force on foraging herons: Effects of plumage color and flocking. *Auk*, **103**, 494–505.

Caldwell, H. R. and Caldwell, J. C. (1931). *South China birds*. Vanderburgh, Shanghai.

Camici, M. and Zimmerli, M. (1993). Graureiher frisst seefrosch im winter. *Ornithologische Beobachter*, **90**, 73.

Campos, F. (1990). Confirmed case of hybridisation between the Grey Heron *Ardea cinerea* and Purple Heron *Ardea purpurea* in Valladolid, Spain. *Gerfaut*, **80**, 57–9.

Campos, F. and Fernandez-Cruz, M. (1989). La poblacion reproductora de garza real (*Ardea cinerea*) en la cuenca del Duero (España), 1988. *Ardeola*, **36**, 102–6.

Campos, F. and Fernandez-Cruz, M. (1991). The breeding biology of the Grey Heron (*Ardea cinerea*) in the Duero River Basin in Spain. *Colonial Waterbirds*, **14**, 57–60.

Campos, F. and Fraile, B. (1990). Les parametres reproducteurs du heron cendre (*Ardea cinerea*) dans le nord-ouest de l'Espagne. *Oiseau et la Revue Francaise d'Ornithologie*, **60**, 212–23.

Campos, F. and Lekuona, J. M. (1997). Temporal variations in the feeding habits of the Purple Heron *Ardea purpurea* during the breeding season. *Ibis*, **139**, 447–51.

Campos, F. and Lekuona, J. M. (2000). Benficios que las garzas imperials reproductoras obtienen de los peces. *Ardeola*, **47**, 105–7.

Campos, F. and Lekuona, J. M. (2001). Are rice fields suitable foraging habitat for Purple Herons during the breeding season? *Waterbirds*, **25**, 450–2.

Campos, R. G. (1988). Chestnut-bellied Heron sighted at Lomas Barbudal: range extension of the Chestnut-bellied Heron (*Agamia agami*) in Costa Rica. *Bee Line News Bulletin*, **2**, 10.

Cardillo, R., Forbes-Watson, A., and Ridgely, R. (1983). The Western Reef-Heron (*Egretta gularis*) at Nantucket Island, Massachusetts. *American Birds*, **37**, 827–9.

Carlson, B. A. (1995). Nest site characteristics of Great Blue Herons (*Ardea herodias*) in northeast Ohio. *Ohio Journal of Science*, **95**, 312–15.

Carlson, B. A. and Mclean, E. B. (1996). Buffer zones and disturbance types as predictors of fledging success in Great Blue Herons, *Ardea herodias*. *Colonial Waterbirds*, **19**, 124–7.

Carmi, N., Weihs, D., and Katzir, G. (1995). Head stabilization during body movement in heron: a quantitative description of the phenomenon and its possible mechansisms and function. *Israel Journal of Zoology*, **41**, 81.

Carpenter, G., and de Jong, M. (2001). Recent records of the Little Bittern (*Ixobrychus minutus*) north of the River Murray. *South Australian Ornithologist*, **33**, 173–4.

Carpegna, F., Alessandria, G., and Della Toffola, M. D. (2000). Status dell'airone bianco maggiore, *Casmerodius albus*, in Italia nord-occidentale. *Rivista Italiana di Ornitologia*, **69**, 169–79.

Carpegna, F., Della Toffola, M. D., and Alessandria, G. (1990). Nidificazione di *Phalacrocorax carbo sinensis* in Piemonte. *Rivista Italiana di Ornitologia*, **60**, 205–7.

Carroll, A. L. K. (1970). The Whitefaced Heron in New Zealand. *Notornis*, **17**, 3–24.

Carruette, P. (1995). Predation du heron cendre (*Ardea cinerea*) sur le grebe castagneux (*Tachybaptus ruficollis*) et d'autres proies inhabituelles. *Nos Oiseaux*, **43**, 239.

Carruth, H. (1989). *Tell me again how the white heron rises and flies across the nacreous River at twilight toward the distant islands.* New Directions, New York.

Carss, D. N. (1993). Grey Heron, *Ardea cinerea* L., predation at cage fish farms in Argyll, western Scotland. *Aquaculture and Fisheries Management*, **24**, 29–45.

Carss, D. N. (1994). Killing of piscivorous birds at Scottish fin fish farms, 1984–87. *Biological Conservation*, **68**, 181–8.

Carss, D. N. and Marquiss, M. (1991). Avian predation at farmed and natural fisheries. *Institute of Fisheries Management Annual Study Course Proceedings*, **22**, 179–96.

Carss, D. N. and Marquiss, M. (1996). The influence of a fish farm on Grey Heron *Ardea cinerea* breeding performance. In *Aquatic predators and their prey*. Proceedings of a Royal Society of Edinburgh conference held at the University of Aberdeen, August 30–September 1, 1994. (eds. S. P. R. Greenstreet and M. L. Tasker), pp. 133–41. Blackwell Scientific Publications, Oxford and Cambridge.

Carter, M. (2003). Cinnamon Bittern *Ixobrychus cinnamomeus* on Christmas Island: a new bird for Australian territory. *Australian Field Ornithology*, **20**, 55–8.

Carter, S. (1992). The heronries census 1928–1991: latest update. *BTO News*, **181**, 8–9.

Casanas, H., Klavins, J., and Cargnelutti, R. (2002). Registro del al garza azul *Egretta caerulea* en la provincia de Tucuman, Argentina. *Cotinga*, **18**, 106–7.

Casement, M. (1995). Great Blue Herons in the North Atlantic. *Birding World*, **8**, 424–5.

Castrale, J. S. (1994). Survey of Great Blue Heron nesting colonies in Indiana, 1983–1993. *Indiana Audubon Quarterly*, **72**, 133–41.

Catley, G. P. (1983). Unusual feeding action of Great White Egret. *British Birds*, **76**, 82–3.

Caupenne, M. (1993). Importance des colonies d'ardéidés arboricoles des marais de Brouage-Seudre et premier cas de nidification du crabier chevelu *Ardeola ralloides* en Charente-Maritime. *Alauda*, **61**, 118–19.

Cavé, A. J. (1983). Purple Heron survival and drought in tropical West Africa. *Ardea*, **71**, 217–24.

Celmins, A. (1992). *List of Latvian bird species*. Ganders Ltd., Riga.

Cempulik, P. (1994). Bestandsentwicklung, brutbiologie und okologie der zwergdommel *Ixobrychus minutus* an fisch- und industrieteichen Oberschlesiens. *Vogelwelt*, **115**, 19–27.

Cezilly, F. (1992). Turbidity as an ecological solution to reduce the impact of fish-eating colonial waterbirds on fish farms. *Colonial Waterbirds*, **15**, 249–52.

Cezilly, F. and Boy, V. (1988). Age related differences in foraging Little Egrets, *Egretta garzetta*. *Colonial Waterbirds*, **11**, 100–6.

Cezilly, F. and Boy, V. (1989). Postural adjustment and foraging efficiency in the Little Egret *Egretta garzetta*. *Revue D'ecologie La Terra et la Vie*, **44**, 95–102.

Cezilly, F., Boy, V., and Hafner, H. (1990). Group foraging in Little Egrets (*Egretta garzetta*): from field evidence to experimental investigation. *Behavioural Processes*, **21**, 69–80.

Chambers, S. (1989). *Birds of New Zealand: locality guide*. Arun Books, Hamilton, New Zealand.

Chang, C.-C. (2000). Malayan Night Heron *Gorsachius melanolophus* breeding in immature plumage. *Forktail*, **16**, 167–8.

Chang, Q., Xhang, B. W., Jin, H., Zhu, L. F., and Zhau, K. Y. (2003). Phylogenetic relationships among 13 species of herons inferred from mitochondrial 12S rRNA gene sequences. *Acta Zoologica Sinica*, **49**, 205–10.

Chapman, B. A. and Forbes, L. S. (1984). Observations on detrimental effects of Great Blue Herons on breeding Black Terns. *Journal of Field Ornithology*, **55**, 251–2.

Chapman, B. R., Grantland, T. L., and Ricklefs, R. E. (1981). Growth and development of temperature regulation in nesting Black-crowned Night Herons. *Colonial Waterbirds*, **4**, 114–19.

Chatto, R. (2000). *Waterbird breeding colonies in the Top End of the Northern Territory*. Technical Report 69, Parks and Wildlife Commission of the Northern Territory, Palmerson.

Chavez Ramirez, F. and Slack, R. D. (1995). Differential use of coastal marsh habitats by nonbreeding wading birds. *Colonial Waterbirds*, **18**, 166–71.

Chebez, J. C. and Gomez, D. (1988). [Zoogeographic notes on some birds of Tierra del Fuego.] *Hornero*, **13**, 75–8.

Cheke, A. S. and Diamond, A. W. (1986). Birds on Moheli and Grande Comore (Comoro Islands) in February 1975. *Bulletin of the British Ornithologists' Club*, **106**, 138–48.

Chen, X., Chen, Z., and Fang, W. (2000a). New record of the White-faced Heron *Egretta novaehollandiae* in Xiamen (Amoy), China. *Emu*, **100**, 340.

Chen, X., Chen, Z., Fang, W., and Lu, L. (2000b). The new record of White-faced Heron (*Egretta novaehollandiea [novaehollandiae]*) in China. *Journal of Xiamen University Natural Science*, **39**, 724–7.

Cherrie, G. K. (1916). A contribution to the ornithology of the Orinoco. *Museum of the Brooklyn Institute of Arts and Science. Science Bulletin*, **2**, 133a–374.

Choudhury, A. (1992). Sighting of the Great White-bellied Heron in Jamjing Reserve Forest, Assam. *Newsletter for Birdwatchers*, **32**, 17.

Choudhury, A. (2000). The birds of Assam. Gauhati: Gibbon Books and World Wide Fund for Nature, India.

Christmas, S. (1994). Greater Manchester ringing report, 1994. *Birds in Greater Manchester*, **19**, 124–8.

Chu, Y. L. and Wang, Y. (1988). [Feeding behaviour and maintenance of feeding territority in Little Egret *Egretta garzetta garzetta*.] *Biological Bulletin National Taiwan Normal University*, **23**, 167–79.

Ciaccio, A. and Siracusa, M. (1989). Nidificazione di airone cenerino, *Ardea cinerea*, e sgarza ciuffetto, *Ardeola ralloides*, e status degli Ardeidi in Sicilia. *Rivista Italiana di Ornitologia*, **59**, 75–9.

Cintra, R. and Yamashita, C. (1990). *Habitats, abundancia e ocorrencia des especies de aves do Pantanal de Poconé, Mato Grosso, Brasil*. Papeis Avulsos de Zoologia 37, article 1. Museu de Zoologia de Universidade de Sao Paulo, Sao Paulo.

Cizek, A. (2001). Black-headed Heron taking Red-chested Flufftail. *Honeyguide*, **47**, 173–4.

Claeys, G., de Bruyker H, and Geers, V. (1989). First record of an Indian pond heron in Belgium and western Europe. *Oriolus*, **55**, 73–5.

Clancey, P. A. (1976). Interesting feeding behavior in the Little Egret and Brown-hooded Kingfisher. *Ostrich*, **47**, 131.

Clapp, R. B., Klimkiewicz, M. K., and Kennard, J. H. (1982). Longevity records of North American birds:

Gaviidae through Alcidae. *Journal of Field Ornithology*, **53**, 81–124.

Clark, W. S., Bloom, P. H., and Oliphant, L. W. (1989). Aplomado Falcon steals prey from Little Blue Heron. *Journal of Field Ornithology*, **60**, 380–1.

Clarke, T. (1999). The Great Blue Heron on Tenerife. *Birding World*, **12**, 158–61.

Clayton, D. H. (1985). Nocturnal foraging of Yellow-crowned Night Herons in the Bahamas. *Florida Field Naturalist*, **13**, 34–5.

Coates, B. J. (1986). *The birds of Papua New Guinea*. Vol. 1. Dove Publications, Alderley, Queensland, Australia.

Coates, B. J., Bishop, K. D., and Gardner, D. (1997). *A guide to the birds of Wallacea, Sulawesi, the Moluccas, and the Lesser Sunda Islands, Indonesia*. Dove Publications, Alderley, Queensland, Australia.

Collar, N. J., Crosby, M. J., and Strattesfield, A. J. (1994). *Birds to watch: The world list of threatened birds.* BirdLife Conservation Series 4. BirdLife International, Cambridge.

Collazo, J. A. (1985). Food habits of nesting Great Blue Herons at Heyburn State Park, Idaho (USA). *Northwest Science*, **59**, 144–6.

Combridge, P. (2000). Great Skua apparently killing Little Egret. *British Birds*, **93**, 144.

Combridge, P. and Parr, C. (1992). Influx of Little Egrets in Britain and Ireland in 1989. *British Birds*, **85**, 16–21.

Commecy, X. (1989). Nidifications du heron bihoreau en Picardie. *Oiseau et la Revue Francaise d'Ornithologie*, **59**, 84–6.

Contreras-Balderas, A. J. (1988). New records of birds from Nuevo Leon, Mexico. *Southwestern Naturalist*, **33**, 251–2.

Contreras, J. R. (1993). Acerca de algunas especies de aves del extremo sudeste de la provincia de Formosa, republica Argentina. *Notulas Faunisticas*, **47**, 1–8.

Cooper, J. (1984). Grey Heron *Ardea cinerea* kleptoparasitizes Cape Cormorants *Phalacrocorax capensis*. *Cormorant*, **12**, 94.

Cooper, J. and Marshall, B. E. (1970). Observations on the nestlings of the Goliath Heron, *Ardea goliath*, in Rhodesia. *Bulletin of the British Ornithologists' Club*, **90**, 148–53.

Cordonnier, P. (1985). Analyse des reprises de bagues d'Ardeides dans la region Rhone-Alpes, statuts et protection des populations de la Dombes. *Bievre*, **7**, 105–13.

Corrick, A. H. (1995). Ownership of wetlands in Victoria and the occurrence of three species of waterbirds. In *People and nature conservation: perspectives on private land use and endangered species recovery* (eds. A. Bennett, G. Backhouse, and T. Clark), pp. 149–55. Surrey Beatty & Sons, Chipping Norton, UK.

Cory, C. B. (1886). Another speciman of *Ardea wuerdmanni*? *Auk*, **3**, 408.

Cory, C. B. (1887). More news of *Ardea wuerdemanni*. *Auk*, **4**, 159.

Cotton, D. and Murphy, N. (1985). Great White Egret in Co. Sligo—a species new to Ireland. *Irish Birds*, **3**, 96–7.

Cottrille, W. P. and Cottrille, B. D. (1958). Great Blue Heron: behavior at the nest. *Miscellaneous Publications of the Museum of Zoology, University of Michigan*, **102**, 3–15.

Crouch, S., Paquette, C., and Vilas, D. (2002). Relocation of a large Black-Crowned Night Heron colony in Southern California. *Waterbirds*, **25**, 474–8.

Courtens, W. (1996). Een nieuwe waarneming van pleisterende koereigers *Bubulcus ibis* in het Blankaartgebied te Woumen in mei 1995. *Mergus*, **10**, 61–7.

Cowles, G. S. (1994). A new genus, three new species and two new records of extinct Holocene birds from Réunion Island, Indian Ocean. *Geobios (Lyon)*, **27**, 87–93.

Cracraft, J. (1967). On the systematic position of the Boat-billed Heron. *Auk*, **84**.

Cramp, S. and Simmons, K. E. L. (1977). *Handbook of the birds of Europe, the Middle East, and North Africa. The birds of the Western Palearctic.* Vol. 1. Oxford University Press, Oxford.

Craney, E., Duquet, M., and Pinston, H. (1989). Comportement de peche particulier chez le heron cendre, *Ardea cinerea*. *Nos Oiseaux*, **40**, 123–4.

Crawford, R. L. and Neel, L. (1997). Apparent longevity record for the Great Blue Heron. *Oriole*, **62**, 6.

Creighton, J. C. and Schnell, G. D. (1996). Proximate control of siblicide in Cattle Egrets: a test of the food-amount hypothesis. *Behavioral Ecology and Sociobiology*, **38**, 371–7.

Crossland, A. C. (1992). First record of white phase Reef Heron (*Egretta sacra*) in New Zealand. *Notornis*, **39**, 233–4.

Crous, R. (1990). Greenbacked Heron fly-fishing. *Babbler*, **20**, 27.

Crous, R. (1994). An incidence of bait-fishing in the Squacco Heron *Ardeola ralloides. Babbler (Gaborone)*, **26–27**, 32.

Cuenoud, P. (1994). Un heron cendre (*Ardea cinerea*) capture un grebe castagneux (*Tachybaptus ruficollis*). *Nos Oiseaux*, **42**, 480.

Cummins, C. L. and D. G. Homberger (1986). The morphology of the gular region of the Snowy Egret, *Egretta thula* (Molina). *American Zoologist*, **26**, 66A.

Curry, R. and Bryant, G. D. (1987). Snowy Egret a new breeding species for Ontario and Canada. *Ontario Birds*, **5**, 64–7.

Custer, C. M. and Galli, J. (2002). Feeding habitat selection by Great Blue Herons and Great Egrets in east central Minnesota. *Waterbirds*, **25**, 115–24.

Custer, T. W. (1991). Green coloration of recently laid Black-crowned Night-heron eggs. *Colonial Waterbirds*, **14**, 184–5.

Custer, T. W., Bunck, C. M., and Kaiser, T. E. (1983b). Organochlorine residues in Atlantic coast Black-crowned Night-heron eggs, 1979. *Colonial Waterbirds*, **6**, 160–7.

Custer, T. W. and Custer, C. M. (1995). Transfer and accumulation of organochlorines from Black-crowned Night-Heron eggs to chicks. *Environmental Toxicology and Chemistry*, **14**, 533–6.

Custer, T. W. and Davis, W. E., Jr. (1982). Nesting by one-year-old Black-crowned Night Herons on Hope Island, Rhode Island. *Auk*, **99**, 784–6.

Custer, T. W. and Frederick, P. C. (1990). Egg size and laying order of Snowy Egrets, Great Egrets, and Black-crowned Night-herons. *Condor*, **92**, 772–5.

Custer, T. W., Hensler, G. L., and Kaiser, T. E. (1983a). Clutch size reproductive success and organochlorine contaminants in Atlantic coast Black-crowned Night Herons. *Auk*, **100**, 699–710.

Custer, T. W., Ohlendorf, H. M., Rigney, M., and Rowe, R. (1984). Reproductive success and organochlorine contaminants in San Francisco Bay Black-crowned Night-herons. *Colonial Waterbird Group Newsletter*, **8**, 45.

Custer, T. W. and Osborn, R. G. (1978a). Feeding habitat use by colonially-breeding herons, egrets, and ibises in North Carolina. *Auk*, **95**, 733–43.

Custer, T. W. and Osborn, R. G. (1978b). Feeding-site description of three heron species near Beaufort, North Carolina. In *Wading birds* (eds. A. Sprunt, IV, J. C. Ogden, and S. Winckler), pp. 355–60. National Audubon Society Research Report No. 7. National Audubon Society, New York.

Custer, T. W. and Peterson, D. W., Jr. (1991). Growth rates of Great Egret, Snowy Egret and Black-crowned Night-heron chicks. *Colonial Waterbirds*, **14**, 46–50.

Custer, T. W., Rattner, B. A., Ohlendorf, H. M., and Melancon, M. J. (1991). Herons and egrets as proposed indicators of estuarine contamination in the United States. In *Acta 20 Congressus Internationalis Ornithologici*. Christchurch, New Zealand 2–9 December 1990. Volume 4 (eds. B. D. Bell, R. O. Cossee, J. E. C. Flux, B. D. Heather, R. A. Hitchmough, C. J. R. Robertson, and M. J. Williams), pp. 2474–9. New Zealand Ornithological Congress Trust Board, Wellington.

Czapulak, A. and Adamski, A. (2002). Biologia rozrodu czapli siwej Ardea cinerea gniazdujacej w szuwarah tricinowch. *Notatki Ornitologiczne*, **43**, 207–17, 2002.

Dahlberg, C. (1998). Great-billed Heron *Ardea sumatrana. Bird Observer*, **791**, 8.

Dami, L., Bennetts, R. E., and Hafner, H. (in prep.). Does occupation by Cattle Egrets influence the spatial distribution of Little Egrets within mixed-species colonies?

Danielson, F. and Skov, H. (1985). South-east Sumatra, July-August 1985. *Interwader*, **6**, 6–7.

Darakchiev, A., Nankinov, D., and Hadjiev, S. (1984). [Investigation on the number and nest biology of Little Bittern (*Ixobrychus minutus* L.) in Plovdiv district.] *Nauchni Trudove Plovdivski Universitet "Paisii Khilendarski"*, **22**, 179–88.

Darnell, J. C. and Darnell, M. H. (1995). Great Egret feeding on birds. *Western Australian Naturalist*, **20**, 119–20.

Darrieu, C. A. and Camperi, A. R. (1997). [Study of a collection of birds from Corrientes Province, Argentine Republic (Rheidae to Picidae).] *Physis Seccion a Los Oceanos y Sus Organismos*, **55**, 5–15.

David, S. and Berrill, M. (1987). Siblicidal attacks by Great Blue Heron, *Ardea herodias*, chicks in a southern Ontario heronry. *Canadian Field-Naturalist*, **101**, 105–7.

Davis, E. M. (1937). Reddish Egret at Matecumbe. *Florida Naturalist*, **10**, 87.

Davis, G. (1982). Rudd for breakfast. *Colchester and District Natural History Society and Field Club Annual Reports and Records*, **1982**, 102.

Davis, W. E., Jr. (1983). Observations of foraging behaviour of a Mangrove Heron. *Australian Birds*, **18**, 9.

Davis, W. E., Jr. (1984). The White-faced Heron. The supreme generalist of Australia's herons. *Australian Natural History*, **21**, 308–10.

Davis, W. E., Jr. (1985a). Foraging White-faced Herons follow Australian White Ibises. *Colonial Waterbirds*, **8**, 129–34.

Davis, W. E., Jr. (1985b). Foraging behavior of a Western Reef Heron in North America. *Colonial Waterbirds*, **8**, 70–3.

Davis, W. E., Jr. (1986a). Effects of old nests on nest-site selection in Black-crowned Night-Herons and Snowy Egrets. *Wilson Bulletin*, **98**, 300–3.

Davis, W. E., Jr. (1986b). A White-faced Heron *Ardea novaehollandiae* uses possible anti-glare foraging behaviour. *Corella*, **10**, 65.

Davis, W. E., Jr., (1993). Black-crowned Night-Heron (*Nycticorax nycticorax*). In *The birds of North America*. No. 74. (eds. A. Poole and F. Gill). Pp 1–19. The Academy of Natural Sciences, Philadelphia, and American Ornithologists' Union, Washington.

Davis, W. E., Jr. (2001). "Practice" foraging by a sub-adult Great Blue Heron. *Florida Field Naturalist*, **29**, 88–9.

Davis, W. E., Jr., Donahue, P. K., and Perkins, E. G. (1980). Observations of the behavior of the Zigzag Heron. *Condor*, **82**, 460–1.

Davis, W. E., Jr., and Jackson, J. A. (2001). Aerial foraging by Tricolored Herons, Snowy and Great Egrets. *Florida Field Naturalist*, **28**, 192–4.

Davis, W. E., Jr. and Kushlan, J. A. (1994). Green Heron, *Butorides virescens*. In *The Birds of North America*. No. 129. (eds. A. Poole and F. Gill), pp. 1–23. The Academy of Natural Sciences, Philadelphia, Pennsylvania and The American Ornithologists' Union, Washington.

Davis, W. E., Jr. and Zickefoose, J. (1988). Bait fishing by birds: a fascinating example of tool use. *Bird Observer*, **27**, 12–8.

Dawn, W. (1964). Nest and eggs of the Cubanis Tiger Heron in Chiapas, Mexico. *Auk*, **81**, 230–1.

Dawson, E. W. (1974). Sight records of Grey Heron (*Ardea cinerea*) in New Zealand, an elucidation. *Notornis*, **21**, 124–8.

Day, J. C. U. and Wilson, J. (1978). Breeding Bitterns in Britain. *British Birds*, **71**, 285–300.

de Heer, P. and Bos, E. (1989). Snowy Egrets on the Azores—a new western Palaearctic bird. *Birding World*, **2**, 58–60.

de Naurois, R. (1966). Le Héron Pourpre de l'Archipel du Cap Vert, *Ardea purpurea bourni*, esp. nov. *Oiseau et la Revue Francaise d'Ornithologie*, **36**, 89–94.

de Naurois, R. (1988a). *Ardea purpurea bournei* endemique de l'Ile de Santiago (Archipel du Cap Vert). *Alauda*, **56**, 261–8.

de Naurois, R. (1988b). [The Phasianidae of the Cape Verde Archipelago (Portugal): *Coturnix coturnix inopinata* Hartert, 1917 and *Numida meleagris bannermani* Frade, 1966.] *Boletim do Museu Municipal do Funchal*, **40**, 215–34.

de Swardt, D. H. (1992). Green-backed Herons in the Orange Free State. *Mirafra*, **9**, 3–4.

de Swardt, D. H. (1998). Interesting observations at Soetdoring Nature Reserve, Bloemfontein District. *Mirafra*, **15**, 29–30.

Dean, W. R. J. (1988). Breeding of the Goliath Heron at Barberspan, Transvaal. *Ostrich*, **59**, 75–6.

Deerenberg, C. and Hafner, H. (1999). Fluctuation in population size and colony dynamics in the Purple Heron *Ardea purpurea* in Mediterranean France. *Ardea*, **87**, 217–26.

Dejohnge, F., Perrin de Brichambaut, J., and Nicolau-Guillaumet, P. (1992). *Oiseaux de Guyane*. Societe d'Etudes Ornithologiques, Brunoy, France.

De Juanas, A. E. and Ferrer, X. (1996). El status de la garceta grande *Egretta alba* en la peninsula Iberica y las isles Baleares. *Ardeola*, **43**, 225–9.

Del Hoyo, J., Elliott, A., and Sargatal, J. editors. (1992). *Handbook of the birds of the world. Volume 1: ostrich to ducks.* Lynx Edicions, Barcelona.

Dell, J. (1985). Irruption of White-necked Heron (*Ardea-pacifica*) into south-western Australia in 1975. *Western Australian Naturalist*, **16**, 28–30.

Delord, K., Kayser, Y., Barbeaud, C. and Hafner, H. (2003). Variability in breeding parameters of the Squacco Heron *Ardeola ralloides. Bird Study*, **50**, 300–5.

Delov, V. G. (1993). A new heronry on the Sofia plain. *Ardea*, **81**, 53–4.

Demey, R. (1986). Little Egrets and Reef Herons hovering. *Malimbus*, **8**, 43–4.

Demy, R. (2003). Recent reports. *Bulletin of the African Bird Club*, **10**, 129–41.

Denis, D., Rodriguez, A., Rodriguez, P., and Jimenez, A. (2003). Reproduction de la Garza Ganadera (*Bubulcus ibis*) en la Cienga de Biramas, Cuba. *Journal of Caribbean Ornithology*, **16**, 45–54.

Denis, D., Torrella, L., Rodriguez, P., and Rodriguez, A. (1999). Nuevo reporte de nidificacion de garza rojiza (*Egretta rufescens*) en Cuba. *Pitirre*, **12**, 96–7.

Desholm, M. and Wegeberg, A. M. (1997). A survey of the avifauna in Similan National Park, Thailand. *Natural History Bulletin of the Siam Society*, **45**, 127–41.

Desmond, T. C. (1939). Reddish Egret nesting near Tavernier, Florida. *Auk*, **56**, 329.

Devine, A., Welch, J. and Smith, D. G. (2000). Behavior exhibited by an American Bittern. *Connecticut Warbler*, **20**, 64–5.

Di Giacomo, A. G. (1988). Nidificacion del chiflon (*Syrigma sibilatrix*) en Salto, Buenos Aires, Argentina. *Hornero*, **13**, 1–7.

Di Giacomo, A. G. (1997). Hallazgo de un nido de *Botaurus pinnatus* en Argentina. *Hornero*, **14**, 262–3.

Dias, P. C. (1991). Les ardéidés nicheurs au Portugal Distribution, biologie, conservation. *Alauda*, **59**, 23–6.

Dickerman, R. W. (1961). A new subspecies of the Pinnated Bittern. *Wilson Bulletin*, **73**, 333–5.

Dickerman, R. W. (1972). Further notes on the Pinnated Bittern in Mexico and Central America. *Wilson Bulletin*, **57**, 84–90.

Dickerman, R. W. (1973a). A review of the Boat-billed Heron *Cochlearius cochlearius*. *Bulletin of the British Ornithologists' Club*, **93**, 111–14.

Dickerman, R. W. (1973b). The Least Bittern in Mexico and Central America. *Auk*, **90**, 689–91.

Dickerman, R. W. (1992). Northeastern records of *Ardea herodias wardi* from the southeastern United States. *Kingbird*, **42**, 10–13.

Dickerman, R. W. (in press). A review of the North American subspecies of the Great Blue Heron (*Ardea herodias*). *Proceedings of the Biological Society of Washington*.

Dickerman, R. W. and Gavino T. G. (1969). Studies of a nesting colony of Green Herons at San Blas, Nayarit, Mexico. *Living Bird*, **18**, 95–111.

Dickerman, R. W. and Juarez L. C. (1971). Nesting studies of the Boat-billed Heron *Cochlearius cochlearius* at San Blas, Nayarit, Mexico. *Ardea*, **59**, 1–16.

Dickerman, R. W. and Parkes, K. C. (1968). Notes on the plumages and generic status of the Little Blue Heron. *Auk*, **85**, 437–40.

Dickerman, R. W., Parkes, K. C., and Bell, J. (1982). Notes on the plumages of the Boat-billed Heron. *Living Bird*, **19**, 115–20.

Dickey, D. R. and van Rossem, A. J. (1938). *The birds of El Salvador*. Publication 406 and Fieldiana Zoology 23 Field Museum of Natural History, Chicago.

Dickinson, E. C., Kennedy, R. S., and Parkes, K. C. (1991). *The birds of the Philippines: an annotated check-list*. BOU Checklist No. 12. British Ornithologists' Union, Hertfordshire.

Dies, J. I. (1999a). Large concentrations of Grey Herons feeding on dead fish. *British Birds*, **92**, 679.

Dies, J. I. (1999b). Grey Heron following fishing boats and diving for discarded fish. *British Birds*, **92**, 679.

Dies, J. I., Prosper, J. and Des, B. (2001). Occasional breeding by Western Reef Egret in eastern Spain, *British Birds*, **94**, 382–6.

Dijkstra, K.-D. (1997). Striated Heron *Butorides striatus* breeding in the Egyptian Nile Valley. *Sandgrouse*, **19**, 140.

Dimalexis, A., Pyrovetsi, M., and Sgardelis, S. (1997). Foraging ecology of the Grey Heron (*Ardea cinerea*), Great Egret (*Ardea alba*) and Little Egret (*Egretta garzetta*) in response to habitat, at 2 Greek wetlands. *Colonial Waterbirds*, **20**, 261–72.

Dindo, J. J. and Marion, K. R. (1992). A comparison of mortality rates due to short term weather events on a mixed-species heronry off coastal Alabama. *Journal of the Alabama Academy of Science*, **63**, 54.

do Nascimento, J. L. X. (1990). Reprodução de *Agamia agami* na usina hidrelétrica Balbina, Amazonas, Brasil. *Ararajuba*, **1**, 79–83.

do Nascimento, J. L. X. (1999). Registros de ninhais de garcas (Ciconiiformes, Ardeidae) na Paraiba, Brasil. *Airo*, **10**, 39–42.

Dobson, A. (1996). More Great Blue Herons. *Birding World*, **9**, 62.

Dodd, M. G. and Murphy, T. M. (1995). Accuracy and precision of techniques for counting Great Blue Heron nests. *Journal of Wildlife Management*, **59**, 667–73.

Dolton, P. (1991). Great White Egret—'fourth' for Suffolk. *Suffolk Birds*, **40**, 138–9.

Dombe, A. H. and McFarlane, R. W. (1978). Feeding ecology of Little Blue Herons as a radionuclide-contaminated reservoir. In *Wading birds* (eds. A. Sprunt, IV, J. C. Ogden, and S. Winckler), pp. 361–4. National Audubon Society Research Report No. 7. National Audubon Society, New York.

Domm, S. and Recher, H. F. (1973). The birds of One Tree Island with notes on their yearly cycle and breeding ecology. *Sunbird*, **4**, 63–86.

Donovan, J. (1988). Highlights 1988. *Pembrokeshire Bird Report*, **1988**, 3–4.

Dowd, E. M. and Flake, L. D. (1984). Great Blue Heron nesting biology on the James River in South Dakota. *Prairie Naturalist*, **16**, 159–66.

Dowd, E. M. and Flake, L. D. (1985). Foraging habitats and movements of nesting Great Blue Herons (*Ardea herodias*) in a prairie river ecosystem, South Dakota. *Journal of Field Ornithology*, **56**, 379–87.

Dowsett, R. J. and de Vos, A. (1964). The ecology and numbers of aquatic birds on the Kajue Flats, Zambia. *Wildfowl Trust Annual Report*, **4**, 67–73.

Dowsett, R. J. and Dowsett-Lemaire, F. (1993). A contribution to the distribution and taxonomy of Afrotropical and Malagasy birds. *Tauraco Research Report*, **5**, 325.

Draulans, D. (1986). On the distribution and foraging behaviour of the Malagasy Heron *Ardea humbloti*. *Ostrich*, **57**, 249–51.

Draulans, D. (1987). The effect of prey density on foraging behaviour and success of adult and first-year Grey Herons (*Ardea cinerea*). *Journal of Animal Ecology*, **56**, 479–93.

Draulans, D. (1988a). Effects of fish-eating birds on freshwater fish stocks: An evaluation. *Biological Conservation*, **44**, 251–64.

Draulans, D. and Hannon, J. (1988). Distribution and foraging behavior of Grey Herons *Ardea cinerea* in adjacent tidal and non-tidal areas. *Ornis Scandinavica*, **19**, 297–304.

Draulans, D., Perremans, K., van Vessem, J., and Pollet, M. (1987). Analysis of pellets of the Grey Heron, *Ardea cinerea*, from colonies in Belgium. *Journal of Zoology (London)*, **211**, 695–708.

Draulans, D. and van Vessem, J. (1984a). The use of radio-tracking in studying foraging ecology of Grey Herons (*Ardea cinerea*). In *Biotelemetry VIII* (eds. H. P. Kimmich and H. J. Klewe), pp. 371–4.

Draulans, D. and van Vessem, J. (1984b). [Distribution and habitat choice of Grey Herons *Ardea cinerea* wintering in Flanders (Belgium).] *De Wielewaal*, **50**, 253–60.

Draulans, D. and van Vessem, J. (1985a). The effect of disturbance on nocturnal abundance and behavior of Grey Herons (*Ardea cinerea*) at a fish-farm in winter. *Journal of Applied Ecology*, **22**, 19–28.

Draulans, D. and van Vessem, J. (1985b). Age-related differences in the use of time and space by radio-tagged Grey Herons (*Ardea cinerea*) in winter. *Journal of Animal Ecology*, **54**, 771–80.

Draulans, D. and van Vessem, J. (1987a). Factors affecting abundance, distribution and behavior of Grey Herons (*Ardea cinerea*) at a fish-farm. *Gerfaut*, **77**, 43–61.

Draulans, D. and van Vessem, J. (1987b). Some aspects of population dynamics and habitat choice of Grey Herons (*Ardea cinerea*) in fish-pond areas. *Gerfaut*, **77**, 389–404.

Draulans, D., van Vessem, J., and De Bont, A. F. (1986a). Origin, age and mortality of Grey Herons (*Ardea cinerea*) wintering in Belgium. *Gerfaut*, **76**, 161–76.

Draulans, D., van Vessem, J., and Symons, F. (1986b). Multivariate analysis of factors affecting the foraging success of Grey Herons (*Ardea cinerea*) at a fish-farm in winter. *Annales de la Societe Royale Zoologique de Belgique*, **116**, 211–26.

Drennan, M. P. and Bowman, R. S. (1993). Ground-nesting Snowy Egrets in Maine—a new northernmost breeding record. *American Birds*, **47**, 376–7.

Drinkwater, H. (1958). Black-crowned Night Herons using bill motion to lure prey. *Wilson Bulletin*, **70**, 201–2.

DuBowy, P. J. (1996). Effects of water levels and weather on wintering herons and egrets. *The Southwestern Naturalist* **41**, 341–347.

Dubois, P. J. and Yesou, P. (1995). Identification of Western Reef Egrets and dark Little Egrets. *British Birds*, **88**, 307–19.

Duebbert, H. F. and Lokemoen, J. T. (1977). Upland nesting of American Bitterns, Marsh Hawks, and Short-eared Owls. *Prairie Naturalist*, **9**, 33–40.

Duhautois, L. (1984). Heron pourpre, butors: le declin. *Courrier de la Nature*, **92**, 21–9.

Duncanson, S. and Paxton, J. R. (1994). Spoonbill poaching under Black Egret canopy. *Honeyguide*, **40**, 174.

du Plessis, S. S. (1963). The feeding behavior of the Black Heron *Metanophoyx ardesiaca*. *Ostrich*, **34**, 111–12.

Duquet, M. (1987). Mise au point sur l'avifaune nicheuse du lac de Remoray (Doubs, France). *Nos Oiseaux*, **39**, 97–111.

Dusi, J. L. (1967). Migration in the Little Blue Heron. *Wilson Bulletin*, **79**, 222–35.

Dusi, J. L. (1968). The competition between Cattle Egrets and Little Blue Herons. *Alabama Birdlife*, **16**, 4–6.

Dusi, J. L. (1979). Heron colony effects on man. *Proceedings of the Colonial Waterbird Group*, **3**, 143–4.

Dusi, J. L. (1982). Cattle Egret management. *Colonial Waterbirds*, **5**, 177.

Dusi, J. L. (1983). Cattle Egret management in Alabama. *Alabama Birdlife*, **30**, 47.

Dusi, J. L. and Dusi, R. T. (1970). Nesting success and mortality of nestlings in a Cattle Egret colony. *Wilson Bulletin*, **82**, 458–60.

Dusi, J. L. and Dusi, R. D. (1995). A classical example of the Cattle Egret (*Bubulcus ibis*) pirating Little Blue Heron (*Egretta caerulea*) nests in Alabama. *Alabama Birdlife*, **42**, 3–5.

Dusi, J. L. and Dusi, R. D. (1996). Heron colonies near Tuskegee, Alabama. *Journal of the Alabama Academy of Science*, **67**, 61.

Dutson, G. (2001). New distributional ranges for Melanesian birds. *Emu*, **101**, 237–48.

Dvorak, M., Nemeth, E., Tebbich, S., Rossler, M., and Busse, K. (1997). Verbreitung, Bestand und Habitatwahl schilfbewohnender Vogelarten in der Naturzone des Nationalparks Neusiedler See—Seewinkel. *BFB-Bericht*, **86**, 1–69.

Dyrcz, A. (1999). Nesting of the Rufescent Tiger-heron *Tigrisoma lineatum* on Barro Colorado Island (Republic of Panama). *Ornitologia Neotropical*, **10**, 105–6.

Eccles, S. D. (1988). The birds of Sao Tome—record of a visit, April 1987 with notes on the rediscovery of Bocage's Longbill. *Malimbus*, **10**, 207–17.

Ecsedi, Z., Olah, J., Jr., and Tar, J. (2000). The Chinese Pond Heron in Hungary. *Birding World*, **13**, 466–7.

Edgar, A. T. (1978). The Reef Heron (*Egretta sacra*) in New Zealand. *Notornis*, **25**, 25–58.

Edwards, S. J. (1988). White-backed Night Herons nesting in rocks. *Honeyguide*, **34**, 119.

Eigenhuis, K. J. (1984). Mogelijke hybriden grote zilverreiger x blauwe reiger in Flevoland in 1981–83. *Dutch Birding*, **6**, 93–4.

Eisenmann, E. (1965). (*Tigrisoma fasciatum*). *Hornero*, **10**, 225–34.

El Din, S. M. B. (1992). Notes on recent changes in the status of breeding herons in the Egyptian Nile Valley and Delta. *Ornithological Society of the Middle East Bulletin*, **29**, 12–15.

Elkin, J. A. (1993). Japanese Night-Heron *Gorsachius goisagi*, a species new to Borneo. *Forktail*, **8**, 154–5.

Elliott, C. C. H. (1983). Unusual breeding records made from a helicopter in Tanzania. *Scopus*, **7**, 33–6.

Ellis, H. I. (1980). Metabolism and solar radiation in dark and white herons in hot climates. *Physiological Zoology*, **53**, 358–72.

Elphick, C. S. (2000). Functional equivalency between rice fields and seminatural wetland habitats. *Conservation Biology*, **14**, 181–91.

Elphick, C. S. and Oring, L. W. (1998). Winter management of Californian rice fields for waterbirds. *Journal of Applied Ecology*, **35**, 95–108.

Ely, A. (1976). *Egretta picata*. *Sunbird*, **7**, 77–8.

Eminov, A. (1993). New habitat of *Phalacrocorax carbo* and Ciconiformes (sic) in Turkmenistan. *Izvestiya Akademii Nauk Turkmenistana Seriya Biologicheskikh Nauk*, **4**, 59–61.

Emond, E. and Medeiros, M. (1999). Is siblicide a possibility in Black-crowned Night Herons? *Journal of the Minnesota Academy of Science*, **63**, 23–4.

Endo, N. and Sawara, Y. (2000). Diel rhythmic activity and foraging site utilization of Black-crowned Night

Heron (*Nycticorax nycticorax*) in its breeding season. *Japanese Journal of Ornithology*, **48**, 183–96.

Engilis, A., Jr. and Pratt, T. K. (1993). Status and population trends of Hawaii's native waterbirds, 1977–1987. *Wilson Bulletin*, **105**, 142–58.

English, M. (1987). More on fishing Greenbacked Herons. *Bokmakierie* **39**, 124–5.

English, P. and Bodenhorst, C. (1991). The voice and first nesting records of the Zigzag Heron in Ecuador. *Wilson Bulletin*, **103**, 661–4.

Erard, C., Guillou, J. J., and Mayaud, N. (1986). Le héron blanc du Banc d'Arguin *Ardea monicae:* Ses affinites, morphologiques: son histoirie. *Alauda*, **54**, 161–9.

Erftmeijer, P. (1989). The occurrence of nests of Nankeen Night Heron *Nycticorax caledonicus* in East Java. *Kukila*, **4**, 146–7.

Eerhert, F. C. and Kurstjens, G. (2000). Aantalsontwikkeling van de kwak *Nycticorax nycticorax* als broedvogel in Nederland in de twintigste eeuw. *Limosa*, **73**, 41–52.

Erikson, H. and Erikson, J. (1999). The first records of the Yellow Bittern *Ixobrychus sinensis* in Oman and Arabia. *Sandgrouse*, **21**, 178–9.

Erikson, J. (1999). New breeding birds in Oman. *Phoenix*, **16**, 22.

Ern, H. (1990). Autumn and winter observations of Great White Herons (*Egretta alba*) on fishponds in Croatia. *Troglodytes*, **3**, 11–12.

Erskine, A. J. (1992). *Atlas of breeding birds of the Maritime Provinces*. Nimus Publications and Nova Scotia Museum, Halifax, Nova Scotia.

Erwee, H. P. (1989). Feeding assembly of Black-headed Herons. *Honeyguide*, **35**, 112.

Erwin, R. M. (1983). Feeding habitats of nesting wading birds spatial use and social influences. *Auk*, **100**, 960–70.

Erwin, R. M. (1985). Foraging decisions, patch use and seasonality in egrets (Aves: Ciconiiformes). *Ecology*, **66**, 837–44.

Erwin, R. M. (1989). Predator–prey interactions, resource depression and patch revisitation. *Behavioural Processes*, **18**, 1–16.

Erwin, R. M., Haig, J. G., Stotts, D. B., and Hatfield, J. S. (1996a). Dispersal and habitat use by post-fledging juvenile Snowy Egrets and Black-crowned Night-herons. *Wilson Bulletin*, **108**, 342–56.

Erwin, R. M., Haig, J. G., Stotts, D. B., and Hatfield, J. S. (1996b). Reproductive success, growth and survival of Black-crowned Night-Heron (*Nycticorax nycticorax*) and Snowy Egret (*Egretta thula*) chicks in coastal Virginia. *Auk*, **113**, 119–30.

Esterhuizen, J. R. (1994). [Green-backed Heron near Bloemfontein.] *Mirafra*, **11**, 34–5.

Etcheberry, R. (1990). The nesting season (1990) in St-Pierre and Miquelon Islands. *Osprey*, **21**, 129–38.

Euston-Brown, M. (1993). Orange River birds. *Promerops*, **209**, 11.

Evans, S. (1986). White-faced Heron *Ardea novaehollandiae* as prey of Powerful Owl *Ninox strenua*. *Australian Bird Watcher*, **11**, 169.

Ewins, P. J., Bazely, D. R., and Recher, H. F. (1990). Communal roosting of Eastern Reef Egrets *Egretta sacra*. *Corella*, **14**, 29.

Ewins, P. J. and Hennessey, B. (1992). Great Blue Herons *Ardea herodias*, feeding at a fishing vessel offshore in Lake Erie. *Canadian Field-Naturalist*, **106**, 521–2.

Farmer, M. (1992). Seasonal foraging habitat for the Reddish Egret in a hypersaline lagoon. *Colonial Waterbird Society Bulletin*, **16**, 60.

Fasola, M. (1982). Feeding dispersion in the Night Heron *Nycticorax nycticorax* and Little Egret *Egretta garzetta* and the information centre hypothesis. *Bollettino di Zoologia*, **49**, 177–86.

Fasola, M. (1983). Nesting populations of herons in Italy depending on feeding habitats. *Bollettino di Zoologia*, **50**, 21–4.

Fasola, M. (1984). Activity rhythm and feeding success of nesting Night Herons *Nycticorax nycticorax*. *Ardea*, **72**, 217–22.

Fasola, M. (1986). Resource use of foraging herons in agricultural and nonagricultural habitats in Italy. *Colonial Waterbirds*, **9**, 139–48.

Fasola, M. (1994). Opportunistic use of foraging resources by heron communities in southern Europe. *Ecography*, **17**, 113–23.

Fasola, M. (1998). Optimal clutch size in herons: Observational and experimental approaches. *Ethology, Ecology and Evolution*, **10**, 33–46.

Fasola, M. and Alieri, R. (1991). Andamento della popolazioni di Ardeidae nidificanti in Italia. *Supplemento Alle Ricerche di Biologia Della Selvaggina*, **16**, 337–40.

Fasola, M. and Alieri, R. (1992a). Nest site characteristics in relation to body size in herons in Italy. *Colonial Waterbirds*, **15**, 185–91.

Fasola, M. and Alieri, R. (1992b). Conservation of heronry Ardeidae sites in North Italian agricultural landscapes. *Biological Conservation*, **62**, 219–28.

Fasola, M., Alieri, R., and Zandonella, N. D. (1992). Strategia per la conservazione delle colonie di Ardeinae e modello per la gestione di specifiche riserve naturali. *Ricerche di Biologia della Selvaggina*, **1992**, 1–49.

Fasola, M. and Barbieri, F. (1978). Factors affecting the distribution of heronries in northern Italy. *Ibis*, **120**, 537–40.

Fasola, M., Barbieri, F., Prigioni, C., and Bogliani, G. (1981). Le garzaie in Italia. *Avocetta*, **7**, 29–36.

Fasola, M., Canova, L., and Saino, N. (1996). Rice fields support a large portion of herons breeding in the Mediterranean region. *Colonial Waterbirds*, **19**, **Special Publication 1**, 129–34.

Fasola, M. and Ghidini, M. (1984). Use of feeding habitat by breeding Night Heron and Little Egret. *Avocetta*, **7**, 29–36.

Fasola, M. and Hafner, H. (1997). *Nycticorax nycticorax* Night Heron. *BWP Update*, **1**, 157–65.

Fasola, M., Hafner, H., Kayser, Y., Bennets, R.E., and Cezilly, F. (2002). Individual dispersal among colonies of Little Egrets *Egretta garzetta*. *Ibis*, **144**, 192–9.

Fasola, M., Hafner, H., Prosper, J., Van Der Kooij, H., and Schogolev, I.V. (2000). Population changes in European herons in relation to African climate. *Ostrich*, **71**, 52–5.

Fasola, M. and Hudec, K. (1997). Night-Heron. In *The EBCC atlas of European breeding birds: their distribution and abundance* (eds. E. J. M. Hagemeijer and M. J. Blair), pp. 44–5. T. & A. D. Poyser, London.

Fasola, M., Movalli, P.A., and Gandini, C. (1998). Heavy metal, organochlorine pesticide, and PCB residues in eggs and feathers of herons breeding in Northern Italy. *Archives of Environmental Contamination and Toxicology*, **34**, 87–93.

Fasola, M., Rosa, P., and Canova, L. (1993). The diets of Squacco Herons, Little Egrets, Night, Purple and Grey Herons in their Italian breeding ranges. *Revue d'Ecologie la Terre et la Vie*, **48**, 35–47.

Fasola, M. and Ruiz, X. (1996). The value of rice fields as substitutes for natural wetlands for waterbirds in the Mediterranean region. *Colonial Waterbirds*, **19**, 108–14.

Fasola, M. and Ruiz, X. (1996). Rice farming and waterbirds: integrated management in an artificial landscape. In *Farming and birds in Europe: the common agricultural policy and its implications for bird conservation* (eds. D. J. Pain and M. W. Pienkowski), pp. 210–35. Academic Press, San Diego.

Fasola, N., Zhang, Y., Zhao, D., Dong, Y. and Wang, H. (2001). Age-assortive mating related to reproductive success in Black-crowned Night Herons. *Waterbirds*, **24**, 272–6.

Feare, C. J. (1975). Further migrant birds in the Seychelles. *Bulletin of the British Ornithologists' Club*, **95**, 48–50.

Fellowes, J. R., Zhou, F., Lee, K. S., Hau, B. C. H., Lau, M. W. N., Lam, V. W. Y., Young, L., and Hafner, H. (2001). Status update on White-eared Night Heron (*Gorsachius magnificus*) in south China. *Bird Conservation International*, **11**, 101–11.

Fellows, D. P. and Paton, P. W. C. (1988). Behavior response of Cattle Egrets to population control measures in Hawaii. *Vertebrate Pest Conference*, **13**, 315–18.

Fenyvesi, L. (1992). Grey Heron (*Ardea cinerea*) × Purple Heron (*Ardea purpurea*) hybridizations at Dinnyes (County Fejer, Hungary). *Aquila A Madartani Intezet Evkonyve*, **99**, 170–1, 183–4.

Fernandez-Cruz, M. and Campos, F. (1993). The breeding of Grey Herons (*Ardea cinerea*) in western Spain: The influence of age. *Colonial Waterbirds*, **16**, 53–8.

Fernandez-Cruz, M., Fernandez Alcazar, G., and Campos, F. (1993). Principales problemas de conservacion de las ardeidas coloniales espanolas. *Alytes (Merida)*, **6**, 453–60.

Fernandez-Cruz, M., Fernandez Alcazar, G., Campos, F., and Dias, P. C. (1992). Colonies of ardeids in Spain and Portugal. In *Managing Mediterranean wetlands and their birds* (eds. M. Finlayson, T. Hollis, and T. Davis). pp. 76–8. IWRB Special Publication No. 20. International Waterfowl and Wetlands Research Bureau, Slimbridge.

Fesenko, G.V. (1984). [The nesting of the Little Egret in the Kiev region (*Egretta garzetta*).] *Vestnik Zoologii*, **1984**, 88.

Festetics, A. and Leisler, B. (1999). The breeding colonies of herons and spoonbills at Lake Neusiedi—Population trends, nestsite selection, and conservation. *Ökologie der Vögel*, **21**, 269–329.

Feunteun, E. and Marion, L. (1994). Assessment of Grey Heron predation on fish communities: the case of the largest European colony. *Hydrobiologia*, **279/280**, 327–44.

ffrench, R. (1973). *A guide to the birds of Trinidad and Tobago*. Livingston Publishing Company, Wynnewood.

Figueroa, R. A. and Corales Stappung, E. S. (2003). Food of breeding Great White Egrets in an agricultural area of southern Chile. *Waterbirds*, **26**, 370–5.

Finch, D. W. (2002). First site record of Grey Heron (*Ardea cinerea*) for Tobago. *Occasional Papers of the Department of Life Sciences, University of the West Indies*, **11**, 199–200.

Findholt, S. L. (1984a). Organochlorine residues eggshell thickness and reproductive success of Snowy Egrets nesting in Idaho. *Condor*, **86**, 163–9.

Findholt, S. L. (1984b). Status and distribution of herons, egrets, ibises, and related species in Wyoming. *Colonial Waterbirds*, **7**, 55–62.

Findholt, S. L. and Berner, K. L. (1988). Current status and distribution of the Ciconiiforms nesting in Wyoming. *Great Basin Naturalist*, **48**, 290–7.

Fjeldsa, J. and Krabbe, N. (1990). *Birds of the High Andes*. Zoological Museum of the University of Copenhagen and Apollo Books, Svendborg.

Fluery, B. E. and Sherry, T. W. (1995). Long-term population trends of colonial wading birds in the southern United States: the impact of crawfish aquaculture on Louisiana populations. *Auk*, **112**, 613–32.

Fogarty, M. J. and Hetrick, W. M. (1973). Summer foods of Cattle Egrets in north central Florida. *Auk*, **90**, 268–80.

Fontanelli, A., Perfetti, A., Puglisi, L., and Taverni, M. (1995). Attività di canto in una popolazione di tarabuso (*Botaurus stellaris*) della toscana. *Supplemento Alle Ricerche di Biologia Della Selvaggina*, **22**, 631–3.

Forbes, L. S. (1982). Prey manipulation in the Great Blue Heron. *Murrelet*, **63**, 89–91.

Forbes, L. S. (1987a). Feeding behavior of Great Blue Herons at Creston, British Columbia (Canada). *Canadian Journal of Zoology*, **65**, 3062–7.

Forbes, L. S. (1987b). Predation on adult Great Blue Herons: Is it important? *Colonial Waterbirds*, **10**, 120–6.

Forbes, L. S. (1989). Prey defences and predator handling behaviour: the dangerous prey hypothesis. *Oikos*, **55**, 155–8.

Forbes, L. S. and McMackin, E. (1984). Extreme aggression in Great Blue Herons. *Wilson Bulletin*, **96**, 318–19.

Forster, R. A. (1989). First United States sight record of Little Egret (*Egretta garzetta*). *Bird Observer*, **17**, 239–44.

Forsyth, E. (1984). It's rough to eat a Ruff. *WBC News*, **124**, 4.

Fox, J. B. (1989a). Grey Heron following a plough. *Irish Birds*, **4**, 71.

Fox, J. B. (1989b). Grey Herons taking roadside rats. *Irish Birds*, **4**, 70–1.

Foxall, C. D. and Drury, D. (1987). Green-backed Heron 'bait-fishing' in Nairobi National Park. *East African Natural History Society Bulletin*, **17**, 11.

Franchimont, J. (1986a). Les causes de l'expansion geographique mondiale du heron garde-boeuf (*Bubulcus ibis L.*). *Cahiers d'Ethologie Appliquee*, **6**, 373–88.

Franchimont, J. (1986b). Les lieux d'alimentation du heron gardeboeuf (*Bubulcus ibis*) dans la nord-ouest Marocain. *Aves*, **23**, 216–24.

Fraser, W. (1974). Feeding association between Little Egret and Reed Cormorant. *Ostrich*, **45**, 262.

Fraser, P. A., Lansdown, P. G., and Rogers, M. J. (1999). Report on scarce migrant birds in Britain in 1996. *British Birds*, **92**, 3–35.

Frederick, P. C. (1997). Tricolored Heron, *Egretta tricolor*. In *The Birds of North America*. No. 306 (eds. A. Poole and F. Gill), pp. 1–28. The Academy of Natural Sciences, Philadelphia, Pennsylvania and The American Ornithologists' Union, Washington.

Frederick, P. C., Bjork, R., Bancroft, G. T., and Powell, G. V. N. (1992). Reproductive success of three species of herons relative to habitat in southern Florida. *Colonial Waterbirds*, **15**, 192–201.

Frederick, P. C. and Collopy, M. W. (1989a). Nesting success of five ciconiiform species in relation to water conditions in the Florida Everglades (USA). *Auk*, **106**, 625–34.

Frederick, P. C. and Collopy, M. W. (1989b). Researcher disturbance in colonies of wading birds: Effects of frequency of visit and egg-marking on reproductive parameters. *Colonial Waterbirds*, **12**, 152–7.

Frederick, P. C. and Collopy, M. W. (1989c). The role of predation in determining the reproductive success of colonially nesting wading birds in the Florida Everglades. *Condor*, **91**, 860–7.

Frederick, P. C., Dwyer, N., Fitzgerald, S., and Bennetts, R. E. (1990). Relative abundance and habitat preferences of Least Bitterns (*Ixobrychus exilis*) in the Everglades. *Florida Field Naturalist*, **18**, 1–9.

Frederick, P. C., Sandoval, J. C., Luthin, C., and Spalding, M. (1997). The importance of the Caribbean coastal wetlands of Nicaragua and Honduras to Central American populations of waterbirds and Jabiru Storks (*Jabiru mycteria*). *Journal of Field Ornithology*, **68**, 287–95.

Freidmanm, H. and Smith, F. D., Jr. (1950). A contribution to the ornithology of northeastern Venezuela. *Proceedings of the United States National Museum*, **100**, 411–538.

Fry, C. H., Hosken, J. H., and Skinner, D. (1986). Further observations on the breeding of Slaty Egrets *Egretta vinaceigula* and Rufous-bellied Herons *Ardeola rufiventris*. *Ostrich*, **57**, 61–4.

Fujioka, M. (1984). Asynchronous hatching, growth and survival of chicks of the Cattle Egret *Bubulcus ibis*. *Tori Bulletin of the Ornithological Society of Japan*, **33**, 1–12.

Fujioka, M. (1985a). Food delivery and sibling competition in experimentally even-aged broods of the Cattle Egret. *Behavioral Ecology and Sociobiology*, **17**, 67–74.

Fujioka, M. (1985b). Sibling competition and siblicide in asynchronously-hatching broods of the Cattle Egret *Bubulcus ibis*. *Animal Behaviour*, **33**, 1228–42.

Fujioka, M. (1987). Mechanisms and factors of brood reduction in the Cattle Egret *Bubulcus ibis*. In *Animal societies: theories and facts* (eds. Y. Ito, J. L. Brown, and J. Kikkawa), pp. 115–23. Japan Scientific Societies Press, Tokyo.

Fujioka, M. (1989). Mate and nestling desertion in colonial Little Egrets. *Auk*, **106**, 292–302.

Fujioka, M. and Yamagishi, S. (1981). Extramarital and pair copulations in the Cattle Egret. *Auk*, **98**, 134–44.

Fujioka, M., Armacost, J. W., Jr., Yoshida, H., and Maeda, T. (2001). Value of fallow farmlands as summer habitat for waterbirds in a Japanese rural area. *Ecological Research*, **16**, 555–67.

Fukumoto, Y., Daimaru, H., Egusa, S., and Yamane K-I (1992). [The second record of the Black Bittern for Japan.] *Japanese Journal of Ornithology*, **39**, 63–4.

Gaasbeek, H. A. (1982). Great White Egret with geranium-red legs. *Dutch Birding*, **4**, 53.

Gaasbeek, H. A. (1985). [Great White Egrets with red legs.] *Vogeljaar*, **33**, 177.

Galbraith, C. A., Pierce, G. J., Spray, C. J., and Robinson, I. H. (1999). The diurnal movement pattern of waterbirds in the Kukut area of Lake Songkla, southern Thailand. *Wildfowl*, **50**, 163–79.

Galeotti, P. (1982). L'allevamento della prole nella Nitticora *Nycticorax nycticorax*. *Avocetta*, **6**, 11–22.

Gantlett, S. (1998). Identification of Great Blue Heron and Grey Heron. *Birding World*, **11**, 12–20.

Gao, Y. R., Xiao, R. G., and Bi, X. F. (2000). [Discovery of endangered White-eared Night Heron in Guandong Province.] *Chinese Journal of Zoology*, **35**, 39–41.

Garnett, S. (1992). *Threatened and extinct birds of Australia*. Birds Australia, Melbourne, Australia. RAOU Report 82.

Garnett, S. (1993). Do White-faced Herons really eat dead cows? *Australian Bird Watcher*, **15**, 40–1.

Garrett, K. L. (1994). A white Green Heron in southern California. *Western Birds*, **25**, 198–200.

Garrido, J.R., Sarasa, C.G., and Fernandez-Cruz, M. (2002). Intraspecific kleptoparasitism in the Cattle Egret. *Journal of Field Ornithology*, **73**, 185–90.

Garrido, O. and Kirkconnell, A. (1990). [The Reddish Egret *Egretta rufescens rufescens* in Cuba.] *Volante Migratorio*, **14**, 20–1.

Gassett, J. W., Folk, T. H., Alexy, K. J., Miller, K. V., Chapman, B. R., Boyd, F. L., and Hall, D. I. (2000). Food habits of Cattle Egrets on St. Croix, U.S. Virgin Islands. *Wilson Bulletin*, **112**, 268–71.

Gast, S. E. and King, B. (1985). Notes on Philippine birds, 7. Recent records of the Chinese Egret *Egretta eulophotes* from Luzon, Mindoro and Palawan, Philippines. *Bulletin of the British Ornithologists' Club*, **105**, 139–41.

Gatter, W. (1988). [Midwinter counts of some groups of Palearctic water birds in Liberia (West Africa).] *Verhandlungen der Ornithologischen Gesellschaft in Bayern*, **24**, 659–88.

Gauggel, K. F. (1999). Regelmassige Uberwinterung der Rohrdommel in einem suddeutschen Baggerseegebiet. *Falke*, **46**, 20–1.

Gavino, T. G. and Dickerman, R. W. (1972). Nesting development of Green Herons at San Blas, Nayarit, Mexico. *Condor*, **74**, 72–9.

Gavris', G. G. (1994). [The occurrence of Large Egret and Little Egret in the left bank Ukraine.] *Vestnik Zoologii*, **1**, 80–3.

Gawlik, D. E. (2002). The effects of prey availability on the numerical response of wading birds. *Ecological Monographs*, **72**, 329–46.

Gebauer, M. N. and Moul, I. E. (2001). Status of the Great Blue Heron in British Columbia. *Wildlife Working Report*, 2001.

Geering, D. J. (1988). Marsh Harrier *Circus aeruginosus* attacks White faced Heron *Ardea novaehollandiae*. *Australian Bird Watcher*, **12**, 164.

Geering, D. J. (1993). The effect of drought-breaking rain on the re-establishment of egret colonies in north coastal New South Wales. *Corella*, **17**, 47–51.

Geering, D. J., Maddock, M., Cam, G. R., Ireland, C., Halse, S. A., and Pearson, G. B. (1998). Movement patterns of Great, Intermediate and Little Egrets from Australian breeding colonies. *Corella*, **22**, 37–46.

Geldenhuys, J. N. (1984). Status of the Fish Eagle and Goliath Heron in the Orange Free State, South Africa. In *Proceedings of the fifth Pan African Ornithological Congress held at Lilongwe, Malawi in 1980* (ed. J. Ledger), pp. 577–87. Southern African Ornithological Society, Johannesburg.

George, C. J. and Moore, A. (1990). Overwintering of the Great Blue Heron at Collins Lake, Schenectady Co., New York. *Kingbird*, **40**, 75–82.

George, J. L. (1941). Great Blue Heron spearing fish. *Wilson Bulletin*, **53**, 123.

Georgiewa, U. (1999). Erstbeobachtung eines Küstenreihers *Egretta gularis* in Bulgarien. *Ornithologische Mitteilungen*, **51**, 351.

Gerard, C. (1986). Nidification du blongios nain dans l'Eure en 1985. *Cormoran*, **5**, 415–6.

Gerhardt, F. and Taliaferro, E.H. (2003). Density-dependent patch selection by foraging Cattle Egrets. *Waterbirds*, **26**, 364–9.

Gerlach, J. and Skerrett, A. (2002). The distribution, ecology and status of the Yellow Bittern *Ixobrychus sinesis* in Seychelles. *African Journal of Ecology*, **40**, 194–6.

Geroudet, P. (1993a). Quand le heron veut avaler le gros poisson. *Nos Oiseaux*, **42**, 226–7.

Geroudet, P. (1993b). Deux "nouvelles" heronnieres lemaniques. *Nos Oiseaux*, **42**, 226.

Gibbs, J. P. (1991). Spatial relationships between nesting colonies and foraging areas of Great Blue Herons. *Auk*, **108**, 764–70.

Gibbs, J. P. and Kinkel, L. K. (1997). Determinants of the size and location of Great Blue Heron colonies. *Colonial Waterbirds*, **20**, 1–7.

Gibbs, J. P. and Melvin, S. M. (1992). American Bittern *Botaurus lentiginosus*. In *Migratory nongame birds of management concern in the Northeast* (eds. K. J. Schneider and D. M. Pence), pp. 51–70. U.S. Fish and Wildlife Service, Newton Corner.

Gibbs, J. P., Melvin, S. M., and Reid, F. A. (1992a). American Bittern, *Botaurus lentiginosus*. In *The birds of North America*. No. 18 (eds. A. Poole, P. Stettenheim, and F. Gill), pp. 1–12. The Academy of Natural Sciences, Philadelphia and The American Ornithologists' Union, Washington.

Gibbs, J. P., Reid, F. A., and Melvin, S. M. (1992b). Least Bittern, *Ixobrychus exilis*. In *The birds of North America*. No. 17 (eds. A. Poole, P. Stettenheim, and F. Gill), pp. 1–11. The Academy of Natural Sciences, Philadelphia and The American Ornithologists' Union, Washington.

Gibbs, J. P., Woodward, S., Hunter, M. L., and Hutchinson, A. E. (1987). Determinants of Great Blue Heron colony distribution in coastal Maine. *Auk*, **104**, 38–47.

Gibbs, J. P., Woodward, S., Hunter, M. L., and Hutchinson, A. E. (1988). Comparison of techniques for censusing Great Blue Heron nests. *Journal of Field Ornithology*, **59**, 130–4.

Gibson, D. D., Tobish, T. G., Jr., and Isleib, M. E. (1987a). The spring migration, March 1–May 31, 1987. Alaska Region. *American Birds*, **41**, 476–7.

Gibson, D. D., Tobish, T. G., Jr., and Isleib, M. E. (1987b). The nesting season. June 1–July 31, 1987. Alaska Region. *American Birds*, **41**, 1475–7.

Gibson, D. D., Tobish, T. G., Jr., and Isleib, M. E. (1988). The nesting season June 1–July 31, 1988. Alaska region. *American Birds*, **42**, 1329–31.

Gilbert, G., McGregor, P. K., and Tyler, G. (1994). Vocal individuality as a census tool: Practical considerations illustrated by a study of two rare species. *Journal of Field Ornithology*, **65**, 335–48.

Gilbert, G., Tyler, G.A., and Smith, K.W. (2002). Local annual survivial of booming male Great Bittern *Botaurus stellaris* in Britain,in the period 1990–1999. *Ibis*, **144**, 51–61.

Gill, B. J. (1988). Records of birds and reptiles from Tonga. *Records of the Auckland Institute and Museum*, **25**, 87–94.

Gill, B. J. (1990). Records of wildlife from Tonga, especially Vava'u. *Records of the Auckland Institute and Museum*, **27**, 165–74.

Gill, H. B. (1979). Display of the Great-Billed Heron *Ardea sumatrana*. *Sunbird*, **10**, 46.

Gillson, G. (1994). Tricolored Heron. *Oregon Birds*, **20**, 44.

Gladstone D. E. (1979). Promiscuity in monogamous colonial birds. *American Naturalist*, **114**, 545–77.

Glahn, J. F., Dorr, B., and Tobin, M. E. (2000). Captive Great Blue Heron predation on farmed channel catfish fingerlings. *North American Journal of Aquaculture*, **62**, 149–56.

Glahn, J. F., Reinhold, D. S., and Smith, P. (1999a). Wading bird depredations on channel catfish *Ictalurus punctatus* in northwest Mississippi. *Journal of the World Aquaculture Society*, **30**, 107–14.

Glahn, J. F., Tomsa, T., and Preusser, K. J. (1999b). Impact of Great Blue Heron predation at trout-rearing facilities in the northeastern United States. *North American Journal of Aquaculture*, **61**, 349–54.

Glahn, J. F., Dorr, B., Harrel, J. B., and Khoo, L. (2002). Foraging ecology and depredation management of Great Blue Herons at Mississippi catfish farms. *Journal of Wildlife Management*, **66**, 194–201.

Gochfeld, M. (1986). Aerial fishing by Pied Herons. *Wilson Bulletin*, **88**, 356–7.

Gochfeld, M., Buckley, P. A., and Buckley, F. G. (1982). Plumage 'atavism' in a Black-crowned Night Heron *Nycticorax nycticorax*. *Bulletin of the British Ornithologists' Club*, **102**, 92–4.

Godfray, H. C. J. (1986). Brood reduction and desertion in herons and egrets. *Trends in Ecology Evolution*, **1**, 33.

Goenka, D. and Pandit, H. (1986). Intimidation among waterbirds at Bharatpur. *Journal of the Bombay Natural History Society*, **83**, 219–20.

Gomez de Silva, H. (2003). Mexico. *North American Birds*, **57**, 408–13.

Gonzalez, J. A. (1996). Kleptoparasitism in mixed-species foraging flocks of wading birds during the late dry season in the llanos of Venezuela. *Colonial Waterbirds*, **19**, 226–31.

Gonzalez, J. A. (1997). Seasonal variation in the foraging ecology of the Wood Stork in the southern Llanos of Venezuela. *Condor*, **99**, 671–80.

Gonzalez-Martin, M. and Ruiz, X. (1996). Brood parasitism in herons. *Colonial Waterbirds*, **19**, 31–8.

Gonzalez-Martin, M., Ruiz, X., and Llorente, G. A. (1992). Breeding parameters, feeding habits and nestling growth in a recovering population of Purple Herons from the Ebro Delta, Spain. *Miscellania Zoologica (Barcelona)*, **16**, 147–60.

Goodman, S. M. and Storer, R. W. (1987). The seabirds of the Egyptian Red Sea and adjacent waters, with notes on selected Ciconiiformes. *Gerfaut*, **77**, 109–45.

Gopakumar, G. (1991). Effect of drought on the breeding activities of Cattle Egret, *Bubulcus ibis coromandus* (Boddaert), in a semi-arid to arid region. *Geobios (Jodhpur)*, **18**, 10–5.

Gordienko, N. S. (1987). [New and rare bird species for Naurzum Nature Reserve.] *Ornitologiya*, **22**, 177–9.

Gordon, P. R. (1986). Squacco Heron eating passerines. *British Birds*, **79**, 590.

Gordon, P. R. (1988). Green Heron at Tyninghame. *Lothian Bird Report*, **1987**, 80.

Gore, M. E. J. and Won, P.-O. (1971). *The birds of Korea*. Royal Asiatic Society, Seoul, Korea.

Gorman, G. (1995). Display of Great Bittern. *British Birds*, **88**, 47.

Goutner, V., Jerrentrup, H., Kazantzidis, S., and Nazirides, T. (1991). Occurrence of the Cattle Egret, *Bubulcus ibis*, in Greece. *Rivista Italiana di Ornitologia*, **61**, 107–12.

Graber, J. W., Graber, R. R., and Kirk, E. L. (1978). *Illinois Birds: Ciconiiformes*. Biological notes 109. Illinois Natural History Survey, Urbana.

Grant, M. and Bennett, D. (1986). Unusual sighting of an Australasian Bittern near Ballarat, Vic. *Bird Observer*, **656**, 85.

Greason, J. (1998). Wiltshire bird report 1997: species new to Wiltshire. Squacco Heron at the Cotswold Water Park: the first 20th Century record. *Hobby*, **24**, 94–5.

Greaves, C. (1991). Grey Heron catching, killing and swallowing Hoopoe. *British Birds*, **84**, 57–8.

Gretch, M. (1987). Great Horned Owl predation of Great Blue Heron. *Kingbird*, **37**, 6.

Gretch, M. (1989). Attrition of nests from an abandoned Great Blue Heron. *Kingbird*, **39**, 157.

Grieef, G. (1991). Bird news. *Manitoba Naturalists Society Bulletin*, **9**, 10.

Griem, J. N. and Martin, K. L. M. (1997). Predatory birds are present on the beach before and during California grunion runs. *American Zoologist*, **37**, 82A.

Grimwood, I. R. and Brocklehurst, M. J. C. (1984). Unusual feeding behavior in the paddybird or Indian Pond Heron *Ardeola grayii*. *Journal of the Bombay Natural History Society*, **81**, 696–7.

Grishchenko, V. N. and Serebryakov, V. V. (1993). Grey Heron migration in the Ukraine according to phonological observation results. *Byulletin' Moskovskogo Obshchestva Ispytalelei Prirody otdel Biologischeskii*, **98**, 33–7.

Gross, A. O. (1923). The Black-crowned Night Herons (*Nycticorax nycticorax*) of Sandy Neck. *Auk*, **40**, 1–30, 191–214.

Grove, R. A. and Henny, C. J. (1990). First nest record of the Least Bittern in Wyoming. *Northwestern Naturalist*, **71**, 12–4.

Gruell, A., Kohler, B., and Rauer, G. (1988). Zum Einsatz weisser Schmuckfedern beim Ausdrucksverhalten der Rohrdommel *Botaurus stellaris*. *Limicola*, **2**, 217–21.

Gruell, A. and Ranner, A. (1998). Populations of the Great Egret and Purple Heron in relation to ecological factors in the reed belt of the Neusiedler See. *Colonial Waterbirds*, **21**, 328–34.

Gruell, A. (1998). Veränderungen in der wahl der nahrungshabitate beim silberreiher (*Casmerodius albus*) am Neusiedler See. *Egretta*, **41**, 1–14.

Grussu, M. (1987). Nidificazione e svernamento del mignattaio, *Plegadis falcinellus*, e nidificazione della sgarza ciuffetto, *Ardeola ralloides*, in Sardegna. *Rivista Italiana di Ornitologia*, **57**, 62–8.

Grussu, M. and Poddesu, G. (1989). Considerazioni sulla presenza dell'airone Schistaceo, *Egretta gularis*, in Europa i sui problemi dell'dentificazione in natura. *Rivista Italiana di Ornitologia*, **59**, 172–82.

Guermeur, Y., ed. (1996). L'annee ornithologique 1996. *Bulletin du Centre Ornithologique d'Ouessant*, **13**, 1–32.

Gunawardene, W. T. T. P. and Wijesinghe, D. P. (1986). The Eurasian Bittern (*Botaurus stellaris*): an addition to the list of Sri Lanka birds. *Loris*, **17**, 52–3.

Guntert, M. (1986). Im Flug fischende Graurecher *Ardea cinerea*. *Ornithologische Beobachter*, **83**, 313–14.

Gustin, M. and Pizzari, T. (1992). First record of Little Egret, *Egretta garzetta*, breeding in a Sardinian marsh. *Avocetta*, **16**, 47.

Gustin, M., Arcamone, E., Corso, A., Passarella, M., and Pizzari, T. (2001). Recent increase in wintering and breeding Cattle Egret *Bubulcus ibis* population in Italy. *Alauda*, **69**, 311–8.

Gynther, I. C. (1995). A Striated Heron *Butorides striatus* inhabiting a freshwater section of the Brisbane River. *Australian Bird Watcher*, **16**, 145–51.

Haas, W. E. (1983). Random notes: Great White Heron, Cooper's Hawk. *Belize Audubon Society Bulletin*, **15**, 4–6.

Hafner, H. (1977). *Contribution à l'étude écologique de quatre espèces de hérons (Egretta g. garzetta L., Ardeola r. ralloides Scop., Ardeola i. ibis L., Nycticorax n. nycticorax L.).* Unpublished thesis, Université Paul Sabatier de Toulouse, France.

Hafner, H. (1978). Le succés de reporduction de quartre espéces d'Ardeides *Egretta g. garzetta* L., *Ardeola r. ralloides* Scop., *Ardeola i. ibis* L., *Nycticorax n. nycticorax* L., en Camargue. *Terre et la Vie*, **32**, 279–89.

Hafner, H. (2000). Herons in the Mediterranean. In *Heron conservation* (eds. J. A. Kushlan and H. Hafner), pp. 33–54. Academic Press, San Diego.

Hafner, H., Bennetts, R. E., and Kayser, Y. (2000). Changes in clutch size, brood size and numbers of nesting Squacco Herons *Ardeola ralloides* over a 32-year period in the Camargue, southern France. *Ibis*, **143**, 11–6.

Hafner, H., Boy, V., and Gory, G. (1982). Feeding methods, flock size and feeding success in the Little Egret

Egretta garzetta and the Squacco Heron *Ardeola ralloides* in the Camargue, southern France. *Ardea*, **70**, 45–54.

Hafner, H. and Britton, R. H. (1983). Changes of foraging sites by nesting Little Egrets (*Egretta garzetta* L.) in relation to food supply. *Colonial Waterbirds*, **6**, 24–30.

Hafner, H. and Didner, E. (1997). *Ardeola ralloides* Squacco Heron. *BWP Update*, **1**, 166–74.

Hafner, H., Dugan, P. J., and Boy, V. (1986). Use of artificial and natural wetlands as feeding sites by Little Egrets (*Egretta garzetta* L.) in the Camarge, southern France. *Colonial Waterbirds*, **9**, 149–54.

Hafner, H., Dugan, P. J., Kersten, M., Pineau, O., and Wallace, J. P. (1993). Flock feeding and food intake in Little Egrets *Egretta garzetta* and their effects on food provisioning and reproductive success. *Ibis*, **135**, 25–32.

Hafner, H. and Fasola, M. (1992). The relationship between feeding habitat and colonially nesting Ardeidae. In *Managing Mediterranean wetlands and their birds* (eds. M. Finlayson, T. Hollis, and T. Davis), pp. 194–201. IWRB Special Publication No. 20. International Waterfowl and Wetlands Research Bureau, Slimbridge.

Hafner, H., Fasola, M., Voisin, C., and Kayser, Y. (2002). Little Egret *Egretta garzetta*. *BWP Update*, **4**, 1–19.

Hafner, H., Kayser, Y., Boy, V., Fasola, M., Julliard, A.-C., Pradel, R., and Cézilly, F. (1998a). Local survival, natal dispersal, and recruitment in Little Egrets *Egretta garzetta*. *Journal of Avian Biology*, **29**, 216–27.

Hafner, H., Lansdown, R. V., Kushlan, J. A., Butler, R. W., Custer, T. W., Davidson, I. J., Erwin, R. M., Hancock, J. A., Lyles, A. M., Maddock, M., Marion, L., Morales, G., Mundkur, T., Perennou, C., Pineau, O., Turner, D., Ulenaers, P., van Vessem, J., and Young, L. (2000). Conservation of herons. In *Heron conservation* (eds. J. A. Kushlan and H. Hafner), pp. 343–75. Academic Press, San Diego.

Hafner, H., Pineau, O., and Kayser, Y. (1994). Ecological determinants of annual fluctuations in numbers of breeding Little Egrets (*Egretta garzetta* L.) in the Camargue, southern France. *Revue d'Ecologie la Terre et la Vie*, **49**, 53–62.

Hafner, H., Pineau, O., Kayser, Y., and Bennetts, R. (1999). A re-analysis of the influence of reproduction on the following year's breeding population of Little Egrets in the Camargue, southern France. *Revue d'Ecologie la Terre et la Vie*, **54**, 187–8.

Hafner, H., Pineau, O., Kayser, Y., Sall, M. A., and Lamarche, B. (1998b). Monitoring of colonial waterbird colonies in the Banc d'Arguin, Islamic Republic of Mauritania. In *2nd International conference on wetlands and development, Dakar, Senegal*. Abstract.

Hafner, H., Pineau, O., and Wallace, J. P. (1992). The effects of winter climate on the size of the Cattle Egret (*Bubulcus ibis* L.) population in the Camargue. *Revue d'Ecologie la Terre et la Vie*, **47**, 403–10.

Hale, S. R. and Homberger, D. G. (1993). Morphology and biological role of the Occipito–Atlanto–Dental joint in the Snowy Egret (*Egretta thula*). *American Zoologist*, **33**, 27A.

Halse, S. A., Pearson, G. B., Jaensch, R. P., Kulmoi, P., Gregory, P., Kay, W. R., and Storey, A. W. (1996). Waterbird surveys of the Middle Fly Floodplain, Papua New Guinea. *Wildlife Research*, **23**, 557–69.

Ham, I. (1986). The successive overwintering of night heron groups is not a random phenomenon. *Larus*, **36–37**, 324–6.

Hammatt, R. B. (1981). Reproductive biology in a Louisiana estuarine heronry. (MS) Thesis, Louisiana State University, Baton Rouge.

Hancock, J. (1984a). Aerial stretch display of the eastern race of the Great White Egret *Egretta alba*. *Ibis*, **126**, 92–4.

Hancock, J. (1984b). Great White Egrets with red legs and black bills. *British Birds*, **77**, 269.

Hancock, J. A. (1999). *Herons and egrets of the world, a photographic journey*. Academic Press, London.

Hancock, J. A. (2000). Herons of North American. Their world in focus. Academic Press, London.

Hancock, J. and Kushlan, J. A. (1984). *The herons handbook*. Croom Helm, London, and Sydney, and Harper and Row, New York.

Hancock, J. A., Kushlan, J. A., and Kahl, M. P. (1992). *Storks, ibises and spoonbills of the world*. Academic Press, London.

Hands, H. M., Drobney, R. D., and Ryan, M. R. (1989a). *Status of the Least Bittern in the northcentral United States*. U.S. Fish and Wildlife Service, Missouri Cooperative Fish and Wildlife Research Unit.

Hands, H. M., Drobney, R. D., and Ryan, M. R. (1989b). *Status of the American Bittern in the northcentral*

United States. U.S. Fish and Wildlife Service, Missouri Cooperative Fish and Wildlife Research Unit.

Harebottle, D. (2003). Foraging behaviour of two egret (*Egretta*) species. *Bird Numbers*, **12**, 9–10.

Hartmann, F. (1995). Heron and ibis 'tread on each others' toes'. *Kenya Birds*, **4**, 30.

Harvey, R. (1999). Tool use by Green Heron. *Connecticut Warbler*, **19**, 172.

Harvey, R. (2000). Tool use by Green Heron. *Connecticut Warbler*, **20**, 29.

Harvey, W. G. (1975). The habitat preferences of different colour morphs of *Egretta garzetta* on the Tasmanian coast. *Bulletin of the British Ornithologists' Club*, **95**, 171–2.

Harvey, W. G. (1990). *Birds in Bangladesh.* University Press Ltd., Dhaka.

Havens, K. E., Bull, L. A., Warren, G. L., Crisman, T. L., Phlips, E. J., and Smith, J. P. (1996). Food web structure in a subtropical lake ecosystem. *Oikos*, **75**, 20–32.

Haverschmidt, F. (1968). Birds of Surinam. Oliver & Boyd, Edinburgh.

Haverschmidt, F. (1983). First record of the Little Egret (*Egretta garzetta*) in Suriname. *Wilson Bulletin*, **95**, 315.

Haverschmidt, F. and Mees, G. F. (1994). *Birds of Suriname.* Vaco Press, Paramaribo.

Hayes, F. E. (2002). Geographic variation, taxonomy and identification of New World *Butorides* herons. *North American Birds*, **56**, 4–10.

Hayes, F. E. and Hardy, D. R. (2001). Nocturnal foraging by the Green Heron (*Butorides virescens*) under artificial illumination in Tobago. *El Pitirre*, **14**, 1–2.

Hayes, F. E. and Rooks, C. (2001). Bait-fishing by the Striated Heron (*Butorides striatus*) in Trinidad. El Pitirre, **14**, 3–4.

Hayes, F. E. and White, G. L. (2001). Status of the Little Egret (*Egretta garzetta*) in Trinidad and Tobago. *El Pitirre*, **14**, 54–58.

Hazevoet, C. J. (1992). A review of the Santiago Purple Heron *Ardea purpurea bournei*, with a report of a new colony. *Bird Conservation International*, **2**, 15–23.

Hazevoet, C. J. (1999). Fourth report on birds from the Cape Verde Islands, including notes on conservation and records of 11 taxa new to the archipelago. *Bulletin Zoologisch Museum Universiteit Van Amsterdam*, **17**, 19–32.

Heather, B. D. (1983). Albinistic White-faced Herons. *Notornis*, **30**, 341–3.

Heather, B. D. and Robertson, H. A. (1996). *The field guide to the birds of New Zealand.* Viking, Aukland.

Heatwole, H. (1965). Some aspects of the association of Cattle Egrets with cattle. *Animal Behaviour*, **13**, 79–83.

Heinz, G. H., Erdman, T. C., Haseltine, S. D., and Stafford, C. (1985). Contaminant levels in colonial waterbirds from Green Bay and Lake Michigan, 1975–1980. *Environmental Monitoring and Assessment*, **5**, 223–36.

Heitmeyer, M. E. (1986). Postbreeding distribution and habitat use of wading birds in Oklahoma, USA. *Colonial Waterbirds*, **9**, 163–70.

Helbig, A. J. (1987). Records of Javan Pond-heron *Ardeola speciosa* and dusky warbler *Phylloscopus fuscatus* in Peninsular Malaysia in March 1986. *Forktail*, **3**, 57–9.

Held, J. J. D. (1981). Population changes in the Purple Heron in relation to drought in the wintering area. *Ardea*, **69**, 185–91.

Helm, G. (1996). Ungewohnliche niststandorte des graureihers (*Ardea cinerea*) in Hamburg. *Hamburger Avifaunistische Beitrage*, **28**, 165–6.

Hemmings, A. D. and Chappell, R. G. (1988). Nesting of White-faced Herons at the Chatham Islands. *Notornis*, **35**, 245–7.

Henderson, C. R. (1985). A Yellow-crowned Night-heron colony in Rutherford County, Tennessee. *Migrant*, **56**, 13.

Henny, C. J. (1972). An analysis of the population dynamics of selected avian species—with special references to changes during the modern pesticide era. *U.S. Fish and Wildlife Service, Wildlife Research Report*, **1**, 1–99.

Henny, C. J. and Bethers, M. R. (1971). Population ecology of the Great Blue Heron with special reference to western Oregon. *Canadian Field-Naturalist*, **85**, 205–9.

Henny, C. J. and Blus, L. J. (1986). Radiotelemetry locates wintering grounds of DDE-contaminated Black-crowned Night-herons. *Wildlife Society Bulletin*, **14**, 236–41.

Henny, C. J., Blus, L. J., and Hulse, C. S. (1985). Trends and effects of organochlorine residues on Oregon and Nevada wading birds, 1979–1983. *Colonial Waterbirds*, **8**, 117–28.

Henny, C. J. and Kurtz, J. E. (1978). Great Blue Herons respond to nesting habitat loss. *Wildlife Society Bulletin*, **6**, 35–7.

Henry, G. M. (1971). *A guide to the birds of Ceylon.* OUP, London.

Herkenrath, P. (2003). Another melanistic Cattle Egret. *British Birds*, **95**, 531.

Herklots, G. A. C. (1961). *The birds of Trinidad and Tobago.* Collins, London.

Hernández-Vazquez, S. and Fernández-Aceves, G. J. (1999). Reproducción de *Cochlearius cochlearius* (Garza Cucharón) y *Butorides virescens* (Garza Verde) en La Manzanilla, Jalisco, México. *Ciencias Marinas*, **25**, 277–91.

Herroelen, P. (1995). A Great Blue Heron in the North Atlantic. *Birding World*, **8**, 356.

Hewson, R. (1985). Scavenging of salmon carcases by birds. *Scottish Birds*, **13**, 179–82.

Hewson, R. (1995). Use of salmonid carcasses by vertebrate scavengers. *Journal of Zoology (London).* **235**, 53–65.

Higgins, P. J. and Smith, J. D. B. (1999). Nocturnal feeding by a White-faced Heron. *Australian Bird Watcher*, **18**, 123–5.

Higuchi, H. (1986). Bait-fishing by the Green-backed Heron *Ardeola striata* in Japan. *Ibis*, **128**, 285–90.

Higuchi, H. (1988a). Bait-fishing by Green-backed Herons in south Florida. *Florida Field Naturalist*, **16**, 8–9.

Higuchi, H. (1988b). Individual differences in bait-fishing by the Green-backed Heron *Ardeola striata* associated with territory quality. *Ibis*, **130**, 39–44.

Hill, B. (1988). Grey Heron ingesting junk. *British Birds*, **81**, 599.

Hilton-Taylor, C. (2000). *2000 IUCN Red List of threatened species.* IUCN—The World Conservation Union, Gland, Switzerland.

Hilty, S. L. and Brown, W. L. (1986). *A guide to the birds of Colombia.* Princeton University Press, Princeton.

Hines, C. J. H. (1985–1987). The birds of eastern Kavango, SWA/Namibia. *Journal SWA (South West Africa) Scientific Society*, **40–41**, 115–47.

Hines, C. J. H. (1992). Observations on the Slaty Egret *Egretta vinaceigula* in northern Namibia. *Ostrich*, **63**, 118–22.

Hiraldo, F., Blanco, J. C., and Bustiamente, J. (1991). Unspecialized exploitation of small animal carcasses by birds. *Bird Study*, **38**, 200–7.

Hirschfeld, E. (1991). Night Herons and Little Egrets preying on fish while flying. *British Birds*, **84**, 57.

Hjartarson, G. and Petursson, G. (2003). [Great White Egret, new for Iceland]. *Bliki*, **24**, 63–4.

Hobbs, J. N. (1986). White-faced Heron *Ardea novaehollandiae* feeding at garbage tip. *Australian Bird Watcher*, **11**, 209–10.

Hobbs, R. (1995). Boomers on the brink. *Natural World (Lincoln)*, **43**, 26–8.

Hoch, E. (1977). Reflections on prehistoric life at Umm anNar (Trucial Oman) based on faunal remains from the third millenium BC. *South Asian Archaeology*, **1977**, 589–638.

Hoffman, D. J., Rattner, B. A., Bunck, C. M., Krynitsky, A., Ohlendorf, H. M., and Lowe, R. W. (1996). Association between PCBs and lower embryonic weight in Black-crowned Night Herons in San Francisco Bay. *Journal of Toxicology and Environmental Health*, **19**, 383–91.

Hoffmann, T. W. (1996). Ceylon Bird Club notes 1995. *Loris*, **21**, 16–18.

Hoffmann, W. H. (1997). Richardson's ground squirrels are the prey of choice for a Great Blue Heron. *Alberta Naturalist*, **27**, 8–9.

Holmes, D. A. and Burton, K. (1987). Recent notes on the avifauna of Kalimantran. *Kukila*, **3**, 2–32.

Holmes, P. R. and Hatchwell, B. J. (1991). Notes on the ecology of the Little Bittern *Ixobrychus minutus* at Haigam Rakh, Kashmir, India. *Forktail*, **6**, 25–33.

Holmstrom, N. (1989a). [Great White Egret found at Utlangen.] *Faglar I Blekinge*, **25**, 126–7.

Holmstrom, N. (1989b). [Little Egret found at Utlangan.] *Faglar I Blekinge*, **25**, 53.

Holt, P. (1991). The American Bittern in Lancashire. *Birding World*, **4**, 49–51.

Holyoak, D. T. (1973). Significance of colour dimorphism in Polynesian populations of *Egretta sacra*. *Ibis*, **115**, 419–20.

Hom, C. W. (1983). Foraging ecology of herons in a southern San Francisco Bay salt marsh. *Colonial Waterbirds*, **6**, 37–44.

Homberger, D. G. (1989). Correlation between morphology of the lingual apparatus and feeding mechanics in birds. *Fortschritte der Zoologie*, **35**, 148–50.

Hoogerwerf, A. (1949–1950). *Birds of Java*. Limosa vol. 22–23.

Horstman, A. J., Nawrot, J. R., and Woolf, A. (1998). Mine-associated wetlands as avian habitat. *Wetlands*, **18**, 298–304.

Horvath, E. G. and Moholt, R. K. (1986). Diurnal feeding cycle at an inland Great Blue Heron colony. *Murrelet*, **67**, 27–8.

Hothem, R. L. and Hatch, D. (2004). Reproductive success of the Black-crowned Night-heron at Alcatraz Island, San Francisco Bay, California, 1990–2002. *Waterbirds*, **27**, 112–25.

Hothem, R. L., Roster, D. L., King, K. A., Keldsen, T. K., Marois, K. C., and Wainwright, S. E. (1995). Spatial and temporal trends of contaminants in eggs of wading birds from San Francisco Bay, California. *Environmental Toxicology and Chemistry*, **14**, 1319–31.

Houston, C. J. (1974). Longevity record for Black-crowned Night Heron:16-1/2 years. *Blue Jay*, **32**, 179.

Howe, S. (1989). Western Reef Heron eating *Phylloscopus* Warblers. *British Birds*, **82**, 564.

Howell, S. N. G. and Webb, S. (1994). *A guide to the birds of Mexico and northern Central America*. Oxford University Press, Oxford.

Howes, J. (1986). New birds in the region. *Interwader*, **7**, 9.

Hoy, M. D. (1994). Depredations by herons and egrets at bait fish farms in Arkansas. *Aquaculture Magazine*, **20**, 52–6.

Hoyer, E. (1991). Zwergrohrdommel—Paarung. *Falke*, **38**, 6–10.

Hoyer, R. C. and Smith, S. D. (1997). Chinese Pond-Heron in Alaska. *Field Notes*, **51**, 953–6.

Hubbard, J. P. (1976). Status of the Night Herons (*Nycticorax*) of the Philippines and vicinity. *Nemouria*, **19**, 1–10.

Hudson, J. W., Dawson, W. R., and Hill, R. W. (1974). Growth and development of temperature regulation in nestling Cattle Egrets. *Comparative Biochemistry and Physiology*, **49**, 717–41.

Hull, A. J. (1999). Green-backed Heron cannibalism. *Honeyguide*, **45**, 17.

Hume, R. A. (1992). Food piracy by Grey Herons on Great Crested Grebes. *British Birds*, **85**, 238.

Humphrey, P. S. and Parkes, K. C. (1963). Plumages and systematics of the Whistling Heron. In *Proceedings of the XIII International Ornithological Congress*. Ithaca 17–24, 1962 (ed. C. G. Sibley), pp. 84–90. American Ornithologists' Union, Baton Rouge.

Hunt, K. A., Hooper, M. J., and Littrell, E. E. (1995). Carbofuran poisoning in herons: Diagnosis using cholinesterase reactivation techniques. *Journal of Wildlife Diseases*, **31**, 186–92.

Husband, C. and Siering, M. (1991). Stosstauchender Graureiher *Ardea cinerea* uber Tiefwasserbereich. *Ornithologische Verhandlungen*, **25**, 76–7.

Hustler, K. (1995). A note on the fishing behaviour and plumage types of the Slaty Egret. *Honeyguide*, **41**, 21.

Hustler, K. and Games, I. (1987). Goliath Herons feeding at night. *Honeyguide*, **33**, 58.

Hustler, K. and Williamson, C. (1985). The Rail Heron in the Hwange National Park. *Honeyguide*, **31**, 145–7.

Huxley, T. H. (1867). On the classification of birds; and on the taxonomic value of the modifications of certain of the cranial bones observable in that class. *Proceedings Zoological Society of London*, **1867**, 415–72.

Igarashi, H. (1996). [Communal roosting of the night heron *Nycticorax nycticorax*.] *Strix*, **14**, 81–94.

Igarashi, H. (1997). [The roosting behaviour of Night Herons in winter.] *Strix*, **15**, 89–94.

Iijima, Y. (1984). Sight records for *Bubulcus ibis*, *Egretta alba*, *E. intermedia* and *E. garzetta* in Taiki, Hokkaido. *Tori Bulletin of the Ornithological Society of Japan*, **33**, 44–6.

Ilani, G. and Shalmon, B. (1984). Will Grey Herons return to nest? *Israel—Land and Nature*, **9**, 125–6.

Ingels, J., Cleere, N. and Pelletier, V. (2003). Noteworthy observation on some French Guianan birds. *Alauda*, **71**, 59–67.

Ingram, R. (1998). Rondevlei sightings. *Promerops*, **234**, 8–9.

Innes, D. W. (1993). Little Blue Heron at Courtenay: second record for British Columbia. *British Columbia Birds*, **3**, 20–1.

Inoue, Y. (1985). The process of asynchronous hatching and sibling competition in the Little Egret *Egretta garzetta*. *Colonial Waterbirds*, **8**, 1–12.

Inskipp, C. (1989). *Nepal's forest birds. Their status and conservation.* ICBP Monograph 4. Cambridge.

Inskipp, C. and Inskipp, P. (1993a). Birds recorded during a visit to Bhutan in Autumn 1991. *Forktail*, **8**, 97–112.

Inskipp, C. and Inskipp, P. (1993b). Birds recorded during a visit to Bhutan in Spring 1993. *Forktail*, **9**, 121–42.

Inskipp, C., Inskipp, T. P., and Grimmett, R. (1999). *Birds of Bhutan*. A. & C. Black / Christopher Helm, London.

Iredale, T. (1956). *Birds of New Guinea*. Georgian House, Melbourne.

Irwin. M. P. S. (1975). Adaptive morphology of the Black and Slaty Egrets *Egretta ardesiaca* and *Egretta vinaceigula* and relationship within the genus *Egretta* (Aves: Ardeidae). *Bonner Zoologische Beitraege*, **26**, 155–63.

Itoh, S. (1984a). The effect of weather on the roosting behaviour of the Little Egret *Egretta garzetta* in winter. *Tori Bulletin of the Ornithological Society of Japan*, **33**, 51–65.

Itoh, S. (1984b). Pre-roosting assemblies in the Little Egret *Egretta garzetta*. *Tori Bulletin of the Ornithological Society of Japan*, **33**, 13–28.

Itoh, S. (1988). *The white egret*. Chronicle Books, San Francisco.

Itoh, S. (1991). Geographical variation of the plumage polymorphism in the Eastern Reef Heron (*Egretta sacra*). *Condor*, **93**, 383–9.

Itzkowitz, M. (1984). Foraging behavior of the Snowy Egret (*Egretta thula*). *Biology of Behaviour*, **9**, 357–70.

Itzkowitz, M. and Makie, D. (1986). Food provisioning and foraging behavior of the Snowy Egret, *Egretta thula*. *Biology of Behaviour*, **11**, 111–15.

IUCN (2003). 2003 *IUCN Red list of threatened species*. IUCN, www.redlist.org.

Iverson, G. C. (1993). Status and distribution of the Great Blue Heron in Indiana. *Proceedings of the Indiana Academy of Science*, **102**, 47–58.

Jackson, B. (2001). The Snowy Egret in Argyll—A new British bird. *Birding World*, **14**, 460–4.

Jacobs, R. (1994). [Communal fishing of Goosander (*Mergus merganser*) and Grey Heron (*Ardea cinerea*) at the Wiehl Valley Dam]. *Charadrius*, **30**, 58–9.

Jaensch, R. P. (1988). Little Bitterns breeding in northern Australia. *Australian Bird Watcher*, **12**, 217–21.

Jaensch, R. P. (1989). Little Bittern *Ixobrychus minutus* breeding at Bool Lagoon, 1984–1986. *South Australian Ornithologist*, **30**, 205–9.

Jaensch, R. P. (1994). An inventory of the sub-humid tropics of the Northern Territory. Report to Australian Nature Conservation Agency and Conservation Commission of Northern Territory.

Jaensch, R. P. (1995). Little Bitterns (*Ixobrychus minutus*) in the Middle Fly wetlands. *Muruk*, **7**, 117–18.

Jaensch, R. P. (1996). A Little Bittern (*Ixobrychus minutus*) at Chambri Lake, Middle Sepik wetlands. *Muruk*, **8**, 27–8.

Jaensch, R. P. (2003). Records of northern Waterbirds in the Barkly wetlands, Northern Territory, 1993–2002. *Northern Territory Naturalist*, **17**, 20–30.

Jaensch, R. P. and Bellchambers, K. (1997). Waterbird conservation values of ephemeral wetlands of the Barkly Tableland, Northern Territory. Report to Australian Heritage Commission and Parks & Wildlife Commission of the Northern Territory.

Jaensch, R. P., Vervest, R. M. and Hewish, M. J. (1988). Waterbirds in nature reserves of south-western Australia: Reserve accounts. Royal Australasian Ornithologists' Union *Report* **30**.

Jamdar, N. (1984). Unusual plumage in a Cattle Egret *Bubulcus ibis coromandus* (Boddaert). *Journal of the Bombay Natural History Society*, **81**, 187.

Jamdar, N. and Srivastava, K. (1990). Black Bittern *Ixobrychus flavicollis* (Latham), an addition to the avifauna of Maharashtra. *Journal of the Bombay Natural History Society*, **87**, 289.

James, D. (1991). First breeding record of Squacco and Night Herons in Arabia and other interesting breeding records for near Riyadh, Saudi Arabia. *Phoenix*, **8**, 3–4.

James, R. D. (1984). Ontario Bird Records Committee Report for 1983. *Ontario Birds*, **2**, 53–65.

James, S. L. (1988). Vagrant White-backed and Rufous-bellied Heron in eastern Zimbabwe. *Honeyguide*, **34**, 120.

Janzen, P., Seibel, D., Rintoul, D., Rader, M., Flowers, T., Rucker, R., McHugh, M., Patti, S., and Gress, B. (2000). 1999 report of the Kansas Bird Records Committee. *Kansas Ornithological Society Bulletin*, **51**, 22–4.

Jeanmonod, J. and Roulin, A. (1989). Nidification terrestre d'un couple de herons cendres. *Nos Oiseaux*, **40**, 173–8.

Jebekk, R. (1996). Siste nytt—uvanlige fugleobservasjoner. *Var Fuglefauna*, **19**, 40–1.

Jenni, D. A. (1969). A study of four species of herons during the breeding season at Lake Alice, Alachua County, Florida. *Ecological Monographs*, **39**, 245–70.

Jepson, P. (1987). Recent reports. *Oriental Bird Club Bulletin*, **6**, 36–40.

Jewett, S. O. (1997). *A white heron*. Candlewick Press, Cambridge.

Jobanek, G. A. (1988). Early history of the Green-backed Heron in Oregon. *Oregon Birds*, **14**, 43–7.

Johnson, R. and Eddie, C. (2000). A Pied Heron at Roma, south central Queensland. *Sunbird*, **30**, 42–4.

Johnson, W. P., Garrettson, P. R., and Rohwer, F. C. (1996). Confirmation of kleptoparasitism by American White Pelicans on Great Blue Herons. *Colonial Waterbirds*, **19**, 262–3.

Johnston, S. N. (1989). First sight record of the common bittern in Malawi. *Nyala*, **14**, 49–50.

Jonasi, G. (1994). White-backed Night Heron breeds at Lake Chivero. *Honeyguide*, **40**, 173.

Jones, H. L. (2001a). Central America. *North American Birds*, **55**, 238–40.

Jones, H. A. (2001b). Central America. *North American Birds*, **55**, 112–5.

Jones, H. L. (2002). Central America. *North American Birds*, **56**, 693–710.

Jongejan, W. (1986). [Traditional reaping activities are a threat to the Purple Heron.] *Vogeljaar*, **34**, 289–90.

Juarez L. C. and Dickerman, R. W. (1972). Nestling development of Boat-billed Herons (*Cochlearius cochlearius*) at San Blas, Nayarit, Mexico. *Wilson Bulletin*, **84**, 456–68.

Julin, K. R. (1986). Decline of second-growth Douglas-fir in relation to Great Blue Heron nesting. *Northwest Science*, **60**, 201–5.

Kahindi, O. (1996). Black-headed Heron devours Laughing Dove. *Kenya Birds*, **5**, 38.

Kahl, M. P. (1971). Some observations on the behavior of Whistling Herons. *Wilson Bulletin*, **83**, 302–3.

Kaiser, M. S. and Fritzell, E. K. (1984). Effects of river recreationists on Green-backed Heron behavior. *Journal of Wildlife Management*, **48**, 561–7.

Kaiser, M. S., Peter, H. U., and Gebauer, A. (1988). Kuhreiher, *Ardeola ibis* (L.) in der Antarktis. *Beitraege Zur Vogelkunde*, **34**, 202–3.

Kaiser, M. S. and Reid, F. A. (1987). A comparison of Green-backed Heron nesting in two freshwater ecosystems. *Colonial Waterbirds*, **10**, 78–83.

Kanai, Y. (1990). [A colony of Great Egrets *Egretta alba* in the Port of Tokyo.] *Strix*, **9**, 218–20.

Kasoma, P. M. B. (1995). Kleptoparasitic attacks on three heron species (Ardeidae) in Queen Elizabeth National Park, Uganda. *African Journal of Ecology*, **33**, 291–3.

Kasoma, P. M. B. (2000). Diurnal activity patterns of three heron species in Queen Elizabeth National Park, Uganda. *Ostrich*, **71**, 127–30.

Kasparek, M. (1986). On the occurrence of the Bittern, *Botaurus stellaris*, in Turkey. *Zoology in the Middle East*, **1**, 33–41.

Kasparek, M., Robel, D., and Siering, M. (1994). Bemerkungen su einigen vogelarten des Jemen. *Zoology In the Middle East*, **10**, 17–22.

Katzir, G. and Intrator, N. (1987). Striking underwater prey by a Reef Heron, *Egretta gularis schistacea*. *Journal of Comparative Physiology A*, **160**, 517–23.

Katzir, G. and Martin, G. R. (1998). Visual field in the Black-Crowned Night Heron *Nycticorax nycticorax*—nocturnality does not result in owl-like features. *Ibis*, **140**, 157–62.

Katzir, G., Strod, T., Schechtman, E., Hareli, S., and Arad, Z. (1999). Cattle Egrets are less able to cope with light refraction than are other herons. *Animal Behaviour*, **57**, 157–62.

Kayser, Y. (1994). Premier cas d'hivernage du heron crabier *Ardeola ralloides* en Camargue. *Faune de Provence*, **15**, 80.

Kayser, Y. (1995). L'epervier d'Europe (*Accipiter nisus*) predateur de l'aigrette garzette (*Egretta garzetta*) et du heron crabier (*Ardeola ralloides*) en Camargue, sud de la France. *Nos Oiseaux*, **43**, 28–9.

Kayser, Y. (1996). Presence du grand-duc d'Europe (*Bubo bubo*) et predation sur une colonie d'ardeides arboricoles en 1994 et 1995 en Camargue (France). *Nos Oiseaux*, **43**, 445–52.

Kayser, Y., Hafner, H., and Massez, G. (1998). Survey of the number of booming males of Eurasian Bitterns *Botaurus stellaris* in the Camargue in 1996. *Alauda*, **66**, 97–102.

Kayser, Y., Pineau, O., Hafner, H., and Walmsley, J. (1994). La nidification de la grande aigrette *Egretta alba* en Camargue. *Ornithos*, **1**, 81–2.

Kazantzidis, S. and Goutner, V. (1996). Foraging ecology and conservation of feeding habitats of Little Egrets (*Egretta garzetta*) in the Axios River Delta, Macedonia, Greece. *Colonial Waterbirds*, **19**, **Special Publication 1**, 115–21.

Kazantzidis, S., Hafner, H., and Goutner, V. (1996). Comparative breeding ecology of the Little Egret (*Egretta g. garzetta*) in the Axios Delta (Greece) and the Camargue (France). *Revue d'Ecologie la Terre et la Vie*, **51**, 313–27.

Kazmierczak, K. (2000). *A field guide to the birds of India, Sri Lanka, Pakistan, Nepal, Bhutan, Bangladesh and the Maldives*. OM Book Service. New Delhi, India.

Keighley, M. and Hall, A. (1995). Grey Heron capturing hirundine in flight. *British Birds*, **88**, 423.

Kelly, J. F., Gawlik, D. E., and Kieckbusch, D. K. (2003). An updated account of wading bird foraging behavior. *Wilson Bulletin*, **115**, 105–7.

Kelly, J. P., Pratt, H. M., and Greene, P. L. (1993). The distribution, reproductive success, and habitat characteristics of heron and egret breeding colonies in the San Francisco Bay area. *Colonial Waterbirds*, **16**, 18–27.

Kennedy, R. S., Mayer, S., and Fisher, T. H. (1984). Notes on Philippine birds, 3. First sight records of the Javan Pond-Heron *Ardeola speciosa* from the Philippines. *Bulletin of the British Ornithologists' Club*, **104**, 102–3.

Kennedy, R. S., Gonzales, P. C., Dickinson, E. C., Miranda, Jr., H. C., and Fisher, Y. H. (2000). *A guide to the birds of the Philippines.* Oxford University Press, Oxford.

Kent, D. M. (1983). In-place hopping of a feeding Little Blue Heron. *Florida Field Naturalist*, **11**, 54–5.

Kent, D. M. (1986a). Behavior, habitat use, and food of three egrets in a marine habitat. *Colonial Waterbirds*, **9**, 25–30.

Kent, D. M. (1986b). Flushing lores of a male Least Bittern. *Florida Field Naturalist*, **14**, 49–51.

Kent, D. M. (1986c). Foraging efficiency of sympatric egrets. *Colonial Waterbirds*, **9**, 81–5.

Kent, D. M. (1987). Effects of varying behavior and habitat on the striking efficiency of egrets. *Colonial Waterbirds*, **10**, 115–19.

Kent, D. M. (1991). Differences in foraging ability of adult and juvenile Snowy Egrets observed in the Saugus River marsh. *Bird Observer*, **19**, 133–6.

Kerr, J. G. (1950). *A naturalist on the Gran Chaco*. Cambridge University Press, Cambridge, UK.

Kersten, M., Britton, R. H., Dugan, P. J., and Hafner, H. (1991). Flock feeding and food intake in Little Egrets: The effects of prey distribution and behaviour. *Journal of Animal Ecology*, **60**, 241–52.

Kibbe, D. P. and Boise, C. M. (ed.) (1984). The spring migration. March 1–May 31, 1984. Niagara-Champlain region. *American Birds*, **38**, 907–9.

Kilham, L. (1980). Association of Great Egret and White Ibis. *Journal of Field Ornithology*, **51**, 73–4.

Kilham, L. (1984). American Crows robbing Great Egrets and White Ibis of large, eel-like salamanders. *Colonial Waterbirds*, **7**, 143–5.

Kim, J. S., Lee, D. P., and Koo, T. H. (1997). Present wintering situation and ecology of the Night Heron *Nycticorax nycticorax*. *Korean Journal of Ornithology*, **4**, 7–16.

King, B. (1985). Green-backed Herons feeding on decaying acorns. *Florida Field Naturalist*, **13**, 99.

King, D. T. and Leblanc, D. (1995). Foraging behaviors of Snowy Egrets (*Egretta thula*) and Yellow-crowned Night-herons (*Nyctanassa violacea*) in south Louisiana. *Colonial Waterbirds*, **18**, 224–5.

Kingery, H. E. (1983a). The nesting season. June 1–July 31, 1983. Mountain west. *American Birds*, **37**, 1010–13.

Kingery, H. E. (1983b). The Spring migration. March 1–May 31, 1983. Mountain west. *American Birds*, **37**, 895–8.

Kingery, H. E. (1991). The Autumn migration. August 1–November 30, 1990. Mountain West Region. *American Birds*, **45**, 131–4.

Kingery, H. E. (1993). The Autumn migration. August 1–November 30, 1992. Mountain West region. *American Birds*, **47**, 123–7.

Kingston, B. L. (1991). Grey Herons stealing fish from cormorants and from bittern. *British Birds*, **84**, 58–9.

Kinzelbach, R. K. (1986). New records of Goliath Heron, Yellow-billed Stork, Blyth's Reed Warbler and Clamorous Reed Warbler in the Middle East. *Ornithological Society of the Middle East Bulletin*, **17**, 13–17.

Kirk, A. (1995). First confirmed breeding of the Least Bittern in Inyo County, California. *Western Birds*, **26**, 165–6.

Kirsch, B. (1998). Grey Heron (62) on the high seas. *Promerops*, **235**, 18.

Kirtland, C. A. E. and Rogers, E. P. (1997). First record and nesting of White-crested Tiger-Heron *Tigriornis leucolophus* in The Gambia. *Bulletin of the African Bird Club*, **4**, 105–6.

Klaptste, J. (1976). Unusual feeding method of Plumed Egrets. *Australian Bird Watcher*, **6**, 179–80.

Klaptste, J. (1987). Carrion eating by herons (Ardeidae). Australian Bird Watcher 14,

Klemm, W. and Pankratius, U. (1987). Purpurreiherbrut in Siebenburgen/Rumanien. *Falke*, **34**, 219–23.

Knief, W. (1986). Zur brutbestandsentwicklung des Graureihers (*Ardea cinerea*) in Schleswig-Holstein von 1984–1986. *Corax*, **12**, 47–53.

Knief, W. and Drenckhahn, D. (1984). Die bestandsentwicklung des graureihers (*Ardea cinerea*) in Schleswig-Holstein von 1974–1983, mit einem nachtrag zum brutvorkommen vor 1974. *Corax*, **10**, 334–54.

Knight, R. L. (1994). Green Herons feed by diving into deep water. *Migrant*, **65**, 47–8.

Knysh, N. P. and Sypko, A. V. (1997). [Materials on distribution and ecology of the Purple Heron in Sumy Region.] *Berkut*, **6**, 23–8.

Kobayashi, S. (2000). [Records of Black-crowned Night Heron, Little Egret and Grey Heron nesting on the ground.] *Strix*, **18**, 121–6.

Kobayashi, S. and Hironaka, T. (1992). Nesting by the Green-backed Heron *Butorides striatus* on a TV-antenna. *Strix*, **11**, 345–7.

Koes, R. F. and Taylor, P. (1991). The Summer season. June 1–July 31, 1991. Prairie Provinces region. *American Birds*, **45**, 1128–30.

Kolbe, U. and Neumann, J. (1990). Der rallenreiher in der DDR. *Falke*, **37**, 182–7.

Kolland, H. (1996). Uberwinterung einer grossen Rohrdommel (*Botaurus stellaris*) bei Hohentauern (Niedere Tauern, Steiermark). *Egretta*, **38**, 132–3.

Kolomiitsev, N. P. (1986). [Some remarkable accidental birds in the environs of Lazovsky Reserve.] *Trudy Zoologicheskogo Instituta*, **150**, 84–5.

Komarov, Y. E. and Lipkovich, A. D. (1986). [Breeding of the Night Heron (*Nycticorax nycticorax*) in central Caucasus.] *Ornitologiya*, **21**, 133–4.

Kopij, G. (1996a). A mixed heronry at Willem Pretorius Game Reserve, Free State Province. *Mirafra*, **13**, 63–7.

Kopij, G. (1996b). Location and structure of Cattle Egret *Bubulcus ibis* nests in a heronry at Wolwekop, Free State Province. *Mirafra*, **13**, 68–70.

Kopij, G. (1997). Timing of colony occupation, clutch size and breeding success in the Cattle Egret *Bubulcus ibis* related to nest location in a South African heronry. *Acta Ornithologica*, **32**, 169–74.

Kopij, G. (1999a). Food of the Cattle Egret (*Bubulcus ibis*) in South African grassland. *Vogelwarte*, **40**, 98–109.

Kopij, G. (1999b). Breeding success in the Cattle Egret in relation to clutch size. *South African Journal of Wildlife Research*, **29**, 112.

Korvin, J. (1987). [Solitary nesting of *Ardea cinerea* near Hlohoveo (Trnava District).] *Tichodroma*, **1**, 137.

Kosugi, K. (1993). [Records of egrets and herons from Rishiri Island, Northern Hokkaido.] *Japanese Journal of Ornithology*, **41**, 49–51.

Kovacs, G. (2002). [Unusual foraging technique of Little Egret (*Egretta garazetta*) in lake bed]. *Aquila*, **107–8**, 104–5, 119.

Kral, B. and Figala, J. (1966). Breeding biology of the Purple Heron, *Ardea purpurea*, in the Velky and Maly Tisy Reserve. *Zoologicke Listy*, **15**, 33–46.

Kramer, A. (1984). Zum einfluss des graureihers *Ardea cinerea* an kleineren fliessgewassern. *Ornithologische Beobachter*, **81**, 149–58.

Kramer, D. L., Manley, D., and Bourgeois, R. (1983). The effects of respiratory mode and oxygen concentration on the risk of aerial predation in fishes. *Canadian Journal of Zoology*, **61**, 653–65.

Krebs, E. A. (1991). Reproduction in the Cattle Egret (*Bubulcus ibis ibis*): The function of breeding plumes. MS Thesis, McGill University, Canada.

Krebs, J. (1974). Colonial nesting and social feeding as strategy for exploiting food resources in the Great Blue Heron (*Ardea herodias*). *Behaviour*, **51**, 99–134.

Krebs, J. (1978). Colonial nesting in birds, with special reference to the ciconiiformes. In *Wading birds* (eds. A. Sprunt, A., J. Ogden, and S. Winkler), pp. 299–314. National Audubon Society, New York.

KKieuziger, J. and Achenbach, E. L. (1998). [Grey Heron (*Ardea cinerea*) swallows a juvenile Great Crested Grebe (*Podiceps cristatus*).] *Vogelwarte*, **39**, 301–2.

Kriedemann, K. (1991). Die Flugaktivitat des Graureihers (*Ardea cinerea*) zwischen Kolonie und Nahrungshabitaten im Tagesverlauf—ein Beitrag zum Raum-Zeit-System. *Ornithologische Mitteilungen*, **43**, 159–65.

Krotoski, T. (1983). Melanistyczny okaz czapli siwej (*Ardea Cinerea*). *Notatki Ornitologiczne*, **24**, 102–3.

Kreuziger. J. and Achenbach, E. L. (1998). Grey Heron (*Ardea cinerea*) swallows a Great Crested Grebe (*Podiceps cristatus*). *Vogelwarte* **39**, 301– 302.

Kruuk, H., Kanchanasaka, B., O'Sullivan, S., and Wanghongsa, S. (1993). Kingfishers *Halcyon capensis* and *Alcedo atthis* and Pond Heron *Ardeola bacchus* associating with otters *Lutra perspicillata*. *Natural History Bulletin of the Siam Society*, **41**, 67–8.

Kurosawa, R. and Higuchi, H. (1993). [Bait-fishing of the Green-backed Heron *Ardeola striata* in different areas of Japan and other countries.] *Strix*, **12**, 1–21.

Kurz, A. and Schmidt, K. (1994). Die Entwicklung des Graureiher (*Ardea cinerea*)—Bestandes in Sudthuringen von 1973 bis 1994. *Anzeiger des Vereins Thueringer Ornithologen*, **2**, 73–7.

Kushlan, J. A. (1972). Aerial feeding in the Snowy Egret. *Wilson Bulletin*, **84**, 199–200.

Kushlan, J. A. (1973a). Bill-vibrating: A prey attracting behavior of the Snowy Egret, *Leucophoyx thula*. *American Midland Naturalist*, **89**, 509–12.

Kushlan, J. A. (1973b). Black-crowned Night Heron diving for prey. *Florida Field Naturalist*, **1**, 27–8.

Kushlan, J. A. (1973c). Least Bittern nesting colonially. *Auk*, **90**, 685–6.

Kushlan, J. A. (1976a). Feeding behavior of North American herons. *Auk*, **93**, 86–94.

Kushlan, J. A. (1976b). Wading bird predation in a seasonally fluctuating pond. *Auk*, **93**, 464–76.

Kushlan, J. A. (1977). The significance of plumage color in the formation of feeding aggregations of ciconiiforms. *Ibis*, **119**, 361–4.

Kushlan, J. A. (1978a). Commensalism in the Little Blue Heron. *Auk*, **95**, 677–81.

Kushlan, J. A. (1978b). Feeding ecology of wading birds. In *Wading birds* (eds. A. Sprunt, IV, J. C. Ogden, and S. Winckler), pp. 249–97. National Audubon Society Research Report No. 7. National Audubon Society, New York.

Kushlan, J. A. (1978c). Nonrigorous foraging by robbing egrets. *Ecology*, **59**, 649–53.

Kushlan, J. A. (1983). Pair formation behavior of the Galapagos Lava Heron. *Wilson Bulletin*, **95**, 118–21.

Kushlan, J. A. and Hafner, H. (2000). *Heron conservation*. Academic Press, London, UK.

Kushlan, J. A., Hancock, J. A., Pinkowski, J., and Pinkowska, B. (1982). Behavior of Whistling and Capped Herons in the seasonal savannas of Venezuela and Argentina. *Condor*, **84**, 255–60.

Kushlan, J. A., Morales, G., and Frohring, P. C. (1985). Foraging niche relations of wading birds in tropical wet savannas. In *Neotropical ornithology* (eds. P. A. Buckley, M. S. Foster, E. S. Morton, R. S. Ridgely, and F. G. Buckley), pp. 663–81. Ornithological Monographs 36. American Ornithologists' Union, Washington.

L'Arrivee, L. P. and Blokpoel, H. (1990). Seasonal distribution and site tenacity of Black-crowned Night-herons, *Nycticorax nycticorax*, banded in Canada. *Canadian Field-Naturalist*, **104**, 534–9.

Labzyuk, V. I. and Glushenko, Y. N. (1984). The Cinnamon Bittern (*Ixobrychus cinnamomeus* Ciconiiformes, Ardeidae), new species in the USSR fauna. *Ornitologiya*, **19**, 181.

Lancaster, D. A. (1970). Breeding behavior of the Cattle Egret in Colombia. *Living Bird*, **9**, 167–94.

Land, H. C. (1970). Birds of Guatemala. Livingston, Wynnewood.

Lane, S. J. and Fujioka, M. (1998). The impact of changes in irrigation practices on the distribution of foraging egrets and herons (Ardeidae) in the rice fields of central Japan. *Biological Conservation*, **83**, 221–30.

Lange, M. (1995). *Texas coastal waterbird colonies: 1973–1990 census summary, atlas, and trends*. U.S. Fish and Wildlife Service and Texas Parks and Wildlife Department, Arlington.

Langley, C. H. (1983). Biology of the Little Bittern in the southwestern Cape. *Ostrich*, **54**, 83–94.

Langrand, O. (1990). *Guide to the birds of Madagascar*. Yale University Press, New Haven.

Lansdown, P. (1985). Identification pitfalls and assessment problems. 8. Purple Heron *Ardea purpurea*. *British Birds*, **78**, 97–102.

Lansdown, P. (1991). Identification pitfalls and assessment problems. 10. Night Heron *Nycticorax nycticorax*. *British Birds*, **84**, 145–8.

Lansdown, P. (2000). Separation of American and Great Bitterns. *British Birds*, **93**, 132–4.

Lansdown, R.V. (1986). Heron studies in Malaysia & Singapore. *Interwader*, **7**, 10.

Lansdown, R.V. (1987a). *The feeding ecology of the larger herons in the Kota Belud Bird Sanctuary. With reference to preparation of management proposals.* World Wildlife Fund Malaysia, Kuala Lumpur.

Lansdown, R.V. (1987b). Recent extensions in breeding range of the Yellow Bittern *Ixobrychus sinensis*. *Forktail*, **3**, 61–3.

Lansdown, R.V. (1988). Some calls, displays and associated morphology of the Cinnamon Bittern (*Ixobrychus cinnamomeus*) and their possible functions. *Colonial Waterbirds*, **11**, 308–10.

Lansdown, R.V. (1989). Displays of the Sumatran Heron *Ardea sumatrana*. *Colonial Waterbirds*, **12**, 113–14.

Lansdown, R.V. (1990). Little-known oriental bird: Chinese Egret. *Bulletin of the Oriental Bird Club*, **11**, 27–30.

Lansdown, R.V. (1992). The status and conservation of the Sumatran Heron *Ardea sumatrana*. *Heron Conservation Newsletter*, **5**, 11–12.

Lansdown, R.V., Mundkur, T., and Young, L. (2000). Herons in east and southeast Asia. In *Heron conservation* (eds. J. A. Kushlan and H. Hafner), pp. 73–98. Academic Press, San Diego.

Lansdown, R.V. and Rajanathan, R. (1993). Some aspects of the ecology of *Ixobrychus* bitterns nesting in Malaysia ricefields. *Colonial Waterbirds*, **16**, 98–101.

Laubhan, M. K. and Reid, F. A. (1991). Characteristics of Yellow-crowned Night-Heron nests in lowland hardwood forests of Missouri. *Wilson Bulletin*, **103**, 486–91.

Laubhan, M. K., Rundle, W. D., Swartz, B. I., and Reid, F. A. (1991). Diurnal activity patterns and foraging success of Yellow-crowned Night-Herons in seasonally flooded wetlands. *Wilson Bulletin*, **103**, 272–7.

Laurenti, S. (1986). Sul comportamento alimentare di airone cenerino *Ardea cinerea*. *Gli Uccelli d'Italia*, **11**, 109–10.

Laurenti, S. (1990). Collaborazione interspecifica tra falco pescatore *Pandion haliaetus* e airone cenerino *Ardea cinerea*. *Gli Uccelli d'Italia*, **15**, 104–5.

le Mair, W., Waaire, E. K., and van Rijck H (1995). Bittern along the motorway. *Vogeljaar*, **43**, 135.

Leader, P. J. (1995). Crested Mynah kleptoparasitising Chinese Ponds Heron chicks. *Hong Kong Bird Report*, **1995**, 253–4.

Leader, P. J. (1996). Field identification of Yellow, Schrenck's and Chestnut Bitterns. *Hong Kong Bird Report*, **1995**, 218–23.

Leader, P. J. (1999). Social and migratory behaviour in Great Bittern. *Hong Kong Bird Report*, **1997**, 156–8.

Lee, K. S. (1998). White-eared Night Heron *Gorsachius magnificus* on sale in a Nanning market. *Bulletin of the Oriental Bird Club*, **28**, 13, 42.

Leibl, F. and Hagemann, P. (1997). Erfolgrieche Bruten von Seiden- (*Egretta garzetta*), Nacht- (*Nycticorax nycticorax*) und Purpurreiher (*Ardea purpurea*) im ostbayerischen Donautal. *Ornithologischer Anzeiger*, **36**, 45–9.

Lekuona, J. M. (1999). Food and foraging activites of Grey Herons (*Ardea cinerea*) in a coastal area during the breeding season. *Foila Zoologica*, **48**, 123–30.

Lekuona, J. M. (2002a). Food intake, feeding behaviour and stock losses of cormorants, and Grey Herons, *Ardea cinerea*, at a fish farm in Arcachon Bay (Southwest France) during breeding and non-breeding season. *Folia Zoologica*, **51**, 23–4.

Lekuona, J. M. (2002b). Kleptoparasitism in wintering Grey Heron *Ardea cinerea*. *Folia Zoologica*, **51**, 215–220.

Lekuona, J. M. and Artazcoz, A. (2000). Ecologia trofica y distribucion de Aves y dormideros de garcilla bueyera (*Bubulcus ibis*) en Navarra durante el periodo invernal. *Anuario Ornitologico De Navarra*, **5**, 8–15.

Lekuona, J. M. and Campos, F. (1993). Agressions et cannibalisme chez des poussins de heron cendre *Ardea cinerea*. *Alauda*, **61**, 198.

Lekuona, J. M. and Campos, F. (1995). Le succès de reproduction du Héron Cendré *Ardea cinerea* dans le Bassin d'Arcachon. *Alauda*, **63**, 179–83.

Lekuona, J. M. and Campos, F. (1996a). Descripcion de un caso de hibridacion entre la garza real *Ardea cinerea* y la garza imperial *Ardea purpurea* en los arrozales de Arguedas (Navarra). *Anuario Ornitologico de Navarra*, **3**, 64.

Lekuona, J. M. and Campos, F. (1996b). Composicion de la dieta y comportamiento trofico de la garza real (*Ardea cinerea*) y de la garza imperial (*Ardea purpurea*) en el Rio Aragon durante la epoca reproductora. *Anuario Ornitologico de Navarra*, **3**, 20–8.

Lekuona, J. M. and Campos, F. (1996c). Seasonal dynamics of Grey Herons on a large river in Spain. *Colonial Waterbirds*, **19**, 214–19.

Lekuona, J. M. and Campos, F. (1998). Extra-pair copulations of Grey Heron *Ardea cinerea* in Arcachon Bay (SW France). *Miscellania Zoologica (Barcelona)*, **21**, 73–80.

Lewis, S. E. and Timm, R. M. (1991). Predation on nestling Bare-throated Tiger-Herons by a Great Black-Hawk. *Ornitologia Neotropical*, **2**, 37.

Ley, A. J. (1991). Pacific heron taking prey from hovering flight. *Australian Bird Watcher*, **14**, 34.

Li, X., Ma, Y., Ma, Y., and Gao, Z. (1989). Study on the band and migration of Grey Heron. *Chinese Wildlife*, **1989**, 51–3.

Lindblad, T. (1995). Simmande rördrom. *Anser*, **34**, 54.

Linnaeus, C. (1956). *Caroli Linnaei Systema naturae: Regnum animale*. Photographic facsimile of the first volume of the tenth edition (1758). British Museum of Natural History, London.

Lippins, L. and Burggraeve, G. (1983). Hybridization de l'aigrette garzette (*Egretta garzetta*) et du heron cendre (*Ardea cinerea*) en Belgique. *Gerfaut*, **73**, 303–311.

Lippins, L. and Burggraeve, G. (1985). Gemendg broedgeval van een Klein Zilverreiger *Egretta garzetta* en een Blauwe eiger *Ardea cinerea* in het Zwin te Knokke. *Wielewaal*, **51**, 310–314.

Lish, J. W. (1983). Comparative eye morphology of ciconiifrom birds. *Dissertation Oklahoma State University, Abstracts, B, Science and Engineering*, **43**, 3157–8.

Litvinenko, N. M. (1983). Nesting of Gray Heron *Ardea cinerea* on sea islands of South Primorye USSR. *Journal of the Yamashina Institute for Ornithology*, **14**, 220–31.

Litvinenko, N. M. and Shibaev, Y. V. (1999). The Chinese Egret *Egretta eulophotes*—a new breeding species for avifauna of Russia. *Russkii Ornitologicheskii Zhurnal Ekspress Vypusk*, **70**, 7–9.

Litvinenko, N. M. and Shibaev, U. V. (2000). Importance of Furugelm Island in the Sea of Japan for wetland birds: the first record of a breeding colony of the Chinese Egret *Egretta eulophotes*. *Oryx*, **34**, 335–7.

Lo, P. L. (1984). Breeding of the White-faced Heron in coastal Manawatu. *Notornis*, **31**, 95.

Lo, P. L. (1991). Diet of the White-faced Heron on Manawatu pastures. *Notornis*, **38**, 63–71.

Lo, P. L. and Fordham, R. A. (1986). Seasonal and diurnal time budgets and feeding intensity of the White-faced Heron in pasture. *Notornis*, **33**, 233–45.

Lock, L. and Cook, K. (1998). The Little Egret in Britain: A successful colonist. *British Birds*, **91**, 273–80.

Lockhart, P. (1996). First sighting of a probable Rufous-bellied Heron in the western Cape. *Promerops*, **223**, 13.

Loison, M. (1991). Comportement alimentaire d'un grand butor (*Botaurus stellaris*). *Aves*, **28**, 39.

Lombardini, K., Bennetts, R. E., and Tourenq, C. (2001). Foraging success and foraging habitat use by Cattle Egrets and Little Egrets in the Camargue, France. *Condor*, **103**, 38–44.

Long, T., *et al.* (1999). Investigation on egret dwelling on the mottled bamboo garden in compact groups in Chengdu City. *Sichuan Journal of Zoology*, **18**, 84–6.

Lopatin, V. V., Carpov, F. F., and Kleymenov, S. V. (1992). Biology of the Little Bittern (*Ixobrychus minutus*) in southeastern Kazakhstan. *Russkii Ornitologicheskii Zhurnal*, **1**, 227–34.

Lopez-Ornat, A. and Ramo, C. (1992). Colonial water-bird populations in the Sian Ka'an Biosphere Reserve (Quintana Roo, Mexico). *Wilson Bulletin*, **104**, 501–15.

Lorenz, K. (1938). A contribution to the comparative sociology of colonial-nesting birds. In *Proceedings of the eighth International Ornithological Congress*. Oxford, July, 1934 (ed. F. C. R. Jourdain), pp. 207–18. Oxford University Press, Oxford.

Lotem, A., Schechtman, E., and Katzir, G. (1991). Capture of submerged prey by Little Egrets (*Egretta garzetta garzetta*): strike depth, strike angle, and the problem of light refraction. *Animal Behaviour*, **42**, 341–6.

Lovell, H. B. (1959). Baiting of fish by a Green Heron. *Wilson Bulletin*, **70**, 280–1.

Lowe, F. A. (1954). *The Heron*. Collins, London, UK.

Lowe, F. (1954). Grey Heron parents attending grounded nestlings. *British Birds*, **77**, 614.

Lowe, K. W. (1983). Feeding behaviour and diet of the White-faced Heron *Ardea novaehollandiae* in Westernport Bay, Victoria. *Corella*, **7**, 101–8.

Lowe, K. W. (1989). Notes on the breeding of the Pacific Heron *Ardea pacifica* near Balranald, New South Wales. *Corella*, **13**, 88–9.

Lowen, J. C., Clay, R. P., Mazar Barnett, J., Madrono, N. A., Pearman, M., Lopez Lanus, B., Tobias, J. A., Liley, D. C., Brooks, T. M., Esquivel, E. Z., and Reid, J. M. (1997). New and noteworthy observations on the Paraguayan avifauna. *Bulletin of the British Ornithologists' Club*, **117**, 275–93.

Lowther, P. E., and Paul, R. T. (2002). Reddish Egret, *Egretta rufescens*. In *The birds of North America*. No. 633. (eds. A. Poole and F. Gill), pp. 1–18, The Birds of North America, Inc. Philadelphia, PA.

Loyn, R. H. (1985). Hovering, aerial feeding and swimming by the Eastern Reef Egret *Egretta sacra* and other herons. *Australian Bird Watcher*, **11**, 133–5.

Loyn, R. H. (1987). Foraging behaviour of herons—a correction. *Australian Bird Watcher*, **12**, 68.

Lu, X. (1987). [The thermoregulation of nestlings of Yellow Bittern.] *Sichuan Journal of Zoology*, **6**, 22–3.

Lucking, R. (1995). Cinnamon Bittern *Ixobrychus cinnamomeus* in Seychelles, first for the Afro-Malagasy region. *Bulletin of the African Bird Club*, **2**, 107–8.

Lunn, J. (1992). Bittern perching and adopting protective stance in a tree. *Magpie*, **4**, 49.

Lyles, A. M. (2000). Captive populations. In *Heron conservation* (eds. J. A. Kushlan and H. Hafner), pp. 293–310. Academic Press, San Diego.

Ma, S. (1990). [Study on the distribution pattern of population of the Yellow Bittern in breeding season.] *Acta Ecologica Sinica*, **10**, 362–6.

Ma, M., Cheung, H. F., Kwok, H. K., and Chow, L. K. (2001). A new record of bird in Xinjian—Chinese Pond Heron *Ardeola bacchus* (Bonaparte). *Journal of Arid Zone Research*, **18**, 76–7.

Maccarone, A. D. and Brzorad, J. N. (2002). Foraging patterns of breeding egrets at coastal and interior locations. *Waterbirds*, **24**, 1–7.

Maccarone, A. D. and Gress, R. (1993). A case of colonial nesting in the Green-backed Herons. *Transactions of the Kansas Academy of Science*, **96**, 225–6.

Maccarone, A. D. and Parsons, K. C. (1994). Factors affecting the use of a freshwater and an estuarine foraging site by egrets and ibises during the breeding season in New York City. *Colonial Waterbirds*, **17**, 60–8.

Mackereth, B. (1992). Sightings of White Heron at Suva. *Notornis*, **39**, 70–1.

MacKinnon, J. and Phillipps, K. (1993). *A field guide to the birds of Borneo, Sumatra, Java, and Bali, the Greater Sunda Islands*. Oxford University Press, Oxford.

MacKinnon, J. and Phillips, K. (2000). A field guide to the birds of China. Oxford University Press, Oxford.

Mackrill, E. J. (1987). Winter breeding of Night Heron in Europe. *Dutch Birding*, **9**, 124.

Maddock, M. (1986). Fledging success of egrets in dry and wet seasons. *Corella*, **10**, 101–7.

Maddock, M. (1988a). Seasonal climatic factors and egret breeding success. In *Proceedings International Symposium on Wetlands* (eds. B. Gilligan, M. Maddock, and K. McDonald), pp. 265–71. Shortland Wetlands Centre, Newcastle.

Maddock, M. (1988b). Field identification characteristics of nestling egrets. In *Proceedings International Symposium on Wetlands* (eds. B. Gilligan, M. Maddock, and K. McDonald), pp. 273–84. Shortland Wetlands Centre, Newcastle.

Maddock, M. (1989a). Colour and first age of breeding in Cattle Egrets as determined from wing-tagged birds. *Corella*, **13**, 1–8.

Maddock, M. (1989b). Bird in hand: Identification of nestling egrets (*Egretta* sp. and *Ardeola ibis*). *Corella* **13**, 24–6.

Maddock, M. (1990). Cattle Egret: south to Tasmania and New Zealand for the winter. *Notornis*, **37**, 1–23.

Maddock, M. (1991). Observations on the biology of the White-faced Heron *Ardea novaehollandiae*. *Corella*, **15**, 79–86.

Maddock, M. (1992). Foraging association between Royal Spoonbills, egrets and the White-faced Heron. *Australian Birds*, **25**, 61–4.

Maddock, M. (1993). Cycle of colour changes in Cattle Egrets *Ardeola ibis (coromandus)* in Australia determined from field observations of marked birds. *Corella*, **17**, 93–9.

Maddock, M. (1995). Response of Cattle Egrets to flood. *Australian Birds*, **29**, 1–10.

Maddock, M. (1999). Egret information sheet: Hunter Valley colonies (revised 1999). Shortland Wetlands Centre, Newcastle.

Maddock, M. (2000). Herons in Australasia and Oceana. In *Heron conservation*. (eds. J. A. Kushlan and H. Hafner), pp. 123–149. Academic Press, San Diego.

Maddock, M. and Baxter, G. S. (1991). Breeding success of egrets related to rainfall: A six-year Australian study. *Colonial Waterbirds*, **14**, 133–9.

Maddock, M. and Bridgeman, H. (1992). Cattle Egret migration and meteorological conditions. *Notornis*, **39**, 73–86.

Maddock, M. and Geering, D. (1993). Cattle Egret migration in south-eastern Australia and New Zealand: An update. *Notornis*, **40**, 109–22.

Maddock, M. and Geering, D. (1994). Range expansion and migration of the Cattle Egret. *Ostrich*, **65**, 191–203.

Madge, S. (1995). A record of Goliath Heron in Assam. *Bulletin of the Oriental Bird Club*, **21**, 47.

Madge, S. C. (1983). Comments on two bittern records from Caradon district. *Birds in Cornwall*, **53**, 96.

Mallord, J. W., Tyler, G. A., Gilbert, G., and Smith, K. W. (2000). The first case of successful double brooding in the Great Bittern *Botaurus stellaris*. *Ibis*, **142**, 672–5.

Mann, C. F. (1987). Notable bird observations from Brunei, Borneo. *Forktail*, **3**, 51–6.

Mann, C. F. (1989). More notable bird observations from Brunei, Borneo. *Forktail*, **5**, 17–22.

Manson, S. A. M. (1985). Systematic list. *Caithness Bird Report*, **1985**, 3–41.

Marchant, S. (1988). Nesting behaviour of the Pacific Heron *Ardea pacifica*. *Australian Birds*, **21**, 61–5.

Marchant, S. and Higgins, P. J. (1990). *Handbook of Australian, New Zealand, and Antarctic birds. Vol. 1., Ratites to Ducks*. Oxford University Press, Oxford.

Marin A. M. (1989). Notes on the breeding of Chestnut-bellied Herons (*Agamia agami*) in Costa Rica. *Condor*, **91**, 215–17.

Marin A. M. and Sheldon, F. H. (1987). Some new nesting records of padi-dwelling birds in Sabah, East Malaysia (North Borneo). *Bulletin of the British Ornithologists' Club*, **107**, 23–5.

Marion, L. (1981). Dynamique d'une population de héron cendré, *Ardea cinerea*, example de la plus grande colonie d'Europe, le Lac de Grand-Lieu. *Oiseau et la Revue Française d'Ornithologie*, **50**, 219–61.

Marion, L. (1984). Mise en evidence par biotelemetrie de territoires alimentaires individuels chez un oiseau colonial, le heron cendre *Ardea cinerea*. Mecanisme de repartition et de regulation des effectifs des colonies de herons. *Oiseau et la Revue Francaise d'Ornithologie*, **54**, 1–78.

Marion, L. (1988). Evolution des strategies démographiques, alimentaires et d'utilsation de l'espece chex le Héron cendré en France: importance des contraines énergétiques et humaines. Thése doctorat d'etat, University de Rennes, France.

Marion, L. (1989). Territorial feeding and colonial breeding are not mutually exclusive: The case of the Grey Heron (*Ardea cinerea*). *Journal of Animal Ecology*, **58**, 693–710.

Marion, L. (1997). Évolution des effectifs nicheurs et de la répartition des hérons coloniaux en France entre 1974 et 1994. *Alauda*, **65**, 86–8.

Marion, L. (2000). Aquaculture. In *Heron conservation* (eds. J. A. Kushlan and H. Hafner), pp. 269–92. Academic Press, San Diego.

Marion, L., Brugiere, D., and Grisser, P. (1993). Invasion de hérons garde-boeufs *Bubulcus ibis* nicheurs en France in 1992. *Alauda*, **61**, 129–36.

Marion, L. and Marion, P. (1987a). Conséquences de la protection du héron cendré sur sa dynamique de population et sur ses strategies d'occupation de l'espace en France. *Revue d'Ecologie: la Terre et la Vie*, **Supplement 4**, 261–70.

Marion, L. and Marion, P. (1994). Premières nidifications réussies de la grande aigrette *Egretta alba* en France, au Lac de Grand-Lieu. *Alauda*, **62**, 149–52.

Marion, L., Ulenaers, P., and van Vessem, J. (2000). Herons in Europe. In *Heron conservation* (eds. J. A. Kushlan and H. Hafner), pp. 1–31. Academic Press, San Diego.

Marion, P. and Marion, L. (1987b). Impact de la prédation exercée par le héron cendré sur une pisciculture. *Revue d'Ecologie: la Terre et la Vie*, **Supplement 4**, 115–16.

Marquiss, M. and Leitch, A. F. (1990). The diet of Gray Herons *Ardea cinerea* breeding at Loch Leven, Scotland, and the importance of their predation on ducklings. *Ibis*, **132**, 535–49.

Marquiss, M., Nicoll, M., and Brockie, K. (1983). Scottish herons and the 1981/82 cold winter. *BTO News*, **125**, 4–5.

Marsh, N. (1995). Nankeen night herons (*Nycticorax caledonicus*) on the Whanganui River. *Notornis*, **42**, 282–3.

Marsh, N. and Lövei, G. L. (1997). The first confirmed breeding by the Nankeen Night Heron (*Nycticorax caledonicus*) in New Zealand. *Notornis*, **44**, 152–5.

Martin, G. R. and Katzir, G. (1994). Visual fields and eye movements in herons (Ardeidae). *Bird Behaviour and Evolution*, **44**, 74–85.

Martin, R. P. and Lester, G. D. (1991). *Atlas and census of wading birds and seabird colonies in Louisiana 1990*. Special Publication 3. Louisiana Department of Wildlife and Fisheries, Lafayette.

Martinez-Abrain, A. (1994). Nota sobre la biologia de *Ixobrychus m. minutus* (L.) durante el periodo de reproduccion en Valencia (E. Espana). *Ardeola*, **41**, 169–71.

Martinez-Abrain, A. (1998). Asociacion trofica de ardeidas en el arrozal de la Albufera de Valencia. *Ardeola*, **45**, 29–34.

Martinez C. (2004). Food and niche overlap of the Scarlet Ibis and the Yellow-crowned Night-Heron in a tropical mangrove swamp. *Waterbirds*, **27**, 1–8.

Martinez, I., Elliott, V. A., and Field, G. D. (1996). Yellow Billed Egret *Egretta intermedia* on the coast of Cameroon. *Malimbus*, **18**, 58.

Mason, V. (1994). An occurrence of Rufous Night-Heron and other interesting observations from Bali. *Kukila*, **7**, 78–9.

Massey, B. W. and Palacios, E. (1994). Avifauna of the wetlands of Baja California, Mexico: current status. In *A century of avifaunal change in western North America. Proceedings of an International Symposium at the Centennial Meeting of the Cooper Ornithological Society, Sacramento, California, April 17, 1993* (eds. J. R. Jehl, Jr. and N. K. Johnson), pp. 45–57. Studies in Avian Biology 15. Cooper Ornithological Society, Camarillo.

Massiah, E. (1997). Identification of Snowy Egret and Little Egret. *Birding World*, **9**, 434–44.

Massiah, E. and Frost, M. (1998). The Little Egrets of Graeme Hall Swamp. *Journal of the Barbados Museum and Historical Society*, **44**, 65–7.

Master, T. L. (1991). Use of tongue-flicking behavior by the Snowy Egret. *Journal of Field Ornithology*, **62**, 399–402.

Master, T. L. (1992). Composition, structure, and dynamics of mixed-species foraging aggregations in a southern New Jersey salt marsh. *Colonial Waterbirds*, **15**, 66–74.

Master, T. L., Frankel, M., and Russell, M. (1993). Benefits of foraging in mixed-species wader aggregations in a southern New Jersey saltmarsh. *Colonial Waterbirds*, **16**, 149–57.

Mathew, D. N., Narendran, T. C., and Zacharias, V. J. (1980). A comparative study of the feeding habits of certain species of Indian birds affecting agriculture. *Journal of the Bombay Natural History Society*, **75**, 1178–97.

Mathew, K. L., Pethani, K.V., and Yadav, D. N. (1997). Cattle Egret *Bubulcus ibis* feeding on baby rats. *Journal of the Bombay Natural History Society*, **94**, 396–7.

Mathews, N. and McQuaid, C. D. (1983). The feeding ecology of the Slatey Egret *Egretta vinaceigula*. *African Journal of Ecology*, **21**, 235–40.

Mathews, N. J. C. and Brooke, R. K. (1998). Notes on the foraging behavior of the Zigzag Heron. *Wilson Bulletin*, **100**, 147–8.

Mathews, N. (1978). In search of a Slaty: U.C.T. Wildlife Society Botswana Expedition. U.C.T. Wildlife Society, 1–93.

Matiukas, G. (1990). [Rare and little investigated bird species of the Plateliai Nature Preserve.] *Acta Ornithologica (Lituanica)*, **3**, 113–17.

Matsunaga, K. (2000). Effects of tidal cycle on the feeding activity and behavior of Grey Herons in a tidal flat in Notsuke Bay, northern Japan. *Waterbirds*, **23**, 226–35.

Matsunaga, K., Matsuda, A., and Fukuda, H. (2000). Changing trends in distribution and status of Grey Heron colonies in Hokkaido, Japan, 1960–1999. *Japanese Journal of Ornithology*, **49**, 9–16, 63.

Mauchamp, A., Chauvelon, P., and Grillas, P. (2002). Restoration of floodplain wetlands: Opening poldes along a coastal river in Mediterranean France, Vistre marshes. *Ecological Engineering*, **18**, 619–32.

Maxwell, G. R., II and Kale, H. W., II (1977). Breeding biology of five species of herons in coastal Florida. *Auk*, **94**, 689–700.

Mayr, E. (1945). Notes on the birds of northern Melanesia. *American Museum Novitates*, **1294**.

McBride, F. (1993). Hammonasset: the bittern and the sparrow. *Connecticut Warbler*, **13**, 56–7.

McCall, R. (1995). A novel foraging association between southern river otters *Lutra longicaudis* and Great Egrets *Casmerodius albus*. *Bulletin of the British Ornithologists' Club*, **116**, 199–200.

McCanch, N. (2003). Grey Heron choking on Little Grebe. *British Birds*, **96**, 86.

McCanch, N.V. and McCanch, M. (1991). Alert posture of Yellow-billed and Great White Egrets. *British Birds*, **84**, 341–2.

McClure, E. (1974). *Migration and survival of the birds of Asia.* US Army Medical Component, SEATO Medical Project, Bangkok, Thailand.

McConnell, P. J. and McKilligan, N. G. (1999). Vertebrates found in Cattle Egret chick regurgitates. *Corella*, **23**, 83–4.

McCracken, K. G. and Sheldon, F. H. (1997). Avian vocalizations and phylogenetic signal. *Proceedings of the National Academy of Sciences of the United States of America*, **94**, 3833–6.

McCracken, K. G. and Sheldon, F. H. (1998). Molecular and osteological heron phylogenies: Sources of incongruence. *Auk*, **115**, 127–41.

McCrimmon, D. A. Jr. (1974). Stretch and snap displays in the Great Egret. *Wilson Bulletin*, **86**, 165–7.

McCrimmon, D. A., Jr., Ogden, J. C., and Bancroft, G. T. (2001). Great Egret *Ardea alba*. In *The birds of North America*. No. 570. (eds. A. Poole and F. Gill), pp. 1–18, The Birds of North America, Inc. Philadelphia, PA.

McCulloch, E. M. (1967). Behavior of White-necked Heron. *Emu*, **67**, 22.

McGregor, P. K. and Byle, P. (1992). Individually distinctive bittern booms: potential as a census tool. *Bioacoustics*, **4**, 93–109.

McKilligan, N. G. (1984). The food and feeding ecology of the Cattle Egret *Ardeola ibis* when nesting in south-east Queensland. *Australian Wildlife Research*, **11**, 133–44.

McKilligan, N. G. (1985). The breeding success of the Indian Cattle Egret *Ardeola ibis* in eastern Australia. *Ibis*, **127**, 530–6.

McKilligan, N. G. (1987). Causes of nesting losses in the Cattle Egret *Ardeola ibis* in eastern Australia with special reference to the pathogenicity of the tick *Argas robertsi* to nestlings. *Australian Journal of Ecology*, **12**, 9–16.

McKilligan, N. G. (1990a). The breeding biology of the Intermediate Egret. Part 1: the physical and behavioural development of the chick, with special reference to sibling aggression and food intake. *Corella*, **14**, 162–9.

McKilligan, N. G. (1990b). Promiscuity in the Cattle Egret (*Bubulcus ibis*). *Auk*, **107**, 334–41.

McKilligan, N. G. (1991). The breeding biology of the Intermediate Egret. Part 2: parental behaviour and nesting investment by the male and female. *Corella*, **15**, 8–12.

McKilligan, N. G. (1995). The breeding success of the Indian Cattle Egret *Ardeola ibis* in eastern Australia. *Ibis*, **127**, 530–6.

McKilligan, N. G. (1996). Field experiments on the effect of ticks on breeding success and chick health of Cattle Egrets. *Australian Journal of Ecology*, **21**, 442–9.

McKilligan, N. G. (1997). A long term study of factors influencing the breeding success of the Cattle Egret in Australia. *Colonial Waterbirds*, **20**, 419–28.

McKilligan, N. (2002). Population dynamics of the Cattle Egret (*Ardea ibis*) in south-east Queensland: a 20-year study. *Emu*, **101**, 1–5.

McKilligan, N. (2002). Aspects of the breeding ecology of the Reef Heron *Egretta sacra*. *Corella*, **26**, 13–8.

McKilligan, N. G. and McConnell, P. (1989). Evidence suggesting a case of bigamy in the Intermediate Egret *Egretta intermedia*. *Australian Bird Watcher*, **13**, 98–9.

McKilligan, N. G., Reimer, D. S., Seton, D. H. C., Davidson, D. H. C., and Willows, J. T. (1993). Survival and seasonal movements of the Cattle Egret in eastern Australia. *Emu*, **93**, 79–87.

McLaren, I. (1989). Thoughts on the Little Egrets. *Nova Scotia Birds*, **31**, 42–3.

McMahon, J. (1984). Great Egret in Muskogee County, Oklahoma, in winter. *Bulletin of the Oklahoma Ornithological Society*, **17**, 29–30.

McMurry, S. L. (1971). Nesting and development of the Reddish Egret (*Dichromanassa rufescens* Gmelen) on a spoil bank chain in the Laguna Madre. MS Thesis. Texas A&I University, Kingsville.

McNeil, R., Benoit, R., and Desgranges, J.-L. (1993). Daytime and nighttime activity at a breeding colony of Great Blue Herons in a nontidal environment. *Canadian Journal of Zoology*, **71**, 1075–8.

McVaugh, W., Jr. (1972). The development of four North American herons. *Living Bird*, **11**, 155–73.

McVaugh, W., Jr. (1975). The development of four North American herons. Pt.2. *Living Bird*, **14**, 163–83.

Meanley, B. (1955). A nesting study of the Little Blue Heron in eastern Arkansas. *Wilson Bulletin*, **67**, 84–99.

Medeiros, M. J., Emond, E. E., and Ploger, B. J. (2000). An unusual type of sibling aggression in Black-crowned Night Herons. *Condor*, **102**, 438–40.

Medland, R. D. (1989). Malagasy Squacco Heron at Blantyre—first record for Malawi. *Nyala*, **13**, 73–4.

Meeks, W. A., Naugle, D. E., Johnson, R. R., and Higgins, K. F. (1996). Interbreeding of a Tricolored Heron and a Snowy Egret in South dakaota. *Auk*, **113**, 955–957.

Mellink, E., Palacios, and Gonzalez, S. (1997). Breeding birds of the delta of the Rio Colorado, Mexico. *Journal of Field Ornithology*, **68**, 153–9

Melville, D. S. (1986). Chinese Pond Herons *Ardeola bacchus* eating flies. *Hong Kong Bird Report*, **1987**, 104–6.

Melville, D. S. (1990). Cattle Egrets and Chinese Pond Herons feeding in trees. *Hong Kong Bird Report*, **1989**, 140–1.

Melville, D. S. (1995). Chinese Pond Heron eating a swallow. *Hong Kong Bird Report*, **1994**, 216.

Melvin, S. L., Gawlik, D. E., and Scharff, T. (1999). Long-term movement patterns for seven species of wading birds. *Waterbirds*, **22**, 411–16.

Merrifield, K. J. (1992). Two unusual foraging behaviors of Great Blue Herons in Lincoln County, Oregon. *Northwestern Naturalist*, **73**, 25–6.

Metz, K. J., Prior, K. A., and Mallory, M. L. (1991). Do Cattle Egrets gain information from conspecifics when foraging? *Oecologia (Heidelberg)*, **86**, 57–61.

Meyerriecks, A. J. (1960). *Comparative breeding behavior of four species of North American herons*. Publications of the Nuttall Ornithological Club 2.

Michelutti, P. L. (1992). [The Rufescent Tiger-Heron (*Tigrisoma lineatum*) in Cordoba.] *Nuestras Aves*, **26**, 27.

Michner, M. C., Weke, J. S., and Clapp, R. B. (1964). A breeding colony of Agami Herons in Veracruz. *Condor*, **66**, 77–8.

Mienis, H. K. (1991). [Predation on the freshwater mussel *Anodonta cygnea* by Grey Herons, *Ardea cinerea*.] *Correspondentieblad van de Nederlandse Malacologische Vereniging*, **263**, 907–8.

Mikuska, T., Kushlan, J. A., and Hartley, S. (1998). Key areas for wintering North American herons. *Colonial Waterbirds*, **21**, 125–34.

Milewski, A. V. (1976). Feeding ecology of the Slaty Egret, *Egretta vinaceigula*. *Ostrich*, **47**, 132–4.

Miller, C. (2001). Long-term monitoring of a breeding colony of White Herons (*Egretta alba*) on the Waitangiroto River, South Westland, New Zealand. *Notornis*, **48**, 157–163.

Miller, J. C. (1988–1989). Unusual behavior of Black-crowned Night-Herons. *Cassinia*, **63**, 72–3.

Miller, P.J. (2001). Long-term monitoring of a breeding colony of White Herons (*Egretta alba*) on the Waitangiroto River, South Westland, New Zealand. *Notornis*, **48**, 157–63.

Miller, P. and Miller, K. (1991). Bitterns using mangroves. *Notornis*, **38**, 79.

Milon, P., Petter, J.-J., and Randrianasolo, G. (1973). *Oiseaux*. Faune de Madagascar 35. ORSTOM, Tananarive and CNRS, Paris.

Milstein, P. le S. and Hunter, H. C. (1974). The spectacular Black Heron. *Bokmakierie*, **26**, 93–7.

Milstein, P. le S., Prestt, I., and Bell, A. A. (1970). The breeding cycle of the Grey Heron. *Ardea*, **58**, 171–257.

Min, B.Y., Honda, K., Shiraishi, S., and Tatsukawa, R. (1984a). Biometry of growth and food habits of young of the eastern White Egret, *Egretta alba modesta*, in Korea. *Journal of the Faculty of Agriculture Kyushu University*, **29**, 23–34.

Min, B.Y. and Shiraishi, S. (1984a). [Breeding ecology of the eastern Great White Egret, *Egretta alba modesta* in Korea: 1. Diurnal activity in the breeding season.] *Science Bulletin of the Faculty of Agriculture Kyushu University*, **39**, 85–92.

Min, B.Y. and Shiraishi, S. (1984b). [Breeding ecology of the eastern Great White Egret *Egretta alba modesta*, in Korea: 2. Nest building behavior and protection procedure.] *Science Bulletin of the Faculty of Agriculture Kyushu University*, **39**, 93–8.

Min, B.Y., Shiraishi, S., and Uchida, T. A. (1984b). [Breeding ecology of the eastern Great White Egret, *Egretta alba modesta*, in Korea: III. Egg-laying, incubation, brooding and growth of the chick.] *Science Bulletin of the Faculty of Agriculture Kyushu University*, **39**, 99–106.

Mirick, S. (2003). "Bubble feeding" behavior in a Snowy Egret. *Bird Observer*, **31**, 291.

Mitchell, P. (1996). Bird reports series 116 (Part 1). *Bird Observer*, **764**,15–16.

Mitsunga, H. and Imamura, K. (1996). [Breeding record of the Chinese Pond Heron *Ardeola bacchus* in Kumamoto Prefecture.] *Strix*, **14**, 190–3.

Mock, D.W. (1974). Aerial hunting by Little Blue Herons. *Wilson Bulletin*, **86**, 280–2.

Mock, D.W. (1975a). Social behavior of the Boat-billed Heron. *Living Bird*, **14**, 185–214.

Mock, D.W. (1975b). Feeding methods of the Boat-Billed Heron, a deductive hypothesis. *Auk*, **92**, 590–2.

Mock, D.W. (1976). Pair-formation displays of the Great Blue Heron. *Wilson Bulletin*, **88**, 185–230.

Mock, D.W. (1979). Display repertoire shifts and "extramarital" courtships in herons. *Behaviour*, **69**, 57–71.

Mock, D.W. (1980a). White-dark polymorphism in herons. In *Proceedings of the first Welder Wildlife Foundation Symposium* (ed. D. L. Drawe), pp. 145–61. Welder Wildlife Foundation Contribution B-7. Rob and Bessie Welder Wildlife Foundation, Sinton.

Mock, D.W. (1980b). Communication strategies of Great Blue Herons and Great Egrets. *Behaviour*, **72**, 156–70.

Mock, D.W. (1985a). Siblicidal brood reduction: The prey-size hypothesis. *American Naturalist*, **125**, 327–43.

Mock, D.W. (1985b). Biology of the Goliath Heron in Zululand. *National Geographic Society Research Report*, **18**, 537–50.

Mock, D.W. (1987). Siblicide, parent-offspring conflict, and unequal parental investment by egrets and herons. *Behavioral Ecology and Sociobiology*, **20**, 247–56.

Mock, D.W., Lamey, T. C., and Ploger, B. J. (1987). Proximate and ultimate roles of food amount in regulating egret sibling aggression. *Ecology*, **68**, 1760–72.

Mock, D.W., Lamey, T. C., and Thompson, B. A. (1988). Falsifiability and the Information Centre Hypothesis. *Ornis Scandinavia*, **19**, 231–48.

Mock, D.W., Lamey, T. C., Williams, C. F., and Pelletier, A. (1987). Flexibility in the development of heron sibling aggression: An intraspecific test of the prey-size hypothesis. *Animal Behaviour*, **35**, 1386–93.

Mock, D.W. and Mock, K. C. (1980). Feeding behavior and ecology of the Goliath Heron. *Auk*, **97**, 433–48.

Mock, D.W. and Ploger, B. J. (1987). Parental manipulation of optimal hatch asynchrony in Cattle Egrets: An experimental study. *Animal Behaviour*, **35**, 150–60.

Möenke, R. and Bick, A. (1990). Vorkommen des Kuhreihers, *Bubulcus ibis* (L.), in der Antarktis. *Mitteilungen aus dem Zoologischen Museum in Berlin*, **66**, **Supplement 14**, 69–79.

Moller, N. W. and Olesen, N. S. (1983). The Grey Heron (*Ardea cinerea*) in Denmark. *Danske Vildtundersogelser*, **36**, 1–23.

Molnar, G. (1990). On food composition of four heron species. *Tiscia (Szeged)*, **25**, 75–82.

Monnet, C. and Varney, A. (1998). Notes on the breeding of the Striated Heron *Butorides striatus patruelis* in Tahiti, French Polynesia. *Emu*, **98**, 132–6.

Monroe, B. L., Jr. and Browning, M. R. (1992). A re-analysis of *Butorides*. *Bulletin of the British Ornithologists' Club*, **112**, 81–5.

Moon, A.V., (ed.) (1997). London bird report for 1997. *London Bird Report*, **62**, 1–239.

Moore, P. J. (1984). Foraging and social behavior of the White-faced Heron at Pauatahanui Inlet. *Notornis*, **31**, 285–99.

Mora, M. A. (1992). Habitat use by foraging Cattle Egrets in the Mexicali Valley, Baja California, Mexico. *Wilson Bulletin*, **104**, 142–8.

Mora, M. A. (1996). Organochlorines and trace elements in four colonial waterbird species nesting in the lower Laguna Madre, Texas. *Archives of Environmental Contamination and Toxicology*, **31**, 533–7.

Mora, M. A. (1997). Transboundary pollution: persistent organochlorine pesticides in migrant birds of the southwestern United States and Mexico. *Environmental Toxicology and Chemistry*, **16**, 3–11.

Morales, G. (2000). Herons in South America. In *Heron conservation* (eds. J. A. Kushlan and H. Hafner), pp. 177–99. Academic Press, San Diego.

Morel, G. J. and Morel, M. Y. (1989). Une heronniere mixte sur le lac de Guier (Senegal) avec reference speciale a *Ixobrychus m. minutus* et *Platalea leucorodia*. *Oiseau et la Revue Francaise d'Ornithologie*, **59**, 290–5.

Morel, G. J. and Morel, M. Y. (1990). *Les oiseaux de Sénégambie: notices et cartes de distribution* . Editions de l'ORSTOM, Paris.

Morey, S. R. and Smits, A. (1987). Aerial plunge foraging by a Great Blue Heron. *Wilson Bulletin*, **99**, 292–3.

Morris, A. (1979). The spread of the Cattle Egret in New South Wales. *Australian Birds*, **13**, 72–4.

Morris, A. K. (1971). Distribution of the Black Bittern in New South Wales. *Emu*, **71**, 175–6.

Morton, S. R., Brennan, K. G., and Armstrong, M. D. (1993). Distribution and abundance of herons, egrets, ibises and spoonbills in the Alligator Rivers region, Northern Territory. *Wildlife Research*, **20**, 23–43.

Moser, M. A. (1984). Resource partitioning in colonial herons with particular reference to the Grey Heron *Ardea cinerea* L. and the Purple Heron *Ardea purpurea* L. in the Camargue, S. France. Doctoral Dissertation, University of Durham, Durham.

Moser, M. E. (1986a). Breeding strategies of Purple Herons (*Ardea purpurea*) in the Camargue, France. *Ardea*, **74**, 91–100.

Moser, M. E. (1986b). Prey profitability for adult Grey Herons *Ardea cinerea* and the constraints on prey size when feeding young nestlings. *Ibis*, **128**, 392–405.

Mosso, E. D. and Beltzer, A. H. (1992). Nuevos aportes a la biologica reproductive de la Garita Azulada *Butorides striatus* (Aves: Ardeidae). *Hornero*, **13**, 236–7.

Mott, D. F. and Flynt, R. D. (1995). Evaluation of an electric fence system for excluding wading birds at catfish ponds. *Progressive Fish-Culturist*, **57**, 88–90.

Movalli, P. A., Vigorita, V., and Sangiorgi, E. (1995). Residui di pesticidi organo-clorurati nelle uova di 10 specie di uccelli nidificanti in Lombardia. *Rivista Italiana di Ornitologia*, **64**, 121–30.

Mugica, L., Acosta Cruz, M., and Denis, D. (1998). Food habits of the Black-crowned Night Heron *Nycticorax nycticorax* in rice fields of the southern Jibaro, Cuba. Society Caribbean Ornithology Meeting Abstracts, 1998.

Mugica Valdes, L., Torres Fundora, O., and Llanes Sosa, A. (1987). Morfometria de la garza ganadera (*Bubulcus ibis*) en algunas regiones de Cuba. *Poeyana*, **334**, 1–8.

Mukherjee, A. and Parasharya, B. M. (1998). Little Green Heron, *Butorides striatus*, and Black Bittern, *Ixobrychus flavicollis*, in Gir forest: Gujarat. *Pavo*, **36**, 123.

Mukherjee, A. and Parasharya, B. M. (2000). Purple Heron *Ardea pupurea* (Linn.) nesting on water hyacinth *Eichhornia crassipes* (Pontederiaceae). *Journal of the Bombay Natural History Society*, **97**, 424–5.

Mundkur, T. (1985). Yellow Bittern nesting, a new record for Poona. *Journal of the Bombay Natural History Society*, **82**, 653–4.

Mundkur, T. (1993). [First sighting of a colour banded Chinese Egret in Singapore.] *Bangkok Bird Club Bulletin*, **10**, 12.

Mundt, J. (1996). Beobachtungen zum Verhalten einer uberwinternden Rohrdommel *Botaurus stellaris*. *Limicola*, **10**, 114–22.

Murphy, W. L. (1992). Notes on the occurrence of the Little Egret (*Egretta garzetta*) in the Americas, with reference to other Palearctic vagrants. *Colonial Waterbirds*, **15**, 113–23.

Murphy, W. L. and Nanan, W. (1987). First confirmed record of Western Reef-Heron (*Egretta gularis*) for South America. *American Birds*, **41**, 392–4.

Murton, R. K. (1971). Polymorphism in Ardeidae. *Ibis*, **113**, 97–9.

Murton, R. K. (1972). The ecology and status of Swinhoe's Egret, with notes on other herons in southeastern China. *Biological Conservation*, **4**, 89–96.

Musgrove, A. J. (2002). The non-breeding status of the Little Egret in Britain. *British Birds*, **95**, 62–80.

Mustoe, G. E. (1993). Eocene bird tracks from the Chuckanut Formation, northwest Washington. *Canadian Journal of Earth Sciences*, **30**, 1205–8.

Nacinovic, J. B., Tavaras, M. S., and Teixeira, D. M. (1986). Sobre a reprodução de "*Botaurus pinnatus*" (Wagler, 1829). *Anais da Sociedade Sul-Riograndensede Ornitologia*, **7**, 36.

Nacinovic, J. B. and Teixeira, D. M. (1989). As aves de Fernando de Noronha: uma lista sistematica anotada. *Revista Brasileira de Biologia*, **49**, 709–29.

Naik, S. and Parasharya, B. M. (1987). Impact of food availability, nesting-habitat destruction and cultural variations of human settlement on the nesting distribution of a coastal bird, *Egretta gularis*, in western India. *Journal of the Bombay Natural History Society*, **84**, 350–360.

Naik, R. M., Parasharya, B. M., Patel, B. H., and Mansuri, A. P. (1981). The timing of breeding season and interbreeding between the colour phases in the Indian Reef Heron, *Egretta gularis* (Bosc). *Journal of the Bombay Natural History Society*, **78**, 494–7.

Nankinov, D. and Kantarzhiev, D. (1988). [On overwintering in the Little Egret (*Egretta garzetta* L.) in Bulgaria.] *Ornitologicheski Informatsionen Byuletin*, **23–24**, 77–81.

Nankinov, D. N. (1991). [General pattern of the bittern migrations in Europe.] *Vestnik Zoologii*, **1991**, 41–8.

Nankinov, D. N. (1999). [On the question of distribution and migrations of the Little Bittern.] *Berkut*, **8**, 15–20.

Naoroji, R. (1990). Predation by *Aquila* eagles on nestling storks and herons in Keoladeo National Park, Bharatpur. *Journal of the Bombay Natural History Society*, **87**, 37–46.

Narusue, M. (1992). [Changes in the distribution and extent of breeding colonies of egrets in Saitama Prefecture.] *Strix*, **11**, 189–209.

Narusue, M. and Uchida, H. (1993). [The effect of structural changes of paddy fields on foraging egrets.] *Strix*, **12**, 121–30.

Naugle, D. E., Johnson, R. R., Meeks, W. A., and Higgins, K. F. (1996). Colonization and growth of a mixed-species heronry in South Dakota. *Colonial Waterbirds*, **19**, 199–206.

Nehls, H. (1992). Extreme aggressive interaction among Great Blue Herons. *Oregon Birds*, **18**, 105.

Nesbitt, S. A. (1984). Dangle feeding by the Green-backed Heron. *Florida Field Naturalist*, **12**, 44.

Newman, K. and Holtshausen, G. (1990). Identifying White Egrets. *Birding in Southern Africa*, **42**, 111–13.

Newton, I., Wyllie, I., and Asher, A. (1994). Pollutants in Great Bitterns. *British Birds*, **87**, 22–5.

Newton, P. N. (1986). Leapfrogging in Cattle Egrets (*Bubulcus ibis*). *Journal of the Bombay Natural History Society*, **83**, 432–3.

Niethammer, K. R. and Kaiser, M. S. (1983). Late summer food habits of three heron species in northeastern Louisiana. *Colonial Waterbirds*, **6**, 148–53.

Niethammer, K. R., Kaiser, M. S., Atkinson, R. D., and Baskett, T. S. (1983). Foods of the Green-backed Heron in the eastern Missouri Ozarks. *Transactions of the Missouri Academy of Science*, **17**, 117–28.

Niikura, M. and Nakamura, K. (1987). [Distribution of the Chinese Pond Heron, *Ardeola bacchus* (Bonaparte) and its present status in Japan.] *Strix*, **6**, 50–65.

Niikura, M. and Takehiro, I. (1986). [Chinese Pond Heron (*Ardeola bacchus*) wintering in Chigasaki City.] *Strix*, **5**, 69–70.

Nikolaus, G. (1983). A bird cemetery in the Nubian Desert, Sudan. *Scopus*, **7**, 48.

Noble, G. K., Wurm, M., and Schmidt, A. (1938). Social behavior of the Black-crowned Night Heron. *Auk*, **55**, 7–40.

Nores, M. (1986). Argentina. In *A directory of neotropical wetlands* (eds. D. A. Scott and M. Carbonell), pp. 1–38. IUCN and IWRB, Slimbridge.

Norman, D. M., Breault, A. M., and Moul, I. E. (1989). Bald Eagle incursions and predation at Great Blue Heron colonies. *Colonial Waterbirds*, **12**, 215–17.

North, P. M. (1978). Relating Grey Heron survival rates to winter weather conditions. *Bird Study*, **26**, 23–8.

Norton, R. L. (1993). West Indies region. Winter 1991–1992. *American Birds*, **47**, 304–5.

Norton, R. L., White, A., and Dobson, A. (2002). West Indies. *North American Birds*, **56**, 372–374.

Norton, R. L., White, A., and Dobson, A. (2003). West Indies and Bermuda. *North American Birds*, **57**, 417–419.

Noske, R. A. and Ueda, K. (1996). First record of Cinnamon Bittern for Timor. *Kukila*, **8**, 158–9.

Nota, Y. (2000). Sexual size dimorphism of the Little Egret *Egretta garzetta*. *Japanese Journal of Ornithology*, **49**, 51–4.

Nota, Y. (2003). Effects of body size and sex on foraging territoriality of the Little Egret (*Egretta garzetta*) in Japan. *Auk*, **120**, 791–8.

NRE (2000). *Threatened vertebrate fauna of Victoria*. Department of Natural Resources and Environment, Melbourne.

Nyirenda, R. G. (1992). First breeding record of White-backed Night Heron in Malawi. *Nyala*, **16**, 30.

Oberhouser, H. C. (1912). A revision of the forms of the Great Blue Heron (*Ardea herodias* Linnaeus). *Proceedings of the United States National Museum*, **43**, 533–59.

Obradovic, R. (1988). Great White Egret. *Acrocephalus (Ljubljana)*, **9**, 69.

O'Connor, T. (1984). A note on the diet of nestling Blackheaded Herons. *Ostrich*, **55**, 221–2.

O'Donnell, C. and Dilks, P. (1988). First record of the Australian Little Bittern (*Ixobrychus minutus*) in New Zealand. *Notornis*, **35**, 153–7.

Ohashi, T. J. and Kimizuka, K. K. (1988). First record of nesting by Cattle Egrets on Maui from observations made during birdstrike control operations. *Elepaio*, **48**, 111–12.

Ohlendorf, H. M. and Marois, K. C. (1990). Organochlorines and selenium in California night-heron and egret eggs. *Environmental Monitoring and Assessment*, **15**, 91–104.

Ólafsson, E., Pétursson, G., and Nielsen, O. K. (1992). Bjarthegrar heimsækja Ísland. *Bliki (Reykjavik)*, **11**, 26–30.

Olioso, G. (1991). Comportement predateur du blongios nain *Ixobrichus minutus*. *Faune de Provence*, **12**, 94.

Oliver, P. J. (1995). Feeding habits of the Green Heron in Mauritius. *Bulletin of the British Ornithologists' Club*, **115**, 69–70.

Olmos, F. (1989). An ornithological survey of the Baixada of the industrial pole of Cubatao (Brazil). *Revista Brasileira de Biologia*, **49**, 373–80.

Olmos, F. and Silva e Silva, R. (2002). Breeding biology of the Little Blue Heron (*Egretta caerulea*) in southeast Brazil. *Ornitologia Neotropical*, **13**, 17–30.

Olmos, F., Silva e Silva, R., and Prado, A. (2001). Breeding season diet of Scarlet Ibises and Little Bllue Herons in a Brazilian mangrove swamp. *Waterbirds*, **24**, 50–7.

Olmos, F. and Souza, M. F. B. (1988). A new record of the Streaked Bittern from northeastern Brazil. *Wilson Bulletin*, **100**, 510–11.

Olson, S. L. (1978). Multiple origins of the Ciconiiformers. *Proceedings of the Colonial Waterbird Group*, **2**, 157–62.

Oren, D. C. (1984). Resultados de uma nova expedica zoologica a Fernando de Noronha. *Boletim do Museu Paraense Emilio Goeldi. Zoologia*, **1**, 19–44.

Orgeira, J. L. (1995). Prencia de garcita bueyera *Bubulcus ibis* en el Oceano Atlantico Sur, otono de 1993. *Hornero*, **14**, 53–4.

Orgeira, J. L. (1996). Cattle Egrets *Bubulcus ibis* at sea in the South Atlantic Ocean. *Marine Ornithology*, **24**, 57–8.

Ortvad, T. E. Pederson, K., and Thorup, K. (2002). Sjaeldne fugle i Danmark og Gronland i 2001. *Dansk Ornitologisk Forenings Tidsskriff*, **96**, 147–60.

Ottema, O. (2002). A possible population of Fasciated Tiger-heron *Tigrisoma fasciatum* on the upper Coppename River, Surinam. *Cotinga*, **18**, 103.

Owen, D. F. (1959). Mortality of the Great Blue Heron as shown by banding recoveries. *Auk*, **76**, 464–70.

Owre, O. T. (1967). The Reef Heron (*Egretta schistacea*) in interior East Africa. *Journal of the East African Natural History Society*, **26**, 61–3.

Palacios, C.-J. and Barone, R. (2001). The Grey Heron Ardea cinerea, new species nesting on Cape Verde Islands. *Alauda*, **69**, 18.

Palacios, E. and Mellink, E. (1983). Additional records of breeding birds from Montague Island, northern Gulf of California. *Western Birds*, **24**, 259–62.

Palacios, E. and Mellink, E. (1995). Breeding birds of esteros Tobari and San Jose, southern Sonora. *Western Birds*, **26**, 99–103.

Palmer, R. S. (1962). *Handbook of North American birds. Vol. 1. Loons through flamingos*. Yale University Press, New Haven.

Pandey, D. N. (1991). Nesting of the Pond Heron *Ardeola grayii* (Sykes) on Eucalyptus trees. *Journal of the Bombay Natural History Society*, **88**, 281.

Parasharya, B. M. (1982). Swimming feeding by the Night Heron, *Nycticorax nycticorax nycticorax* (Linnaeus). *Pavo*, **20**, 42–3.

Parasharya, B. M. (1983). An erithristic Pond Heron. *Pavo*, **21**, 107–8.

Parasharya, B. M. (1985). Pairing between Pond Heron and Intermediate Egret. *Pavo*, **23**, 103–4.

Parasharya, B. M. and Bhat, H. R. (1987). Unusual feeding strategies of the Little Egret and Pond Heron. *Pavo*, **25**, 13–15.

Parasharya, B. M. and Naik, R. M. (1984). The juvenile plumage of the Little Egret compared with that of the white-phase Indian Reef Heron. *Journal of the Bombay Natural History Society*, **81**, 693–5.

Parasharya, B. M. and Naik, R. M. (1987). Changes in the soft part colouration of the Indian Reef Heron, *Egretta gularis*, related to age and breeding status. *Journal of the Bombay Natural History Society*, **84**, 1–6.

Parasharya, B. M. and Naik, R. M. (1988). Breeding biology of the Indian Reef Heron. *Journal of the Bombay Natural History Society*, **85**, 251–62.

Parasharya, B. M. and Naik, R. M. (1990). Ciconiiform birds breeding in Bhavnagar City, Gujarat: a study of their nesting and plea for conservation. In *Conservation in developing countries: Problems and prospects. Proceedings of the Centenary Seminar of the Bombay Natural History Society* (eds. J. C. Daniel and J. S. Serrao), pp. 429–45. Bombay Natural History Society and Oxford University Press, Bombay and New York.

Parejo, D. (1997). Black-crowned Night-Herons *Nycticorax nycticorax* feeding on carcasses. *Butlleti del Grup Catala d'Anellament*, **14**, 59–61.

Parejo, D., Sanchez-Guzman, J. M., and Aviles, J. M. (2001a). Breeding biology of the Cattle Egret *Bubulcus ibis* in southwest Spain. *Bird Study*, **48**, 367–72.

Parejo, D., Sanchez, J. M. and Aviles, J. M. (2001b). Breeding biology of the Little Egret (*Egretta garzetta*) in southwestern Spain. *Ornis Fennica*, **78**, 31–8.

Parera, A. and Bosso, A. (1996). Description of a mixed waterbird colony at Laguna Ibera, Corrientes Province, Argentina. *Hornero*, **14**, 81–4.

Parish, D. and Melville, D. (1985). Deep frozen waterbirds. *Interwader*, **6**, 8–9.

Parkes, K. C. (1955). Systematic notes on North American birds. 1. The herons and ibises (Ciconiiformes). *Annuals of the Carnegie Museum*, **33**, 287–93.

Parkes, K. C. (1974). Buller's New Zealand specimen of Grey Heron. *Notornis*, **21**, 121–3.

Parkes, K. C. (1998). First record of the Great Blue Heron for Brazil. *Colonial Waterbirds*, **21**, 89–90.

Parkhurst, J. A., Brooks, R. P., and Arnold, D. E. (1992). Assessment of predation at trout hatcheries in central Pennsylvania. *Wildlife Society Bulletin*, **20**, 411–19.

Parrini, R. and Pacheco, J. F. (1997). Seis novos registros de aves para o Estado de Minas Gerais. *Atualidades Ornitologicas*, **82**, 6.

Parrott, J., Phillips, J., and Wood, V. (1987). Tricolored Heron on Azores in October 1985. *Dutch Birding*, **9**, 17–19.

Parsons, K. C. (1995). Heron nesting at Pea Patch Island, upper Delaware Bay, USA: abundance and reproductive success. *Colonial Waterbirds*, **18**, 69–78.

Parsons, K. C. (2003). Reproductive success of wading birds using *Phragmites* marsh and upland nesting habitats. *Estuaries*, **2**, 596–601.

Parsons, K. C. and Burger, J. (1982). Human disturbance and nestling behavior in Black-crowned Night Herons. *Condor*, **84**, 184–7.

Parsons, K. C. and McColpin, A. C. (1995). Great Blue Heron reproductive success in upper Delaware Bay. *Journal of Field Ornithology*, **66**, 184–91.

Parsons, K. C. and Master T. L. (2000). Snowy Egret, *Egretta thula*. In *The birds of North America*. No. 489 (eds. A. Poole and F. Gill), pp. 1–24. The birds of North America, The Academy of Natural Sciences, Philadelphia and The American Ornithologists' Union, Washington.

Parsons, K. C., Schmidt, S. R., and Matz, A. C. (2001). Regional pattens of wading bird productivity in Northeastern U. S. estuaries. *Waterbirds*, **24**, 323–30.

Partridge, J. K. (1986). Results of ringing herons in Connemara. *Irish Birds*, **3**, 267–72.

Passarella, M., Altieri, E., and Verza, E. (1999). Prima osservazione di ibrido airone cenerino x airone rosso *Ardea cinerea* x *purpurea* in Italia (Delta del Po, Rovigo). *Avocetta*, **23**, 60.

Paton, P. W. C., Fellows, D. P., and Tomich, P. Q. (1986). Distribution of Cattle Egret roosts in Hawaii with notes on the problems egrets pose to airports. *Elepaio*, **46**, 143–7.

Paton, P.W. C. and Kneedy, C. (1993). Great Blue Herons and Double-crested Cormorants nesting on powerline towers. *Utah Birds*, **9**, 33–8.

Paton, P.W. C. and MacIvor, L. H. (1983). Green Heron in Hawaii. *American Birds*, **37**, 232–3.

Paul, R.T. (1991). *Status report—Egretta rufescens (Gmelen) Reddish Egret*. U.S. Fish and Wildlife Service, Houston.

Paul, R.T. (1996). Reddish Egret, In *Rare and endangered species of Florida. Vol. V. Birds* (eds. J. A. Rodgers, Jr., Herbert Kale, II, and H.T. Smith) University Presses of Florida, Gainesville.

Paul, R.T., Kale, H.W., II, and Nelson, D.A. (1979). Reddish Egret nesting on Florida's east coast. *Florida Field Naturalist*, **7**, 24–25.

Paul, R.T., Meyerriecks, A. J., and Dunstan, F. M. (1975). Return of Reddish Egrets to Tampa Bay. *Florida Field Naturlist*, **3**, 9–10.

Pauw, R. (1991). Goliath Heron. *Promerops*, **199**, 16.

Paxton, R. O., Burgiel, J. C., and Cutler, D.A. (1999). Fall migration: August through November 1998. Hudson–Delaware Region. *North American Birds*, **53**, 34–8.

Payne, R. B. (1974). Species limits and variation of the New World Green Herons *Butorides virescens* and Striated Herons *B. striatus*. *Bulletin of the British Ornithologists' Club*, **94**, 81–6.

Payne, R. B. (1979). Family Ardeidae. In *Check-list of the birds of the world*. Vol. 1, 2nd ed. (eds. E. Mayr and G.W. Cottrell), pp. 193–244. Museum of Comparative Zoology, Cambridge.

Payne, R. B. and Risley, C. J. (1976). *Systematics and evolutionary relationships among the herons*. Miscellaneous Publications, Museum of Zoology, University of Michigan, 150.

Paz, U. (1987). *The birds of Israel*. Stephen Greene Press, Lexington.

Pedersen, A., Nielsen, S. S., Thuy, L. D., and Trai, L.T. (1998). The status and conservation of threatened and near-threatened species of birds in the Red River Delta, Vietnam. *Bird Conservation International*, **8**, 31–51.

Pelletier, F.-X. (1999). The bird people of Manchar. *GEO Australasia*, **21**, 69–73.

Penny, M. (1974). *The birds of the Seychelles*. Collins, London.

Perennou, C. (1987). L'impact du heron cendre *Ardea cinerea* en bassin d'alevinage. *Oiseau et la Revue Francaise d'Ornithologie*, **57**, 262–5.

Perennou, C., Lansdown, R.V., and Hancock, J.A. (2000). Herons in south and west Asia. In *Heron conservation* (eds. J. A. Kushlan and H. Hafner), pp. 44–72. Academic Press.

Perennou, C., Mundkur, T., Scott, D. A., Follestad, A., and Kvenild, L. (1994). *The Asian waterfowl census 1987–91: distribution and status of Asian waterfowl*. AWB Publication No. 86 and IWRB Special Publication 24. Asian Wetland Bureau, Kuala Lumpur and International Waterfowl and Wetlands Research Bureau, Slimbridge.

Peris, S. J., Briz, F. J., and Campos, F. (1995). Shifts in the diet of Grey Heron (*Ardea cinerea*) in the Duero Basin, central-west Spain, following the introduction of exotic fish species. *Folia Zoologica*, **44**, 97–102.

Peris, S. J., Briz, F. J., and Campos, F. (1996). Prey selection by the Grey Heron *Ardea cinerea* in relation to prey availability. *Airo*, **7**, 15–20.

Perkins, S. (1992). The Spring season. March 1–May 31, 1992. New England Region. *American Birds*, **46**, 393–6.

Perkins, S. (1995). New England Region. *National Audubon Society Field Notes*, **49**, 227–30.

Perlmutter, G. B. (1992). Environmental factors influencing roost arrival of Black-crowned Night-Herons. *Journal of Field Ornithology*, **63**, 462–5.

Peter, J. M. (1993). A feeding interaction between Great Egret *Ardea alba* and Fairy Tern *Sterna nereis*. *Australian Bird Watcher*, **15**, 228.

Peter, J. M. (1995). The Eastern Reef Egret in Victoria and Tasmania: past and present. *Australian Bird Watcher*, **16**, 120–5.

Peters, J. L. (1931). *Check-list of birds of the world. Vol. 1*. Harvard University Press, Cambridge.

Petersen, A. (1985). [New bird species for the Icelandic checklist.] *Bliki (Reykjavik)*, **4**, 57–66.

Peterson, W. R. (1992). The Summer season. June 1–July 31, 1992. New England region. *American Birds*, **46**, 1119–22.

Petry, M.V. and Hoffmann, G. R. (2002). Ocupacao e construcao de ninhos em um ninhal misto de garcas e macaricos (Ciconiiformes) no Rio Grande do Sul, Brazil. *Biociencias*, **10**, 55–63.

Petursson, G. [Green Heron, new for Iceland]. *Bliki*, **24**, 69–70.

Petzer, R. (1990). Predation by Wahlberg's Eagles on a heron/egret colony. *Gabar*, **5**, 25.

Phalan, B. (1997). Foot-stirring behaviour of Little Egret. *British Birds*, **90**, 524.

Phillips, B. (1983). Confrontation at Great Meadows. *Bird Observer of Eastern Massachusetts*, **11**, 285.

Phillips, N. J. (1984). Migrant species new to Seychelles. *Bulletin of the British Ornithologists' Club*, **104**, 9–10.

Piacentini, D. (1993). Prima nidificazione accertata di airone bianco maggiore, *Egretta alba*, in Italia. *Rivista Italiana di Ornitologia*, **63**, 107.

Piskorski, M. (1999). Rozmieszczenie, liczebnosc i ochrona baka *Botaurus stellaris* na Pojezierzu Leczynsko-Wlodawskim. *Chronmy Przyrode Ojczysta*, **55**, 52–64.

Pitelka, F. A. (1938). Melanism in the Black-crowned Night Heron. *Auk*, **55**, 518–19.

Pitt, W. C., Beauchamp, D. A., and Conover, M. R. (1998). Evaluation of bioenergetics models for predicting Great Blue Heron consumption of rainbow trout at hatcheries. *North American Journal of Fisheries Management*, **18**, 52–65.

Pitt, W. C. and Conover, M. R. (1996). Predation at intermountain west fish hatcheries. *Journal of Wildlife Management*, **60**, 616–24.

Pittaway, R. and Burke, P. (1996). Cory's Least Bittern. *Ontario Birds*, **14**, 26–40.

Pleshak, T.V. (1999). Vagrant Ciconiiformes birds in Arkhangelsk region. *Russkii Ornitologicheskii Zhurnal Ekspress Vypusk*, **58**, 14–15.

Ploger, B. J. and Mock, D. W. (1986). Role of sibling aggression in food distribution to nestling Cattle Egrets (*Bubulcus ibis*). *Auk*, **103**, 768–76.

Polasek, Z. (1991). The Grey Heron nests on marsh plant growth in Czechoslovakia. *Ziva*, **39**, 279–80.

Pollock, A.B. and Noden, P. (2002). Notes on the Black Bittern (Ixobrychus flavicollis) in far eastern Victoria. *Australian Bird Watcher*, **19**, 173–7

Poole, C. M., Park, J.-Y., and Moores, N. (1999). The identification of Chinese Egret and Pacific Reef Egret. *Bulletin of the Oriental Bird Club*, **30**, 38–41.

Porter, R. F. and Al-Saghier, O. (1998). The birds of some of Yemen's Red Sea islands. *Sandgrouse*, **20**, 66–7.

Post, W. (1990). Nest survival in a large ibis-heron colony during a three-year decline to extinction. *Colonial Waterbirds*, **13**, 50–61.

Post, W. (1998). Reproduction of Least Bitterns in a managed wetland. *Colonial Waterbirds*, **21**, 268–73.

Post, W. and Seals, C. A. (1993). Nesting associations of Least Bitterns and boat-tailed grackles. *Condor*, **95**, 139–44.

Potter, E. F., Brown, S., Barkley, N. P., and Alderink, R. (1997). Nesting ecology of Great Blue Herons at three previously unreported sites in Wake, Franklin, and Johnston counties, North Carolina. *Chat*, **61**, 183–99.

Poulin, B. and Lefebvre, G. (2003). Optimal sampling of booming Bitterns (*Botaurus stellaris*). *Ornis Fennica*, **80**, 11–20.

Powell, G.V. N. (1983). Food availability and reproduction by Geat White Herons, *Ardea herodias*: a food addition study. *Colonial Waterbirds*, **6**, 139–47.

Powell, G.V. N. (1987). Habitat use by wading birds in a subtropical estuary: Implications of hydrography. *Auk*, **104**, 740–9.

Powell, G.V. N., Bjork, R. D., Ogden, J. C., Paul, R. T., Powell, A. H., and Robertson, W. B., Jr. (1989). Population trends in some Florida Bay wading birds. *Wilson Bulletin*, **101**, 436–57.

Powell, G.V. N. and Powell, A. H. (1986). Reproduction by Great White Herons *Ardea herodias* in Florida Bay as an indicator of habitat quality. *Biological Conservation*, **36**, 101–13.

Prashant, J. J., Vasudeva Rao, V., and Nagulu, V. (1994). Nesting, egg size, incubation and factors affecting clutch size in Little Egret, *Egretta garzetta* at Nellore Andhra Pradesh. *Pavo*, **32**, 67–72.

Prasad, J. N. and Hemanth, J. (1992). Pond Heron *Ardeola grayii* (Sykes) feeding on bees. *Journal of the Bombay Natural History Society*, **89**, 246.

Pratt, H. D. and Bruner, P. L. (1992). A probable Little Egret (*Egretta garzetta*) on O'ahu. *Elepaio*, **52**, 15, 17.

Pratt, H. M. (1970). Breeding biology of Great Blue Herons and Common Egrets in central California. *Condor*, **72**, 407–16.

Pratt, H. M. (1972). Nesting success of Common Egrets and Great Blue Herons in the San Francisco Bay region. *Condor*, **74**, 447–53.

Pratt, H. M. (1983). Marin County California heron colonies: 1967–1981. *Western Birds*, **14**, 169–84.

Pratt, H. M. and Winkler, D.W. (1985). Clutch size, timing of laying and reproductive success in a colony of Great Blue Herons (*Ardea herodias*) and Great Egrets (*Casmerodius albus*). *Auk*, **102**, 49–63.

Preston, C. R., Moseley, H., and Moseley, C. (1986). Green-backed Heron baits fish with insects. *Wilson Bulletin*, **98**, 613–14.

Prigioni, C., Galeotti, P., and Fasola, M. (1985). Accrescimento dei pulli e riuscita della riproduzione nella nitticora *Nycticorax nycticorax*. *Avocetta*, **9**, 127–33.

Prosper, J. and Hafner, H. (1996). Breeding aspects of the colonial Ardeidae in the Albufera de Valencia, Spain: population changes, phenology, and reproductive success of the three most abundant species. *Colonial Waterbirds*, **19**, **Special Publication 1**, 98–107.

Prytherch, R. (1980). Squacco Heron possibly using insects as bait. *British Birds*, **73**, 183–4.

Pugacewicz, E. and Kowalski, J. (1997). Pierwsze w 20. wieku legi czapli bialej *Egretta alba* w Polsce. *Notatki Ornitologiczne*, **38**, 323–5.

Puglisi, L. and Baldaccini, N. E. (2000). Nocturnal flights during spring migration in the Bittern (*Botaurus stellaris*). *Vogelwarte*, **40**, 235–8.

Puglisi, L., Cima, O., and Baldaccini, N. E. (1997). A study of the seasonal booming activity of the Bittern *Botaurus stellaris*: what is the biological significance of the booms? *Ibis*, **139**, 638–45.

Puglisi, L., Fontanelli, A., and Baldaccini, N. E. (1995a). [The avifauna of Diaccia Botrona Marshes (Italy): present status and recent modifications.] *Ricerche di Biologia della Selvaggina*, **95**, 1–50.

Puglisi, L., Fontanelli, A., Perfetti, A., and Taverni, M. (1995b). The population of Bittern *Botaurus stellaris* in the Diaccia Botrona marsh, central Italy: four years of census (1991–94). *Avocetta*, **19**, 182–8.

Puglisi, L., Franceschini, S., Balestri, P. C., and Baldaccini, N. E. (1999). Comportamento spaziale di maschi di tarabuso Botaurus stellaris studiato mediante il radiotracking. *Avocetta*, **23**, 70.

Pullin, B. P. (1987). Restoring Great Egrets to the Tennessee Valley. *Proceedings of the Southeastern Nongame Endangered Wildlife Symposium*, **3**, 11–13.

Pullin, B. P. (1990). Size and trends of wading bird populations in Tennessee during 1977–1988. *Migrant*, **61**, 95–104.

Pyle, J. (2000). Fish-baiting behaviour observed in Striated Heron *Butorides striatus*. *South Australian Ornithologist*, **33**, 85–6.

Pyle, R. L. (1984). The Winter season. Hawaiian Islands region. *American Birds*, **38**, 359–61.

Pyle, R. L. (1999). Hawaiian Islands Region. North American Birds, 53, 435–7.

Pyrovetsi, M. and Crivelli, A. (1988). Habitat use by water-birds in Prespa National Park, Greece. *Biological Conservation*, **45**, 135–54.

Quinney, T. E. (1982). Growth, diet, and mortality of nestling Great Blue Herons. *Wilson Bulletin*, **94**, 571–7.

Quinney, T. E. (1983). Comparison of Great Blue Heron *Ardea herodias* reproduction at Boot Island and other Nova Scotia colonies. *Canadian Field-Naturalist*, **97**, 275–8.

Quinney, T. E. and Smith, P. C. (1980). Comparative foraging behavior and efficiency of adult and juvenile Great Blue Herons. *Canadian Journal of Zoology*, **58**, 1168–73.

Raffaele, H., Wiley, J., Garrido, O., Keith, A., and Raffaele, J. (1998). *A guide to the birds of the West Indies*. Princeton University Press, Princeton.

Raine, C. L. (1996). Some factors influencing roosting choice of the Cattle Egret, *Ardeola ibis*, during the out-migration period at five sites in the Lower Hudson Valley. Honours Thesis University of Newcastle, Newcastle.

Rajchard, J. and Novak, R. (1998). Hnizdeni volavky stribrite (*Egretta garzetta*) a kolpika bileho (*Platalea leucorodia*) v kolonii kvakose nocniho (*Nycticorax nycticorax*) na Ceskobudejovicku. *Sylvia*, **34**, 93–6.

Ramo, C. (1993). Extra-pair copulations of Grey Herons nesting at high densities. *Ardea*, **81**, 115–20.

Ramo, C. and Busto, B. (1982). Notes on the breeding of the Chestnut-bellied Heron (*Agamia agami*) in Venezuela. *Auk*, **99**, 784.

Ramo, C. and Busto, B. (1993). Resource use by herons in a Yucatan wetland during the breeding season. *Wilson Bulletin*, **105**, 573–86.

Rand, A. L. (1936). The distribution and habits of Madagascar birds: summary of the field notes of the Mission Zoologique Franco–Anglo–Américaine à Madagascar. *Bulletin of the American Museum of Natural History*, **72**, 143–499.

Rand, A. L. and Gilliard, E. T. (1967). *Handbook of New Guinea birds*. Weidenfeld and Nicolson, London.

Rand, A. L. and Rabor, D. A. (1960). Birds from the Philippines Islands: Siquijor, Mt. Malindang, Bohol and Samar. *Fieldiana Zoology*, **35**, 223–441.

Randall, R. and Heeremans, M. (1994). Breeding of the Slaty Egret *Egretta vinaceigula* along the Boro River in the central Okavango Delta (Botswana). *Ostrich*, **65**, 39–43.

Ranglack, G. S., Angus, R. A., and Marion, K. R. (1991). Physical and temporal factors influencing breeding success of Cattle Egrets (*Bubulcus ibis*) in a west Alabama colony. *Colonial Waterbirds*, **14**, 140–9.

Ranglack, G. S. and Marion, K. R. (1988). Hatching success and nestling survivorship studies in an Alabama Cattle Egret rookery. *Journal of the Alabama Academy of Science*, **59**, 129.

Rao, V. V., Anjaneyulu, M., Raza, S. H., and Nagulu, V. (1996). Comparative ethological studies of Little Egret (*Egretta garzetta*) and Cattle Egret (*Bubulcus ibis*) in Andhra Pradesh. *Acta Ecologica (Talvandi)*, **18**, 91–106.

Rasmussen, D. T., Olson, S. L., and Simons, E. L. (1987). Fossil birds from the Oligocene Jebel Qatrani Formation, Fayum Province, Egypt. *Smithsonian Contributions to Paleobiology*, **62**, 1–18.

Rattner, B. A., Hoffman, D. J., Melancon, M. J., Olsen, G. H., Schmidt, S. R., and Parsons, K. C. (2000). Organochlorine and metal contaminant exposure and effects in hatching Black-crowned Night Herons (*Nycticorax nycticorax*) in Delaware Bay. *Archives of Environmental Contamination and Toxicology*, **39**, 38–45.

Rault, G. (1995). Premiere mention Francaise du heron vert *Butorides virescens*. *Ornithos*, **2**, 90–2.

Raymakers, S. (1992). White-backed Night Heron breeding at Vanderbijlpark. *WBC News*, **158**, 14.

Raza, R. H. (1993). Pond Heron foraging on earthworms. *Newsletter for Birdwatchers*, **33**, 52–3.

Recher, H. F. (1972). Territorial and agnostic behavior of the Reef Heron. *Emu*, **72**, 126–30.

Recher, H. F. and Holmes, R. T. (1982). *The foraging behavior of herons and egrets on the Magela Creek flood plain, Northern Territory*. Australian Government Publishing Service, Canberra.

Recher, H. F., Holmes, R. T., Davis, W. E., Jr., and Morton, S. (1983). Foraging behaviour of Australian herons. *Colonial Waterbirds*, **6**, 1–10.

Recher, H. F. and Recher, J. A. (1969). Comparative foraging efficiency of adult and immature Little Blue Herons (*Florida caerulea*). *Animal Behaviour*, **17**, 320–2.

Recher, H. F. and Recher, J. A. (1972). The foraging behavior of the Reef Heron. *Emu*, **72**, 85–90.

Recher, H. F. and Recher, J. A. (1980). Why are there different kinds of herons? *Transactions of the Linnaean Society of New York*, **9**, 135–58.

Reichholf, J. (1983). Flugfischen von silber *Casmerodius albus* und schneereihern *Egretta thula* am mittleren Amazonas. *Verhandlungen der Ornithologischen Gesellschaft In Bayern*, **23**, 530.

Relton, A. (1996). Is red-legged pond heron a female? *Newsletter for Birdwatchers*, **36**, 59.

Reynaud, P. A. and Kushlan, J. A. in press. Nesting of Agami Herons in French Guiana.

Richards, D. (1992). Short notes on herons from Botswana. *Birding in Southern Africa*, **44**, 96.

Richardson. A. J. and Taylor, I. R. (2003). Are rice fields in southeastern Australia an adequate substitute for natural wetlands as foraging areas for egrets? *Waterbirds*, **26**, 353–63.

Richardson, C. T. and Bannon, J. K. (1991). Notes on identification of non-breeding Indian Pond Heron *Ardeola grayii* in the UAE. *Emirates Bird Report*, **15**, 34–5.

Richardson, S. (1999). Intermediate Egret at Midway Atoll. *North American Birds*, **53**, 441–3.

Richner, H. (1985). One adult Grey Heron killing another on feeding territory. *British Birds*, **78**, 297.

Richner, H. (1986). Intraspezifische Verhaltensvariationen: Das Beispiel des Graureihers *Ardea cinerea*. *Ornithologische Beobachter*, **83**, 291–4.

Richter, R. (1985). Red-tailed Hawk drowns Yellow-crowned Night Heron. *Florida Field Naturalist*, **13**, 12–13.

Riddell, I. C. (1987). Aggression between Dwarf Bitterns. *Honeyguide*, **33**, 14.

Ridgely, R. S. and Gwynne, J. A., Jr. (1989). *A guide to the birds of Panama: with Costa Rica, Nicaragua, and Honduras*. 2nd ed. Princeton University Press, Princeton.

Ridgely, R. S. and Gwynne, J. A., Jr. (1993). *Guia de las aves de Panamá*. Asociación Nacional para la Conservación de la Naturaleza, Ciudad de Panamá.

Riegner, M. F. (1982a). Prey handling in Yellow-crowned Night Herons. *Auk*, **99**, 380–1.

Riegner, M. F. (1982b). The diet of Yellow-crowned Night-Herons in the Eastern and Southern United States. *Colonial Waterbirds*, **5**, 173–6.

Riehl, C. (2001). Black-crowned Night Heron fishes with bait. *Waterbirds*, **24**, 285–6.

Rimbert, P. (1997). Premier point sur l'evolution de l'effectif nicheur du heron gardeboeufs *Bubulcus ibis* en plaine du Forez, Loire, France. *Nos Oiseaux*, **44**, 241–2.

Ripley, S. D. (1982). *A synopsis of the birds of India and Pakistan together with those of Nepal, Bhutan, Bangladesh and Sri Lanka*. Bombay Natural History Society, Bombay.

Robbins, M. B., Capparella, A. P., Ridgely, R. S., and Cardiff, S. W. (1991). Avifauna of the Rio Maniti and Quebrada Vainilla, Peru. *Proceedings of the Academy of Natural Sciences of Philadelphia*, **143**.

Roberson, D. (1980). *Rare birds of the West Coast*. Woodstock Publishing Company, Pacific Grove.

Roberts, T. J (1991). *The birds of Pakistan. Vol. 1*. Oxford University Press, Oxford.

Robertson, D. (1987). Rufous-bellied Heron at Dullstroom. *WBC News*, **139**, 13.

Robertson, H. A. and Heather, B. D. (1999). Effect of water levels on the seasonal use of Lake Wairarapa by waders. *Notornis*, **46**, 79–88.

Robertson, I. (1993). Unusual records from Cameroon. *Malimbus*, **14**, 62–3.

Robertson, I. (1995). Recent reports. *Bulletin of the African Bird Club*, **2**, 61–4.

Robinson, J. C. (1990). An annotated checklist of the birds of Tennessee. University of Tennessee Press, Knoxville.

Robinson, S. K. (1994). Use of bait and lures by Green-backed Herons in Amazonian Peru. *Wilson Bulletin*, **106**, 567–9.

Robinson, W. L., Groebner, D. J., and Schultz, R. N. (1991). Foraging of gray wolves at a Great Blue Heron rookery. *American Midland Naturalist*, **126**, 196–7.

Rockingham-Gill, D.V. (1998). African waterfowl census—and other. *Honeyguide*, **44**, 166–7.

Rodgers, J. A., Jr. (1974). Aerial feeding by Snowy and Great Egrets in Louisiana waters. *Wilson Bulletin*, **86**, 70–1.

Rodgers, J. A., Jr. (1975). Additional observations on hover-feeding by North American ardeids. *Wilson Bulletin*, **87**, 420.

Rodgers, J. A., Jr. (1977). Breeding displays of the Louisiana Heron. *Wilson Bulletin*, **89**, 266–85.

Rodgers, J. A., Jr. (1978a). Breeding behavior of the Louisiana Heron. *Wilson Bulletin*, **90**, 45–59.

Rodgers, J. A., Jr. (1978b). Egg removal by Little Blue Heron. *Florida Field Naturalist*, **6**, 48.

Rodgers, J. A., Jr. (1980a). Little Blue Heron breeding bare-part coloration and plumage characteristics. *Florida Field Naturalist*, **8**, 46–7.

Rodgers, J. A., Jr. (1980b). Breeding ecology of the Little Blue Heron on the west coast of Florida. *Condor*, **82**, 164–9.

Rodgers, J. A., Jr. (1980c). Little Blue Heron breeding behavior. *Auk*, **97**, 371–84.

Rodgers, J. A., Jr. (1980d). Reproductive success of three heron species on the west coast of Florida. *Florida Field Naturalist*, **8**, 37–40.

Rodgers, J. A., Jr. (1982). Food of nestling Little Blue Herons on the west coast of Florida. *Florida Field Naturalist*, **10**, 25–30.

Rodgers, J. A., Jr. (1983). Foraging behavior of seven species of herons in Tampa Bay, Florida. *Colonial Waterbirds*, **6**, 11–23.

Rodgers, J. A., Jr. (1987). Breeding chronology and reproductive success of Cattle Egrets and Little Blue Herons on the west coast of Florida, USA. *Colonial Waterbirds*, **10**, 38–44.

Rodgers, J. A., Jr. and Schwikert, S. T. (1986). Recolonization of Pelican Island by Reddish Egrets. *Florida Field Naturalist*, **14**, 76–7.

Rodgers, J. A., Jr. and Schwikert, S. T. (1999). Breeding ecology of the Least Bittern in Central Florida. *Florida Field Naturalist*, **27**, 141–9.

Rodgers, J. A., Jr. and Smith, H. T. (1995a). Little Blue Heron, *Egretta caerulea*. In *The birds of North America*. No. 145 (eds. A. Poole and F. Gill), pp. 1–32. The Academy of Natural Sciences, Philadelphia, Pennsylvania and The American Ornithologists' Union, Washington.

Rodgers, J. A., Jr. and Smith, H. T. (1995b). Set-back distances to protect nesting bird colonies from human disturbance in Florida. *Conservation Biology*, **9**, 89–99.

Rodriguez de los Santos, M. and Canavate, J. P. (1985). Selection des proies par le heron pourpre *Ardea purpurea* pendant la periode de reproduction dans les marais du Guadalquivir (Espagne). *Oiseau et la Revue Francaise d'Ornithologie*, **55**, 195–203.

Rodriguez-Mojica, R. (2002). A Reddish Egret (*Egretta rufescens*) in the western coast of Puerto Rico. *El Pitirre*, **15**, 40–1.

Rodwell, S. P. (1996). Notes on the distribution and abundance of birds observed in Guinea-Bissau, 21 February to 3 April 1992. *Malimbus*, **18**, 25–43.

Rogers, M. J. (2002). Report on rare birds in Great Britain in 2001. *British Birds*, **95**, 476–528.

Rohwer, S. (1990). Foraging differences between white and dark morphs of the Pacific Reef Heron *Egretta sacra*. *Ibis*, **132**, 21–6.

Rojas, L. M., McNeil, R., Cabana, T., and Lachapelle, P. (1999). Behavioral, morphological and physiological correlates of diurnal and nocturnal vision in selected wading bird species. *Brain Behavior and Evolution*, **53**, 227–42.

Root, T. (1988). *Atlas of wintering North American birds, an analysis of Christmas Count Data*. University of Chicago Press, Chicago.

Rose, P. M. and Scott, D. A. (1994). *Waterfowl population estimates. International Waterfowl and Wetlands Research Bureau*, IWRB Publication No. 29, Slimbridge.

Rose, P. M. and Scott, D. A. (1997). *Waterfowl population estimates*. 2nd ed. Wetlands International Publication 44. Wetlands International, Wageningen.

Rosenberg, G. H. and Stejskal, D. (1992). The spring season. March 1–May 31, 1992. Southwest Region Arizona. *American Birds*, **46**, 457–9.

Ross, A. G. and Bell, P. T. (1985). Green Backed Heron in Humberside. *British Birds*, **78**, 106–7.

Rossi, J. A. H. (1958). Contribución al conocimento de la Biologia de la Garza Roja. *Universidad de Buenos Aires Facultad de Ciencias Exactas y Naturalis, ser. Zoo*, **1**, (2).

Round, P. D. (1992a). Recent reports. February–April 1992. *Bangkok Bird Club Bulletin*, **9**, 11–12.

Round, P. D. (1992b). Recent reports June–July 1992. *Bangkok Bird Club Bulletin*, **9**, 11.

Round, P. D. (1994). Recent reports April–August 1994. *Bird Conservation Society of Thailand Bulletin*, **11**, 12–13.

Round, P. D. (1995a). Ringing recovery. *Bird Conservation Society of Thailand Bulletin*, **12**, 16.

Round, P. D. (1995b). Recent reports March–May 1995. *Bird Conservation Society of Thailand Bulletin*, **12**, 14.

Round, P. D. (1995c). Recent reports April–June 1995. *Bird Conservation Society of Thailand Bulletin*, **12**, 14–15.

Rowe, S. and Rowe, J. (1998). Falcon chases heron. *OSNZ (Ornithological Society of New Zealand) News*, **87**, 12.

Rowlands, B. W. (1987). Seabirds and shorebirds observed on Rodrigues and its lagoon islets in July 1986. *Cormorant*, **14**, 31–8.

RSPN. (2003). *White-bellied Heron study project, first quarter report, April–Aug 2003.*, Royal Society for Protection of Nature, Bhutan.

Ruiz, X. (1985). An analysis of the diet of Cattle Egrets in the Ebro Delta, Spain. *Ardea*, **73**, 49–60.

Rusev, I. T. (1984). Wintering of the Night Heron (*Nycticorax nycticorax*) in the Dnestr River Delta. *Ornitologiya*, **19**, 185–6.

Ryan, P. (1999). River dance. *Africa Birds & Birding*, **4**, 48–51.

Ryan, R. C. (1997). First record of the Little Egret Egretta garzetta for Guyana. *Cotinga*, **8**, 92.

Ryder, R. A. (1978). Breeding, distribution, movements, and mortality of Snowy Egrets in North America. In *Wading birds* (eds. A. Sprunt, IV, J. C. Ogden, and S. Winckler), pp. 197–205. National Audubon Society Research Report No. 7. National Audubon Society, New York.

Ryder, R. A. (1998). Snowy Egret. In *Colorado breeding bird atlas* (eds. H. E. Kingery), pp. 56–7. Colorado Bird Atlas Partnership and Colorado Division of Wildlife and Parks, Denver.

Ryder, R. A., Schnaderbeck, R. W., and Jeske, C. W. (1989). The Little Blue Heron, a new breeding bird for Colorado. *C.F.O. Journal*, **23**, 102–3.

Sage, B. (1994). Great Blue Herons *Ardea herodias* and Magnificent Frigatebirds *Fregata magnificens* feeding on young turtles. *Bulletin of the British Ornithologists' Club*, **115**, 68.

Saikia, P. and Bhattacharjee, P. C. (1990). The present status of waders and other water birds of Brahmaputra Valley, Assam (India). *Stilt*, **17**, 65–70.

Saleh, M. A. and Mohamed, S. A. (1990). The Western Reef Heron (*Egretta gularis* Hemprich & Ehrenberg, 1833) in Bahrain. *Fauna of Saudi Arabia*, **11**, 306–17.

Salewski, V. and Grafe, T. U. (1999). New tape recordings of three West African birds. *Malimbus*, **21**, 117–21.

Samusenko, I. E. (1993). Distribution and population dynamics of Grey Heron (*Ardea cinerea*) in Belorussia. *Ring*, **15**, 115–20.

Samusenko, I. E. and Pinchuk, P.V. (1999). [The first record of breeding Black-crowned Night-Heron (*Nycticorax nycticorax*) in Belarus.] *Subbuteo*, **2**, 51.

Sanders, E. and Ebels, E. B. (1998). Melanistische blauwe reigers in Nederland. *Dutch Birding*, **20**, 113–15.

Sandilands, A. P. (1977). Green Heron nesting in a Wood Duck box. *Auk*, **94**, 390.

Sanger, M. B. (1967). *World of the Great White Heron*. Devin-Adair Company, New York.

Sarasa, C. G., Bartolome, J., Fernandez-Cruz, M., and Farinha, J. C. (1993). Segundo censo de Ardeidas invernantes en la Peninsula Ibérica y Baleares (1992–93). *Airo*, **4**, 41–50.

Sasaki, M. (2001). Distribution of breeding colonies of herons and egrets and their protection in Kyoto Prefecture. *Strix*, **19**, 149–60.

Sato, N. and Maruyama, N. (1996). Foraging site preference of Intermediate Egrets *Egretta intermedia* during the breeding season in the eastern part of the Kanto Plain, Japan. *Journal of the Yamashina Institute for Ornithology*, **28**, 19–34.

Sauer, J. (1991). The status and distribution of Great Egrets in Wisconsin. *Passenger Pigeon*, **53**, 303–13.

Sawara, Y., Azuma, N., Hino, K., Fukui, K., Demachi, G., and Sakuyama, M. (1992). Feeding activity of the Grey Heron *Ardea cinerea* in tidal and non-tidal environments. *Japanese Journal of Ornithology*, **39**, 45–52.

Sawara, Y., Sakuyama, M., and Demachi, G. (1994). [Diets and foraging site utilization of the Grey Heron, *Adrea cinerea*, in the breeding season.] *Japanese Journal of Ornithology*, **43**, 61–71.

Schaldach, W., Jr., Escalante, P. B. P., and Winker, K. (1997). Further notes on the avifauna of Oaxaca, Mexico. *Anales del Instituto de Biologia Universidad Nacional Autonoma de Mexico Serie Zoologia*, **68**, 91–135.

Schantz, T. (1996). Reddish Egret in Jasper County. *Iowa Bird Life*, **66**, 63–4.

Scharf, W. C. (1989). Coastal Great Blue Heron and Great Egret colonies of the Michigan Great Lakes. *Jack-Pine Warbler*, **67**, 52–65.

Schels, C. (1993). The second breeding record of the Goliath Heron *Ardea goliath* in Somalia. *Scopus*, **17**, 59–60.

Scherner, E. R. (2001). Ein weisser Reiher mit Merkmalen von *Egretta gularis* in Nieders. *Ornithologsche Mitteilungen*, **53**, 379–84.

Schipper, W. L. (1985). Observations of birds on Kwajalein Atoll 1978–1983. *Elepaio*, **46**, 27–32.

Schlumprecht, H. (1999). Die grosse Rohrdommel als Zielart. Leitbilder, Ziele und Massnahmen am Beispiel eines LIFE-Projektgebiets. *Naturschutz und Landschaftsplanung*, **31**, 247–52.

Schodde, R., Mason, I. J., Dudzinski, M. L., and McKean, J. L. (1980). Variation in the Striated Heron *Butorides striatus* in Australasia. *Emu*, **80**, 203–12.

Schogolev, I.V. (1996). Fluctuations and trends in breeding populations of colonial waterbirds in the Dnestr Delta, Ukraine, Black Sea. *Colonial Waterbirds*, **19**, **Special Publication 1**, 91–7.

Schonwetter, M. (1967). *Handbuch der Oölogie*. AkademieVerlag, Berlin.

Schuster, A., Nemeth, E., Grull, A., and Rossler, M. (1998). The Little Egret (*Egretta garzetta*): A new breeding bird for Austria. *Egretta*, **41**, 61–6.

Schwabl, H., Mock, D. W., and Gieg, J. A. (1997). A hormonal mechanism for parental favouritism. *Nature*, **386**, 231.

Scientific Advisory Committee. (1997). *Final submitted recommendation on a nomination for listing: Little Bittern Ixobrychus minutus.* Scientific Advisory Committee for

Flora and Fauna Nomination no. 439. Department of Natural Resources and Environment, Melbourne.

Scott, D. A. (1984). The feeding success of Cattle Egrets in flocks. *Animal Behaviour*, **32**, 1089–100.

Scott, D. A. (1989). *A directory of Asian wetlands*. IUCN, The World Conservation Union, Gland.

Scott, D. A. and Carbonell, M. (1986). *A directory of neotropical wetlands*. IUCN and IWRB, Slimbridge, U.K.

Scott, J. M., Pyle, R., and Coleman, R. (1983). Records of small white egrets in Hawaii and Samoa with notes on identification. *Elepaio*, **43**, 79–82.

Scott, M. D. and Powell, J. A. (1982). Commensal feeding of Little Blue Herons with manatees. *Wilson Bulletin*, **94**, 215–16.

Seabolt, T. (1984). The aggressive responses of a nesting Osprey in a Great Blue Heron colony. *Loon*, **56**, 214–15.

Selby, A. D. (1997). Some further notes on albino birds. *Bird Observer*, **668**, 107.

Sellin, D. (1986). Zam Nahrungserwerb uberwinternder Graureiher (*Ardea cinerea*). *Oekologie der Voegel*, **8**, 110–12.

Seton, D. H. C. (1972). Observations on breeding of the Great-billed Heron in Northern Queensland. *Emu*, **73**, 9–11.

Severinghaus, L. L. (1989). Endangered birds in Taiwan. In *Proceedings of the joint meeting of ICBP Asian Section and East Asian Bird Protection Conference, Bangkok, Thailand, April 1989*. ICBP, Cambridge.

Shanley, E., Jr. (1984). Snowy Egrets and Great Egrets using foot-dragging behavior in aerial feeding. *Bulletin of the Texas Ornithological Society*, **17**, 17–18.

Sheldon, F. H. (1987). Phylogeny of herons estimated from DNA-DNA hybridization data. *Auk*, **104**, 97–108.

Sheldon, F. H. and Marin A, M. (1984). The sympatry of Night Herons in Borneo. *Bulletin of the British Ornithologists' Club*, **105**, 76–8.

Sheldon, F. H., Jones, C. E., and K. G. McCracken. (2000). Relative patterns and rates of evolution in heron nucler and mitochondrial DNA. *Molecular Biology and Evolution*, **17**, 437–450.

Sheldon, F. H., McCracken, K. G., and Stuebing, K. D. (1995). Phylogenetic relationships of the Zigzag Heron (*Zebrilus undulatus*) and White-crested Bittern (*Tigriornis leucolophus*) estimated by DNA–DNA hybridization. *Auk*, **112**, 672–9.

Sheldon, F. H., Moyle, R. G., and Kennard, J. (2001). Ornithology of Sabbah: History, gazetteer, annotated checklist and bibliography. *Ornithological Monographs* **52**, American Ornithologists' Union, Washington.

Sheldon, F. H. and Slikas, B. (1997). Advances in Ciconiiform systematics 1976–1996. *Colonial Waterbirds*, **20**, 106–14.

Sheldon, F. H. and Whittingham, L. A. (1997). Phylogeny in studies of bird ecology, behavior, and morphology. In *Avian molecular evolution and systematics* (ed. D. P. Mindell), pp. 279–99. Academic Press.

Shen, R. C. and Chen, L. J. (1996). A preliminary study of the habitat, distribution and breeding behavior of the tiger bittern (*Gorsachius melanolophus*) in Taiwan. *Annual of Taiwan Museum*, **39**, 183–99.

Shen, Y., Hu, X., and Liu, X. (1987). [A study of the breeding ecology of Chinese Pond Heron in Changsha.] *Natural Science Journal of Hunan Normal University*, **10**, 65–73.

Shernazarov, E. (1995). Distribution and number of Great egret (*Egretta alba*) in Uzbekistan. *Ornitologiya*, **26**, 220–2.

Shimada, T., Kuwabara, K., Minowa, Y., Kaneda, H., and Suzuki, Y. (1994). [Seasonal fluctuations of heron populations at the mouth of the Tama River in Tokyo and Kanagawa Prefecture.] *Strix*, **13**, 85–92.

Shiquan, M (1990). Study of the distribution pattern of population of the Yellow Bittern in breeding season. *Acta Ecolgica Sinica*, **10**, 362–6.

Short, L. L. (1969). Agonistic display of the Whistling Heron *Syrigma sybilatrix*. *Wilson Bulletin*, **81**, 330–1.

Shrubb, M. (2003). Farming and birds: An historic perspective. *British Birds*, **96**, 158–77.

Si Bachi, A., Hafner, H., Tourenq, J. N., and Doumandji, S. (2000). Structure de l'habitat et biologie de reproduction du heron garde-boeuf, *Bubulcus ibis*, dans une colonie de la vallee de la Soummam (Petite Kabylie, Algerie). *Revue d'Ecologie la Terre et la Vie*, **55**, 33–43.

Sibley, C. G. and Monroe, B. L. (1990). *Distribution and taxonomy of birds of the world*. Yale University Press, New Haven.

Sick, H. and Belton, W. translator. (1993). *Birds in Brazil*. Princeton University Press, Princeton.

Siegfried, W. R. (1966a). Age at which Cattle Egrets first breed. *Ostrich*, **37**, 198.

Siegfried, W. R. (1966b). The status of the Cattle Egret in South Africa with notes on neighbouring territories. *Ostrich*, **37**, 157–69.

Siegfried, W. R. (1966c). On the food of nestling Cattle Egrets. *Ostrich*, **37**, 219–20.

Siegfried, W. R. (1969). Energy metabolism of Cattle Egrets. *Zoologica Africana*, **4**, 265–73.

Siegfried, W. R. (1970). Mortality and dispersal of ringed Cattle Egrets. *Ostrich*, **41**, 122–35.

Siegfried, W. R. (1971a). The food of the Cattle Egret. *Journal of Applied Ecology*, **8**, 447–68.

Siegfried, W. R. (1971b). Communal roosting of the Cattle Egret. *Tansactions of the Royal Society of South Africa*, **39**, 419–43.

Siegfried, W. R. (1971c). Feeding activity of the Cattle Egret. *Ardea*, **59**, 38–46.

Siegfried, W. R. (1971d). Plumage and moult of the Cattle Egret. *Ostrich*, **Supplement 9**, 153–64.

Siegfried, W. R. (1972a). Aspects of the feeding ecology of Cattle Egrets (*Ardeola ibis*) in South Africa. *Journal Animal Ecology*, **41**, 71–8.

Siegfried, W. R. (1972b). Breeding success and reproductive output of the Cattle Egret. *Ostrich*, **43**, 43–55.

Siegfried, W. R. (1973). Food requirements and growth of Cattle Egrets in South America. *Living Bird*, **11**, 193–206.

Siegfried, W. R. (1978). Habitat and modern range expansion of the Cattle Egret. In National Audubon Society Research Report 7 (eds. A. Sprunt, J. Ogden, and S. Winckler), pp. 315–24, National Audubon Society, New York.

Silcock, R. (1984). First nesting of Cattle Egret & Little Blue Heron for Iowa. *Iowa Bird Life*, **54**, 101–3.

Silvius, M. (1986). South-east Sumatra. *Interwader*, **7**, 7.

Silvius, M. J. (1991). Chinese Egret notes. *Bulletin of the Oriental Bird Club*, **13**, 33.

Simpson, K., Smith, J. N. M., and Kelsall, J. P. (1987). Correlates and consequences of coloniality in Great Blue Herons. *Canadian Journal of Zoology*, **65**, 572–7.

Sindar, K. S. G. (in prep). Abundance and seasonality of Pond Herons *Ardeola grayii* with red legs in Etawah and Mainpuri districts, Uttah Pradesh.

Singh, J. P. and Roy, S. P. (1995). Territorial behaviour in wader population (Ardeidae: Ciconiiformes) of Kawar Lake, Begusarai, North-Bihar. *Journal of Freshwater Biology*, **7**, 269–72.

Sirois, J., Low, G., and McCollum, J. T. (1991). First Great Egret recorded in the Northwest Territories. *Blue Jay*, **49**, 36–7.

Sivasubramanian, C. (1988). Aerial feeding by Median Egret (*Egretta intermedia*), Little Egret (*Egretta garzetta*) and Pond Heron (*Ardeola grayii*). *Journal of the Bombay Natural History Society*, **85**, 611–12.

Skeen, R. Q. (1988). Unusual escape flight of Grey Heron. *British Birds*, **81**, 236–7.

Skinner, J., Wallace, J. P., Altenburg, W., and Fofana, B. (1987). The status of heron colonies in the Inner Niger Delta Mali. *Malimbus*, **9**, 65–82.

Skipnes, K. and Folvik, A. (1997). Forstegangsfunn av fugler i Norge, i fugleskinnsamlingen pa Stavanger Museum. *Stavanger Museum Arbok*, **107**, 91–102.

Small, A. (1994). *California birds: their status and distribution*. Ibis Publishing Co., Vista, California.

Smiddy, P. and Duffy, B. (1997). Little Egret *Egretta garzetta*: a new breeding bird for Ireland. *Irish Birds*, **6**, 55–6.

Smiddy, P. and O'Sullivan, O. (1998). The status of the Little Egret *Egretta garzetta* in Ireland. *Irish Birds*, **6**, 201–6.

Smith, J. P. (1995a). Foraging flights and habitat use of nesting wading birds (Ciconiiformes) at Lake Okeechobee, Florida. *Colonial Waterbirds*, **18**, 139–58.

Smith, J. P. (1995b). Foraging sociability of nesting wading birds (Ciconiiformes) at Lake Okeechobee, Florida. *Wilson Bulletin*, **107**, 437–51.

Smith, J. P. (1997). Nesting season food habits of 4 species of herons and egrets at Lake Okeechobee, Florida. *Colonial Waterbirds*, **20**, 198–220.

Smith, J. P. and Collopy, M. W. (1995). Colony turnover, nest success and productivity, and causes of nest failure among wading birds (Ciconiiformes) at Lake Okeechobee, Florida (1989–1992). *Ergebnisse der Limnologie*, **45**, 287–316.

Smith, J. P., Richardson, J. P. and Collopy, M. W. (1995). Foraging habitat selection among wading birds (Ciconiiformes) at Lake Okeechobee, Florida, in relation to hydrology and vegetative cover. *Archives Hydrobiologia*, **45**, 247–85.

Smith, K. W. and Tyler, G. A. (1993). Trends in the numbers of breeding bitterns in the UK. In *Britain's birds in 1990–91: the conservation and monitoring review* (eds. J. Andrews and S. P. Carter), pp. 139–40. British Trust for Ornithology & Joint Nature Conservation Committee, Thetford & Peterborough, UK.

Smith, K., Welch, G., Tyler, G., Gilbert, G., Hawkins, I., and Hirons, G. (2000). Management of RSPB Minsmere Reserve reedbeds and its impact on breeding bitterns. *British Wildlife*, **12**, 16–21.

Smith, P. H. (1989). Great White Egrets feeding in wake and robbing Black-headed Gulls. *British Birds*, **82**, 27.

Smith, P. W. and Hutt, M. B. (1984). First sight record of Western Reef-Herons for Barbados. Are we about to see another transatlantic colonization? *American Birds*, **38**, 254–6.

Smith, P. W. and Smith, S. A. (1990). Gray Herons on Barbados. Has this species been overlooked on the East coast? *American Birds*, **44**, 1096.

Smythies, B. E. (1986). *The birds of Burma*. World of Ornithology Series. Nimrod Press Ltd., Liss and Silvio Mattacchione & Co., Pickering.

Soderholm, S. (2001). Flying Bitterns *Botaurus stellaris* with unusual behavior. *Ornis Svecica*, **11**, 103–8.

Sodhi, N. S. (1985). Food of Black-crowned Night Heron (*Nycticorax nycticorax nycticorax*) nestlings. *Pavo*, **23**, 47–52.

Sodhi, N. S. (1986). Feeding ecology of Indian Pond Heron and its comparison with that of Little Egret. *Pavo*, **24**, 97–112.

Sodhi, N. S. (1992). Food-niche relationships of five sympatric north Indian herons. *Forktail*, **7**, 125–30.

Sodhi, N. S. and Khera, S. (1984). Food, food requirement during growth, and feeding behaviour of nestling *Bubulcus ibis coromandus* (Boddaert). *Pavo*, **22**, 21–9.

Sodhi, N. S. and Khera, S. (1986). Feeding habits of the Median Egret. *Research Bulletin of the Panjab University, Science*, **37**, 9–12.

Soto-Largo, E., Velasco, T., and Sanz-Zuasti, J. (1998). La población reproductora del avetoro común (*Botaurus stellaris*) en la Ribera Navarra. *Anuario Ornitologico de Navarra*, **5**, 35–41.

Spaans, A. L. and de Jong, B. H. J. (1982). Present status of some colonial waterbirds in Suriname, South America. *Journal of Field Ornithology*, **53**, 269–72.

Spahn, S. A. and Sherry, T. W. (1999). Cadmium and lead exposure associated with reduced growth rates, poorer fledging success of Little Blue Heron chicks (*Egretta caerulea*) in south Louisiana wetlands. *Archives of Environmental Contamination and Toxicology*, **37**, 377–84.

Spalding, M. G., Frederick, P. C., McGill, H. C., Bouton, S. N., and McDowell, L. R. (2000a). Methylmercury accumulation in tissues and its effects on growth and appetite in captive Great Egrets. *Journal of Wildlife Diseases*, **36**, 411–22.

Spendelow, J. A. and Patton, S. R. (1988). *National atlas of coastal waterbird colonies in the contiguous United States: 1976–82*. Biological Report 88(5). U.S. Fish and Wildlife Service, Washington.

Squires, J. R. (1998). Attempted kleptoparasitism of Ospreys by Great Blue Herons. *Wilson Bulletin*, **110**, 560.

Stagg, A. J. (1984). Birds of Saudi Arabia. A note on the breed birds of Kutambil Island on the Red Sea coast of Saudi Arabia. *Fauna of Saudi Arabia*, **6**, 546–8.

Stanworth, A. (2001). The Green Heron in Lincolnshire. *Birding World*, **14**, 370–1.

Stawarczyk, T. (1984). [Occurrence of Great White Egret in Poland in post-war period.] *Notatki Ornitologiczne*, **25**, 3–13.

Steadman D. W., Norton, R. K., Browning, M. R., and Arendt, W. J. (1997). The birds of St. Kitts, Lesser Antilles. *Caribbean Journal of Science*, **33**, 1–20.

Steadman, D. W., Tellkamp, M. P., and Wake. T. A. (2003). Prehistoric exploitation of birds on the Pacific Coast of Chiapas, Mexico. *Condor*, **105**, 572–9.

Steadman, D. W., Worthy, T. H., Anderson, A. J., and Walter, R. (2000). New species and records of birds from prehistoric sites on Niue, southwest Pacific. *Wilson Bulletin*, **112**, 165–86.

Stewart, J. R. (2004). Wetland birds in the recent fossil record of Britain and northwest Europe. *British Birds*, **97**, 33–43.

Stickley, A. R., Jr., Glahn, J. F., King, J. O., and King, D. T. (1995). Impact of Great Blue Heron depredations on channel catfish farms. *Journal of the World Aquaculture Society*, **26**, 194–9.

Stiles, F. G. and Skutch, A. F. (1989). *A guide to the birds of Costa Rica*. Comstock Publishing Associates, Ithaca.

Stival, E. (1987). Un caso de cleptoparassitismo di sterna comune *Sterna hirundo* nei confronti de sgarza ciuffetto *Ardeola ralloides*. *Avocetta*, **11**, 67.

Stocker, R. (1994). Green-backed Heron preying on Quelea. *Honeyguide*, **40**, 247.

Stokes, T. (1983). Bird casualties in 1975–76 at the Booby Island lightstation, Torres Strait. *Sunbird*, **13**, 53–8.

Stokes, T. (1990). Confirmation of Yellow Bittern record from Christmas Island, Indian Ocean. *Australian Bird Watcher*, **13**, 165.

Stokes, T., Sheils, W., and Dunn, K. (1984). Birds of the Cocos (Keeling) Islands, Indian Ocean. *Emu*, **84**, 23–8.

Stolen, E. D. (2001). Great Blue Heron eating a Pied-billed Grebe. *Florida Field Naturalist*, **29**, 87.

Straube, F. C. (1991). Novos registros de duas aves raras no Estado do Paraná: *Crypturellus noctivagus* (Tinamiformes: Tinamidae) e *Tigrisoma fasciatum* (Ciconiiformes: Ardeidae). *Ararajuba*, **2**, 93–4.

Strong, A. M., Bancroft, G. T., and Jewell, S. D. (1997). Hydrological constraints on Tricolored Heron and Snowy Egret resource use. *Condor*, **99**, 894–905.

Stuart, C. T. and Dürk, T. E. (1984). Food of the Blackheaded Heron at a breeding colony. *Ostrich*, **55**, 103–4.

Subramanya, S. (1996). Distribution, status and conservation of Indian heronries. *Journal of the Bombay Natural History Society*, **93**, 459–86.

Sueur, F. (1998). Commensalisme de trois espèces d'ardéidés vis-à-vis du grand cormoran *Phalacrocorax carbo*. *Alauda*, **66**, 168.

Sugathan, R., Melville, D. S., and Alagar Rajan S. (1987). Further additions to the avifauna of Point Calimere. *Journal of the Bombay Natural History Society*, **84**, 206–7.

Sullivan, J. P. (1988). Effects of provisioning rates and number fledged on nestling aggression in Great Blue Herons. *Colonial Waterbirds*, **11**, 220–6.

Summers-Smith, D. (1984). Bird notes from the Cape Verde Islands. *Bulletin of the British Ornithologists' Club*, **104**, 148–9.

Sutton, G. M., Lea, R. B., and Edwards, E. P. (1950). Notes on the ranges and breeding habits of certain Mexican birds. *Bird-Banding*, **21**, 45–59.

Svedarsky, W. D. (1992). Observations on nesting of the American Bittern in northwest Minnesota. *Prairie Naturalist*, **24**, 241–50.

Swennen, C. and Won, P. O. (1993). Study of egrets in the Shin Islet colony, Korea, 1991. *Bulletin of the Institute of Ornithology, Kyung Hee University*, **4**, 106–11.

Sykes, B. R. (1996). The status and identification of *Gorsachius* herons in southern Taiwan. *Bulletin of the Oriental Bird Club*, **23**, 57–8.

Takumi, Y. and Ezaki, Y. (1998). Habitat segregation among herons and egrets in a pond-paddy complex area in Harima, Hyogo Prefecture. *Japanese Journal of Ecology Tokyo*, **48**, 17–26.

Tallone, G. (1991). Una garzaia di airone cenerino (*Ardea cinerea*) in ambiente suburbano: studio e gestione (Aves, Ardeidae). *Rivista Piemontese di Storia Naturale*, **12**, 89–99.

Tarboton, W. R. (1980). Notes on Dwarf Bittern. *WBC News*, **108**, 35.

Tarboton, W. (1996). Slaty Egret. First breeding record for South Africa. *Africa Birds & Birding*, **1**, 8.

Taylor, P. S. and Carss, D. N. (1990). An observation of a heron plunge-dive. *Scottish Birds*, **16**, 44.

Teixeira, D. M. (1991). O uso de instrumentos por algumas aves brasileiras. *Ararajuba*, **2**, 89–90.

Teixeira, D. M. and Alvarenga, H. M. F. (1985). The first recorded Cory's Bittern (*Ixobrychus 'neoxenus'*) from South America. *Auk*, **102**, 413.

Teixeira, D. M. and Carvalho, M. C. S. (1985). Notas sobre a garcea-real, *Pilherodius pileatus* (Boddaert, 1783). *Anais da Sociedade Sul-Riograndense de Ornitologia*, **3**, 13–15.

Teixeira, D. M. and Nacinovic, J. B. (1985). O soco-boi-baio, *Botaurus pinnatus* (Wagler, 1829), no Rio de Janeiro. *Anais da Sociedade Sul-Riograndense de Ornitologia*, **3**, 9–12.

Teixeira, D. M., Nacinovic, J. B., and Pontual, F. B. (1987). Notes on some birds of northeastern Brazil (2). *Bulletin of the British Ornithologists' Club*, **107**, 151–7.

Teixeira, D. M., Otoch, R., Luigi, G., Raposo, M. A., and de Almeida, A. C. C. (1993). Notes on the birds of northeastern Brazil. *Bulletin of the British Ornithologists' Club*, **113**, 48–52.

Telfair, R. C., II. (1983). *The Cattle Egret: a Texas focus and world view.* Texas Agricultural Experimental Station, Texas A & M University, College Station.

Telfair, R. C., II. (1994). Cattle Egret, *Bubulcus ibis.* In *The birds of North America* No. 113 (eds. A. Poole and F. Gill), pp. 1–32. The Academy of Natural Sciences, Philadelphia and The American Ornithologists' Union, Washington.

Telfair, R. C. III and Bister, T. J. (2004). Long-term breeding success of Cattle Egrets in Texas. *Waterbirds*, **27**, 69–78.

Telfair, R. C., II., McCrimmon, D. A., Jr., and Fryska, S. T. (2000). Population dynamics of the Cattle Egret in Texas, 1954–1999. *Waterbirds*, **23**, 187–95.

Terbough, J. W., Fitzpatrick, J. W., and Emmons, L. (1984). Annotated checklist of bird and mammal species of Cocha-Cashu Biological Station, Manu National Park, Peru. *Fieldiana Zoology*, **21**, 1–29.

Thomas, C. M. and Anthony, R. G. (2003). Environmental contaminants and nesting behavior of Great Blue Herons from the Columbia and Williamette Rivers, Oregon and Washington. *Northwest Science*, **77**, 320–30.

Thomas, F., Deerenberg, C., Lepley, M., and Hafner, H. (1999). Do breeding site characteristics influence breeding performance of the purple heron *Ardea purpurea* in the Camargue? *Revue d'Ecologie la Terre et la Vie*, **54**, 269–281.

Thomas, F. and Hafner, H. (2000). Breeding habitat expansion in the Grey Heron (*Ardea cinerea*). *Acta Oecologica*, **21**, 91–5.

Thomas, F., Kayser, Y., and Hafner, H. (1999). Nesting size rank in the Little Egret (*Egretta garzetta*) influences subsequent breeding success of offspring. *Behavioral Ecology and Sociobiology*, **45**, 466–70.

Thompson, C. F., Lanyin, S. M., and Thompson, K. M. (1982). The influence of foraging benefits on association of Cattle Egrets (*Bubulcus ibis*) with cattle. *Oceologia*, **52**, 167–70.

Thompson, P. M., Harvey, W. C., Johnson, D. L., Millin, D. J., Rashid, S. M. A., Scott, D. A., Stanford, C., and Woolner, J. D. (1993). Recent notable bird records from Bangladesh. *Forktail*, **9**, 13–44.

Thuy, L. D. (1994). [Counting number of two species: Little Egret and Little Cormorant in breeding season in Bac Lieu colony—Minh Hai Province.] *Tap Chi Sinh Hoc*, **16**, 5–8.

Tkachenko, A. A. and Shakula, V. F. (1983). Some behavioral features of interspecies relationships in a mixed colony of ciconiiformes and corvidae. In *Proceedings of the 3rd All-Union Conference On Animal Behavior, Vol. 2. Behavior of Animals In Communities* (ed. A. A. Zakharov), pp. 190–2. Moscow.

Tobish, T. G., Jr. and Isleib, M. E. (1992). The Summer season. June 1–July 31, 1992. Alaska region. *American Birds*, **46**, 1165–7.

Todhunter, J. F. (1987). Carrion Crow versus Grey Heron. *SOG (Suffolk Ornithologists' Group) Bulletin*, **76**, 31–2.

Tojo, H. (1996). Habitat selection, foraging behaviour and prey of five heron species in Japan. *Japanese Journal of Ornithology*, **45**, 141–58.

Toland, B. (1991). Successful nesting by Reddish Egrets at Oslo Island, Indian River County, Florida. *Florida Field Naturalist*, **19**, 51–3.

Toland, B. (1999). Mid-air capture of fish by Tricolored Herons and Snowy Egrets in southeastern Florida. *Florida Field Naturalist*, **27**, 171–2.

Tomlinson, D. (1992). Rise of the grey fishers. The latest heron figures. *Country Life (London)*, **186**, 54.

Tomlinson, D. N. S. (1974). Studies of the Purple Heron. *Ostrich*, **45**, 149–60, 209–23.

Tomlinson, D. N. S. (1975). Studies of the Purple Heron. *Ostrich*, **46**, 157–65.

Tomlinson, D. N. S. (1976). Breeding behaviour of the Great White Heron. *Ostrich*, **47**, 61–78.

Tomlinson, D. N. S. (1979). Interspecific relations in a mixed heronry. *Ostrich*, **50**, 193–98.

Toquenaga, Y., Kajitani, I., and Hoshino, T. C. (1995). Egrets of a feather flock together. *Artificial Life*, **1**, 391–411.

Tosi, G. and Toso, S. (1979). Night Herons (*Nycticorax nyxticorax*) wintering in the Po River Valley. *Ibis*, **121**, 336.

Tourenq, C., Barbraud, S., Sadoul, N., Sandoz, A., Lombardini, K.Kayser, Y., and Martin, J.-L. (2001). Does foraging habitat quality affect reproductive performance in the Little Egret, *Egretta garzetta*? *Animal Biodiversity and Conservation*, **24**, 107–16.

Traube, F. and Grylle, M. (2001). The Intermediate Egret in Italy: A new European bird. *Birding World*, **14**, 247–9.

Tremblay, J. and Ellison, L. N. (1979). Effects of human disturbance on breeding of Black-crowned Night Herons. *Auk*, **96**, 364–9.

Tsao, M. H. (1998). A record of Malayan Night-Heron. *Yuhina*, **58**, 25–8.

Tsuboshima, Y. (1994). [Little Egrets using their own beaks as baits to lure fish.] *Strix*, **13**, 221–3.

Tucker, G. M. and Heath, M. F. (1994). *Birds in Europe: their conservation status.* BirdLife Conservation Series 3. BirdLife International, Cambridge.

Turner, D. (2000). Herons in Africa and the Malagassy region. In *Heron conservation* (eds. J. A. Kushlan and H. Hafner), pp. 99–122. Academic Press, San Diego.

Tyler, G. (1994). Management of reedbeds for bitterns and opportunities for reedbed creation. *RSPB Conservation Review*, **8**, 57–62.

Tyler, G. A., Smith, K. W., and Burges, D. J. (1998). Reedbed management and breeding bitterns *Botaurus stellaris* in the U.K. *Biological Conservation*, **86**, 257–66.

Tyler, S. J. (1979). Underwing feeding by Night Heron. *British Birds*, **72**, 475.

Tyler, S. J. (1993). A commensal feeding association between African Spoonbills *Platalea alba* and Great White Egrets *Egretta alba* in Tanzania. *Scopus*, **16**, 109–10.

Tyler, S. J. (2003). Day and roost counts of Waterbirds in the Okavango Delta, January to March 2003. *Babbler*, **42**, 14–37.

Tyler, S. J. (in prep.). Foraging techniques of the Chinese Pond Heron.

Tyler, S. J. and Tyler, L. (2001). Black-headed Heron *Ardea melanocephalus* eating Lesser Moorhen *Gal angulata*. *Babbler*, **39**, 50.

Uchida, H. and Matsuda, T. (1992). Colonial breeding of the Yellow Bittern *Ixobrychus sinensis*. *Japanese Journal of Ornithology*, **39**, 53–61.

Ueda, K. (1992). [Food of the Yellow Bittern *Ixobrychus sinensis* in boluses regurgitated by nestlings.] *Journal of the Yamashina Institute for Ornithology*, **24**, 13–17.

Ueda, K. (1993). [A case of inter-specific brood parasitism in the moorhen *Gallinula chloropus*.] *Strix*, **12**, 224–6.

Ueda, K. (1994). [The population of egrets at a breeding colony of Inba-gun, Chiba Prefecture.] *Strix*, **13**, 65–72.

Ueda, K. (1996). [Nest site preference and coloniality in the Yellow Bittern *Ixobrychus sinensis*.] *Strix*, **14**, 55–63.

Ullman, M. (1985). [Some flight-characteristics of Purple Heron, *Ardea purpurea*.] *Var Fagelvarld*, **44**, 81–2.

Utschick, H. (1983). Die Brutbestandsentwicklung des Graureihers (Ardea cinerea) in Bayem. *Journal fur Ornithologie*, **124**, 233–250.

Utschick, H. (1984a). Okologische untersuchungen zur rolle des graureihers *Ardea cinerea* in de sportfischerei. *Verhandlungen der Ornitholgischen Gessellschaft in Bayern*, **24**, 111–124.

Utschick, H. (1984b). Untersuchungen zur rolle des graureihers *Ardea cinerea* in der teichwirtschaff. *Verhandlungen der Ornitholgischen Gessellschaft in Bayern*, **24**, 87–110.

Uys, C. J. and Underhill, G. D. (1995). Black-headed Heron catches adult common quail. *Promerops*, **217**, 13.

Van Balen, S. (1999). First record of the Grey Heron in Kalimantan. *Kukila*, **10**, 156–7.

Van Balen, S., Adikerana, A., and Boeadi (1994). Schrenck's Bittern in Barito Ulu, central Kalimantan. *Kukila*, **7**, 77.

Van den Berg, A. B. (2000). Reports. *Dutch Birding*, **22**, 114–73.

van der Ham, N. F. (1984). Squacco Heron attempting to swallow Wood Sandpiper. *British Birds*, **77**, 72.

van der Kelen, G. (1993). Observation d'un heron cendre *Ardea cinerea* nageur se nourrissant d'un cadavre de canard colvert *Anas platyrhynchos* mort de botulisme. *Oiseau et la Revue Francaise d'Ornithologie*, **63**, 133–5.

van der Kooij, H. (1991). Nest-habitat of the purple heron *Ardea purpurea* in the Netherlands. *Limosa*, **64**, 103–12.

van der Kooij, H. (1997). Wordt het broedresultaat van purperreigers *Ardea purpurea* beinvloed door de nesthoogte? *Limosa*, **70**, 145–50.

van der Kooij, H. and Voslamber, B. (1997). Aantalsontwikkeling van de grote zilverreiger *Egretta alba* in Nederland sinds 1970 in een Europees perspectief. *Limosa*, **70**, 119–25.

van Dijk, G. and Ledant, J.-P. (1983). La valeur ornithologique des zones humides de l'est Algerien. *Biological Conservation*, **26**, 215–26.

van Gass, F. (1989). Korsman's Bird Sanctuary, Lakefield, Benoni. *Bokmakierie*, **41**, 3, 16–17.

van Nierkerk, D. J. (1992). [Green-backed Herons near Petrusburg.] *Mirafra*, **9**, 45–6.

van Vessem, J. (1991a). Timing of egg-laying, clutch size and breeding success of the Grey Heron, *Ardea cinerea*, in the north of Belgium. *Gerfaut*, **81**, 177–93.

Van Vessem, J. (1991b). Populaqtion dynamics of breeding Gray Herons (*Ardea cinerea*) in Flanders Belgium during the period 1981–1989. *Ibis*, **133**, 135.

van Vessem, J. and Draulans, D. (1986a). The adaptive significance of colonial breeding in the Grey Heron *Ardea cinerea*: inter- and intra-colony variability in breeding. *Ornis Scandinavica*, **17**, 356–62.

van Vessem, J. and Draulans, D. (1986b). Factors affecting the length of the breeding cycle and the frequency of nest attendance by Grey Herons *Ardea cinerea*. *Bird Study*, **33**, 98–104.

van Vessem, J. and Draulans, D. (1986c). Nest attendance by male and female Gray Herons. *Journal of Field Ornithology*, **57**, 34–41.

van Vessem, J. and Draulans, D. (1987a). Spatial distribution and time budget of radio-tagged Grey Herons, *Ardea cinerea*, during the breeding season. *Journal of Zoology (London)*, **213**, 507–34.

van Vessem, J. and Draulans, D. (1987b). Patterns of arrival and departure of Grey Herons *Ardea cinerea* at two breeding colonies. *Ibis*, **129**, 353–63.

Vanden Berge, J. C. (1970). A comparative study of the appendicular musculature of the order Ciconiiformes. *American Midland Naturalist*, **84**, 289–364.

Vandewalle, F. J. (1985). Combined canopying and foot-stirring functions in the Black Egret. *Bokmakierie*, **37**, 73–5.

Vaniman, M. and Luna, C. (1998). Great Blue Heron observed attempting to ingest an American Coot. *Prairie Naturalist*, **30**, 189–90.

Vargas, H. (1996). First record of the Green Heron (*Butorides viriscens*) in the Galapagos Islands. *Noticias de Galapagos*, **57**, 5–6.

Vaurie, C. (1965). *Birds of the Palaearctic fauna, nonpasserines*. Witherby, London.

Velmala, W., Clarke, T., Lindroos, R. and Nikkinen, L. (2002). A Dwarf Bittern on the Canary Islands—The fourth Western Palearctic record. *Birding World*, **15**, 392–3.

Venchi, A., Camann, S., Tallone, G., Bari, A., Gonella, P., and Piazza, R. (1994). Individual recognition in the Black-crowned Night Heron *Nycticorax n. nycticorax*: the role of the greeting call. *Atti Convegno Italiano di Ornitologia*, **6**, 215–26.

Verbruggen, M. (1989). First breeding by a wild pair of Night Heron in Flanders this century. *Oriolus*, **55**, 148–9.

Verheugt, W. J. M., Skov, H., and Danielsen, F. (1993). Notes on the birds of the tidal lowlands and floodplains of South Sumatra Province, Indonesia. *Kukila*, **6**, 53–84.

Vernon, C. (1993). The bittern in the Cape Province. *Bee-eater*, **44**, 38–43.

Vernon, C. J. (1971). Observations on *Egretta vinaceigula*. *Bulletin of the British Ornithologists' Club*, **91**, 157–9.

Vicens, P. (1997). Situacio de la queca *Botaurus stellaris* a s'Albufera de Mallorca. *Anuari Ornitologic de les Balears*, **12**, 95–104.

Vides-Almonacid, R. (1988). Aportes sobre la presencia en Tucuman del hoco oscuro *Tigrisoma fasciatum pallescens* Olrog. *Neotropica (La Plata)*, **34**, 123–4.

Vincent, A. (1990). Interesting recent sightings. *WBC News*, **149**, 22.

Viney, C. (1988). Great Egret swimming. *Hong Kong Bird Report*, **1988**, 114.

Vogrin, M. (1999). The Great White Egret *Egretta alba* during autumn and winter on Dravsko Polje in northeastern Slovenia. *Ornis Hungarica*, **8–9**, 1–2.

Voisin, C. (1970). Observations sut le comportement du héron bihoreau (*Nycticorax nycticorax*) en période de reproduction. *Oiseau et la Revue Francaise d'Ornithologie*, **40**, 307–39.

Voisin, C. (1976). Etude du comportement de l'aigrette garzette (*Egretta garzetta*) en periode de reproduction. *Oiseau et la Revue Française d'Ornithologie*, **46**, 387–425.

Voisin, C. (1977). Etude du comportement de l'aigrette garzette (*Egretta garzetta*) en periode de reproduction. *Oiseau et la Revue Française d'Ornithologie*, **47**, 65–103.

Voisin, C. (1978). Utilisation des zones humides du Delta Rhodanien par les Ardeides (a suivre). *Oiseau et la Revue Française d'Ornithologie*, **48**, 217–61.

Voisin, C. (1983). Les ardeides du delta du flueve Senegal. *Oiseau Et La Revue Francaise D'ornithologia*, **40**, 307–39.

Voisin, C. (1996). The migration routes of Purple Herons (*Ardea purpurea*) ringed in France. *Vogelwarte*, **38**, 155–68.

Voisin, C. (1991). *The herons of Europe*. T. & A.D. Poyser, London.

Volponi, S. and Emiliani, D. (1995). Nidificazione di airone bianco maggiore, *Egretta alba* (L.) nel biotopo di Punte Alberete (Ravenna). *Supplemento Alle Ricerche di Biologia della Selvaggina*, **22**, 719–22.

Voous, K. H. (1986). Striated or Green Herons in the south Caribbean Islands? *Annalen des Naturhistorischen Museums in Wien Serie B Botanik und Zoologie*, **88–89**, 101–6.

Voous, K. H. and van Marle, J. G. (1988). *The birds of Sumatra: an annotated check-list*. B.O.U. Check-list 10. British Ornithologists' Union, London.

Voskamp, P. and Zoetebier, D. (1999). De kwak *Nycticorax nycticorax* als broedvogel in de uiterwaarden van de IJssel: territoria, nesten, habitat en gedrag. *Limosa*, **72**, 131–42.

Voslamber, B. (1992). Egrets *Egretta* sp. in Oostvaardersplassen in 1991. *Limosa*, **65**, 89–92.

Wakisaka, H. (2002). The first record of the Cinnamon Bittern in Shimane Prefecture. *Strix*, **20**, 185–7.

Wall, L. E. (1986). Strange nest site for White-faced Heron. *Australian Bird Watcher*, **11**, 237.

Wallace, J., Skinner, J., Altenburg, W., and Fofana, B. (1992). Breeding success of three species of egret in the Inner Niger Delta, Mali. *Tauraco*, **2**, 37–45.

Walsh, J. F. (1987). Inland records of Western Reef Heron *Egretta gularis*. *Malimbus*, **9**, 58.

Walsh, J. F., Grunewald, J., and Grunewald, B. (1985). Green-backed Herons (*Butorides striatus*) possibly using a lure and using apparent bait. *Journal für Ornithologie*, **126**, 439–42.

Walters, G. J. (1983). Carrion Crows causing death of Grey Heron. *British Birds*, **76**, 459.

Walters. M. (1976). Some observations of the eggs of the Great White Bellied Heron *Ardea insignis*. *Journal of the Bombay Natural History Society*, **73**, 213–14.

Walters, M. (2001). The correct scientific name of the White-bellied Heron. *Bulletin of the British Ornithologists' Club*, b, 234–6.

Warner, D. (1986). Herring Gull attacks on Grey Herons. *Scottish Birds*, **14**, 50.

Wassink, A. (1979). Some additional field characters of the Western Reef Heron (*Egretta gularis*). *Ardea*, **66**, 123–4.

Watmough, B. R. (1978). Observations on nocturnal feeding by Night Herons *Nycticorax nycticorax*. *Ibis*, **129**, 356–8.

Watts, B. D. (1988). Foraging implications of food usage patterns in Yellow-crowned Night-Herons. *Condor*, **90**, 860–5.

Watts, B. D. (1995a). Yellow-crowned Night-heron, *Nyctanassa violacea*. In *The birds of North America*. No. 161 (eds. A. Poole and F. Gill), pp. 1–24. The Academy of Natural Sciences, Philadelphia and The American Ornithologists' Union, Washington.

Watts, B. D. (1995b). Incidence of breeding by Yellow-crowned Night-Herons in juvenal plumage. *Colonial Waterbirds*, **18**, 222–3.

Webster, J. D. (1988). Some bird specimens from Sitka, Alaska. *Murrelet*, **69**, 46–8.

Weeks, H. P. (1976). Breeding of the Yellow-crowned Night Heron in Indiana. *Indiana Audubon Quarterly*, **54**, 83–6.

Wei, G., Chen, X., Lin, Q., Hu, H. J., and Chen, J. R. (2002). The breeding behaviour and reproductive capacity of the Little Egret. *Journal of Xiamen University Natural Science*, **41**, 647–52.

Wie, G., Chen, X., and Hu, H. (2003). Observation on some activities of reproduction in Little Egret (*Egretta garzetta*). At Jiyu Island in Xiamen. *Zoological Research*, **24**, 343–347.

Weir, R. D. (1986). The Autumn migration. Ontario region. *American Birds*, **38**, 195–9.

Weir, R. D. (1986). The nesting season. June 1–July 31, 1986. Ontario region. *American Birds*, **40**, 1197–201.

Welch, G. (1995). Digging for bitterns— reshaping Minsmere's scrape. *Suffolk Birds*, **44**, 7–9.

Weller, M. W. (1961). Breeding biology of the Least Bittern. *Wilson Bulletin*, **73**, 11–35.

Wen, Z. Z. (1997). [The breeding biology of the Chinese Egret.] *Acta Zoologica Sinica*, **43**, 216–17.

Wen, Z. Z. and Sun, R.Y. (1993). [The breeding, growth and development of homeothermy in Cattle Egret (*Bubulcus ibis*).] *Acta Zoologica Sinica*, **39**, 263–71.

Werner, S. J., Tobin, M. E., and Fioranelli, P. B. (2001). Great Egret preferences for catfish size classes. *Wetlands*, **24**, 381–5.

Werschkul, D. F. (1979). Nestling mortality and the adaptive significance of early locomotion in the Little Blue Heron. *Auk*, **96**, 116–30.

Werschkul, D. F. (1982). Nesting ecology of the Little Blue Heron: promiscuous behavior. *Condor*, **84**, 381–4.

Wesley, H. D. (1993). Genetics of the red tarsi and feet in the Pond Heron. *Newsletter for Birdwatchers*, **33**, 74.

Wesley, H. D. (1996). More red-legged Pond Herons. *Newsletter for Birdwatchers*, **36**, 5–6.

Wetmore, A. (1960). *A classification of the birds of the world.* Smithsonian Miscellaneous Collection 139, No. 11. Smithsonian Institution, Washington.

Wetmore, A. (1965). *The birds of the Republic of Panama.* Part 1, Tinamidae (tinamous) to Rynchopidae (skimmers). Smithsonian Miscellaneous Collections 150. Smithsonian Institution, Washington.

Wheelwright, N. T. (1978). Snowy Egret in the Strait of Magellan. *Auk*, **95**, 590.

White, C. M. N. (1973). Night Herons in Wallacea. *Bulletin of the British Ornithologists' Club*, **93**.

White, C. M. N. and Bruce, M. D. (1986). *The birds of Wallacea (Sulawesi, the Moluccas, and Lesser Sunda Islands, Indonesia): An annotated checklist.* BOU Checklist 7. British Ornithologists' Union, London.

Whitehall, B. (1999). Feeding relationship between Little Egrets and Red-breasted Mergansers. *British Birds*, **92**, 365.

Whiteside, A. J. (1989). The behavior of bitterns and their use of habitat. *Notornis*, **36**, 89–95.

Whitfield, A. K. and Blaber, S. J. M. (1978). Feeding ecology of piscivorous birds at Lake Sta Lucia, Part 2. Wading birds. *Ostrich*, **Supplement 9**, 19.

Whitmore, R. C., Whitmore, R. C., and Whitmore, M. M. (1999). A previously unreported nesting colony of the Yellow-crowned Night-Heron near Mulege, Baja California Sur. *Western Birds*, **30**, 52–3.

Wiese, J. H. (1976). Courtship and pair formation in the Great Egret. *Auk*, **93**, 709–24.

Wiggins, D. A. (1991). Foraging success and aggression in solitary and group-feeding Great Egrets (*Casmerodius albus*). *Colonial Waterbirds*, **14**, 176–9.

Wild Bird Society of Japan. (1982). *A field guide to the birds of Japan.* Wild Bird Society of Japan, Tokyo.

Wiley, J. W. (2001). Green Heron (*Butorides virescens*) predation at Village Weaver (*Ploceus cucullatus*) nests. *El Pitirre*, **14**, 130–3.

Willard, D. E. (1977). The feeding ecology and behavior of five species of herons in southeastern New Jersey. *Condor*, **79**, 462–70.

Willard, W. E. (1979). Comments on the feeding behavior of the Boat-billed Heron (*Cochlearius cochlearius*). *Biotropica*, **11**, 158.

Williams, D. B. (1978). Yellow-crowned Night Heron; a new breeding species for the Great Smoky Mountains National Park. *Migrant*, **49**, 81–82.

Willoughby, P. J. (2001). Melanistic Cattle Egret. *British Birds*, **94**, 390–1.

Windisch, R. (1999). Mausejagd einer uberwinternden Rohrdommel. *Falke*, **46**, 56.

Wingate, D. B. (1982). Successful reintroduction of the Yellow-crowned Night-heron as a nesting resident on Bermuda. *Colonial Waterbirds*, **5**, 104–15.

Witherby, H. F., Jourdain, F. C. R., Ticehurst, N. F., and Tucker, B. W. (1939). *The handbook of British birds. Vol. 3, Hawks to ducks.* Witherby, London.

Wittenberger, J. and Hunt, G. (1995). The adaptive significance of coloniality in birds. In *Avian Biology Vol. 3* (eds. D. Farner, J. King, and K. Parkes), pp. 1–78.

Wolf, B. O. and Jones, S. L. (1989). Great Blue Heron deaths caused by predation on Pacific lamprey. *Condor*, **91**, 482–4.

Wolford, J. W. and Boag, D. A. (1971). Food habits of Black-crowned Night Herons in southern Alberta. *Auk*, **88**, 435–7.

Wolinski, R. A. (1988). Status of the Yellow-crowned Night-Heron in Michigan. *Jack-Pine Warbler*, **66**, 117–19.

Wolinski, R. A. and Pike, E. A. (1987). First sight record of Yellow-crowned Night-heron on north Manitou Island. *Jack-Pine Warbler*, **65**, 16.

Wood, D. S. (1986). On iris color in *Botaurus* bitterns. *Colonial Waterbirds*, **9**, 108–9.

Woods, M. C. (1994). Little Egret at Chincoteague. *Raven*, **65**, 88.

Wooldridge, G. E. (1998). Little Egrets perching on horse. *British Birds*, **91**, 60.

Wooten, C. W. and Cox, S. (1986). Reddish Egret in Anderson County, S.C.: first inland record for state. *Chat*, **50**, 46–9.

Worthy, T. H., Walter, R., and Anderson, A. J. (1998). Fossil and archaeological avifauna of Niue Island, Pacific Ocean. *Notornis*, **45**, 177–90.

Wüst, W. (1983). Kustenreiher (*Egretta gularis schistacea*) aus Pakistan nach Mittelfranken importiert. *Ornithologische Mitteilungen*, **35**, 132–3.

Yamada, K. (1994). Feeding behaviour and territoriality according to the prey types and habitats in the Little Egret *Egretta garzetta*. *Japanese Journal of Ornithology*, **42**, 61–75.

Yamashita, C. and de Paula Valle, M. (1990). Ocorrência de duas aves raras no Brasil Central: *Mergus octosetaceus* e *Tigrisoma fasciatum fasciatum*. *Ararajuba*, **1**, 107–9.

Yap, B. (1981). *Birds in medieval manuscripts*. The British Library, London.

Yao, C. T, Yo, S. P. and Lin, L.K. (2000). Preliminary study on the reproductive chronology and behavior of the Malaysian Night Heron. *Chinese Bioscience*, **43**, 11–7.

Yao, C. T. (2002). Breeding ecology of the Malaysian Night Heron (*Gorsachius melanophus*) in Central Taiwan. Master's Thesis, Tunghai University, Taichung, Taiwan.

Yen, C. W. (1991). Food of nestling egrets and Night Herons in the western lowlands of Central Taiwan. *Journal of Taiwan Museum*, **44**, 309–20.

Yesou, P. (1984). Little Egrets with uncommon bare-parts coloration. *British Birds*, **77**, 315–17.

Yin, Z. H. and Lei, F. M. (2002). A preliminary study of the breeding biology of the Chinese Egret. *Acta Zoologica Sinica*, **48**, 824–7.

Yosef, R., Rydberg-Hedaen, M., Mitchell, L., and Smit, H. (2002). The first breeding record of Striated Heron (*Butorides striatus*) in Israel. *Sandgrouse*, **24**,

Young, L. (1998). The importance to ardeids of the Deep Bay fish ponds, Hong Kong. *Biological Conservation*, **84**, 293–300.

Young, L. and Chan, G. (1997). The significance of drained fish ponds for wintering waterbirds at the Mai Po Marshes, Hong Kong. *Ibis*, **139**, 694–8.

Yiraberg, N.-J. (1992). White-tailed Eagle *Haliaeetus albicilla* hunting in a Grey Heron *Ardea cinerea* colony in Sunnmore. *Fauna Norvegica Series C, Cinclus*, **15**, 25–30.

Yu, J.-P. and Hahm, K.-H. (1997). Breeding ecology of the Black-crowned Night Heron in Korea. *Acta Zoologica Cracoviensia*, **40**, 269–78.

Zhang L. and Liu Z. (1991). Ecological study on the breeding habits of Little Egret. *Shanxi Daxue Xuebao Ziran Kexue Ban*, **14**, 202–8.

Zhang, L., Liu, Z., and Zhang, F. (1994). [Breeding ecology and biology of four species of egret and heron.] *Acta Ecologica Sinica*, **14**, 80–3.

Zhang, Y. M., Ruan, L. Z., Dong, Y. H., Gong, Z. M., Li, Y. D., Wang, H., and Fasola, M. (2000). Breeding biology of Night Heron (*Nycticorax nycticorax*) and Little Egret (*Egretta garzetta*) in Thaihu Lake of Wuxi, China. *Zoological Research*, **21**, 275–8.

Zhou, F. (1996). White-eared Night-Heron in Guangxi. *Oriental Bird Club Bulletin*, **23**, 89.

Zhou, F. (2003). Guangxi will set up two heron nature reserves. *China Crane News* **7(1)**:28–29.

Zhou, L., Song, Y., and Ma, Y. (1998). Breeding biology of the three herons in Zipeng Mountains. *Chinese Journal of Zoology*, **33**, 34–8.

Zhou, F. W., Chen, Z. H., Lin, Q. X. and Chen, X. L. (2002). The avian fauna of the Xiamen coastline wetlands, southeast China (1999–2000). *Journal of Xiamen University Natural Science*, **41**, 653–8.

Zhu, X., Ma, S., Dai, Y., Zhong, X., and Zhou, L. (1994). [Study on the activities and biomass production of breeding population of Chinese Pond Heron *Ardeola bacchus*.] *Acta Ecologica Sinica*, **14**, 75–9.

Zhu, X., Yang, S. D., Zou, X. P., Chen, W. Z., and Ren, J. B. (2000). [The breeding habits and growth of the nestling Black-crowned Night Heron, *Nycticorax nycticorax*.] *Zoological Research*, **21**, 58–64.

Zickefoose, J. and Davis, W. E., Jr. (1998). Great Blue Heron (*Ardea herodias*) uses bread as bait for fish. *Colonial Waterbirds*, **21**, 87–8.

Zimmerman, D. A., Turner, D. A., and Pearson, D. J. (1996). *Birds of Kenya and northern Tanzania*. Princeton University Press, Princeton.

Index

Page numbers in **bold** indicate main species accounts and numbers in *italic* indicate illustrations.

I

J

K, L

M

The manufacturer's authorised representative in the EU for product safety is Oxford University Press España S.A. of El Parque Empresarial San Fernando de Henares, Avenida de Castilla, 2 - 28830 Madrid (www.oup.es/en or product.safety@oup.com). OUP España S.A. also acts as importer into Spain of products made by the manufacturer.
Printed and bound by CPI Group (UK) Ltd, Croydon, CR0 4YY
14/09/2026
02216284-0019